Nanometer CMOS ICs

Harry Veendrick

Nanometer CMOS ICs

From Basics to ASICs

Third Edition

Springer

BITS ON CHIPS

Harry Veendrick
Bits on Chips
Heeze, The Netherlands

ISBN 978-3-031-64248-7 ISBN 978-3-031-64249-4 (eBook)
https://doi.org/10.1007/978-3-031-64249-4

All Illustrations are created by Kim Veendrick.

This Springer imprint is published by the registered company Springer Nature Switzerland AG
The registered company address is: Gewerbestrasse 11, 6330 Cham, Switzerland

Foreword

Any sufficiently advanced technology is indistinguishable from magic (Clarke's third law, 1973, Profiles of the future). This book leads the reader into the magic of semiconductor nano-scale integrated devices, circuits and systems. It originates from the authors own rich research experience in IC design and decades of training and teaching the details of semiconductors. The two preceding editions of this book have already inspired tens of thousands of engineers and researchers from a multitude of companies and academic teams.

From white goods to healthcare, on the road or in the theme park, ordering pizza or electricity from the grid, reading an e-book or creating a speech with generative AI, we all rely insatiably on semiconductors. Like with any addiction we often are unaware of the underlying chemistry that is ruling, the processes that are involved and the actors that play. This book will provide—after you took a deep breath and worked it through—a solid understanding and colourful nuances in the complex world of semiconductor technology and the nano-scale integrated circuits.

The reach of the book, from the basics of different semiconductor device families all the way up to large circuit design, is ambitious. But it reflects the importance for the innovators and engineers in the state-of-the-art semiconductor technology to have a fundamental appreciation of the full stack of disciplines that are required to build a complex system-on-chip or a specialty circuit tuned to a specific application. The fundamental understanding of the device physics and the parameters that govern the CMOS scaling into the sub-2 nm regime leads to a better understanding of reliability and variability that play at circuit and systems level. How Artificial Intelligence workloads are optimally ported on a chiplet-based heterogeneous compute architecture will be the result of an optimisation based on elevated insight in technology options for compute, memory—or compute-in-memory—, interconnect and packaging. Indeed, in advanced digital applications one can find numerous examples that illustrate the need for System-Technology Co-Optimisation. Where in the 'happy scaling' era the design community could make abstraction of the technology hocus-pocus in the clean rooms for every new generation, today the dialog is getting much more intense. And in that multi-disciplinarity the most exciting innovations happen.

Materials scientist, solid-state electronics engineer, lithography expert, circuit designer, digital system architect, or just curious... I trust this book will inspire the reader to fully embrace the magic of semiconductors.

EVP and Chief Strategy Officer IMEC
Professor KU Leuven, Belgium
Leuven, Belgium
February, 2024

Jo De Boeck

Preface

An *integrated circuit* (*IC*) is a piece of semiconductor material, on which a number of electronic components are interconnected. These 'chip' components implement a specific function. The semiconductor material is usually silicon but alternatives include gallium arsenide (for RF devices), silicon carbide (for power electronics) and indium phosphite (for fibre optics and lasers). ICs are essential in all modern electronic products.

Figure 1 illustrates the progress in IC complexity. This figure shows the numbers of components for advanced ICs and the year in which these ICs were first presented. Originally, this doubling in complexity every 18 months was predicted by Moore (Intel 1964). The slope in the line gradually saturates and as such we can conclude that we are slowly reaching the limits of scaling. The complexity doubling of certain memories, for instance, now happens at a three-to-four year cycle. Every dot on the line represents the maximum number of transistors on the 'champion chip' in the given year. In the first decade small logic chips contained the largest number of transistors, while roughly from the 1970s to the 1990s DRAM memory chips had taken over the lead. After that period, NAND flash chips are the true champions in this respect.

Figure 2 shows the relative semiconductor revenue per IC category. CMOS ICs represent 75–80% of the total semiconductor market. They can be subdivided into three main categories: logic, microprocessors and memories. Less than 10% of the CMOS ICs are of an analog nature. Today's logic ICs may contain several hundreds of millions to more than a hundred billion transistors on a single die of 1–8 cm^2. Figures 3, 4, 5, 6, 7, 9 illustrate the evolution in IC technology. The first IC was created by Jack Kilby in 1959. Photographs of this device and the inventor are shown in Fig. 3. Figure 4 shows the interior of a discrete BC107 transistor. The digital filter shown in Fig. 5 comprises a few thousand transistors while the Digital Audio Broadcasting (DAB) chip in Fig. 6 contains more than six million transistors. The Intel Haswell-E/EP eight-core processor of the Xeon family in Fig. 7.30 (Sect. 7.6), contains 2.6 billion transistors on a 355 mm^2 die, fabricated in a 22 nm process with a maximum power consumption of 140 W *thermal design power (TDP)*. This is the maximum amount of heat that the chip's cooling system can dissipate. Another strong workhorse, shown in Fig. 7, is the AMD Zen-based Opteron, made in 14 nm FinFET technology, which packs 32 cores on two dies, containing about five billion transistors, mounted in one package which will function as a single 32 core chip.

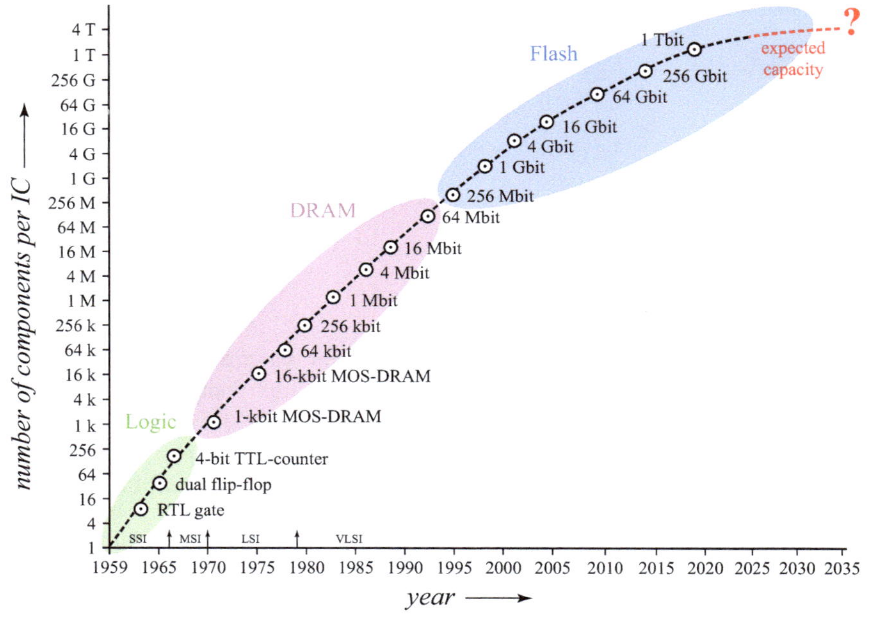

Fig. 1 Growth in the number of components per IC

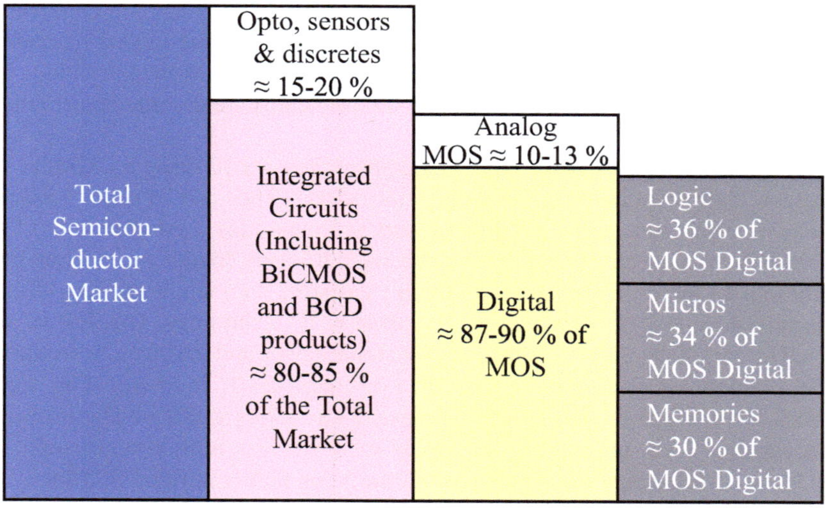

Fig. 2 Relative semiconductor revenue by IC category (Source: IC Insights; 2022)

Fig. 3 The development of the first IC: in 1958 Jack Kilby demonstrated the feasibility of resistors and capacitors, in addition to transistors, based on semiconductor technology. Kilby, an employee of Texas Instruments, submitted the patent request entitled 'Miniaturized Electronic Circuits' in 1959. His request was honoured. Recognition by a number of Japanese companies in 1990 means that Texas Instruments is still benefiting from Kilby's patent (Photo: Texas Instruments / Koning and Hartman, reprinted with permission; 1959).

Figure 8 shows the Nvidia B200 (2024) streaming multi-processor chip of more than 800 mm^2. It contains 104 billion transistors and is fabricated in a custom TSMC 4N process, which is somewhat different from the generic 4 nm TSMC process. It is one of the dies used in the Nvidia Grace Blackwell 200 (GB200) 'superchip', which is built from three *chiplets* (see Sect. 10.3.3): two B200 dies and a Grace-CPU die and consumes more than 1000 W. That includes the power consumption of the high-bandwidth memory (HBM). It can perform a huge number of matrix operations necessary in today's artificial intelligence and neural-network applications and requires liquid cooling to exploit the full potential of the GB200 chipset.

A 1 Tb 3D NAND flash memory with 232 memory cell layers is shown in Fig. 9. It is implemented with TLC memory cells (3 bit/cell), with a density of 14.6 Gb/mm^2. The main target applications are computer products, mobiles, etc. Finally, Fig. 10 illustrates the sizes of various semiconductor components, such as a silicon atom, a single transistor and an integrated circuit, in perspective. The sizes of an individual MOS transistor are already similar to those of a virus.

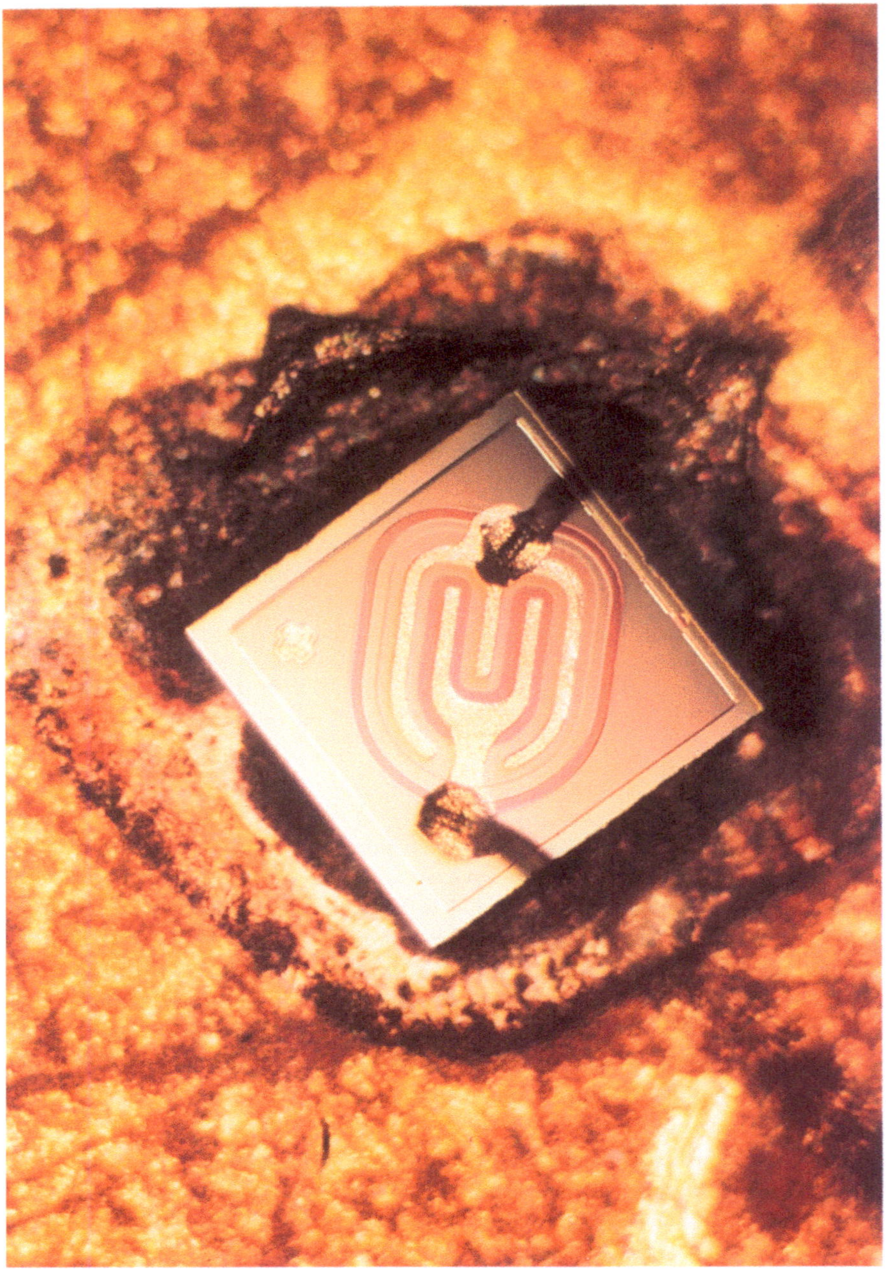

Fig. 4 A single BC107 bipolar transistor (Photo: NXP Semiconductors, reprinted with permission; early 1960s)

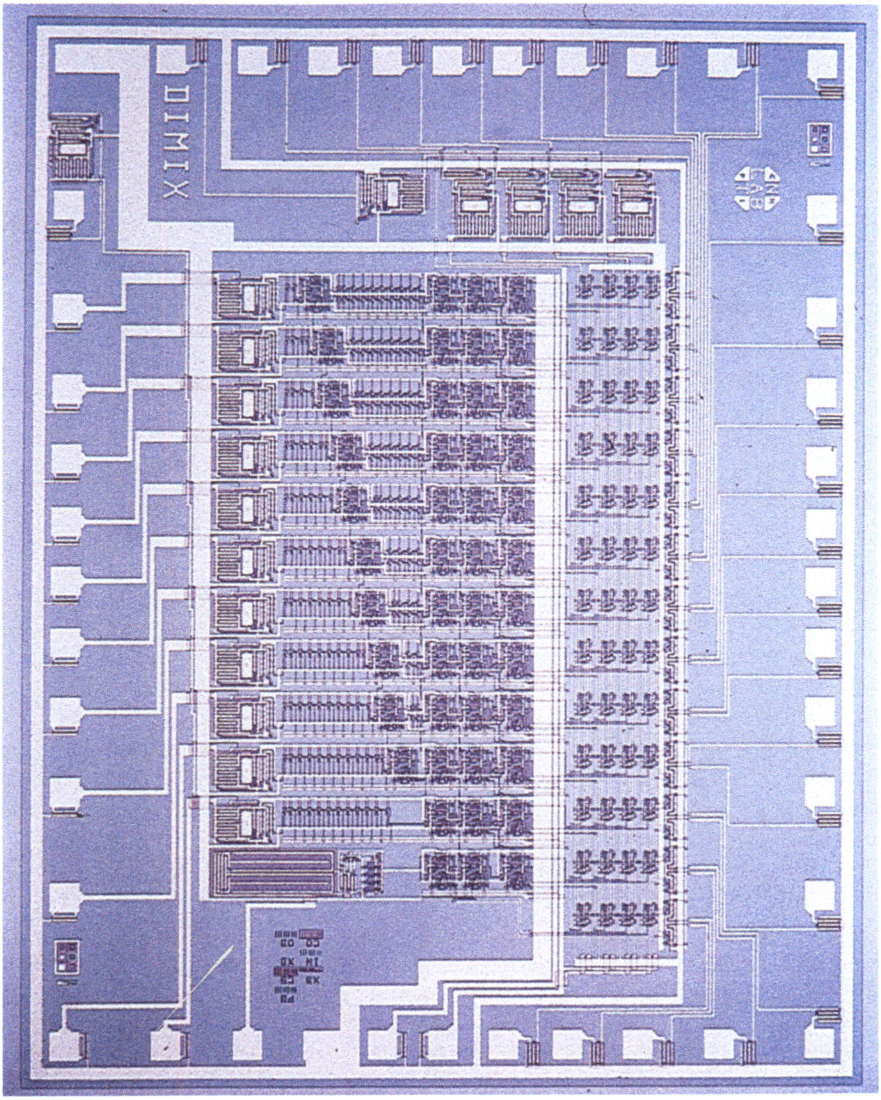

Fig. 5 A digital filter which comprises a few thousand transistors (Photo: NXP Semiconductors, reprinted with permission; 1978)

Fig. 6 A Digital Audio Broadcasting (DAB) chip, which comprises more than six million transistors (Photo: NXP Semiconductors, reprinted with permission; 1997)

This book provides an insight into all aspects associated with CMOS ICs. The topics presented include relevant fundamental physics. Process technology, lithography, design and implementation aspects are also explained and applications are discussed. CAD tools used for the realisation of ICs are described while current and expected developments also receive attention.

The contents of this book are based on a CMOS ICs course as part of an industry-oriented course program entitled 'An introduction to IC techniques'. This CMOS course has been given for more than four decades at various companies, which include Philips, NXP, TSMC, ASML, Applied Materials, IMEC, ST-Microelectronics, Infineon, Nexperia and several universities in China and India. Continuous revision and expansion of the course material ensures that this

Fig. 7 The 14 nm FinFET-based AMD Opteron 6300 packs 8 core 'Piledriver' CPUs on a single die. The Zen-based Opteron will pack 32 cores on two dies, containing about five billion transistors, mounted in one package which will function as a single 32 core Opteron chip (Photo: Advanced Micro Devices, Inc., reprinted with permission; 2014)

Fig. 8 The Nvidia Grace Blackwell 200 (GB200) 'superchip' combines three chiplets: two B200 dies and a Grace-CPU die. This image shows such a B200 *chiplet* (see Sect. 10.3.3). It contains 104 billion transistors, has a die area of over 800 mm^2 and is made in TSMC N4 process, which is somewhat different from the generic 4 nm TSMC process. Each chiplet also has a 4096-bit memory bus connected to 96 GB of DRAM memory. This 96 GB is divided over four 24 GB high-bandwidth memories (HBM3e), each built from 16 memory dies, stacked on each other. The B200 package features a 1.8 TB/s NVLink interface for host connectivity and connectivity to another B200 chip. It consumes around 1000 W, but that includes the power consumption of the high-bandwidth memory. GPU architectures are built for performing many simpler tasks in parallel. A single task is then split into several sub-tasks which run simultaneously on multiple simple processors. It is designed for use in advanced applications like artificial intelligence (AI) and high-performance computing. For more information on this subject, refer to the following link: https://www.techpowerup.com/320542/unwrapping-the-nvidia-b200-and-gb200-ai-gpu-announcements (Photo: Nvidia, reprinted with permission; 2024)

Fig. 9 This 3D NAND flash memory with 232 memory cell layers contains 1 Tb. It is implemented with TLC memory cells (3 bit/cell), with a density of 14.6 Gb/mm^2. The main target applications are computer products, mobiles, etc. (Photo: Micron Technology, reprinted with permission; 2023)

book is highly relevant to the IC industry. The level of the discussions makes this book a suitable introduction for designers; technologists; CAD developers; test-, packaging-, failure analysis- and reliability engineers; technical-commercial personnel; and IC applicants. The text is also suitable for both graduates and undergraduates in related engineering courses.

Considerable effort has been made to enhance the readability of this book and only essential formulae are included. The large number of diagrams and photographs should reinforce the explanations. The design and application examples

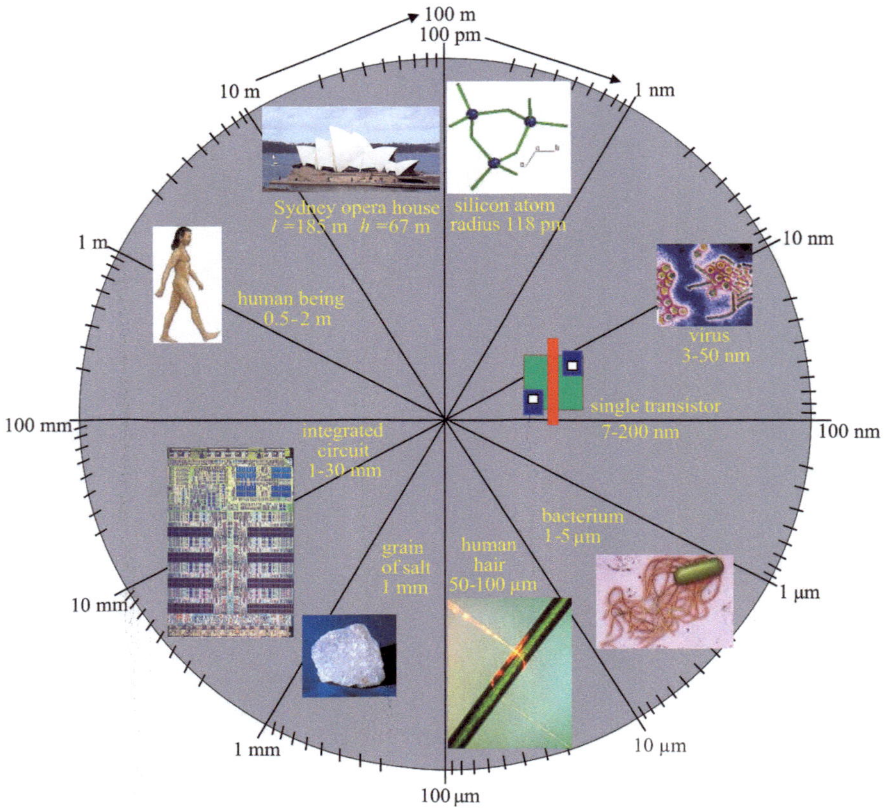

Fig. 10 Various semiconductor component sizes (e.g., atom, transistor, integrated circuit) in perspective

are mainly digital. This reflects the fact that more than 90% of all modern CMOS ICs are digital circuits. However, the material presented will also provide the analogue designer with a basic understanding of the physics, manufacture and operation of nanometer CMOS circuits. The chapters are summarised below. For educational purposes the first four chapters each start with a discussion on nMOS physics, nMOS transistor operation, nMOS circuit behaviour, nMOS manufacturing process, etc. Because the pMOS transistor operation is fully complementary to that of the nMOS transistor, it is then easier to understand the operation and fabrication of complementary MOS (CMOS) circuits. The subjects per chapter are chosen in a very organised and logical sequence so as to gradually built the knowledge: *from Basics to ASICs*. These are ICs which are realised for a single end-user and dedicated to a particular application. ASICs therefore implement customer-specified functions. In fact, most logic ICs fall in this category.

The knowledge gathered from each chapter is required to understand the information presented in the next chapter(s). Each chapter ends with a reference list and exercises. The exercises summarise the important topics of the chapter and form an important part of the complete learning process.

Chapter 1 contains detailed discussions on the basic principles and fundamental physics of the MOS transistor. The derivation of simple current-voltage equations for MOS devices and the explanation of their characteristics illustrates the relationship between process parameters and circuit performance.

The continuous reduction of transistor dimensions led to increased deviations between the performance predicted by the simple MOS formulae and actual transistor behaviour. The effects of temperature and the impact of the continuous scaling of the geometry on this behaviour are explained in Chap. 2. In addition to their influence on transistor and circuit performance, these effects can also reduce device lifetime and reliability.

The various technologies for the manufacture of CMOS ICs are examined in Chap. 3. After a summary on the available different substrates (wafers) used as starting material, an explanation of the most important associated photolithography and processing steps is provided. This precedes the discussion of advanced CMOS processes in the sub-10 nm nodes for the manufacture of modern VLSI circuits.

The design of CMOS circuits is treated in Chap. 4. An introduction to the performance aspects of nMOS circuits provides an extremely useful background for the explanation of the CMOS design and layout procedures. This chapter also includes discussions and examples of FinFET layouts.

MOS technologies and their derivatives are used to realise the special devices discussed in Chap. 5. Charge-coupled devices (CCDs), CMOS imagers, power MOS transistors and IGBTs are among the special devices. Chapter 5 concludes the presentation of the fundamental concepts behind BICMOS circuit operation.

Stand-alone memories currently represent about 20–25% of the total semiconductor market revenue. However, also in logic and microprocessor ICs embedded memories represent close to 80% of the total transistor count. So, of all transistors produced in the world, today, more than 99.5% end up in either a stand-alone, or in an embedded memory. This share is expected to stay at this level or to increase. The majority of available memory types are therefore examined in Chap. 6. The basic structures and the operating principles of the various types are explained. In addition, the relationships between their respective properties and application areas are made clear.

Developments in IC technology now facilitate the integration of complete systems on a logic chip, which contain several hundreds of millions to more than a hundred billion transistors. The various IC design and realisation techniques used for these VLSI ICs are presented in Chap. 7. The advantages and disadvantages of the techniques and the associated CAD tools are examined. Various modern technologies are used to realise a separate class of VLSI ICs, which are specified by applicants rather than manufacturers. These application-specific ICs (ASICs) are examined in this chapter as well. Motives for their use are also discussed.

As a result of the continuous increase of power consumption, the maximum level that can be sustained by cheap plastic packages has been reached. Therefore, all CMOS designers must have a 'less-power attitude'. Chapter 8 presents a complete overview of less-power and less-leakage options for CMOS technologies, as well as for the different levels of design hierarchy.

Increased VLSI design complexities, combined with higher frequencies at reduced operating voltages, create a higher sensitivity to physical effects. These effects dominate the reliability and signal integrity of nanometer CMOS ICs. Chapter 9 extensively discusses these effects and the design measures to be taken to maintain both reliability and signal integrity at a sufficiently high level.

Finally, testing, yield, packaging, debug and failure analysis are important factors that contribute to the ultimate costs of an IC. Chapter 10 presents an overview of the state-of-the-art techniques that support testing, debugging and failure analysis. It also includes a rather detailed summary on available packaging technologies and gives an insight into their future trends. Essential factors related to IC production are also examined; these factors include quality and reliability and failure analysis.

The continuous reduction of transistor dimensions associated with successive process generations is the subject of the final chapter (Chap. 11). This scaling has various consequences for transistor behaviour and IC performance. The resulting increase of physical effects and the associated effects on reliability and signal integrity are important topics of attention. The expected consequences of and road blocks for further miniaturisation are described. This provides an insight into the challenges facing the IC industry in the race towards real nanometer devices. The major drive for continuation of Moore's law has been the reduction of the cost per transistor. However, sub-10 nm nodes require an increasing number of extremely complex process steps that makes further cost reduction questionable for certain IC categories. This final chapter therefore also includes some of the cost trends related to both the design and manufacturing of integrated circuits.

Not all data in this book is completely sprout from my mind. A lot of books and papers contributed to make the presented material state-of-the-art. Considerable effort has been made to make the reference list complete and correct. I apologise for possible imperfections.

Heeze, The Netherlands Harry Veendrick

Acknowledgements

I wish to express my gratitude to all those who contributed to the realisation of this book; it is impossible to include all their names. I greatly value my professional environment: Philips Research labs, of which the semiconductor research department is now part of NXP Semiconductors. It offered me the opportunity to work with many internationally highly valued colleagues who were all real specialists in their field of semiconductor expertise. Their contributions included fruitful discussions, relevant texts and manuscript reviews. I would like to make an exception, here, for my colleagues Marcel Pelgrom and Maarten Vertregt, who greatly contributed to the discussions held on trends in MOS transistor currents and variability matters throughout this book and Roger Cuppens, Maurits Storms and Roelof Salters for the discussions on non-volatile and random-access memories, respectively.

I would especially like to thank Andries Scholten and Ronald van Langevelde for reviewing Chap. 2 and for the discussions on leakage mechanisms in this chapter, Casper Juffermans and Johannes van Wingerden (both NXP) and Ewoud Vreugdenhil and Hoite Tolsma (both ASML) for their inputs to and review of the lithography section in Chap. 3. I would also like to sincerely thank Robert Lander for his detailed review of the section on CMOS process technologies and future trends in CMOS devices and Gerben Doornbos for the correct sizes and doping levels used in the manufacture of state-of-the-art CMOS devices. I appreciate the many circuit simulations that Octavio Santana has done to create the tapering-factor table in Chap. 4. I am grateful for the review of Chap. 5 on special circuits and devices based on MOS transistor operation: Albert Theuwissen (Harvest Imaging) for the section on CCD and image sensors, Johan Donkers and Erwin Hijzen for the BICMOS section and Jan Sonsky for the high voltage section. I also appreciate their willingness to supply me with great photographic material. Toby Doorn and Ewoud Vreugdenhil are thanked for their review of the memory chapter (Chap. 6). The latter person has also contributed a lot to the lithography section. I appreciate Paul Wielage's work on statistical simulations with respect to memory yield loss. I thank Ad Peeters for information on and reviewing the part on asynchronous design in the low-power chapter (Chap. 8). Reliability is an important part of Chap. 9, which discusses the robustness of ICs. In this respect I want to thank Andrea Scarpa for reviewing the hot-carrier and NBTI subjects, Frans List and Theo Smedes for the ESD and latch-up subjects and Yuang Li for the part on electromigration. I also greatly value the work of Bram Kruseman, Henk Thoonen and Frank Zachariasse for reviewing the sections on testing, packaging and failure analysis, respectively. I also like to express them my appreciation for supplying me with a lot of figures and photographs, which support and enrich the discussions on these subjects in Chap. 10. Finally, I want to thank Chris Wyland and John Janssen, for their remarks and additions on electrical and thermal aspects of IC packages, respectively.

I am very grateful to all those who attended the course, because their feedback on educational aspects, their corrections and constructive criticism contributed to the quality and completeness of this book.

In addition, I want to thank Philips Research and NXP Semiconductors, in general, for the co-operation I was afforded. I thank my son Bram for the layout of the cover and the layout diagrams in Chap. 4, and Ron Salfrais for the correctness of a large part of the English text.

I would also like to express my gratitude to Henny Alblas for the many hours he has spent on the creation of excellent and colourful art work.

Finally, I wish to thank Harold Benten and Dré van den Elshout for their conscientious editing and typesetting work. Their efforts to ensure high quality should not go unnoticed by the reader.

However, the most important appreciation and gratitude must go to my family, again, and in particular to my wife, for her years of exceptional tolerance, patience and understanding. The year 2007 was very demanding. Lost hours can never be regained, but I hope that I can give her now a lot more free time in return.

First Edition Harry J. M. Veendrick
Heeze, The Netherlands
February, 2008

This **second full-colour edition** covers the same subjects, but then they are completely revised and updated with the most recent state-of-the-art material. It covers all subjects, related to nanometer CMOS ICs: physics, technologies, design, testing, packaging and failure analysis. The contents include substantially new material along with extended discussions on existing topics, which leads to a more detailed and complete description of all semiconductor disciplines. The result is a better self-contained book which makes it perfectly accessible to semiconductor professionals, academic staff, PhD and (under) graduate students.

Finally, I wish to thank Harold Benten and Kim Veendrick, again, for their conscientious text-editing and excellent art work, respectively.

Second Edition Harry J. M. Veendrick
Heeze, The Netherlands
October, 2016

As all editions of this book describe the full development chain of a chip, from basics to ASICs, this **third edition**, again, covers the same subjects. It particularly contains updates in the lithography, fabrication and memory sections as these are very susceptible to the rapid technology changes. In this respect, it describes lithography solutions and new transistor architectures that are needed for sub-5 nm CMOS process nodes. In 3D NAND flash memories it seems that 'the sky is the limit', where the number of stacked memory-cell layers have already exceeded three hundred, today. Also the yield section has been extended. It now includes discussions on the cleanliness in a cleanroom and yield spread across the wafer. Moreover, the increasing interest in chiplets, which are seen as an alternative to a monolithic chip realisation to reduce cost, is explained as well. Because of the increasing complexity of the power and signal distribution across an advanced chip, several semiconductor companies and research institutes are looking for alternative solutions. Back-side power distribution seems to be a promising alternative to improve the robustness and reliability of chip operation.

Finally, I especially want to thank my daughter Kim, who has done both the text editing and the artwork for this third edition. She spent many hours on creating new drawings and tables. This, in combination with her dedication to consistently use defined colours for all semiconductor cross sections throughout this book, has contributed a lot to its quality and clarity.

Third Edition
Heeze, The Netherlands
February, 2024

Harry J. M. Veendrick

Overview of Symbols

α	channel-shortening factor or clustering factor
A	area
A	aspect ratio
a	activity factor
β	MOS transistor gain factor
$\beta_{\square}$	gain factor for MOS transistor with square channel
β_{n}	nMOS transistor gain factor
β_{p}	pMOS transistor gain factor
β_{total}	equivalent gain factor for a combination of transistors
BV	breakdown voltage
C	capacitance
C_{b}	bitline capacitance
C_{d}	depletion layer capacitance
C_{db}	drain-substrate capacitance
C_{g}	gate capacitance
C_{gb}	gate-substrate capacitance
C_{gd}	gate-drain capacitance
C_{gs}	gate-source capacitance
C_{gdo}	voltage-independent gate-drain capacitance
C_{gso}	voltage-independent gate-source capacitance
C_{par}	parasitic capacitance
C_{min}	minimum capacitance
C_{s}	scaled capacitance
C_{ox}	oxide capacitance
C_{s}	silicon surface-interior capacitance
C_{sb}	source-substrate (source-bulk) voltage
C_{t}	total capacitance
CD	critical dimension
ΔL	difference between drawn and effective channel length
ΔV_{T}	threshold voltage variation
D_0	defect density for uniformly distributed errors (dust particles)
D_{l}	threshold-voltage channel-length dependence factor
D_{w}	threshold-voltage channel-width dependence factor
ϵ	dielectric constant

ϵ_0	absolute permittivity
ϵ_{ox}	relative permittivity of oxide
ϵ_r	relative permittivity
ϵ_{si}	relative permittivity of silicon
E	electric field strength
E_c	conduction band energy level
E_f	Fermi energy level
E_i	intrinsic (Fermi) energy level
E_{mx}	maximum horizontal electric field strength
E_{ox}	electric field across an oxide layer
E_v	valence band energy level
E_x	horizontal electric field strength
E_{x_c}	critical horizontal field strength
E_z	vertical electric field strength
ϕ	electric potential
ϕ_f	Fermi potential
ϕ_s	surface potential of silicon w.r.t. the substrate interior
ϕ_{MS}	contact potential between gate and substrate
F	feature size (= size of a half pitch used for stand-alone memories)
f	clock frequency
f_{max}	maximum clock frequency
γ	factor which expresses relationship between drain-source voltage and threshold-voltage variation
g_m	transconductance
I	current
I_b	substrate current
I_{ds}	drain-source current
I_{ds0}	characteristic subthreshold current for gate-substrate voltage of 0 V
I_{dsD}	driver transistor drain-source current
I_{dsL}	load transistor drain-source current
$I_{ds\,sat}$	saturated transistor drain-source current
$I_{ds\,sub}$	subthreshold drain-source current
I_{max}	maximum current
I_{on}	on current
I_R	current through resistance
$i(t)$	time-dependent current
j	current density
k	Boltzman's constant
K	K-factor; expresses relationship between source-substrate voltage and threshold voltage
K	amplification factor
λ	wavelength of light
L	effective transistor channel length and inductance
L_{CLM}	channel length reduction due to channel length modulation
L_{eff}	effective channel length

L_{ref}	effective channel length of reference transistor
M	yield model parameter
μ_0	substrate carrier mobility
μ_n	channel electron mobility
μ_p	channel hole mobility
n	number of electrons in a material
n_i	intrinsic carrier concentration
N_A	substrate doping concentration
NA	numeric aperture
ρ	charge density
P	power dissipation
P_{dyn}	dynamic power dissipation
P_{stat}	static power dissipation
p	voltage scaling factor
p	also represents the number of holes in a material, in related expressions
Q	charge
q	elementary charge of a single electron
Q_d	depletion layer charge
Q_g	gate charge
Q_m	total mobile charge in the inversion layer
Q_n	mobile charge per unit area in the channel
Q_{ox}	oxide charge
Q_s	total charge in the semiconductor
R	resistance
R_{JA}	junction-to-air thermal resistance
R_{JC}	junction-to-case thermal resistance
R_L	load resistance
R_{out}	output resistance or channel resistance
R_{therm}	thermal resistance of a package
r	tapering factor
s	scale factor
s_{subthr}	subthreshold slope
σ	conductivity of a semiconductor material
τ	delay time
τ_f	fall time
τ_r	rise time
τ_R	dielectric relaxation time
T	clock period
T_{min}	minimum clock period
$Temp$	temperature
$Temp_A$	ambient temperature
$Temp_C$	case temperature
$Temp_J$	junction temperature
T_{lf}	transistor lifetime
t	time

t_{cond}	conductor thickness
t_{d}	depletion layer thickness
$t_{\mathrm{dielectric}}$	dielectric thickness
t_{ox}	gate-oxide thickness
t_{is}	isolator thickness
U	computing power
v	carrier velocity
v_{sat}	carrier saturation velocity
V	voltage
V_{B}	breakdown voltage
V_{r}	scaled voltage
V_0	depletion layer voltage
V_{bb}	substrate voltage
V_{dd}	supply voltage
V_{c}	voltage at silicon surface
V_{ds}	drain-source voltage
$V_{\mathrm{ds\,sat}}$	drain-source voltage of saturated transistor
V_{E}	Early voltage
V_{fb}	flat-band voltage
V_{g}	gate voltage
V_{gg}	extra supply voltage
V_{gs}	gate-source voltage
$V_{\mathrm{gs_L}}$	load transistor gate-source voltage
V_{H}	high voltage level
V_{in}	input voltage
V_{j}	junction voltage
V_{L}	low voltage level
V_{PT}	transistor punch-through voltage
V_{sb}	source-substrate (back-bias) voltage
V_{ss}	ground voltage
V_{ws}	well-source voltage
V_{T}	threshold voltage
$V_{\mathrm{T_D}}$	driver transistor threshold voltage
$V_{\mathrm{T_{dep}}}$	depletion transistor threshold voltage
$V_{\mathrm{T_{enh}}}$	enhancement transistor threshold voltage
$V_{\mathrm{T_L}}$	load transistor threshold voltage
$V_{\mathrm{T_n}}$	nMOS transistor threshold voltage
$V_{\mathrm{T_p}}$	pMOS transistor threshold voltage
$V_{\mathrm{T\,par}}$	parasitic transistor threshold voltage
V_{out}	output voltage
$V(x)$	potential at position x
V_{x}	process-dependent threshold voltage term
$V_{\mathrm{X_L}}$	process-dependent threshold voltage term for load transistor
$V_{\mathrm{X_D}}$	process-dependent threshold voltage term for driver transistor
W	transistor channel width
W_{n}	nMOS transistor channel width

W_{p}	pMOS transistor channel width
W_{ref}	reference transistor channel width
$\frac{W}{L}$	transistor aspect ratio
$(\frac{W}{L})_{\mathrm{n}}$	nMOS transistor aspect ratio
$(\frac{W}{L})_{\mathrm{p}}$	pMOS transistor aspect ratio
x	distance w.r.t. specific reference point
Y	yield
Z_{i}	input impedance

Explanation of Atomic-Scale Terms

- *Electron:* an elementary particle, meaning that it is not built from substructures. They can be fixed bound to the nucleus of an atom, or freely moving around. When free electrons move through vacuum or a conductor, they create a flow of charge. This is called electrical current, which, by definition, flow in the opposite direction of the negatively charged electrons. Electrons have a mass of $9.11 \cdot 10^{-31}$ kg ($\approx$1/1836 the mass of a proton) and a negative charge $1.6 \cdot 10^{-19}$ C. Electrons play a primary role in electronic, magnetic, electromagnetic, chemistry and nuclear physics. In semiconductor circuits, their main role is charging or discharging analog, logic and memory nodes.
- *Proton:* a subatomic particle with a positive charge of $1.6 \cdot 10^{-19}$ C and a mass of $1.67 \cdot 10^{-27}$ kg. Protons form, together with neutrons, the basic elements from which all atomic nuclei are built and are held together by a strong nuclear force.
- *Neutron:* a subatomic particle with no charge, with a mass which is about equal to that of a proton.
- *Atom:* an atom is the smallest unit of any material in the periodic system of elements. It consists of a nucleus with a fixed number of protons and neutrons, surrounded by one or more shells, which each contain a certain number of electrons. Since an atom is electrically neutral, the total number of electrons in these shells (one or more; hydrogen has only one electron) is identical to the number of protons in the nucleus, since neutrons have no net electrical charge. The number of protons in the nucleus defines the atomic number of the element in the periodic table of elements and determines their physical and chemical properties and behaviour. Most of the CMOS circuits are fabricated on silicon wafers. Silicon is in group IV, which means that it has four electrons in the outer shell. In a monocrystalline silicon substrate, each of these four electrons can form bonding pairs with corresponding electrons from four neighbour silicon atoms, meaning that each silicon atom is directly surrounded by four others. By replacing some of the silicon atoms by boron or phosphorous, one can change the conductivity of the substrate material. In this way nMOS or pMOS transistors can be created. Atom sizes are of the order of $0.1-0.4$ nm. In a monocrystalline silicon substrate the atom to atom spacing is 0.222 nm, meaning that there are between 4–5 silicon atoms in one nanometer.
- *Molecule:* a molecule is the smallest part of a substance that still incorporates the chemical properties of that substance. It is built from an electrically neutral

group of atoms, which are bound to each other in a fixed order. The mass of a molecule is the sum of the masses of the individual atoms, from which it is built. A simple hydrogen molecule (H_2), for example, only consists of two hydrogen atoms which are bound by one electron pair. A water molecule (H_2O) consists of two light hydrogen atoms and one (about 16 times heavier) oxygen atom.

- *Ion:* an ion is an electrically charged atom or molecule or other group of bound atoms, created by the removal or addition of electrons by radiation effects or chemical reactions. It can be positively or negatively charged by, respectively, the shortage or surplus of one or more electrons.

List of Physical Constants

$\epsilon_0 = 8.85 \times 10^{-12}$ F/m
$\epsilon_{ox} = 3.9$ for silicon dioxide
$\epsilon_{si} = 11.7$
$\phi_f \approx 0.35$ V for silicon substrate
$k = 1.4 \times 10^{-23}$ Joule/K
$q = 1.6 \times 10^{-19}$ Coulomb

bandgap for Si:	1.12 eV
bandgap for SiO_2:	9 eV

Contents

1 Basic Principles ... 1
1.1 Introduction ... 1
1.2 The Field-Effect Principle ... 1
1.3 The Inversion-Layer MOS Transistor ... 4
1.3.1 The Metal-Oxide-Semiconductor (MOS) Capacitor ... 8
1.3.2 The Inversion-Layer MOS Transistor ... 11
1.4 Derivation of Simple MOS Formulae ... 18
1.5 The Back-Bias Effect (The Influence of the Threshold Voltage on Transistor and Chip Behaviour) ... 21
1.6 Factors Which Characterise the Behaviour of the MOS Transistor ... 25
1.7 Different Types of MOS Transistors ... 27
1.8 Parasitic MOS Transistors ... 28
1.9 MOS Transistor Symbols ... 29
1.10 Capacitances in MOS Structures ... 31
1.11 Conclusions ... 38
1.12 Exercises ... 39
References ... 43
General Basic Physics ... 43
MOS Capacitances ... 43

2 Geometrical, Physical and Field-Scaling Impact on MOS Transistor Behaviour ... 45
2.1 Introduction ... 45
2.2 The Zero-Field Mobility ... 45
2.3 Carrier Mobility Reduction ... 46
2.3.1 Vertical and Lateral Field Carrier Mobility Reduction ... 47
2.3.2 Stress-Induced Carrier Mobility Effects ... 51
2.4 Channel Length Modulation ... 51
2.5 Short- and Narrow-Channel Effects ... 53
2.5.1 Short-Channel Effects ... 53
2.5.2 Narrow-Channel Effect ... 55
2.6 Temperature Influence on Carrier Mobility and Threshold Voltage ... 57

2.7 MOS Transistor Leakage Mechanisms 59
2.7.1 Weak-Inversion (Subthreshold) Behaviour of the MOS Transistor 60
2.7.2 Gate-Oxide Tunnelling 63
2.7.3 Reverse-Bias Junction Leakage 64
2.7.4 Gate-Induced Drain Leakage (GIDL) 65
2.7.5 Hot-Carrier Injection and Impact Ionisation 66
2.7.6 Overall Leakage Interactions and Considerations 67
2.8 MOS Transistor Models and Simulation 69
2.8.1 Worst-Case (Slow), Typical and Best-Case (Fast) Process Parameters and Operating Conditions 69
2.9 Conclusions 71
2.10 Exercises 71
References 72

3 Manufacture of MOS Devices 75
3.1 Introduction 75
3.2 Different Substrates (Wafers) as Starting Material 76
3.2.1 Wafer Sizes 80
3.2.2 Standard CMOS Epi 81
3.2.3 Crystalline Orientation of the Silicon Wafer 83
3.2.4 Silicon-on-Insulator (SOI) 84
3.3 Lithography in MOS Processes 89
3.3.1 Lithography Basics 89
3.3.2 Lithographic Extensions Beyond 30 nm 103
3.3.3 Accuracy Measurement of Lithographic Systems 115
3.3.4 Mask Cost Reduction Techniques for Low-Volume Production 116
3.3.5 Pattern Imaging 118
3.4 Oxidation 122
3.5 Deposition 126
3.6 Etching 131
3.7 Diffusion and Ion Implantation 134
3.8 Planarisation 137
3.9 Basic MOS Technologies 140
3.9.1 The Basic Silicon-Gate nMOS Process 141
3.9.2 The Basic Complementary MOS (CMOS) Process 145
3.9.3 Sub-100 nm CMOS Processes 147
3.9.4 Further Transistor Scaling: From Planar to FinFETs 155
3.9.5 CMOS Technologies Beyond 10 nm 167
3.9.6 Further Interconnect (BEOL and MEOL) Scaling 169
3.10 Conclusions 172
3.11 Exercises 173
References 174

4 CMOS Circuit, Layout and Library Design 179
4.1 Introduction 179
4.2 The Basic nMOS Inverter 180
4.2.1 Introduction 180
4.2.2 The DC Behaviour 181
4.2.3 Comparison of the Different nMOS Inverters 188
4.2.4 Transforming a Logic Function Into an nMOS Transistor Circuit 189
4.3 Electrical Design of CMOS Circuits 192
4.3.1 Introduction 192
4.3.2 The CMOS Inverter 193
4.4 Digital CMOS Circuits 206
4.4.1 Introduction 206
4.4.2 Static CMOS Circuits 206
4.4.3 Clocked Static CMOS Circuits 211
4.4.4 Dynamic CMOS Circuits 214
4.4.5 Other Types of CMOS Circuits 218
4.4.6 Choosing a CMOS Implementation 219
4.4.7 Clocking Strategies 220
4.5 CMOS Input and Output (I/O) Circuits 220
4.5.1 CMOS Input Circuits 221
4.5.2 CMOS Output Buffers (Drivers) 222
4.6 The Layout Process 223
4.6.1 Introduction 223
4.6.2 Layout Design Rules 224
4.6.3 Stick Diagram 227
4.6.4 Example of the Layout Procedure 230
4.6.5 Guidelines for Layout Design 232
4.7 Libraries and Library Design 234
4.8 FinFET Layout 237
4.9 Conclusions 239
4.10 Exercises 240
References and Further Reading 243

5 Special Circuits, Devices and Technologies 245
5.1 Introduction 245
5.2 CCD and CMOS Image Sensors 245
5.2.1 Introduction 245
5.2.2 Basic CCD Operation 246
5.2.3 CMOS Image Sensors 250
5.2.4 Conclusions on Image Sensors 252
5.3 BICMOS Circuits 253
5.3.1 Introduction 253
5.3.2 BICMOS Technology 253
5.3.3 BICMOS Characteristics 255

5.3.4 BICMOS Circuit Performance ... 256
5.3.5 Future Expectations and Market Trends ... 258
5.4 Power MOSFETs ... 258
5.4.1 Introduction ... 258
5.4.2 Technology and Operation ... 259
5.4.3 Applications ... 261
5.5 Bipolar-CMOS-DMOS (BCD) Processes ... 262
5.6 Conclusions ... 266
5.7 Exercises ... 266
References ... 267

6 Memories ... 269
6.1 Introduction ... 269
6.2 Serial Memories ... 272
6.3 Content-Addressable Memories (CAM) ... 273
6.4 Random-Access Memories (RAM) ... 273
6.4.1 Introduction ... 273
6.4.2 Static RAMs (SRAM) ... 273
6.4.3 Dynamic RAMs (DRAM) ... 285
6.4.4 High-Performance DRAMs ... 297
6.4.5 Single- and Dual Port Memories ... 304
6.4.6 Error Sensitivity ... 305
6.5 Non-volatile Memories ... 305
6.5.1 Introduction ... 305
6.5.2 Read-Only Memories (ROM) ... 305
6.5.3 Programmable Read-Only Memories ... 309
6.5.4 EEPROMs and Flash Memories ... 311
6.5.5 Non-volatile RAM (NVRAM) ... 329
6.5.6 BRAM (Battery RAM) ... 329
6.5.7 FRAM, MRAM, PRAM (PCM) and RRAM ... 330
6.6 Embedded Memories ... 334
6.7 In-memory Computing ... 338
6.8 Redundancy in Memories ... 339
6.9 Classification of the Various Memories ... 341
6.10 Conclusions ... 341
6.11 Exercises ... 343
References ... 344
Further Reading ... 348

7 Very-Large-Scale Integration (VLSI) and ASICs ... 349
7.1 Introduction ... 349
7.2 Digital ICs ... 351
7.3 Abstraction Levels for VLSI ... 355
7.3.1 Introduction ... 355
7.3.2 System Level ... 359
7.3.3 Functional Level ... 360

7.3.4 RTL Level ... 362
7.3.5 Logic-Gate Level ... 364
7.3.6 Transistor Level ... 365
7.3.7 Layout Level ... 366
7.3.8 Conclusions ... 367
7.4 Digital VLSI Design ... 369
7.4.1 Introduction ... 369
7.4.2 The Design Trajectory and Flow ... 369
7.4.3 Example of Synthesis from VHDL Description to Layout ... 374
7.4.4 Floor-Planning ... 379
7.5 The Use of ASICs ... 381
7.6 Silicon Realisation of VLSI and ASICs ... 382
7.6.1 Introduction ... 382
7.6.2 Handcrafted Layout Implementation ... 384
7.6.3 Bit-Slice Layout Implementation ... 384
7.6.4 ROM, PAL and PLA Layout Implementations ... 386
7.6.5 Cell-Based Layout Implementation ... 389
7.6.6 (Mask-Programmable) Gate Array Layout Implementation ... 390
7.6.7 Programmable Logic Devices (PLDs) ... 394
7.6.8 Embedded Arrays, Structured ASICs and Platform ASICs ... 400
7.6.9 Hierarchical Design Approach ... 403
7.6.10 The Choice of a Layout Implementation Form ... 405
7.7 Conclusions ... 407
7.8 Exercises ... 408
References ... 408

8 Less Power: A Hot Topic in IC Design ... 411
8.1 Introduction ... 411
8.2 Battery Technology Summary ... 412
8.3 Sources of CMOS Power Consumption ... 414
8.4 Technology Options for Low Power ... 415
8.4.1 Reduction of P_{leak} by Technological Measures ... 415
8.4.2 Reduction of P_{dyn} by Technology Measures ... 420
8.4.3 Reduction of P_{dyn} by Reduced Voltage Processes ... 421
8.5 Design Options for Power Reduction ... 424
8.5.1 Reduction of P_{short} by Design Measures ... 424
8.5.2 Reduction/Elimination of P_{stat} by Design Measures ... 425
8.5.3 Reduction of P_{dyn} by Design Measures ... 426
8.6 Computing Power Versus Chip Power: A Scaling Perspective ... 455
8.7 Conclusions ... 457
8.8 Exercises ... 458
References ... 459

9 Robustness of Nanometer CMOS Designs: Signal Integrity, Variability and Reliability 461
9.1 Introduction 461
9.2 Clock Generation, Clock Distribution and Critical Timing 462
9.2.1 Introduction 462
9.2.2 Clock Distribution and Critical Timing Issues 463
9.2.3 Clock Generation and Synchronisation in Different (Clock) Domains on a Chip 470
9.3 Signal Integrity 474
9.3.1 Cross Talk and Signal Propagation 474
9.3.2 Power Integrity, Supply and Ground Bounce 480
9.3.3 Substrate Bounce 485
9.3.4 EMC 487
9.3.5 Soft Errors 488
9.3.6 Signal Integrity Summary and Trends 492
9.4 Variability 494
9.4.1 Spatial vs. Time-Based Variations 494
9.4.2 Global vs. Local Variations 494
9.4.3 Transistor Matching 499
9.4.4 From Deterministic to Probabilistic Design 501
9.4.5 Can the Variability Problem Be Solved? 502
9.5 Reliability 504
9.5.1 Punch-Through 504
9.5.2 Electromigration 505
9.5.3 Hot-Carrier Injection (HCI) 507
9.5.4 Bias Temperature Instability (BTI, NBTI and PBTI) 511
9.5.5 Latch-Up 513
9.5.6 Electrostatic Discharge (ESD) 516
9.5.7 The Use of Guard Rings 522
9.5.8 Charge Injection During the Fabrication Process 523
9.5.9 Reliability Summary and Trends 523
9.6 Design Organisation 524
9.7 Conclusions 525
9.8 Exercises 526
References 527

10 Testing, Yield, Packaging, Debug and Failure Analysis 531
10.1 Introduction 531
10.2 Testing 532
10.2.1 Basic IC Tests 535
10.2.2 Design for Testability 547
10.3 Yield 548
10.3.1 A Simple Yield Model and Yield Control 551
10.3.2 Design for Manufacturability 557

10.3.3 Towards Chiplet Design to Reduce Manufacturing Costs 560
10.3.4 Redundancy in Logic 563
10.4 Packaging 563
10.4.1 Introduction 563
10.4.2 Package Categories 564
10.4.3 Packaging Process Flow 567
10.4.4 Electrical Aspects of Packaging 574
10.4.5 Thermal Aspects of Packaging 575
10.4.6 Reliability Aspects of Packaging 577
10.4.7 Future Trends in Packaging Technology 578
10.4.8 System on a Chip (SoC) Versus System in a Package (SiP) 580
10.4.9 Quality and Reliability of Packaged Dies 583
10.4.10 Conclusions 588
10.5 Potential First Silicon Problems 589
10.5.1 Problems with Testing 589
10.5.2 Problems Caused by Marginal or Out-of-Specification Processing 591
10.5.3 Problems Caused by Marginal Design 593
10.6 First-Silicon Debug and Failure Analysis 593
10.6.1 Introduction 593
10.6.2 I_{ddq} and ΔI_{ddq} Testing 594
10.6.3 Traditional Debug, Diagnosis and Failure Analysis (FA) Techniques 594
10.6.4 More Recent Debug and Failure Analysis Techniques 601
10.6.5 Observing the Failure 611
10.6.6 Circuit Editing Techniques 613
10.6.7 Design for Debug and Design for Failure Analysis 616
10.7 Conclusions 617
10.8 Exercises 618
References 619

11 Effects of Scaling on MOS IC Design and Consequences for the Roadmap 621
11.1 Introduction 621
11.2 Transistor Scaling Effects 623
11.3 Interconnection Scaling Effects 624
11.4 Scaling Consequences for Overall Chip Performance and Robustness 627
11.5 Potential Limitations of the Pace of Scaling 633
11.6 Conclusions 643
11.7 Exercises 645
References 645

Index 647

About the Author

Harry Veendrick graduated from the Technical University Eindhoven, the Netherlands, in 1977. In the same year he joined Philips Research Laboratories, also in Eindhoven, where he has been involved in the design of memories, gate arrays and complex digital video-signal processors. He holds 30 US and many more European patents in the area of CMOS circuit design. He is the (co-)author of more than 30 publications on robust, high-performance and low-power CMOS IC design. In this respect, he has contributed to many conferences and workshops, as reviewer, speaker, invited speaker, panelist, organizer, guest editor and programme committee member. In addition, he is the author of MOS ICs (VCH 1992), Deep-Submicron CMOS ICs, from Basics to ASICs (Kluwer Academic Publishers: 1st edition 1998, 2nd edition 2000) and Nanometer CMOS ICs, from Basics to ASICs (Springer: 1st edition June 2008, 2nd edition 2017 and 3rd edition 2024). He is a co-author of Low-Power Electronics Design (CRCPress, 2004). His principle research interests included the design of low-power and high-speed complex digital CMOS circuits, with an emphasis on nanometer-scale physical effects and scaling aspects. From 2002 to 2009 he has been project leader of the Deep-Submicron/Nanometer CMOS Electrical Design Cluster within the Mixed-Signal Circuits and Systems research group at NXP Semiconductors, which is the 2006 spin-off of the former Philips Semiconductors business. Complementary to this is his interest in IC technology, which allowed him to act as an interface between digital IC design and IC process technology. From 1980 onwards, next to his research activities, he has been actively involved in the training of more than 7000 semiconductor design, process, test, product and CAD tools engineers. In 2002, he received the PhD degree in electronic engineering from the Technical University of Eindhoven, the Netherlands. He was a research fellow at Philips Research Laboratories and NXP Research and has been a Visiting Professor to the Department of Electronic and Electrical Engineering of the University of Strathclyde, Glasgow, Scotland, UK. In May 2009, he left NXP Research and started his own training activity, teaching 1-day, 3-day and 5-day courses for different target audiences (www.bitsonchips.com).

Basic Principles

1

1.1 Introduction

The majority of current VLSI (very-large-scale integration) circuits are manufactured in CMOS technologies. Familiar examples are DRAM memories (1 Gb, 4 Gb and 16 Gb), NAND flash memories (128 Gb, 512 Gb and 1 Tb), mobile phone chips, microprocessors, graphic processors and signal processors. A good fundamental treatment of basic MOS devices is therefore essential for an understanding of the design and manufacture of modern *VLSI* circuits. This chapter describes the operation and characteristics of MOS devices. The material requirements for their realisation are discussed and equations that predict their behaviour are derived.

The acronym MOS represents the *M*etal, *O*xide and *S*emiconductor materials used to realise early versions of the MOS transistor. The fundamental basis for the operation of MOS transistors is the field-effect principle. This principle is quite old, with related publications first appearing in the 1930s. These include a patent application filed by J.E. Lilienfeld in Canada and the USA in 1930 and one filed by O. Heil, independently of Lilienfeld, in England in 1935. At that time, however, insufficient knowledge of material properties resulted in devices which were unfit for use. The rapid development of electronic valves probably also hindered the development of the MOS transistor by largely fulfilling the transistor's envisaged role.

1.2 The Field-Effect Principle

The field-effect principle is explained with the aid of Fig. 1.1. This figure shows a rectangular conductor, called a channel, with length L, width W and thickness t_{cond}. The free electrons present in the channel are the mobile charge carriers. There are n electrons per m^3 and the charge q per electron equals -1.602×10^{-19} C(coulomb). The application of a horizontal electric field of magnitude E to the channel causes the electrons to acquire an average velocity $v = -\mu_{\mathrm{n}} \cdot E$. The electron mobility μ_{n}

H. Veendrick, *Nanometer CMOS ICs*,
https://doi.org/10.1007/978-3-031-64249-4_1

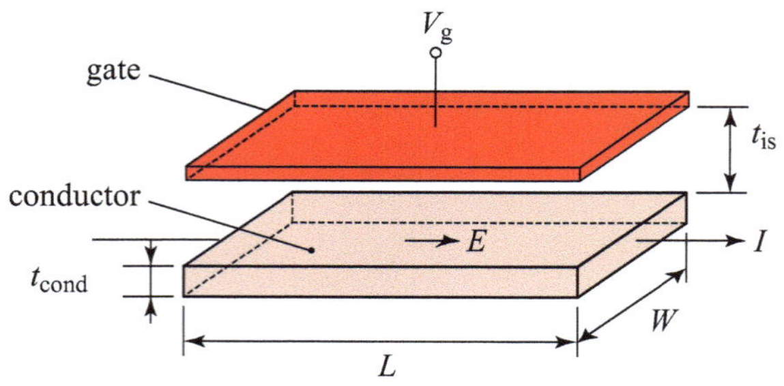

Fig. 1.1 The field-effect principle

is positive. The direction of v therefore opposes the direction of E. The resulting current density j is the product of the average electron velocity and the mobile charge density ρ:

$$j = \rho \cdot v = -n \cdot q \cdot \mu_n \cdot E \tag{1.1}$$

A gate electrode situated above the channel is separated from it by an insulator of thickness t_{is}. A change in the gate voltage V_g influences the charge density ρ in the channel. The current density j is therefore determined by V_g.

Example. Suppose the insulator is silicon dioxide (SiO_2) with a thickness of 2 nm ($t_{is} = 2 \times 10^{-9}$m). The gate capacitance will then be about 17 mF/m^2. The total gate capacitance C_g is therefore expressed as follows:

$$C_g = 17 \times 10^{-3} \cdot W \cdot L \text{ [F]}$$

A change in gate charge $\Delta Q_g = C_g \cdot \Delta V_g$ causes the following change in channel charge:

$$+ C_g \cdot \Delta V_g = 17 \times 10^{-3} \cdot W \cdot L \cdot \Delta V_g = W \cdot L \cdot t_{cond} \cdot \Delta\rho$$

Thus

$$\Delta\rho = \frac{17 \times 10^{-3} \cdot \Delta V_g}{t_{cond}} \text{ C/m}^3$$

and

$$\mid \Delta n \mid = \mid \frac{\Delta\rho}{q} \mid = \frac{10.6 \times 10^{16} \cdot \Delta V_g}{t_{cond}} \text{ electrons/m}^3$$

If we assume that a 0.5 V change in gate voltage is to cause a 10,000 times increase in current density j, which value is representative for a MOS transistor today, then the following must apply:

$$\frac{\Delta j}{j} = \frac{\Delta \rho}{\rho} = \frac{\Delta n}{n} = \frac{10.6 \times 10^{16} \cdot 0.5}{t_{\text{cond}} \cdot n} = 10{,}000 \Rightarrow t_{\text{cond}} = \frac{5.3 \times 10^{12}}{n}$$

Examination of two materials reveals the implications of this expression for t_{cond}:

Case a The channel material is copper.
This has $n \approx 10^{28}$ electrons/m^3 and hence $t_{\text{cond}} \approx 5.3 \times 10^{-16}$ m.
The required channel thickness is thus less than the size of one atom ($\approx 3 \times 10^{-10}$ m). This is impossible to realise and its excessive number of free carriers renders copper unsuitable as channel material.

Case b The channel material is 5Ωcm n-type silicon.
This has $n \approx 10^{21}$ electrons/m^3 and hence $t_{\text{cond}} \approx 5.3$ nm.
Today, the channel dope level is often in the range of 10^{22} to 10^{23}, leading to a t_{cond} much less than 1 nm.

From the above example, it is clear that field-effect devices can only be realised with semiconductor materials. Aware of this fact, Lilienfeld used copper sulphide as a semiconductor in 1930. Germanium was used during the early 1950s. Until 1960, however, usable MOS transistors could not be manufactured. Unlike the transistor channel, which comprised a manufactured thin layer, the channel in these *inversion-layer transistors* is a thin conductive layer, which is realised electrically. The breakthrough for the fast development of MOS transistors came with advances in planar silicon technology and the accompanying research into the physical phenomena in the semiconductor surface.

Generally, circuits are integrated in silicon because widely accepted military specifications can be met with this material. These specifications require products to function correctly at a maximum operating temperature of 125 °C. The maximum operating temperature of germanium is only 70 °C, while that of silicon is 150 °C. This does not mean that silicon ICs cannot operate above 125 °C, but it means that the transistor leakage currents become extremely high and, as such, chip designers need to incorporate that in the respective designs. A comparison of a few other germanium (Ge) and silicon (Si) material constants is presented below:

Material constant	Germanium	Silicon
Melting point [°C]	937	1415
Breakdown field [V/μm]	8	30
Relative expansion coeff. [°C]$^{-1}$	5.8×10^{-6}	2.5×10^{-6}
ϵ_r	16.8	11.7
Max. operating temp. [°C]	70	150

From the above, it is clear that almost all MOS ICs are fabricated on silicon substrates (wafers). Like all semiconductor materials, silicon substrates come in two versions: undoped (intrinsic) and doped silicon. At room temperature, silicon has an intrinsic carrier concentration of $n_{\mathrm{i}} = 1.45 \times 10^{10}$ [cm^{-3}]. Thermal energy creates electron-hole pairs in this intrinsic silicon. With n being the number of electrons and p the number of holes, it always holds in undoped and doped silicon that

$$p \cdot n = n_{\mathrm{i}}^2$$

This is called the *mass action law*. As will be explained in Sect. 1.3, we can dope silicon with acceptor atoms (e.g. boron) with a concentration N_{A} [cm^{-3}] or with donor atoms (e.g. phosphorous or arsenic) with a concentration N_{D} [cm^{-3}]. For n-type silicon, with $n_{\mathrm{n}} = N_{\mathrm{D}}$ representing the electron density in n-type silicon and p_{n} representing the hole density in n-type silicon, it holds that

$$N_{\mathrm{D}} \cdot p_{\mathrm{n}} = n_{\mathrm{i}}^2$$

Complementary, for p-type silicon it holds that

$$N_{\mathrm{A}} \cdot n_{\mathrm{p}} = n_{\mathrm{i}}^2$$

with $p_{\mathrm{p}} = N_{\mathrm{A}}$ representing the hole density in p-type silicon and n_{p} representing the electron density in p-type silicon. In fact, by doping the silicon, the number of free carriers and thus the *conductivity* σ is changed:

$$\sigma = q(\mu_{\mathrm{n}} n + \mu_{\mathrm{p}} p)$$

For n-type silicon, $N_{\mathrm{D}} = n_{\mathrm{n}} \gg n_{\mathrm{p}}$, and thus, $\sigma = q\mu_{\mathrm{n}} n_{\mathrm{n}}$. Complementary, for p-type silicon, $N_{\mathrm{A}} = n_{\mathrm{p}} \gg n_{\mathrm{a}}$, and thus, $\sigma = q\mu_{\mathrm{p}} n_{\mathrm{p}}$. Since the mobility of electrons (μ_{n}) is larger than the mobility of holes (μ_{p}), nMOS transistors show larger conductivities than pMOS transistors (see also Sect. 2.3).

1.3 The Inversion-Layer MOS Transistor

A schematic drawing of the inversion-layer nMOS transistor, or simply 'nMOSt', is shown in Fig. 1.2, which is used to explain its structure and operation. The two n^+ areas in the p-type substrate are called the *source* and *drain*. The *gate* electrode is situated above the p area between them. The '+' and '−', respectively, used in the n^+ source/drain and p^- substrate areas, refer to the level of doping concentration, e.g. an n^+ region refers to a high-doped n region; a p^- region refers to a low-doped p region. This gate electrode is either a metal plate, e.g. aluminium or molybdenum, a heavily doped and thus low-ohmic polycrystalline silicon layer, or a combination of both. Normally, the source and drain areas are also heavily doped to minimise series resistance. The resistance R of a 10 µm long and 2 µm wide track is $\frac{10}{2} \cdot R_{\square}$, where $R_{\square}$ is the sheet resistance of the track material. The sheet resistance of the source and drain areas usually ranges from 3 to 100 $\Omega/\square$ with doping levels upto $5 \cdot 10^{19}$ to $2 \cdot 10^{20}$ atoms per cm^3. The dope concentration in the p-type substrate is approximately 10^{14}–10^{16} atoms per cm^3, while the channel dope (by threshold

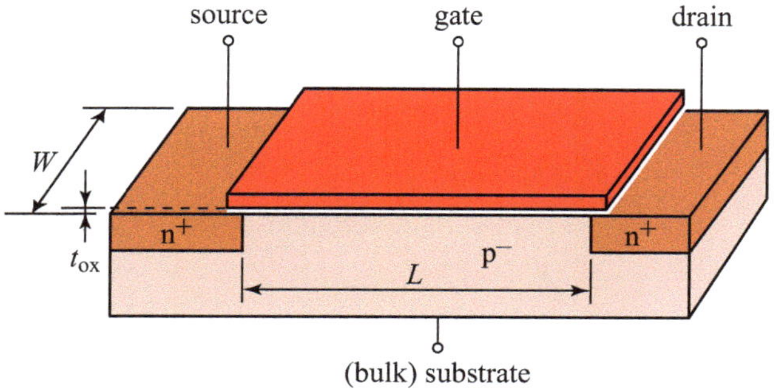

Fig. 1.2 Cross section of an inversion-layer nMOS transistor

Fig. 1.3 The structure of a free silicon atom

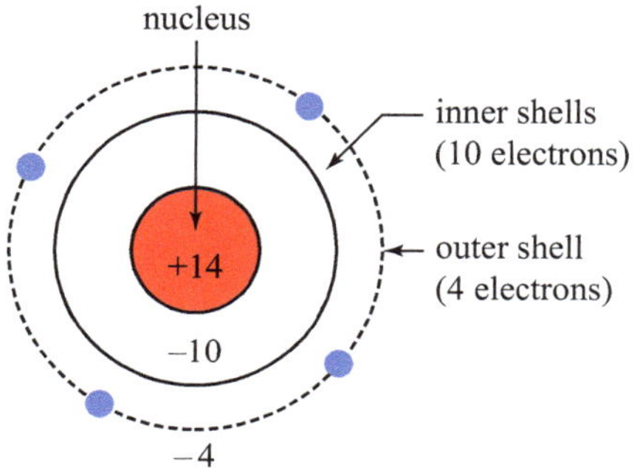

adjustment implantation, etc.) is between 10^{17}–10^{18} atoms per cm^3. A p-channel transistor differs from the above n-channel type in that it contains a p^+ source and drain in an n^--type substrate.

Characteristic parameters of a MOS transistor are indicated in Fig. 1.2. These include the width W and length L of the channel and the thickness t_{ox} of the insulating oxide which separates the gate and channel. In current CMOS VLSI circuits, the minimum values of L (real channel length) range from 12 nm to 250 nm, depending on the process node, while the minimum W is often close to 1.5 times the minimum in L in that node. t_{ox} ranges from 1 nm to 5 nm. Continuous development will reduce these values in the future. The depth of the source and drain junctions are in the order of 20% to 30% of the transistor channel length [1].

The energy band theory and its application to the MOS transistor are briefly summarised below. An understanding of this summary is a prerequisite for a detailed discussion of the behaviour of the MOS transistor.

The structure of a free silicon atom is shown in Fig. 1.3. This atom comprises a nucleus, inner shells and an outer shell. The nucleus contains 14 protons and 14 neutrons, while the shells contain 14 electrons. Ten of the electrons are in the inner shells and four are in the outer shell. The positive charge of the protons and the

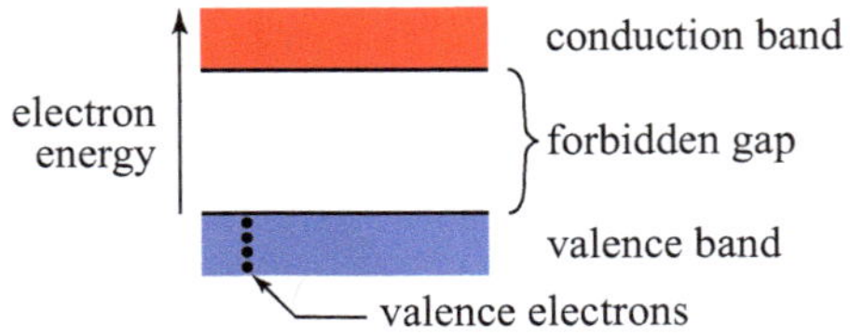

Fig. 1.4 Schematic representation of electron energy bands in a typical solid material

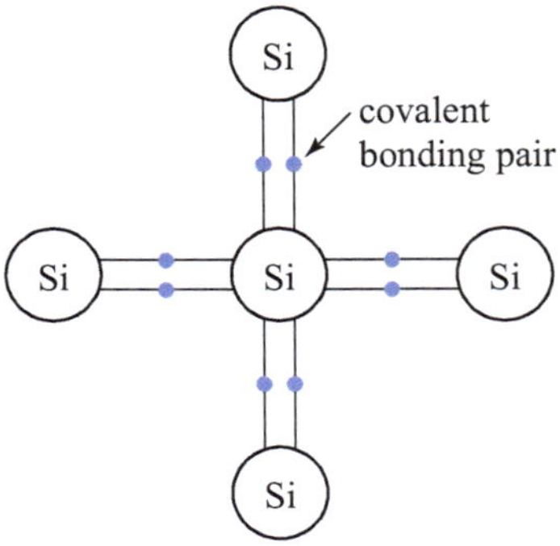

Fig. 1.5 Silicon crystal

negative charge of the electrons compensate each other to produce an atom with a net neutral charge.

The electrons in an atom may possess certain energy levels. These energy levels are grouped into energy bands, which are separated by energy gaps. An energy gap represents impossible levels of electron energy. The energy bands that apply to the electrons in an atom's outer shell are *valence* and *conduction* bands. Figure 1.4 shows these bands and the energy gap for a typical solid material. The valence electrons determine the physical and chemical properties of a material.

The four electrons in the outer shell of a silicon atom are in the material's valence band. Figure 1.5 shows the bonds that these electrons form with neighbouring atoms to yield a silicon crystal.

The required energy to break a covalent bond is called the *band gap*. A broken bond results in a free electron that can freely move through the crystal and a hole. When an electron from a neighbouring bond fills the hole, the hole has moved to that neighbouring bond. In a conductor, the electrons can easily go from the valence band to the conduction band. Therefore, the conduction and valence bands in a conductor partly overlap, as shown in Fig. 1.6a. In an insulator, however, none of the valence electrons can reach the conduction band. Figure 1.6b shows the large band gap generally associated with insulators. A semiconductor lies somewhere between a conductor and an insulator. The associated small band gap is shown in Fig. 1.6c. Valence electrons may acquire sufficient thermal energy to reach the conduction band and therefore leave an equal number of positively-charged ions, or 'holes', in the valence band. This produces a limited conduction mechanism in semiconductors.

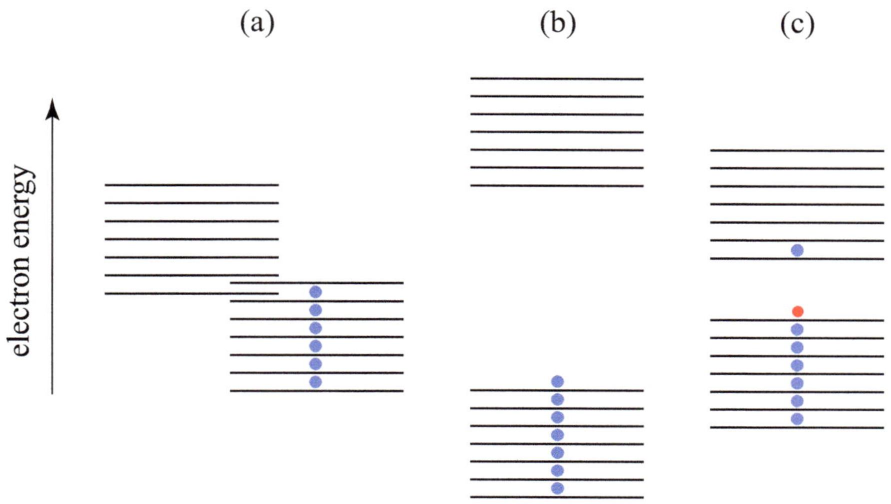

Fig. 1.6 Energy bands of a conductor, an insulator and an intrinsic semiconductor

Semiconductor materials are located in group IV of this system. The introduction of an element from group III or V in a semiconductor crystal produces an 'acceptor' or a 'donor' atom. This *semiconductor doping process* dramatically changes the crystal properties. The following table shows the location of semiconductor materials in the *periodic system of elements*.

Group		
III (Acceptors)	IV	V (Donors)
Boron	Carbon	Nitrogen
Aluminium	Silicon	Phosphorous
Gallium	Germanium	Arsenic
Indium	Stannic (tin)	Stibnite

The presence of a group III atom in a silicon crystal lattice is considered first. The situation for boron (B) is illustrated in Fig. 1.7a. Boron has one electron less than silicon and cannot therefore provide an electron required for a bond with one of its four neighbouring silicon atoms. The hole in the resulting p-type semiconductor is a willing '*acceptor*' for an electron from an alternative source. This hole can be removed relatively easily with the ionisation energy of approximately 0.045 eV shown in the energy band diagram of Fig. 1.7a.

Similar reasoning applies when a group V atom, such as phosphorus (P), is present in the silicon lattice. This situation is illustrated in Fig. 1.7c. The extra electron in the phosphorus atom cannot be accommodated in the regular bonding structure of the silicon lattice. It is therefore easy to remove this '*donor*' electron in

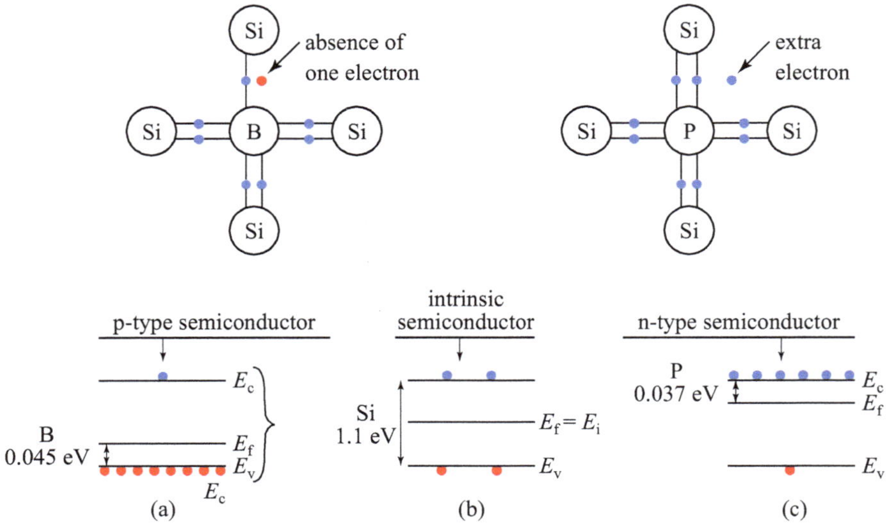

Fig. 1.7 Energy band diagrams for p-type, intrinsic, and n-type semiconductor materials

the resulting n-type semiconductor. The mere 0.037 eV ionisation energy required is much lower than the 1.11 eV band gap energy of silicon. Figure 1.7b shows the energy band diagram of an intrinsic silicon lattice, which contains no donor or acceptor ‘impurity’ atoms.

The energy level indicated by E_f in Fig. 1.7 is called the *Fermi level*. An electron with this energy has an equal probability of location in the valence band and the conduction band. This probability is exactly 0.5.

The Fermi level of an intrinsic semiconductor is often referred to as the intrinsic Fermi level E_i. The Fermi level E_f in a p-type semiconductor is situated near the valence band E_v, while it is close to the conduction band E_c in an n-type semiconductor. The above theory concerning the different types of semiconductors and their respective energy band diagrams will now be used to explain the behaviour of the MOS transistor. This explanation is preceded by a description of the structure and operation of the MOS capacitor.

1.3.1 The Metal-Oxide-Semiconductor (MOS) Capacitor

Figure 1.8 shows a cross section of a basic MOS capacitor. This structure is identical to a MOS transistor except that the source and drain diffusion regions are omitted. The p-type substrate is made with an acceptor dope material, e.g. boron. The substrate is assumed to behave as a normal conductor and contains many free holes. The situation which occurs when the p-type substrate is grounded and a negative voltage is applied to the gate electrode is shown in Fig. 1.8a. The negative

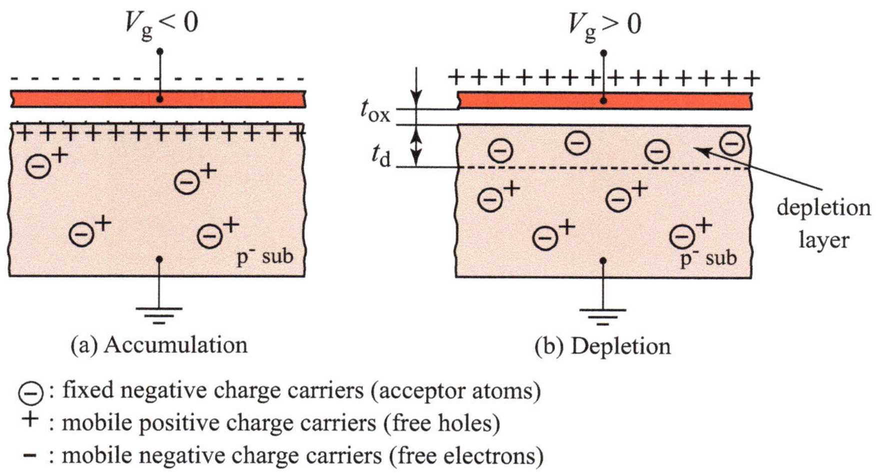

Fig. 1.8 Cross section of a MOS structure without source and drain areas. There is a capacitance between the gate and substrate.

charge on the gate is compensated by an equal but positive charge in the substrate. This is accomplished by positively charged holes which accumulate at the Si-SiO_2 interface. These holes are the majority charge carriers in the substrate. This *'accumulation'* process continues until the positive charge at the substrate surface equals the negative charge on the gate electrode. Extra holes are supplied through the ground contact to the substrate. The resulting accumulation capacitor can be viewed as an ideal parallel plate capacitor.

A different situation occurs when the potential on the gate electrode is made positive with respect to the grounded substrate. This situation is shown in the cross section of Fig. 1.8b. The positive charge which is present on the gate must be counterbalanced by a negative charge at the Si-SiO_2 interface in the substrate. Free positively charged holes are pushed away from the substrate surface to yield a negatively charged depletion layer. This *'depletion'* process stops when the negative charge of the depletion layer equals the positive charge on the gate electrode. Clearly, the thickness t_d of the depletion layer in the equilibrium situation is proportional to the gate voltage. It is important to realise that a depletion layer only contains a fixed charge, i.e. ions fixed in the solid state lattice, and no mobile charge carriers.

Various energy band diagrams are used to explain the behaviour of the inversion layer MOS transistor. To provide a better understanding of these diagrams, Poisson's law is first applied to the different regions of the MOS capacitor. These regions include the gate, the SiO_2 insulator, the depletion layer in silicon and the p-type silicon substrate. Poisson's law is used to investigate the charge distribution $Q(z)$, the electric field $E(z)$ and the electric potential $\phi(z)$ in these regions as a function of the distance z from the Si-SiO_2 interface.

In its one-dimensional form, Poisson's law is formulated as follows:

$$\frac{d^2\phi(z)}{dz^2} = -\frac{\rho}{\epsilon} \tag{1.2}$$

$$\begin{aligned} \text{where } \phi(z) &= \text{electrical potential at position z;} \\ z &= \text{distance from the} \\ &\quad Si - SiO_2 \text{ interface;} \\ \rho &= \text{space charge;} \\ \epsilon &= \text{dielectric constant.} \end{aligned}$$

The situation in which no space charge is present is considered first. This is almost true in the SiO_2 insulator, in which case $\rho = 0$. Integration of Formula (1.2) once gives the electric field:

$$E(z) = C_1, C_1 = \text{integration constant.}$$

Integration of Formula (1.2) twice gives the electric potential in SiO_2:

$$\phi(z) = C_1 \cdot z + C_2$$

The electric field in the insulator is thus constant, and the electric potential is a linear function of the distance z from the Si-SiO_2 interface.
Next, the situation in which a constant space charge is present is considered. This is assumed to be true in the depletion layer, whose width is W_D. In this case

$$\begin{aligned} \rho &= -q \cdot N_A \\ \text{where } q &= \text{the charge of an electron} \\ \text{and } N_A &= \text{the total number of fixed ions} \\ &\quad \text{in the depletion layer of thickness } t_d. \end{aligned}$$

Integrating Formula (1.2) once gives the electric field:

$$E(z) = \frac{q \cdot N_A}{\epsilon} \cdot z + C_1$$

Integrating Formula (1.2) twice gives the electric potential in the depletion layer:

$$\phi(z) = \frac{q \cdot N_A}{2\epsilon} \cdot z^2 + C_1 \cdot z + C_2$$

Therefore, the electric field in a depletion layer with constant space charge is a linear function of z, while the electric potential is a square function of z. The space charge in a depletion layer is only constant when the dope of the substrate has a constant value at all distances z from the Si-SiO_2 interface. In practice, the space charge profile is related to the dope profile which exists in the substrate.

The gate and the substrate region outside the depletion layer are assumed to behave as ideal conductors. The electric potentials in these regions are therefore constant and their electric fields are zero.

The above results of the application of Poisson's law to the MOS capacitor are illustrated in Fig. 1.9. Discontinuities in the diagrams are caused by differences between the dielectric constant of silicon and silicon dioxide. The electric charge, the electric field and potential are zero in the grounded substrate outside the depletion region. The observation that the electric potential is a square function of z in the depletion layer is particularly important.

1.3.2 The Inversion-Layer MOS Transistor

Figure 1.10 shows a cross section of an nMOS transistor with 0 V on all of its terminals. The figure also contains the associated energy band diagram.
It is assumed that the presence of the gate does *not* affect the distribution of holes and electrons in the semiconductor. With the exception of the depletion areas around the n^+ areas, the entire p-substrate is assumed to be homogeneous and devoid of an electric field ($E = 0$). There is no charge on the gate and no surface charge in the silicon. Generally, the electron energies at the Fermi levels of the different materials in the structure will differ. Their work functions (i.e. the energy required to remove an electron from the Fermi level to vacuum) will also differ. When the voltage between the gate and source is zero ($V_{gs} = 0$) and the metal gate is short circuited to the semiconductor, electrons will flow from the metal to the semiconductor or vice versa until a voltage potential is built up between the two materials. This voltage potential counterbalances the difference in their work functions. The Fermi levels in the metal and the semiconductor are then aligned.

Therefore, there will be an electrostatic potential difference between the gate and substrate which will cause the energy bands to bend. The *'flat-band condition'* exists when there is no band bending at the metal-semiconductor interface. The *'flat-band voltage'* V_{fb} is the gate voltage required to produce the flat-band condition. It is the difference between the work functions of the metal (ϕ_M) and the semiconductor (ϕ_S), i.e. $V_{fb} = \phi_{MS} = \phi_M - \phi_S$. Since equilibrium holds, the Fermi level in the semiconductor remains constant regardless of the value of the gate voltage.

A negative charge is induced in the semiconductor surface when a small positive voltage is applied to the gate, while the source, drain and substrate are at 0 V, see also Fig. 1.11. The negative charge is caused by holes being pushed away from the insulator interface. The negatively charged acceptor atoms that are left behind form a negative space charge, i.e. a depletion layer. The thickness of this depletion layer

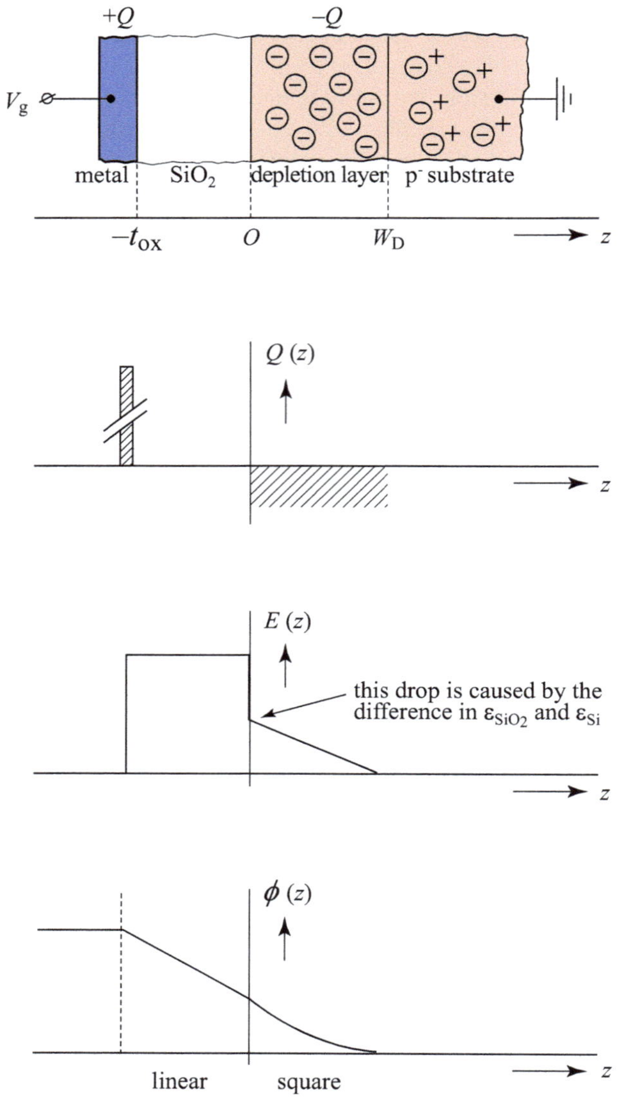

Fig. 1.9 The sections of a MOS capacitor and the associated charge distribution $Q(z)$, electric field $E(z)$ and electric potential $\phi(z)$

is determined by the potential V_c at the silicon surface. The gate voltage V_{gs} now consists of two parts:

a. The voltage across the oxide $V_g - V_c$
b. The voltage across the depletion layer V_c

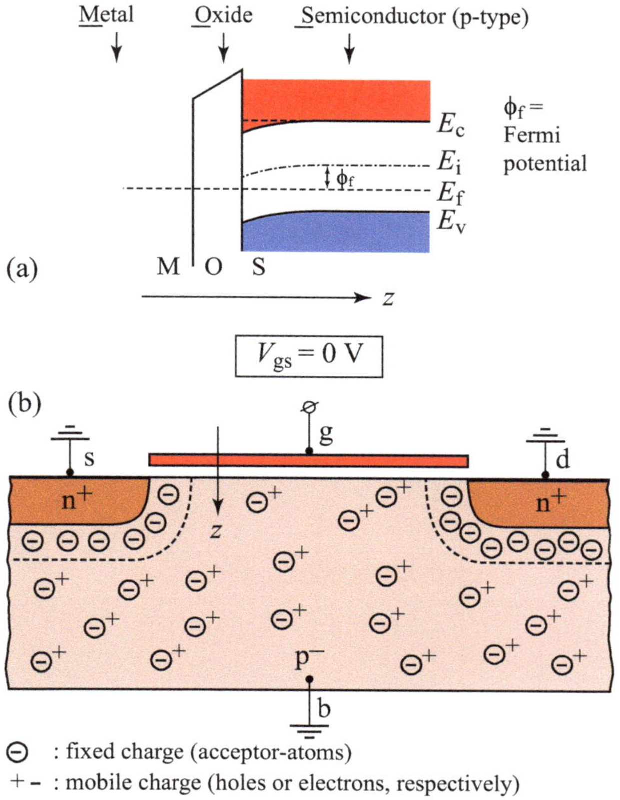

Fig. 1.10 Cross section of a MOS transistor with $V_{gs} = V_{ds} = V_{sb} = 0$ V and the associated energy band diagram

The capacitance between the gate and substrate now consists of the series connection of the oxide capacitance C_{ox} and the depletion-layer capacitance C_d.

The term V_T in Fig. 1.11 represents the *threshold voltage* of the transistor. This is the gate voltage at which the band bending at the silicon surface is exactly $2\phi_f$. At this band bending, the electron concentration at the semiconductor surface becomes equal to the hole concentration in the bulk. This situation is called *(strong) inversion*, and the layer of free electrons created at the surface is called an inversion layer. For the present, V_T is assumed to be positive for an inversion-layer nMOS transistor. This assumption is confirmed later in the text.

If the gate voltage is further increased ($V_{gs} > V_T$), then the band bending at the silicon surface will be larger than $2\phi_f$. This situation is illustrated in Fig. 1.12. A comparison of Figs. 1.12 and 1.7c reveals that the energy band at the silicon surface corresponds to an n-type semiconductor.

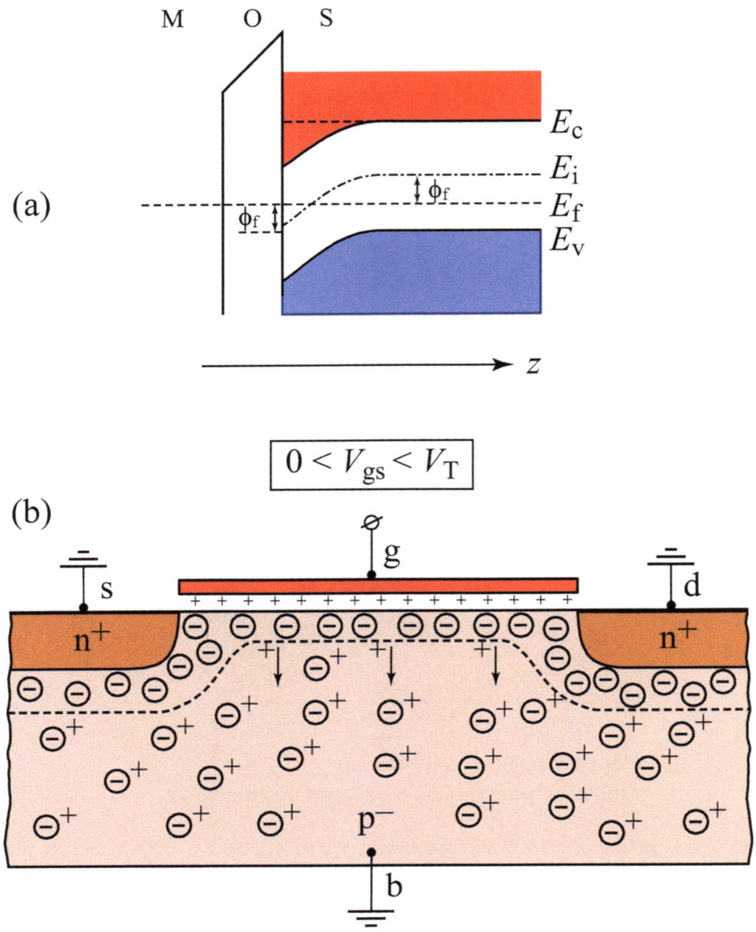

Fig. 1.11 Cross section of a MOS transistor with $0 < V_{gs} < V_T$ and $V_{ds} = V_{sb} = 0$ V and its corresponding energy band diagram

Deep in the substrate, however, the energy band corresponds to a p-type semiconductor. A very narrow n-type layer has therefore been created at the surface of a p-type silicon substrate. In addition to the negative acceptor atoms already present, this *inversion layer* contains electrons which act as mobile negative charge carriers. Conduction in the n-type inversion layer is mainly performed by these electrons, which are minority carriers in the p-type substrate. The inversion layer forms a conducting *channel* between the transistor's source and drain. No current flows in this channel if there is no voltage difference between the drain and source terminals, i.e. $I_{ds} = 0\,\text{A}$ if $V_{ds} = 0\,\text{V}$. The number of electrons in the channel can be controlled by the gate-source voltage V_{gs}.

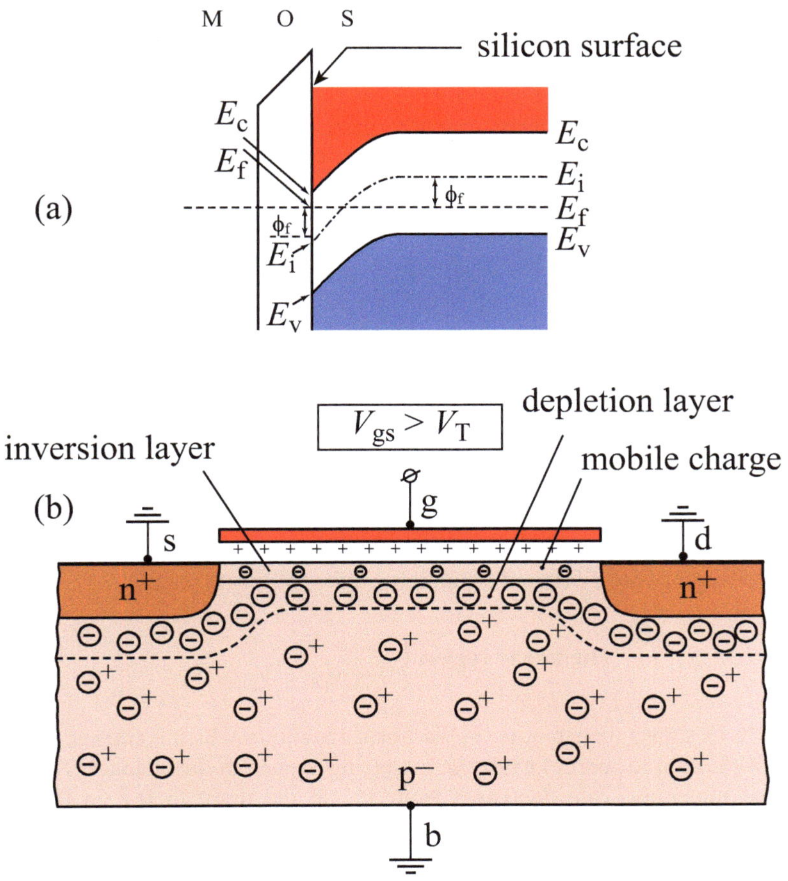

Fig. 1.12 Cross section of a MOS transistor with $V_{gs} > V_T$ ($V_T > 0$) and $V_{ds} = V_{sb} = 0$ V and its corresponding energy band diagram

Assuming that $V_{gs} > V_T$, the effects of increasing V_{ds} from 0 V are divided into the following regions:

1. $0 < V_{ds} < V_{gs} - V_T$.
   This is called the *linear* or *triode* region of the MOS transistor's operating characteristic.
2. $V_{ds} = V_{gs} - V_T$.
   At this point, a transition takes place from the linear to the so-called saturation region.
3. $V_{ds} > V_{gs} - V_T$.
   This is the *saturation* region of the MOS transistor's operating characteristic.

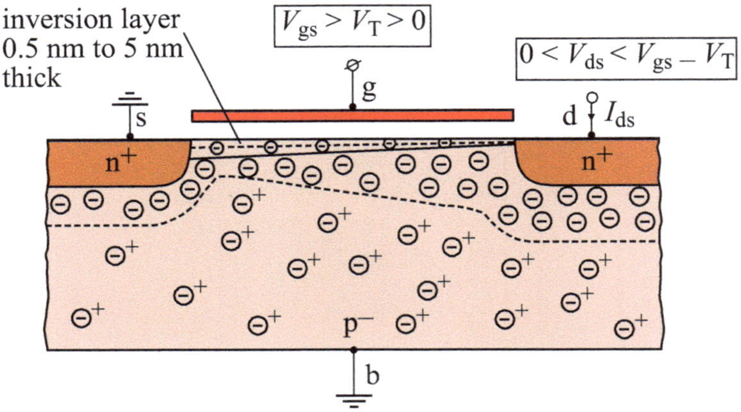

Fig. 1.13 Cross section of a transistor operating in the linear (triode) region

The three regions are discussed separately on the following pages.

$$\textbf{The linear region} \begin{cases} V_{gs} > V_T > 0 \\ 0 < V_{ds} < V_{gs} - V_T \end{cases}$$

Figure 1.13 shows the situation in the linear region, in which a current I_{ds} (which flows from drain to source) causes a voltage difference in the channel. The surface potential under the gate decreases from V_{ds} in the drain to 0 V in the source. The maximum potential difference between the gate and channel is at the source. Therefore, the strongest inversion and the highest concentration of electrons in the inversion layer occur adjacent to the source. The maximum potential difference between the channel and substrate is at the drain. The depletion layer is therefore thickest here. In the linear region, the drain current I_{ds} increases with increasing V_{ds} for a constant V_{gs}.

$$\textbf{The transition region} \begin{cases} V_{gs} > V_T > 0 \\ V_{ds} = V_{gs} - V_T \end{cases}$$

An increase in V_{ds}, with V_{gs} constant, decreases the voltage difference between the gate and channel at the drain. The inversion layer disappears at the drain when the voltage difference between the gate and channel equals the threshold voltage V_T. The channel end then coincides with the drain-substrate junction. This situation occurs when $V_{ds} = V_{gs} - V_T$ and is shown in Fig. 1.14.

$$\textbf{The saturation region} \begin{cases} V_{gs} > V_T > 0 \\ V_{ds} > V_{gs} - V_T \end{cases}$$

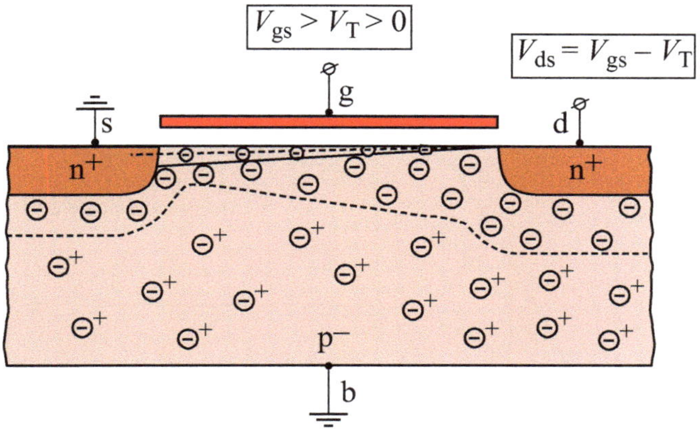

Fig. 1.14 Situation during transition from triode to saturation region, i.e. $V_{ds} = V_{gs} - V_T$

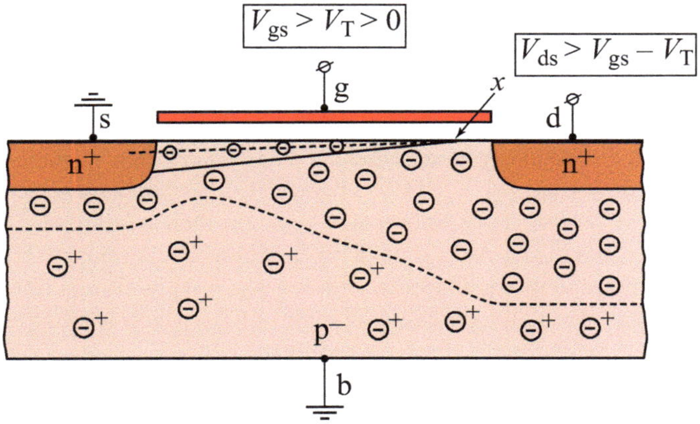

Fig. 1.15 Situation in the saturation region, i.e. $V_{ds} > V_{gs} - V_T$

The channel end no longer coincides with the drain when V_{ds} is larger than $V_{gs} - V_T$. This situation is shown in Fig. 1.15.
The voltage V_x at the end point x of the inversion layer equals $V_{gs} - V_T$. Therefore, V_T is the voltage difference between the gate and channel at position x. If this *pinch-off point* is considered to be the *virtual drain* of the transistor, then I_{ds} is determined by the voltage $V_x = V_{gs} - V_T$. In other words, the drain current in the saturation region equals the drain current at the transition point between the linear and saturation regions. The value of the *saturation current* is clearly proportional to V_{gs}. Electrons are emitted from the inversion layer into the depletion layer at the

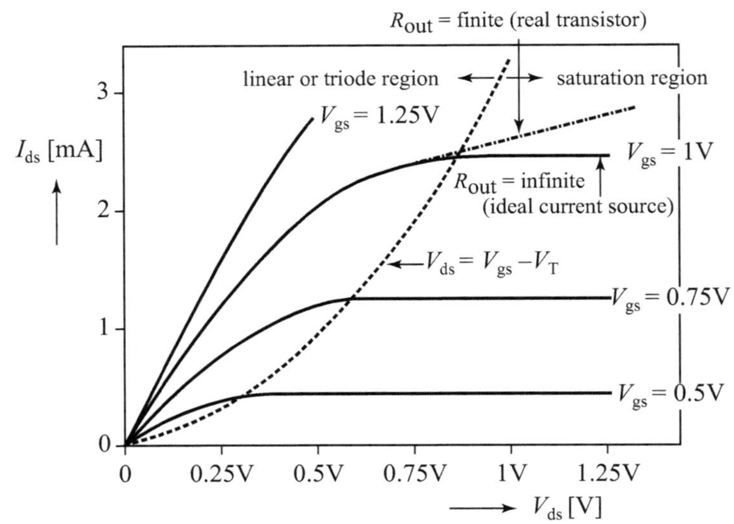

Fig. 1.16 The $I_{ds} = f(V_{ds})$ characteristic for various values of V_{gs}

pinch-off point. These electrons will be attracted and collected by the drain because $V_{ds} > V_x$, which builds a large electric field across the very narrow pinch-off region.

Figure 1.16 shows the $I_{ds} = f(V_{ds})$ characteristic for various gate voltages. If $V_{ds} = 0\,\text{V}$, then $I_{ds} = 0\,\text{A}$. If V_{ds} is less than $V_{gs} - V_T$, then the transistor operates in the triode region, and the current I_{ds} displays an almost linear relationship with V_{ds}. Current I_{ds} increases to its saturation value when $V_{ds} = V_{gs} - V_T$. Further increases of V_{ds} above $V_{gs} - V_T$ no longer cause increases in I_{ds}. The transition between the triode and saturation regions is characterised by the curve $V_{ds} = V_{gs} - V_T$.

1.4 Derivation of Simple MOS Formulae

The inversion-layer nMOS transistor shown in Fig. 1.17 has a width W perpendicular to the plane of the page and an oxide capacitance C_{ox} per unit area. A commonly used unit for C_{ox} is fF/μm^2, where $1\,\text{fF} = 10^{-15}\,\text{F}$.

Based on the *law for conservation of charge*, the following equality must hold at any position x between the source and drain:

$$Q_g + Q_{ox} + Q_n + Q_d = 0. \tag{1.3}$$

The components in this equation are charges per unit area, specified as follows:

Q_g = the gate charge [C/m^3]
Q_{ox}= primarily a small fixed charge which in practice always

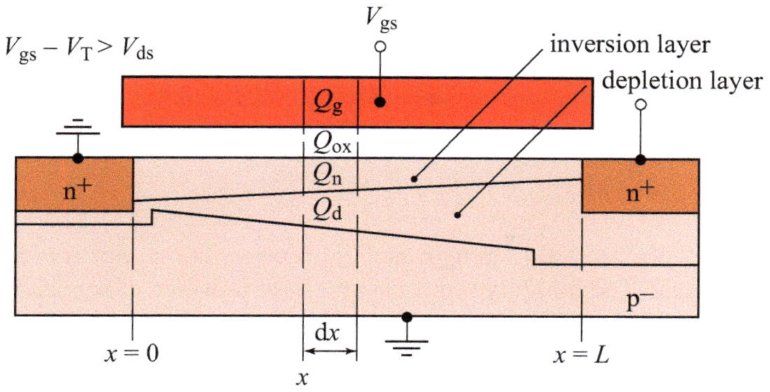

Fig. 1.17 Charges in a MOS transistor operating in the linear region

appears to be present in the thin gate oxide [C/m^3]
Q_n = the mobile charge in the inversion layer [C/m^3]
Q_d = the fixed charge in the depletion layer [C/m^3]

For gate voltages larger than V_T, the inversion layer shields the depletion layer from the gate. The charge in the depletion layer can then be considered constant:

$$Q_{ox} + Q_d = -C_{ox} \cdot V_T \tag{1.4}$$

The threshold voltage V_T is assumed to be constant. The potential in the channel at a position x is $V(x)$. With $Q_g = C_{ox}[V_{gs} - V(x)]$ and substituting (1.4) into (1.3) yields

$$Q_n = -C_{ox}[V_{gs} - V_T - V(x)]$$

The total mobile charge $\mathrm{d}Q_m$ in a section of the channel with length $\mathrm{d}x$ is defined as

$$\mathrm{d}Q_m = Q_n \cdot W \cdot \mathrm{d}x = -W \cdot C_{ox}[V_{gs} - V_T - V(x)] \cdot \mathrm{d}x \tag{1.5}$$

$$\Rightarrow \frac{\mathrm{d}Q_m}{\mathrm{d}x} = -W \cdot C_{ox}[V_{gs} - V_T - V(x)] \tag{1.6}$$

The drain current I_{ds} is expressed as

$$I_{ds} = \frac{\mathrm{d}Q_m}{\mathrm{d}t} = \frac{\mathrm{d}Q_m}{\mathrm{d}x} \cdot \frac{\mathrm{d}x}{\mathrm{d}t} \tag{1.7}$$

where $\frac{\mathrm{d}Q_m}{\mathrm{d}x}$ is defined in Eq. (1.6) and

$\frac{dx}{dt}$ is the velocity v at which the charge Q_m moves from the source to the drain region.
This is the velocity of the electrons in the inversion layer and is expressed as

$$v = -\mu_n \cdot E = -\mu_n \cdot \frac{dV(x)}{dx} \tag{1.8}$$

where E is the electric field strength and μ_n represents the electron mobility in the inversion layer. The mobility represents the ease in which charge carriers move within a semiconductor.

In practice, the effective mobility appears to be less than one third of the electron mobility in the substrate (see Sect. 2.3). Combining Eqs. (1.6), (1.7) and (1.8) yields

$$I_{ds} = \mu_n \cdot C_{ox} \cdot W \cdot [V_{gs} - V_T - V(x)] \cdot \frac{dV(x)}{dx} \tag{1.9}$$

Substituting $\beta_\square = \mu_n \cdot C_{ox}$ yields

$$I_{ds} \cdot dx = \beta_\square \cdot W \cdot [V_{gs} - V_T - V(x)] \cdot dV(x) \tag{1.10}$$

Integrating the left-hand side from 0 to L and the right-hand side from 0 to V_{ds} yields

$$I_{ds} = \frac{W}{L} \cdot \beta_\square \cdot (V_{gs} - V_T - \frac{1}{2} V_{ds}) \cdot V_{ds} \tag{1.11}$$

Equation (1.11) has a maximum value when $V_{ds} = V_{gs} - V_T$. In this case, the current I_{ds} is expressed as

$$I_{ds} = \frac{1}{2} \cdot \frac{W}{L} \cdot \beta_\square \cdot (V_{gs} - V_T)^2 \tag{1.12}$$

If $V_{gs} = V_T$, then $I_{ds} = 0\,A$. This clearly agrees with the earlier assumption that V_T is positive for an inversion-layer nMOS transistor. The term β is usually used to represent $\frac{W}{L} \cdot \beta_\square$. This factor is called the transistor *gain factor* and depends on geometry. The gain term $\beta_\square$ is a process parameter which depends on such things as the oxide thickness t_{ox}:

$$\beta_\square = \mu_n \cdot C_{ox} = \mu_n \cdot \frac{\epsilon_0 \epsilon_{ox}}{t_{ox}} \tag{1.13}$$

The unit of measurement for both β and $\beta_\square$ is A/V^2. However, $\mu A/V^2$ and mA/V^2 are the most commonly used units. For an n-channel MOS transistor, $\beta_\square$ varies from $360\,\mu A/V^2$ to $750\,\mu A/V^2$ for oxide thicknesses of 3.2 nm and 1.6 nm, respectively. Note that these values for $\beta_\square$ resemble the zero-field mobility in the substrate. The effective mobility in the channel, and so the effective gain factor, is much lower (often more than one order of magnitude) due to several second-order effects as discussed in Chap. 2.

According to Eq. (1.11), I_{ds} would reach a maximum value and then decrease for increasing V_{ds}. In the discussion concerning Figs. 1.15 and 1.16, however, it was stated that the current remains constant for an increasing V_{ds} once $V_{ds} > V_{gs} - V_T$. The transistor has two operating regions which are characterised by corresponding expressions for I_{ds}. These regions and their I_{ds} expressions are defined as follows:

1. The linear or triode region. $0 < V_{ds} < V_{gs} - V_T$.

$$I_{ds} = \beta \cdot (V_{gs} - V_T - \frac{1}{2}V_{ds}) \cdot V_{ds} \tag{1.14}$$

2. The saturation region. $V_{ds} \geq V_{gs} - V_T$.

$$I_{ds} = \frac{\beta}{2} \cdot (V_{gs} - V_T)^2 \tag{1.15}$$

These two expressions above are derived from the Shockley first-order transistor model, which gives a basic inside into transistor currents in both operating regions. The transistor models used in today's circuit simulations, however, are much more complex and include all parasitic components and physical effects that result from transistor scaling. Several of them are discussed in Chap. 2. Still, we use the above simple expressions throughout the rest of this book, as they are very well suited to explain these physical effects.

According to Eq. (1.15), I_{ds} is independent of V_{ds} in the saturation region. The output impedance dV_{ds}/dI_{ds} should then be infinite, and the transistor should behave like an ideal current source. In practice, however, MOS transistors show a finite output impedance which is dependent on geometry. This is explained in Chap. 2. Figure 1.16 shows both the ideal (theoretical) and the real current-voltage characteristics of a transistor with a threshold voltage $V_T = 0.25$ V.

The $I_{ds} = f(V_{ds})|_{V_{gs}=\text{constant}}$ curves in Fig. 1.16 are joined by the dotted curve $V_{ds} = V_{gs} - V_T$ at the points where Eq. (1.14) yields maximum values for I_{ds}. This curve divides the $I_{ds} - V_{ds}$ plane into two regions:

1. Left of the dotted curve: the triode or linear region, which is defined by Eq. (1.14)
2. Right of the dotted curve: the saturation region, which is defined by Eq. (1.15)

1.5 The Back-Bias Effect (The Influence of the Threshold Voltage on Transistor and Chip Behaviour)

The simple MOS formulae derived in Sect. 1.4 appear to be reasonably satisfactory in most cases. The very important *back-bias effect* is, however, not included in these formulae. This effect accounts for the modulation of the threshold voltage by the substrate bias and the subsequent effects on the drain current.

During normal operation (when $V_{gs} > V_T$ and $V_{ds} > V_{gs} - V_T$), a depletion layer is formed, as shown in Fig. 1.15. However, the thickness of the depletion region under the channel increases when a negative *back-bias voltage* (positive V_{sb}) is applied to the bulk (b) with respect to the source. This is caused by the increased reverse-bias voltage across the fictive channel-substrate junction. The increased depletion layer requires additional charge. The channel charge therefore decreases if V_{gs} is held constant. The channel conductivity can only be maintained if V_{gs} is increased. The threshold voltage is therefore related to the back-bias voltage. This dependence is expressed as follows:

$$V_T = V_x + K\sqrt{V_{sb} + 2\phi_f} \tag{1.16}$$

$$V_{T0} = V_x + K\sqrt{2\phi_f} \tag{1.17}$$

The terms in these formulae are as follows:

V_x = process-related constant threshold voltage term; this term depends on the *work function* difference between the gate and the substrate (bulk) and is also somewhat proportional with the gate-oxide thickness (t_{ox}). The work function of a material is the amount of work (energy) that is required to extract an electron completely from its surface. For most materials, this energy is between 2 and 5 eV. The combination of the gate-stack materials is therefore very important in the determination of the transistor threshold voltage.

V_{T0} = $V_T|_{V_{sb}=0V}$.

K = process parameter equal to $\frac{1}{C_{ox}}\sqrt{2N_A q \epsilon_0 \epsilon_{si}}$; also known as the *'body factor'* or K-factor.

N_A = substrate (bulk) dope concentration.

V_{sb} = source-bulk (back-bias) voltage.

ϕ_f = the Fermi potential; this parameter represents the difference between Fermi level and intrinsic energy level (see Fig. 1.10a), often with values between 0.35 V and 0.5 V, depending on the process node.

$2\phi_f$ = band bending at which strong inversion in the top of the silicon channel occurs.

As can be seen from Expression 1.16, the threshold voltage of a transistor is determined by the following:

1. The materials in the transistor stack (V_x)
2. The transistor channel dope N_A, which influences K in the term

The required $2\phi_f$ band bending in Fig. 1.12 to achieve full channel inversion is built from these two components. In a p-type doped substrate, as used in standard nMOS transistors, the Fermi potential ϕ_f is determined by the amount of p dope in the substrate. So, the first part of the applied gate voltage is used to create this 'one ϕ_f' band bending. The rest of the gate voltage, until it reaches the threshold voltage, is used to create further band bending until the channel region is fully depleted. So, this part of the V_T depends on the previously discussed work function. What happens in the channel is that, when V_{gs} has reached the threshold voltage V_T, at $2\phi_f$ band bending, all holes in the channel region are filled. When V_{gs} becomes larger than V_T, free electrons are extracted from the source into the channel, which then create the inversion layer, and the transistor starts conducting.

Many CMOS processes beyond the 32 nm node processes use transistors (e.g. high-ϵ, metal-gate and FinFETs) with (almost) undoped channel regions (intrinsic silicon), meaning that their threshold voltages are almost completely determined by the transistor stack (work function). Certain applications, as explained extensively in Chap. 8, require ICs with multiple threshold voltages (V_T's), which can be created by the deposition of different work-function layers. In other words, a high-V_T transistor in such a *dual-V_T process* contains another work-function layer than a low-V_T transistor. This makes these processes additionally complex. Many CMOS processes above the 32 nm node with planar silicon-gate MOS transistors use p-type and n-type substrates for their nMOS and pMOS transistors, respectively. The threshold voltages of these devices can be increased by changing the substrate dope N_A in the channel, as discussed above. A dual-V_T process based on this effect requires transistors with different channel dope levels. The threshold voltage in circuits built in these processes can also be increased by by adding a negative substrate (bulk) voltage (positive V_{sb} in expression 1.16) for an nMOS transistor, or positive substrate (n-well) voltage for a pMOS transistor. The influence of this substrate voltage on the threshold voltage is called *back-gate effect*, due to the fact that it can control the transistor current, although to a lesser extent, similar to the normal transistor (top) gate.

This back-bias effect causes MOS transistors of the same type and dimensions to have different threshold voltages. Assume the nMOS inverter of Fig. 1.18 is designed in a 1.2 V 65 nm technology. Applying Eq. (1.16) yields the following

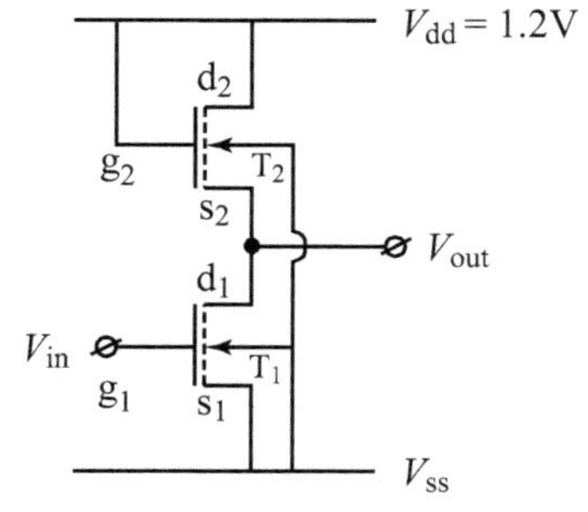

Fig. 1.18 nMOS-inverter with enhancement load

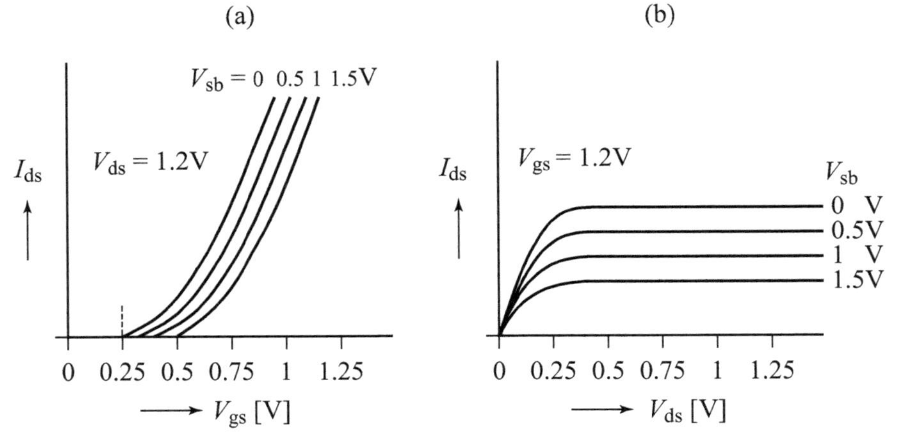

Fig. 1.19 Back-bias effect on MOS transistor characteristics: (**a**) $I_{ds} = f(V_{gs})|_{V_{ds}=\text{const}}$. (**b**) $I_{ds} = f(V_{ds})|_{V_{gs}=\text{const}}$

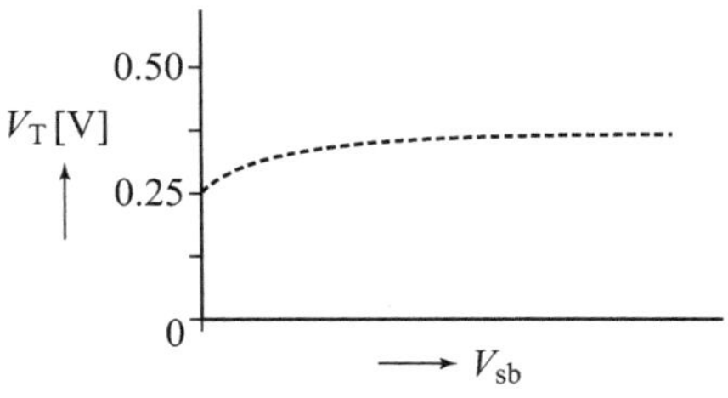

Fig. 1.20 $V_T = f(V_{sb})$: Threshold voltage as a function of source-bulk voltage

equations for transistors T_1 and T_2, respectively:

$$V_{T1} = V_x + K\sqrt{V_{s_1b} + 2\phi_f}$$

$$V_{T2} = V_x + K\sqrt{V_{s_2b} + 2\phi_f}$$

If the output is 'high' ($\approx 0.7\,\text{V}$), the source-bulk voltages of T_1 and T_2 are $V_{s_1b} = V_{ss} - V_{bb} = 0\,\text{V}$ and $V_{s_2b} = V_{out} - V_{bb} = 0.7\,\text{V}$, respectively. The source-bulk voltage can therefore cause the threshold voltage V_{T2} of the upper transistor to be considerably larger than the threshold voltage V_{T1} of the lower transistor.

Figure 1.19 shows the influence of the back-bias effect on different transistor characteristics. Formula (1.16) clearly shows that the threshold voltage V_T increases with an increasing back-gate voltage V_{sb}. For a constant V_{gs}, the drain-source current therefore decreases for an increasing V_{sb}. This is illustrated in Fig. 1.19b.

Figure 1.20 shows the dependence of V_T on V_{sb}. The starting point of this graph is determined by V_{T0} in Eq. (1.17), while its curve depends on the K-factor.

The back-bias effect must be accurately included in the design of MOS circuits. The most important reasons for using a back-bias voltage are as follows:

- Normally, the K-factor in Eqs. (1.16) and (1.17) spreads more than the term V_x. The influence of the K-factor on the threshold voltage is larger when a back-bias voltage is applied. This results in a more stable threshold voltage.
- The depletion layer around the source and drain junctions of the MOS transistor becomes thicker as a result of the increased reverse voltage across these p-n junctions. This reduces the parasitic capacitances of the source and drain.
- Negative voltage pulses which may occur in dynamic MOS logic circuits may forward-bias the p-n diode between the substrate and a source or drain. Application of a negative voltage to the substrate virtually removes this possibility.
- Because an additional back-bias voltage increases the V_T, it reduces the subthreshold leakage current. This usage is described in Chap. 8.

Next to the back-bias effect, there also exists a *forward-bias effect*. This effect is sometimes used to reduce the threshold voltage in order to improve the performance of certain logic or memory cores on a chip. In such cases, the source-to-substrate junction is put in forward bias but with a voltage level below the junction voltage, which is equal to about 0.6 V.
The MOS transistor formulae are summarised as follows:

$$\begin{array}{ll} \text{– linear region} & : I_{ds} = \beta(V_{gs} - V_T - \frac{V_{ds}}{2})V_{ds} \\ \text{– saturation region} & : I_{ds} = I_{ds_{sat}} = \frac{\beta}{2}(V_{gs} - V_T)^2 \\ \text{where} & V_T = V_x + K\sqrt{V_{sb} + 2\phi_f} \\ \text{and} & V_{T0} = V_x + K\sqrt{2\phi_f} \end{array} \quad (1.18)$$

1.6 Factors Which Characterise the Behaviour of the MOS Transistor

The previously discussed current-voltage characteristics represent the relationship between a transistor's current (I_{ds}) and its various applied voltages (V_{gs}, V_{ds} and V_{sb}). A number of important parameters which are frequently used to describe the behaviour of a transistor are explained below.
The *transconductance* g_m describes the relationship between the change δI_{ds} in the transistor current caused by a change δV_{gs} in the gate voltage:

$$g_m = \frac{\delta I_{ds}}{\delta V_{gs}}|_{V_{ds} = \text{const}} \quad (1.19)$$

Referring to Fig. 1.16, it is clear that the value of g_m depends on the transistor's operating region:

$$\text{Linear region}: g_m = \beta \cdot V_{ds} \quad (1.20)$$

$$\text{Saturation region}: g_{m\text{sat.}} = \beta \cdot (V_{gs} - V_T) \quad (1.21)$$

Another parameter that characterises conduction in a transistor is its *output conductance*. In the transistor's linear operating region, this conductance (which is also called the channel conductance) is defined as

$$g_{ds} = (\frac{\delta I_{ds}}{\delta V_{ds}}) = \{\beta(V_{gs} - V_T) - \beta V_{ds}\} \quad (1.22)$$

If V_{ds} is small, then

$$g_{ds} = \beta(V_{gs} - V_T) \quad (1.23)$$

For an ideal MOS transistor operating in the saturation region, we have $\frac{\delta I_{ds}}{\delta V_{ds}} = 0$. The transistor current is then independent of V_{ds}. The output resistance is therefore infinite and the transistor acts as an ideal current source. In practice, however, the MOS transistor always has a finite output resistance, and its current remains dependent on V_{ds}. This is illustrated in Fig. 1.16 and is treated in Sect. 2.4.

We will now briefly discuss two other figure of merits, which represent the frequency response of a MOS transistor. If we ignore parasitic effects, the average transit time τ of a carrier across the channel L of an intrinsic device, operating in its linear region, is equal to

$$\tau = \frac{L}{\mu E} = \frac{L^2}{\mu V_{dd}} \quad (1.24)$$

and

$$\tau = \frac{L}{v_{sat}} \quad (1.25)$$

when the device is in velocity saturation (see Chap. 2) and where v_{sat} equals the saturation speed of the carriers in the channel.

In most traditional CMOS technologies, $v_{sat} \approx 10^7$ cm/s, so that $\tau \approx 0.2$ ps for a transistor with a channel length $L = 20$ nm. This leads to the so-called cut-off (or threshold) frequency f_T at which the device can fill and empty the channel as a response to an ac signal:

$$f_T = \frac{1}{2\pi\tau} = \frac{v_{sat}}{2\pi L} = \frac{g_m}{2\pi C_{gs}} \quad (1.26)$$

with the average carrier velocity in the channel equal to $g_m/C_{ox} \approx g_m/C_{gs}$.

In saturation, it is clear that $v_{average} = v_{sat}$. Scaling thus leads to a reduction of the transit time and to an increase of the transconductance and cut-off frequency.

1.7 Different Types of MOS Transistors

1. The previous discussions are all related to *n-channel* MOS transistors. The substrate material of these nMOS transistors is p-type, and the drain and gate voltages are *positive* with respect to the source during normal operation. The substrate is the most *negative* electrode of an nMOS transistor.
2. *P-channel* MOS transistors are produced on an n-type substrate. The voltages at the gate and drain of these pMOS transistors are *negative* with respect to the source during normal operation. The substrate is the most *positive* electrode.

Generally, nMOS circuits are faster than those with pMOS transistors. The *power-delay* (τD) *product* of a logic gate is the product of its delay τ and dissipation D. The τD products of nMOS logic gates are lower than those of pMOS logic gates. This is because of the difference between the *mobility* of electrons and holes. Electron mobility is a factor of about three times higher than hole mobility in both the bulk silicon and inversion layers of the respective devices. Figure 2.1 illustrates this relationship. Typically, at room temperature (300°K) and at low doping levels, electron mobility for silicon is close to 1400 cm^2/Vs, while the hole mobility is close to 450 cm^2/Vs. The following relationship then follows from Eq. (1.13):

$$\beta_{\square n} \approx 3 \cdot \beta_{\square p}$$

In conventional process nodes, an nMOS transistor therefore conducts approximately two and a half times as much current as a pMOS transistor of equal dimensions and with equal absolute voltages. Advanced CMOS technologies, today, apply different channel-stress and crystal-orientation techniques to improve the transistor drive currents and which also drives the performance of the pMOSts closer to that of the nMOSts. These are discussed in Chap. 3.

Figure 1.21 shows a schematic overview of transistors which are distinguished on the basis of threshold voltage V_T. This distinction applies to both pMOS and nMOS transistors and results in the following types:

- *Enhancement* or *normally off* transistors:
  No current flows through an enhancement transistor when $V_{gs} = 0$.
  $V_T > 0$ for an nMOS enhancement transistor and $V_T < 0$ for a pMOS enhancement transistor.
- *Depletion* or *normally on* transistors:
  Current flows through a depletion transistor when $V_{gs} = 0$.
  $V_T < 0$ for an nMOS depletion transistor and $V_T > 0$ for a pMOS depletion transistor.

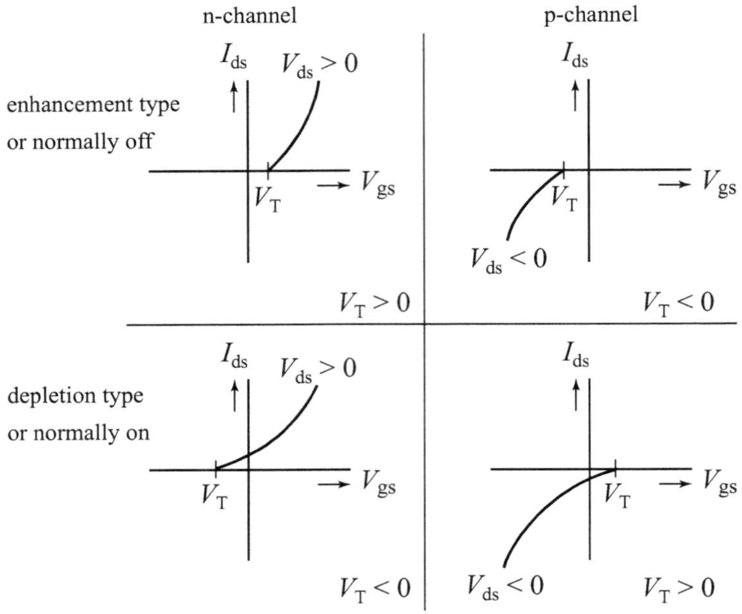

Fig. 1.21 Schematic overview of the different types of MOS transistors

1.8 Parasitic MOS Transistors

MOS (V)LSI circuits comprise many closely packed transistors. This leads to the presence of parasitic MOS transistors, as illustrated in Fig. 1.22.

Transistors T_1 and T_2 are separated by the *field oxide*. Parasitic MOS transistor T_3 is formed by a metal interconnection track on the field oxide and the n^+ areas of transistors T_1 and T_2. This field oxide is thick in comparison with the gate oxide, which ensures that the threshold voltage $V_{T_{par}}$ of transistor T_3 is larger than the threshold voltages of transistors T_1 and T_2. The field strength at the silicon surface in T_3 is therefore lower than in T_1 and T_2. Transistor T_3 will never conduct if its gate voltage never exceeds $V_{T_{par}}$.

Many MOS production processes use an extra diffusion or ion implantation to artificially increase the threshold voltage $V_{T_{par}}$ of parasitic transistors. For this purpose, boron is used to create a p-type layer beneath the thick oxide in processes that use p^--type substrates. This makes it much more difficult to create an n-type inversion layer in these areas.

Processes that use n^--type substrates use phosphorus to increase $|V_{T_{par}}|$. The terms *channel stopper implant* is used to refer to these boron and phosphorous implantations.

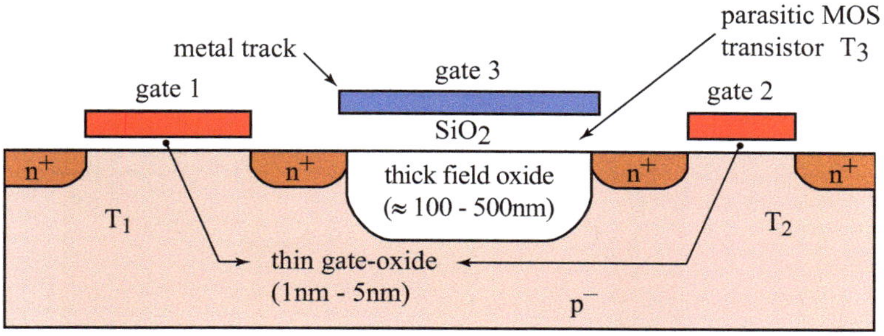

Fig. 1.22 Example of a parasitic MOS transistor

Note Parasitic MOS transistors also appear in bipolar circuits. The absolute value of parasitic threshold voltages is always higher in n-type substrates than in p-type substrates. This is one of the reasons why planar IC technologies were mainly developed on n-epi layers.

1.9 MOS Transistor Symbols

Figure 1.23 shows various symbols used in literature to represent MOS transistors. Their meanings are as follows:

(a) The inward pointing arrow indicates that the transistor is n-channel, and the broken line between s and d indicates that it is an enhancement transistor.
(b) The solid line from s to d indicates that this n-channel transistor is a depletion device.
(c) The outward pointing arrow indicates that the transistor is p-channel, and the broken line between s and d indicates that it is an enhancement transistor.
(d) The solid line from s to d indicates that this p-channel transistor is a depletion device.
(e) This symbol for an n-channel enhancement transistor is analogous to the npn transistor symbol.
(f) This p-channel transistor is by definition not necessarily an enhancement type.
(g) This general symbol represents a MOS transistor of any type.

Adaptations of the above symbols are also used. MOS symbols must therefore be interpreted with caution. The following rules are generally applied:

1. A transistor symbol with a broken line between its source and drain is always an enhancement or normally off type.
2. Arrows indicate the forward directions of the substrate-channel 'junctions'.

The symbols in Fig. 1.24 are used throughout this book.

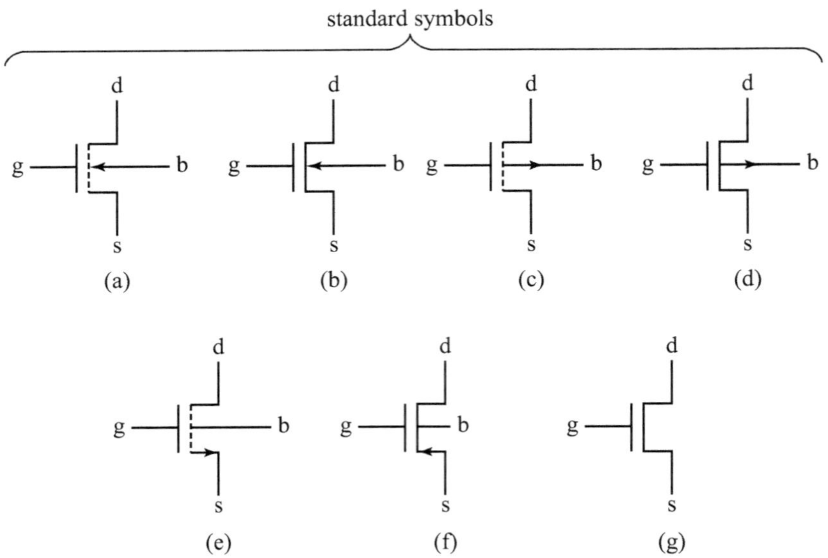

Fig. 1.23 Various transistor symbols

Fig. 1.24 Transistor symbols used throughout this book

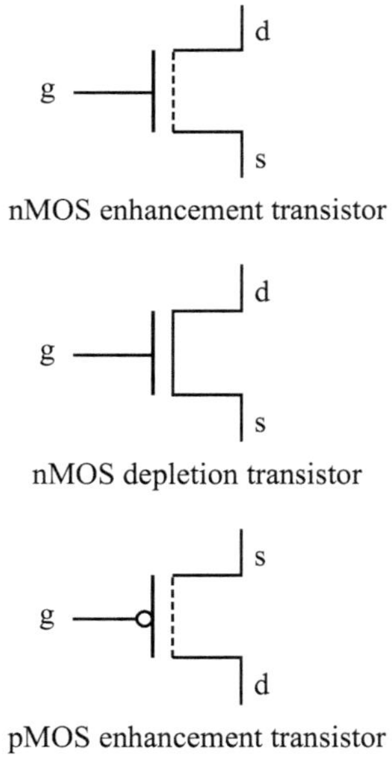

1.10 Capacitances in MOS Structures

Figure 1.25 illustrates the MOS capacitance, whose value depends on such things as V_g and the frequency at which it varies. Section 1.3.1 describes the MOS capacitance and presents a qualitative discussion of its related charges, fields and voltages. Figure 1.26 shows a plot of the total capacitance C_t between the gate and ground terminals as a function of their voltage difference.

The various regions of the *C-V curve* in Fig. 1.26 are explained as follows:

1. $V_g \ll V_T$ for a p-type substrate; $V_g \gg V_T$ for an n-type substrate.
   Here, the surface potential ϕ_s is highly negative and majority carriers in the p-type substrate will form a surface layer of holes. This accumulation layer is thin in comparison with the oxide thickness and exists as long as V_g is much smaller than V_T. Now, the silicon behaves like a metal plate, and the MOS capacitance is equal

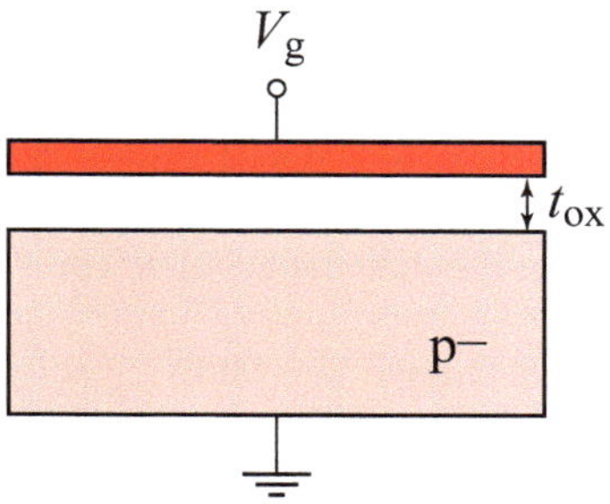

Fig. 1.25 The MOS capacitance

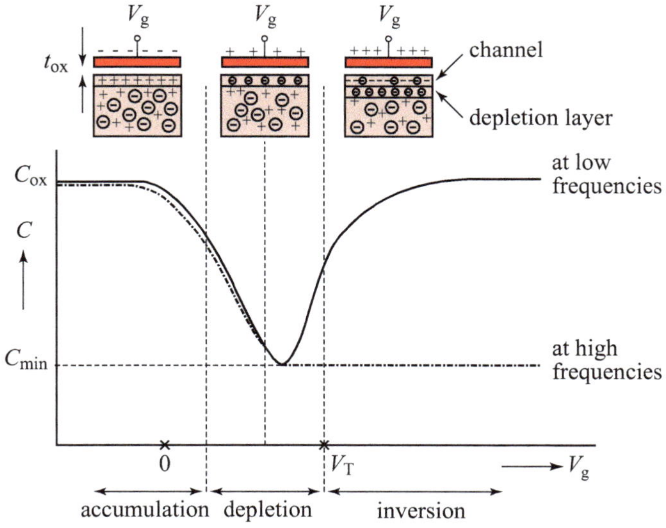

Fig. 1.26 Capacitance behaviour of a MOS structure

to the oxide capacitance C_{ox}. Deviations only appear at very high frequencies ($>$ 1 GHz), where the *dielectric relaxation time* τ_R is important. For the 10Ωcm silicon, $\tau_R \approx 10\,\text{ps}$ ($=10^{-11}$ s).

2. $V_g \approx V_T$; thus, $\phi_s \approx 0 \ldots 2\phi_f$.
   As V_g gradually becomes more positive, the accumulation layer decreases for a p-type substrate. A depletion layer is created under the gate when $\phi_s > 0$. A voltage change ΔV at the gate causes a change ΔQ in the charge at the edge of the depletion layer. In fact, the total capacitance is now determined by the series connection of the gate capacitance and the depletion-layer capacitance. The capacitance therefore decreases.
3. $V_g \gg V_T$ for a p-type substrate; $V_g \ll V_T$ for an n-type substrate.
   Now, ϕ_s is highly positive and an inversion layer is created. This layer is thin compared to the oxide thickness. At low frequencies ($<$ 100 kHz), the capacitance will again be equal to the oxide capacitance C_{ox}. However, the inversion layer for a p-type substrate consists of electrons that are supplied and absorbed by the substrate. This relies on the process of thermal generation and recombination of minorities, i.e. the electrons. With a constant temperature, the speed of the generation/recombination process is limited. This accounts for the lower capacitance shown in Fig. 1.26 at higher frequencies ($>$ 1 MHz). At these high frequencies, the capacitance C_t will be about equal to the series connection of the gate capacitance and the depletion-layer capacitance.

As discussed, the MOS capacitance can be considered as a series connection of two capacitances: the oxide capacitance C_{ox} between the gate and the silicon surface and a capacitance C_s between the silicon surface and the substrate interior. This is explained below. The voltage V_g can be expressed as follows:

$$V_g = V_{ox} + \phi_{ms} + \phi_s \tag{1.27}$$

The law for conservation of charge yields the following equation:

$$Q_g + Q_{ox} + Q_n + Q_d = 0 \tag{1.28}$$

where

V_{ox} = voltage across the oxide between gate and silicon surfaces
ϕ_{ms} = contact potential between gate and substrate
ϕ_s = surface potential of the silicon with respect to the substrate interior
Q_g = charge on the gate
Q_{ox} = charge in the oxide
Q_n = charge in the inversion layer
Q_d = charge in the depletion layer

The following expression for a change ΔV_g in gate voltage can be derived from Eq. (1.27):

$$\Delta V_g = \Delta V_{ox} + \Delta\phi_s (\phi_{ms} \text{is constant, thus } \Delta\phi_{ms} = 0) \tag{1.29}$$

Substituting $Q_n + Q_d = Q_s$ in Eq. (1.28) yields

$$\Delta Q_g = -\Delta Q_{ox} - \Delta Q_s \tag{1.30}$$

If Q_{ox} is considered constant, then

$$\Delta Q_g = -\Delta Q_s \tag{1.31}$$

Equations (1.29) and (1.31) yield the following expressions:

$$\frac{\Delta V_g}{\Delta Q_g} = \frac{\Delta V_{ox}}{\Delta Q_g} + \frac{\Delta\phi_s}{\Delta Q_g} = \frac{\Delta V_{ox}}{\Delta Q_g} - \frac{\Delta\phi_s}{\Delta Q_s}$$

where

$\frac{\Delta Q_g}{\Delta V_g} = C_t =$ the total capacitance of the MOS structure

$\frac{\Delta Q_g}{\Delta V_{ox}} = C_{ox} =$ oxide capacitance

$-\frac{\Delta Q_s}{\Delta\phi_s} = C_s =$ capacitance between the silicon surface and the semiconductor interior (depletion-layer capacitance)

C_t can now be expressed as follows:

$$C_t = (\frac{1}{C_{ox}} + \frac{1}{C_s})^{-1} \tag{1.32}$$

Capacitance C_s is responsible for the drop in the $C - V$ curve. The value of C_s is determined by the substrate doping concentration and the potential difference across the depletion layer. The minimum value C_{min} in the $C - V$ curve is also determined by C_{ox}. A smaller C_{ox} leads to a larger $\frac{1}{C_{ox}}$ and a smaller C_{min}. C_{min} can be as low as $0.1C_{ox}$.

The $C - V$ curve is often used during MOS manufacturing processes to get a quick impression of the value of V_T.

Figure 1.27 shows a MOS capacitance with an additional n^+ area, which causes significant changes in the capacitance behaviour. The structure is in fact equivalent to a MOS transistor without a drain or to a MOS transistor with an external short circuit between its drain and source. This structure is generally called a *MOS*

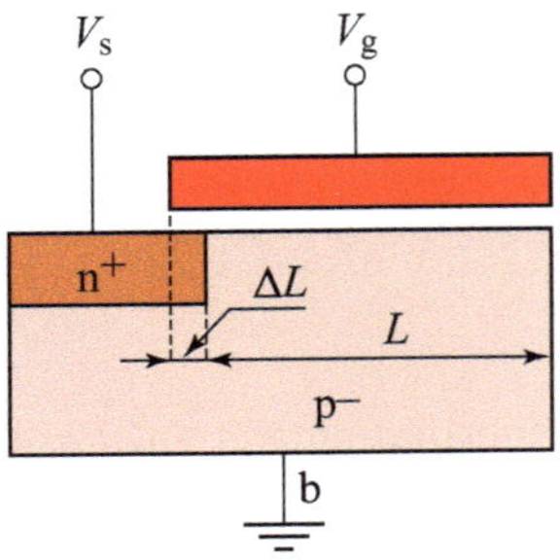

Fig. 1.27 MOS capacitance with source and/or drain area

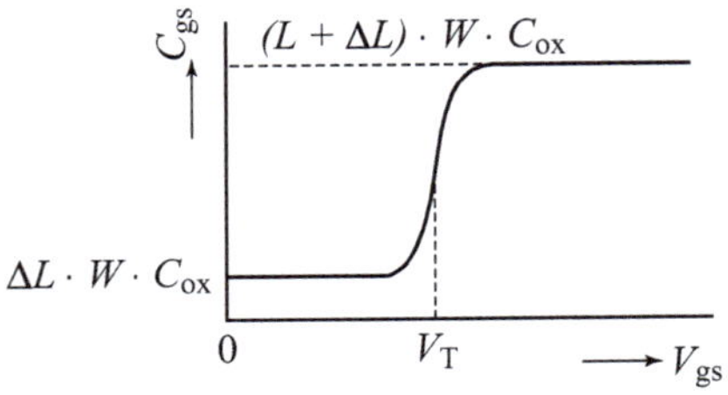

Fig. 1.28 Non-linear behaviour of a MOS capacitance

capacitance or a *MOS varactor*. Dynamic MOS circuits, in particular, use this device very often.

While $V_{gs} < V_T$, there is no inversion layer in a MOS capacitance, and the behaviour of the gate capacitance is unchanged. However, an inversion layer is created when $V_{gs} > V_T$. The electrons in this inversion layer are supplied by the n^+ area instead of by thermal generation/recombination processes of minorities in the substrate. This n^+ area can generate and absorb electrons at very high frequencies (> 1 GHz). Therefore, C_t will now equal C_{ox} under all normal operating conditions. In this case, C_t represents the capacitance between the gate and source, i.e. $C_t = C_{gs} = C_{ox}(L + \Delta L) \cdot W$.

The dependence of the capacitance C_{gs} on the applied voltage V_{gs} is summarised as follows:

- When $V_{gs} < V_T$, there is no inversion layer. Here, the value of C_{gs} is determined by the channel width W and the gate overlap ΔL on the source/drain area: $C_{gs} = \Delta L \cdot W \cdot C_{ox}$.
- When $V_{gs} > V_T$, there is an inversion layer. Here, C_{gs} is determined by the channel length $L : C_{gs} = (L + \Delta L) \cdot W \cdot C_{ox}$.

The above non-linear behaviour of $C_{gs} = f(V_{gs})$ is shown in Fig. 1.28.

Note There is no inversion layer when $V_{gs} < V_T$. Figure 1.26 shows how the gate-substrate capacitance then behaves.

Figure 1.29 shows the large number of capacitances in a real MOS transistor. These capacitances, which are largely non-linear, are defined as follows:

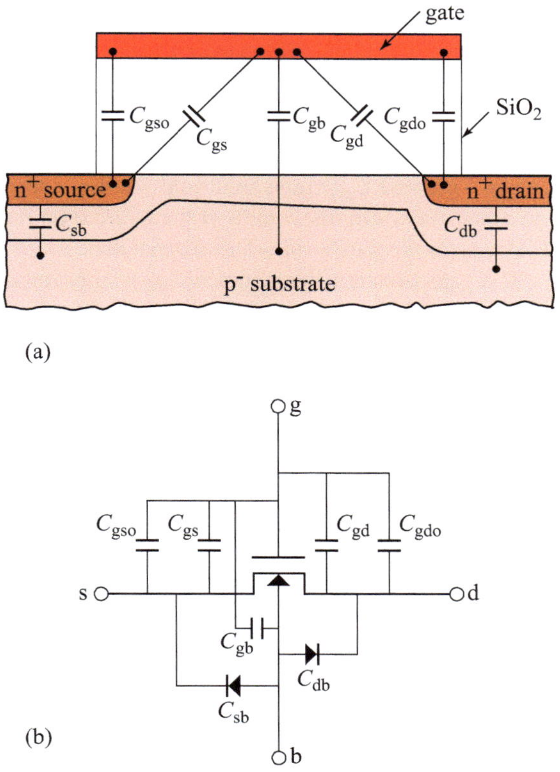

Fig. 1.29 Capacitances in a MOS transistor

C_{db}, C_{sb} : drain-substrate and source-substrate capacitances, which are non-linearly dependent on V_{db} and V_{sb}, respectively

C_{gdo}, C_{gso}: gate-drain and gate-source capacitances, which are voltage-independent

C_{gd}, C_{gs} : gate-drain and gate-source capacitances (via the inversion layer), which are non-linearly dependent on V_{gs}, V_{gd} and V_{gb}

C_{gb} : gate-substrate capacitance, which is non-linearly dependent on V_{gb}

The values of the C_{db} and C_{sb} diode capacitances in Fig. 1.29 are expressed as follows:

$$C(V) = \frac{C_{\mathrm{o}}}{(1 + \frac{V}{V_{\mathrm{j}}})^{1/m}} \tag{1.33}$$

where

C_o = capacitance when $V = 0$
V_j = junction voltage(0.6 V to 0.9 V)
m = grading factor, $2 \leq m \leq 3$: $m = 2$ for an abrupt junction and $m = 3$ for a linear junction

Terms C_{gdo} and C_{gso} represent gate overlap capacitances that are determined by the transistor width, the length of the overlap on the drain and source areas and the thickness of the gate oxide. These capacitances are clearly voltage independent.

The gate-substrate capacitance C_{gb} is only important if $V_{gs} \ll V_T$. Now, C_{gb} is often expressed as $C_{gb} \approx (0.12 \text{ to } 0.2) \cdot W \cdot L \cdot C_{ox}$. The inversion layer shields the substrate from the gate and $C_{gb} = 0$ when $V_{gs} \geq V_T$.

Terms C_{gd} and C_{gs} represent gate-drain and gate-source capacitances, respectively, which are present via the inversion layer (Fig. 1.28). The values of these capacitances depend strongly on the bias voltage on the terminals of the MOS transistor. The following cases are distinguished:

Case a $V_{gs} < V_T$; no inversion layer, thus $C_{gd} = C_{gs} = 0$.
Case b $V_{gs} > V_T$ and $V_{ds} = 0$.
For reasons of symmetry, $C_{gs} = C_{gd} = \frac{1}{2} \cdot W \cdot L \cdot C_{ox}$.
Case c $V_{gs} > V_T$ and $V_{ds} > V_{d_{sat}} (V_{d_{sat}} = V_{gs} - V_T)$.
The transistor is in saturation and there is no inversion layer at the drain: $C_{gd} = 0$ and $C_{gs} = \frac{2}{3} \cdot W \cdot L \cdot C_{ox}$.
This expression for C_{gs} is derived below.
Case d $V_{gs} > V_T$ and $0 < V_{ds} < V_{d_{sat}}$.
In this case, a linear interpolation between the values in cases b and c closely corresponds to the actual values, which are shown in Fig. 1.30.

The above expression in case c for the gate-source capacitance C_{gs} of a saturated MOS transistor is explained with the aid of Fig. 1.31. This figure shows a cross section of a MOS transistor biased in the saturated region. The channel does not reach the drain area but stops at a point where the channel potential is exactly $V_{gs} - V_T$.

Equation (1.5) leads to the following expression for the charge dQ in a channel section of length dx at position x:

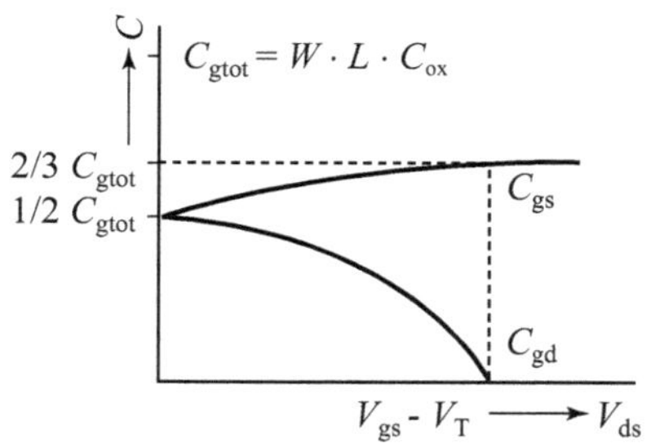

Fig. 1.30 C_{gs} and C_{gd} dependence on V_{ds} for $V_{gs} > V_T$

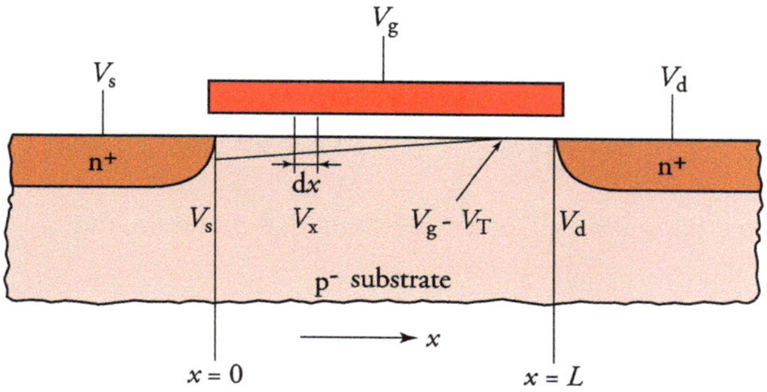

Fig. 1.31 Cross section of a saturated MOS transistor. $C_{gd} = 0$ and $C_{gs} = \frac{2}{3} \cdot W \cdot L \cdot C_{ox}$

$$dQ(x) = Q_n \cdot W \cdot dx = -W \cdot C_{ox}[V_{gs} - V_T - V(x)] \cdot dx \tag{1.34}$$

The following expression for dx is derived from Eq. (1.9):

$$dx = \mu_n \cdot C_{ox} \cdot W \cdot [V_{gs} - V_T - V(x)] \cdot \frac{dV(x)}{I_{ds}} \tag{1.35}$$

Combining Eqs. (1.34) and (1.35) yields the following expression for $dQ(x)$:

$$dQ(x) = \frac{\mu_n \cdot C_{ox}^2 \cdot W^2 \cdot [V_{gs} - V_T - V(x)]^2}{I_{ds}} \cdot dV(x) \tag{1.36}$$

Equation (1.15) yields the following expression for the drain current I_{ds} in a saturated MOS transistor:

$$I_{ds} = \frac{\beta}{2} \cdot (V_{gs} - V_T)^2 = \frac{\mu_n \cdot C_{ox}}{2} \cdot \frac{W}{L} \cdot (V_{gs} - V_T)^2 \tag{1.37}$$

Substituting Eq. (1.37) in Eq. (1.36) yields

$$dQ(x) = \frac{C_{ox} \cdot W \cdot L \cdot 2 \cdot [V_{gs} - V_T - V(x)]^2}{(V_{gs} - V_T)^2} \cdot dV(x) \tag{1.38}$$

Integrating Eq. (1.38) from the source to the imaginary drain gives

$$Q = \int_{V_s}^{V_{gs} - V_T} \frac{C_{ox} \cdot W \cdot L \cdot 2 \cdot [V_{gs} - V_T - V(x)]^2}{(V_{gs} - V_T)^2} \cdot dV(x)$$

$$= \frac{C_{\text{ox}} \cdot W \cdot L \cdot 2}{(V_{\text{gs}} - V_{\text{T}})^2} \cdot \left[-\frac{1}{3} \cdot [V_{\text{gs}} - V_{\text{T}} - V(x)]^3 \right|_{V_s}^{V_{\text{gs}} - V_{\text{T}}}$$

$$\Rightarrow Q = \frac{2}{3} \cdot W \cdot L \cdot C_{\text{ox}} \cdot (V_{\text{gs}} - V_{\text{T}}) \tag{1.39}$$

The gate-source capacitance C_{gs} can be found by differentiating Q in Eq. (1.39) with respect to V_{gs}

$$C_{\text{gs}} = \frac{\mathrm{d}Q}{\mathrm{d}V_{\text{gs}}} = \frac{2}{3} \cdot W \cdot L \cdot C_{\text{ox}} \tag{1.40}$$

The C_{gs} of a saturated MOS transistor is therefore only two thirds of the total value, while the gate-drain capacitance is zero.

In summary:

Most capacitances in a MOS transistor are non-linearly dependent on the terminal voltages. For each capacitance, these dependencies are as follows:

1. The diode capacitances C_{db} and C_{sb} :
   $C(V) = \frac{C_0}{(1+\frac{V}{V_{\text{j}}})^{1/m}}$, where $V_{\text{j}} \approx 0.6 \ldots 0.9\,\text{V}$ and $2 \leq m \leq 3$.
2. Figure 1.28 shows the voltage dependence of gate-channel capacitances C_{gd} and C_{gs} when the drain and source are short circuited, as is the case in a MOS capacitance. Figure 1.30 shows the voltage dependence of C_{gd} and C_{gs} when the drain and source are at different voltages, i.e. during normal transistor operation.
3. The gate-substrate capacitance C_{gb} is 0 when $V_{\text{gs}} > V_{\text{T}}$ and
   $C_{\text{gb}} = 0.2 \cdot W \cdot L \cdot C_{\text{ox}}$ if $V_{\text{gs}} < V_{\text{T}}$.
4. The overlap capacitances C_{gdo} and C_{gso} are the only capacitances which are not dependent on the terminal voltages.

1.11 Conclusions

The basic principles of the operation of the MOS transistor can be explained in different ways. The fairly simple approach adopted in this chapter should provide a good fundamental understanding of this operation. The current-voltage characteristics presented are derived by means of the simplest mathematical expressions for MOS transistor behaviour.

Second-order and parasitic effects are not essential to an understanding of the basic principles of MOS transistor operation. They have therefore been neglected in this chapter. However, these effects should be included in accurate descriptions of MOS transistors and are therefore discussed in Chap. 2. Most of these effects are included in the MOS transistor models used by commonly used compact MOS models in circuit simulation programs.

1.12 Exercises

Note $2\phi_f = 0.7$ V throughout these exercises.

1. What happens to the depletion layer in Fig. 1.12 when the substrate (b) is connected to a negative voltage (≈ -1 V) instead of ground?
   What effect does this have on the threshold voltage V_T ?
2. Current I_{ds} in a transistor ($\frac{W}{L} = 2$) is 100 µA when its gate-source voltage V is 0.8 V. The current is 324 µA when $V = 1.2$ V.

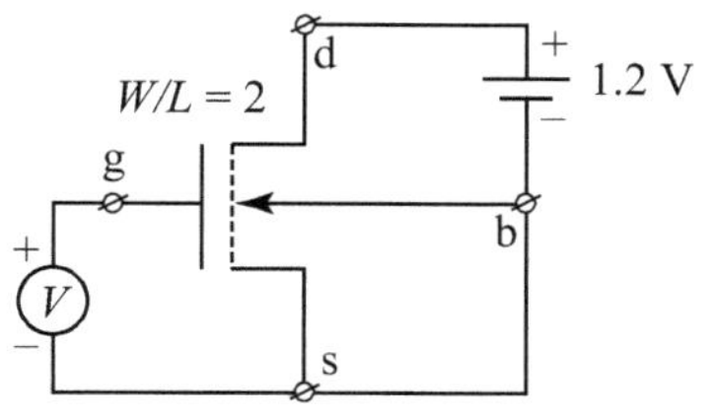

   (a) Which transistor operating regions (linear or saturated) do these values of V correspond to?
   (b) Calculate $\beta_\square$ and V_T for the given transistor.
3. Given:

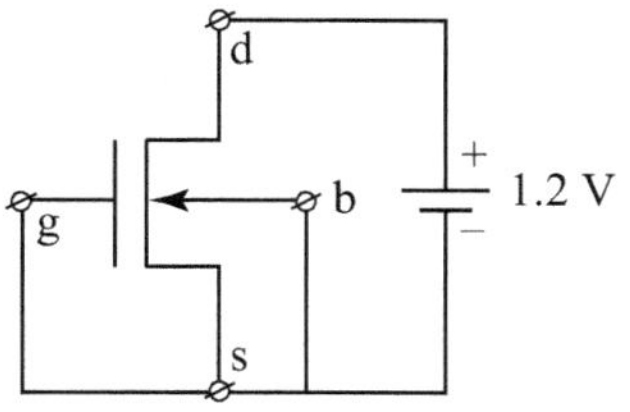

   (a) What type is the transistor shown?
   (b) Calculate I_{ds} when this transistor has the same β as the transistor in Exercise 2 and $V_T = -1$ V.
4. Given:

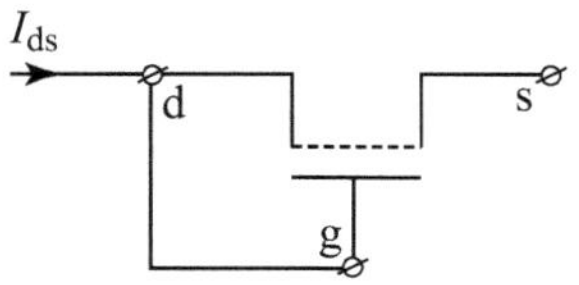

   If this is an n-type enhancement MOS transistor and the current $I_{ds} > 0$, explain the following:
   (a) This transistor is always in its saturation region.
   (b) This connection is often called a MOS diode.

5. For this exercise, the threshold voltage V_T is 0.25 V. There is *no* thermal generation of electron/hole pairs.

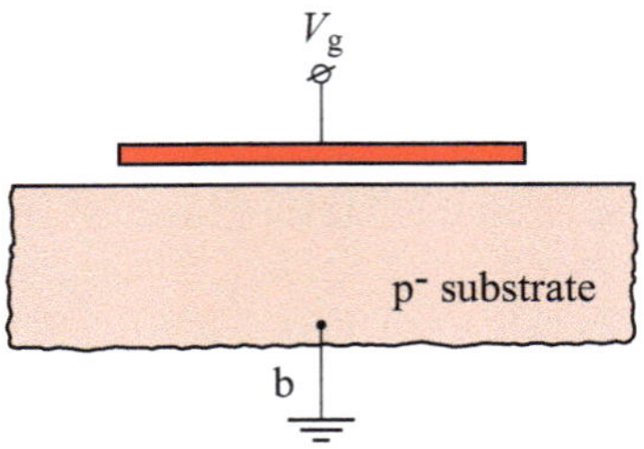

(a) The above structure exists when the source and drain areas of an nMOS transistor are excluded.
Copy this structure and include the possible depletion and inversion layers for the following values of V_g: −0.6 V, 0.1 V, 0.6 V and 1.2 V.

(b) An n^+ area is now added to the structure in Exercise 5a.

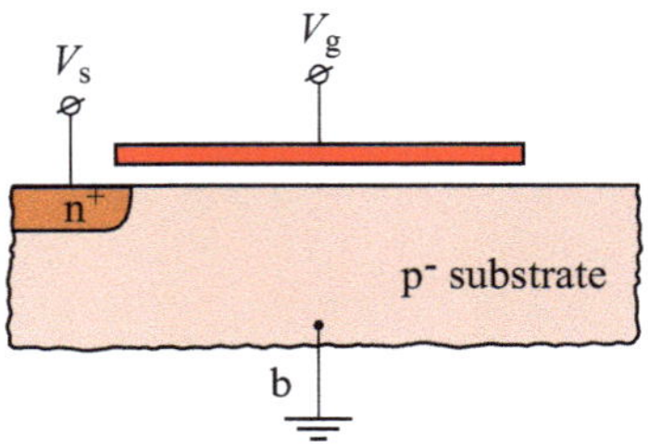

Repeat Exercise 5a for $V_s = 0$ V and for $V_s = 0.5$ V.

(c) The substrate of the structure in Exercise 5b is connected to a negative voltage: $V_{bb} = -1$ V.
What happens to the depletion and inversion layers if $V_s = 0$ V and $V_g = 0.5$ V?

(d) A second n^+ area is added to the structure of Exercise 5b to yield the following structure.

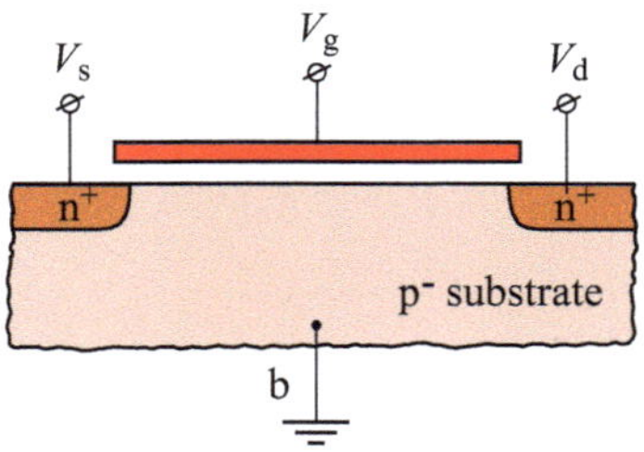

Repeat Exercise 5a for $V_s = 0$ V and $V_d = 0.8$ V.

(e) In practice, there are thermally generated electron hole pairs in the silicon substrate. The resulting free electrons in the depletion layer move in the opposite direction to the applied external electric field.
Draw the direction of movement of the thermally generated electrons and holes for $V_g = 1.2$ V in the structure of Exercise 5a.
If this situation continues for a longer period, a new equilibrium is reached and the electrons and holes accumulate in the structure. Draw this situation.

6. The following values apply in the figure shown:
$V_{dd} = 1.2$ V, $\beta = 1$ mA/V^2, $V_x = -1$ V, $V_{bb} = -1$ V.

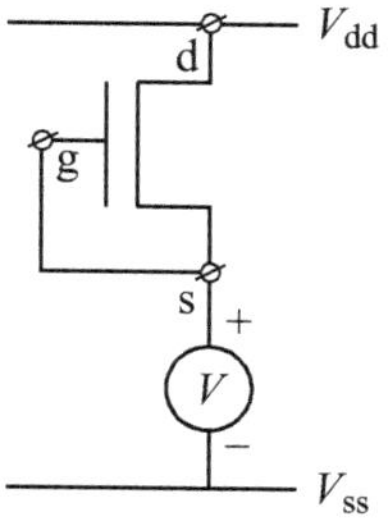

(a) What type is the transistor and why?
(b) Calculate and draw the graph $I_{ds} = f(V_{ds})$ for $K = 0\,\mathrm{V}^{1/2}$ and $V_{ds} = 0$, 0.2, 0.4, 0.6, 0.8, 1.0 and 1.2 V.
(c) Repeat b) for $K = 0.2\,\mathrm{V}^{1/2}$.
(d) Assuming $K = 0.2\,\mathrm{V}^{1/2}$, calculate the output impedance of the transistor for $V_{ds} = 50$ mV and for $V_{ds} = 0.6$ V.
(Note: the drain remains at 1.2 V).

7. The following values apply for the circuit shown:
$V_{dd} = 1.2\,\mathrm{V}$, $V_{bb} = -1\,\mathrm{V}$, $V_{ss} = 0\,\mathrm{V}$, K=$0.2\,\mathrm{V}^{1/2}$, $\beta_\square = 400\,\mu\mathrm{A/V}^2$, $V_{X_L} = -1\,\mathrm{V}$ and $V_{X_D} = 0.2\,\mathrm{V}$.

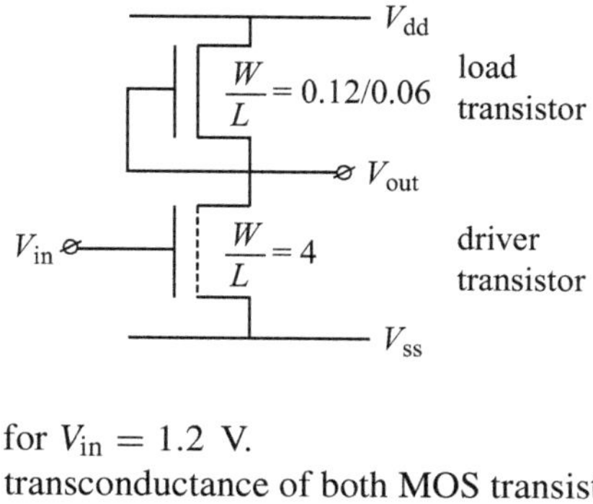

(a) Calculate V_{out} for $V_{in} = 1.2$ V.
(b) Determine the transconductance of both MOS transistors for this situation.
(c) What value does V_{out} reach when $V_{in} = 0.1$ V?
(d) The same low output level must be maintained when the load transistor is replaced by an enhancement-type transistor of the same size and with its

gate at V_{dd}. Does this require a driver transistor with the same $\frac{W}{L}$ and with a smaller or a larger channel width W? Explain your answer.

8. The aspect ratio of this transistor is $W/L = 200\text{nm}/50\text{nm}$. Results of measurements on it are summarised in the following table ($2\phi_f = 1\text{V}$):

	I_{ds} [μA]	
V_{sb} [V]	$V_{gs} = 0.5$ V	$V_{gs} = 1$ V
0	40	360
1.25	10	–

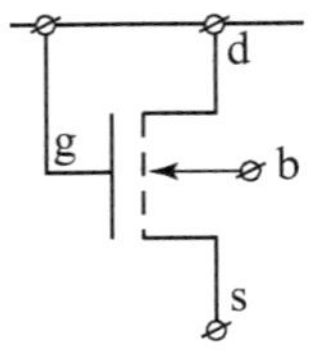

   (a) Determine V_x, K and $\beta_\square$ for this transistor.
   (b) Calculate and draw the graph $V_T = f(V_{sb})$ for at least five V_{sb} values (0 V$< V_{sb} <$ 2 V).

9. Define an expression for the transconductance with respect to the substrate voltage V_{sb} when the transconductance with respect to the normal gate voltage is defined as $g_m = \frac{\delta I_{ds}}{\delta V_{gs}}$.

10. Assume that we build a decoupling capacitor between V_{dd} and V_{ss} using an nMOS transistor with a gate-oxide thickness $t_{ox} = 1.6$ nm.
    (a) Draw how this nMOS transistor is connected between the V_{dd} and V_{ss} lines to form this capacitor.
    (b) What would be its capacitance value per unit area?
    (c) Assuming a pMOS transistor operates fully complementary to an nMOSt, how would you connect such a pMOSt as an additional capacitor in the same circuit as in a).

11. The following values apply in the figure shown: both nMOS transistors are identical, $V_{dd} = 1$ V, $V_T = 0.3$ V when $k = 0\,\text{V}^{1/2}$.

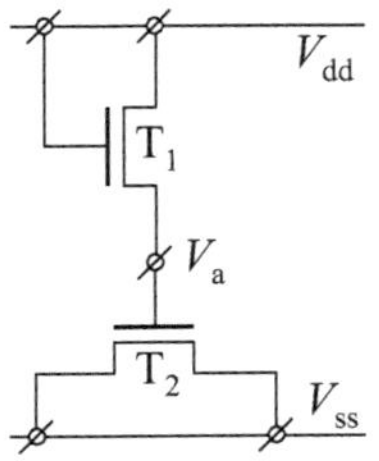

    (a) With what type of device could you compare T_1?

(b) With what type of device could you compare T_2?
(c) What would be the voltage V_a when $k = 0\,V^{1/2}$?
(d) What would be the voltage V_a when $k = 0.2\,V^{1/2}$?
(e) If one of the V_{ss} connections of T_2 would be left open (floating), what would be the result in terms of operation of the device T_2 and of the operation of the total circuit?
(f) Describe what happens with the voltage V_a, when V_{dd} peaks to 1.2V and back.
(g) By what device should T_1 be replaced to have the circuit operating as an efficient decoupling capacitor (with full V_{dd} across it)?

References

1. ITRS Roadmap (2014). www.ITRS.net

General Basic Physics

2. R.S.C. Cobbold, *Theory and Applications of Field Effect Transistors* (Wiley, New York, 1970)
3. S.M. Sze, *Semiconductor Devices: Physics and Technology*, 2nd edn. (Illustrated) (Wiley, New York, 2001)
4. Y.P. Tsividis, *Operation and Modelling of the MOS Transistor* (WCB McGraw-Hill, Boston, 1999)
5. C. Kittel, *Introduction to Solid State Physics*, 7th edn. (Wiley, New York, 1996)
6. D. Bhattacharya, N.K. Jha, FinFETs: From devices to architectures. Adv. Electron. **2014**, Article ID 365689, 21 p. (2014). http://www.hindawi.com/archive/2014/365689/
7. B. Ho, Evolutionary MOSFET Structure and Channel Design for Nanoscale CMOS Technology, dissertation for University of California, Berkeley, Fall 2012

MOS Capacitances

8. E.W. Greenwich, An analytical model for the gate capacity of small-geometry MOS structures. IEEE Trans. Electron Dev. **ED-30**, 1838–1839 (1983)
9. J.J.Paulos, D.A. Antoniadis, Y.P. Tsividis, Measurement of Intrinsic Capacitances of MOS Transistors. ISSCC Digest of technical papers, pp. 238–239 (1982)
10. D.E. Ward, R.W. Dutton, A charge-oriented model for MOS transistor capacitances. IEEE J. Solid State Circuits **sc-13**, 703–707 (1978)
11. H. Kogure, et al., Analysis of CMOS ADC non-linear input capacitance. IEICE Trans. Electron. **E85-C**(5), 1182–1190 (2002)
12. Chenming-Hu, MOS Capacitor, Chapter 5, Hu_ch05v3.fm, p. 194, February 13th (2009)

Geometrical, Physical and Field-Scaling Impact on MOS Transistor Behaviour

2

2.1 Introduction

The simple formulae derived in Sects. 1.4 and 1.5 account for the first-order effects which influence the behaviour of MOS transistors. Until the mid-seventies, Formulae (1.18) appeared quite adequate for predicting the performance of MOS circuits. However, these transistor formulae ignore several physical and geometrical effects which significantly degrade the behaviour of MOS transistors. The results are therefore considerably more optimistic than the actual performance observed in MOS circuits. The deviation becomes more significant as MOS transistor sizes decrease in VLSI circuits.

This chapter contains a brief overview of the most important effects, in nanometer CMOS technologies, which degrade the performance of MOS devices. The chapter concludes with a detailed discussion on transistor leakage mechanisms.

2.2 The Zero-Field Mobility

As discussed in Chap. 1, the MOS transistor current is heavily determined by the *gain factor* β of the transistor:

$$\beta = \frac{W}{L} \cdot \beta_{\square} = \frac{W}{L} \cdot \mu \cdot C_{\text{ox}} \tag{2.1}$$

where W and L represent the transistor channel width and length, respectively, C_{ox} represents the gate oxide capacitance per unit of area and μ represents the actual *mobility* of the carriers in the channel. This mobility can be quite different from the zero-field or substrate mobility μ_0, which depends on the doping concentration in the substrate. Figure 2.1 shows zero-field electron and hole mobilities in silicon at room temperature as a function of the doping concentration. The reason that we

H. Veendrick, *Nanometer CMOS ICs*,
https://doi.org/10.1007/978-3-031-64249-4_2

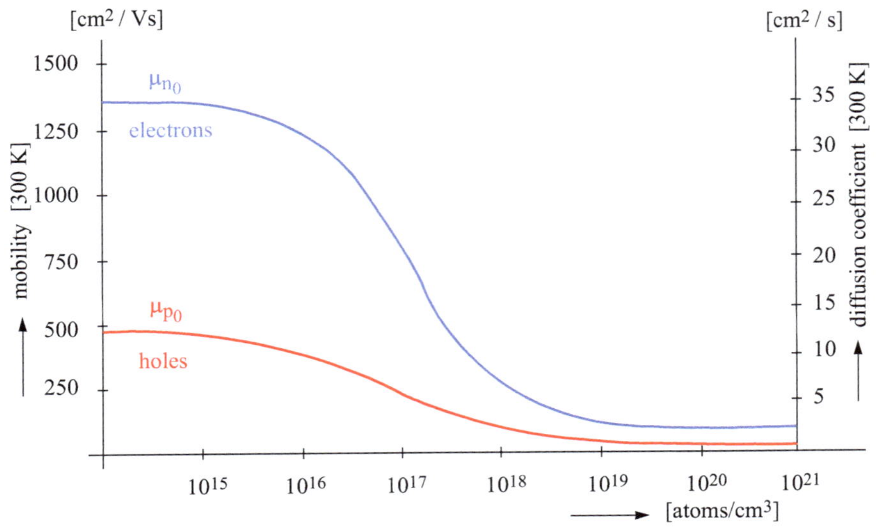

Fig. 2.1 Zero-field carrier mobility and diffusion coefficient as a function of doping concentration in silicon at room temperature

can have two axis representing the same diagram is that there is a linear relationship between the mobility and the diffusion coefficient (Einstein relation: $\mu = D{\cdot}q/kT$).

For a channel doping concentration of 10^{17} atoms/ cm^3, the mobility of electrons (μ_{n0}) is about three times that of holes (μ_{p0}), in the absence of an electric field. This is the major reason that the I_{on} current (which is the saturation current when $V_{gs} = V_{dd}$) of an nMOS transistor is about two to three times higher than the I_{on} of an equally sized pMOS transistor, depending on the technology node. It also depends on the transistor stress engineering and crystal orientation. However, several other effects dramatically reduce the mobility of the carriers in the channel. These are discussed in Sect. 2.3.

2.3 Carrier Mobility Reduction

During normal transistor operation, electrical fields are applied in both the lateral (horizontal) and transversal (vertical) directions, which influence the mobility of the carriers in the channel. Moreover, when the chip temperature is increased, either by an increase of the ambient temperature or by the chip's own dissipation, this will have a negative effect on the carrier mobility and thus on the β of each transistor.

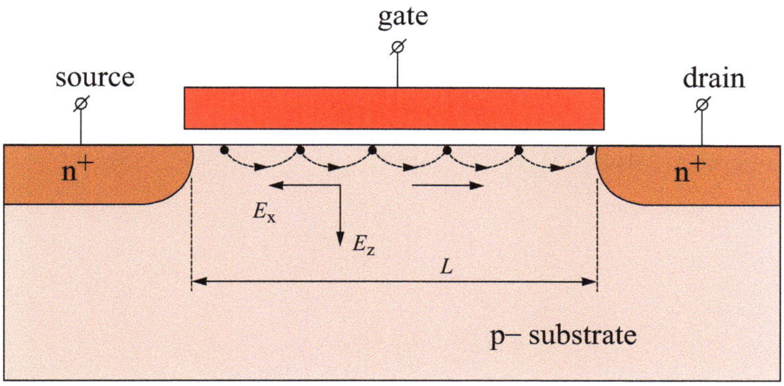

Fig. 2.2 Components which affect carrier mobility in MOS transistors

2.3.1 Vertical and Lateral Field Carrier Mobility Reduction

During normal operation, the effective mobility μ of the carriers in the transistor channel is degraded by the mechanisms indicated in Fig. 2.2. These include the *vertical electric field* E_z, the lateral electric field E_x and the carrier velocity v.

When the vertical electric field E_z is high, the minority carriers in an n-channel device are strongly attracted to the silicon surface, where they rebound. The resulting *'surface scattering'* is indicated by the dashed lines in Fig. 2.2. This causes a reduction of the recombination time and of carrier mobility μ with increasing E_z.

In [1], some experimental results are presented with respect to the vertical field carrier mobility degradation. The vertical electric field depends on the gate voltage and on the substrate voltage. The relationship between these voltages and the mobility can be expressed as follows:

$$\mu = \frac{\mu_0}{1 + \theta_1(V_{gs} - V_T) + \theta_2(\sqrt{V_{sb} + 2\phi_F} - \sqrt{2\phi_F})} \tag{2.2}$$

where μ_0 represents the zero-field substrate mobility, ϕ_F represents the Fermi level in the substrate and θ_1 and θ_2 are technology defined constants.

The carriers in the transistor channel are accelerated to a maximum velocity when the *lateral electric field* E_x becomes high. This means that, above a critical field $E_{x_{sat}}$ (Fig. 2.3), the carrier velocity is no longer related to E_x and reaches a constant level (v_{sat}). A good first-order approximation for this 'velocity saturation' phenomenon is

$$\mu = \frac{\mu_0}{1 + E_x/E_{x_{sat}}} \tag{2.3}$$

where

$$E_{\mathrm{x}} \approx \frac{V_{\mathrm{ds}}}{L} \tag{2.4}$$

Substituting Eq. (2.4) into Eq. (2.3) yields

$$\mu = \frac{\mu_0}{1 + \theta_3 \cdot V_{\mathrm{ds}}} \tag{2.5}$$

$$\text{where} \quad \theta_3 = \frac{1}{L \cdot E_{\mathrm{x_{sat}}}} \tag{2.6}$$

The above effects are included in the following expression for carrier mobility:

$$\mu = \frac{\mu_0}{(1 + \theta_1(V_{\mathrm{gs}} - V_{\mathrm{T}}) + \theta_2(\sqrt{V_{\mathrm{sb}} + 2\phi_{\mathrm{F}}} - \sqrt{2\phi_{\mathrm{F}}}))(1 + \theta_3 V_{\mathrm{ds}})} \tag{2.7}$$

At high gate voltages, the vertical field influence (represented by the voltage terms containing V_{gs} and V_{sb}) may reduce the transistor current by about 50%. The lateral field influence may be of the same order of magnitude. Note that this lateral field close to the source dominates the drain-source current. At a level of about 1 V/μm, this lateral field also reduces the electron mobility in the channel of an nMOS transistor by almost 50%. Thus, the total field-dependent mobility reduction can amount to a factor four.

The actual mobility is equal to the substrate mobility when $E_{\mathrm{z}} = 0$. Some transistor models include the series resistance of the source (R_s) and the drain (R_d) in the surface scattering factor θ_1 and in the velocity-saturation factor θ_3. Moreover, these resistances are weakly dependent on the terminal voltages and are therefore included in the device equations. Their influence can be incorporated in Eq. (2.7) by replacing θ_1 and θ_3 by $\theta_1{}'$ and $\theta_3{}'$, respectively, where

$$\theta_1' = \theta_1 + \beta \cdot (R_s + R_d) \tag{2.8}$$

and

$$\theta_3' = \theta_3 - \beta \cdot R_d \tag{2.9}$$

Because these resistances are in series with the transistor terminals, they have a reducing effect on the transistor current.

Due to the ultrashort-channel lengths, today, transistors show electric fields that exceed 10 V/μm. Because $E_{\mathrm{x_{sat}}}$ is smaller for electrons than for holes, the performance of nMOS transistors (conduction by electrons) is more degraded by velocity saturation than that of pMOS transistors (conduction by holes). Figure 2.3 shows the carrier velocity v as a function of the electric field E_{x} in the channel. The

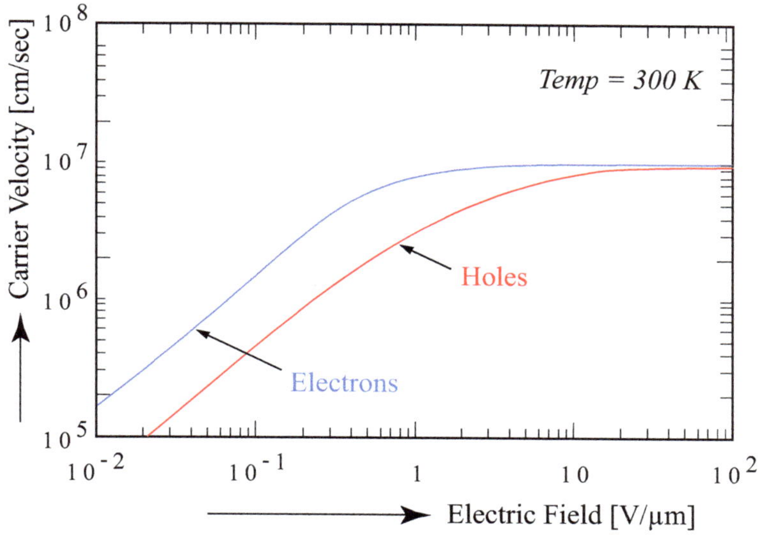

Fig. 2.3 Carrier velocity as a function of the lateral electric field in the channel

critical field $E_{x_{sat}}$ to reach velocity saturation depends on the doping level and is about 3 V/µm for electrons and about 10 V/µm for holes. In fact, holes may reach a saturation velocity comparable to the v_{sat} of electrons.
This carrier velocity is defined by

$$v = \mu \cdot E_x \approx \mu \cdot V_{ds}/L \tag{2.10}$$

So, in the derivation of the current expression (1.11), with $\beta = \mu \cdot C_{ox}$, we can replace $\mu \cdot V_{ds}/L$ by v. Particularly in 90 nm CMOS processes and beyond, most carriers in the channel travel at a maximum *saturation velocity* v_{sat}. This would lead to a saturation current equal to

$$I_{ds} = \frac{v_{sat} C_{ox} W}{2} \cdot (V_{dd} - V_T) \tag{2.11}$$

This reduces the channel length's influence on the current, which is one of the reasons that the transistor's *drive current* has shown negligible increase over the last couple of technology generations. This is not expected to change for future generations; for almost all CMOS processes from 180 to 32 nm, the I_{on} for the nMOS and pMOS transistors has almost remained constant at about 650 µA/µm and 270 µA/µm, respectively. As a result, the effective gain factor $\beta_{\square eff}$ of a transistor with minimum drawn gate length (L_{drawn}) almost remains at a constant value, which is close to (or today even below) 50 µA/V^2. Figure 2.4 shows this gain factor as a function of the drawn channel length for various technology generations [2]. *LSTP*

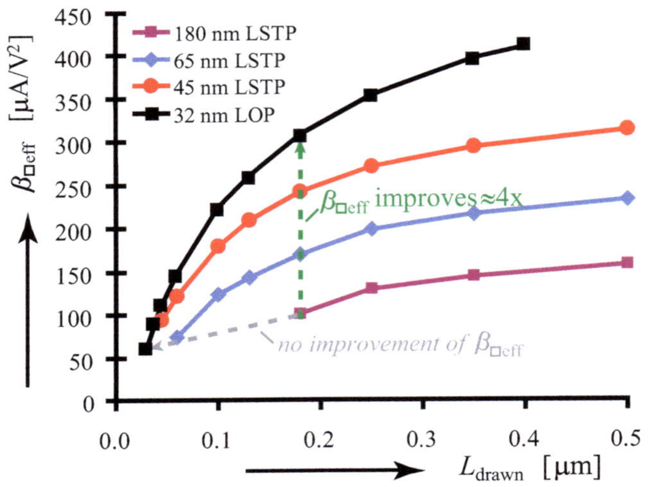

Fig. 2.4 The effective gain factor versus the drawn channel length for various technology generations [2]

and *LOP* refer to a *low standby power* and a *low operating power*, respectively. Table 11.5 (Chap. 11) shows values for and trends in the effective mobility (μ_{eff}) and the transistor on-current I_{on} for advances process nodes beyond 5 nm. Also here, μ_{eff} is expected to reduce from 125 to 40 cm^2/Vs over five process nodes, while the trend in I_{on} remains more or less flat.

It also shows that $\beta_{\square\text{eff}}$ improves dramatically with increasing L_{drawn}. For example, an increase of L_{drawn} in the 45 nm CMOS node from minimum length to about 100 nm will double the value of $\beta_{\square\text{eff}}$. For digital circuits, this channel length increase is certainly not an appropriate solution for performance increase, because it also increases the chip area. However, if we compare a minimum drawn transistor ($L_{\text{drawn}} = 0.18\,\mu\text{m}$) in 0.18 μm technology with a transistor with the same $L_{\text{drawn}} = 0.18\,\mu\text{m}$ in 32 nm technology, the performance improvement is close to a factor of four. So, for analogue circuits, which usually take a small portion of the chip area on a mixed analog/digital design, a larger channel length serves two goals. First it improves the transconductance ($g_m \sim \beta_{\square\text{eff}}(V_{\text{gs}} - V_{\text{T}})$) and, secondly, it reduces the threshold voltage variation, which improves the transistor *matching* properties and enables a better performance prediction. These variability-related topics are discussed in Chap. 9.

Chapter 3 presents a few mobility enhancement techniques which will improve the performance of both the digital and analogue circuits in advanced and future nanometer CMOS ICs.

2.3.2 Stress-Induced Carrier Mobility Effects

The mechanical stress (*STI stress*) induced in the wafer by shallow-trench isolation has an increasing effect on the carrier mobility of the nMOS and pMOS devices. It degrades the mobility of nMOS devices, while it slightly improves it of the pMOS devices. Chapter 3 discusses mobility enhancement techniques to improve transistor device performance in advanced CMOS technologies.

2.4 Channel Length Modulation

The ideal I_{ds} vs V_{ds} characteristics illustrated in Fig. 1.16 do not show the influence of V_{ds} on I_{ds} in the saturation region. In practice, an increase in V_{ds} in the saturation region causes an increase in I_{ds}. This phenomenon is particularly obvious in short-channel devices and is caused by *channel length modulation*.

The distribution of carriers in an nMOS transistor operating in the saturation region ($V_{\mathrm{ds}} > V_{\mathrm{dssat}} = V_{\mathrm{gs}} - V_{\mathrm{T}}$) is illustrated in Fig. 2.5. The operation of the basic MOS transistor in this region is discussed in Sect. 1.3. Clearly, the end of the inversion layer (which is called the virtual drain) does not reach the actual drain. The effective channel length therefore equals $L - \Delta L_{\mathrm{CLM}}$.

The saturation current specified in Eq. (1.15) must be changed to account for the effective channel length. The modified expression is as shown in Eq. (2.12):

$$I_{\mathrm{ds_{sat}}} = \frac{W}{L - \Delta L_{\mathrm{CLM}}} \cdot \frac{\beta_{\square}}{2} \cdot (V_{\mathrm{gs}} - V_{\mathrm{T}})^2 \tag{2.12}$$

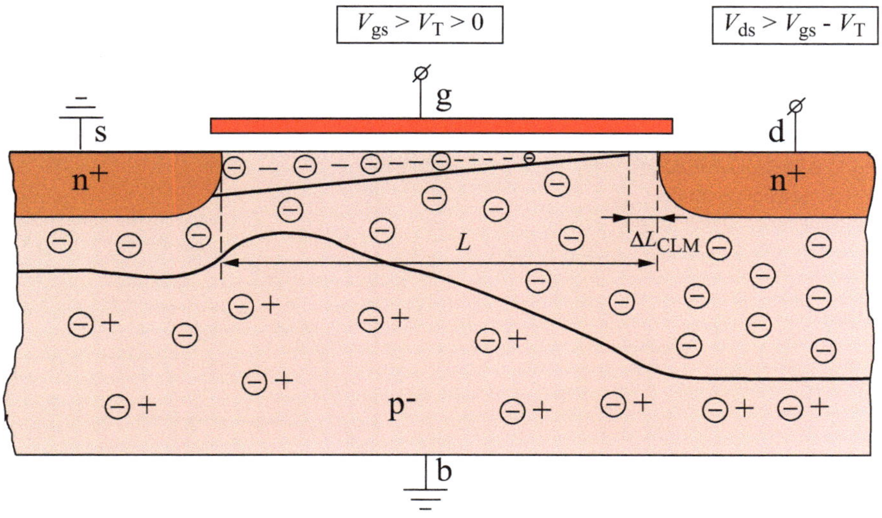

Fig. 2.5 A MOS transistor in the saturation region ($V_{\mathrm{ds}} > V_{\mathrm{gs}} - V_{\mathrm{T}}$)

where ΔL is the length of the depletion region at the silicon surface between the inversion layer and the drain. In the above expression, the total field-dependent mobility degradation, as discussed before, is not included.

The voltage $V_{ds} - V_{dssat}$ across this *'pinch-off' region* modulates ΔL_{CLM}. This effect can be modelled by

$$\frac{\Delta L_{CLM}}{L} = \alpha \ln{(1 + \frac{V_{ds} - V_{dssat}}{V_P})} \tag{2.13}$$

where α and V_P are constants, which may vary with the transistor geometry. The expression clearly shows the relation between ΔL_{CLM} and the amount of V_{ds} voltage above V_{dssat}.

The above discussions show that the additional contribution to the drain current of a MOS transistor operating in the saturation region is proportional to $V_{ds} - V_{dssat}$. This effect is sometimes approximated by the following modified current expression:

$$I_{ds} = (1 + \lambda V_{ds}) \cdot I_{ds0} \tag{2.14}$$

where I_{ds0} is the transistor current when the channel length modulation is ignored and λ is a semi-empirical channel length modulation parameter, whose reciprocal value $(1/\lambda)$ is analogous to the BJT Early voltage. The effect of this channel length modulation on the $I_{ds} = f(V_{ds})$ characteristics is shown in Fig. 2.6, where the extrapolation of the curves in the saturation region would all intersect the x-axis closely to the point $1/\lambda$.

Channel length modulation is an undesired effect, which is particularly an issue for analogue design. Because channel length modulation is a short-channel effect which rapidly decreases with longer channels, analogue designs typically require larger than minimum transistor channel lengths to improve the performance and operating margins.

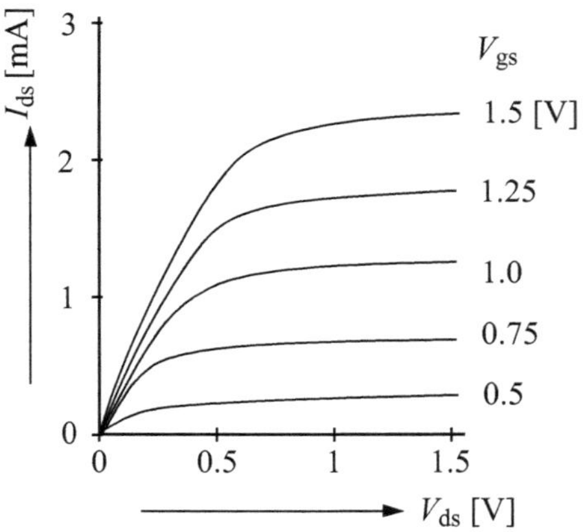

Fig. 2.6 Effect of channel length modulation on the MOS transistor characteristic

2.5 Short- and Narrow-Channel Effects

The electrical behaviour of a MOS transistor is primarily determined by its gain factor β, its threshold voltage V_T and its body factor K. Generally, the values of these parameters are largely dependent on the width W and length L of a transistor. The influence of these dependencies increases as transistor dimensions decrease. These small-channel effects, which are discussed below, are particularly significant in deep-submicron and nanometer MOS processes.

2.5.1 Short-Channel Effects

The cross section presented in Fig. 2.7 is used to explain short-channel effects.

Even in the absence of a gate voltage, the regions under the gate close to the source and drain are inherently depleted of majority carriers, i.e. holes and electrons in nMOS and pMOS transistors, respectively. In a short-channel transistor, the distance between these depletion regions is small. The creation of a complete depletion area under the gate therefore requires a relatively small gate voltage. In other words, the threshold voltage is reduced. This is a typical two-dimensional effect, which can be reduced by shallow source and drain diffusions. However, the associated smaller diffusion edge radii cause a higher electric field near the drain edge in the channel when $V_{ds} > V_{gs} > V_T$. One way to overcome this problem is to reduce the supply voltage. This short-channel effect on the threshold voltage occurs at shorter gate lengths and causes *threshold voltage roll-off* (see Fig. 2.8).

The use of shallow source and drain extensions (see Fig. 3.51), with less doping than the real source and drain implants, in combination with local higher doped channel regions (so-called *halo* or *pocket implants*) suppresses the depletion-layer

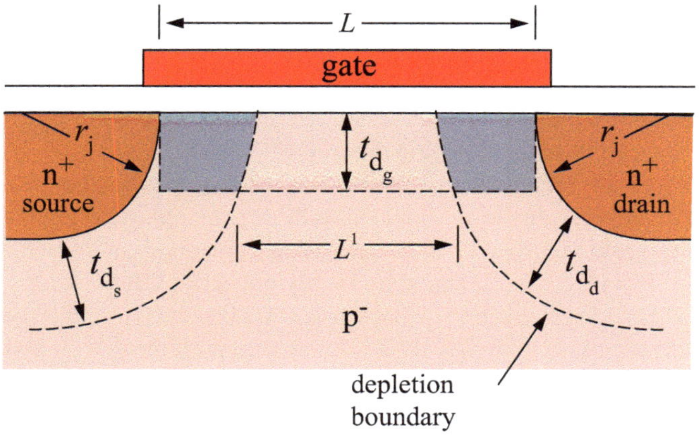

Fig. 2.7 Cross section of a short-channel transistor, showing several depletion areas that affect each other

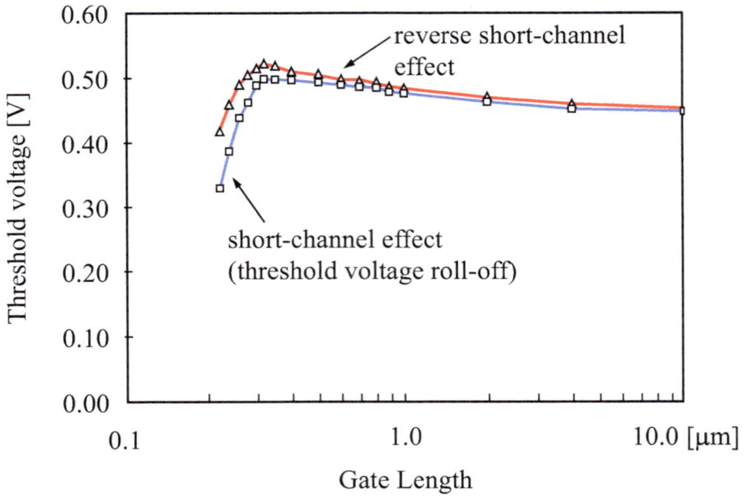

Fig. 2.8 Short-channel and reverse short-channel effect on the threshold voltage V_T of an nMOS transistor

width in the channel and contributes to a reduction of the *short-channel effect* (SCE). Usually, the *halos* are implanted after the gate formation (self-aligned with the gate) at a tilt angle with peak dope just below the active channel, adjacent to the source and drain. During operation, a positive voltage is applied to the drain. This brings the drain junction into reverse bias, causing expansion of the depletion region surrounding the drain, also in the direction of the source. This depletion layer may then merge with that of the source. It can short the channel, i.e. even at zero gate-source voltage (*punch-through*), or it increases the subthreshold leakage current. This effect is closely related to the *DIBL*, as discussed in Sect. 2.7.1. The halo implant is optimised for transistors with the smallest channel lengths in a given process. These transistors will have the nominal threshold voltage, while transistors with longer channels will have higher threshold voltages. The use of source and drain extensions, as well as of halos in the architecture of both nMOS and pMOS transistors, is clearly illustrated in Fig. 3.51.

A second effect that depends on the channel length is the *reverse short-channel effect (RSCE)*. In conventional CMOS devices, this effect, which involves increasing threshold voltages V_T with decreasing gate length, is caused by a lateral non-uniform channel doping induced by locally enhanced diffusion.

As described before, current devices use so-called halo implants to suppress short-channel effects. Figure 2.9 shows a possible dope profile in a device with halos. In devices with relatively long channels, these halos occupy a smaller region of the channel. When the channel becomes shorter, these halos get closer to one another and will also cause V_T roll-up. In 180 nm CMOS technologies and beyond, these halos intentionally cause roll-up and suppress the onset of roll-off.

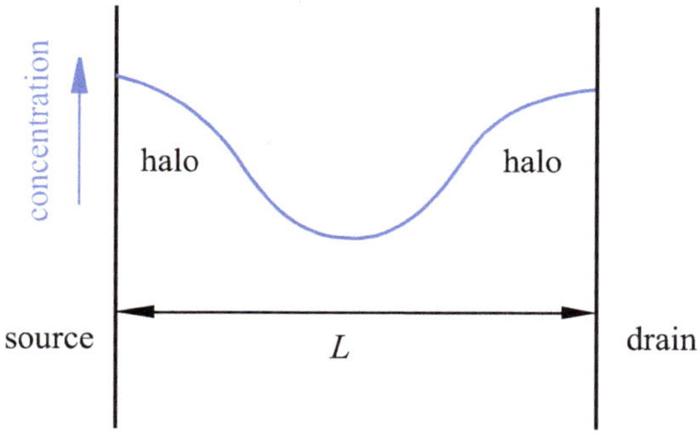

Fig. 2.9 Potential doping profile in the channel of a MOS device including the halos

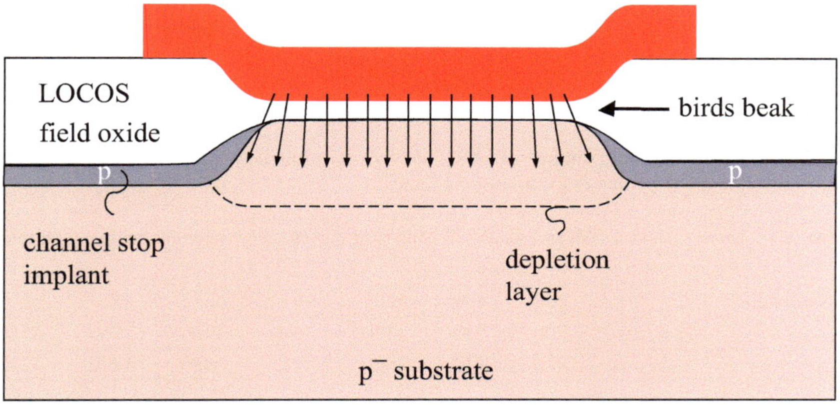

Fig. 2.10 Cross section of a narrow-channel transistor showing the distribution of electric field lines under the gate

2.5.2 Narrow-Channel Effect

Also, the width of an active device influences the threshold voltage. The depletion layer extends under the edges of the gate, where the gate electrode crosses the field oxide. With a LOCOS type of field isolation (see Fig. 2.10), this effect is primarily caused by the encroachment of the channel stop dopant at the edge of the field isolation.

The additional depletion region charge has to be compensated by an additional gate voltage. This results in an increase of the threshold voltage at reduced width of the device. The encroachment of channel stop dopant is especially pronounced for a conventional diffused well technology. The channel stop dopants are implanted prior to the high-temperature LOCOS oxidation and cause a large shift in V_T. In a retrograde implanted well process, the field oxidation is performed prior to the well implants, and less encroachment of dopant atoms occurs under the gate edge. However, the threshold voltage is still increased as a result of the bird's beak and two-dimensional spreading of the field lines at the edge.

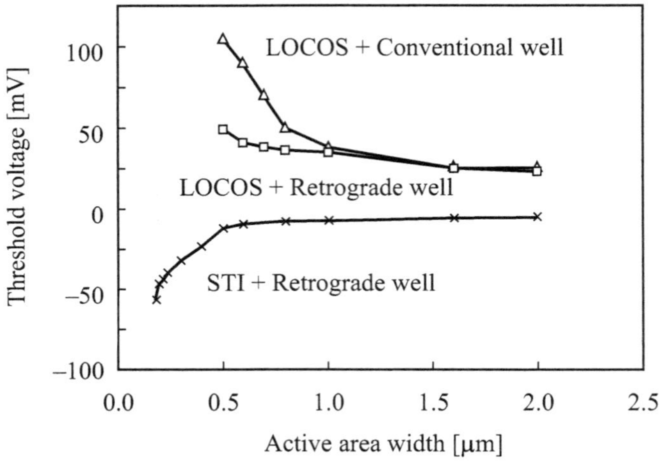

Fig. 2.11 Shift of threshold voltage of nMOS devices as a function of the active area width for different well technology and field isolation schemes

Table 2.1 An evolution of nMOS and pMOS transistor current drive strength against technology node [3]

Process node	nm	250	180	130	90	65	45	32
$I_{\text{dsat,n}}$	$\mu A/\mu\text{m}$	700	940	1170	1449	1460	1360	1550
$I_{\text{dsat,p}}$	$\mu A/\mu\text{m}$	320	420	600	725	880	1070	1210

Figure 2.11 shows this *narrow-channel effect*, together with the influence of the channel width on the threshold voltage in a shallow-trench isolation (see Chap. 3) scheme. In contrast to the conventional narrow-width effect, the threshold voltage is even decreased at very narrow-channel widths of around 0.2 μm. This *inverse narrow-width effect (INWE)* is attributed to a sharp corner at the top of the shallow-trench isolation. The fringing field at this corner results in an increased electrical field strength and reduces the threshold voltage. Also, the quality of the oxide used to fill the trench is not as good as the thermally grown LOCOS field oxide. A positive fixed oxide charge is present in the oxide, and, in nMOS devices, it contributes to the decreased threshold voltage. This contribution of the fixed oxide charge is less severe than the fringing field component and depends also on the deposition method used to fill the trench.

The additional process steps to compensate mobility reduction and short-channel effects resulted in a much closer current drive strength of nMOS and pMOS transistors. Table 2.1 shows the Intel current drive strengths for high-performance devices in various technology nodes.

For smaller process nodes, the nMOS and pMOS transistor current efficiencies have reduced per node, due to increasing surface scattering and other carrier mobility reduction effects. By the introduction of FinFET devices, the effective current per unit area has increased due to the use of vertical channels. However, carrier mobility reduction effects have also reduced the $I_{\text{dsat,n}}$ and $I_{\text{dsat,p}}$ of FinFETs over several process nodes. According to the 2022 IRDS More Moore (see Reference [90] in Chap. 3), these values are about 450 and 200 μA/μm for the nMOS and pMOS, respectively, and are expected to remain at those levels until the year 2035.

2.6 Temperature Influence on Carrier Mobility and Threshold Voltage

An increase in the operating temperature of a MOS transistor affects its behaviour in two different ways:

1. The mobility of the majority carriers, e.g. electrons in an nMOS transistor, in the channel decreases. Consequently, the transistor gain factor $\beta_\square$ also decreases. Its temperature dependence is expressed as follows [4]:

$$\beta_\square\,(Temp) = \beta_\square\,(298\,\text{K}) \cdot \left(\frac{298}{Temp}\right)^m \tag{2.15}$$

   From physical considerations, the value of the exponent m is close to 1.5, depending on the technology. For process nodes in the order of 10 nm and below, this value is close to 1.
2. The threshold voltage V_T of both nMOS and pMOS transistors decreases slightly. The magnitude of the influence of temperature change on threshold voltage variation ΔV_T depends on the substrate doping level. A variation between −0.6 mV/°C and −1 mV/°C is quite typical, with the smaller values being more applicable for the sub-10 nm nodes.

Both effects have different consequences for the speed of an IC. This speed is determined by the delay τ of a logic gate, which is defined as

$$\tau = \frac{CV}{I} = 2 \cdot \frac{CV}{\beta(V_\text{gs} - V_\text{T})^2} \tag{2.16}$$

In conventional CMOS processes, the overall circuit performance reduces with increasing temperature, because its effect on the mobility reduction in the transistor current was traditionally larger than the effect on the reduction of the threshold voltage. This was one of the reasons to keep high-speed processors cool, by using a fan. Also, worst-case corner simulations were usually done at high temperatures. However, today's CMOS technologies offer several different threshold voltages to support both high-speed and low-leakage applications. For general-purpose and high-speed processes, V_T is relatively low, and a further reduction with 1 mV/°C has less influence on this speed than the reduction in the β. For low-leakage processes, with a relatively large V_T, both effects partly compensate each other,

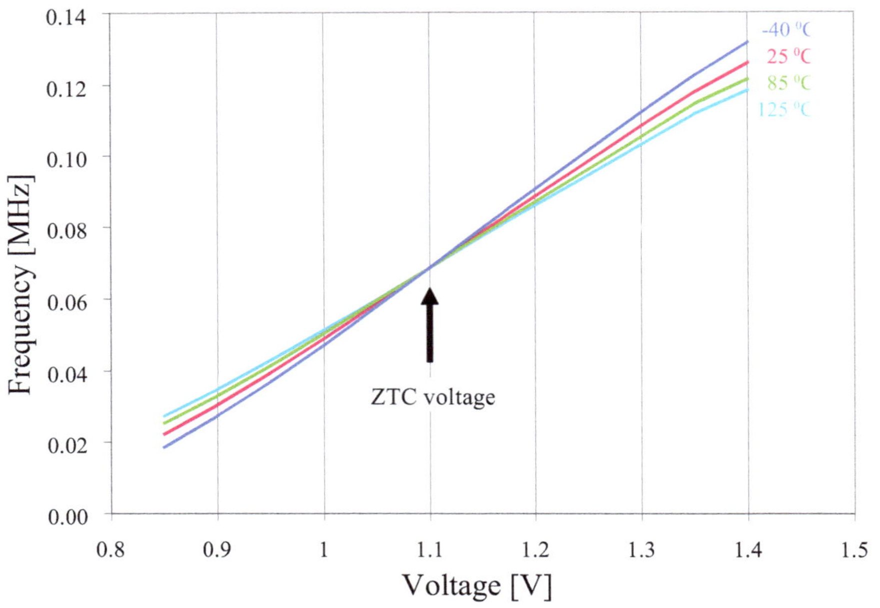

Fig. 2.12 Ring oscillator frequency response as a function of the supply voltage at different temperatures

because of the increasing competition between mobility and threshold voltage, so that there is a reduced influence on the speed. At a certain supply voltage, the above two mechanisms fully cancel each other's contribution to the transistor current, such that the circuit speed has no longer a relation with the temperature. This is the so-called *zero-temperature-coefficient (ZTC) voltage* [5, 6]. This reducing temperature dependence, which is expected to continue with further scaling of the supply voltage, has serious consequences for the static timing analysis, as it may invalidate the approach of defining *PVT* (process, voltage and temperature) corners, by independently varying voltage and temperature [7]. Figure 2.12 shows the frequency response of a high-V_T ring oscillator as a function of the supply voltage, for different operating temperatures.

Above the ZTC voltage of 1.1 V, which is close to the nominal supply voltage of 1.2 V in this technology, the frequency reduces with increasing temperature, while below this voltage, the effect is opposite. For the same ring oscillator fabricated with a standard V_T, this ZTC is reduced to 0.95 V.

As a result of this varying temperature behaviour, the worst-case and best-case corners for simulation need to be reconsidered, since for modern CMOS technologies a higher temperature not automatically corresponds to a lower performance! For the 45 nm technology node and beyond, the temperature effect will diminish further, because of an increasing compensation of the β and V_T contributions to the transistor current [8]. Also, due to the combined use of high-V_T and standard-V_T

or low-V_T cores on the same chip, some cores may run faster, while other cores may run slower at increasing temperatures. This must accurately and extensively be simulated in order to prevent timing problems in the application. Since different companies/foundries use different transistor architectures in the smaller process nodes (from 20 to 1 nm), the ZTC phenomenon must be extensively simulated in order to find out which parts of the chip must be simulated at what worst-case temperature. Reference [9] gives an example of ZTC behaviour for a nano-sheet nMOS transistor.

ZTC also has consequences for certain failure analysis methods (see Chap. 10) that use local heating to detect changes in circuit behaviour, because these changes will become smaller and less visible in modern technologies.

2.7 MOS Transistor Leakage Mechanisms

Due to the continuous scaling of the physical MOS device dimensions, such as channel length and gate-oxide thickness, and increasing doping levels to suppress short-channel effects (SCEs), MOS devices will increasingly drift away from an ideal switching behaviour. As a result, an increasing number of leakage mechanisms is influencing their performance, particularly during off-state. Figure 2.13 shows the major contributions to the total transistor leakage current.

The reduction of the supply voltage (V_{dd}) for a series of technology generations has caused the threshold voltage (V_T) to reduce accordingly, in order to achieve sufficient performance increase. A lower V_T leads to a larger off-current: a larger

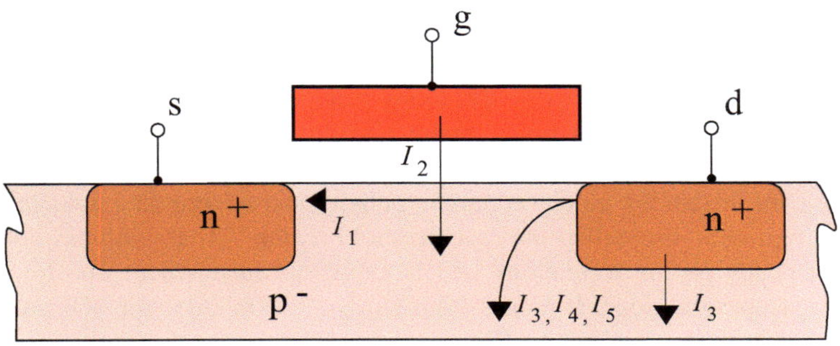

I_1 = sub-threshold leakage (incl. drain-induced barrier lowering (DIBL))
I_2 = gate oxide tunneling
I_3 = reverse bias junction leakage
I_4 = gate-induced drain leakage (GIDL)
I_5 = impact ionisation current

Fig. 2.13 Leakage mechanisms in nanometer CMOS technologies drain-to-source leakage current when the gate-to-source voltage is zero, subthreshold leakage current.

The reduction of the gate-oxide thickness below 2.5 nm has resulted in charge carriers tunnelling through the oxide from the channel to the gate or vice versa and causes a gate-leakage current. This tunnelling is not caused by the field only, but mainly by the penetration of the carriers into the oxide. A potential difference of 2 V across an oxide thickness of 2 nm will cause tunnelling, while a 5 V potential difference across 5 nm oxide won't.

Finally, the increasing number of SCEs, in particular the threshold voltage roll-off effect, requires additional local pocket implants in the channel region close to the source and drain. This leads to a larger reverse-bias drain junction leakage current.

A few other leakage mechanisms contribute to the total leakage, such as gate-induced drain leakage (GIDL) and impact ionisation current. The next subsections will discuss the mechanisms behind these leakage currents in more detail. Particularly, during the standby mode of most portable electronic devices, these leakage currents cause a relatively large standby power, thereby limiting the battery lifetime. Design methods to reduce these leakage power components are presented in Chap. 8.

2.7.1 Weak-Inversion (Subthreshold) Behaviour of the MOS Transistor

An nMOS transistor operates in the '*weak-inversion*' region when its gate-source voltage (V_{gs}) is below its threshold voltage (V_T); see Fig. 2.14. Below the threshold voltage, the current decreases exponentially. On a logarithmic scale, the slope (for gate voltages below the threshold voltage, this slope is called the *subthreshold slope*) is inversely proportional to the thermal energy kT. Some electrons still have enough thermal energy to cross the gate-controlled potential barrier (Fig. 2.15) and travel to the drain. At (very) low threshold voltages, the resulting *subthreshold leakage current* may lead to an unacceptably high power consumption. This leakage current should not exceed a few tens of nano-amperes for a one hundred million transistor chip in standby mode (no circuit activity and at zero gate voltage). This operating region is also called the '*subthreshold region*'. The *subthreshold slope* (s_{subthr}) depends on the technology and is expressed in mV/dec. The lower the value, the steeper the slope. It ranges from $s_{subthr} \approx 63$ mV/dec for SOI processes to $s_{subthr} \approx 80$ mV/dec for bulk CMOS processes. This means that the subthreshold leakage current increases about 18 times ($\approx 10^{100/80}$) for every 100 mV reduction in V_T, for bulk CMOS transistors.

The normal strong-inversion equations (that apply above the threshold voltage) do not apply to the weak-inversion region. The drain-source current in a transistor with a long channel and a constant drain-source voltage operating in the weak-inversion region is expressed as follows:

$$I_{dssub} = \frac{W}{L} \cdot C\, I_{ds0}\, e^{V_{gs}/mU_T} \tag{2.17}$$

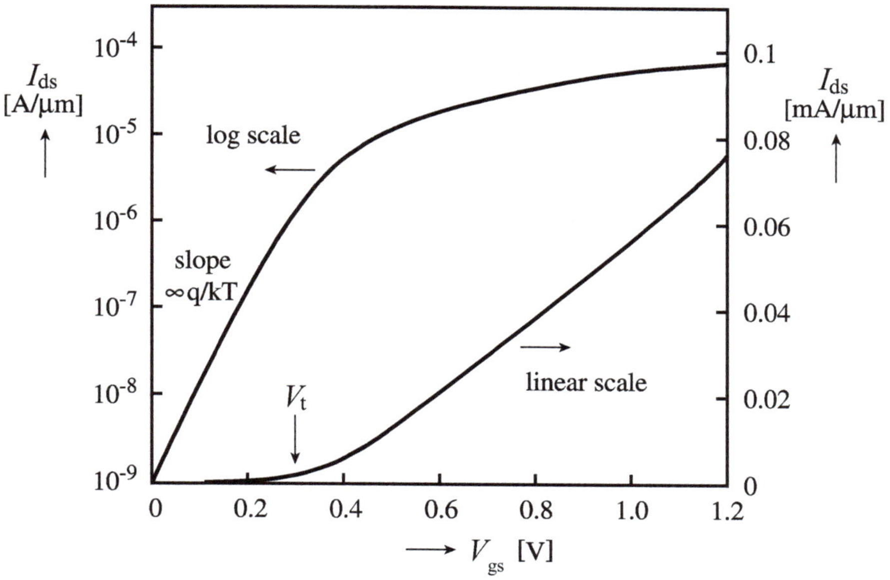

Fig. 2.14 MOSFET current representation on linear scale (right axis) and log scale (left axis)

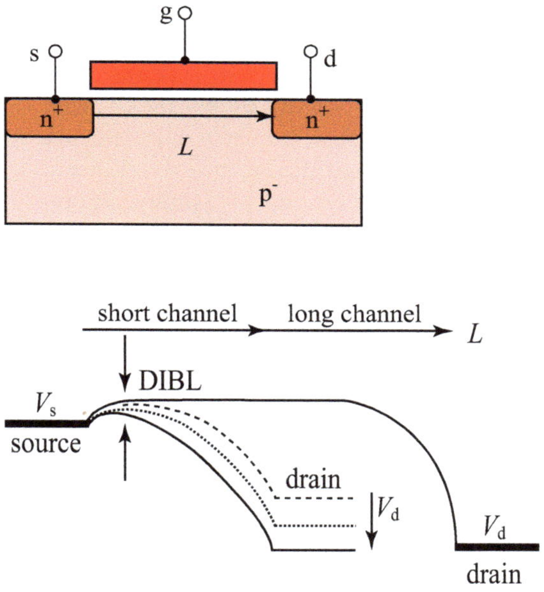

Fig. 2.15 The effect of the drain voltage on lowering the barrier

The terms in Eq. (2.17) are defined as follows:

$$C = 1 - e^{-V_{ds}/U_T}$$

$$U_T = \frac{kT}{q} \approx 25\,\text{mV at room temperature}$$

$$I_{ds0} = \text{characteristic current at } V_{gs} = 0\,\text{V}; I_{ds0} \propto e^{-V_T/mU_T}$$

$$m = \text{slope} \approx 1.5 \text{ and depends on the channel length}$$

Equation (2.17) applies when V_{gs} is not larger than a few U_T below V_T. The subthreshold transistor current I_{dssub} can lead to a considerable *standby current* in transistors that are supposedly inactive. Expression (1.16) in Chap. 1 shows that when we apply a negative (positive) voltage to the substrate or back gate of an nMOS (pMOS) transistor, we can increase the threshold voltage, thereby reducing the subthreshold leakage. Methods to use this back-bias effect in reducing the standby power of CMOS ICs are discussed in Chap. 8.

An accurate description of the behaviour of a transistor operating in the weak-inversion region is contained in References [10, 11]. The following statements briefly summarise this operating region:

1. At low V_T, there is a significant subthreshold current when $V_{gs} = 0$ V. This has the following consequences:
   (a) There is a considerable standby current in (C)MOS VLSI and memory circuits.
   (b) The minimum required clock frequency of dynamic circuits is increased as a result of leakage currents. DRAMs are among the circuits affected.
2. In long-channel transistors, the source and drain depletion regions have a limited influence on the electric field in the channel. These devices show a much lower threshold-voltage dependence on channel length and drain voltage. In short-channel transistors, on the contrary, the source and drain depletion layers cover a much larger part of the channel (see Fig. 2.7). They interact with each other, such that the drain voltage reduces the potential barrier at the source. Figure 2.15 shows the influence of the drain voltage and channel length on the barrier height. An increase of the drain-source voltage in short-channel transistors will reduce the barrier height. This *drain-induced barrier lowering effect (DIBL)* leads to a reduction of the threshold voltage V_T of the transistor.

   In a 65 nm CMOS technology, for example, a transistor with a channel length of 60 nm can carry a six times larger subthreshold leakage current than one with a channel length of 90 nm at the same operating voltages! The total subthreshold leakage current is also dependent on the temperature: it almost doubles for every 25 °C increase in temperature. It is clear that these effects are very important in the development and the modelling of deep-submicron technologies.

3. Analogue circuit techniques use weak-inversion behaviour in low-current applications. The voltage gain of a MOS transistor operating in the weak-inversion region is relatively high and comparable to the voltage gain of bipolar transistors.

2.7.2 Gate-Oxide Tunnelling

The continuous scaling of MOS devices over the past four decades has caused a reduction of their lateral dimensions with an average scaling factor s $\approx$ 0.7. In order to achieve a higher speed, the transistor current needs to be increased. From the current expressions, derived in Chap. 1, it can be seen that we need to increase the transistor gain factor β, in order to achieve a higher current drive. Because β is inversely proportional to the gate-oxide thickness t_{ox}, this thickness was required to be reduced with the same factor, leading to an increasing probability of direct electron tunnelling through it. However, when t_{ox} becomes less than 2.5 nm, this tunnelling becomes visible in the total transistor leakage picture. Both this probability and the resulting leakage current are strong exponential functions of t_{ox}, as well as a function of the voltage across the gate oxide [13]. It is not only the electrical field across the oxide that determines the amount of tunnelling but also, and even more dominantly, the penetration of carriers into the oxide, e.g. 2 V across 2 nm oxide causes tunnelling, while 5 V across 5 nm does not. Figure 2.16 shows these relationships [10]. The gate leakage increases roughly by a factor of ten for every 0.2 nm reduction in oxide thickness. It also increases by about a factor of ten with the doubling of the voltage across the oxide.

The total *gate-oxide leakage* is dependent on the state of the transistor. In the transistor on-state, the whole channel area contributes to the gate-oxide leakage current. In the off-state, only the gate-to-source and gate-to-drain overlap area contribute, depending on the voltage on these terminals.

It should be noted that gate-oxide (SiO_2) leakage will be dominated by the n-MOS device, because the p-MOS device shows an increased barrier for hole

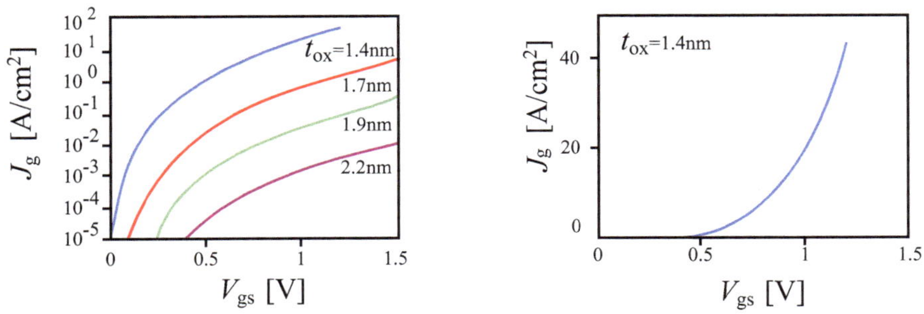

Fig. 2.16 Gate-oxide leakage current density as a function of the oxide thickness and oxide voltage

tunneling. As a result, a pMOS transistor will exhibit roughly an order of magnitude lower gate-oxide leakage current than its nMOS counterpart [11].

The gate-oxide leakage can be reduced by applying an isolating material with a relatively high dielectric constant, so that the dielectric thickness can be increased and the gate-oxide leakage reduced. Some semiconductor vendors have already replaced SiO_2 by a hafnium-based dielectric in their 45 nm node, while others started to use it in the 32 nm or 28 nm node. Hafnium-based dielectrics show dielectric constants close to 20 or higher compared to 4 of SiO_2. Section 3.9.4 presents more details on the use of high-ϵ dielectrics in combination with metal gate.

2.7.3 Reverse-Bias Junction Leakage

In commonly used CMOS circuits, when the source and drain junctions are biased, they are reverse biased. The larger reverse bias across the drain junction is particularly responsible for the increasing junction leakage. There are several mechanisms that contribute to this junction leakage. Two of them, which are commonly known from the basic conventional diode operation, are diffusion and drift of minority carriers and electron-hole pair (Shockley-Read-Hall; SRH) generation in the depletion region of the reverse biased junction. For the lower temperature range, the generation mechanism dominates and is proportional to the intrinsic carrier concentration n_i. The diffusion mechanism is dependent on the thermal energy kT, which causes the carriers to move at random even when no field is applied. At high temperatures, this dominates the leakage contribution, which is then more proportional to ${n_i}^2$.

If the reverse electrical field increases but is still in the relatively low region, the reverse leakage current is mainly assisted by interface traps. This so-called *trap-assisted tunnelling (TAT)* increases with the density of traps and can be increased by electrical stress.

If no special technology measures had been taken, four decades of transistor channel length scaling would have resulted in the source and drain depletion layers touching each other, causing the SCE as discussed in Sect. 2.5.1. To suppress SCE, close to the source and drain regions in the channel, halos (pockets) are implanted with increasing peak doping levels ($\approx 1 - 2 \cdot 10^{19}$ atoms/cm^3), depending on the technology node. Because of this, the depletion layers become so thin that also here direct tunnelling of carriers will occur, just like through the thin gate oxide.

A high field across the reverse-biased p-n junction at the drain causes significant tunnelling of electrons through this junction, from the valence band of the p-halo to the conduction band of the n-drain (Fig. 2.17). This so-called *sub-surface band-to-band tunnelling* (sub-surface BTBT) is a major contribution to the total *reverse-bias junction leakage current* at high reverse electrical fields.

This BTBT-leakage current is exponentially dependent on the reverse-bias electric field across the junction and on the doping levels at both sides of the junction.

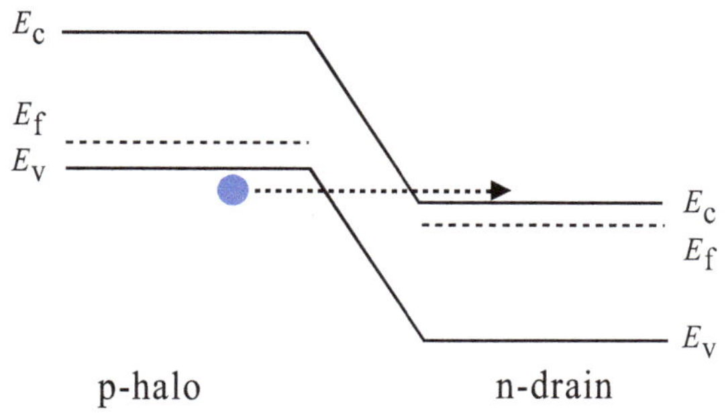

Fig. 2.17 Sub-surface BTBT through an nMOS drain junction

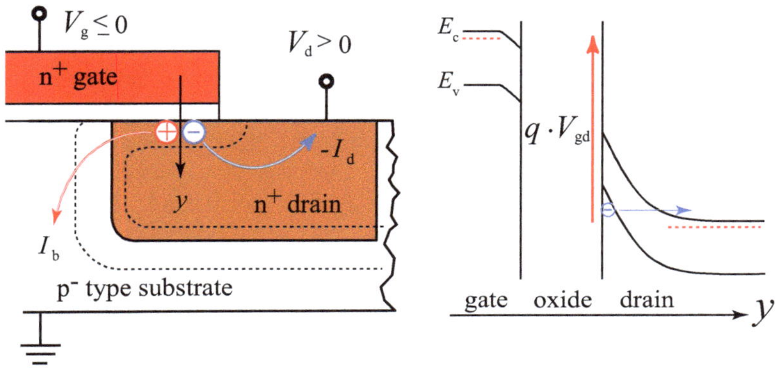

Fig. 2.18 Cross section and energy-band diagram used to explain the GIDL effect

The use of a negative back-bias voltage increases the reverse-bias potential and may lead to a dramatic increase of this BTBT leakage. So, the use of a reverse body-bias to reduce the overall transistor leakage becomes questionable, since it influences the various leakage components in a different way: it will reduce the subthreshold leakage, but it will increase the reverse-bias junction leakage. Section 2.7.6 presents a discussion on the combined leakage behaviour in relation with the temperature and the use of a reverse body bias.

2.7.4 Gate-Induced Drain Leakage (GIDL)

When a large drain-to-gate bias is applied, a deep depletion condition is created in the drain region underneath the gate. This can create a sufficient energy-band bending (greater than the silicon band gap) at the Si-SiO_2 interface in the drain for valence-band electrons to tunnel into the conduction band. In other words, when, in case of an nMOS transistor, $V_{gs} \leq 0\,\text{V}$ while $V_{ds} = V_{dd}$, electrons may tunnel through this deep-depletion layer and flow to the drain terminal, while the remaining holes flow to the substrate. Figure 2.18 shows the effect by the energy-band diagram.

This *surface BTBT* together with trap-assisted tunnelling current at the interface is the main cause of the so-called *gate-induced drain leakage (GIDL)* current [12]. This current only depends upon the conditions near the gate-to-drain overlap. A decreasing gate-oxide thickness causes an increase of the drain-to-gate electrical field and results in an increase of the GIDL current. The resulting drain and substrate currents increase exponentially with V_{dg}. The effects of GIDL remain almost constant with constant-field scaling (see Chap. 11). It is not expected that the use of high-ϵ dielectrics will have much impact on GIDL.

2.7.5 Hot-Carrier Injection and Impact Ionisation

As a result of the continuous scaling of the devices, the electric field near the drain has reached extraordinary large values in short-channel devices. Consequently, a carrier can acquire sufficient energy that it can cause *impact ionisation* upon impact with an atom, e.g. such a *hot carrier* can cause the transition of an electron from the valence band to the conduction band, leaving a free hole behind. This produces an extra conduction electron and a hole. These generated carriers face the same large electric field that has caused their formation. Therefore, in an nMOS device, both electrons will flow to the drain, while the hole drifts to the substrate. In other words, impact ionisation occurs, when the excess of energy that a carrier has collected, from passing a high electric field, is used to create other carriers. This gives rise to an increase in the drain-source current I_{ds} and to a substrate current I_{b}. The degree to which impact ionisation occurs and the magnitude of I_{b} are directly proportional to the square of the maximum electric field in the pinch-off region near the drain. Every measure to reduce this maximum causes a reduction in I_{b}.

In an nMOS device, a carrier impact may generate high-energy electrons which may be scattered towards the substrate-to-gate oxide interface. Electrons with the highest energy can even be injected into the gate oxide. This so-called *hot-carrier injection* will degrade the device performance, and it may lead to reliability problems (see Chap. 9). Impact ionisation and the occurrence of hot carriers are more pronounced in short-channel devices operating at the maximum drain-source voltage (equal to the supply voltage), when the gate-source voltage is equal to half of the supply voltage.

Because analogue circuits tend to be designed with non-minimum channel length devices, they will suffer less from these effects. Because the band gap of silicon is 1.15 V, the trend of reducing the supply voltages further below 1.2 V will cause a reduction of occurrence of impact ionisation in future CMOS devices. However, a change in the transistor gate stack will have a severe impact on the hot-carrier behaviour and reliability of the transistor.

2.7.6 Overall Leakage Interactions and Considerations

The previous subsections presented the most important individual leakage components in an MOS transistor. However, these leakage components are interdependent and respond differently to a change in one or more of the parameters (e.g. oxide thickness, channel length, channel doping, source and drain doping, gate material/doping, terminal voltages, chip temperature).

Figure 2.19 (top) shows an example of the gate, drain and bulk currents of an nMOS transistor in a 65 nm low-leakage (LSTP) CMOS technology as a function of the gate voltage at 25 °C and 125 °C. Figure 2.19 (bottom) shows the total

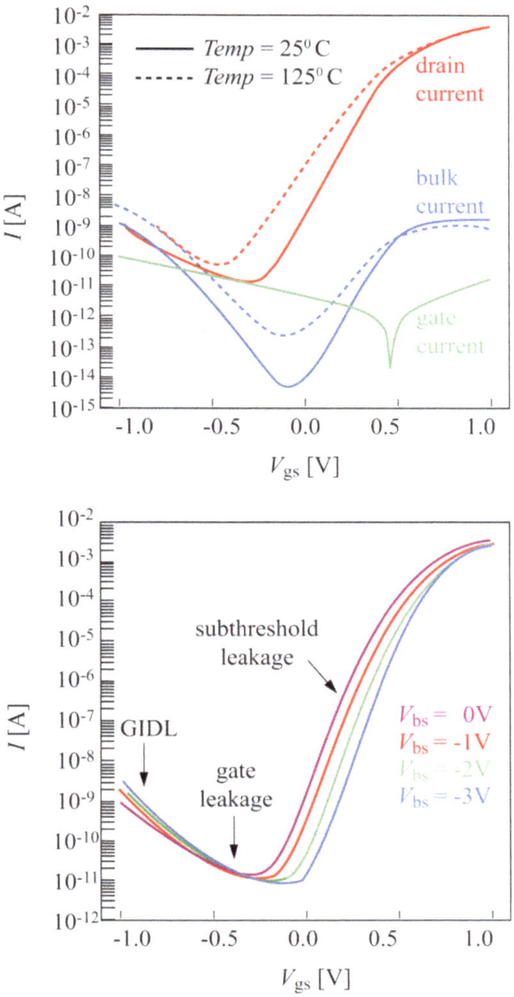

Fig. 2.19 Most dominant contributions to the total leakage current in an nMOS transistor in a low-leakage 65 nm CMOS technology at 25 °C and 125 °C (top) and at different back-bias voltages (bottom)

subthreshold leakage current characteristic as a function of the gate voltage. At higher gate voltages, the subthreshold leakage currents are most dominant, while for the negative gate voltages, the GIDL component becomes most dominant.

The position of this current-voltage characteristic with respect to the gate voltage depends on the transistor's threshold voltage. An additional reverse substrate bias voltage increases this threshold voltage and shifts the current voltage characteristic to the right, as can be seen in the lower diagram.

Leakage currents will therefore vary dramatically between different technology nodes and between *low standby power (LSTP)* and general purpose CMOS processes. Even in the same technology node, the subthreshold leakage current may almost vary an order of magnitude between wafers from different batches.

Advanced technologies using metal gates and high-k dielectrics show reduced subthreshold- and gate-leakage currents.

The subthreshold current is exponentially proportional with the temperature, while the gate tunnelling current is almost completely independent of it. For process nodes beyond 20 nm, the contributions of the gate and subthreshold currents will dramatically increase, because transistors in these nodes have a much thinner gate oxide and a smaller V_T. However, these contributions may vary dramatically between different technology nodes and between low-power and general-purpose processes from different foundries. Figure 2.20 [15] shows the exponential increase in leakage current over various process generations.

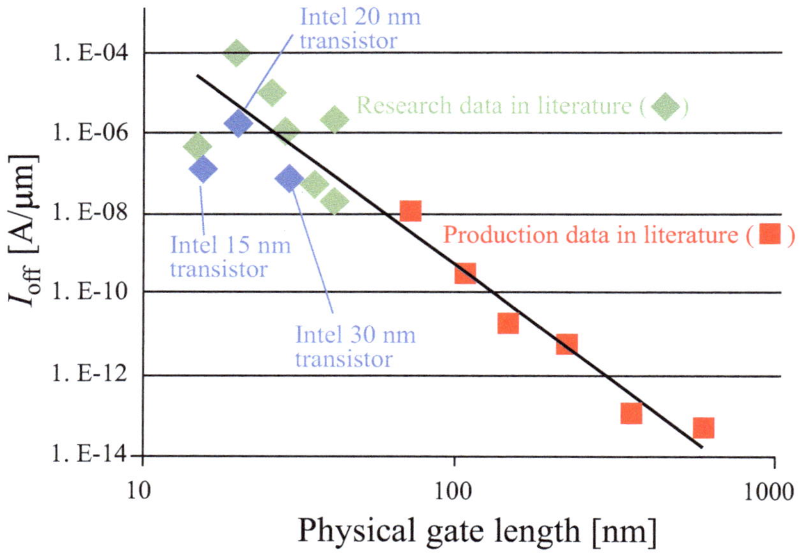

Fig. 2.20 Exponential increase of the total leakage current vs. channel length (technology generation) [15]

More on leakage current mechanisms can be found in [16]. Alternative technology and design solutions to reduce leakage currents are presented in Chap. 8.

2.8 MOS Transistor Models and Simulation

All previously discussed physical mechanisms, combined with an accurate description of the (overlap) capacitances, are included in today's (compact) MOS models. A MOS model gives a description of the physical behaviour of a MOS device from weak-inversion to strong-inversion operation and calculates the device currents, charges and noise as a function of the terminal voltages. Particularly, the analogue and RF applications of the MOS transistor require an accurate description of their operation over the full range of operating voltages. A compact model is based on only a limited number of parameters. In combination with a circuit simulator, a compact model allows full simulation of the electrical behaviour of rather complex analogue and RF circuits and predicts their behaviour, before they are integrated on real silicon.

The majority of recently developed MOS models are based on so-called surface-potential (SP) models, in which the surface potential equations at both ends of the channel are solved. In December 2005, the Compact MOS Council (CMC) has selected the PSP model, which was a merge of two existing SP models, as a standard, next to the BSIM MOSFET model for future CMOS technologies [12, 14, 17]. Most process nodes of 16 nm and beyond are FinFET processes, while GAA transistors are expected to be used in 3 nm nodes and beyond. FinFETs, GAA and nano-wire transistors use the third dimension to increase the transistor channel width. Therefore, the models for these transistors are quite different from models for planar transistors [18, 19]. More details about the physical and mathematical basics of MOS modelling are beyond the scope of this book but are largely available through the internet.

2.8.1 Worst-Case (Slow), Typical and Best-Case (Fast) Process Parameters and Operating Conditions

In the semiconductor industry, it is common practice to include the statistical variations of various *process parameters* in the circuit simulator models. These variations are usually referred to as inter-die (global) and intra-die (local) variations. Section 9.4 presents more details on the possible causes of these variations. In this chapter, we will restrict ourselves to the influence they may have on the electrical performance of the circuits. Therefore, we return to the simple current expression for a saturated nMOS transistor (Expression 1.18):

$$I_{\text{ds}} = \frac{\beta}{2} \cdot (V_{\text{gs}} - V_{\text{T}})^n \text{ with } \beta = \frac{W}{L} \cdot \mu_{\text{n}} \cdot C_{\text{ox}} = \frac{W}{L} \cdot \mu_{\text{n}} \cdot \frac{\epsilon_0 \epsilon_{\text{ox}}}{t_{\text{ox}}}$$

Table 2.2 Example of worst-case (slow) and best-case (fast) process parameters (for 16 nm CMOS process) and operating conditions for a given circuit

Condition	Parameter	Worst-case (slow)	Best-case (fast)	Potential causes
Application	Supply voltage [V]	0.9 V_{dd}	1.1 V_{dd}	Application spec
	Temperature [T]	125 °C[a]	−55 °C[a]	Application spec; process node (ZTC !!!)
Process and litho	Channel length	L_{max}	L_{min}	LER (random var.)
				litho; poly etching (syst. var.)
	Channel width	W_{min}	W_{max}	STI etching
	Threshold voltage	V_{Tmax}	V_{Tmin}	Work function (gate stack; syst. var.)
				Channel dope (random & syst. var.)
	Gate-oxide thickness	$t_{ox_{max}}$	$t_{ox_{min}}$	ALD deposition (random & syst. var.)

[a]The real worst-case and best-case temperatures depend on the process node, the supply voltage and the threshold voltage as discussed further in this subsection

In Chap. 1, we started the derivation of the simple current expressions assuming ideal MOS transistor operation.

Therefore, the exponent n in the above current expression was equal to 2. Due to second-order effects, today's MOS transistors often show exponents in the range between 1.5 and 2. The dielectric parameters ϵ_0 and ϵ_{ox} are rather constant across and between dies. The mobility μ_n may vary across and between dies due to a variation in stress-inducing doping (profiles) and proximity effects (Sect. 9.4). Table 2.2 shows the influence of the process and the lithography on the most important process parameters: V_T and t_{ox} as well as on the transistor dimensions, W and L.

Looking at the above current expression, it will be clear that, in the best-case (fast) situation, the current is larger (i.e. the circuit is faster) for transistors with the shortest channel length (L_{min}), the largest channel width (W_{max}), the minimum threshold voltage ($V_{T,min}$) and the thinnest gate-oxide thickness ($t_{ox,min}$). The right column of the table shows some causes for the different variations of these process parameters. Next to the variation in process parameters, also the operating conditions may have a strong impact in circuit behaviour. The customer is usually allowed to operate the circuit at a 10% higher or at a 10% lower supply voltage. Next to this, he may operate the chip in different temperature environments, such that the junction temperature of the chip is always between −55 °C and +125°C (military spec). For most processes above the 45 nm node, the best-case temperature is the lowest one (−55 °C). However, for circuits around or below the 45 nm node, it depends on the value of the threshold voltage whether the circuit becomes slower or faster at increasing temperatures. This is a result of the so-called ZTC voltage, as discussed in Sect. 2.6. Some design houses simulate both the worst-case and best-case situation at 125°C and −55 °C. In summary, for the circuit according to Table 2.2, the worst-case (slowest) circuit operation is achieved, when all parameters are at the worst-case *PVT* (process, voltage, temperature) corner: $0.9V_{dd}$, 125 °C, L_{max}, W_{min}, $V_{T,max}$ and $t_{ox,max}$. This means that even under these worst-case conditions, all the circuits on the chip must still operate according to the speed requirements dictated by the application. These worst-case parameters must therefore be compensated by larger transistor widths. In fact, for all ICs that are better than worst case, these increased channel widths lead to less dies on a wafer and to reducing process yields. This so-called traditional PVT corner/case analysis is often too pessimistic and requires too many simulation runs. There is an increasing effort to base circuit analysis on parameter distributions and statistical timing analysis. Details can be found on the Internet and are beyond the scope of this book.

2.9 Conclusions

The formulae derived in Chap. 1 provide a good insight into the fundamental behaviour of MOS devices. These formulae were used to predict circuit behaviour with reasonable accuracy until the mid 1980s. The continuous drive for higher circuit densities with smaller transistors, however, has given rise to an increased contribution from physical and geometrical effects. These effects cause deviations from the ideal transistor behaviour assumed in Chap. 1. In addition, the magnitude of these deviations increases as transistor dimensions shrink. These effects combine to reduce the ideal transistor current by more than a factor ten for channel lengths below 90 nm. There are also effects that permanently degrade the performance of a MOS transistor. Particularly, its behaviour in off-state has a great impact on the standby power consumption of CMOS ICs. Therefore, weak-inversion behaviour and transistor leakage-current mechanisms have been discussed in detail, to allow better understanding of the low standby power solutions presented in Chap. 8. The continuous scaling of both devices and interconnects also has severe consequences for the reliability of the IC and may dramatically reduce its lifetime. These consequences and potential technology and design solutions are discussed in Chap. 9.

2.10 Exercises

1. At room temperature (25 °C), the magnitude of an nMOS transistor's gain factor β is 240 μA/V^2 in a 5 nm process, and its threshold voltage V_T is 0.4 V. Assume that the exponent m in Eq. (2.15) is equal to 1 and that the threshold voltage reduces 0.6 mV/°C.
   (a) Calculate the gain factor β when the transistor is operating at 125 °C.
   (b) Calculate the threshold voltage for the temperature in (a).

(c) What would be the consequences of this reduced threshold voltage for the standby current in an SRAM, for instance?

2. (a) Explain in detail what ZTC means and what consequences it may have for a sub-32 nm design that is realised in a multi-V_T process?
   (b) Assume the transistor in Exercise 1 is saturated with its gate connected to its drain. At what V_{gs} would the influence of the temperature difference on the gain factor β and on the threshold voltage V_T fully compensate each other (in other words, what would be the ZTC voltage value of V_{gs})?
3. (a) What is the effect on the gain factor β of a pMOS transistor with $L = 45$ nm when the mobility is only influenced by velocity saturation caused by a very large horizontal electric field, $E_x = 0.95 \cdot E_{x\,sat}$?
   (b) Calculate the drain-source voltage at which the relevant reduction in mobility occurs if $\theta_3 = 0.86\,\mathrm{V}^{-1}$.
4. Assume Eq. (2.7) can be approximated by

$$\mu = \frac{\mu_0}{(1 + \theta_1'(V_{gs} - V_T) + \theta_3' \cdot V_{ds})}$$

What can you say about the relation between the mobility and the series resistance of the drain (R_d), when the transistor operates in the saturation region?

References

1. A.J. Walker, P.H. Woerlee, A mobility model for MOSFET device simulations. Journal de Physique, colloque C4, **49**(9), 256 (1988)
2. M. Vertregt, The Analog Challenge in Nanometer CMOS, IEDM Digest of Technical Papers (2006), pp. 11–18
3. N. Weste, D.M. Harris, *CMOS VLSI Design, a Systems Perspective*, 4th edn. (Addison-Wesley, Boston, 2011)
4. R.S.C. Cobbold, *Theory and Applications of Field Effect Transistors* (Wiley, New York, 1970)
5. I.M. Filanovsky, A. Allam, Mutual compensation of mobility and threshold voltage temperature effects with applications in CMOS circuits. IEEE Trans. Circuits Syst. Fundam. Theory Appl. **48**(7), 876–884 (2001)
6. E. Long et al., Detection of temperature sensitive defects using ZTC, in *Proceedings of 22nd IEEE VLSI Test Symposium* (VTS 2004)
7. A. Dasnan et al., Handling inverted temperature dependance in static timing analysis. ACM Trans. Des. Autom. Electron. Syst. **11**(2), 306–324 (2006)
8. R. Kumar et al., Reversed temperature-dependent propagation delay characteristics in nanometer CMOS circuits. IEEE Trans. Circuits Syst.-II: Express Briefs **53**(10), 1078–1082 (2006)
9. C.H.S. Coelho, Analysis of zero-temperature coefficient behavior on vertically stacked double nanosheet nMOS devices. Microelectron. J. **117**, 105277 (2021)
10. R. van Langevelde et al., Gate current: modelling, ΔL extraction and impact on RF performance, in *IEDM Technical Digest* (2001), pp. 289–292
11. F. Hamzaoglu et al., Circuit-level techniques to control gate leakage for sub-100 nm CMOS, in *Proceedings of the 2002 ISLPED Symposium*, pp. 60–63 (2002)
12. G. Gildenblat et al., PSP: an advanced surface-potential-based MOSFET model for circuit simulation. IEEE Trans. Electron Devices **53**(9), 1979–1993 (2006)

13. D. Lee et al., Gate oxide leakage current analysis and reduction for VLSI circuits. IEEE Trans. VLSI Syst. **12**(2), 155–166 (2004)
14. J. Assenmacher, BSIM4 modelling and parameter extraction. http://www.ieee.org/r5/denver/sscs/References/2003_03_Assenmacher.pdf
15. G. Marcyk et al., New transistors for 2005 and beyond. http://www.eas.asu.edu/~vasilesk/EEE531/TeraHertzlong.pdf
16. A. Scholten et al., The physical background of JUNCAP2. IEEE Trans. Electron Devices **53**(9), 2098–2107 (2006)
17. R. Woltjer et al., An industrial view on compact modeling, in *Proceedings of the 36th European Solid-State Device Research Conference* (2006), pp. 41–48
18. J. Duarte, Mathematical Compact Models of Advanced Transistors for Numerical Simulation and Hardware Design, Technical Report No. UCB/EECS-2018-24 (2018)
19. K.O. Petrosyants, Comparative characterization of NWFET and FinFET transistor structures using TCAD modeling. Micromachines **13**(8), 1293 (2022).

General Basic Physics Reference

20. S.M. Sze, *Very Large Scale Integration Technology*, 2nd edn. (Mc Graw-Hill, New York, 1998)

Manufacture of MOS Devices

3

3.1 Introduction

Until the mid-1980s, the nMOS silicon-gate process was the most commonly used process for MOS LSI and VLSI circuits. However, nearly all modern VLSI and memory circuits are made in CMOS processes. CMOS circuits are explained in Chap. 4; the technology used for their manufacture is discussed in this chapter.

Modern sub-10 nm CMOS processes have emerged from the numerous manufacturing processes which have evolved since the introduction of the MOS transistor in integrated circuits. Differences between the processes were mainly characterised by the following features:

- The minimum feature sizes that can be produced.
- The gate-oxide thickness.
- The number of interconnection levels.
- The type of substrate material. Alternatives include n-type and p-type, high-resistive or low-resistive, bulk silicon, epitaxial or SOI wafers.
- The choice of the gate material. Initially, the gate material was the aluminium implied in the acronym MOS (metal oxide semiconductor). Molybdenum has also been used. From 6 µm until and including 120 nm MOS processes, however, nearly all use *polycrystalline* silicon (polysilicon) as gate material. One of the main reasons is that a polysilicon gate facilitates the creation of self-aligned source and drain areas. Another reason for using polysilicon as gate material is that it allows accurate control of the formation of the gate oxide. From 90 nm onwards, some companies used a stack of W-WN-polysilicon and SiO_xN_y. A combination of a metal gate with *high-ϵ dielectrics* (= *high-k dielectrics*) was first introduced in the 45 nm node by Intel. Other companies have introduced high-ϵ metal-gate devices in their 32 or 28 nm CMOS nodes.

H. Veendrick, *Nanometer CMOS ICs*,
https://doi.org/10.1007/978-3-031-64249-4_3

- The method to isolate transistors. Conventional CMOS processes used the so-called LOCOS isolation while most of today's processes use shallow-trench isolation (STI); see Sect. 3.4.
- The type of transistors used: nMOS, pMOS, enhancement and/or depletion, etc.

Many of the transistor parameters, in terms of performance, power consumption, and reliability, are determined by the substrate that is used as starting material. A short summary on the properties and use of the different substrate materials will therefore be presented first.

Modern manufacturing processes consist of numerous photolithographic, etching, oxidation, deposition, implantation, diffusion and planarisation steps. These steps are frequently repeated throughout the process, and they currently may exceed a thousand steps. The IC fabrication discussion starts with a brief description of each step. Most processes use masks to define the required patterns in all or most of the IC diffusion and interconnect layers. Modern CMOS manufacturing processes may use 40 or more masks. For products that need double or quadruple patterning, the number of masks can easily exceed 70 in the most advanced process nodes. However, the initial discussion of IC manufacturing processes in this chapter focuses on a basic nMOS process with just five masks.

Subsequently, a basic CMOS process flow is briefly examined. Fundamental differences between various CMOS processes are then highlighted.

Finally, a sample nanometer CMOS process is explained. Many of the associated additional processing steps are an extension of those in the basic CMOS process flow. Therefore, only the most fundamental deviations from the conventional steps are explained. The quality and reliability of packaged dies are important issues in the IC manufacture industry. An insight into the associated tests is included in Chap. 10.

3.2 Different Substrates (Wafers) as Starting Material

To create silicon wafers, first, pure silicon is heated at temperatures up to 1500 °C in a huge furnace. Then, a seed of single silicon crystal, with the right (110) crystal orientation (see Sect. 3.2.3), is mounted on a shaft and is dipped into the molten silicon. This seed is then slowly rotated and raised upwards out of the melt just fast enough to pull the molten silicon with it by cohesion, thereby physically growing the silicon crystal. In this so-called *Czochralski* process, the *crystal growth* is a continuous process of forming new thin films of the silicon melt on the bottom of the cooled previous films, roughly at about 25 mm an hour. This process is named after a polish scientist, who, by accident in 1915, dipped his pen into molten tin instead of in his inkwell. The tin that he drew out of the melt turned out to be a monocrystalline tin filament. The diameter of the grown mono crystalline silicon bar (also called *ingot*) varies over the length, and a grinder is used to create a bar (Fig. 3.1) with a homogeneous diameter.

A 300 mm crystal ingot can be as long as 2 metres and may weight several hundred kilograms. Next, wafers are sawn by a diamond-coated saw. Because

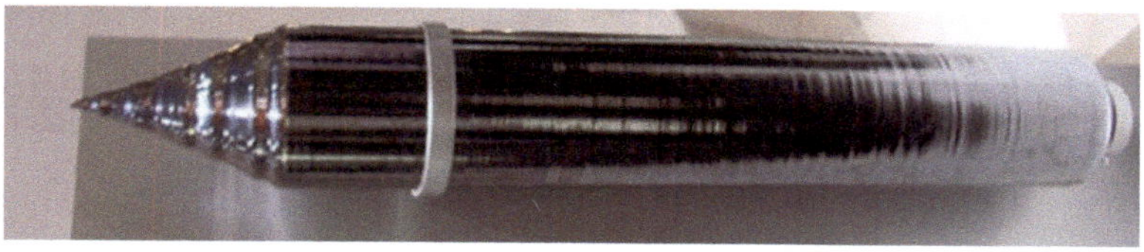

Fig. 3.1 A 300 mm silicon ingot from which 300 mm wafers are sawn. (Courtesy of Smartalix)

the transistors are fabricated close to the silicon surface, their performance and reliability are very much dependent on the flatness and crystal integrity of the silicon surface. Theoretically, for good MOS transistor operation, the wafers could be as thin as a micron, but with this thickness, a wafer would easily break during handling. Therefore, most wafers have a thickness between 400 μm and 1 mm. Most 300 mm wafers used for CMOS processes have a thickness of about 775 μm $\pm$ 20 μm.

A very critical element in the operation of an integrated circuit is the electrical isolation between the individual devices. Unintended electrical interference can dramatically affect their performance. Smaller minimum feature sizes reduce the distance between devices and increase their sensitivity at the same time. An important factor in the isolation properties is the substrate on which the devices are built. In all discussions, so far, we have assumed a *bulk silicon* substrate (wafer) as the starting material for our (C)MOS processes. However, CMOS technologies used epitaxial wafers in the past, because the Czochralski (CZ) process could not deliver high-quality wafer surfaces. One of the main problems was the formation of defects and impurities in the crystal structure of the wafer during the growth process. This was caused by the introduction of impurities from the container, which can dissolve into the melt and become incorporated into the crystal structure. Impurities such as oxygen, carbon and metals can negatively impact device performance and yield. To mitigate this problem, manufacturers needed to use high-purity materials and carefully control the process conditions to minimise impurity introduction. Another problem was the formation of dislocations in the crystal structure during the growth process. Dislocations are defects in the crystal structure that can negatively impact device performance and yield. Dislocations can be caused by factors such as temperature gradients, thermal stresses and crystal growth rate variations. To minimise dislocation formation, manufacturers needed to carefully control the process conditions and optimise the crystal growth parameters. This required the growth of an accurately controlled epitaxial layer on top of the wafer. Advances in crystal growth technology and process control have helped to address many of these problems, and the CZ process remains a widely used method for producing high-quality silicon wafers. As said, the ingots are sliced into thin wafers using wire saws with many parallel wires to cut many wafers at once. The slicing process creates a rough and irregular edge on the wafers, which needs to be smoothed and rounded. The edge of a semiconductor wafer has therefore been etched at an angle to create a sort of slanting edge or is rounded by a polishing step. This region is called *wafer bevel* area (Fig. 3.2). The slanting or slightly rounded edge helps in preventing

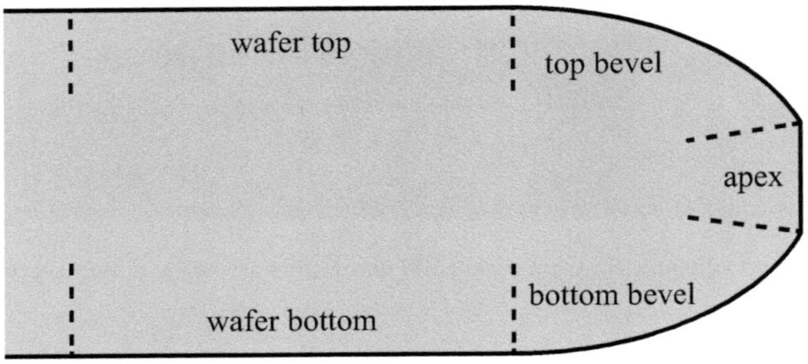

Fig. 3.2 Various area definitions of a wafer edge [1]

chipping or cracking of the wafer during handling and subsequent processing. The bevel essentially reduces the risk of damage to the wafer and increases yield during manufacturing processes. The bevel also facilitates automatic handling of the wafer during fabrication, transportation and packaging and thus minimises the risk of damage to the sensitive products.

Another concern to produce ICs with high accuracy, performance, yield and reliability is the control of the *wafer flatness*. With respect to lithography, wafer flatness variations must be smaller than the *depth of focus (DOF)* of the lithography exposure tools over the illuminated region of the top surface of the wafer including deposited films (see Sect. 3.3.1). There are several parameters that describe the wafer flatness. The so-called *total thickness variation (TTV)* represents the maximum difference between the thinnest and thickest position across the entire wafer surface (see Fig. 3.3a), generally calculated by measuring the thickness in five locations. TTV can occur during various steps in the wafer manufacturing process, such as slicing, grinding and polishing. *Double-sided polishing* can create wafers with the lowest TTV values, even below 1 μm, necessary in advanced process nodes to create the required flatness. This is often followed by an extra single-sided polishing step to reduce the surface roughness and particle count even further.

Particularly in multilayer *3D die stacking* [14], the wafer backside grinding is a very critical process step to achieve ultra-thin wafers of only several tens of micrometres (sometimes 30 to 50 μm with TTV values smaller than 0.2 μm) thickness. These are then used in the 3D stacking of dies, e.g. in *HBM* (stacking of DRAM dies) or in USB sticks, (stacking of *3D NAND flash* memory dies) or in high-speed logic applications (stacking *chiplets*). These applications are discussed in further chapters in this book. There are two other parameters that affect the flatness of a wafer: bow and warp(age). *Bow* is the deviation of the median wafer surface centre of an un-clamped wafer, from the reference plain (Fig. 3.3b), and *warp* (or *warpage*) is the difference between the maximum and minimum deviation of the median surface from the reference plane, again for a free un-clamped wafer plain (Fig. 3.3c). Typical ranges for these parameters for 300 mm (single-side polished) wafers used in advanced CMOS processes are TTV, $<3\,\mu$m and bow and warp, $<10\,\mu$m. A 3D NAND flash memory can even show a warp of more than 300 μm.

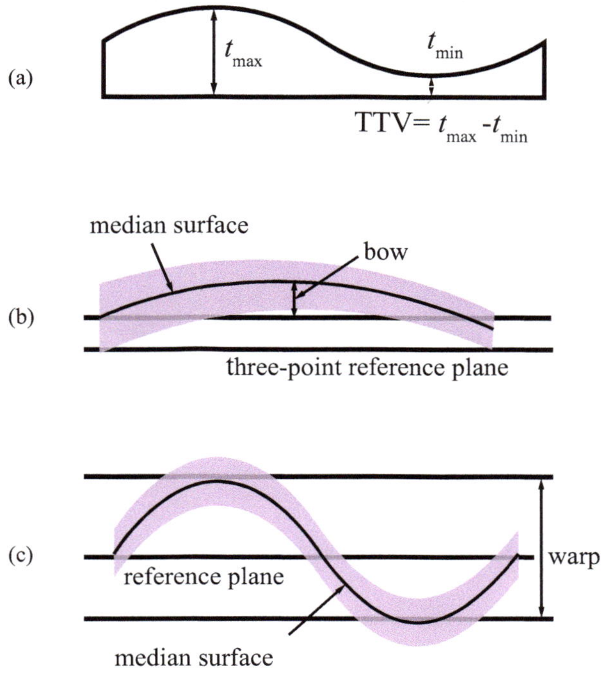

Fig. 3.3 Definition of TTV, bow and warp (age) [15]

TTV, bow and warp may be the result of the wafer creation process, but bow and warp may also vary during further IC process steps in the cleanroom. Certain process steps, like implantation of different doping ions inside the wafer or deposition of different materials into or on top of the wafer, may create mechanical stress in the wafer leading to different values for these three parameters. This is particularly an issue with processes that involves the deposition of many layers: a large number of metal layers in a logic process or many memory cell layers in a 3D NAND flash, for example. It affects the performance, reliability and yield of integrated circuits (ICs) fabricated on the wafer.

TTV, bow and warp need to be controlled to tight tolerances which require effective metrology solutions. These parameters are typically measured using non-contact techniques. Some are based on capacitance measurements between a fixed probe point above the wafer and the top wafer surface and a fixed probe point below the wafer and the bottom wafer surface. The wafer is then scanned between these fixed probe points. Another method depends on the accurate detection of colours from light that is reflected back from the wafer surface.

Fig. 3.4 The increase in wafer size from 1 inch in the 1970s through 2, 3, 4, 5, 6 and 8 inch to 12 inch today. (Photo: Joop van Lammeren)

3.2.1 Wafer Sizes

From an economical perspective, larger wafers have lead to reduced IC manufacturing costs per transistor. This rule drove the wafer diameter from about 1 inch (≈25 mm), about five decades ago, to 300 mm (≈12 inch) today (Fig. 3.4). This has put severe pressure on maintaining the wafer flatness, its resistivity and low crystal defect density homogeneous across a rapidly increasing wafer area. The figure shows that wafers with a diameter up to and including 6 inch have a flat, while 8 and 12 inch wafers have a *notch* for aligning the wafers in the lithography tool.

However, the introduction of a new wafer diameter generation requires a huge amount of development costs. This has put the transition from 300 to 450 mm wafers on hold, and it still needs billions of dollar investment before volume production can take off. In this respect, around the year 2010, big semiconductor houses such as Intel, TSMC, IBM, Samsung and Globalfoundries joined R&D forces [2] in a five-company consortium (G450C), in a partnership with the College of Nanoscale Science and Engineering at the State University of New York, to develop the next generation wafer technology. Figure 3.5 shows the 300 mm *fab cost* evolution as a function of the feature size.

Upgrading a 300 mm fab from 32 to 22 nm would cost around $2B. However, the transition from 300 mm production to 450 mm production would require about $7B, because all fab equipment must be upgraded. Another disadvantage is the more

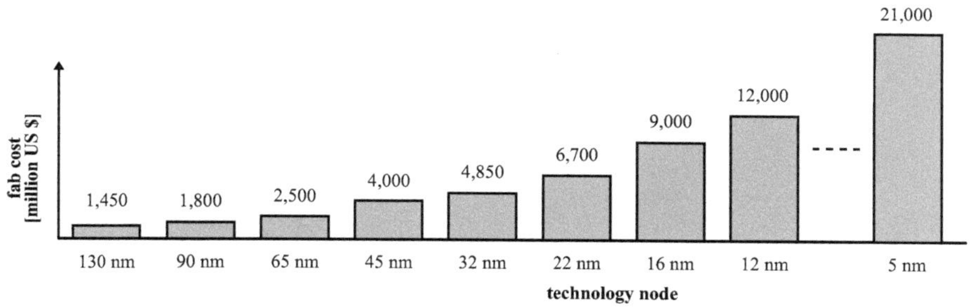

Fig. 3.5 300 mm logic fab cost as a function of feature size. (Courtesy: Globalfoundries and Fabricated Knowledge)

complex fabrication of the 450 mm diameter silicon bar (*crystal ingot*). It will be three times heavier (around 1000 kg). This, combined with the much larger time required for cooling, will almost double the process time. As stated before, the development of 450 mm technology is almost completely put on hold [4]. Currently, there are no signs of any further 450 mm wafer development

3.2.2 Standard CMOS Epi

Epitaxial wafers (epitaxial from Greek: epi = above; taxis = to arrange) consist of a thin, monocrystalline silicon layer grown on the polished surface of a bulk silicon substrate [3]. This so-called *epi-layer* is defined to meet the specific requirements of the devices in terms of performance, isolation and reliability. This layer must be free of surface imperfections to guarantee a low defect density and limit the number of device failures. Since the carriers in a transistor channel only travel in the surface region of the device, the epi-layer thickness is only defined by the transistor architecture (source/drain and STI depths) and ranges from one to a few microns. Usually, the total thickness of a 300 mm wafer is typically 775 μm but may range between 400 μm and 1 mm, depending on the wafer size and technology node. It means that the top epi-layer forms only less than one percent of the total wafer and that the major part of the wafer mainly serves as a substrate carrier for the ICs made on it.

Although the resistance of this substrate hardly affects the performance of digital circuits, it has influence on the robustness of the ICs built on it. Most conventional CMOS processes, including the 180 nm node, use/used low -resistivity (5–10 mΩcm at doping levels between $5 \cdot 10^{18}$ and $1 \cdot 10^{19}$ atoms/cm^3) wafers, in order to reduce the chance of latch-up occurrence (see Chap. 9). With reducing supply voltages, the chance for triggering the parasitic transistor to initiate latch-up is also diminishing. This, combined with the increasing integration of GHz RF functions, has made the use of high-resistivity (10–50 Ωcm at doping levels between $1 \cdot 10^{15}$ and $1.5 \cdot 10^{15}$ atoms/cm^3) substrates very popular from the 120 nm CMOS

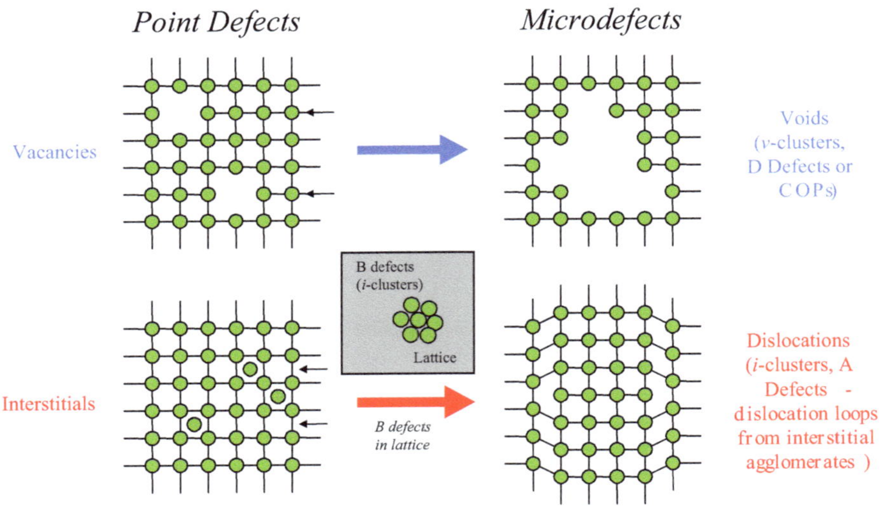

Fig. 3.6 Defects in silicon. (Source: MEMC)

node downwards. It leads to performance increase of passive components, such as inductors, but also to a better electrical isolation between the noisy digital circuits and the sensitive RF and analogue ones (less substrate noise; Chap. 9).

Because the full device operation occurs within this thin top epi layer, it puts severe demands on the homogeneity of the layer thickness, of the resistivity and of the crystal defectivity. When growing single crystal silicon, either for creating bulk silicon wafers or for creating thin epi-layers, a few typical defects in the silicon may show up. *Point defects* may originate from single empty locations (*vacancies*) in the monocrystalline atomic structure (Fig. 3.6), while *micro defects* or *crystal-oriented particles (COP)* can be the result of a cluster of voids.

Interstitials are atoms located in between the atoms of the crystal, while *dislocations* may be caused by clusters of interstitials. The average atomic spacing is also dependent on the covalent atomic radius of the specific material: silicon (Si) 1.17 Å, boron (B) 0.88 Å, phosphorous (P) 1.10 Å, arsenic (As) 1.18 Å and stibnite (Sn) 1.36 Å. So, B is a smaller atom than Si. Doping Si with B (or P) reduces the average atomic spacing of the Si crystal. Another result of this is that the average atomic spacing in the p^- epi-layer is larger than that in the p^+ substrate, because the substrate contains a higher concentration of smaller atoms. Large differences in the atomic spacing of different layers may lead to so-called misfit dislocations. To prevent *misfit dislocations* in a thin epi-layer on a resistive substrate, a simple rule of thumb is applied [6, 7]:

$Epi\ Thickness\ in\ \mu m \leq Substrate\ Resistivity\ in\ m\Omega$

Today, the quality of the Czochralski process of creating the silicon has improved such that it results in extremely pure (purity of 99.9999999%) monocrystalline silicon, which is almost defect-free. The defectivity level of these bulk silicon wafers is certainly comparable to or even better than that of wafers with an epitaxial layer. It is no longer needed to compensate bulk defects. Some semiconductor applications, however, still require epitaxial wafers, but then, it is for better control and adjustment of the dopant and the resistivity of the layer.

These examples show that not all ICs can be made on the same substrate. The following subjects discuss substrates that enhance the device performance.

3.2.3 Crystalline Orientation of the Silicon Wafer

As discussed in Chap. 2, the effective mobility of the carriers in the channel has reduced dramatically over time, due to the continuous scaling of the transistors. Suppressing short-channel effects by increasing the channel doping has led to an increased density of charged impurity scattering sites, thereby reducing the mobility of the carriers in the channel. The intrinsic speed of a logic gate, in first approximation, is proportional to the mobility. Therefore, a lot of research is currently performed in a variety of ways to improve carrier mobility. In this respect, also the crystalline orientation of the silicon substrate plays an important role.

Traditionally, CMOS has been fabricated on wafers with a (100) crystalline orientation (Fig. 3.7), mainly due to the high electron mobility and low interface trap density. However, the pMOS transistors on this substrate suffer from a reduced mobility. By moving away from the (100) orientation, electron mobility is degraded, while hole mobility is improved. Compared to a traditional (100) wafer, a (110) wafer shows a higher atom density in the plane of conduction, as shown in the figure, and so, the electrons suffer from an increased number of atom collisions.

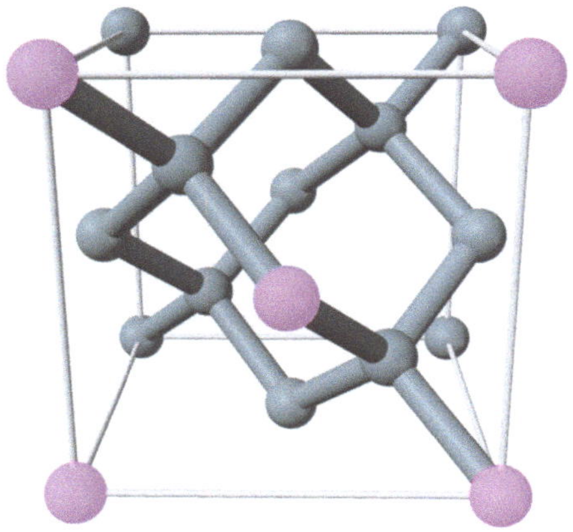

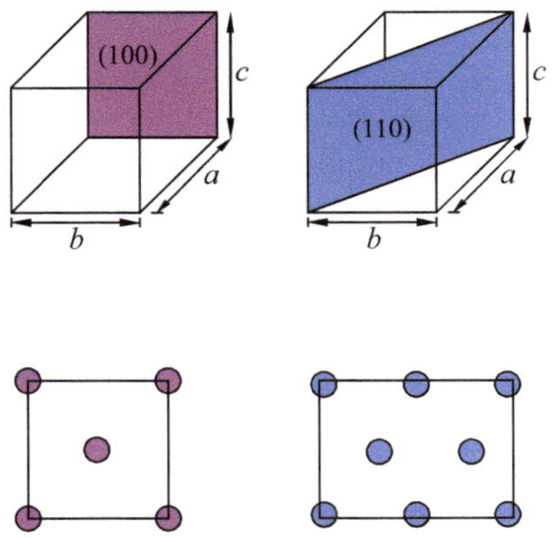

(a) (b)

Fig. 3.7 (**a**) The diamond cubic crystal structure of monocrystalline silicon and (**b**) cross sections through the front (100) plane and diagonal (110) plane

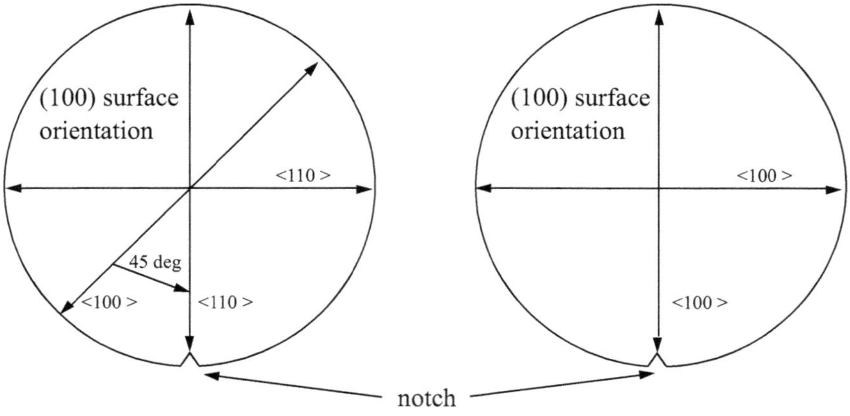

Fig. 3.8 (**a**) traditional notch grinding and (**b**) grinding the notch in the <100> direction. (Source: MEMC)

This may reduce the electron mobility by about 5–10%, while hole mobility shows improvements up to 30% in practice.

An optimum technology, with a much better balance between nMOS and pMOS device performance, would be a hybrid-orientation technology: the (100) plane for nMOSts and the (110) plane for the pMOSts [8, 9]; see also Sect. 3.9.4.

If the pMOS channel is oriented along the <100> direction on a (100) wafer, its mobility and performance may be increased by about 15%, with almost no degradation of the nMOSt performance. Another advantage is that the pMOS transistor will also exhibit a reduced variability. This is only a minor change in the starting wafer, with no further consequences for the device technology and layout (Fig. 3.8).

The only difference is that the wafer flat alignment or notch is changed from the standard <110> direction to the <100> direction. Traditionally, the *notch* is cut during crystal grinding in the <110> direction (Fig. 3.8a). To orient the channel direction along <100> requires a crystal rotation of 45° to grind the notch in <100> direction (Fig. 3.8b). This orientation change is a low-cost solution to enhance the pMOS device, logic gate and memory cell performance with no risk or consequences for the integration process. This wafer option is already in use in high volume production since the 120 nm node.

3.2.4 Silicon-on-Insulator (SOI)

Bulk-CMOS devices show relatively large source/drain capacitances. This can be avoided with the *SOI-CMOS* devices illustrated in Fig. 3.9. The complete isolation of nMOS and pMOS transistors associated with this process also completely removes the possibility of latch-up.

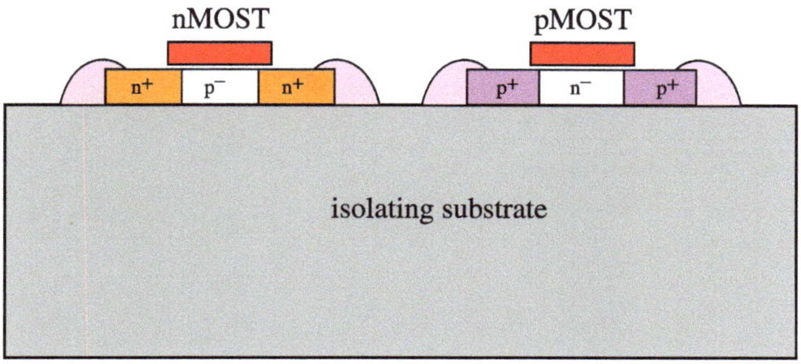

Fig. 3.9 Cross section of a basic SOI-CMOS process

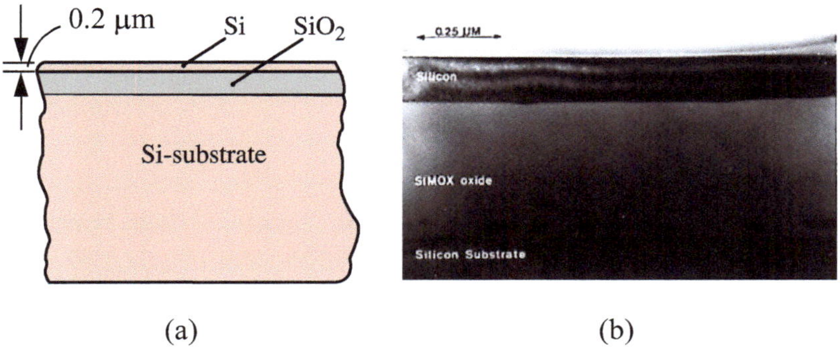

Fig. 3.10 (**a**) Cross section of a SIMOX wafer and (**b**) SEM photograph of such a cross section

Neither the nMOS nor pMOS transistor channels require overcompensating impurity dopes. Very small body effects and source/drain capacitances are therefore possible for both types of transistor. In addition, the n^+ and p^+ source and drain regions do not have bottom junctions. Consequently, the parasitic capacitances are much less than those of the bulk-CMOS processes. This makes the SOI-CMOS process particularly suitable for high-speed and/or low-power circuits. Murphy's law, however, ensures that there are also several disadvantages associated with SOI-CMOS processes. The absence of substrate diodes, for example, complicates the protection of inputs and outputs against the *ESD* pulses discussed in Chap. 9.

Sapphire was originally used as the isolating substrate in SOI circuits, despite the fact that it is substantially more expensive than silicon. The *SIMOX* (*'Separation by IMplantation of OXygen'*) process provides a cheap alternative for these *silicon-on-sapphire* or 'SOS-CMOS' processes. Several modern SOI-CMOS processes were based on SIMOX. These processes use a retrograde implantation of oxide atoms to obtain a highly concentrated oxygen layer beneath the surface of a bare silicon wafer. The resulting damage to the wafer's crystalline structure is corrected in an annealing step. The result is shown in Fig. 3.10.

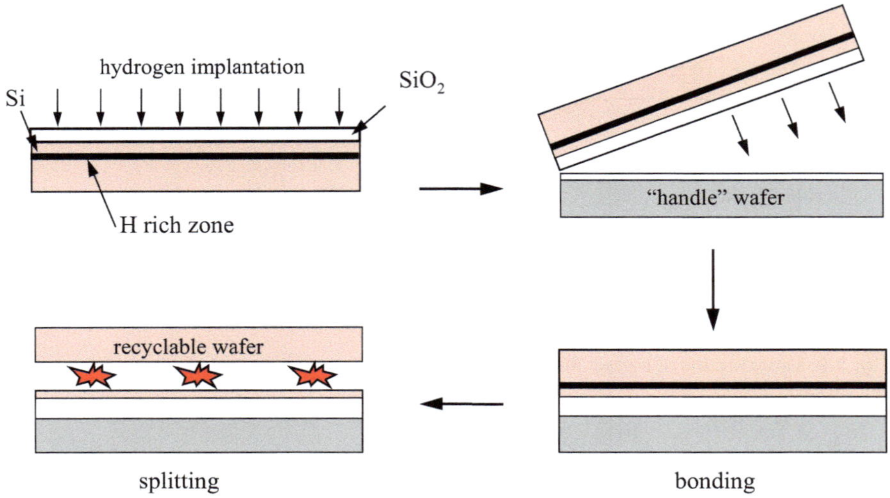

Fig. 3.11 Smart-Cut process flow. (Source: SOITEC)

SIMOX wafers were delivered with a *buried-oxide layer (BOX layer)* (SiO2) varying from less than 20 to 150 nm, with a top silicon layer varying from less than 10 to 100 nm. This is done to reduce the consequences of damage on the wafer surface. Fully depleted devices can be realised by reducing the thickness of the top layer to below 50 nm, for example, during processing. An alternative to the SIMOX process flow to create SOI is the Smart-Cut process flow (Fig. 3.11).

After the original wafer is first oxidised to create an isolating layer, H^+ ions are implanted to form a 'weak' layer at a certain distance below the surface. The thickness of the top layer is determined by the implantation energy. Next, the wafer is cleaned and bonded upside down to another wafer for further handling. During the 'smart-cut' step, the wafer is heated, such that the wafer is split exactly at the implanted weak H^+ layer. The remaining part of the wafer is reused again as original wafer or as carrier for a new SOI wafer, and the process cycle starts again. Finally, the SOI wafer needs an annealing step to recover the atomic structure, which was damaged during the implantation step. After a CMP planarisation step, the SOI wafer is ready. This smart-cut technology can be used for a wide range of SOI and BOX thickness.

In an SOI device with a thick top silicon layer (Fig. 3.12.a), this layer can only become partially depleted *(PD-SOI)* during operation, showing such parasitic effects as the floating-body and kink effect. A thin-body device (<50 nm) (Fig. 3.12b) will become fully depleted *(FD-SOI)* and does not show these effects.

For advanced low-voltage CMOS ($\leq$1 V) system-on-chip designs with digital, analogue and RF parts, SOI is expected to offer a better performance than bulk-CMOS technology [11, 12]. SOI is said to deliver more speed at the same power consumption or to consume less power at the same speed. Furthermore, SOI realises

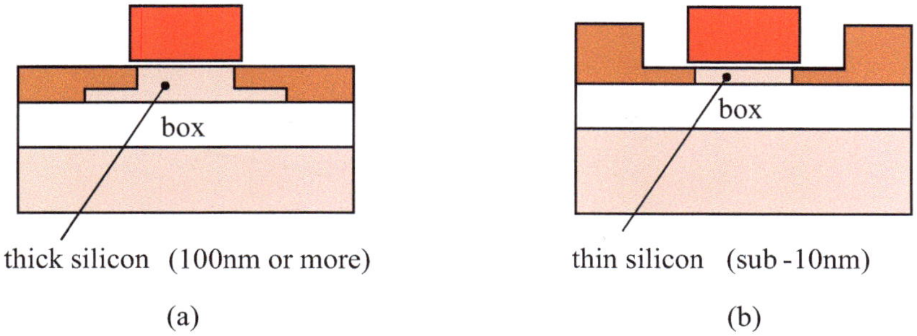

Fig. 3.12 Cross section of a (**a**) partially depleted SOI device and (**b**) a fully depleted SOI device

better isolation between digital, analogue and RF parts on the IC. Those circuits will therefore be less affected by substrate noise. Additionally, an SOI transistor has lower parasitic capacitances and consequently exhibits a better RF performance. SOI devices are thermally insulated from the substrate by the buried-oxide layer. This leads to a substantial elevation of temperature *(self-heating)* within the SOI device, which consequently modifies the output *IV* characteristics of the device, showing negative conductance. These effects, which are considerably larger in SOI than in bulk devices under similar conditions, must be taken into account by device technology engineers, model builders and designers. Since the body is isolated, SOI circuits show several advantages, compared to bulk CMOS:

- Smaller junction capacitances
- No deep well required (this is especially an advantage for FD-SOI)
- Less n^+ to p^+ spacing, due to absence of wells
- Significant reduction in substrate noise (questionable at high frequencies >1.5 GHz)
- No manifestation of latch-up
- Reduced soft-error rate (SER), because the electron-hole pairs generated in the substrate cannot reach the transistors
- Steeper subthreshold slope, which can be close to the theoretical limit of 63 mV/decade, compared to around 80 mV/decade for bulk CMOS devices

The future for planar partially depleted SOI devices is not completely clear. The relative performance benefit due to the smaller junction capacitances of SOI will gradually reduce because this advantage diminishes with scaling. Junction area capacitance decreases with the square of the scaling factor, while gate and perimeter capacitances decrease only linearly. Next to this, the increasing impacts of interconnect capacitances and delays will also reduce the performance benefits of SOI.

For the 45 nm node, most semiconductor manufacturers still use bulk CMOS as their main process technology. However, beyond this node, FD-SOI may become

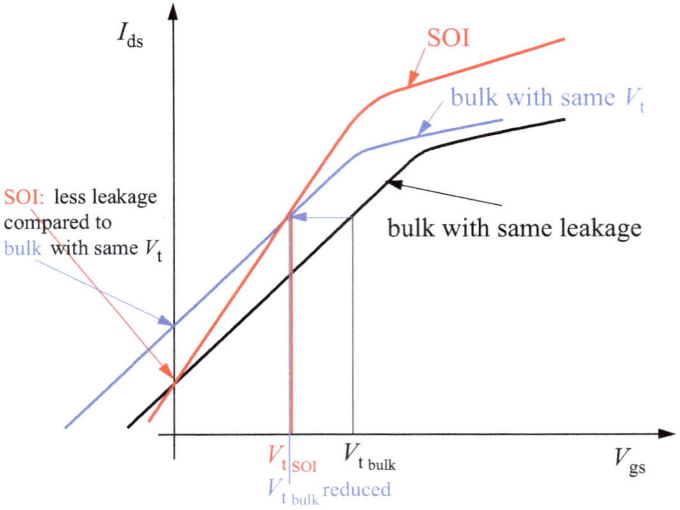

Fig. 3.13 Schematic illustration of current characteristics and subthreshold behaviour of bulk CMOS and FD-SOI

a good alternative to bulk CMOS. Since the channel region is fully depleted, it largely eliminates the neutral body. It therefore hardly exhibits the floating-body, history and kink effects. Moreover, it is expected to show improved short-channel effects (SCE) and drain-induced barrier lowering (DIBL). FD-SOI requires a reduced channel-doping concentration, leading to a higher mobility and a much steeper *subthreshold slope*, which almost matches the ideal value of ≈63 mV/decade (Fig. 3.13), compared to the ≈80 mV/decade for a bulk-CMOS process.

The diagram shows that in an SOI process, a transistor may have a lower V_T than in a bulk-CMOS process while carrying the same subthreshold leakage current. This advantage can either be used for speed improvement, when running SOI at the same supply voltage as bulk CMOS, or for power reduction, when running SOI at a lower supply voltage but at the same speed. FD-SOI allows sub-1V RF circuits, with improved F_t and F_{max} and reduced noise levels.

The transistors in such a nanometer FD-SOI process are fabricated in a thin film, with a thickness ≈5 to 20 nm on a box thickness between 5 and 50 nm [10]. Because the body between source and drain is fully depleted, the V_T-spread in these devices is much less dominated by the doping levels. Instead, it now depends heavily on the film thickness, whose uniformity across an 8 inch or 12 inch wafer has become a major criterion in the success of FD-SOI. This uniformity is therefore likely to have a more global (inter-chip) than local (intra-chip) impact on the variability in device operation. Below the 22 nm node planar SOI devices are expected to show device current degradation due to 'quantum confinement' [13].

Many other alternative device and process options have been applied in technologies beyond the 45 nm node. A flavour of these technology options in both the devices and interconnects is presented in Sect. 3.9.4.

3.3 Lithography in MOS Processes

3.3.1 Lithography Basics

The integration of a circuit requires a translation of its specifications into a description of the layers necessary for IC manufacture. Usually, these layers are represented in a *layout*. The generation of such a layout is usually done via an interactive graphics display for handcrafted layouts or by means of synthesis and place-and-route tools, as discussed in Chap. 7. Figure 3.14 shows an example of a complex IC containing several synthesized functional blocks.

A complete design is subjected to functional, electrical and layout design rule checks. If these checks prove satisfactory, then the layout is stored in a computer file (gds2 file). This *database* is now ready for *tape-out*. This term originates from the past when the complete database was loaded in gds2 format onto a magnetic

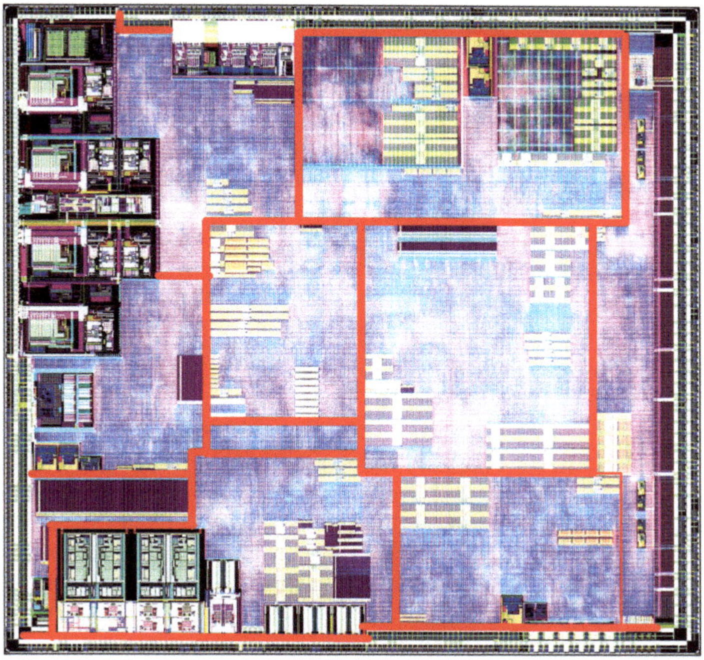

Fig. 3.14 Example of a complex signal processor chip, containing several existing IP cores with newly synthesised cores. (Source: NXP Semiconductors)

tape. Tape-out is the final design description which will be used for manufacture. The first activity in the manufacturing process is the creation of the physical masks. A software program (post-processor) is used to convert this database to a series of commands. These commands control an *electron-beam pattern generator (EBPG)* or a *laser-beam pattern generator (LBPG)*, which creates an image of each mask on a photographic plate called a *reticle* (Fig. 3.16). Such a reticle contains a magnified copy of the mask patterns. The reticle pattern is thus demagnified as it passes through the projection optics. Usually, a reticle contains four times the physical sizes of the patterns. The sizes of image field of 26 × 33 mm are the physical sizes on the wafer. On reticle level, these sizes are four times larger. The grey areas at the left and right side of the image field contain barcodes, reticle ID and pre-alignment marks. The *alignment marks* consist of an X-mark and a Y-mark, respectively, above and left from the image field. On the wafer, they will become imaged in the scribe lanes, which are typically between 30 and 100 μm wide to enable separation of the individual dies by mechanical sawing or laser cutting. For wafer alignment, about 16 X/Y pairs, distributed over the wafer, are being measured. Because of wafer warpage and bow, and because of topology differences between the wafer centre and wafer edge, more alignment marks are measured near the wafer edge. Litho systems use *alignment sensors* that determine the wafer positions based on the reflected light pattern from the features in the alignment mark. After copying the first mask to the wafer, alignment marks in the (inter-field) *scribe lanes* on the wafer are left behind. Subsequent mask layers need then to be aligned with previous ones. This is called *alignment sequence*. Often, several masks are aligned with respect to the first one. During the corresponding process steps, the alignment marks of the first layer are covered with several deposited layers. This requires alignment sensors with different colours of light that can ‘look’ through these layers to position the new mask. However, each colour will lead to a different reflection pattern and, so, to a different position measurement. The solution to these problems is currently included in today’s alignment sensors. In this approach, the sensor can ‘look’ through a limited number of layers. After that, the next mask includes a new set of alignment marks, which are then used for the alignment of again a series of subsequent masks, and so on. This results in a so-called *alignment tree*. The amount of mask layers, before a new alignment mark is required, depends on the process and the applied litho tool: DUV or EUV. An important disadvantage of this alignment tree is the accumulation of alignment tolerances: one for each new set of marks.

An alternative technique is to use a so-called etch-clearance approach, in which, after a limited number of ‘look-through’ layers on top of the first-layer alignment mark, these layers are etched away. Next, a subsequent number of mask layers is again aligned to that first alignment mask layer, followed by another etch-clearance step, etc. Although this requires additional process complexity and cost, it leads to smaller overlay numbers.

The vast majority of dies are much smaller than the reticle field size. Therefore, more dies fit on one reticle. In DRAM design, for example, the die sizes are often chosen such that an array of an integer number of dies fit on one reticle, e.g. in case of a 4 Gb DRAM, there may be 20 of these memory dies on one reticle field, which

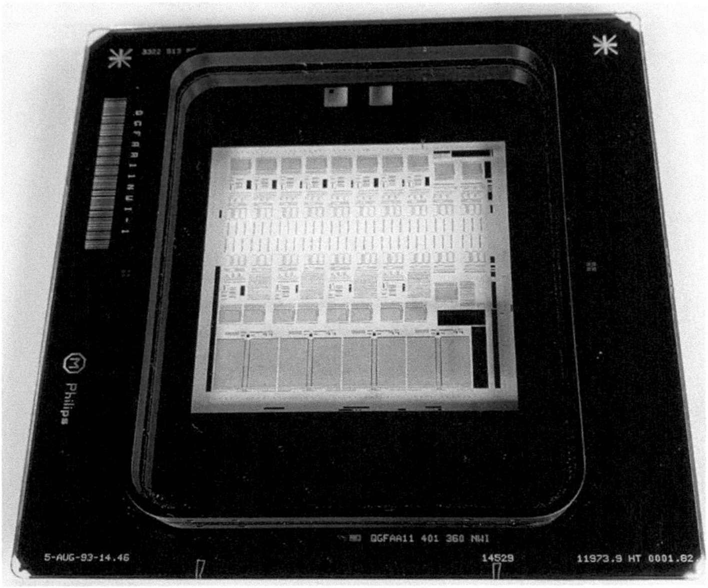

Fig. 3.15 Example of the second-metal mask patterns of a video processor chip of the 1990s

greatly reduces the cost. Many of these multi-die reticles also include alignment marks in the (intra-field) scribe lanes between these dies.

In conclusion, the used alignment method and sequence is very much dependent on the fabrication plant, the process node and the used litho tool [17]. An additional remark to be made here is that the *scribe lanes* are also used by the foundry to position their *process control modules* (*PCM*s), which contain many transistor and process structures that are measured during different stages in the process and yields (early) feedback on the margins and quality of the process. Traditionally, on 6-inch and smaller wafers, there are three to five chip positions assigned to PCMs, more or less evenly distributed across the wafer. On the one- and two-inch wafers in Fig. 3.4, the reader can still distinguish a few of these PCMs. Today, foundries do no longer waste five potential chip positions for PCM usage, and they are therefore located in the scribe lanes. However, as the processes become extremely complex, foundries require an increasing number of devices and structures in their PCMs for a good coverage of the process parameters and quality. Sometimes, foundries complain about the intense use of alignment marks, so that there is not enough space left in the scribe lanes for their PCMs. Section 10.3.1 presents more details on PCMs.

During the printing process, often pellicles are used to protect the reticle from harmful particles. The reticle in Fig. 3.15 shows the second-metal mask patterns of a video processor chip of the 1990s.

On both sides of this reticle, the chip patterns are surrounded by a kind of rectangular metal frame with a heigth between 5 and 10 nm. This is covered with an

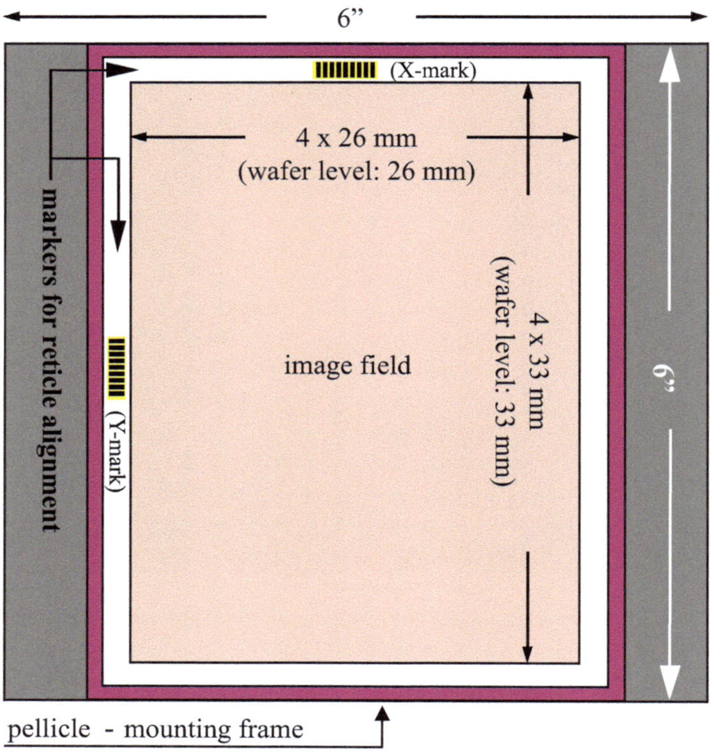

Fig. 3.16 Schematic layout of a 4x reduction reticle for step and scan systems. (Source: ASML)

extremely transparant polymer membrane called a *pellicle*. It acts as a dust cover, which keeps particles out of focus during the lithographic process, so these will not image onto the wafer and reduces the possibility of printing defects. Particularly, with the introduction of 193 nm lithography systems, the light transmission loss in the pellicles increases with the number of exposures. The major cause of this so-called *haze effect* results from chemical residuals from mask cleaning but also from contaminants from the scanner ambient such that they frequently need to be replaced. The cost of a mask set is subject of discussion in Sect. 3.3.4.

The process of copying a pattern from a mask (reticle) onto a wafer is known as photolithography. The wafer is first coated with a photosensitive layer, also called *photoresist layer*. Then, a light beam shines through the mask onto this layer. The open areas in the mask become exposed areas in the photoresist layer. This exposure step changes the chemical and physical properties of the photoresist material in these exposed areas, which can then selectively be removed during the development step. The process steps related to the deposition, bake and development of the resist material are usually performed in a so-called *track system*, which is directly coupled to the wafer scanner.

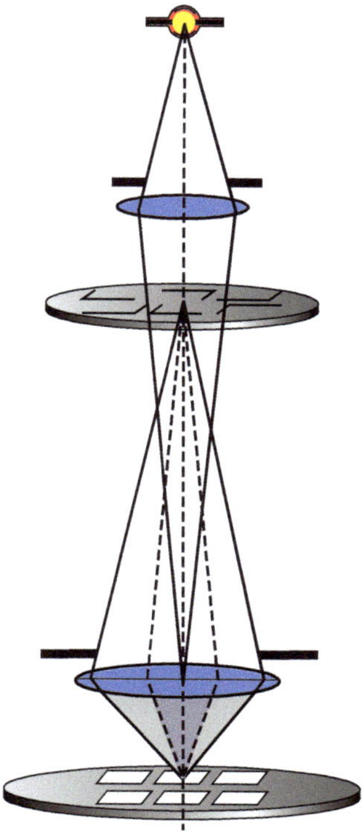

Fig. 3.17 Basic schematic of generic optical projection system

Small feature sizes, such as currently required in deep-submicron (<0.25 μm channel lengths) and nanometer (<100 nm) CMOS processes, are obtained by using reduction steppers or scanners. Current reduction steppers and scanners use *four-to-one* (4:1) reduction *step-and-repeat* or *step-and-scan*. A traditional step-and-repeat system only moves the wafer rapidly to the next die (or reticle field) position and holds while the whole reticle field is exposed during a single exposure. In a step-and-scan system, both the wafer and the reticle move synchronous in opposite directions at (four times) different speed controlled by a high-precision tracking system. It scans a narrow image field across the total reticle field. After the total field has been scanned, it steps to the next field position on the wafer. The reduction is achieved by means of a system of (very) complex lenses. Figure 3.17 shows a basic schematic of a generic optical projection system. In a real photolithographic system, both the illuminator path (light source to reticle) and the projection path (reticle to wafer) consist of a couple of lenses.

Limitations of these projection lithography techniques are not only determined by the wavelength λ of the applied light source and the numerical aperture NA. A stepper/scanner also needs to create clear, high-contrast images, while it must offer a sufficient *depth of focus (DOF)* to accommodate system and process-focus (height) variations which also lead to critical dimension (CD) variation. The combination of a large number of metal layers and extremely large-area designs create significant topographies across these designs and put stringent demands to the *DOF*. Current CMP planarisation technology limits topology variations to below 40 nm.

The resolution of the resulting projections is limited by diffraction and also depends on the properties of the photoresist. Better photoresists allow smaller minimum feature sizes. There are two expressions, developed by Ernst Abbe around 1867, which describe the most important characteristics of a lithographic imaging system. Firstly, the feature size F (≡ half pitch for memories, often also referred to as *critical dimension (CD)*), which refers to the minimum width of the printed patterns, is defined by

$$F = CD = k_1 \cdot \frac{\lambda}{NA} = k_1 \cdot \frac{\lambda}{n \sin \alpha} \tag{3.1}$$

where k_1 is a constant, which is a function of the resist, the mask, illumination and resolution enhancement techniques *(RET)*, which will be discussed later. With 'conventional' *three-beam imaging*, where only the zero- (0) and first-diffraction order rays (-1 and $+1$) pass the lens, the value of k_1 is restricted to: $k_1 \geq \frac{1}{2}$. When phase-shift masks *(PSM)* or off-axis illumination (both techniques are explained later in this section) are applied, only two diffraction orders pass through the lens, and k_1 can be further reduced to $\frac{1}{4} \leq k_1 \leq \frac{1}{2}$. These techniques are usually also referred to as *two-beam imaging*. NA represents the *numerical aperture* and n the refraction index of the medium between the lens and the wafer (1 for an air-based system) and α is the collection half angle as shown in Fig. 3.24. Secondly, *depth of focus (DOF)*, which refers to the distance along the optical axis over which features of illuminated surface are in focus and can be exposed with sufficient accuracy, is defined by

$$DOF = k_2 \cdot \frac{n \cdot \lambda}{NA^2} \tag{3.2}$$

where k_2 represents another lithographic constant, determining the allowable image blur from defocus. Current values for k_2 are around 0.5. Needless to say, F should be minimised and DOF should be maximised. In fact, a trade-off has to be made. Whereas the resolution of the imaging system is improving (reducing F) by increasing NA, its depth of focus will be reduced. For 193 nm ArF immersion scanners with an NA equal to 1.35, the DOF over the illuminated region may vary between 50 and 100 nm, depending on the photoresist material and its thickness [16]. Variations in CD, which are specified by the *CD uniformity (CDU)*, depend on the following:

- The actual light energy
- The pattern on the reticle (isolated or dense lines)
- The depth of focus (DOF)

In extreme cases, focus errors cause blurring of the lines to be printed. The DOF depends on several parameters:

- Illumination mode of the system (NA and σ (of k_2))
- Substrate flatness (planarisation) and substrate reflectivity
- Minimum feature size
- Pattern structure (again, isolated or dense lines)

For many technology generations in the past, the values for k_1 and NA were about the same, resulting in minimum feature sizes, which were about equal to the wavelength of the used light source. 0.35 μm feature sizes were mostly printed on i-line (365 nm) steppers. From a cost perspective, there is a strong drive to extend the wavelength of the light source to smaller technologies. The 248 nm deep-UV (DUV) steppers, with a krypton-fluoride (KrF) light source, are even used for 90 nm feature sizes, while the argon-fluoride (ArF) 193 nm DUV can potentially be used for feature sizes until 60 nm with dry lithography and until 30 nm with immersion lithography. Steppers (scanners) with shorter wavelengths will become very expensive and need many workarounds, as traditional optical lithography will no longer be viable at much shorter wavelengths.

When creating smaller feature sizes with the same wavelength, we need to compensate for non-ideal patterning, such as lens aberrations, variations in exposure dose, photoresist, pattern sensitivity, die distribution across the reticle and the field (reticle) size. The extension of the use of the 193 nm wavelength to sub-100 nm technologies cannot be done without the use of several additional *resolution enhancement techniques (RET)*: optical-proximity correction (OPC), off-axis illumination (OAI), phase-shift masks (PSM), better resist technologies, immersion lithography and design support. In the following, these techniques are discussed in some detail to present the reader a flavour of the increasing complexity and costs of the lithographic process, starting with the basic conventional binary mask.

The conventional binary mask is used in combination with the 193 nm light source to depict features with half pitch (HP) sizes as small as 90 nm. A *binary (photo) mask* is composed of quartz and chrome features (Fig. 3.18). Light passes through the clear quartz areas and is blocked by the opaque chrome areas. Where the light reaches the wafer, the photoresist is exposed, and those areas are later removed in the develop process, leaving the unexposed areas as features on the wafer. Binary masks are relatively cheap and they show long lifetimes, because they can be cleaned an almost infinite number of times. Moreover, they use the lowest exposure dose and enable high throughput rates. Preferably, all masks should be binary masks since it would reduce the overall production costs.

As feature sizes and pitches shrink, the resolution of the projection optics begins to limit the quality of the resist image. In the example above, there is significant energy (and intensity, which is proportional to the square of the energy) even below the opaque chrome areas, due to the very close proximity of the neighbouring clear quartz areas. This 'unwanted' energy influences the quality of the resist profiles, which are ideally vertical.

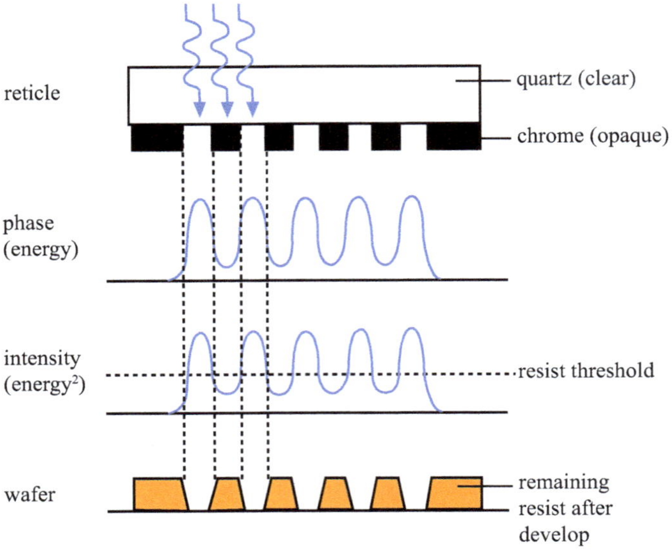

Fig. 3.18 Basic use of a binary photomask. (Source: ASML)

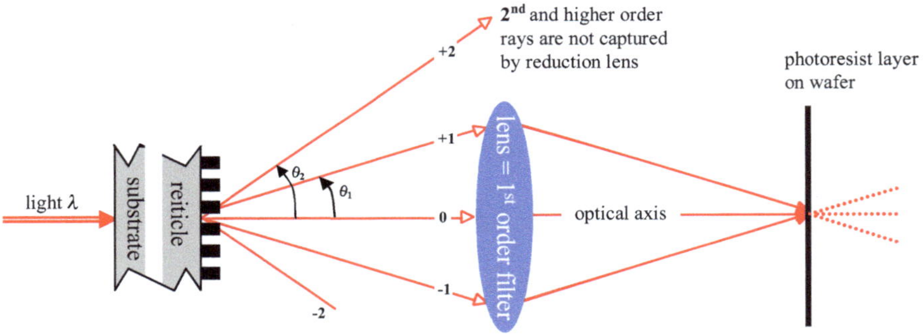

Fig. 3.19 Three-beam imaging concept

A conventional binary mask with a dense pattern of lines will produce a pattern of discrete light diffraction orders (-n, −(n−1),.., −2, −1, 0, 1, 2,.., n−1, n). The example in Fig. 3.19 shows a so-called *three-beam imaging* system. Here, a binary mask is used in combination with a projection lens that acts as a first-order ray filter. This prevents the capture of higher-order rays.

The interference of the zero-order diffracted light beam with the two first-order diffracted light beams produces a reduced (4:1) image of the pattern. If the line pitch in the pattern becomes smaller, the first-order light beam diffracts with an angle, which is too large to be captured by the lens, which is then incapable of producing the right image. Therefore phase-shift techniques, such as off-axis illumination and

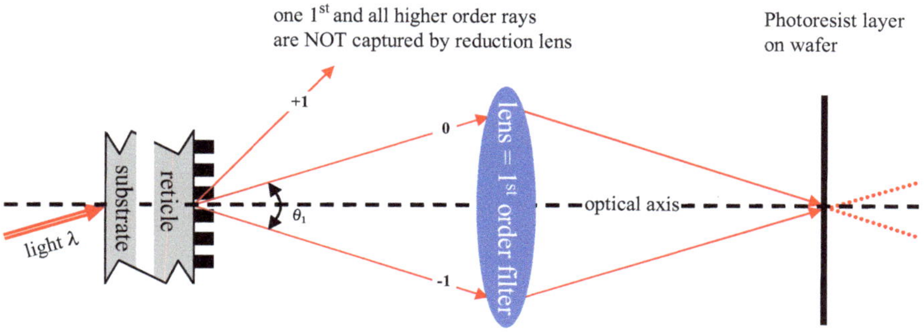

Fig. 3.20 Off-axis illumination (two-beam imaging concept)

PSM, are designed to 'sharpen' the intensity profile and thus the resist profile, which allows smaller features to be printed. When a binary mask is illuminated at a different from normal angle, this angle can be chosen such that one of the first-order diffracted light beams can no longer be captured by the lens and the image is produced by only two diffracted beams (the zero and remaining first order). This so-called *off-axis illumination (OAI)* technique (Fig. 3.20) is therefore an example of *two-beam imaging*. A further optimisation of this imaging technique can be achieved by choosing the angle of illumination such that the remaining beams are symmetric with respect to the centre of the lens. An OAI system can improve the resolution limit of a dense line pattern with a factor of two.

However, another benefit from a two-beam imaging system comes from the enhanced *depth of focus (DOF)*. It can be seen that in a three-beam imaging system (Fig. 3.19), the first-order diffracted beams travel across a different path than the zero-order beam, before arriving at the wafer surface. It can therefore provide only a very narrow range, in which the zero and first diffraction orders remain in phase (basically only in the focal plane), limiting its depth of focus. Outside this range, it creates a phase error. A minor displacement of the wafer out of the focal plane causes an increase of this phase error and leads to a degraded image at the wafer surface. In a two-beam imaging system (Fig. 3.20), assuming full spatial symmetry, the diffraction patterns are in phase and will interfere properly. The same wafer displacement in such a system will result in a satisfactory image over a longer range, thereby increasing its depth of focus.

An alternative to off-axis illumination is the *phase-shift mask (PSM)* technology, which has been pioneered in recent years to extend the limits of optical lithography. PSM technology is divided into two categories: attenuated PSM and alternating PSM.

Attenuated phase-shift masks (AttPSM) form their patterns through adjacent areas of quartz and, for example, molybdenum silicide (MoSi). Unlike chrome, MoSi allows a small percentage of the light to pass through (typically 6% or 18%). However, the thickness of the MoSi is chosen so that the transmitted light is 180° out of phase with the light that passes through the neighbouring clear quartz areas

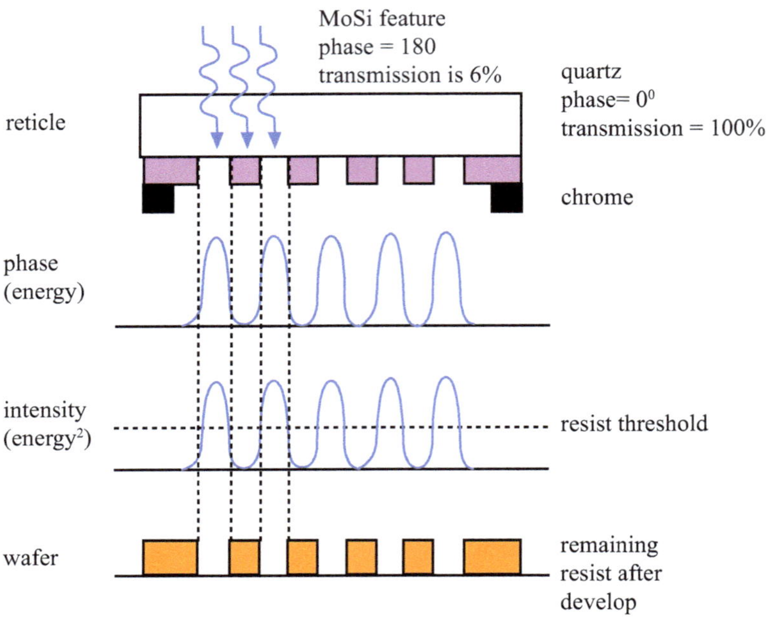

Fig. 3.21 Basic use of an attenuated phase-shift mask (attPSM). (Source: ASML)

(Fig. 3.21). The light that passes through the MoSi areas is too weak to expose the resist, and its 180° phase shift reduces the intensity in these areas such that they appear to be 'darker' than similar features in chrome. The result is a sharper intensity profile which allows smaller features to be printed on the wafer. The 180° phase shift is only achieved for light at a given fixed wavelength. AttPSM masks can therefore only be used for one type of scanners only, while binary masks can be used for scanners with different wavelengths. In fact, the use of attPSM filters one of the first-order diffracted light beams of a three-beam imaging system (Fig. 3.19), which makes it a two-beam imaging system, similar to OAI imaging (Fig. 3.20). Figure 3.22 shows a comparison of the three different imaging systems. It clearly shows the improvement of the DOF in the two-beam imaging systems.

OAI systems and attenuated phase-shift masks are used for critical patterns that require higher resolution than photolithography systems that employ binary masks only. An alternative powerful but complex two-beam illumination system is the *alternating phase-shift mask (altPSM)* concept (Fig. 3.23). Such masks employ alternating areas of chrome, 0° phase quartz and 180° phase-shifted quartz to form features on the wafer. The pattern is etched into the quartz on the reticle causing a 180° phase shift compared to the unetched areas (0° phase). As the phase goes from positive to negative, it passes through 0. The intensity (proportional to the square of the phase) also goes through 0, making a very dark and sharp line on the wafer. The process of manufacturing the mask is considerably more

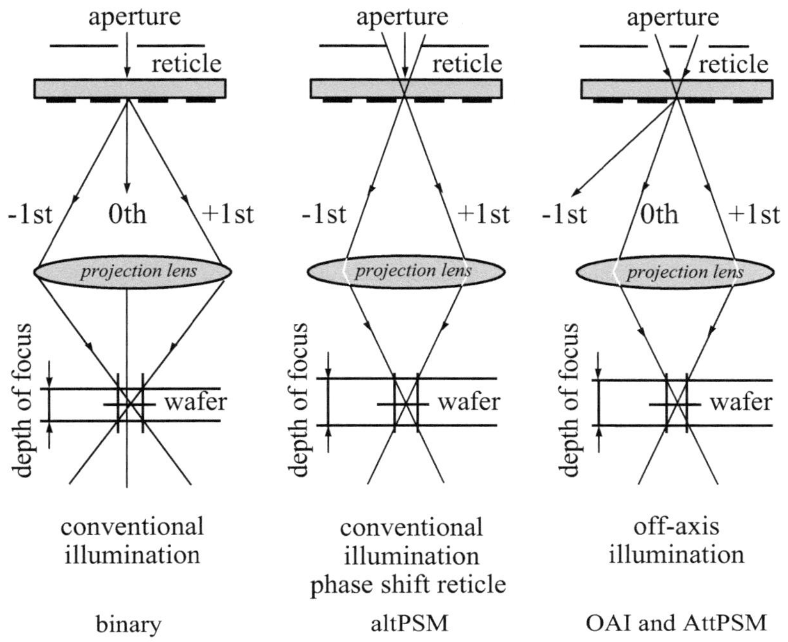

Fig. 3.22 Comparison of the three different imaging systems. (Source: ASML)

demanding and expensive than that for binary masks. Furthermore, the AltPSM requires an additional binary 'trim' mask and exposure step, resulting in extra costs and decreased stepper/scanner throughput; however, it enables excellent CD control.

AltPSM is used for the production of high-performance ICs that only allow extremely limited variations in line width, such as high-speed microprocessors.

As explained, the above-presented lithographic techniques are basically applied to increase the resolution and/or depth of focus of the total illumination system. Another technique, which is currently already applied to enhance the lithographic properties, is called *immersion lithography*. If we immerse the photolithographic process in water ($n = 1.43$) and if we assume that $\sin\alpha$ in Expression (3.1) can reach a maximum value of 0.95, then this 'water-immersion lithography' can yield an NA close to 1.37. Only the lower part of the optics is immersed in water (Fig. 3.24).

The left half in the figure shows the diffraction of the light beams in air, with a diffraction index $n = 1$ and some of the beams being reflected. The right half uses an immersion liquid with $n > 1$, which reduces the amount of reflected light, increasing the resolving power and allowing finer feature sizes. Immersion lithography also improves the DOF, which may resolve some of the related topography problems.

Compared to an air-based system, immersion lithography shows a number of additional problems. To achieve a high throughput, the stage has to step quickly

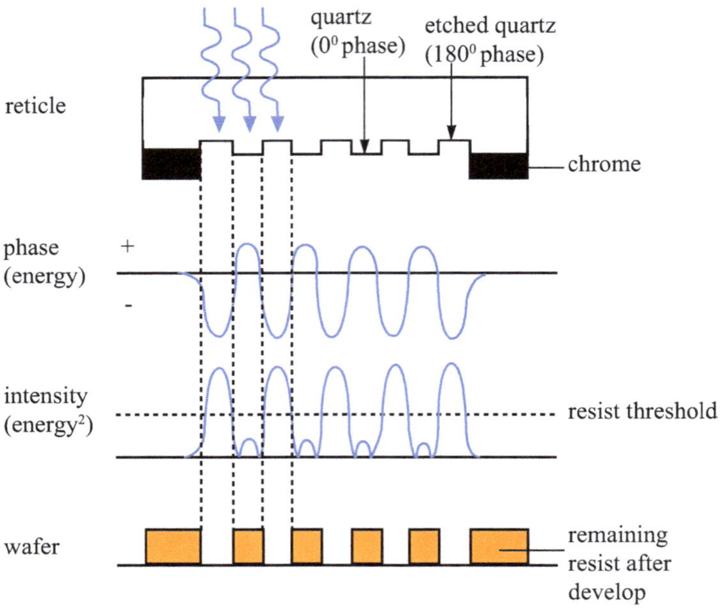

Fig. 3.23 Basic use of an alternating phase-shift mask (altPSM). (Source: ASML)

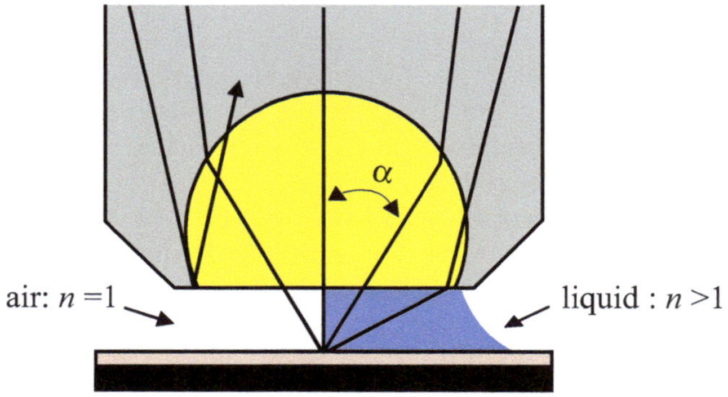

Fig. 3.24 Basic principle of immersion lithography. (Source: ASML)

from one chip position to the next, which may create *bubbles* into the water, deteriorating the imaging capability of the system. There are several solutions to this problem, but these are beyond the scope of this text.

Using one of the above-described resolution enhancement techniques (RETs) is a prerequisite to create lithographic images with a satisfactory resolution and *DOF*. But it is not sufficient. When printing patterns with sub-wavelength resolution, they need to be compensated for the aberrations in the patterning. In other words, the fabricated IC patterns are no longer accurate replica of the originally designed

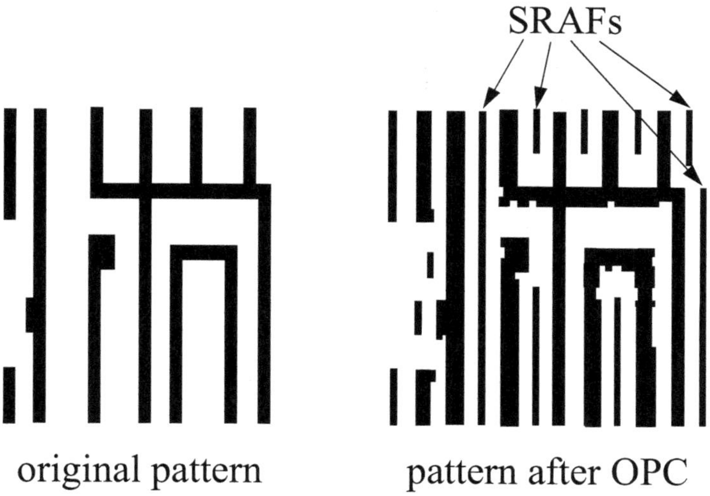

Fig. 3.25 OPC (including SRAFs) applied in the mask-definition process. (Source: ASML)

patterns. So, we need to proactively compensate (make corrections) for these shortcomings in the mask. Figure 3.25 shows how *optical proximity correction (OPC)* is applied in the mask-definition process. The right mask pattern is used during lithography, to get left (original layout) pattern image on the chip. More optimal imaging results can be achieved by using so-called *subresolution assist features (SRAFs)*, such as scattering bars and hammerheads, which are not printed onto the wafer, but help to reduce resolution enhancement variations across the mask.

This has several consequences for the layout designer: he should leave enough space to add OPC features and/or he should draw the patterns with constant proximity and/or he should leave enough space to add SRAFs. It will certainly make the design process more complex.

While the above-described RETs improve the resolution of the imaging system, the use of OPC masks will make them work. Mask costs, however, very much depend on the applied technology. When normalising the costs of a binary mask to 1, then an attPSM (without OPC) mask would cost 1.2 times as much and an attPSM (with OPC) mask 2.5 times. The use of altPSM is much more costly (six times and ten times more for altPSM without and with OPC, respectively), since it requires an additional binary trim mask and thus needs double exposure.

For the time being, we still relied on innovations that extend the use of photolithography beyond the 28 nm node. Support from the design side has already alleviated several problems related to the extended use of 193 nm lithography into the sub-50 nm CMOS technologies. To improve yield, complex design for manufacturability (DfM) design rules have already been used in many technology nodes.

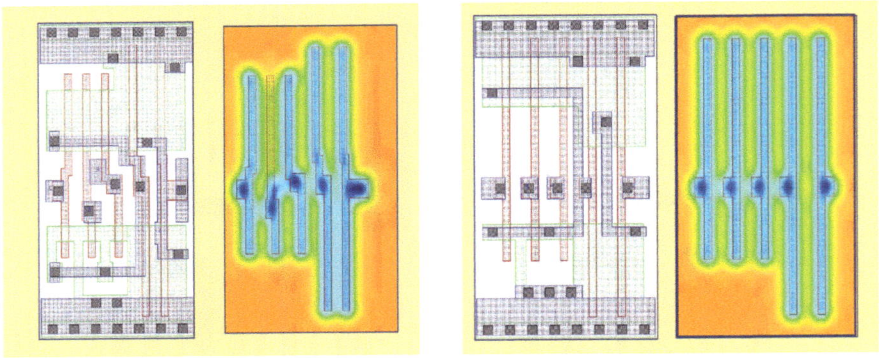

Fig. 3.26 Comparison of an original (left) and a litho-friendly layout (right) with more regularity. (Source: NXP Semiconductors)

For technologies beyond 70 nm, this was certainly not enough. They also required strict *design for lithography (DfL)* design rules. DfL, also called *litho-friendly design*, litho-driven design or litho-centric DfM, is focused on more regular layout structures. It simplifies the lithographic process; it supports SRAFs and might reduce the mask costs. It also leads to a more aggressive scaling and to yield improvement due to a smaller variety of patterns to be printed. Moreover, more regularity in the standard cells results in a better portability to the next technology node. Figure 3.26 shows two layout versions of a standard cell: the original layout with a plot of simulated line widths and the litho-friendly layout with a plot of simulated line widths, showing more regularity. In the litho-friendly layout, all polysilicon lines would be in the projected image on the wafer, while in the original layout, the second-from-left polysilicon line would be missing in the image.

For this particular cell, litho-friendly design shows a relatively large impact on the cell area. For an average library, however, the area increase can be limited to just a few percent. Next to the already discussed implications of RET and DfL for layout design, these techniques are supported by the design flow and got more and more attention from design for yield (DfY) EDA tools and tool vendors. An overview of EDA-vendor DfY activities is presented in [18].

Litho-friendly design usually uses a limited number of poly-pitches. Such a fixed-pitch litho-friendly library design is a step towards a *fully regular library architecture*. Next to the process spread caused by lithographic imperfections, such an architecture may also reduce the influence of other process-spread mechanisms, by using only one size nMOS and one size pMOS transistor. The high-density gate-array architecture shown in Fig. 7.38 is an example of such an architecture, which can also be used as standard-cell template.

Before we continue our discussions, it is good to present some typical sizes and dimensions which are characteristic for a 28 nm CMOS process (Table 3.1). It shows that there are different definitions for critical dimensions (CD), feature sizes and

Table 3.1 Various definitions for critical dimensions (CD), pitches and out diffusion, depending on the lithographic and manufacturing process step and on the type of circuit, for a 28 nm process

	LOGIC process		Stand-alone memory
Dimension	High density [nm]	High performance [nm]	(e.g. planar flash) [nm]
CPP (=contacted poly-pitch)	114	130	56
CD litho print target	50	60	28
CD after resist trim	40	50	–
Pattern transfer etch	35	45	28
Out diffusion	30–35	40–45	28

pitches. They not only depend on the type of circuit but also on the particular phase during the lithographic and manufacturing process.

Let's summarise the individual contributions of the above-described RETs: the combination of PSM and OPC may lead to a minimum k_1 of about 0.25, while water immersion can lead to a maximum NA of approximately 1.37. Using these values, for a 193 nm lithography, in Expression (3.1) for F, leads to a minimum feature size ($\equiv$ half pitch; most common for memories) of around 35 nm. For smaller line widths, the 157 nm DUV (deep ultraviolet)-line (from a fluorine light source) lithography would have been an option. However, it was expected that this lithography would extend the lifetime of photolithography for just one or two process generations. The investments to create sufficiently transparent lenses with a homogeneous light refraction, adequate photoresists and pellicles to build a lithographic for such a short lifetime were too high. Therefore, chip makers decided to stretch the use of 193 nm immersion lithography to its limits, by applying additional techniques to enhance its resolution. Some of them are discussed in the next subparagraph.

3.3.2 Lithographic Extensions Beyond 30 nm

An increased pattern resolution can be achieved by combining immersion lithography with *double-patterning techniques (DPT)*. The most commonly used DPT is the so-called *litho-etch, litho-etch (LELE)* which uses two masks and double exposure. The second DPT, which is also called *self-aligned double patterning (SADP)*, creates an increased pattern density by a specific sequence of process steps. Both techniques will now be explained further.

- *Litho-etch, litho-etch (LELE)*. When the pitch of two lines in a dense pattern is less than 76 nm, it becomes a sub-resolution pitch, which can no longer be imaged correctly, with current lithographic techniques. Therefore, this can be done with an image split: first, image the odd lines with twice the minimum feature pitch (Fig. 3.27), and then, image the even lines, also with twice the pitch.

Fig. 3.27 Example of LELE double patterning

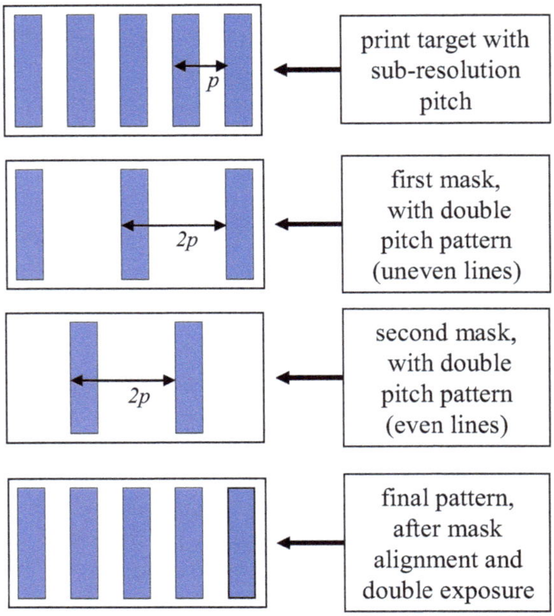

This procedure requires two masks and two exposures. The biggest challenge is the high accuracy of the alignment of the masks during exposure. Another challenge is to effectively decompose the single pattern layer into two individual masks. LELE double-patterning techniques are often used in advanced logic chips because of their non-uniform patterns. Several companies are currently experimenting *triple* and *quadruple patterning* techniques. With a 193 nm immersion lithographic system, *triple patterning* would enable 16 nm features, while quadruple patterning would even enable feature size down to 10 nm [20]. These techniques also contribute to increasing mask and processing cost. All *multi-patterning techniques* require an intelligent split of a single mask pattern into more separated masks, each with a lower-resolution pattern than the original pattern. In standard-cell design, this can be handled by tools; however, in optimised memory and analogue circuit design, the designer faces additional design rules to fulfil the requirements of double (triple or quadruple) patterning. An example of the decomposition of an original layout into two or three masks is shown in Fig. 3.28 [21].

This type of pattern decomposition was used for process nodes between 22 and 14 nm. Actually, LELE lithography has never become very popular. It causes severe problems with overlay and requires doubling of the number of masks and exposures or more in case of multi-patterning.

- Use of *self-aligned double patterning (SADP) (spacer lithography)*. In this technology, the final pattern on the wafer is created by the formation of sub-resolution features during semiconductor process steps, rather than by high-resolution

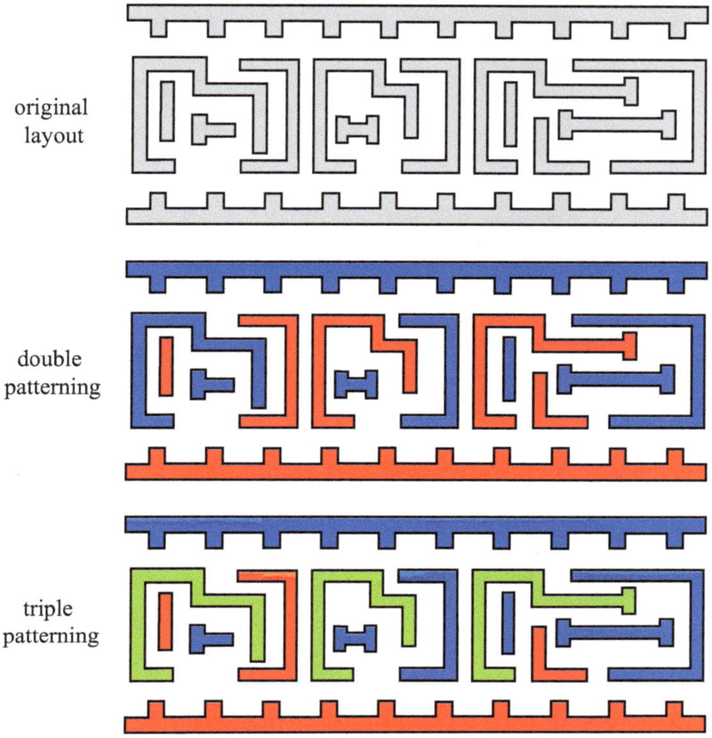

Fig. 3.28 Decomposition of an original layout into two or three individual masks. (Image: David Abercrombie; Mentor Graphics Corp.)

lithography. The process flow in this technology is as follows (Fig. 3.29). The print target is shown in (a). As a first step, a hard mask layer is deposited or grown on the wafer. To support the formation of sub-resolution spacers, a sacrificial polysilicon layer is deposited on the wafer and patterned with a relatively large optical lithography pitch (b). Since many of the layers are deposited with an atomic layer deposition (ALD) step, where no high temperature step is involved, the polysilicon is often replaced by photoresist material.

Next, an oxide (or nitride or other) layer is deposited on top of the structure and then etched back until sub-resolution side-wall spacers are left (c). Then, the sacrificial polysilicon is removed (etched) (d), followed by a pattern transfer from spacer to hard mask (e). Finally, the pattern in the hard mask is used to create the final pattern on the wafer (f). This spacer technology is a convenient approach to achieve sub-resolution patterning with relatively large optical resolution pitches, avoiding problems of, e.g. overlay between successive exposures in a double-patterning technology. Another advantage of this technique is that the printed *critical dimension uniformity (CDU)* is independent of the *line-edge roughness (LER)*. LER is caused by the diffusion of resist during a heat step after the

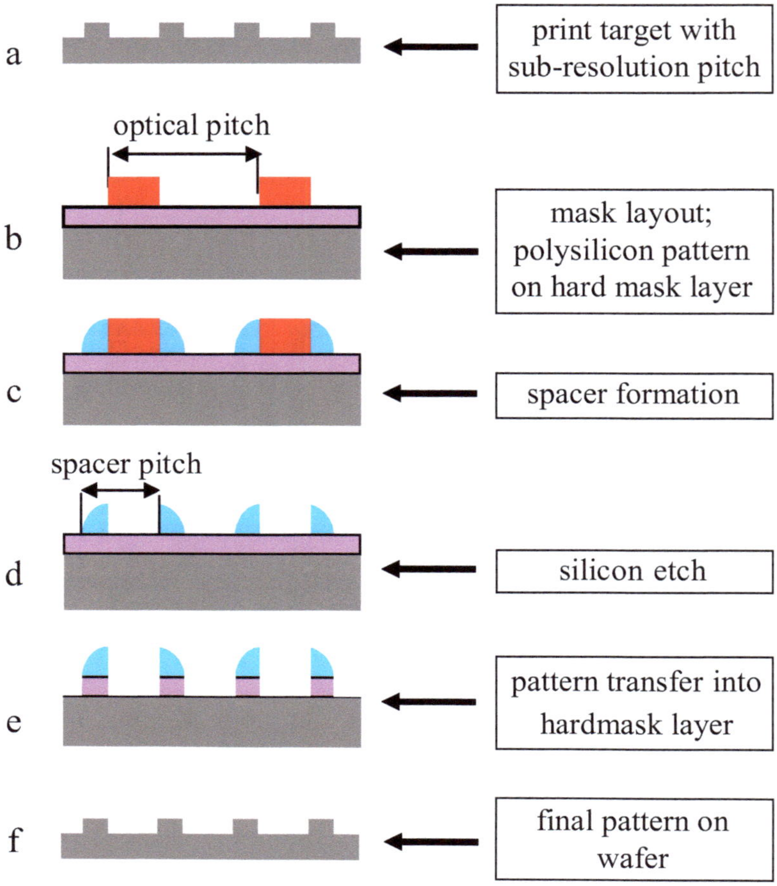

Fig. 3.29 Basic steps in spacer lithography

exposure (post-exposure bake at 200–220 °C) but before the development of the resist. This diffusion is random and may lead to diffusion lengths of 40 nm, which causes intra-line variations leading to frayed lines. In spacer technology, however, the pattern transfer is done through spacers and not through resists, showing almost no LER. A disadvantage of the spacer lithography is that it is only applicable for mono *CD (critical dimension)* which reflects the smallest geometrical features (contacts, metal width, trenches, etc.), so, for patterns with only one width. Patterns with features that also have two times the line width can be produced by the formation of two spacers directly positioned next to each other. SADP [19] is preferably used in the creation of dense regular patterns of parallel fixed-pitched lines in only one direction. In litho-friendly designs, including FinFET designs, which contain fixed-pitched transistor gates, SADP is used for patterning the polysilicon layer (Fig. 3.30). In this case, a pattern of continuous

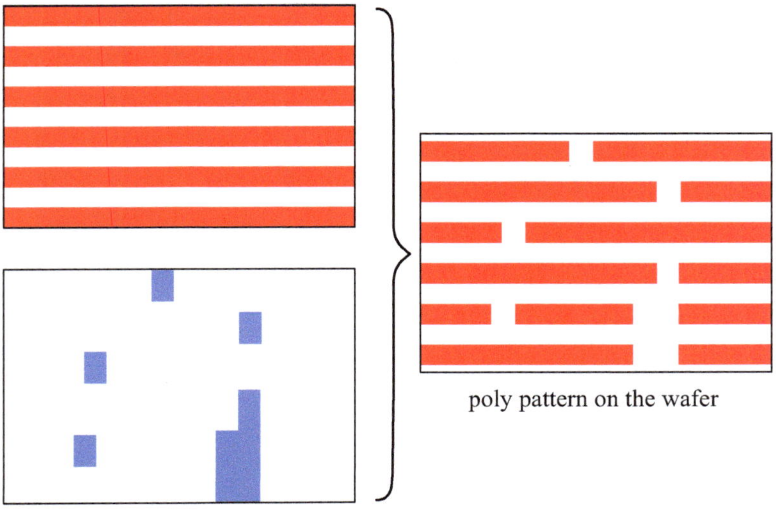

Fig. 3.30 Example of the use of the cut mask to create a much improved poly-CD control

poly-lines at a single pitch is printed first. Then, to create individual gates, the unwanted portions of the polysilicon are etched away by using a *cut mask*. This leads to a much improved CD control and reproducibility of the polysilicon gates, because the final pattern is much less influenced by lithographic aberrations.

The cut mask may contain a dense pattern of high-resolution features, which will make it costly. SADP requires restricted design rules, resulting in patterning one-dimensional lines with fixed spacings. Patterning the critical metal layers in the back end of the process demands a shift in IC design and requires 1D metal patterns. The metal features in one metal layer are then fully orthogonal with respect to ones in the previous layer.

The SADP spacer lithography allows the pitch to be halved with just one single lithographic exposure step. The CD control is then determined by the thickness of the deposited spacer layer, which is very accurately controlled since the formation of this spacer layer is done with an *atomic layer deposition* step (*ALD*). Let us assume that we now use the pattern structure in step (e) in Fig. 3.29 as a starting point for a second SADP iteration and we repeat steps (c) to (f), and then, we have again doubled the number of features. This is often referred to as *self-aligned quadruple patterning (SAQP)* (Fig. 3.31). SADP double patterning is often used in advanced memories, because memories typically consist of uniform pattern distributions. Spacer lithography is also increasingly used in the formation of the fins in FinFET process nodes of 20 nm and beyond. Even logic circuits in advanced FinFET processes are increasingly built from fully regular

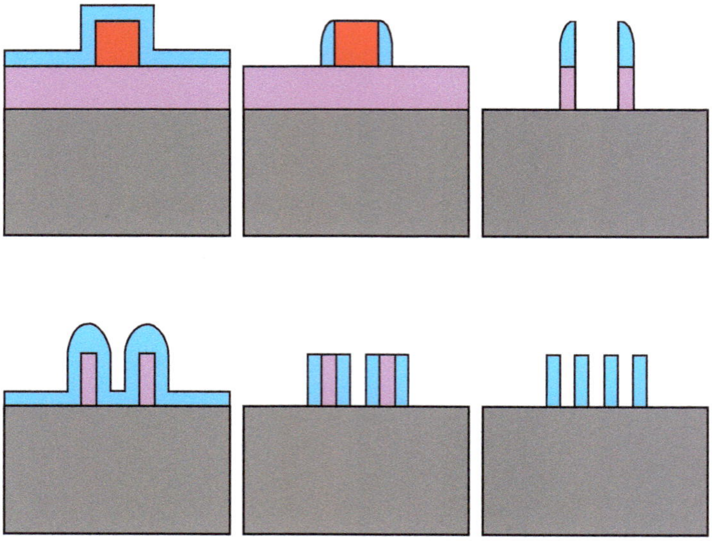

Fig. 3.31 Example of quadruple patterning using two iterations of self-aligned double patterning

layout patterns in the creation of fins and transistor gates. Section 4.8 in the next chapter describes a potential FinFET layout architecture in an example 16 nm CMOS process node.

- *Computational lithography.* In Sect. 3.3.1, several resolution enhancement techniques (RETs), such as optical proximity correction (OPC), phase-shift mask (PSM) and off-axis illumination (OAI), have been discussed. OPC already uses a lot of computational effort to compensate lithographic aberrations by changing the patterns on the reticle. Computational lithography uses complex model-based mathematical algorithms to optimise these RETs. It models and simulates the light, from source to wafer as it travels through the reticle, the lenses and the photoresist. Potential light distortions are pre-corrected by changing the patterns on the reticle. It not only includes the adjustment of OPC geometries but also accounts for variations in exposure time, dose and focus. The growth in the number of mask features combined with the increasing number of RETs has caused an exponential growth in computer simulation time. Many CPU years are required for the complete computational mask fabrication of a state-of-the-art chip. Mask-making companies run several graphic cards and other dedicated hardware accelerators in parallel to timely produce a complete chip mask set. Use of NVIDIA cuLitho, which is a library with optimised tools and algorithms for GPU-accelerating computational lithography, is expected to result in more masks being generated per day. The adoption of a newer mask technology such as *inverse lithography technology (ILT)* which uses *inverse curvilinear masks (ILM)* has been hindered due to their high computational requirements. This is expected

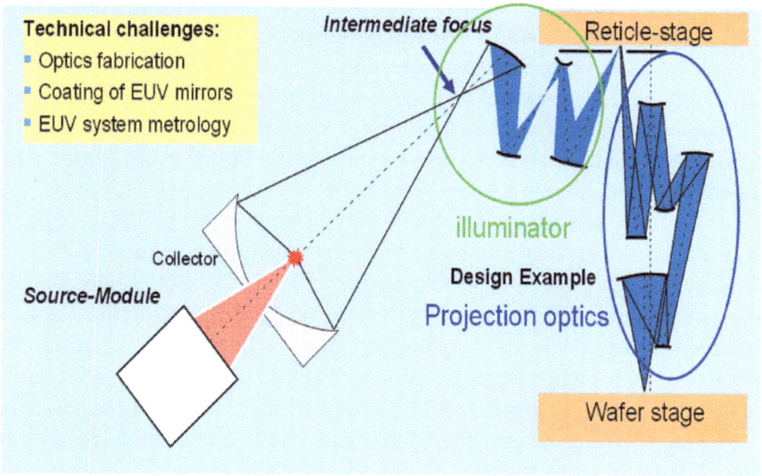

Fig. 3.32 The transmission path of the light in an EUV scanner as it travels from source to wafer. (Courtesy of: Carl Zeiss)

to worsen in the coming years. Curvilinear or 45° mask shapes (see Fig. 6.29) can be used on both optical and EUV masks, which currently use *Manhattan* (orthogonal) *shapes*. The data processing demand for high-NA EUV and ILT masks is expected to increase by a factor of ten. Computational lithography for these advanced lithography technologies will therefore require a massive use of computing power, such as offered in the above-mentioned NVIDIA solution [23]. Computational lithography, in combination with double or triple patterning, will enable the printing of 22 nm half pitch layouts.

- Use of *extreme-UV (EUV) lithography*. With a light source wavelength of 13.5 nm, EUV was often regarded as the most probable potential lithography solution for technology nodes beyond 30 nm. However, EUV 'light' is absorbed by all materials, including air. Therefore, mirrors have to be used in a vacuum-based system with reflective instead of refractive optics and reticles. The light transmission takes place via a large number of mirror lenses (Fig. 3.32). To achieve maximum reflectivity, each mirror lens is built from more than 100 layers of carefully selected materials. An extremely well-controlled polishing process results in a lens area smoothness variation of less than the thickness of one atom. A *laser-produced plasma (LPP)* source is used to deliver the required EUV power [24], by focusing a CO_2 laser beam onto tiny vaporised tin (Sn) drops. A droplet generator creates 50,000 tin drops per second. In fact, each tin drop, with an average diameter of about 10 µm, is hit by two separate laser pulses, creating highly ionised plasmas. Not all tin is transferred into plasma. Some of the tin splashes around inside the EUV source chamber. A hydrogen buffer gas in this chamber reduces the amount of tin debris traveling towards the EUV collector (collector protection) and enables a chemical reaction with the tin debris to remove it from the collector (collector reflectivity restoration) [25]. The highly ionised plasmas, with temperatures up to 400,000 degrees Celsius (about 40 times the temperature of the plasma in the sun), cause an isotropically radiation of EUV photons. These are then gathered by a special coated (≈0.5 m) mirror called collector (Fig. 3.32) and focussed to an *intermediate focus point*, from where they are directed through the illuminator path, the reticle and the projection optics to the wafer. This puts stringent requirements on the EUV light source.

The required sensitivity for EUV lithography is typically defined by the minimum required exposure dose needed to pattern a feature with a certain critical dimension. The current industry target for EUV lithography is to achieve a sensitivity of around 20–30 mJ/cm^2 or lower. Achieving higher sensitivity is important because it allows for shorter exposure times and higher throughput, which is crucial for manufacturing advanced semiconductor devices in sub-10 nm nodes [27]. However, improving sensitivity while maintaining acceptable line-width roughness and pattern fidelity remains challenging due to the complexity of the EUV lithography process. When variations in the width of a resist feature occur quickly over the length of the feature, this variation is called *line-width roughness (LWR)*. When examining these variations along just one edge it is called *line-edge roughness (LER)*. EUV scanners require clean EUV power of 250 W at the intermediate focus point (IF point) to generate about 1 W EUV power at wafer level. This required a laser output power of about 25 kW. With a laser efficiency of only 2.8%, this required (in the last decade) a total laser input electrical power of 0.88 MW [28] (compared to 0.16 MW for an immersion scanner), with instantaneous laser peak power of several MW [29]. Although the laser efficiency has increased to about 5%, a high-NA EUV machine may still consume between 1 and 1.5 MW.

It has been a struggle for almost a decade to combine the best EUV power source with the perfect resist to enable sufficient EUV power at wafer level. This has led to acceptable throughputs towards 200 or more wafer exposures per hour. This explains the need for an improved light-transmission system to improve the throughput time and reduce the power consumption.

In 2006, the first EUV lithography tools have already been shipped. It was not meant for production, but it supported R&D programs at IMEC (Leuven, Belgium) and at CNSE (University of Albany, New York) [26]. Pilot line production took off in 2012, when Intel, TSMC and Samsung start participation in a co-investment program to enable acceleration of the development of key lithography technologies, particularly that of EUV. Although current immersion scanners show throughputs of more than 300 wafers per hour, the effective throughput per mask layer with double, triple or quadruple patterning reduces with a factor of two, three or four, respectively. Currently (2024), EUV wafer throughput is between 175 and 200 wafers per hour, based on a 30 mJ/cm^2 dose, while expected machine upgrades will lead to ≥220 wafers per hour at the same dose [32]. This made EUV very competitive with alternative multi-patterning lithography technologies [33].

What is really important in the operating efficiency of an EUV lithography system is its average throughput. This is a combination of its actual throughput and its availability (uptime). An example of a physical effect that limits the *uptime* is the fact that when a tin drop is hit by a laser (see Fig. 3.32), parts of it 'splash' in all directions. Some land on the collector, which then loses its reflectivity. A solution for that problem is that the tin can be drained from the collector by a hydrogen flow. This does not fully prevent the degradation of the mirrors. The small loss in reflection can then be compensated by increasing the light source power, rather than replacing the mirrors. Currently (2024), about once a year, the collector mirror is taken out of the system, the tin debris is removed and the mirror is then placed back. Today's availability is between 90% and 95%. ASML continuously rolls out new upgrades to increase uptime of their EUV systems. The three leading semiconductor foundries (TSMC, Samsung and Intel) have all installed many EUV tools in their fabs. EUV is then expected to be used only for the most critical layers, while the other layers will still be printed with a combination of multi-patterning and 193 nm (immersion) scanners or even DUV scanners.
EUV masks are different from conventional photomasks. While conventional optical masks transmit light, EUV masks reflect light. Analogous to the optical lenses in an EUV tool, to achieve maximum reflectivity of the EUV mask, it consists of 40 to 50 thin alternating layers of silicon and molybdenum on a substrate. This results in a multilayer stack that is more than 300 nm thick. Identical to traditional lithography, also EUV masks require a *pellicle* to prevent particles to land on the mask. AS EUV lithography is based on reflective optics, the pellicle only needs to be mounted on one side of the reticle and needs to transmit both the incoming and reflected EUV beam. Generally, for larger die sizes, the use of pellicles is crucial to achieve reasonable yield numbers. Using EUV reticles without pellicles would require additional extensive pattern checks on the wafer. ASML developed its first polysilicon-based EUV pellicles in 2016 with a transmission efficiency around 78% on a simulated 175-watt source [34]. EUV pellicles need far more than 90% of transmittance to support low defectivity and high productivity of EUV lithography. Today's (2024) EUV pellicles show transmission efficiencies of over 92% have a transmission non-uniformity spec of less than 2% with less than 0.03% reflectivity. Next to transmission efficiency of the pellicle, also its temperature is an important parameter. When a pellicle is hit by EUV light, it may absorb several hundred watt source power and can heat up to more than 600 °C. This may cause degradation of the pellicle. Current research is therefore focused on materials that allow both a higher transmission efficiency and a higher temperature budget. IMEC has been developing EUV pellicles based on carbon nanotubes.
To improve the resolution of an EUV system, it also requires the use of *phase-shift masks (PSM)*, because the 'dark' areas on the EUV mask still reflect light ($\approx$3%) since the absorber still transmits light through the multilayer underneath. This reflected light, however, is shifted in phase ($\sim$150 degrees). This requires a different EUV mask with other absorber materials [35]. PSM is also here an option to push the resolution limits of EUV lithography.

Compared to 193 nm dry lithography, 193 nm immersion lithography has a much higher *numerical aperture (NA)*, enabling finer feature sizes and improved DOF (Sect. 3.3.1). To achieve comparable improvements in EUV systems, also *high-NA EUV* systems have been developed. Current EUV tools have a numerical aperture of 0.33. As critical dimensions shrink, also 0.33NA EUV systems require multi-patterning to print the smallest features. While 0.33NA EUV systems are able to print features with 26 nm pitches, high NA EUV systems (0.55NA) enable 16 nm pitches. High NA is achieved by a combination of larger light cones (larger angles; see discussions related to Fig. 3.24) close to both reticle and wafer and a change in the photoresist material. A higher NA at wafer level also means a higher NA at reticle level. Because the sizes of the reticle are four times those at wafer level, the larger cone of the incoming EUV light at the reticle overlaps with that of the reflected light [36]. This overlap problem can be solved by increasing the magnification of the patterns on the reticle from four to eight, but this is only needed in one direction. If we call this the y-direction, then the magnification of reticle patterns in this direction is eight times, while the magnification in x-direction is still kept the same (four times). This direction-dependent magnification projection is known as *anamorphic projection* (as opposed to isomorphic for normal projection systems). As a result, the maximum size of a high-NA reticle is limited to half field (26 × 16.5 mm). A high-NA EUV tool requires twice the number of exposures, while the overlay error must be minimised. The reduced field size of high-NA exposure tools requires *stitching* for the fabrication of chips that do not fit within the 26 × 16.5 mm exposure field. Stitching means that we have to merge one part of the chip that is patterned by one mask with another part of the same chip that is patterned by a second mask. Overall, compared to normal EUV systems, high-NA EUV systems show a 70% increase of resolution and can be used for process node of 3 nm and beyond. High-NA systems consume close to 1.5 GW of power! In summary, the adoption of high-NA EUV leads to relaxed design rules to get around EUV double patterning, to reduce process complexity and to increase yields. However, these benefits come with the drawback of a half-sized exposure field and a more complex fabrication process, particularly for larger dies. While there are still challenges to overcome in implementing this technology, we may assume that the ongoing research and development efforts will lead to acceptation of high-NA EUV in the (sub-) 2 nm nodes in the near future.

Also several investigations for a potential successor of high-NA: *hyper-NA EUV* (with an NA between 0.7 and 0.75), have already been conducted. Compared to high-NA EUV, the lenses must be bigger, the reflection layers of the lenses are different, and the depth of focus reduces. According to former ASML CTO Martin van den Brink in an interview in September 2022: 'Theoretically, it can be done. Technologically, it can be done. If the cost of hyper-NA grows as fast as we've seen with high-NA, it will pretty much be economically unfeasible'.

- Use of alternative techniques to fabricate image-critical patterns in sub-10 nm technologies. For many years, *X-ray lithography (XRL)* has been a potential candidate for *next-generation lithography (NGL)*. It uses X-rays, which generate photons with a wavelength often between 1 and 4 nm to expose the resist film deposited on the wafer, enabling much finer features than current optical lithography tools. However, it has some major disadvantages. Generally, at smaller wavelengths, all optical materials become opaque, but at X-ray wavelengths, these materials become transparent again. Moreover, at these wavelengths, the refraction index is almost 1.0 for all materials. Conventional lenses are unable to focus X-rays, and, consequently, XRL tools cannot use a lens to shrink a mask's features. Therefore, its 1:1 pattern transfer methodology requires mask patterns with only one-fourth of the feature sizes used in the 4:1 photolithography masks. In addition, it requires an extremely expensive synchrotron, which converts an electron beam into an X-ray beam. It is therefore expected that the use of XRL will be limited to fabrication processes that create niche devices, such as devices and structures in a (bio)chemistry and medical R&D environment, in bio-sensors (lab on a chip) and MEMs [37].

- An alternative to photolithography is the *nano-imprint lithography (NIL)*@NAND logic D-type flip-flop. This 1:1 technology is based on physically pressing a hard mould (typically identical to the quartz/chrome material commonly used for optical lithography) with a pattern of nano-structures onto a thin blanket of thermal plastic monomer or polymer resist layer on the sample substrate, to which the structure needs to be replicated. This imprinting step is usually done with the resist heated, such that it becomes liquid and can be deformed by the pattern on the mould. After cooling down, the mould is separated from the sample, which now contains a copy of the original pattern. Its major advantage is that it can replicate features with nanometer dimensions [38]. This process is already used in volume production in electrical, optical and biological applications. For semiconductor applications, the 'step-and-flash' imprint (SFIL) seems to be the most viable one. It allows imprinting at room temperature with only a little pressure using a low-viscosity UV curing solution instead of the resist layer. The higher the sensitivity to UV, the less exposure time the solution needs and the higher the throughput. In this imprint technology, some of the wafer process complexity has moved to the fabrication of the mould. Still, a lot of key issues, particularly related to overlay and defects, need to be solved, but the results of this disruptive technology, so far, are promising. A potential barrier for using the imprint lithography is that it requires very advanced lithographic processes to create the patterns on the mould. Because it is a 1:1 pattern transfer process, the pattern dimensions are only one-fourth of those printed on a photomask, which is one of its major challenges. Moreover, low throughput has become the real show stopper for this technology. Reference [38] discusses the process and potentials of nano-imprint in more detail. NIL is also seen as an alternative to photolithography in photonics applications, such as in the fabrication of LEDs and photovoltaic (PV) cells. For the fabrication of ICs, NIL regained interest in 2016 by a 3D NAND flash manufacturer, as a way to reduce the production cost of NAND flash memories [30]. Line widths down to 15 nm were claimed, while the cost could be less than the use of double-/quadruple-patterning techniques or EUV. The mould can be made using e-beam lithography. In 2023, Canon claimed in a press release [39] that its investments in nano-imprint lithography have resulted in a new NIL machine, which is capable of printing patterns with a minimum line width of 14 nm, equivalent to the 5 nm node. With further improvements in this lithography, NIL is expected to enable patterning with line widths of 10 nm, which corresponds to 2 nm node. The work is done in cooperation with KIOXIA Corporation (which is the former Toshiba Memory Corporation). However, still a lot of problems need to be solved before reaching the levels required for mass production. The 3D NAND may be the best point of entry for NIL into the semiconductor business. Canon has also included DRAM and logic in their road map, but these still face many defectivity and overlay issues.

- *E-beam lithography (EBL)* is another alternative to photolithography. For a long time, the most important use of EBL is in the production of photomasks. Today, it is also used as a direct-write lithography in the chip fabrication process. It uses a focused electron beam that directly writes the pattern shapes into the electron-sensitive resist layer on the wafer. The intensity of the electron beam changes the solubility properties of the resist material. Exposed areas, or their inverse, depending on the tone (positive or negative) of the resist, are then removed during a development step. Advanced SoC ICs may contain several billion transistors, connected by wires in about ten different metal layers and patterned by 35 to 40 masks. When all rectangles in each of the masks need to be written by a single e-beam, the throughput time of the total manufacturing process would explode. A solution to this problem is to use many e-beams in parallel. An example of such a maskless lithography tool is based on a massively parallel electron-beam writing system that uses high-speed optical data transport for switching the electron beams [31]. With 13,260 electron beams in parallel, this tool enables a throughput of 10 wph. The amount of data for each 26×33 mm field is 8 TB. One such tool has a footprint of 1.1×1.65 m. Due to its relatively low throughput, e-beam lithography applications are limited to prototype ICs and low-volume specialty products. Also in environments which explore semiconductor (test) circuits and designs, this maskless lithography would avoid the development of an expensive mask set.

  In January 2019, ASML has acquired all mapper IP for different purposes, e.g. e-beam inspection, etc., and the highly skilled employees.

Moore's law is driven by the economical requirements of the semiconductor markets. This means that all semiconductor disciplines (design, litho, diffusion, packaging, testing, etc.) are cost driven. For the lithography, it means that there is a constant drive to make masks cheaper or to use cheaper masks for certain low-resolution process steps. Binary masks are relatively simple and cheap, guarantee high throughput and can be non-destructively cleaned. Attenuated PSM masks suffer from radiation damage. Moreover, they are immersed in a chemical liquid for cleaning, which is a destructive process, such that they can only be cleaned by a limited number of times and are therefore much more expensive. Today, radiation damage is reduced by roughly 40% by using so-called *AID (anti-irradiation damage) PSM*. It also improves cleaning durability.

To minimise mask costs during the fabrication process, the more expensive masks are only used to image those patterns that really need the smallest feature sizes. For the production of one type of memory, for example, different mask categories can be used. To reduce the production costs of a flash memory process of 40 masks, it may use 8 ArF (attPSM + OPC) masks, 20 KrF (about 10 binary and 10 attPSM) and 10 I-line (binary) masks.

Finally, particularly, the semiconductor memory vendors have found a way to increase bit density without the use of very advanced and expensive lithography tools. By using multiple layers of silicon *(3D stacked silicon)*, memory capacity can be increased dramatically, without increasing the footprint of the memory chip. Some SRAM products use cells with three vertically stacked transistors, while some flash memories are currently being fabricated using hundreds of stacked layers of memory cells. The first OTPs built from four memory cell layers were already introduced in 2004. NAND flashes with more than 500 stacked layers are in development. 3D technologies are only economically viable when the complexity of the devices fabricated in these stacked layers is very limited. Because non-volatile memories use only one type of transistor in the cells (see Chap. 6), they are particularly suited for 3D stacking. Therefore, these layers are only used to fabricate arrays of memory cells and require only a very limited number (zero (3D NAND flash) to three) masks per layer, which can be fabricated by existing photolithography tools. These arrays use the peripheral address selection and sense amplifier circuits of the original first memory array located at wafer level.

3.3.3 Accuracy Measurement of Lithographic Systems

As lithography is one of the most critical steps during the creation of integrated circuits, it requires very accurate control of the printed features. To be able to measure the quality of these features, today, two very advanced *metrology* tools are available: diffraction-based optical measurement and e-beam inspection [40]. In a diffraction-based optical system (ASML: Yieldstar™), a light beam is projected on a pattern of lines on the wafer. In case of a very regular pattern, the resulting scattered light pattern can be easily predicted by simulation. In the measurement system, the resulting scattered pattern is captured by a very-high-resolution camera. The comparison of the measured with the predicted images is a measure for the quality of the printed lines. The system is also used in the measurement of alignment accuracy between two and more layers on the wafer (*overlay*). An e-beam inspection tool offers a higher resolution then the diffraction-based optical system, but at the cost of longer measurement times. Particularly, with EUV and high-NA EUV, the photoresist layers get thinner, which causes problems to acquire high-resolution images and accurate sub-nanometer measurements of the patterns. The energy of e-beams may deform these thin photoresist patterns. Therefore, to acquire accurate measurements in the (high-NA) EUV resist, the e-beam metrology tool must produce very narrow e-beams with a 'dose' low enough to not interact with the fragile resist. The highest-resolution e-beam tools currently can detect defects smaller than 5 nm and achieve a precision of less than 0.1 nm CD with 0.05 nm measurement sensitivity (source: ASML). Various metrology/measurement tools are also used during the lithographic process. The previously mentioned alignment sensors are one example of such a tool. Another example is a level sensor or height sensor that can measure the topography of a wafer surface and which is integrated in the litho tool. It measures the height of a wafer as a function of the wafer position. The resulting height map is then used to correct the height position of the wafer during the litho exposure step to ensure proper focus all across the wafer [41].

3.3.4 Mask Cost Reduction Techniques for Low-Volume Production

The amount to which mask cost contribute to the total chip development cost depends on the process node and the application area (volume) of the chip. This has a direct relation with the number of wafer exposures per mask. This may therefore vary from several thousands for ASICs, several 100.000 for microprocessors (e.g. Intel, AMD, etc.) to more than a million for stand-alone memories. As such, the mask cost per chip for high-volume DRAMs and NAND flash memories are neglectable, while they can be more than 10% in low-volume applications as medical and aviation. The cost of a mask set increases with shrinking process nodes. In fact, it doubles when compared at their year of introduction [44]. It then reduces roughly with an average of 20% per year. A mask set for 32 nm could cost close to three million US$. A mask set for a state-of-art semiconductor technology like 5 nm costs around 10–15 million US$. And these *mask costs* are expected to increase to above 20 million US$, when approaching the 1 nm node. Sometimes, two or three extra mask sets are required, because of the chance of potential mask defects during mask handling or mask degradation as a result of many exposures. An important effect in photomask usage is the so-called *haze effect* (also called *hazing*), which refers to the accumulation of particulate contamination on the surface of the photomask, which can degrade the mask's optical properties. This contamination can result in reduced transparency, reduced contrast and decreased resolution in the final printed image. However, when it comes to the pellicle, haze refers to a different phenomenon. A *pellicle* is a thin, transparent film that is placed on the surface of a photomask to protect it from contamination. The haze effect on the pellicle refers to the presence of defects, such as particles or blemishes that can reduce the optical clarity and overall performance of the pellicle. These defects can impact the quality of the projected image during the lithography process in semiconductor manufacturing. It can also limit the number of lithography exposures. To maintain optimal performance, it is essential to periodically inspect and clean the pellicle or, if necessary, replace it. Typically, a lithography mask can be used for several ten thousand wafer exposures, for an attenuated PSM mask, to more than hundred thousand wafer exposures for a binary mask, before it needs to be replaced. However, this number can vary significantly depending on various factors such as the complexity of the design, the process node, the exposure settings and the type of materials used. Regular inspection and maintenance can also help detect any issues like haze or defects that may affect the performance of the mask (or pellicle) and require its replacement. Particularly with EUV masks, the light-source power of 250 W during exposure is expected to heat the pellicle to 500 °C. This would cause deformation and transmission losses of the pellicle. High-NA EUV may require light power sources of over 600 W. Current silicide composite pellicles in EUV systems are only 13 nm thick and can withstand temperatures up to the above-mentioned 500 °C.

Close to two thirds of the masks are non-critical, in that they can be implemented as simple binary masks. The other third can be considered as critical, of which three to six masks fall in the category of extremely critical [5]. The critical masks can be produced with scanning laser equipment with rather good throughput, while e-beam equipment is required for the extremely critical masks which may require 24 hours of e-beam exposure time. An e-beam writes patterns on the mask based on a given chip design. For traditional binary optical masks, the most commonly used e-beam tool uses a single-beam system, based on variable shape beam (VSB) technology, which is mainly used in process nodes down to 5 nm. A transparent substrate (quartz or glass plate) coated with an opaque (chrome) film with an e-beam-sensitive resist film on top is loaded in the system. The electrons hit the resist in form of shots. This, in turn, patterns the resist in rectangular-like shapes, which pattern is then copied into the chrome layer by an etching step. EUV masks have much smaller and more complex features, which require so-called multi-beam *mask writers*, with up to 250,000 or more beams to reduce the write time of the mask [42]. Such a multi-beam system can write an optical mask in about 10 hours. This is often referred to as the turnaround time of the system. For a conventional system, this could be as high as 30 hours [43].

There are several approaches to reduce mask cost. On so-called *multi-project wafers (MPW)*, several products are included on the same mask set to reduce overall mask costs (Fig. 3.33). Another way to share the mask costs is the *multilayer reticle (MLR)*, on which several mask layers of the same product are grouped together to reduce the physical number of masks. These MLRs do not combine designs of different products. Both techniques are particularly used for small-volume designs, for prototyping and for educational purposes. To save mask costs completely, *direct-writing techniques* use an *electron-beam (e-beam)* or *laser-beam* system, which writes the layout pattern directly onto a wafer resist layer, without using a mask. It requires the deposition of an additional conductive layer on the resist layer, to prevent damage by electron charging during the patterning process.

The resolution yielded by an e-beam machine is better than 5 nm, but at a lower throughput, because it writes every feature individually. It is free of wavelength aberration. Laser-beam systems are gaining market share at the cost of e-beam systems, because they are cheaper since they do not require a vacuum environment. Because of their low throughput, both e-beam and laser-beam systems usage, today, is limited to fabricate low-volume samples, such as MPWs, prototyping products and test silicon for process development. Next to that, these techniques are used to fabricate the physical glass-chrome masks (reticles) for use in photolithography processes. These direct-writing techniques are also called *mask-less lithography*

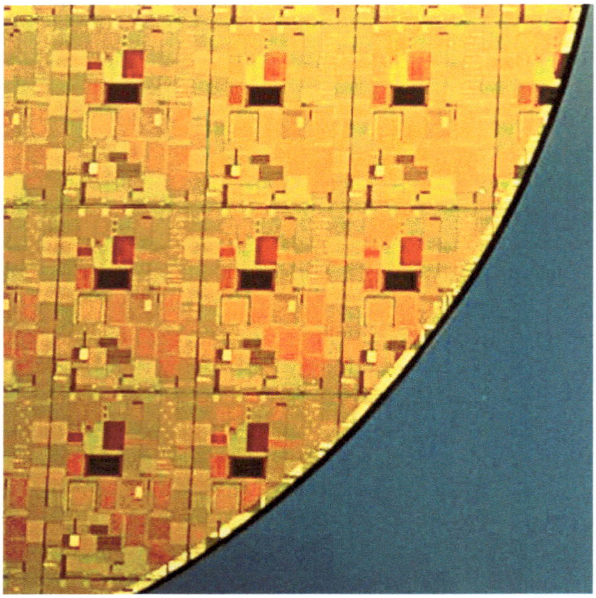

Fig. 3.33 Example of a multi-project wafer (MPW)

(MLL or ML2) and are currently also being explored as an alternative for, or successor of the conventional photolithography, even for high-volume production. The main reason is the rapidly increasing costs of an optical mask set as explained before. Over the last decade, a lot of progress has already been made to improve throughput. The potentials of maskless e-beam lithography are further discussed in [45].

More information on future lithography techniques can be found in numerous publications and also on the Internet and is beyond the scope of this book. To summarise the evolution of the wafer stepper/scanner, Table 3.2 presents several key parameters which reflect the improvements made over different generations of steppers/scanners.

3.3.5 Pattern Imaging

The photolithographic steps involved in the transfer of a mask pattern to a wafer are explained with the aid of Fig. 3.34. Usually, the first step is oxidation and comprises the growth of a 30 to 50 nm thick silicon dioxide (SiO_2) layer on the wafer. Subsequently, a nitride (Si_3N_4) or other *hard mask* layer is deposited (Fig. 3.34a). Today, the nitride is often replaced by a (spin-on) carbon layer, which can be deposited in the track system of the lithography tool. Next, this hard mask layer is covered with a 0.5–2 μm thick *photoresist* layer (Fig. 3.34b). The mask is used to selectively expose the photoresist layer to light (Fig. 3.34c, d). This exposure causes a change in the chemical properties of the resist, so that it can be removed by a

Table 3.2 The evolution of the wafer scanner. (Source: ASML, 2024)

		dry				wet	EUV	
Light source		I-Line	40 W KrF	KrF	ArF	ArFi	Low-NA	High-NA
Litho-tool @ ASML	Step & scan	450	860M	1060K	1470	2050i	3600D	5000
Reduction proj. Lens	Hor/vert	4×/4×	4×/4×	4×/4×	4×/4×	4×/4×	4× isomorpic	4×/8× anamorphic
Wavelength (nm)		365	248	248	193	193	13.5	13.5
Platform	XT/NXT/NXE	XT	XT	XT	NXT	NXT	NXE	EXE
NA_{MAX}	proj.lens	0.65	0,55–0,80	0.93	0.93	1.35	0.33	0.55
Resolution (nm)	Half-pitch	220	110	80	<57	<38 nm	13	8
Productivity	wph	>300	>300	>205	>300	295	100	??
Max. field size	mm × mm	28 × 33	27 × 33	26 × 33	26 × 33	26 × 33	26 × 33	$26 \times 16\frac{1}{2}$
On-product overlay		20	10	3.5	4.5	2.5	≤2	<2
				Same chuck		to EUV		
Min k1	Dipool-mode			0.3	0.27	0.265		
Min k1	Annular		0.35					
Min k1	Conventional	0.40					0.32	

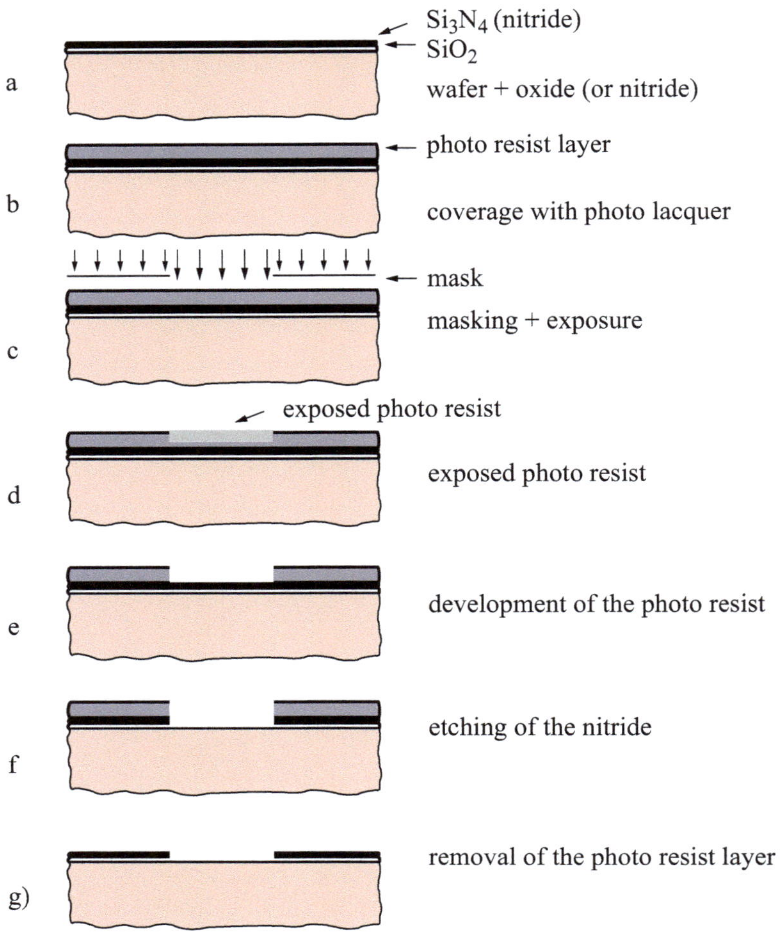

Fig. 3.34 Pattern transfer from mask to wafer

special solution, called *developer*. It removes the exposed areas if the photoresist is positive. The resulting pattern in the resist after development (Fig. 3.34e) acts as an etch barrier in the subsequent hard mask etching step (Fig. 3.34f), in which the unprotected hard mask pattern is removed (stripped). Finally, after the photoresist is no longer needed, it must be removed from the substrate, and an image of the mask pattern remains in the hard mask layer (Fig. 3.34g). This hard mask pattern acts as a barrier for a subsequent processing step. The wafer is then cleaned (Fig. 3.34g) before the real process step (implantation, etching, etc.) is performed. The use of a hard mask, as discussed above, is not required for every litho step. In fact, most photoresists are soft/weak polymers or other organic materials. These resist materials can easily be etched by several process steps that use, e.g. plasma etching,

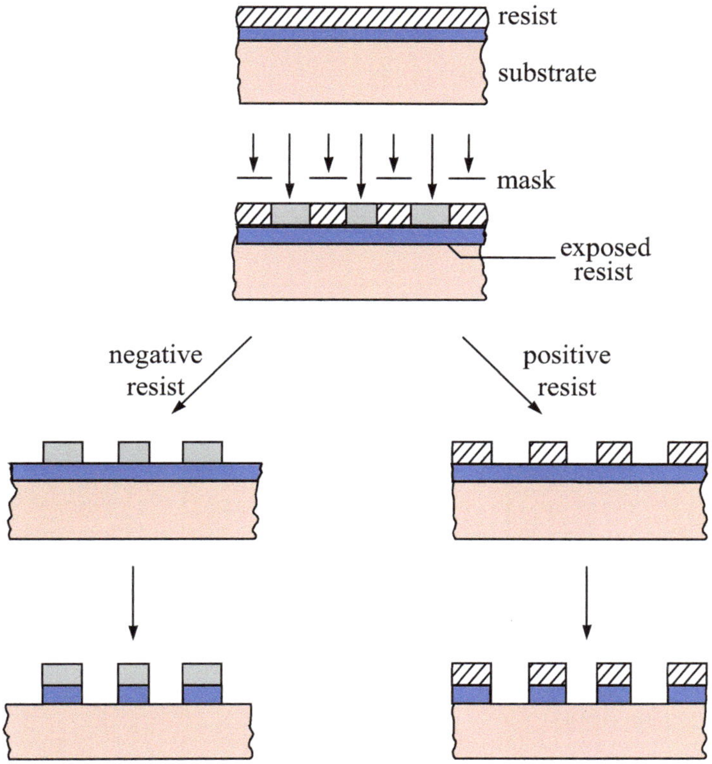

Fig. 3.35 The use of positive and negative resist for pattern imaging

such as oxygen, chlorine and fluorine. Hard masks, which can be dielectrics, or metals, prevent this etching. Also, implantation steps can be aggressive and may require a hard mask as well.

Both *positive* and *negative resists* exist. The differences in physical properties of these resist materials result in inverting images; see Fig. 3.35.

The combination of pattern transfer and one or more processing steps is repeated for all masks required to manufacture the IC. The types of layers used for the pattern transfer may differ from the silicon dioxide and silicon nitride layers described above. The principle, however, remains the same. The processing steps that follow pattern transfer may comprise etching, oxidation, implantation or diffusion and planarisation. Deposition is also an important processing step. These steps are described in detail in the following sections.

3.4 Oxidation

The dielectrics used in the manufacture of nanometer CMOS circuits must fulfil several important requirements [48]:

- High breakdown voltage
- Low dielectric constant of inter metal dielectrics
- High dielectric constant for gate dielectric
- No built-in charge
- Good adhesion to other process materials
- Low defect density (no pinholes)
- Easy to be etched
- Permeable to hydrogen

One of the materials that incorporates most of these properties is silicon dioxide (SiO_2). SiO_2 can be created by different processes: thermal *oxidation* or deposition. A *thermal oxide* was used to isolate the transistor areas in conventional MOS ICs. In these isolation areas, the oxide must be relatively thick to allow low capacitive values for signals (tracks) which cross these areas. This *thick oxide* was created by exposing the monocrystalline silicon substrate to pure oxygen or water vapour at a high temperature of 900 °C to 1200 °C. The oxygen and water vapour molecules can easily diffuse through the resulting silicon dioxide at these temperatures. The following respective chemical reactions occur when the oxygen and water vapour reach the silicon surface:

$$\text{Dry oxidation} : \text{Si (solid)} + O_2 \text{ (vapour)} \longrightarrow SiO_2 \text{ (solid)}$$
$$\text{Wet oxidation} : \text{Si (solid)} + 2H_2O \text{ (vapour)} \longrightarrow SiO_2 \text{ (solid)} + 2\,H_2$$

The *Local Oxidation of Silicon (LOCOS)* process is an oxidation technique which has found universal acceptance in MOS processes with gate lengths down to 0.5 µm. Silicon is substantially consumed at the wafer surface during this process. The resulting silicon dioxide layer extends about 46% below the original wafer surface and about 54% above it. The exact percentages are determined by the concentration of the oxide, which contains about $2.3 \cdot 10^{22}$ atoms/cm^3, while silicon contains about $5 \cdot 10^{22}$ atoms/cm^3. A disadvantage of the LOCOS process is the associated rounded thick oxide edge. This *bird's beak* is shown in Fig. 3.36a.

The formation of the bird's beak causes a loss of geometric control, which becomes considerable as transistor sizes shrink. Intensive research efforts aimed at suppression of bird's beak formation have resulted in lengths of just 0.1–0.15 µm for an oxide thickness of 0.5 µm. Even with a suppressed bird's beak, the use of LOCOS is limited to the isolation of over-0.25 µm process nodes.

An important alternative to these LOCOS techniques, already used in 0.35 µm CMOS technologies and below, is the *shallow-trench isolation (STI)*. STI uses deposited dielectrics to fill trenches which are etched in the silicon between active areas. The use of STI for nanometer technologies is discussed later in this chapter (Sect. 3.9.3).

Fig. 3.36 Comparison of (**a**) a conventional LOCOS process and (**b**) use of shallow-trench isolation (STI) to isolate transistors

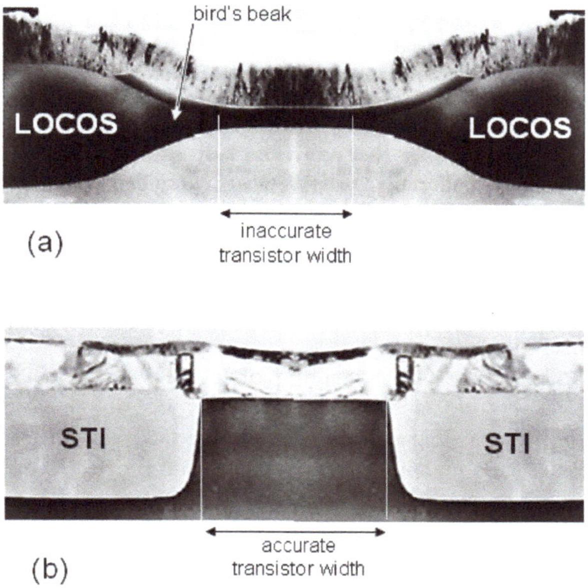

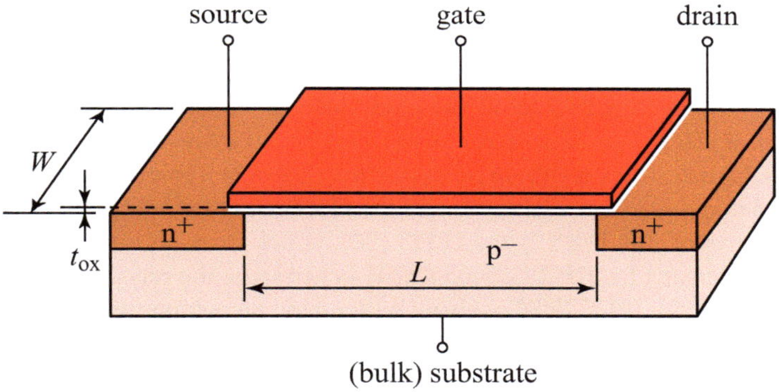

Fig. 3.37 Schematic cross section of a MOS transistor

Another important application of thermally grown oxide was the oxide layer between a transistor gate and the substrate in conventional CMOS processes. This *'gate oxide'* must be of high quality and very reliable. Defects such as pinholes and oxide charges have a negative effect on electrical performance and transistor lifetime. Because the gate oxide is only a few atoms thick, it is particularly a challenge for the industry to scale it further and/or find alternative ways to increase its capacitance. Figure 3.37 shows a cross section of a MOS transistor.

The *gate-oxide thickness* must be sufficiently uniform across the die, from die to die, from wafer to wafer and from run to run. It scales with the technology node

Table 3.3 Trends in gate-oxide thickness and threshold voltage

Technology	L [nm]	t_{ox} [nm]	V_{dd} [V]	V_t [V]
0.35 µm	350	7	3.3	0.6
0.25 µm	250	5	2.5	0.55
0.18 µm	180	3.2	1.8	0.55/0.45
0.13 µm	120	2	1.2	0.45/0.35/0.2
90 nm	80	2.2/1.6	1.2/1.0	0.45/0.4/0.35/0.3/0.2
65 nm	60	1.8/1.2	1.2/1.0	0.5/0.4/0.3/0.2
45 nm	40	1.8/1	1.1/0.9	0.5/0.4/0.3/0.1

according to Table 3.3, which shows representative values for various technology nodes. L represents the physical gate length.

The table also shows the divergence in gate-oxide thicknesses, supply and threshold voltages. This is due to the fact that today's semiconductor technologies must support applications with a wide range of performance requirements: high density, low active power, low standby power, high speed, etc. In each technology node, the input and output (I/O) transistors usually operate at a larger voltage (1.2 V, 1.8 V, 2.5 V and/or 3.3 V) and require an additional oxide thickness and threshold voltage. The simultaneous use of more oxide thicknesses and threshold voltages in one process is of course at the cost of more mask, lithography and processing steps. Each of the processes offers usually only two or three different threshold voltages, to limit the number of masks, lithography and processing steps. Technology nodes from 90 to 14 nm offer different process versions, e.g. a *high performance (HP)*, a *low-operating power (LOP)* and a *low standby power (LSTP) process*. Characteristics for these processes are shown in Table 3.4, according to the ITRS road map [49]. The final version of the ITRS road map was published in 2013. The follow-up of this ITRS is the IRDS [90], which does no longer include three different road maps for HP, LOP and LSTP. I encourage the reader to take a look at this free-accessible IEEE IRDS website with many interesting trends and tables.

When a *multi-*V_T process offers three different threshold voltages (V_T's), these are usually referred to as HV_T (high-V_T), SV_T (standard-V_T) and LV_T (low-V_T). In most planar processes, these different V_T's were created by a different V_T-implant. For certain applications, a design may use two different V_T's (see Sect. 8.4.1) for both nMOS and pMOS transistors, at the cost of four masks and implantation steps. For high-ϵ metal-gate devices, the different V_T's are created by the deposition of different work function layers (see Sect. 1.5), which makes the process much more complex. For process nodes below 7 nm, multi-V_T options become a significant increase in process complexity [50]. As an example, only of the low-V_T transistor, some companies offer already five or more sub-options like LV_T, ULV_T (ultra low-V_T), ELV_T (extreme low-V_T), LV_TLL (low-V_T/low leakage and ULV_TLL (ultra low-V_T/low leakage).

Although most of these processes include high-ϵ metal-gate transistors, the oxide thickness (t_{ox}) is still expressed as if silicon dioxide was used for the gate dielectric. Therefore, the *equivalent oxide thickness (EOT)* refers to an equivalent silicon

Table 3.4 Characteristics for HP, LOP and LSTP processes according to ITRS road map

Technology node	Process	L[nm]	t_{ox}(EOT)[nm]	V_{dd} [V]	V_T [V]
32 nm	HP	22	0.88	0.87	0.3
	LOP	24	0.98	0.7	0.3
	LSTP	27	1.4	0.9	0.48
28 nm	HP	20	0.84	0.85	0.3
	LOP	21	0.94	0.67	0.3
	LSTP	24	1.3	0.87	0.48
22 nm	HP	17	0.8	0.8	0.3
	LOP	18	0.9	0.63	0.3
	LSTP	20	1.2	0.81	0.48
14 nm	HP	12.8	0.68	0.73	0.3
	LOP	13.1	0.78	0.57	0.3
	LSTP	14.1	0.95	0.72	0.48

dioxide thickness with the same capacitance value as the used high-ϵ dielectric stack.

The use of dielectric SiO_2 layers below about 2 nm thickness causes gate-oxide direct *tunnelling*, resulting in currents which may exceed a level of 1 A/cm^2. At these gate-oxide thicknesses, pMOS transistors with heavily boron-doped polysilicon suffer from boron penetration into the gate oxide, causing an undesirable positive threshold voltage shift and a performance and reliability degradation. The quality of the gate oxide is greatly improved with *nitrided gate oxide (SiON)* [51], wherein a conventionally created silicon oxide dielectric is impregnated with a small dose of nitrogen. It reduces boron penetration and improves gate-oxide breakdown characteristics and reliability [53]. It also leads to a minor increase in the dielectric constant. On the other hand, too much nitrogen close to the gate-oxide/Si-substrate interface enhances negative-bias temperature instability (NBTI; see also Chap. 9) [54]. Moreover, the combination of thinner gate oxide and increased channel doping also causes depletion of the bottom region of the gate material, and this effect becomes more pronounced with further scaling of the oxide thickness. This is called *gate depletion*. As a result of these effects, the *double-flavoured polysilicon* (n^+ doped gate for nMOS transistors and p^+ doped gate for pMOS transistors) is replaced by a metal. Other alternatives, which were under research and also prevent gate depletion, include *fully silicided (FUSI)* polysilicon gates. Section 3.9.4 discusses further details on FUSI gates and high-ϵ metal-gate processes. Most advanced CMOS processes use atomic layer deposition (ALD) to fabricate the very thin gate-oxide layer. This is discussed in the next subsection.

3.5 Deposition

The *deposition* of thin layers of dielectrical material, polysilicon and metal is an important aspect of IC production.

The growth of an *epitaxial film* (layer) is the result of a deposition step combined with a chemical reaction between the deposited and substrate material. The term *epitaxy* is based on the Greek words *epi*, which means 'above', and *taxis*, which means 'in ordered manner'. Therefore, epitaxial can be translated as 'in an ordered manner arranged upon'. If the deposited layer is the same material as the substrate, it is called *homo-epitaxy* or epi-layer for short. Silicon on sapphire is an example of *hetero-epitaxy*, in which the deposited and substrate materials differ [55]. Epitaxial deposition is created by a *chemical vapour deposition (CVD)* process. This is a process during which vapour-phase reactants are transported to and react with the substrate surface, thereby creating a film and some by-products. These by-products are then removed from the surface. Normally, the actual film created by a CVD process is the result of a sequence of chemical reactions. However, a different overall reaction can generally be given for each of the silicon sources. The hydrogen reduction of silicon tetrachloride ($SiCl_4$), for example, can be represented as

$$SiCl_4 + 2H_2 \longrightarrow Si + 4\,HCl$$

Several parameters determine the growth rate of a film, including the source material and deposition temperature. Usually, high temperatures ($>1000\,°C$) are used for the depositions because the growth rate is then less dependent on the temperature and thus shows fewer thickness variations. The overall reaction for the deposition of polysilicon is

$$SiH_4\ (\text{vapour}) \longrightarrow Si\ (\text{solid}) + 2\,H_2\ (\text{vapour})$$

This reaction can take place at lower temperatures, because SiH_4 decomposes at a higher rate. The creation of dielectric layers during IC manufacture is also performed by some form of CVD process. The most commonly used dielectric materials are silicon dioxide (SiO_2) and silicon nitride (Si_3N_4). In an *atmospheric pressure CVD (APCVD)* process, the material is deposited by gas-phase reactions. This deposition generally results in overhangs and a poor step coverage (Fig. 3.39). APCVD is currently used to deposit borophosphosilicate glass *(BPSG)* epitaxial layers and form the scratch-protection layer or *passivation layer (PSG)*. PSG is a phosphorus-doped silicon dioxide dielectric which is deposited on top of polysilicon (between polysilicon and first metal) to create a smooth topography that is beneficial for the deposition of the metal layers. BPSG contains boron and phosphorus for a better flow (spread) of the dielectric. The phosphorus also serves to improve internal passivation. The following reactions apply for the deposition of SiO_2 and Si_3N_4, respectively:

LPCVD:	$Si(OC_2H_5)_4$	$\longrightarrow SiO_2$ + by-products
PECVD:	$Si(OC_2H_5)_4 + O_2$	$\longrightarrow SiO_2$ + by-products
LPCVD:	$3SiCl_2H_2 + 4NH_3$	$\longrightarrow Si_3N_4 + 6HCL + 6H_2$

Two versions of CVD have been introduced by the above reactions: LPCVD and PECVD. *LPCVD* is a low-pressure CVD process, usually performed in a vacuum chamber at medium vacuum (0.25–2.0 torr) and at temperatures between 550 °C and 750 °C. Under these conditions, the vapour-phase reactions are suppressed, while the decomposition now occurs at the surface, leading to a much better step coverage. In the previously discussed CVD process, the chemical reactions are initiated and sustained only by thermal energy. *PECVD* is a plasma-enhanced CVD process. A *plasma* is defined to be a partially ionised gas which contains charged particles (ions and electrons) and neutrals. The plasma is generated by applying an RF field to a low-pressure gas, thereby creating free electrons within the discharge regions [55]. The electrons gain sufficient energy so that they collide with gas molecules, thereby causing gas-phase dissociation and ionisation of the reactant gases. At room temperature, a plasma therefore already contains high-energy electrons. Thus, even at low temperatures, a PECVD process can generate reactive particles; it therefore has a higher deposition rate than other CVD processes. The ionised gas flows out of the anode and is attracted to the surface of the substrate (cathode), which has a slightly positive charge. The heat at the cathode removes the by-products by desorption, which are then removed from the process chamber by the vacuum pump.

If we compare the previous reactions to depositing SiO_2, we see that the LPCVD, which occurs at high temperature, therefore needs no additional oxygen, while the PECVD process needs additional oxygen because the oxygen cannot be dissociated from the *TEOS* (tetra-ethyl orthosilicate: $Si(OC_2H_5)_4$) at low temperatures. A *sub-atmospheric CVD (SACVD)* process occurs at temperatures around 700 °C to 800 °C. Because of the high pressure (≈1/2 atmosphere instead of a few torr), the deposition speed will be higher, resulting in a higher throughput. This form of CVD is particularly used for BPSG.

Metal layers are deposited by both physical and chemical methods. In physical vapour deposition (PVD) methods, such as *evaporation* and *sputtering*, the material is physically moved onto the substrate. PVD evaporation is a deposition process, in which a vapour of the material to be deposited is transported to the wafer in a low-pressure environment. After condensation at the wafer surface, it forms a thin film on it. When using the PVD sputtering technique for the deposition of aluminium, for instance, an aluminium target is bombarded with argon ions, which physically dislodge aluminium molecules from the target, causing a flux of aluminium to flow from the target to the wafer surface. The aluminium was alloyed with 0.5% copper to improve electromigration behaviour. After deposition of the aluminium, photolithographic and etching steps are used to create the required metal pattern.

Copper cannot be deposited and etched as easy as aluminium. Potential etching plasmas create non-volatile residuals that remain on the wafer. Moreover, copper diffuses through oxides leading to transistor threshold voltage shifts and reliability

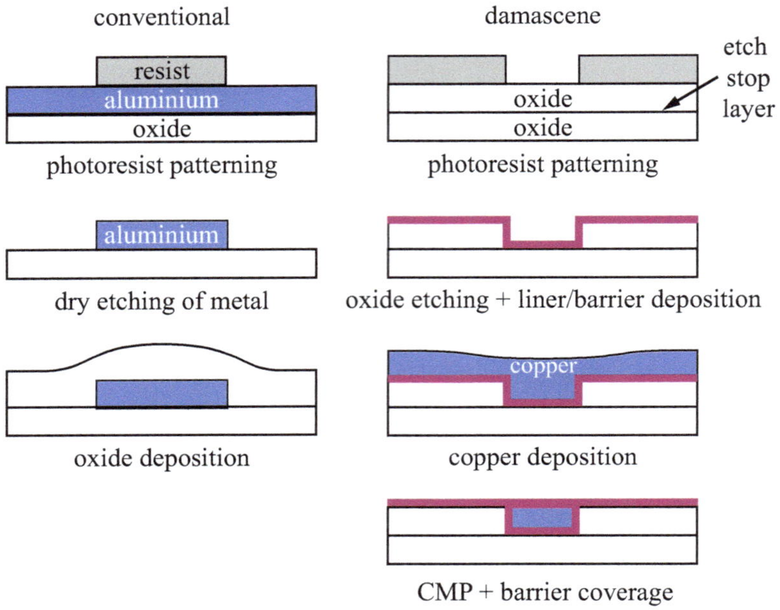

Fig. 3.38 Comparison of conventional and damascene processing

problems. Therefore, a copper back-end technology is quite different from a conventional aluminium one. In the latter, the aluminium deposition step is followed by a dry etching step to etch the metal away according to the mask pattern and then filling the gaps with a dielectric. A copper back-end uses a so-called *damascene process flow*, in which the conventional subtractive metal etching process flow is replaced by a metal inlay process flow. Figure 3.38 shows a comparison of both flows.

In a damascene process, first an oxide layer is deposited, identical to an aluminium back-end process. Then, an etch-stop layer is deposited on top of this oxide layer, followed by the deposition of another oxide layer. These oxide layers are also referred to as *inter-level dielectric (ILD)* layers. Next, an oxide etching step creates trenches in the top oxide layer, according to the pattern in the corresponding metal mask. The etch-stop barrier blocks the etching process, such that it cannot etch the lower oxide layer. Then, a thin liner/barrier layer is deposited by an *atomic layer deposition (ALD)* step on top of the ILD layer and prevents the diffusion of copper. This layer is a combination of Ta and TaN. Current CMOS copper back-end processes include the deposition of several layers. After etching the trenches in the dielectric layer according to the copper interconnect pattern, first a (tantalium nitride) barrier layer is deposited with ALD. This barrier layer prevents copper diffusion into the dielectric. Next, a (tantalium) liner layer, which is also deposited with ALD, enhances adhesion between the barrier and the copper. Next, a seed layer is deposited, also with ALD. This seed layer is a very thin copper layer deposited on the liner layer to provide a conductive surface for the subsequent electroplating process. (Sect. 3.9 presents changes in the back-end materials for the 10 nm node and beyond.) In fact the deposition starts with a Ta of a few atomic layers thick and then gradually increases the amount of N such that the last atomic layers consist of TaN. Next, a seed layer is deposited to provide a conductive layer, which is required for the electroplate deposition process of the copper, to improve copper adhesion and coverage. Then, copper deposition is done, using an electrochemical process: *electroplating*, in which the wafer is immersed in a (salt/acid) solution of copper sulfate (and some other acids and/or additives to enhance the filling capabilities) and connected to a negative terminal of the power supply. The positive supply terminal is connected to a copper body, which creates copper ions into the salt solution. These positively charged copper ions are attracted to the negative wafer surface and form a thick copper blanket across the total wafer. Then, a planarisation step, called CMP (Sect. 3.8), polishes the wafer until it has reached the bottom of the barrier layer (copper and barrier are removed in one step!). Copper tracks are then remaining as a metal inlay in the trenches (damascene processing), similar to the metal inlay in swords, made in ancient times in Damascus, Syria. Then again, a barrier layer is deposited to cover the top of the copper inlays, such that copper is fully encapsulated within the barrier layer. In 20 nm CMOS process, the barrier may consist of a TaN film and a Ta film, each of which is deposited with ALD technology with a thickness of approximately 3 nm each. Because the copper width in this node and smaller nodes will be so narrow, the chance of creating voids (poor copper fill) is rapidly increasing. By depositing a thin cobalt film before the copper deposition greatly improves the copper's fill performance at smaller geometries, leading to less voids and improved electromigration properties [56]. After the copper layer has been deposited on the cobalt barrier layer and planarised thereafter, the top barrier layer is then formed by a *selective cobalt deposition* only on the copper tracks. This is done by exposing the substrate to a cobalt precursor gas to selectively form a cobalt capping layer over the copper surface while leaving exposed the dielectric surface during a vapour deposition process [57].

Today, most fabs use a dual-damascene back end, in which both the *vias* (also called *studs*, or *pillars*, which are contacts between two metal layers) and trenches are simultaneously etched into the ILD layer. Also in the next sequence of deposition steps for the barrier, the seed layer and the electroplate copper, respectively, the vias and tracks are simultaneously filled, thereby reducing processing costs.

Although the resistance of copper is 40% less than that of aluminium, this advantage cannot fully be exploited, because part of the available track volume is occupied by the barrier material, which has a much higher resistance value. The use of copper instead of aluminium for interconnection resulted in only a limited reduction of the effective interconnect resistivity by 25% to 30%. In combination with the use of low-ϵ dielectrics, the interconnect capacitance is reduced and leads to faster or less power circuits. Copper can also withstand higher current densities, resulting in a reduced chance of electromigration (see Chap. 9).

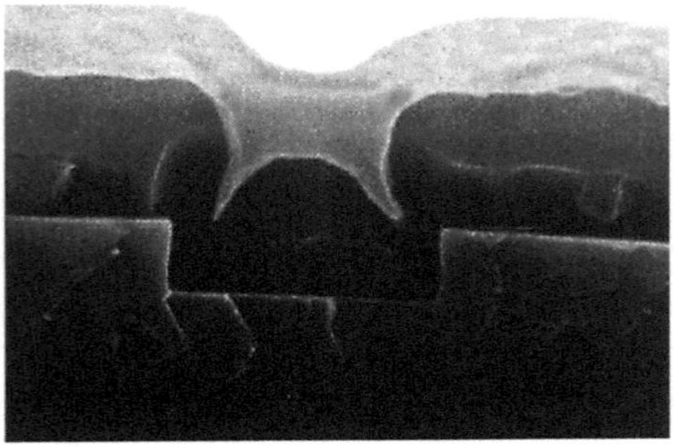

Fig. 3.39 Example of poor step coverage in a conventional CMOS process

CVD methods form the chemical alternative for the deposition of metals. Tungsten (W), for example, may yield the following CVD reaction:

$$WF_6 + 3H_2 \longrightarrow W + 6HF$$

The choice of deposition method is determined by a number of factors, of which *step coverage* is the most important. Figure 3.39 shows an example of bad aluminium step coverage on a contact hole in a conventional CMOS process. Such a step coverage can dramatically reduce the lifetime of an IC. It also causes problems during further processing steps, and the associated temperature variations can lead to voids in the aluminium.

Moreover, the local narrowings cannot withstand high current densities. *Current densities* of $\approx 10^5$ A/cm^2 are not exceptional in modern integrated circuits. Excessive current densities in metal tracks cause *electromigration*. This leads to the physical destruction of metal tracks and is another phenomenon that reduces the reliability of ICs. This topic is examined more closely in Chap. 9.

One deposition step that got a lot of attention over the last decades and which was already mentioned a couple of times before is the so-called *atomic layer deposition (ALD)*, particularly for its potential applications in advanced (high-ϵ) gate dielectrics, DRAM capacitor dielectrics and copper diffusion barriers in advanced CMOS and memory processes. ALD is a special form of chemical vapour deposition. Without going deep into the chemical and physical reactions, ALD basically uses pulses of gas, each creating a unique self-limiting growth mechanism, which results in the deposition of one atomic layer at a time. So, the deposited film thickness is only dependent on the number of deposition cycles (gas pulses) providing extremely high uniformity and thickness control. It is therefore also of interest in all (sub)-nanometer applications that benefit from accurate control of (ultra-) thin films. The drive for ALD development was to replace the thermally grown thin gate-oxide layer creation. ALD is a cyclic process with a low thermal budget. The film deposition consists of a repetition of these cycles, with every single cycle creating a one-atomic-thick layer film. More details on ALD can be found in [52, 58].

3.6 Etching

Many of the deposited layers require an etching step to remove the material where it is not needed. For example, to create a polysilicon gate pattern, first the whole wafer is deposited with polysilicon, and next, it is etched away according to the polysilicon mask pattern, at locations where no polysilicon tracks are needed. The photolithographic steps described in Sect. 3.3.5 produce a pattern in a nitride or equivalent barrier layer. This pattern acts as a protection while its image is duplicated on its underlying layer by means of *etching* processes. There are several different etching techniques. The etching process must fulfil the following requirements: a high degree of anisotropy, good dimensional control, a high etching rate to minimise processing time, a high selectivity for different materials, a perfect homogeneity and reproducibility (e.g. 8 billion trenches in a 8 Gb DRAM) and a limited damage or contamination to satisfy reliability standards. The degree of anisotropy depends on the requirements of the process step, e.g. during the STI etch, an extremely vertical and sharp profile may increase stress and the occurrence of defects.

With *wet etching*, the wafer is immersed in a chemical etching liquid. The wet-etching methods are *isotropic*, i.e. the etching rate is the same in all directions. The associated *'under-etch'* problem illustrated in Fig. 3.40a becomes serious when the minimum line width of the etched layer approaches its thickness.

Dry etching methods may consist of both physical and chemical processes (anisotropic) or of a chemical process only (isotropic). Dry etching methods, which use a plasma, allow *anisotropic* etching, i.e. the etching process is limited to one direction by the perpendicular trajectory of the ions used at the wafer surface. The result, shown in Fig. 3.40b, is an accurate copy of the mask pattern on the underlying layer.

With *plasma etching* techniques [46], the wafers are immersed in a plasma containing chlorine or fluorine ions that etch, e.g. Al and SiO_2, respectively. It comprises a plasma chamber, which contains a certain process gas. To transfer from the gas state into the plasma state, the chamber is pumped to the required pressure, and energy is supplied to produce a glow discharge plasma by a radio-

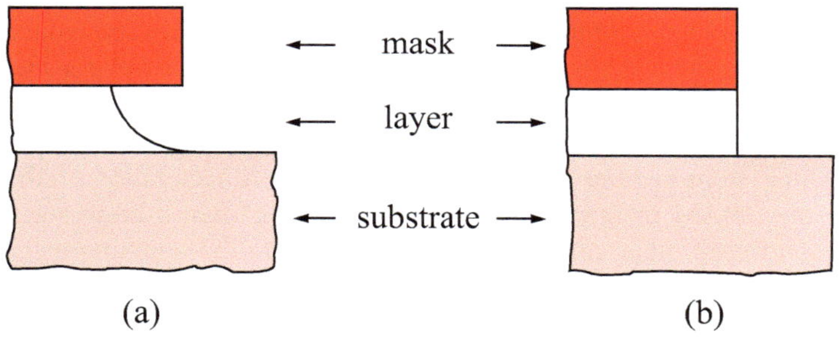

Fig. 3.40 The results of different etching methods. (**a**) Isotropic. (**b**) Anisotropic

frequency (RF) electromagnetic field. This causes ionisation of the low-temperature plasma: after collision with molecules, they create many different gaseous species, free radicals, electrons, ions, neutrals, photons and by-products. These are then accelerated by an electrical field towards the surface material, which can then be etched quickly and selectively. The etching process depends on the gas pressure and flux and on the applied RF field. In *sputter etching* techniques, the wafer is bombarded by gas ions such as argon (Ar^+). As a result, the atoms at the wafer surface are physically dislodged and removed.

Finally, a combination of plasma and sputter etching techniques is used in *reactive ion etching (RIE)*. During RIE, ionised gaseous molecules from the plasma are accelerated by an electric field towards the surface and react with the surface atoms forming new electrically neutral molecules which then floats away.

Satisfactory etching processes have been developed for most materials that are currently used in IC manufacturing processes. New process generations, however, require improved selectivity, uniformity, reproducibility and process control. Selectivity can be improved by the compound of the gaseous plasma or by the creation of polymers at the underlying layer. The use of an additional carbonaceous substance such as CHF_3 during etching enhances its anisotropic properties. The use of this substance creates a thin layer close to the side wall of a contact hole, for example, which improves the anisotropy of the etching process. A second advantage is that carbon reacts with oxygen. It therefore increases the selectivity of the etching process because, when used in the etching of a contact to silicon, the reaction is stopped immediately on arrival at the silicon surface. Carbon does not react with silicon.

For critical anisotropic etching steps, both low-pressure etching techniques and *high-density plasma (HDP)* techniques are used. In HDP, energy is coupled into the plasma inductively to increase the number of free electrons. HDP is operated at low (some mtorr) pressure. This in turn results in a higher plasma density and a higher degree of ionisation. HDP is used to provide high aspect ratios.

During the formation of the transistors, also a combination of anisotropic (dry etching) and isotropic etching (wet etching) is used. Particularly in the formation of STI, the anisotropic etching step is used to create the steep edges of the trench, while an isotropic etching step is used at the end of the STI etching process, to create smooth very round corners in the bottom of these trenches. Rounded corners, both in the top and the bottom of the STI, limit the local electric field and reduce leakage currents. After etching the trenches, better rounded corners can be achieved by a high-temperature thermal oxidation, which reduces stress in the substrate. In the advanced 3D memory architectures, such as DRAMs and NAND flash memories, many etching steps require high to extremely high aspect ratios of the contact holes. In the DRAM memories, the 3rd dimension is often used to stack various dies on top of each other and use *through-silicon via (TSV)* etching techniques to connect the individual dies. 3D NAND flash memories introduce some significant changes to the traditional etching techniques. In these devices, the third dimension is used to produce many different layers of memory cells stacked on top of each other. The related extremely high aspect ratios for contacts (up to 100) require new

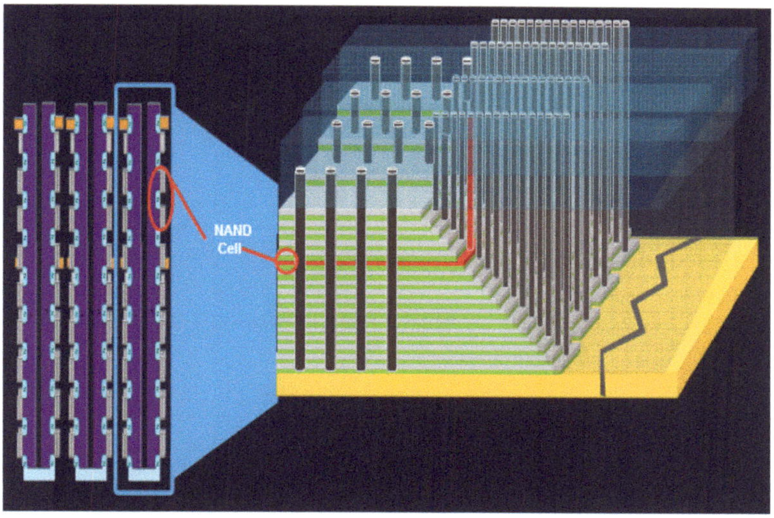

Fig. 3.41 Cross section of the etching requirements in an example 3D NAND flash. (Courtesy of Applied Materials)

etching techniques. Figure 3.41 shows a cross section of a 3D example NAND flash memory [47]. Details about these etching techniques are beyond the scope of this book. Further details on 3D memories can be found in Chap. 6.

Complementary to atomic layer deposition to form extremely thin layers on a wafer, *atomic layer etching (ALE)* enables the etching of layers with atomic precision [52]. ALE is sometimes also called reverse ALD. ALE has already been researched for more than two decades. It enables, with atomic-level precision, both anisotropic and isotropic etching which is based on sequential and self-limiting chemical reactions [59]. The application area of ALE is less than that of ALD, while the process is more complex. It still requires further R&D effort, before it is mature enough to become available in highly accurate etching steps in the high-volume production of semiconductor devices and ICs. The focus on new etching techniques does not preclude further development of existing techniques such as high-pressure etching and RIE.

Many process steps use plasma or sputter etching techniques, in which charged particles are collected on conducting surface materials (polysilicon, metals). Also, during ion implantation, charge can be built up. These techniques can create significant electrical fields across the thin gate oxides; this is called the *antenna effect*. The gate oxide can be stressed to such an extent that it can be damaged (so-called process or *plasma-induced damage: PID*), and the transistor's reliability can no longer be guaranteed. The antenna effect can also cause a V_T-shift, which affects matching of transistors in analogue functions. It is industry practice to introduce additional 'antenna design rules' to limit the ratio of antenna area to gate-oxide area. There are different rules for polysilicon, contact, via and metal-antenna ratios.

These ratios may vary, e.g. from 10 (contact-on-poly area to poly-gate area) to 5000 (accumulated-metal area to poly-gate area). An antenna rule, for example, may limit the maximum wire length in a certain metal layer to several hundred micron, depending on the metal layer and process technology. Also, in some libraries, protection diodes are used to shunt the gate. Each input to a logic gate in a standard-cell library then contains a protection diode.

3.7 Diffusion and Ion Implantation

Diffusion and *ion implantation* are the two most commonly used methods to force impurities or dopants into the silicon.

Diffusion

Diffusion is the process by which the impurities are spread as a result of the existing gradient in the concentration of the chemical. Diffusion is often a two-step process.

The first step is called *pre-deposition* and comprises the deposition of a high concentration of the required impurity. The impurities penetrate some tenths of a micrometre into the silicon, generally at temperatures between 700 °C and 900°C. Assuming that the impurities flow in one direction, then the flux is expressed as

$$J = -D \cdot \frac{\delta C(x,t)}{\delta x}$$

where D represents the *diffusion coefficient* of the impurity in [cm^2/s] and $\frac{\delta C}{\delta x}$ is the impurity concentration gradient.

As the diffusion strongly depends on temperature, each different diffusion process requires individual calibration for different processing conditions. During the diffusion process, silicon atoms in the lattice are then substituted by impurity atoms.

The second step is called *drive-in diffusion*. This high-temperature ($>$1000 °C) step decreases the surface impurity concentration, forces the impurity deeper into the wafer, creates a better homogeneous distribution of the impurities and activates the dopants. This drive-in diffusion also causes an identical lateral diffusion.

As a result of the increased requirements of accurate doping and doping profiles, diffusion techniques are losing favour, and ion implantation has become the most popular method for introducing impurities into silicon.

Ion Implantation

The ion implantation process is quite different from the diffusion process. It takes place in an *ion implanter*, which comprises a vacuum chamber and an ion source that can supply phosphorus, arsenic or boron ions, for example. The silicon wafers are placed in the vacuum chamber and the ions are accelerated towards the silicon under the influence of electric and magnetic fields. The *penetration depth* in the silicon depends on the ion energy. This is determined by the mass and electrical charge of the ion and the value of the accelerating voltage. Ion implanters are equipped with a mass spectrometer (analysing magnet), which ensures that only ions of the correct mass and charge can reach the silicon wafer. Ion implantation is characterised by the following four parameters:

- The type of ion. Generally, this is phosphorus, arsenic or boron. The mass and electrical charge of the ion are important.
- The accelerating voltage (V), which varies from a few kilovolt to several MV.
- The current strength (I), which usually lies between 0.1 μA and 1 mA. High current implanters may use even higher currents.
- The implantation duration (t), which is in the order of tens of seconds per wafer. The total charge $Q = I \cdot t$ determines the number of ions that will enter the silicon. Typical doses range from 10^{11} to 10^{18} atoms/cm^2.

Variables V, I and t can be measured with very high accuracy. This makes ion implantation much more reproducible for doping silicon than classical diffusion techniques. In addition, V and I can be varied as a function of t to produce a large variety of doping profiles that are not possible with diffusion. The maximum impurity concentration is almost always at the surface when diffusion techniques are used.

The ion implantation technique, however, can be used to selectively create profiles with peaks below the wafer surface. The concentration of impurities decreases towards the wafer surface in these *'retrograde profiles'*. The most important material that is used to mask ion implanting is photoresist. Ion implantation causes serious damage (disorder) in the crystal lattice of the target. In addition, only a fraction of the implanted ions occupies a silicon atom location. The other part does not occupy lattice sites. These *interstitial dope atoms* are electrically inactive and do not operate as donors or acceptors. A subsequent thermal step, at temperatures around 650 °C, is used to recover the crystal structure. Another *thermal activation* step of around 850 °C causes the majority (sometimes only less than fifty percent) of the dopants to become electrically active on the lattice sites. A disadvantage of this *anneal step* is that at a high temperature the doping atoms diffuse in all directions thereby increasing the junction depth and reducing the channel length. Often, hydrogen is used to assist by transferring heat uniformly over the surface of the wafer. It also penetrates into the crystal lattice to react with atomic impurities. *Rapid thermal anneal (RTA)* is a short temperature cycle to limit the diffusion. It consists of a constant temperature (e.g. 600 °C) for stabilisation, followed by a temperature spike of just a few seconds up to 1000 °C. Then, a radiative cooling step in the order of 30 seconds takes the temperature back to the normal value.

Because of the high energy involved, the equipment needs to be cooled and the focussed ion beam, which may have a beam size of around 20 cm^2, must be scanned over the wafer to avoid heating. This scan follows a very controlled pattern, to create both a sufficiently high local and global dose uniformity. Ion implantation adds flexibility and increased process control to CMOS manufacture. It is superior to chemical deposition techniques for the control of impurities ranging from 10^{14} to

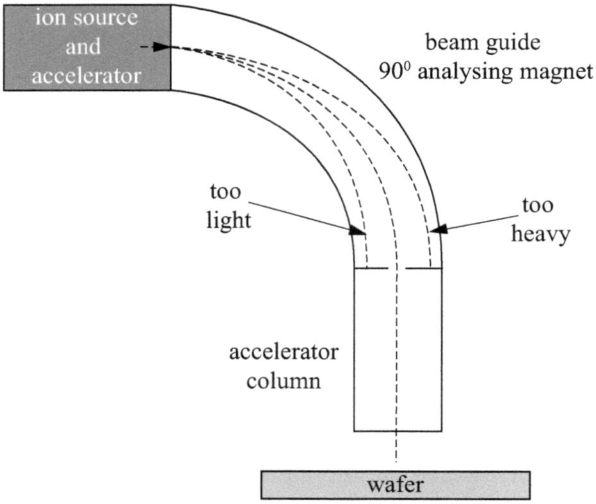

Fig. 3.42 Schematic drawing of an ion implanter

10^{21} atoms/cm^3. The heart of an ion implanter is formed by an ion source, usually an RF-activated plasma, from which ions are extracted by a small electric field, and a 90° analysing magnet. Because the ion beam is a mixture of different fractions of molecules and atoms of the source material, the 90° analysing magnet causes only the selected ions, with exactly the right charge and mass that face equal centrifugal and centripetal forces, to reach the wafer through the accelerator column and the resolving aperture; see Fig. 3.42 and [60]. Lighter ions strike the inner wall; heavier ions strike the outer wall.

Ion implantation is an essential and accurate technology to dope various regions inside, in between and below the transistors. Examples of the use of ion implantation are the following:

- Threshold voltage adjustment (e.g. $1 \cdot 10^{18} - 5 \cdot 10^{18}$ atoms/cm^3, however, this leads to a steep retrograde dope profile, in which the surface (channel) dope concentration is between $1 \cdot 10^{17}$ and $5 \cdot 10^{17}$ atoms/cm^3; see Table 3.3 for the different threshold voltages that are currently applied in the different technology nodes.)
- Retrograde-well implantation
- Channel-stop implantation
- Source/drain formation (including S/D implants and S/D extension implants)
- Halo implant
- Triple-well implant
- Doping of three-dimensional architectures used in memories (DRAM and flash)

Non-ideal aspects of ion implantation include the following:

- Lateral distribution of impurities is not completely zero.
- Throughput is lower than in diffusion process.
- Complex and expensive implanters.
- Initial cost of equipment: 2 to 5 M$.

The depth of the source and drain junctions are often in the same order of magnitude as the transistor channel length. The use of ion implantation in the formation of source/drain extension regions becomes increasingly challenging as these junctions become very shallow (4–20 nm) in advanced processes. Source/drain extension depths are in the order of 1/3 of the deep source/drain depths. The doping concentration does not change much with scaling. Only the energy during implantation must be adjusted to create those shallow junctions. *Silicidation* of sources and drains becomes a problem in that silicide can penetrate through the shallow junctions. This is called *junction spiking*. Unsilicided sources and drains show a five to ten times higher sheet and contact resistance, affecting the electrical properties of the transistors. Because of this, all modern CMOS processes today use silicided sources and drains. More on the creation of sources and drains in advanced CMOS processes can be found in Sect. 3.9.4. During the implantation process, the stationary ion beam is scanned over the wafers, which are positioned with 13 on a rotating disc. The wafer scan follows a controlled pattern to create sufficiently high local and global dose uniformities. The implant equipment must be cooled during use. The implant beam size may be in the order of 20 cm^2. The formation of nMOS and pMOS transistors requires a large number of different implants (see Fig. 3.51) for the wells, the sources and drains, their extension and halo implants, their threshold implants (e.g. low-V_T and high-V_T), etc. Advanced wafer fabs, which produce one to several hundred thousand wafers per month, with processes that use 50 or more different implants, may therefore require 20 different implanters, each with a capacity of more than 200 wafers/hour.

The doping concentration with diffusion is always higher towards the surface of the wafer, with the peak dope at the surface. With ion implantation, we can accurately adjust the ion implant acceleration speed, thereby creating the peak dope at a very well-controlled distance below the surface. Such an implant is also called a *retrograde implant*. Retrograde implant profiles are particularly used in the formation of the wells, as discussed in Sect. 3.9.3.

3.8 Planarisation

The increase in the number of processing steps, combined with a decrease in feature sizes, results in an increasingly uneven surface. For example, after completing the transistors, an isolation layer is deposited before the metal layers are deposited and patterned. The step height of the underlying surface is replicated into this isolation layer. This introduces two potential problems in the fabrication process. When the first metal is directly deposited onto this layer, its thickness can dramatically reduce at these steps, causing an increase in metal resistance and an increase in the occurrence of electromigration. Secondly, as already discussed in the lithography section, new lithography tools allow a smaller depth of focus (DOF), tolerating only very small height variations. During imaging, these variations can introduce focus problems at the high and low areas. Therefore, all current CMOS processes use several planarisation steps. These steps flatten or 'planarise' the surface before the next processing step is performed.

In conventional CMOS processes, *planarisation* was used during the back end of the process, i.e. in between the formation of successive metal layers to flatten the surface before the next metal layer was defined. In such a *spin-on-glass (SOG)* formation, the surface was coated with a liquid at room temperature. After this, the wafer was rotated (spun), such that the liquid flowed all over the wafer to equalise the surface. Next, the wafer undergoes a high-temperature curing process to form a hard silicate or siloxane film. To prevent cracking, phosphorus was often incorporated in the film. The resulting dielectric layer was planarised to a certain extent. An advantage of SOG is that very small gaps are easy to fill. However, with SOG, the surface is locally, but not globally, planarized; see Fig. 3.43. On locally rough areas (A and B), the surface is reasonably planarised.

There is still a global height difference after SOG planarisation, depending on the local pattern densities (area A, B and C). In a multilevel metal chip, this effect would be much worse and would lead to etching problems and problems with the *DOF* of the stepper. In all CMOS technologies below 0.25 µm, a very good alternative planarisation technique is used: *Chemical Mechanical Polishing (CMP)*.

CMP is based on the combination of mechanical action and the simultaneous use of a chemical liquid (slurry) and actually polishes the surface; see Fig. 3.44.

The *slurry* contains polishing particles (e.g. silica or alumina) and an etching substance (KOH or NH_4OH (e.g. ammonia)). A polishing pad together with the slurry planarises the wafer surface. Because CMP is also based on a mechanical action, it is much better suited for the local and global planarisation of rough areas, by offering a reduced topography for a more consistent focus across the field of exposure. It is particularly used for the creation and oxide filling of trenches (STI; Sect. 3.9.3) and during the metallisation (back-end) part of a multilayer metal process.

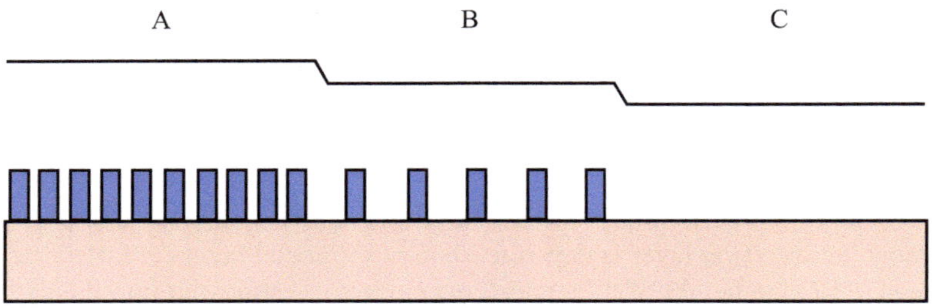

Fig. 3.43 SOG planarisation results

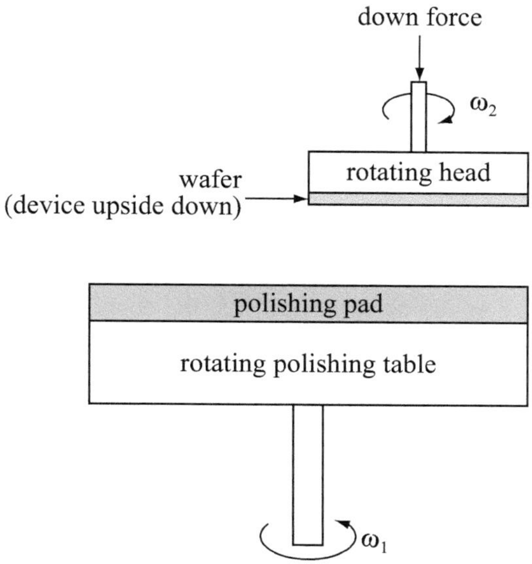

Fig. 3.44 Schematic overview of the CMP polishing process

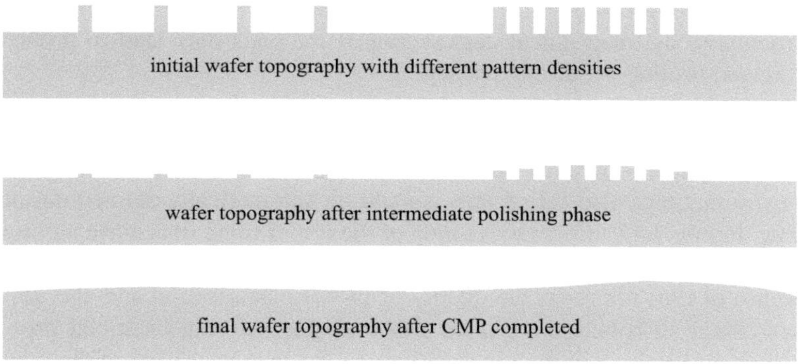

Fig. 3.45 Changing wafer topography after different CMP polishing phases

From the previous text, the reader might conclude that CMP leads to an ideal planarisation result. However, there are several issues related to differences in pattern densities and differences in polishing rates of the various materials. Figure 3.45 shows the polishing results at three different phases of the CMP process.

The forces, exhibited during the polishing process, cause a higher pressure on the individual features in sparsely dense areas than in high-dense areas. As a result, an increased polishing rate is observed on areas with very sparse patterns, compared to areas with the high-density patterns. This may lead to problems with the *DOF* during the lithography process and to reliability problems because of different contact heights.

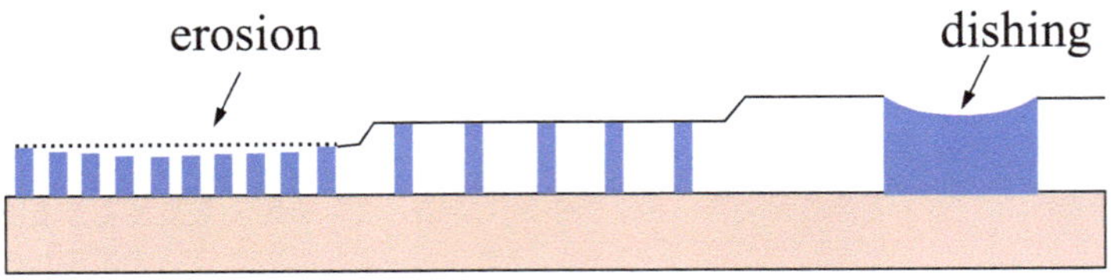

Fig. 3.46 Potential problems of copper CMP

As discussed in Sect. 3.5, the copper CMP process includes the simultaneous removal of copper and barrier. The soft centre of relatively large copper areas (wide copper lines and pads) polishes faster than the barrier/dielectric interface. This so-called *dishing* effect (Fig. 3.46) increases the resistance of these lines and reduces pad reliability. Also due to the difference in polishing rates, areas with dense copper patterns will polish faster than areas with only sparse copper patterns. This so-called *erosion* will also lead to thinner copper lines with higher resistance.

These polishing problems, in combination with the increased porosity of the inter-metal dielectrics, require constant monitoring through test structures for maintaining or improving both yield and reliability.

Particularly, the mechanical degradation of the pads may lead to problems as cracking and peeling-off during packaging.

Measures to prevent planarisation problems in the back-end metallisation process include the creation of dummy metal patterns in scarcely filled areas. The idea is to create metal patterns with as uniform a density as possible. These *dummy metal patterns*, sometimes also called *tiles*, should be automatically defined during chip finishing. Figure 3.47 shows an example of the use of tiling to achieve an improved metal distribution for optimised planarisation.

The use of tiles improves the quality of global planarisation and also results in a better charge distribution (reduced *antenna effect*) during back-end processing (deposition and etching of the successive metal layers). The shape of the individual tiles should be chosen such that it hardly affects the yield, performance and signal integrity of a logic block.

A disadvantage of CMP is the mechanical wear of the polishing pad. As a result, the speed of polishing is reduced, and, sometimes after each wafer, a diamond-brush step is performed to recondition the pad. After about 500 wafers, the polishing pad must be completely replaced. Figure 3.48 shows the result of the CMP planarisation technique in a multi-metal layer process.

3.9 Basic MOS Technologies

Sections 3.3 to 3.8 illustrate that MOS processes mainly consist of several basic actions that are repeated. In modern CMOS processes, the total number of actions has increased to several hundreds.

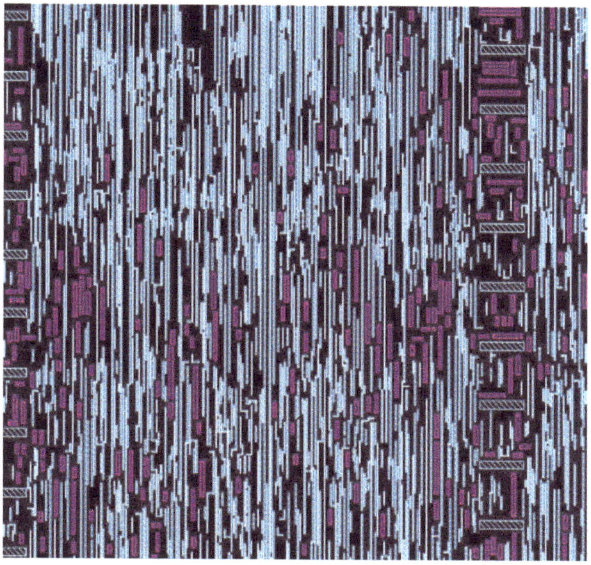

Fig. 3.47 Improved homogenous metal distribution by the use of tiles (purple)

In this section, a basic nMOS process with just five masks is discussed. A good understanding of this *silicon-gate nMOS* process enables a smooth transition to the complex modern CMOS processes. With the exception of some new steps, these CMOS processes are just an extension of the basic nMOS process presented here. A good insight into both technology types is a prerequisite when comparing the advantages and disadvantages of nMOS and CMOS.

Finally, a nanometer CMOS process is presented and the associated fundamentally new steps are discussed. The section is concluded with a quantitative discussion of CMOS technology options beyond 45 nm.

3.9.1 The Basic Silicon-Gate nMOS Process

An *nMOS process* which uses a mere five masks is explained with the aid of Fig. 3.49. First, an oxide is grown on the base silicon wafer. Next, the oxidised silicon wafer is coated with a silicon nitride (Si_3N_4) layer, as shown in Fig. 3.49a.

The first mask is the ACTIVE mask, which is used to define nitride areas corresponding to substrate regions where transistors should be formed. After the nitride is etched, boron is implanted through the resulting holes to produce the channel stopper, discussed in Sect. 1.8 and indicated in Fig. 3.49b. The wafer is then oxidised to produce the LOCOS areas in Fig. 3.49c. The resulting thick oxide only exists at places that were not covered by the nitride. The channel stopper is thus automatically present everywhere beneath the LOCOS oxide. This is a great advantage of the LOCOS process. The removal of the remaining nitride reveals the

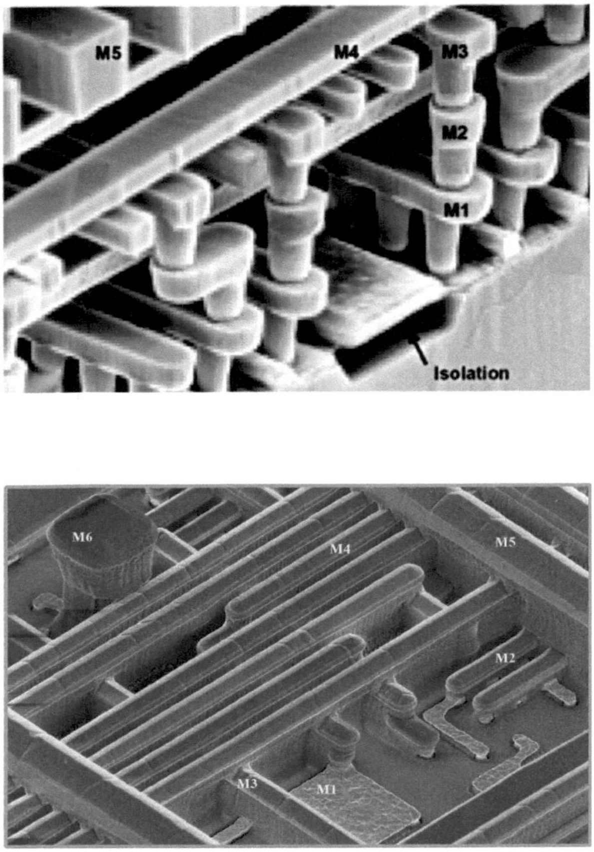

Fig. 3.48 Cross sections of CMOS back end, showing the potentials of CMP planarisation. (Source: NXP Semiconductors)

areas in which transistors will be created. Now, the oxide is removed by a wet HF dip. The next step is the growth of a thin oxide in these areas.

The thickness of this oxide varies from a few to a few tens of nanometers in most MOS processes. The threshold voltage adjustment implantation which follows this oxidation damages the thin oxide. The implantation is therefore done through this *sacrificial pad oxide*. Low-energy impurity atoms such as iron (Fe) and/or copper (Cu) from the ion implanter may be caught in and/or masked by the sacrificial gate oxide during the implantation. This sacrificial pad oxide is subsequently removed and the actual thin gate oxide is grown. The properties of a MOS transistor are largely determined by the gate oxide. Gate oxidation is therefore one of the most critical processing steps. Its thickness is between 1 and 7 nm (see Table 3.3).

After this, a polysilicon layer of about 0.1 to 0.4 μm thickness is deposited. A subsequent phosphorus diffusion, used to dope the polysilicon, is followed by photolithographic and etching steps, which yield polysilicon of the required pattern on the wafer. The POLY mask is the second mask step in this process and is used

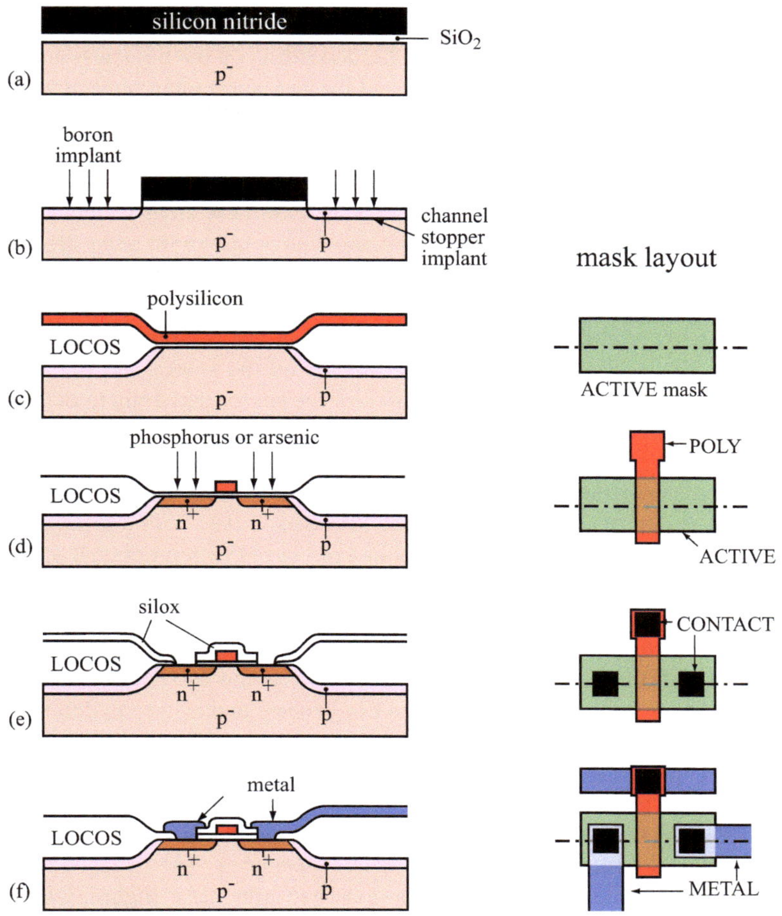

Fig. 3.49 The basic silicon-gate nMOS process with LOCOS isolation

to define the pattern in the polysilicon layer. This step corresponds to Fig. 3.49d. Solid silicon is used in various different phases. The most popular ones used in semiconductor fabrication are amorphous silicon, polycrystalline silicon and monocrystalline silicon. An important parameter for the conductivity and sheet resistance is the intrinsic *carrier mobility*. This varies from 1 cm^2/Vs for *amorphous silicon* to 250 cm^2/Vs for *polycrystalline silicon* (or *polysilicon*) and 1400 cm^2/Vs for *monocrystalline silicon*. CMOS circuits are built on monocrystalline silicon wafers. Polysilicon is used both as MOS transistor gate material, where it lies on thin oxide, and as an interconnection layer, where it lies on thick oxide (LOCOS). The resistance value of a polysilicon film with large grain sizes is comparable to that of monocrystalline silicon with equivalent doping level. However, polysilicon films with small grain sizes may exhibit a ten times larger resistance than monocrystalline silicon with equivalent doping level. The *sheet resistance* of polysilicon interconnections lies between 100 Ω/□ and 1k Ω/□, depending on the thickness and doping level. Polysilicon can therefore only be used for very short interconnections (inside library cells).

Phosphorus (P) or arsenic (As) is mainly used to create the source and drain areas. The source and drain junctions are implanted through the gate oxide which was covering the complete wafer. The sheet resistance of these areas is about the same as that of polysilicon. Today's polysilicon and source and drain areas are silicided to reduce the resistance values to about 8 Ω/□ (see Sect. 3.9.3 and Table 4.2). The edges of the n^+ areas are defined by the LOCOS and the polysilicon gate. Source and drain areas are thus not defined by a mask but are *self-aligned*, according to the location of the gate. The overlap of the gate on the source and drain areas is therefore determined by the *lateral diffusion* of the source and drain under the gate. In the nMOS processes that used diffusion to create sources and drains, the length of the lateral diffusion is about 60% of the diffusion depth of the drain and source.

Currently, lower doped thin drain extensions are used which show a lateral diffusion of about 40% of their depth; see also Sect. 3.9.3. With a *drain extension* of 10 nm, the lateral diffusion is only about 4 nm in a 45 nm process. The *effective transistor channel length* is therefore equal to the polysilicon width minus twice the lateral diffusion.

The wafer is then covered with a new oxide layer, deposited by an LPCVD step. The resulting SILOX layer indicated in Fig. 3.49e is about 200 to 600 nm thick. The CONTACT mask is the third mask step in this process and is used to define contact holes in the SILOX layer; see also Fig. 3.49e. The metal layer is then deposited by means of sputtering; see Sect. 3.5. The METAL mask is the fourth mask in this sample process. It is used to define the pattern in the aluminium layer. Basically, the processing is now completed; see Fig. 3.49f. However, as a final step, the entire wafer is covered with a plasma-nitride *passivation* layer. This *scratch-protection* layer protects the integrated circuit from external influences. Figure 3.49f shows the situation before deposition of the scratch protection. With a final mask step, the scratch protection is etched away at the bonding pad positions to be able to make wiring connections from the chip to the package. This mask and the associated processing steps are not included in the figure.

In summary, the mask sequence for the considered basic silicon-gate nMOS process is as follows:

1. ACTIVE	definition of active areas
2. POLY	polysilicon pattern definition
3. CONTACT	definition of contact holes between aluminium and monocrystalline silicon or polysilicon
4. METAL	interconnection pattern definition in aluminium

Finally, the NITRIDE mask is used to etch openings in the nitride passivation layer, to be able to connect bonding pads with package leads.

Note The temperatures used for the source and drain diffusion exceed 900 °C. Aluminium evaporates at these temperatures. Self-aligned source/drain formation is therefore impossible in an aluminium-gate process. Molybdenum gates have also been experimented with. However, they have never been industrially applied. In current CMOS technologies, the sources and drains are implanted rather than diffused, due to the very high accuracy of the channel length definition.

The silicon-gate nMOS process has the following properties:

- Small gate-source and gate-drain overlap capacitances, caused by the self-aligned implantations.
- A relatively low number of masks, i.e. basically five to six.
- Three interconnection layers, i.e. n^+ diffusion, polysilicon and aluminium. However, intersections of n^+ and polysilicon interconnections are not possible as these result in the formation of a transistor. Chapter 4 presents a basic summary on the properties of nMOS circuits.

3.9.2 The Basic Complementary MOS (CMOS) Process

CMOS circuits and technologies are more complex than their nMOS counterparts. In addition, a static CMOS circuit contains more transistors than its nMOS equivalent and occupies a larger area in the same process generation. However, CMOS circuits dissipate less power than their nMOS equivalents. This is an important consideration when circuit complexity is limited by the 1 to 2 W maximum power dissipation associated with cheap plastic IC packages. In fact, reduced dissipation is the main reason for using CMOS instead of nMOS.

Both n-type and p-type transistors are integrated in CMOS processes. Figure 3.50 illustrates the flow of a simple CMOS process with an *n-well*, or *n-tub*, in which the pMOS transistors are implemented. This process serves as an example for the many existing CMOS technologies.

The basic CMOS process begins with the oxidation, to some tens of nanometers, of a monocrystalline p-type silicon wafer. A layer of silicon nitride (Si_3N_4) is then deposited on the wafer. This is followed by a photoresist layer. A mask is used to produce a pattern in the photoresist layer corresponding to *active areas*. Circuit elements will be created in these areas.

The defined pattern determines which silicon nitride remains during a subsequent etching step. The photoresist is then completely removed, as shown in Fig. 3.50a. LOCOS oxide is then grown by exposing the wafer to oxygen at a high temperature. This oxide will not be grown on the areas that are still covered by the nitride. The LOCOS oxide separates active areas; see Fig. 3.50b for an indication of the result. Instead of LOCOS, STI is used in deep-submicron and nanometer CMOS processes to separate active areas (see next subsection). A new photoresist layer is then deposited, and the p-type transistor areas are 'opened' during photolithographic

Fig. 3.50 The basic CMOS process with LOCOS isolation

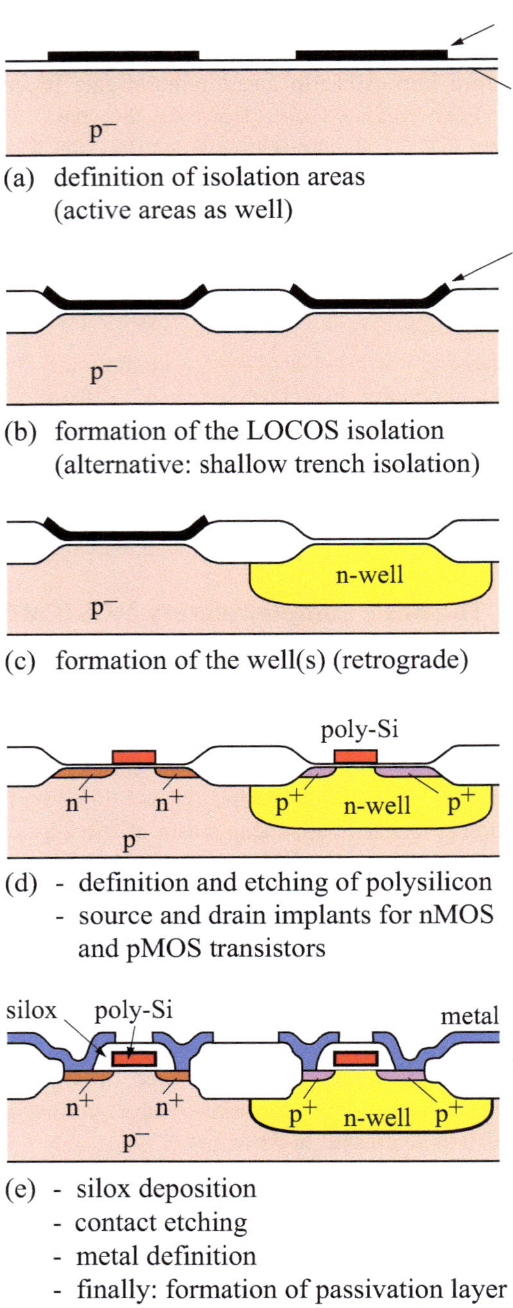

steps. In conventional processes, the n-well was created by depositing a high concentration of donors (mostly phosphorous) in these areas, as shown in Fig. 3.50c. Initially, these ions collect at the silicon surface, but they diffuse more deeply during a subsequent high temperature step. Today, the n-well (and p-well) are implanted (see next subsection). A layer of polysilicon is then deposited on the wafer, which now consists of n-type n-well areas with a limited submicrometre depth and p-type substrate areas.

Polysilicon doping reveals either n-type polysilicon for both nMOS and pMOS transistor gates or *double-flavoured polysilicon* (n-type and p-type polysilicon for nMOS and pMOS transistor gates, respectively). This is also sometimes referred to as n^+/p^+ *dual polysilicon*.

A photolithographic step follows and the polysilicon pattern is etched. The resulting polysilicon is used for short interconnections and for transistor gates.

Separate masks are used for the self-aligned source/drain implantations: nplus and pplus for the nMOS and pMOS transistors in the substrate and n-well, respectively. The result is shown in Fig. 3.50d.

The first step in the creation of interconnections between the different transistor areas is to deposit an SiO_2 (SILOX) layer on the wafer. Contact holes are etched in this layer to allow connections to the gates, drains and sources of the transistors. A metal layer is then deposited, in which the final interconnect pattern is created by means of photolithographic and etching steps. Figure 3.50e shows the final result.

Modern CMOS processes use 25 to 35 masks. Basically, these processes are all extensions of the simple CMOS process described above. VLSI and memory processes now use channel (gate) lengths of 20 nm to 0.25 μm and offer several levels of polysilicon and/or metal. These multiple interconnection layers facilitate higher *circuit densities*. The next section discusses a state-of-the-art nanometer CMOS process.

3.9.3 Sub-100 nm CMOS Processes

Compared to the basic CMOS process discussed before, an advanced nanometer CMOS process, with channel lengths below 100 nm, incorporates several major different processing steps. These differences will now be discussed in some detail (Fig. 3.51).

Shallow-Trench Isolation

Actually, LOCOS is thick SiO_2 that is thermally grown between the active areas. In contrast, *shallow-trench isolation (STI)* is implemented at significantly lower temperatures, preventing many warpage and stress problems associated with a high-temperature step. The STI process starts with a thermally grown oxide with a thickness between 10 and 14 nm. This is followed by an LPCVD deposition of 100 to 160 nm nitride. Next, the active areas are masked, and a dry etch step is applied to create the trenches, which have a typical depth between 250 and 500 nm. The

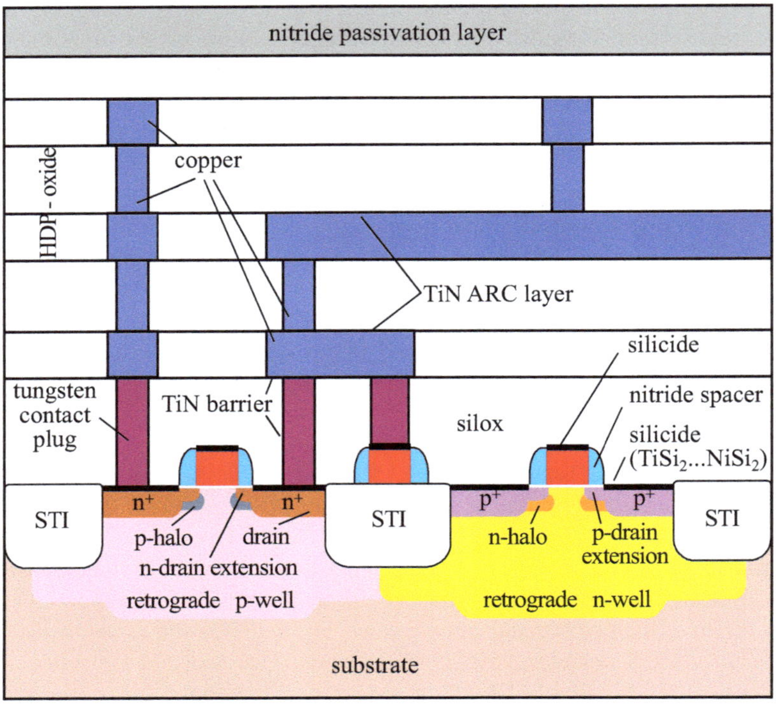

Fig. 3.51 An advanced nanometer CMOS process with STI isolation

corners at the bottom and the top of the trench are rounded by a thermally grown oxide layer (between 20 and 50 nm) along the side walls of the trench; see Fig. 3.52.

After removing the resist, a thick oxide high-density plasma (HDP), typically 700 to 1100 nm, is deposited. HDP is capable of filling the high aspect ratio of the trenches, which includes the pad oxide and nitride layer thicknesses.

In dense areas, the oxide level is well above the silicon nitride, while the oxide thickness equals the deposited oxide thickness in large open areas. The remaining topology is planarised using CMP; see Sect. 3.8. The nitride layer is used as chemical etch stop; see Fig. 3.53.

Next, the nitride masking layer is removed, using a wet etch and subsequently sacrificial oxide, gate oxide (by ALD) and polysilicon is deposited, etc. Figure 3.54 shows a cross section through the width of the device. The gate oxide between the polysilicon layer and the monocrystalline silicon substrate can be as thin as 1 nm in very advanced nanometer CMOS ICs.

In this way, device widths well below 20 nm can be well defined. Figure 3.36 showed already a comparison between LOCOS and STI field isolation techniques. It is clear that the STI is much more accurately defined and enables the creation of high aspect-ratio field-oxide isolation areas to improve the circuit density in nanometer CMOS ICs.

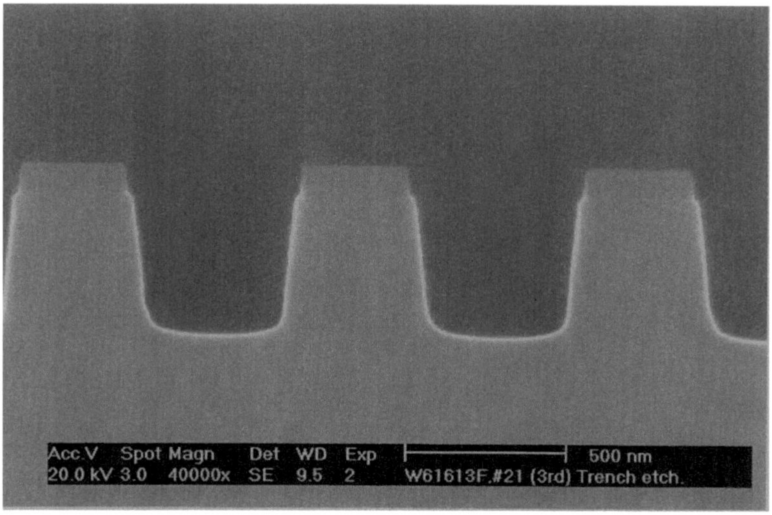

Fig. 3.52 Cross section after etching the trenches in the silicon

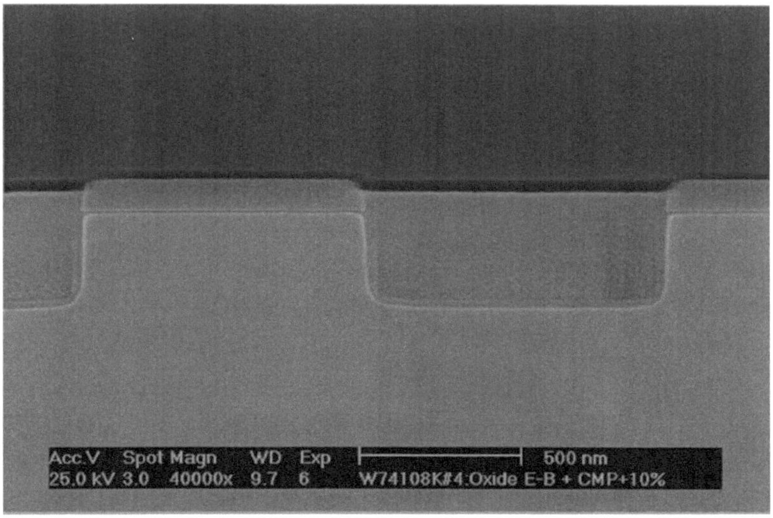

Fig. 3.53 SEM cross section after C MP

Retrograde-Well Formation

A *retrograde-well* process (Fig. 3.51) uses both n-wells and p-wells and is also called a twin-well process. These wells form the substrate for p-type and n-type devices, respectively. High-energy implantation of the wells yields doping profiles with maxima between 250 and 600 nm beneath the wafer surface in active areas. The maximum dope level beneath thick oxide areas (STI areas) is only a short distance

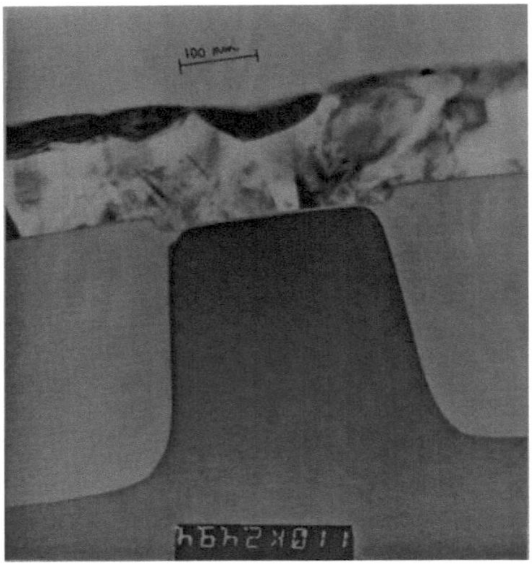

Fig. 3.54 TEM cross section through the width of the device

below the bottom of these oxides. The implantation therefore acts as a very effective *channel stopper* for parasitic devices in these areas.

Only a limited temperature is required to drive the well implants to appropriate depths, which results in limited lateral diffusion. Consequently, the wells can be accurately defined and their separation from source and drain areas of their own type (e.g. n-well to n^+ source/drain regions and p-well to p^+ source/drain regions) can be relatively small. This is the most important reason for applying retrograde-well processing.

Each well can be optimised to yield the highest performance for both types of transistors. This can be done by minimising source/drain junction capacitances and body effect or by using an *'anti-punch-through' (APT)* implant. Another advantage is the associated feasible symmetrical electrical behaviour. In addition, the two wells are usually each other's complement and can be formed by defining only a single mask during the design, while the other one is defined during the post-processing or chip finishing. Also, the throughput time for a retrograde well is shorter than that of a diffused well. Finally, another significant advantage of twin-well CMOS processes is formed by the better scaling properties, which facilitate the rapid transfer of a design from one process generation to another. The consequences of scaling are extensively discussed in Chap. 11.

Optimising technologies for high-speed digital designs generally degrades analogue circuit performance of long-channel devices. Careful optimisation of the front-end process (including the wells) is required to improve mixed analogue/digital circuit performance [61].

Drain Extension and Halo Implant

The *hot-carrier injection*, which will be discussed in Chap. 9, manifests itself more when carriers acquire more kinetic energy than about 3.2 eV. In 1.2 V processes and below, it becomes almost impossible for the charge carriers to penetrate into the gate oxide (energy equals $q \cdot V = 1.2\,\text{eV}$ in a 1.2 V process). Carriers can only acquire such energies after a lot of collisions in the pinch-off region. As the pinch-off regions are very narrow for nanometer CMOS technologies, this is becoming very unlikely to happen.

The LDD (Chap. 9) implants, as used in processes of 0.35 μm and larger to reduce the probability of occurence of hot carriers, are thus replaced by higher doped source/drain extension implants (Fig. 3.51). This source and drain extension is produced similar to the LDD. However, the peak doping concentration ($\approx 1 \cdot 10^{20} - 2 \cdot 10^{20}$ atoms/cm^3), today, is much higher than usually applied in an LDD and almost equals the peak dope in the highly doped source and drain regions. It results in a lower series resistance. Moreover, oxide spacers have been mostly replaced by nitride spacers, and a lot more doping profile engineering has been performed, to create smooth junctions to reduce junction leakage (band-to-band tunnelling) and punch-through. This is achieved by a combination of four different implants. First, halos are implanted after the formation of the gate, at a tilt angle (see Sect. 2.5.1). Next, a very thin off-axis As implant is applied to create the source/drain extension. This implant, in combination with its small lateral diffusion under the gate, reduces the source/drain extension dope concentration at the transistor channel edges with 1 or 2 orders of magnitude ($\approx 10^{18}$/cm^3) to reduce short-channel effects (depletion layer thickness reduction). Next, a much deeper As n^+ implant is used for the source/drain formation, followed by an even deeper phosphorous implant with a reduced doping, to create the smooth junction. The source/drain extension implant is much less deep (4–20 nm) than the actual source/drain junctions, which allows a better control of the channel length and reduces the short-channel effects. Actually, such an extension acts as a hard mini drain. In some cases in literature, only one implant is used to create the drain. This is then without extension implant and called *highly doped drain (HDD)*. The phosphorous halo with increased dope in the channel around the drain reduces the depletion layer thickness and suppresses short-channel effects such as threshold roll-off and punch-through.

Silicides, Polycides and Salicides

Silicidation is the process of creating a surface layer of a refractory metal silicide on silicon. *Silicides* may be formed by the use of $TiSi_2$, WSi_2, $CoSi_2$, NiSi or other metal silicides. When, for example, a titanium film is deposited directly on a silicon surface, after the definition of the polysilicon and the formation of the source/drain junctions, the titanium and the silicon react to form a silicide layer during a subsequent heating step. Titanium (and some other metals) react with exposed polysilicon and source/drain regions to form $TiSi_2$ silicide (or other silicides). A layer of titanium nitride (TiN) is formed simultaneously on the silicon dioxide. This will be selectively etched away. Silicidation yields low-ohmic silicide top layers in polysilicon and source/drain regions to reduce *RC* delays by five to ten times and improve circuit performance. Because the silicidation step is maskless, it is also called *self-aligned silicide* or *salicide*. In a *polycide* process, only the polysilicon is silicided. Sheet resistance values for silicided and unsilicided source, drain and polysilicon regions are presented in Table 4.2 in Chap. 4. $TiSi_2$ was introduced as silicide in the 250 nm technology node. Shrinking of lines has a dramatic effect on the resistivity of $TiSi_2$, and therefore, titanium has been replaced by cobalt (Co) for a couple of process generations. $NiSi_2$ is currently the most popular silicide used, due to a lower thermal budget during processing and its lower Si consumption during the formation.

Ti/TiN Film

Titanium (Ti) is used in the contact holes to remove oxides and to create a better contact with the underlying silicide. A *titanium nitride* (TiN) film is used in the contacts, as well as on top of the PETEOS (plasma-enhanced tetra-ethyl orthosilicate) oxide, because of its good adhesive properties. After the tungsten deposition to fill the contact holes, the tungsten must be removed on top of the oxide areas, which is done through a plasma etching step. The TiN layer is then used as an etch stop. The TiN is also responsible for an increased resistance of the contact plugs.

Anti-reflective Coating (ARC)

Reflections during exposure of a metal mask may cause local narrowing in the resist pattern and, consequently, in the underlying metal pattern, which is to be defined. A titanium nitride film is often deposited on top of the metal layer and serves as an *anti-reflective coating (ARC)*. Today, organic ARC is used during all lithographic steps in nanometer technologies. This film is highly absorbent at the exposure wavelength. It absorbs most ($\approx$75%) of the radiation that penetrates the resist. It also suppresses scattering from topographical features.

Contact (Re)fill

In many processes, particularly those which include planarisation steps, oxide thickness may vary significantly. Deep contact holes with high aspect ratios require special techniques to guarantee good filling of such contacts. This *contact filling* is often done by tungsten, called (tungsten) plugs, pillars or studs. As these aspect ratios become more aggressive with scaling, poor step coverage and voids in the contact plug become apparent. To fill the plugs void-free with tungsten, very thin Ti/TiN films are used as a low resistance glue layer for better adhesion to the dielectric at the contact side walls. The reason for using tungsten instead of copper or aluminium to fill these high aspect ratio contacts is that tungsten material shows excellent adhesion to these Ti/TiN films inside the pillers.

Damascene Metal Patterning

In 0.18 μm CMOS processes, metal patterning is done by depositing an aluminium layer, followed by a dry etching step to etch the aluminium away according to a mask pattern. In the damascene process, copper patterns are created by etching trenches in the dielectric, overfilling these trenches with copper and then polishing the overfill away using CMP, until the polishing pad lands on the dielectric. Damascene copper processing is discussed in some detail in Sect. 3.5.

Damascene patterning is used, particularly in 120 nm and below, to form copper wires. In a *dual-damascene* process, plugs (studs, pillars) and wires are deposited simultaneously. This process replaces the deposition of the plug and its etching, thereby reducing processing costs. The damascene process is mainly used to pattern copper, which cannot be etched like aluminium in plasma reactors. The copper will create too many by-products which remain on the surface and cannot be removed. The use of copper instead of aluminium for interconnection results in a reduction of the interconnection resistivity by 25% to 30%. This advantage is mainly exploited by a reduction of the metal height, so that about the same track resistance is achieved, but at a reduced mutual wire capacitance. This serves two goals: power reduction due to the reduced load capacitance of the driving gate and cross-talk reduction due to the smaller mutual wire capacitance to neighbouring wires. In combination with the use of low-ϵ dielectrics, the speed can be improved even more, or the power can be reduced further. Copper can also withstand higher current densities (reduced chance of electromigration, see also Chap. 9).

Up to 32 nm CMOS, the process flow was characterised by a series of *front-end-of-line (FEOL)* and *back-end-of-line (BEOL)* process steps. FEOL includes all process steps to create the transistors. BEOL include all contact, via and metal layer process steps. To enable small metal widths and spacings in 28 nm processes and beyond, related process flows also contain *mid-end-of-line (MEOL)* process steps. In these processes, the MEOL process steps refer to the creation of the local interconnect layer (LIL) (not drawn in the figure), including the first contact-hole layer, the first metal layer (M_0) and the first via layer (see Fig. 3.55). In general, the MEOL part of a process may include the silicidation steps, the tungsten contact plug formation and the first local interconnect layer(s). MEOL currently represents the most critical part of the process flow, particularly for the lithography tool, as it contains the smallest feature sizes on a CMOS chip (see also Sect. 3.9.6). The BEOL process steps then refer to Mx (this may include six layers M_1–M_6, depending on the technology node and performance), My (which may include layers M_7–M_9) and Mz (which may include layers M_{10}–M_{13} layers, when present). The number of metal layers in each Mx, My and Mz depends on the technology node and the process target, e.g. low power or high performance. The M_1 local interconnect layer may show metal patterns in both directions with somewhat larger pitches, while the other Mx layers often only include one-directional metal lines with small pitches. My layers are a little thicker than Mx layers (Fig. 3.55) and usually come with larger design rules (widths and spacings). On its turn, Mz layer(s) are thicker than My layers, also with larger design rules. These My metal layers are used for global routing, while the top Mz metal layers are normally used for the power distribution network. Some companies may offer even one or more thicker metal layers (between 1 and 3.5 μm thick) for specific applications.

Fig. 3.55 Cross section to show the various metal layers in the back end of the advanced CMOS process

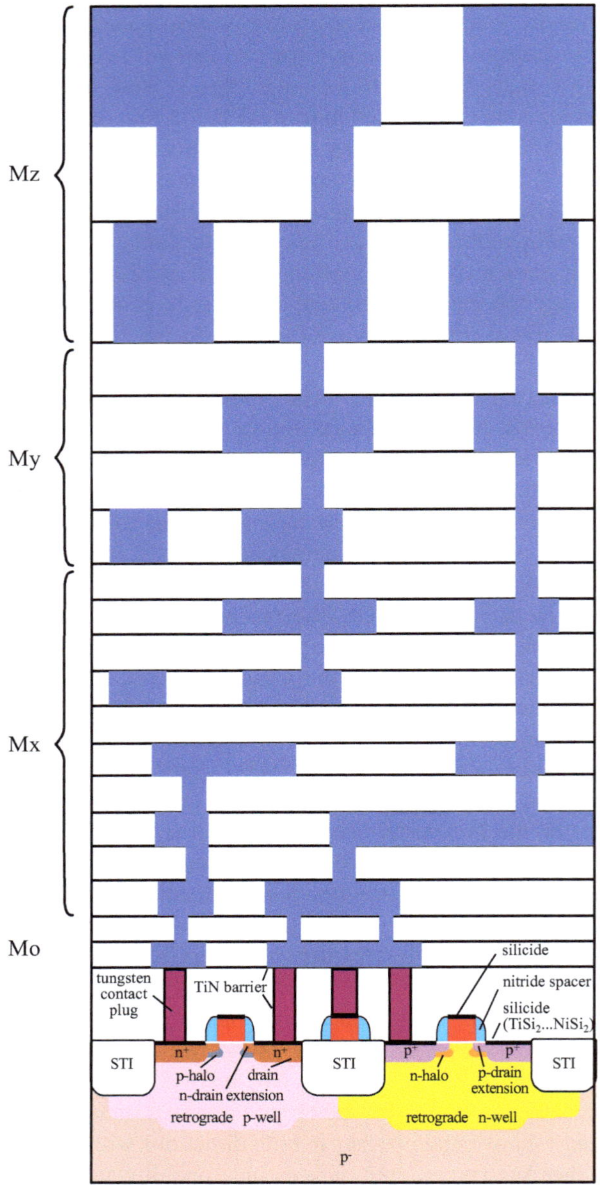

After the top metal layer has been deposited and patterned, the chip is fully covered with a strong passivation layer. Every chip must be connected from the outside, which means that on certain positions (bond pads), the passivation layer must be etched away. On these positions, the top metal layer would be accessible for contacting. Because copper oxidises relatively fast when exposed to an oxygen environment (like air), all bond pads in the top metal layer must be covered with an aluminium cap (please refer to Fig. 10.36), to which the eventual bond wire can be connected.

As will be clear from Fig. 3.55, all layers need to be accurately defined (positioned; aligned) with respect to the previous layer(s). This is done during the photolithography process in which the wafer is aligned with respect to the reticle being exposed. State-of-the-art processes require between 60 and 90 reticles to define all patterns in the individual layers. Each reticle must be aligned with respect to marks on the wafer which were created during process steps that correspond to a pattern image defined by a previous reticle. This leads to a so-called *reticle-alignment sequence* (see Sect. 3.3.1). Processing of the active areas (corresponding with the first mask; ACTIVE mask; Sect. 3.9.1) also leaves *alignment marks* (see Sect. 3.3.1) in the ACTIVE layer on the wafer in the scribe lanes. All successive masks are now being aligned with respect to these ACTIVE marks on the wafer, including the POLY mask. After the POLY mask and corresponding process steps, all IMPLANT masks as well as the CONTACT mask are being aligned with the POLY marks on the wafer. Then, the 1-st METAL mask is aligned to the CONTACT marks on the wafer, while the following VIA and METAL layers are both aligned to the METAL marks in the previously processed metal layer. The above alignment sequence is just serving as an example. It depends on the lithographic tool and on the required accuracy.

3.9.4 Further Transistor Scaling: From Planar to FinFETs

Approaching the end of Moore's law, by reaching the physical limits of scaling planar CMOS devices, has challenged both process and design engineers to create solutions to extend CMOS technology scaling towards 7 nm feature sizes. Local circuit speed is dominated by the devices (transistors' driving currents), while the global speed is dominated by a combination of the devices and interconnects (signal propagation). There are several issues related to the continuous scaling of the devices and interconnects.

Planar Devices

The transistor's driving current depends heavily on its threshold voltage and carrier mobility. Scaling introduces several mechanisms that reduce this mobility, directly or indirectly. First of all, the carrier velocity saturation and surface scattering effects, introduced in Chap. 2, are responsible for a two to six times mobility reduction. Apart from this, there was an increased depletion of the bottom side of the polysilicon gate *(gate depletion; gate inversion)*, due to the increased levels of halo implants for suppression of short-channel effects. Because mainly this bottom side of the gate is responsible for the drive current of the transistor, this gate depletion will dramatically reduce it. Alternatives of polysilicon gates are fully silicided *(FUSI gate)* and metal gates. It has taken many R&D resources to replace polysilicon gates with an appropriate metal-gate material. This is due to the fact that the metal work function (which also determines the V_T) is affected by the metal-gate composition, the gate dielectric and heat cycles. Few (metal) gate stacks have been identified giving a correct V_T after integration in a manufacturable CMOS process flow.

In a FUSI gate, the chemical reaction during silicidation continues until the gate is siliced all the way down to the bottom of the gate. Its operation then resembles that of a metal gate and does not show bottom depletion. Some companies have introduced *metal gates* in their 45 nm node, while others started to use them from the 32 nm node onwards.

The conventional way of increasing the transistor current is to reduce the gate-oxide thickness. But with oxide thickness values (far) below 2 nm, the transistor exhibits relatively large gate leakage currents, which increase with a factor of close to ten for every 0.2 nm further reduction of the oxide thickness. A high-ϵ gate dielectric (hafnium oxide, zirconium oxide and others) was therefore a must to continue device scaling with an affordable leakage budget. The choice of new materials in the transistor gate stack is not only dependent on the target improvement of its characteristics, such as increasing performance and reducing leakage. It also heavily depends on their mechanical (stress/strain), physical (optical, dielectrical), thermal (temperature expansion coefficient, thermal resistance) and chemical (lattice matching, adhesion, chemical affinity with adjacent layers) properties. As a result, the search for the right combination of high-ϵ gate dielectric with the right gate electrode with the right work function and tolerance to high-temperature process steps was very difficult. Therefore, the metal-gate architecture consists of a stack of different materials to fulfil the electrical, mechanical and physical requirements, such as lattice mismatches or adhesion. nMOS and pMOS transistors show complementary behaviour and have different strain requirements. Therefore, nMOS and pMOS transistors require different gate stacks. The different metals and work functions for the nMOS and pMOS transistors, often referred to as *work function metal (WF metal)*, are tuned to control the required nMOS and pMOS threshold voltages (V_T). Figure 3.56 shows example cross sections of the gate stacks for an nMOS and pMOS transistor.

The gate stack must be compatible with these strain requirements and also be able to survive the high-temperature anneal step needed to recover the silicon crystal structure after the source/drain implant and to activate the doping ions. Intel was the first to use high-ϵ dielectrics in combination with a metal gate and fabricated their Penryn dual-core processor in that process [62]. Many metal-gate and FinFET processes include a TiN diffusion barrier layer in between the metal-gate and the work function layer.

To limit short-channel effects, also the depths of the source/drain junctions and the shallow source/drain extensions have been scaled along with the gate length.

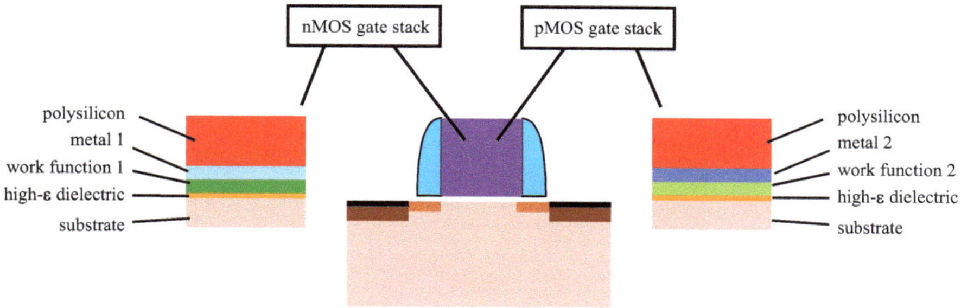

Fig. 3.56 Example gate stacks for an nMOS and pMOS transistor

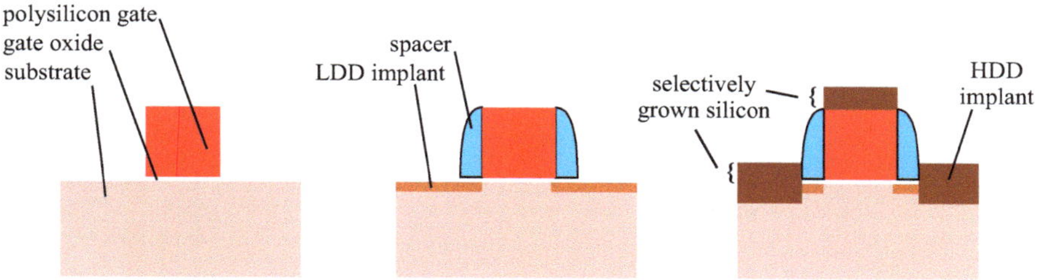

Fig. 3.57 Process flow for raised source/drain process

This caused an increased contact resistance, due to smaller contact areas and a relatively increase of sub-diffusion of the source and drain dopants beneath the side-wall spacers. A solution to this problem is to elevate the sources and drains to above the silicon wafer surface. It allows thinner spacers, which reduces the transistor area. Figure 3.57 shows simple process cross sections explaining these so-called *elevated sources and drains* or *raised sources and drains*. Although not drawn in the cross sections, of course also halo implants are implemented.

The elevated areas are created by a *selective epitaxial growth (SEG)* of silicon on source and drain (and gate) areas. The process reduces their resistances by providing more silicon for the silicide formation on the sources and drains [63].

A CMOS process with polysilicon gates is by definition a *gate-first process*, in that the gate is created before the source and drain implants. In this process, the gate oxide is deposited first, followed by the deposition of the polysilicon layer, which is then etched such that the polysilicon gates remain. Next, the sources and drains are implanted using the polysilicon gate as a barrier. The *channel length* is defined by polysilicon width. Then, the wafer must be annealed (*rapid thermal anneal (RTA)*: short high-temperature step to limit the diffusion) to repair the damage done during implantation, establish the desired doping profile and activate the implanted ions. As explained before, with high-ϵ metal gate, the gate stack consists of layers of oxide, metal and a few other materials which create a kind of sandwich. These thin additional layers need to compensate for lattice mismatches, adhesion or other physical properties that 'glues' it all together. A gate-first process with a *high-ϵ*

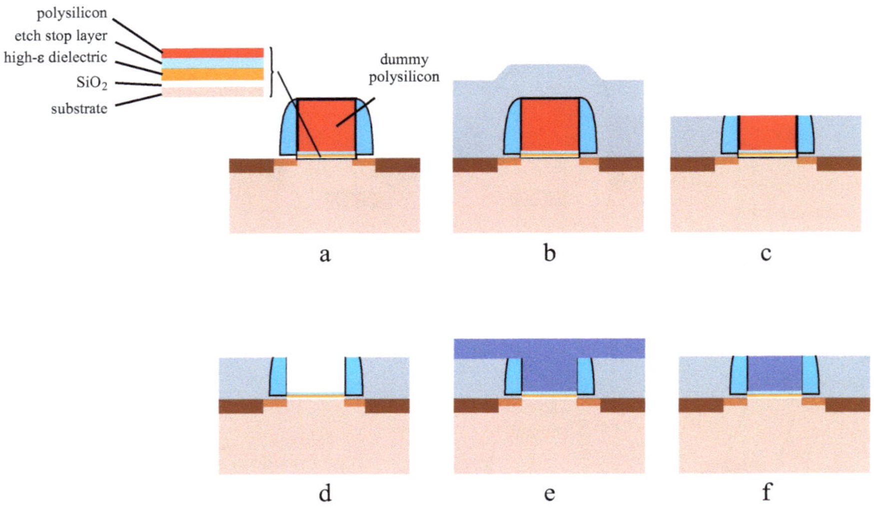

Fig. 3.58 (**a**) creation of S/D and extensions using spacers and dummy polysilicon as a barrier (**b**) deposition of inter-level dielectrics (**c**) CMP (**d**) polysilicon etch (**e**) metal fill (**f**) metal CMP [65]

metal-gate composition is similar to the traditional polysilicon gate CMOS process; however, the stack must withstand the S/D anneal step and maintain leakage and reliability standards. Metal tuning and the incorporation of an additional cap layer (work function layer in Fig. 3.56) are means to achieve the right transistor properties (V_T) [64]. The problem is now that the required anneal (high temperature) step can destroy the reliability of that stack. A solution to this problem is to use a sacrificial polysilicon gate to mask the source and drain implants, then perform the anneal step and remove the sacrificial gate and build a new gate stack after the anneal step. In such a *gate-last process*, this polysilicon gate is often referred to as *replacement gate*. Figure 3.58 shows an example of the creation of a transistor in such a gate-last process [65]. Although the figure shows the formation of one transistor only, nMOS and pMOS transistors require their own gate stacks as explained in Fig. 3.56. Note that, also here, the halos are not drawn in the cross sections in this figure, as they would hardly be visible. Of course in the process, they are there, as they need to suppress, in combination with the S/D extensions, the depletion layers at the channel edges near the sources and drains.

During the metal CMP, wide areas without metal wear down faster, causing an irregular surface (dishing or cupping). The gate-last process therefore requires additional *restrictive design rules (RDRs)*: poly can go in only one direction; no wide capacitors allowed; poly resistors must be replaced by bigger metal resistors. This results in an overall area penalty between 10% and 20% [66].

Although a high-ϵ metal-gate process contains a much thicker gate dielectric, its 'gate-oxide capacitance' is often still expressed as '*equivalent oxide thickness*' (*EOT*. It means that the combination of dielectric layers that together form the high-

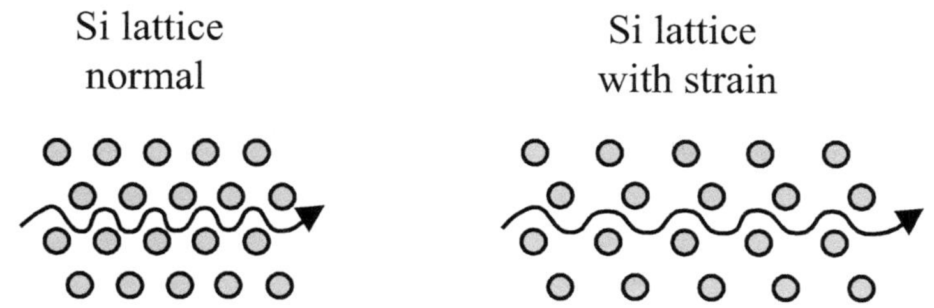

Fig. 3.59 Strained Si shows a reduced atom density, allowing improved carrier mobility

ϵ dielectrics has the same capacitance value as an equivalent transistor with a silicon dioxide dielectric. Today, almost all FinFETs have an EOT around 0.4 nm, which hardly scales.

Another way of increasing the transistor current is to improve the channel mobility. The use of *strained silicon* is one of the alternatives to achieve this. To achieve the best mobility improvements, the strain should be compressive ($\rightarrow\leftarrow$) for the pMOS transistors and tensile ($\longleftarrow\!\!\rightarrow$) for the nMOS transistors. In unstrained nanometer CMOS processes, the average hole mobility in the silicon is about two times lower than the electron mobility. Therefore, in many cases, the improvement of the pMOS transistor mobility has been given more priority.

In a *strain-relaxed buffer (SRB)* technology, a SiGe layer is grown on a silicon substrate. Germanium atoms physically take more space than silicon.

Next, a thin (about 10 nm thick) silicon layer is grown on top of the thicker SiGe layer. This top layer's atomic structure adapts itself to the atomic structure of the SiGe layer below. This creates strain in this silicon top layer (Fig. 3.59), introducing physical (tensile) stress in it, thereby increasing the channel mobility. The left picture in Fig. 3.60 shows a cross section of such a transistor. Experimental SiGe strained silicon showed 20% improvement in channel mobility [67]. To achieve a sufficient improvement in mobility, about 20% to 30% of the silicon atoms must be replaced by germanium. Germanium, however, exhibits a much larger thermal resistance than silicon, leading to self-heating problems comparable to SOI. A second problem related to this type of strained SiGe is the fact that germanium oxide is dissolvent to water, which is used during wafer cleaning to remove residual material from previous processing steps.

A third problem is that the SRB technology implicitly creates threading dislocations from the top of the SiGe layer into the strained silicon top layer [68]. These may have severe impact on the junction leakage and yield. Other SiGe methods have replaced the SRB technology. An alternative means of introducing strain to enhance the mobility is to embed an epitaxially grown strained $Si_{1-x}Ge_x$ (embedded silicon germanium; eSiGe) film in the source and drain areas *(recessed source/drain)*. Germanium atoms are slightly larger than silicon atoms (5.66 Å vs 5.43 Å), which generates a *compressive strain* in the transistor channel, which results in an enhanced hole mobility (Fig. 3.60 right transistor) in pMOS transistors [69].

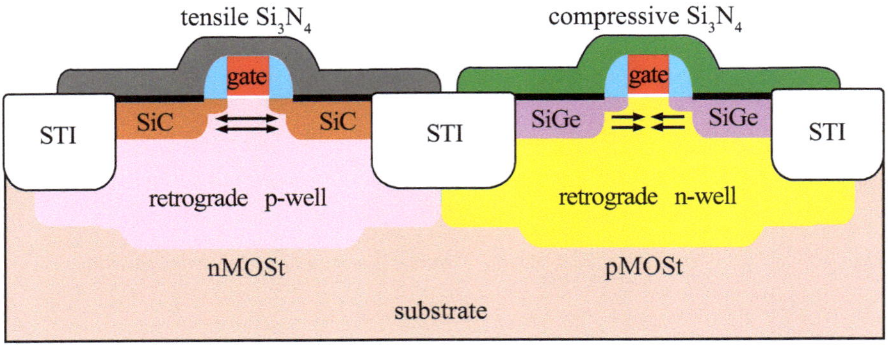

Fig. 3.60 Use of process-induced strain to enhance mobility by creating tensile stress in nMOS transistors and compressive stress in pMOS transistors

However, it puts severe demands to the transistor engineering, in particular with the alignment (overlay) of the gate with respect to the STI isolation areas. In order to fabricate a device with symmetrical behaviour, the self-aligned source and drain must be of equal size to induce the same amount of stress into the channel. *Tensile strain*, as opposed to compressive strain, can be created by using carbon (3.56 Å) which has a smaller lattice constant to substitute some silicon atoms. nMOS and pMOS transistors react differently under the influence of strain. As a result, the introduction of tensile strain improves the performance of nMOS devices while it degrades the performance of pMOS devices and vice versa. nMOS and pMOS devices are therefore built with built-in tensile and compressive strain, respectively (Fig. 3.60). Incorporating TiNi in the gate at a high temperature also introduces strain in the channel after cooling, due to the different temperature expansion coefficients of the various gate materials. The tensile and compressive stress in the nMOSt and pMOSt, respectively, is also enhanced by the deposition of a silicon nitride compound on top of the respective transistor gates. Whether this silicon nitride acts as a tensile or compressive layer depends on the ratio of silicon and nitride in the compound, as well as at the temperature budget during and after the deposition.

The carrier mobility in the channel is also related to their physical crystal orientation (see also Sect. 3.2). It is known that the mobility of holes in a (110) silicon substrate with a current flow along the <110> direction is about two times higher than in conventional (100) silicon. A combination of (110) oriented crystal lattice for the pMOS transistors with a (100) lattice for nMOS provides a much better balance between nMOS and pMOS transistor performance. The (110) orientation for the pMOS could lead to a 45% increase in drive current [70]. Figure 3.61 shows a cross section of a potential nMOS and pMOS device architecture built with different crystal orientations.

Figure 3.62 shows a summary of a potential technology options to boost the intrisic device speed.

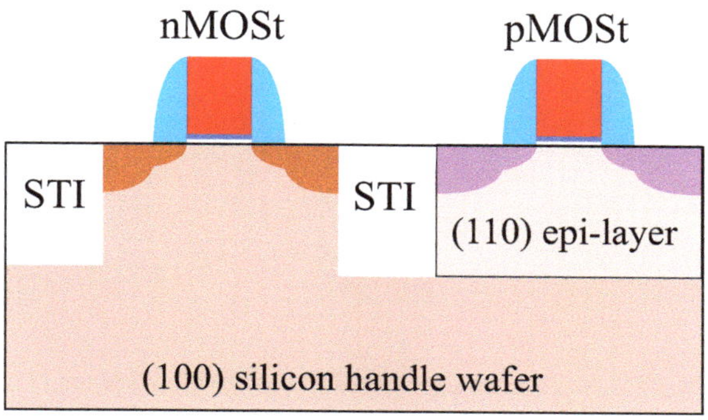

Fig. 3.61 Hybrid substrate architecture with nMOSt on (100) and pMOSt on (110) crystal orientation

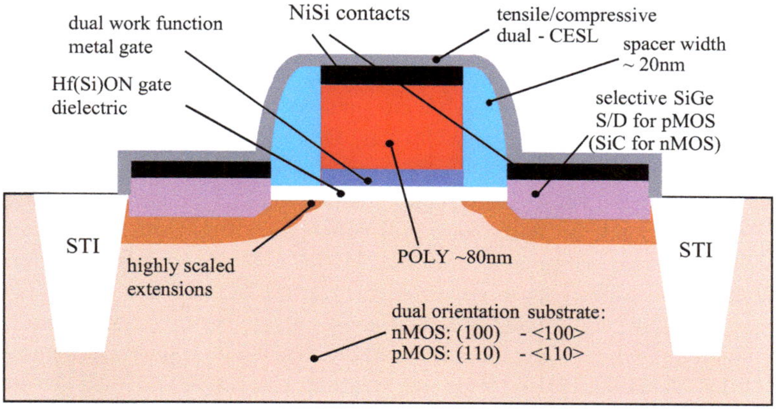

Fig. 3.62 Potential technology options for performance boost of MOS devices. (Source: NXP Semiconductors)

The optimum combination of stress and device orientations has driven and will still further drive the I_{on} current to much higher values than available in today's high-volume CMOS processes as discussed in Chap. 2.3.1. However, it is not only the real value of I_{on} that counts, but it is more the total $I_{\text{ds}} = f(V_{\text{ds}})$ characteristic that counts, because during switching the transistor cycles through the whole current to voltage characteristic.

3D and Alternative Devices

A fourth alternative to increase the transistor current is to use a double-gate or FinFET transistor. In a *double-gate transistor* (Fig. 3.63a), the transistor body is still lateral, but embedded in between two gates, a bottom gate and a top gate. Above a certain thickness of the body, there are two parallel channels contributing to the total current of the device, which now behave as two parallel fully depleted SOI transistors.

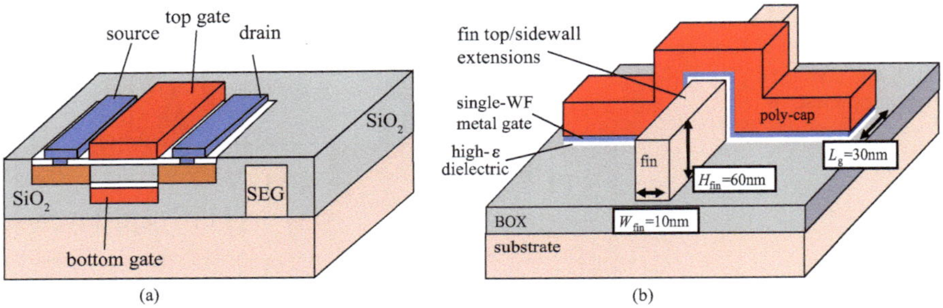

Fig. 3.63 (**a**) Double-gate transistor and (**b**) cross section of a FinFET. (Source: NXP Semiconductors)

Planar MOS devices have been used for more than four decades. Although double-gate transistors also offer better current driving capabilities than planar devices, FinFETs are easier to fabricate, and as such, they have already been accepted as the successor of planar devices. Due to increasing process parameter variations and leakage currents, industry has replaced planar device technologies by FinFET technologies. These are expected to scale to the 3 nm node or even further. FinFETs can be made on bulk or on SOI substrates. The *FinFET* architecture of Fig. 3.63b is created on an SOI substrate. In this example, the substrate consists of a thick silicon wafer with a buried oxide layer (BOX) and thin silicon layer on top. The silicon in the top layer is etched away outside the transistor areas, so that silicon fins remain in the active areas, which are then covered with a thin gate-oxide layer. Order of magnitude for W_{fin} and H_{fin} are, respectively, 10 and 30 nm in a 16 nm FinFET process. Then, a thin metal layer with a polysilicon cap is formed, covering the gate-oxide areas at all sides: left, top and right side. If the fin (or body) is very thin, this device will operate as a fully depleted SOI transistor with a higher driving current, due to the parallel current channels. The width of the transistor is determined by the height of the thin substrate, meaning that only one-size (width) transistors can be fabricated. In this example device, the transistor width is equal to the width of the fin plus two times its height, resulting in a transistor width of 130 nm. The transistor width can only be increased by putting more transistors in parallel (Fig. 3.64), enabling only quantised channel widths. Most FinFETs today, however, are built on bulk silicon wafers for improved compatibility with the planar CMOS process and to reduce cost. In an example of a bulk CMOS 22 nm FinFET process [71], the formation of the fins is similar to the formation of the active areas in a planar CMOS process (Fig. 3.64), by etching trenches (STI) in the silicon wafer and filling them with isolation oxide. Next, the wafer is planarised, and then, the STI oxide is etched back (recessed) so that the fins reveal. The following process steps, to create the gate stack and interconnections, are similar to those in a high-ϵ/metal-gate process.

The expected minimum fin width is around 5 nm, below which electrostatics reduce the FinFET's current drive and both the leakage current and the threshold

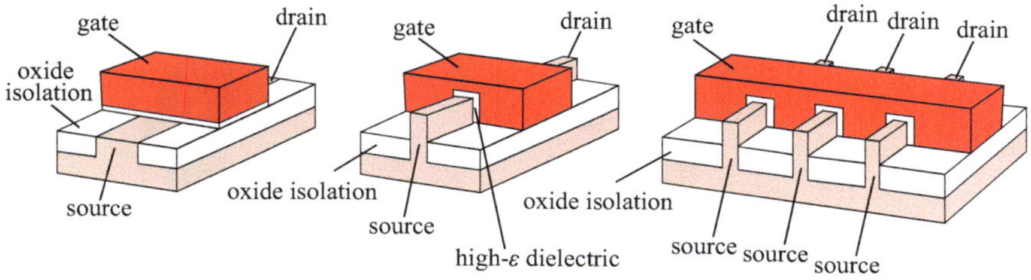

Fig. 3.64 Traditional planar CMOS transistor (left) and FinFET (middle), both on a bulk silicon wafer, and three FinFETs in parallel controlled by the same gate (right)

voltage variation will increase dramatically. Typical values for the *fin height* and *fin width* in the 5 nm process node are 60 and 6 nm, respectively. Fins may be undoped or low doped (10^{15} atoms/cm^3) and usually have a trapezoidal shape. After the gate formation is completed, capping layers are deposited above the gate to induce additional stress in the channel. The *dual stress liner (DSL)* approach uses a selectively deposited tensile silicon nitride film over the nMOS transistor and a compressive silicon nitride film over the pMOSt. The remaining (undoped) source and drain fins would dramatically increase the contact and series resistance of the S/D terminals. To enable low S/D resistance, an additional spacer oxide is deposited on the FinFET gate side walls. Next, a *selective epitaxial growth (SEG)* adds silicon volume (in all directions) to the sources and drains. In an nMOS, the fins are subjected to a mixture of silane and carbon gases in an oven at 500–600 °C, where this SEG adds more silicon volume to the fins and build strain into the nMOS channel at the same time. This creates raised S/D junctions with a SiC (2% carbon) to create tensile stress in the n-channel for improved electron mobility. During the SEG of the pMOS, a SiGe (55% Ge) layer is epitaxially grown on the source and drain of the pMOS transistor, introducing compressive stress in the channel to increase the hole mobility. Epitaxial growth is different along different crystal orientations (e.g. 110 or 110). This leads to a sloped structure for both source and drain *(faceted S/D)* (Fig. 3.65). Appropriate types of dopants are used to bring the S/D resistance to acceptable levels. An anneal step follows to activate the implanted dopants. Now, the stressors (nitride layers above the gates) may be removed before silicidation; however, the stress remains in the channels (*stress memorisation*) because it was transferred from the nitride to the channel during the annealing step. The S/D resistance is further reduced by silicidation. Figure 3.65 shows a 3D representation of a FinFET with raised S/D regions [72].

Assume that the fins in Fig. 3.64 have been processed up to the status of Fig. 3.65. By means of a long rectangular contact strap, also called a local interconnect, multiple parallel fins can be strapped together to create a multi-fin device. Chapter 4, Sect. 4.8, presents a summary on FinFET layout principles.

Dual V_T FinFET transistors are possible, e.g. to reduce subthreshold leakage in embedded SRAM, by fabricating gate stacks with different work functions,

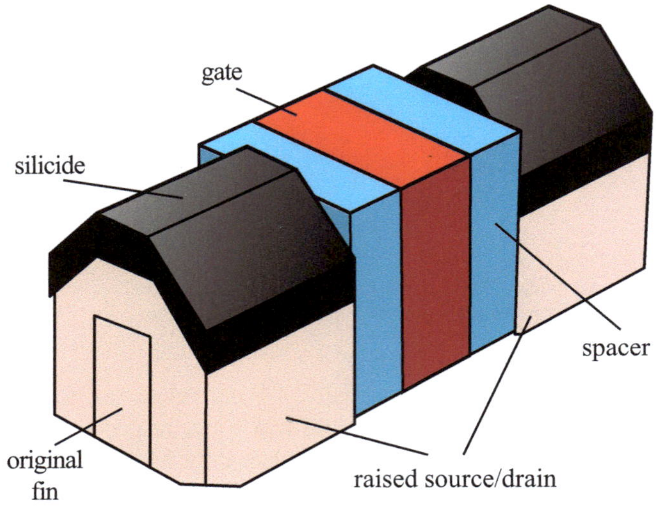

Fig. 3.65 3D representation of a FinFET with a raised source and drain [72]

which requires significant additional process complexity [73]. Raised S/D junctions introduce an increase of the gate-source and gate-drain capacitances. FinFET process spread depends on the uniformity of the fin width and height, the gate *line-edge roughness (LER)* and the gate-length (CD) variation. Fully depleted devices do not exhibit doping fluctuations leading to improved matching characteristics. Because of these properties, FinFETs are expected to scale relatively easy. The way the threshold voltage is defined (combination of gate-stack materials, including work function (WF) layers) forms an additional source for process variations. The raised S/D junctions introduce spread in the S/D series resistance. Because of the different architecture of FinFETs, they suffer from several physical variations in channel length, gate-oxide thickness, fin-thickness and gate underlap [74, 75]. FinFETs require complex 3D modelling to include all profiles and geometries of the fins, the spacer oxides and the gates. FinFET geometries are not ideal, so the usually drawn rectangular shape is in fact a trapezoidal shape on the die as shown in Fig. 3.66 [76].

It is clear that the trapezoidal shape will create threshold voltage variation across the height of the fin. This is symbolically represented by the three different V_T's in the fin. Below the fin, a high p-dope channel-stop implant prevents leakage currents through the bulk from source to drain and improves the short-channel effects of the bulk FinFET. Generally, the control of the gate over the channel is in FinFETs much better than in traditional lateral MOS devices resulting in a reduced subthreshold leakage.

The fin often gets an additional boron implantation step between 10^{15} and 10^{16} dopants/cm^3 for tuning the etching process. Doped fins etch different. Since the V_T

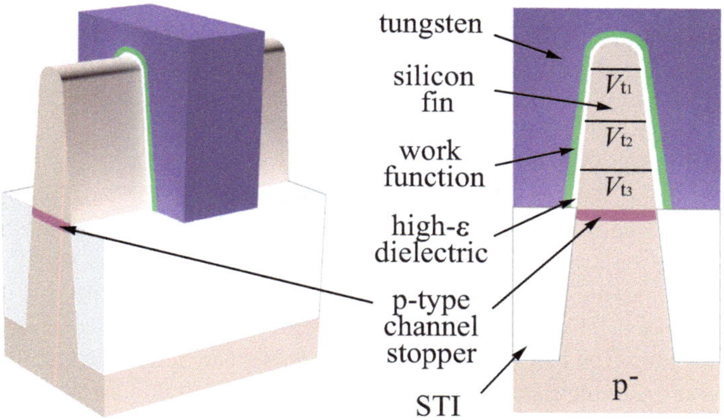

Fig. 3.66 Trapezoidal shape of a FinFET transistor (original FinFET cross section is courtesy of Intel)

is mainly determined by the work function of the gate electrode, random dopant fluctuations have only a minor effect on it. The spread in V_T is primarily caused by the spread in both the grain size and the grain orientation of the work function material(s). More detailed information on design and key characteristics of 14 nm FinFETs can be found in [77].

In planar transistors, the gate is unable to effectively control the leakage currents that are far below the gate oxide. Further reduction of the gate-oxide thickness does no longer help to reduce this leakage. The two major challenges of planar devices at technology nodes of 28 nm and below are transistor parameter variability due to *random dopant fluctuations (RDF)* and the efficiency of the gate to control the channel conductance (electrostatic behaviour of the transistor). As explained before, double-gate and FinFET devices, which are also called *multi-gate FETs* or *MUGFETs*, control the channel from both sides. FinFETs create a channel on both sides of the fin, as well as on top of the fin. These devices are therefore also called *tri-gate devices*. As such, they have a much better control over the leakage current and suffer less from short-channel effects. However, further reduction of the fin width of the FinFET towards 4 to 5 nm introduces channel width variations leading to undesirable variability effects.

In a 16 nm FinFET process, the gate stack may consist of 0.6 nm SiO_2 dielectric layer, followed by a 1.2 nm high-ϵ dielectric layer, a 1.3 nm WF layer and a 7 nm metal gate. The threshold voltage V_T of a FinFET transistor is determined by gate work function engineering rather than by doping of the channel region, as is common in traditional planar CMOS. To understand the difference, refer to Expression 1.16 for the threshold voltage, and the corresponding text regarding the explanation of the parameters that determine the threshold voltage. FinFETs, therefore, hardly suffer from the back-gate effect. Creating dual-V_T FinFETs would require the integration of different WF gates. FinFET technology does not allow the use of back-gate voltage to control the threshold voltage in low-power standby modes (see Chap. 8). Finally compared to bulk FinFET technology, SOI FinFET devices show a higher thermal resistance to the substrate due to the isolating BOX layer reducing their cooling capabilities.

Intel has introduced FinFETs in their 22 nm CMOS node. Next to Intel, also Global Foundries, Samsung and TSMC have developed FinFET technologies in the 16 and 14 nm nodes and beyond. Layout design of FinFET CMOS is not much different from traditional planar CMOS. The main difference is that the FinFET drive strength can only be improved during layout by adding more fins in parallel. The fins are often fabricated using a SADP process flow (see Sect. 3.3.2). With a CUT mask, long fins can be separated into individual ones. Because the fins are created by this double-patterning process flow, the minimum number of fins may be two in a practical manufacturing process. Some circuits, e.g. SRAM bit cells, require isolated fin patterning. To create an isolated fin, its 'spacer companion' must be removed (etched).

A FinFET may provide 80% more drive current for the same silicon area compared to a lateral MOSFET. This advantage can be used in two ways: FinFET circuits can run at lower voltages and consume less power while providing the same performance, or they provide higher performance when running at the same voltage as planar devices. Both are a drive to use FinFET technology.

Normally, the continued scaling according to Moore's law was driven by the cost reduction per logic gate. Below the 28 nm node, however, it looks like this trend has come to an end and that further scaling leads to an increase in cost per gate (see Sect. 11.5 and [78]). Finally, FinFET on SOI is more expensive than FinFET on bulk material, mainly due to the increased wafer cost: \$500 instead of \$120. This is partly compensated by a reduced number of FEOL litho and process steps for the FinFET on SOI process, resulting in an overall cost increase of \$136 [80].

Generally, the thicker the transistor body is, the less it can be controlled by the gate and the larger the subthreshold leakage current will be. This has not only resulted in fully depleted MOS devices and FinFETs with ultra-thin fins but also to alternatives, such as the *Ultra-Thin Body and BOX SOI (UTBBSOI)* [81] and the *Extremely-Thin SOI (ETSOI)* [82]. Figure 3.67 shows a cross section of both technologies.

Both technologies use an extremely thin, undoped fully depleted body, with the threshold voltage determined by work function engineering of the gate stack. Representative values for the body and BOX thicknesses are 7 and 20 nm, respectively. With small source and drain capacitances, reduced short-channel effects, reduced V_T variation and no well-proximity effects, these planar technologies are promising alternatives to the FinFET technology. Moreover, due to the thin BOX, the channel can also be influenced by the substrate (back gate in Fig. 3.67c) voltage enabling V_T control to trade-off leakage and speed (see Chap. 8). In the UTBB-SOI process, the

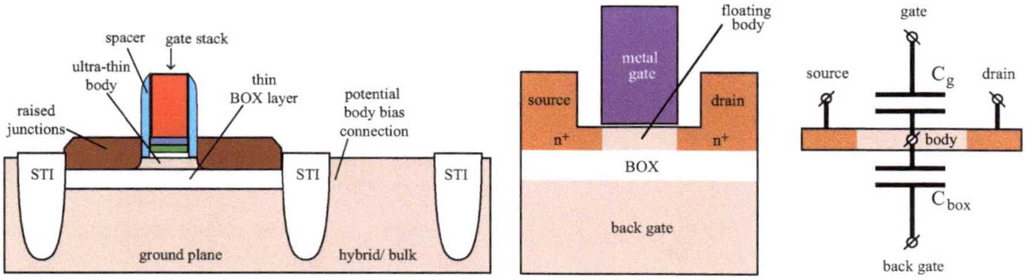

Fig. 3.67 Cross sections of UTBB-SOI (left) and ET-SOI (middle) and an equivalent circuit (right)

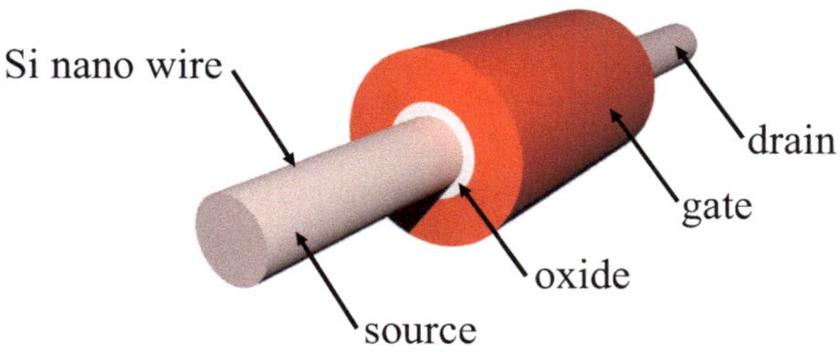

Fig. 3.68 Representation of a cylindrical Gate-All-Around transistor

BOX layer can be removed to enable bulk devices, like resistors, diodes and bipolar transistors, to support analogue circuits and improve reliability (ESD; Chap. 9). More technical details on these SOI technologies can be found in References [81] and [82] but are beyond the scope of this book.

A further evolution of the FinFET is the *Gate-All-Around FinFET (GAA FinFET)*, in which the fin is fully encapsulated by the gate [22]. Figure 3.68 shows a drawing of a cylindrical *gate-all-around transistor* (also called *nano- wire FET*).

The device provides a much better control of the gate over the channel and therefore guarantees optimal electrostatic behaviour. As a result, short-channel effects are suppressed. Today, this cylindrical GAA transistor is already used for more than a decade in the vertical 3D flash memories (Sect. 6.5.4) [83].These flash memories come with relatively long program and erase cycles: tens to hundreds of microseconds, depending on the memory organisation. Also, random reading of individual memory cells takes relatively much more time than in RAM memory. Therefore, 3D NAND memory cells may be stacked on top of each other, and they need not be created in a monocrystalline layer.

3.9.5 CMOS Technologies Beyond 10 nm

Transistors in logic circuits require the highest current efficiency. This will enable either high-speed, or, if the speed is not needed, less-power circuits, because then, the voltage can be reduced. This means that logic transistors must always be

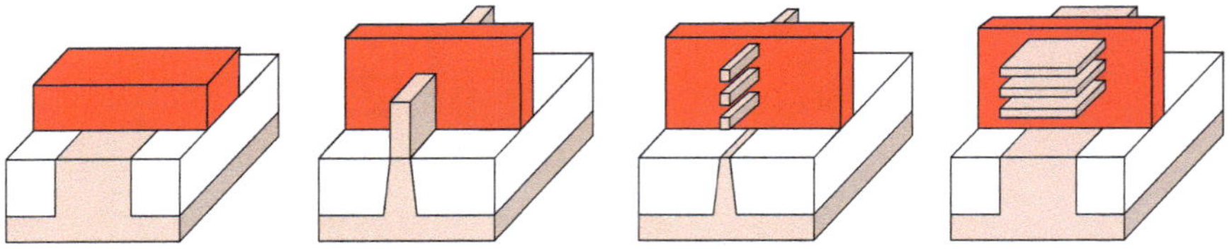

Fig. 3.69 Comparison of alternative MOS transistor architectures. (Source: Semiconductor Engineering)

fabricated in monocrystalline silicon. One cannot stack logic transistors in polycrystalline silicon on top of each other, since the carrier mobility in polycrystalline silicon is more than five times smaller than in monocrystalline (see Sect. 3.9.1). *Gate-All-Around (GAA)* transistors are therefore very complex to be used in logic processes. FinFETs, which were introduced in the 22 nm node, use three sides of the fin for the channel, while the bottom side is still unused. To continue further scaling, the gate length and the track height of the standard cells (see Sect. 4.8) needs to be reduced as well. Single fin devices in process nodes beyond 5 nm cannot deliver sufficient drive current. For this reason, several foundries have developed so-called *nano-ribbon*, *nano-wire*, *nano-sheet* or *fork-sheet* transistors for processes around and beyond the 3 nm node. These transistor architectures are based on the GAA device as shown in Fig. 3.68. Figure 3.69 presents a comparison of a planar, a FinFET and logic GAA transistor architectures [84, 85].

In the example of a nano-sheet transistor, a superlattice is built from alternating epitaxially grown layers of Si and SiGe on top of each other on a (monocrystalline) silicon substrate. Then, the SiGe layers are selectively etched away, leaving the nano-sheets as bridges between the source and drain. Next, atomic layer deposition steps are used to surround the channel region with thin gate dielectric materials and metal gate.

Researchers are currently working on stacks of tens of nano-sheets using indium gallium arsenide for improved current capability, but a simpler solution is to use doped germanium, which shows a higher mobility for both electrons and holes, compared to silicon. Although these alternative 3D transistor architectures for logic chips are promising candidates for the 3 and 2 nm nodes, they are very complex and may still show some unforeseen process and reliability problems [85]. An extension of the nano-sheet transistor, where either a multiple of nMOS devices or pMOS devices are stacked (Fig. 3.69 right) on top of each other, is a *CFET (complementary FET)*, where one or more pMOS nano-sheet transistors are stacked on one or more nMOS nano-sheet transistor (or vice versa). The CFET(s) may become the transistor architecture after the gate-all-around multichannel transistors [86, 87]. Finally, an additional challenge for both GAA and CFET chips is the cost of IP qualification. For a 3 nm node GAA chip, it may be 1.5 times what it costs for a 3 nm FinFET chip [88].

3.9.6 Further Interconnect (BEOL and MEOL) Scaling

There are several reasons why future CMOS ICs still need an increasing number of interconnect layers. Every new technology node offers us more transistors at a two times higher density. This requires more metal resources to support the increasing need for connecting these transistors. Secondly, they require a more dense power distribution network to be able to supply the increasing current needs. Since the introduction of 120 nm CMOS technologies, the aluminium back end has been replaced by a copper back end. Due to the required use of a barrier layer in the copper (Sect. 3.5) formation process, the effective copper metal track resistance has only reduced by about 25% compared to aluminium. This has been exploited by reducing the metal height, so that metal tracks show resistances comparable to aluminium but show less mutual capacitance to neighbouring signals while maintaining the signal propagation across them. However, further reductions of the metal heights are limited by the increasing current densities and the chance of electromigration. There is also an issue in the scaling of the contacts and vias. Since their number and aspect ratio (height/width ratio) increase with scaling, while their sizes decrease, they are becoming a very important part in the determination of the global chip performance, reliability and yield. Because of the increasing currents, the contacts and vias show an increasing amount of voltage drop, particularly when the signal line switches many times from one metal layer to another. Another result of the increasing current is the increased possibility of electromigration occurrence, thereby threatening the reliability. Finally, due to the high aspect ratios, there is an increased chance for bad contacts or opens, which will affect the yield. Already today, but certainly in the future, *design for manufacturabilty (DfM)* becomes an integral part of the design flow to support yield-improving measures (see also Chap. 10). A few examples are (1) *wire spreading*, where wires are routed at larger pitches (spreaded) because there is more area available than needed by minimum pitch routing, and (2) *via doubling*, where more vias are used for the same connection, only at locations where there is sufficient space, to improve yield.

Because the size of the contacts and vias scale, while their number is increasing, it becomes increasingly difficult to position them accurately at the right position between the successive metal layers to achieve sufficient contact area. Until recently, all vias between two successive metal layers were defined by a via mask pattern, in combination with lithographic and etching steps. It requires high alignment accuracies and expensive lithography. In many cases, the minimum via spacing is 20% larger than the minimum metal wire spacing, limiting the metal wire density. *Self-aligned via* interconnections are therefore developed to enable the use of relaxed lithographic steps [79]. The method uses operations performed on the metal patterns in the successive metal layers to be connected to each other and defines a pattern of potential via positions. This accurate potential via position pattern is combined with the more relaxed via mask layer to accurately define the required vias. Although the via-creating process step uses (relaxed) masking and exposure steps, the via position is accurately defined by the location where both to-be-connected metal wires cross each other. Self-aligned vias thus enable accurate via position, guaranteeing sufficient contact area without the need for extremely high-resolution lithographic process steps.

Most of the further improvements of the interconnect network has to come from further reduction of the dielectric constant (low-ϵ dielectrics) of the *inter-level dielectric (ILD)* layers between the metal layers and between the metal lines within one layer. This is realised by using the evaporation of a solvent in the dielectric material, which converts it into a thin porous film with a foam-like structure. During the last three decades, this dielectric constant has gradually reduced from 4 to 2.5. Alternative *low-ϵ dielectrics*, such as SiCOH, for instance, have reduced the dielectric constant to below 2.5 [89]. According to the IRDS road map [90], this dielectric constant is expected to remain between 2.2 and 2.5 (see Table 11.5), but it still needs many innovations to guarantee sufficient reliability. For more than a decade, research [91] has been focussed on *air gaps*, in which the dielectric material between metal lines in the same layer is replaced by air only. This will theoretically reduce the dielectric constant to even below 2 (the effective dielectric constant will not be equal to 1 (of air), because there are also mutual electric field lines from the top and bottom areas to the respective top and bottom areas of neighbouring metal lines. However, in practice, the values are expected to remain in the above-predicted IRDS range. The reliability of these air gaps is very important, since the encapsulation of the copper lines also determines their electromigration properties. This is circumvented by only partly replace the dielectric material by air gaps between metal wires in the same layer. An example of first usage of *air gaps* is in the Intel Broadwell processor family in a 14 nm node, as shown in Fig. 3.70 [91,92]. This chip contains 13 metal layers, of which the lower 8 levels use low-ϵ dielectrics. The air gaps are used in between metal 4 and 6 levels. Air gaps are also used in planar NAND flash memories, to reduce the lateral capacitance between two neighbouring cells to prevent their parasitic programming. The use of air gaps not only reduces the overall metal capacitance, leading to reduced active power consumption, but it also reduces the lateral capacitance between metal wires, thereby reducing the interference between as well as improving the signal propagation across the metal lines.

The combined move from aluminium to copper wiring and silicon dioxide to low-ϵ dielectrics required a change in the bonding process because the adhesion and stability are different. Low-ϵ dielectrics are more porous and include more air, so they become less robust and more sensitive to plasma damage during damascene processing and to pressure during test (probing) and bonding. Particularly, when *bond-over-active* techniques are used, where pads are not only located at the chip's periphery but also on top of circuits, these low-ϵ dielectrics must guarantee sufficient reliability. So, changing pad-related design and technology concepts also influences the reliability of the bonding process. Poor bond pad surface contamination may lead to a bond pad metal peel-off which leads to wedge bond or ball bond lifting.

Finally, the continuous process scaling also affects the copper resistivity level. Further scaling leads to an increase of the copper resistivity, a larger voltage drop and an increased power dissipation in the interconnect layers. The resistivity of a

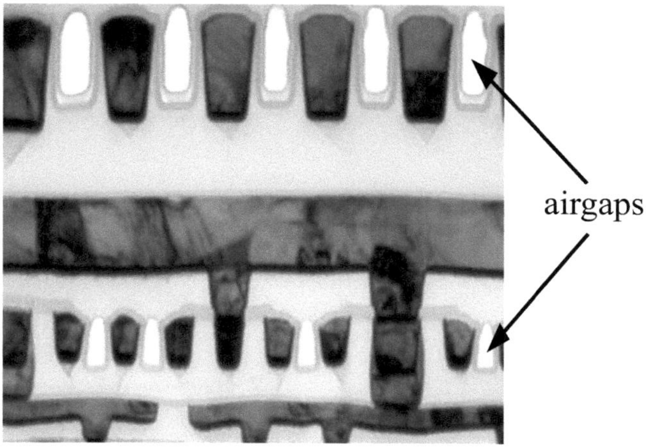

Fig. 3.70 TEM image of the first use of air gaps in the Intel Broadwell processor. (Source: Chipworks)

line is related to its width and to the electron *mean free path* of the conducting material. The mean free path of an electron is defined as the mean distance it travels between two collisions. For copper interconnections, the electron mean free path is close to 40 nm at room temperature and decreases with increasing temperature. So, as the line width has approached that value, increasing side-wall scattering will dramatically increase the resistivity. Scaling requires further track narrowing, which leads to significant increase of the copper track resistance, since the barrier and liner hardly scale. Therefore, the so-called *MEOL* process steps, which includes the silicidation steps of the fins and the formation of the local transistor interconnections, are continuously being optimised for the given process node. Until the 10 nm node, tungsten has been the favorite material for making electrical connections to the transistor's silicided source, drain and gate regions. Recently, tungsten contacts were replaced by cobalt with a thin TiN barrier, because the mean free path of cobalt is only a 10 nm. Therefore, cobalt with the thin barrier has also been introduced in the lower tight-pitch metal wires (often referred to M_0) and vias, where it shows much lower resistivity values than copper. For the advanced process nodes, however, not every waferfab's BEOL process is the same. In the 'Intel 4' process, for example, Intel replaced cobalt by tungsten again in the contact while increasing the tungsten volume for contact resistance reduction [93]. The process also includes an improved copper (Cu) scheme called 'enhanced' Cu. Previously, Intel used doped Cu, but now they are using pure Cu encapsulated in cobalt (Co). First, a barrier layer is deposited, followed by a Co layer which acts as a seed for electroplating. After the electroplating and planarisation steps, a Co cap layer is deposited on the Cu. While this process slightly degrades the electromigration resistance compared to using only Co, it still fulfils reliability requirements, and the resistance of the line is reduced. Even the *RC* value is

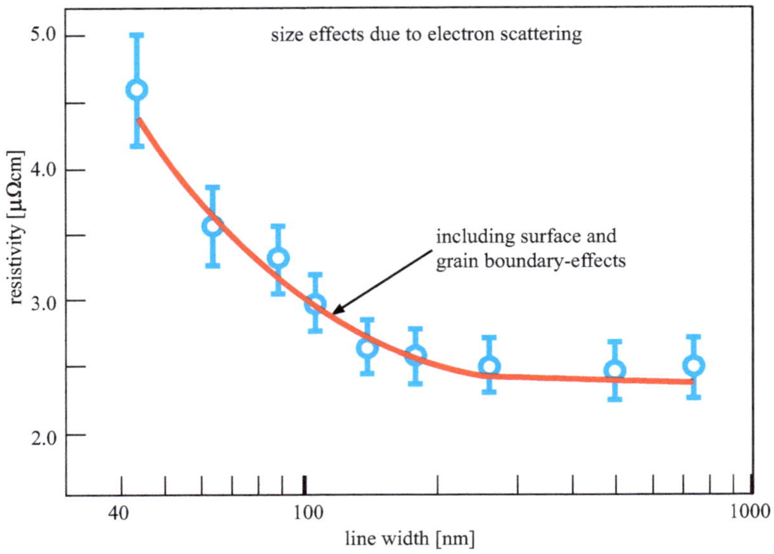

Fig. 3.71 Measured narrow Cu line resistivity (circles) versus simulated results including surface and grain boundary effects [96]

maintained, compared to the 'Intel 7' process. The overall 'Intel 4' process includes five enhanced copper layers, two giant metal layers and eleven 'standard' metal layers, making a total of 18 layers [94]. Further scaling drives the need for ultra-thin, high conductivity barriers and the exploration of 'barrier-less' approaches. An example of this is given in [95], which shows that reducing the barrier/liner to a total of 2 nm (or less) is required for future nodes and that this is can be achieved with a stack of Mn-based/Ru. This barrier/liner is an alternative solution to extend the Cu metallisation to future interconnects nodes. Figure 3.71 shows the trend in copper resistivity increase as a function of the copper line width [96]. From the above, it will be clear that the optimum combination of liner/barrier/conductor/capping material for the lower tight-pitch metal tracks depends on many physical issues: adhesion to the dielectric layers, barrier thickness, track resistance, electromigration issues and electroplating. Therefore, beyond 7 nm, each node may require a different metallisation process for these tracks. A further discussion on copper metallisation can be found in [97].

3.10 Conclusions

It is clear that the realisation of optimum electronic systems is based on a perfect match between the substrate (wafer), the transistors and the interconnections. The increasing number of application areas have leads to a large variety of substrate and technology options to support high-speed and low-power products.

So, the processing requirements for different types of circuits can be quite diverse. RAMs, for example, require a technology that allows very high bit densities. CMOS static RAMs therefore require tight n^+-diffusion to n-well spacings. This can be achieved when a retrograde-well implantation is used to minimise lateral well diffusion.

The discussions have started with a basic description of the most important processing steps that are repeatedly used throughout the fabrication of a CMOS chip. For educational purposes, the complexity of the described processes gradually increased from a simple five-mask nMOS process in the early 80s to a complex over-forty-masks nanometer CMOS process today. Due to the use of certain double- and quadruple-patterning techniques, the real number of masks for certain ICs may increase to close to or even above 70. The *number of process steps* (including measurement steps) has increased from about 1000 for the 32 nm node to more than 1200 for the 5 nm node. In addition, the *throughput time* in a wafer fab, also called *wafer cycle time*, has increased as well: from about one-day-per-mask for the 32 nm node to between 1.5 to 2 days per-mask for the 5 nm node.

Finally, several trends are discussed which focus on state of-the-art and future technology requirements. Chapters 9 and 11 focus on the physical and electrical design consequences of the continuous scaling process.

Finally, the increasing complexity of both the lithographic and manufacturing process is reflected by the growing cost of a fab. To ramp up a fab to volume production in a 5 nm node requires a time frame of about two and a half years and a budget of $ 15 to 22 billion. This has prompted many semiconductor companies to become *'fab-lite'* or maybe even totally *fabless*. As a result, there are only three semiconductor companies left that still invest in LOGIC waferfabs: Intel, Samsung and TSMC. Starting with the 28 nm node, the price per logic gate only gradually reduces with scaling to the next node. This means that cost reduction is no longer the big driver for further scaling. Therefore, the number of products for which scaling to the next process node is still economically attractive is continuously reducing. However, the world needs more powerful chips. Example applications in this respect are deep learning, artificial intelligence, bitcoin mining and weather forecast. More detailed discussions on design and manufacturing costs can be found in Chap. 11.

3.11 Exercises

1. Why is the formation of the gate oxide a very important and accurate process step?
2. Briefly explain the major differences between the diffusion process and the ion-implantation process. What are the corresponding advantages and disadvantages?
3. What are the possible consequences of an aluminium track with a bad step coverage?
4. Describe the main differences between the formation of LOCOS and STI.
5. What are the major advantages of self-aligned sources and drains?

6. Why is planarisation increasingly important in modern deep-submicron technologies?
7. Assume that the ninth metal layer in a 22 nm CMOS process is optional, in which designs would you use the ninth metal and why? What is/are the advantage(s)/disadvantage(s) of using an additional metal layer?
8. Why was copper not used earlier in the metallisation part of a CMOS process?
9. What are the disadvantages of plasma etching?
10. What are 'tiles', as meant in the manufacture of a nanometer CMOS chip? Why may they be needed in such a design?
11. What is meant with CMP and what are its side effects?
12. Mention several solutions to reduce the source, drain and gate resistances of a MOS transistor. What would be the physical consequence for each of these solutions?
13. For which type of circuits would SOI be particularly beneficial in terms of speed and power?
14. Summarise all potential (technological as well as electronic) solutions to increase the I_{on} current of a transistor. Distinguish between nMOS and pMOS solutions.
15. Describe the major differences between a bulk-CMOS planar and a bulk-CMOS FinFET transistor in terms of lithography and in terms of fabrication process.
16. What are the major differences in current drive capability between a planar MOS and a FinFET MOS transistor, regarding (a) the effective channel width, (b) the increase in drive strength, (c) the use in analogue circuits and (d) the effect on self-heating?
17. How can the threshold voltage be controlled?
18. Describe why there are depletion layers around the source and drain junctions. How can its thickness at the channel edges be reduced?

References

1. P. Nanja, *Challenges and Solutions for Silicon Wafer Bevel Defects During 3D NAND Flash Manufacturing* (A Conventor Inc. publication, 2019)
2. J. Hruska, 450 mm silicon wafers aren't happening any time soon as major consortium collapses, 13 Jan 2017, https://www.extremetech.com/computing/242699-450mm-silicon-wafers-arent-happening-time-soon-major-consortium-collapses
3. www.memc.com
4. IC Insights, Companies Maximize 300 mm, 200 mm Wafers; Slow Progress on 450 mm. Design & reuse, 14 Sep 2015, http://www.design-reuse.com/news/38229/global-wafer-capacity-2015-2019-report.html
5. R. Wilson, Chip industry tackles escalating mask costs, *EETIMES*, 17 June 2002
6. M. Porrini, Growing Ingots of Single Crystal Si, in *MEMC Silicon Workshop at IMEC*, Leuven, 22 June 2006
7. G. Vaccari, Silicon Epitaxi for CMOS and power applications, in *MEMC Silicon Workshop at IMEC*, Leueven, 22 June 2006
8. L. Chang et al., CMOS circuit performance enhancement by surface orietation optimization. IEEE Trans. Electron Devices **51**(10), 1621–1627 (2004)

9. M. Yang, et al., Hybrid-orientation technology (HOT): opportunities and challenges. IEEE Trans. Electron Devices **53**(5), 965–978 (2006)
10. S.R. Alla, Ultra-thin body SOI FETs, 20 May 2014, http://www.slideshare.net/sindhureddy14/538-34932218
11. T. Buchholtz et al., A 660 MHz 64b SOI Processor with Cu Interconnects, ISSCC, Digest of Technical Papers (2000)
12. J.L. Pelloie et al., SOI Technology Performance and Modelling, ISSCC, Digest of Technical Papers (1999), pp. 428–429
13. H. Majima et al., Experimental evidence for quantum mechanical narrow channel effect. IEEE Electron Device Lett. **21**, 396–398 (2000)
14. Y. Liu et al., An investigation on the total thickness variation control and optimization in the wafer backside grinding process. Materials **15**, 4230 (2022). https://doi.org/10.3390/ma15124230
15. Measure Wafer Bow, Warp and TTV with Capacitance. https://vitrek.com/mti-instruments/measure-wafer-bow-warp-ttv/
16. J. Song et al., A study on the resolution and depth of focus of ArF immersion photolithography. Micromachines **13**, 1971 (2022)
17. R. van Haren et al., The impact of the reticle and wafer alignment mark placement accuracy on the intra-field mask-to-mask overlay, in *Proceedings of SPIE 11178*, Photomask, 2019
18. T. Lecklider, Yield: the key to nanometer profits. Eval. Eng. (2005). www.evaluationengineering.com/archive/articles/0305/0305yield.asp
19. Y.K. Choi et al., Sublithographic nanofabrication technology for nanocatalysts and DNA chips. J. Vac. Sci. Technol. **B21**(6), 2951–2955 (2003)
20. Michael White, A Look Behind the Mask of Multi-Patterning, 28 Jan 2015, https://blogs.sw.siemens.com/calibre/2015/01/28/a-look-behind-the-mask-of-multi-patterning/
21. J. Kwan, Sign-off lithography simulation and multi-patterning must play well together, 12 Aug 2014, http://www.techdesignforums.com/practice/tag/multi-patterning/
22. Y.-K. Choi, Multiple gate CMOS and beyond. in *Nanotechnology-forum*, Forum_6, Seoel, 5–6 June 2012
23. P. Alcorn, Nvidia tackles chipmaking process, Claims 40X speed up with cuLitho, Tom's Hardware, Mar 2023
24. D.C. Brandt et al., Laser produced plasma EUV sources for device development and HVM, 2012. http://www.cymer.com/files/pdfs/Technology/2012/Laser_Produced_Plasma_EUV_Sources_for_Device_Development_and_HVM.pdf
25. D.T. Elg et al., In situ collector cleaning and extreme ultraviolet reflectivity restoration by hydrogen plasma for extreme ultraviolet sources. J. Vacuum Sci. Technol. A **34**, 021305 (2016)
26. M. LaPedus, ASML ships world's first EUV tool, 28 Aug 2006, www.eetimes.com
27. M. Feldman (ed.), *Nanolithography: The Art of Fabricating Nanoelectronic and Nanophotonic Devices and Systems*. Woodhead Publishing Series in Electronic and Optical Materials (Cambridge, United Kingdom, 2014)
28. H. Mizoguchi et al., Performance of 100-W HVM LPP-EUV source. Adv. Opt. Technol. **4**(4), 297–309 (2015)
29. I. Fomenkov, Status and outlook of LPP light sources for HVM EUV, in *EUVL Workshop 2015*, 18 June 2015
30. P. Clarke, Report: Toshiba adopts imprint litho for NAND production, EETIMES (Analog), 07 June 2016
31. G. de Boer et al., MAPPER: progress toward a high-volume manufacturing system, in *SPIE Proceedings Vol. 8680: Alternative Lithographic Technologies V*, Mar 2013
32. S. Jones, SPIE 2020 – ASML EUV and Inspection Update, SemiWiki, IC Knowledge, 4 Apr 2020
33. E. Korczynski, EUV Resists and Stochastic Processes, Semiconductor Manufacturing & Design Community, 4 Mar 2016, http://semimd.com/blog/tag/euv/
34. M. Lapedus, Yield rises with mask protection; multiple sources will likely reduce costs, Semiconductor Engineering, 22 Mar 2021

35. F. Chen, The Uncertain Phase Shifts of EUV Masks, SemiWike, 13 Mar 2020
36. J. van Schoot et al., Improving the resolution of extreme-UV lithography scanners, Jan 2016, https://spie.org/news/6259-improving-the-resolution-of-extreme-uv-lithography-scanners?SSO=1
37. A. Bharti et al., X-ray lithography for nanofabrication: is there a future? Front. Nanotechnol. **4**, Article 835701 (2022). Sec. Nanofabrication
38. P. Singer, Nanoimprint lithography: a contender for 32 nm? in *Semiconductor International*, 1 Aug 2006
39. Canon Inc., Nanoimprint lithography semiconductor manufacturing system that covers diverse applications with simple patterning mechanism, Press Release, 13 Oct 2023
40. ASML, Measuring accuracy, https://www.asml.com/en/technology/lithography-principles/measuring-accuracy
41. S.R. Huisman et al., Level Sensor and Lithographic Apparatus, United States Patent Application 20210072652, 3 Nov 2021
42. J. Yasuda et al., Recent progress and future of electron multi-beam mask writer. Jpn. J. Appl. Phys. **62**(SG), id.SG0803 (2023)
43. M. Lapedus, Battling Fab Cycle Times, Semiconductor Engineering, Feb 2017
44. K. Jeong et al., New yield-aware mask strategies, in *Proceedings of SPIE*, vol. 8081, 2011 SPIE, p. 80810P
45. H.C. Pfeiffer et al., Microlithography world – the history and potential of maskless E-beam lithography, Solid State Technology, Feb 2005, http://sst.pennnet.com/Articles/Article_Display.cfm?Section=ARTCL&ARTICLE_ID=221612&VERSION_NUM=4&p=28
46. K. Suzuki, N. Itabashi, Future prospects for dry etching. Pure Appl. Chem. **68**(5), 1011–1015 (1996)
47. G. Lee, Flash below 20 nm: what is coming and when. Challenges in 3D NAND, Flash Memory Summit 2013
48. D. Pramanik, *Challenges for Intermetal Dielectrics* (Future Fab International, Palo Alto, USA, 1997)
49. Process Integration, Devices, and Structures (PIDS), Tables, ITRS Roadmap, edition 2011
50. V.S. Chang et al., Enabling multiple-VT device scaling for CMOS technology beyond 7 nm node, in *VLSI Symposium 2020*, Paper TC1.1 (2020)
51. D.-G. Park, X. Wang, High-k gate dielectrics for nanoscale CMOS devices: status, challenges. ECS Trans. **28**(2), 39–50 (2010). The Electrical Chemical Society
52. T. Faraz et al., Atomic layer etching: what can we learn from atomic layer deposition? ECS J. Solid State Sci. Technol. **4**(6), N5023–N5032 (2015)
53. B. Mann, Development of thin gate oxides for advanced CMOS applications, in *22nd Annual Microelectronic Engineering Conference*, May 2004
54. Y. Mitani et al., NBTI Mechanism in Ultra-thin Gate Dielectric-Nitrogen-Originated Mechanism in SiON-, International Electron Devices Meeting Technical Digest (2002), pp. 509–512
55. S. Wolf, R.N. Tauber, Silicon processing for the VLSI era, *Process Technology*, vol. 1 (Lattice Press, San Fransisco, USA, 1986)
56. J. Hruska, How combining cobalt and copper could improve chip yields, boost performance, ExtremeTech, 14 May 2014, http://www.extremetech.com/extreme/182386-how-combining-cobalt-and-copper-could-improve-chip-yields-boost-performance
57. S.-H. Yu et al., Selective cobalt deposition on copper surfaces, US Patent 20090269507 A1, 29 Oct 2008
58. B.S. Lim et al., Atomic layer deposition of transition metals. Nat. Mater. **2** (2003), www.nature.com/naturematerials
59. A. Mackus et al., Etching with atomic-level precision: the emerging field of atomic layer etching. Nevac Blad **58**(2), 40–42 (2020)
60. L. Rubin, J. Poate, Ion implantation in silicon technology, The Industrial Physicist, June/July 2003, pp. 12–15
61. R.F.M. Roes et al., Implications of pocket optimisation on analog performance in deep sub-micron CMOS, ESSDERC, digest of technical papers, 1999, pp. 176–179

62. M. Bohr et al., The High-k Solution, IEEE Spectrum, Oct 2007, pp. 23–29
63. Y. Song, Source/drain technologies for the scaling of nanoscale CMOS device, Solid-State Sci. **13**, 294–305 (2013)
64. S. Kesapragada et al., High-k/metal gate stacks in gate first and replacement gate schemes, in *Advanced Semiconductor Manufacturing Conference (ASMC)* (IEEE/SEMI, 2010), pp. 256–259
65. Gate-last and gate-first high-k metal, IMEC Scientific Report 2010, http://www.imec.be/ScientificReport/SR2010/2010/1159059.html
66. B. Moyer, Gate first vs. last, Electron. Eng. J. Posted on 14 Nov 2011, http://www.eejournal.com/archives/articles/20111114-gate/
67. S. Thompson et al., A 90 nm logic technology featuring 50 nm strained silicon channel transistors, 7 layers of Cu interconnects, low-k ILD, and 1 mm SRAM cell, IEEE International Electron Devices Meeting, 2002
68. G. Eneman et al., N+/P and P+/N junctions in strained Si on strain relaxed SiGe buffers: the effect of defect density and layer structure. Mater. Res. Soc. Symp. Proc. **864**, E3.7.1–E3.7.6 (2005). Materials Research Society
69. P.R. Chidambaram, 35% drive current improvement from recessed-SiGe drain extensions on 37 nm gate length PMOS, in *2004 Symposium on VLSI Technology Digest of Technical Papers*, pp. 48–49
70. M. Yang et al., High Performance CMOS Fabricated on Hybrid Substrate With Different Crystal Orientations, Electron Devices Meeting, 2003. IEDM'03 Technical Digest
71. M.-H. Chi, Challenges in Manufacturing FinFET at 20 nm node and beyond (2012), https://www.semanticscholar.org/paper/Challenges-in-Manufacturing-FinFET-at-20-nm-node-Chi/3788a8c0c3641bb5dc1ac8d4a31094ba22aab34f
72. T. Dillinger, Challenges for FinFET extraction, in *IEEE Electronic Design Process Symposium*, 15 Apr 2013
73. D.R. Muralidher et al., Meeting the challenge of multiple threshold voltages in highly scaled undoped FinFETs. IEEE Trans. Electron Devices **60**(3), 1276–1278 (2013)
74. X. Wang et al., Statistical variability and reliability in nanoscale FinFETs, in *Proceedings of the IEEE International Electron Devices Meeting (IEDM'11)*, Washington, DC, Dec 2011, pp. 541–544
75. S. Chaudhuri, N.K. Jha, 3D vs. 2D analysis of FinFET logic gates under process variations, in *Proceedings of the 29th IEEE International Conference on Computer Design (ICCD'11)*, Amherst, Nov 2011, pp. 435–436
76. P. Clarke, Intel's FinFETs are less fin and more triangle, 17 May 2012, http://www.embedded.com/electronics-news/4373195/Intel-FinFETs-shape-revealed
77. J.-H. Lee, Bulk FinFETs: design at 14 nm node and key characteristics, in *Nano Devices and Circuit Techniques for Low-Energy Applications and Energy Harvesting*, ed. by C.M. Kyung (Springer Science+Business Media, Dordrecht, 2016), pp. 33–64. ISBN: 978-94-017-9989-8
78. LexInnova Technologies LLC, FinFET – extending Moore's law, Report, 2015, http://www.wipo.int/export/sites/www/patentscope/en/programs/patent_landscapes/documents/lexinnova_plr_finfet.pdf
79. Rieger, et al., Self-aligned via interconnect using relaxed patterning exposure, US 2014/0015135 A1, 16 Jan 2014
80. D. Fried et al., Comparison study of FinFETs: SOI vs. bulk, performance, manufacturing variability and cost (SOI Industry Consortium, 2011), https://docplayer.net/21500591-Comparison-study-of-finfets-soi-vs-bulk-performance-manufacturing-variability-and-cost.html
81. M. Haond, FDSOI for low power system on chip, 2011, http://semieurope.omnibooksonline.com/2011/semicon_europa/SEMI_TechARENA_presentations/NewMaterial_05_Michel.Haond_STMicroelectronics.pdf
82. A. Majumdar, Undoped-body extremely thin SOI MOSFETs with back gates. IEEE Trans. Electron Devices **56**(10), 2270–2276 (2009)
83. B. Prince, *Vertical 3D Memory Technologies* (Wiley, 2014), ISBN: 978-1-118-76051-2

84. F.N.A. Hassan Agha et al., Review of nanosheet transistors technology, Tikrit J. Eng. Sci. **28**, 40–48 (2021)
85. P. Ye et al., The last silicon transistor: nanosheet devices could be the final evolutionary step for Moore's Law. IEEE Spectr. **56**(8), 30–35 (2019)
86. M. Lapedus, Transistors reach tipping point at 3 nm, Semiconductor Engineering, 23 Feb 2022
87. P. Clarke, Intel, TSMC both set to report on stacked CFETs at IEDM, eeNews, 3 Oct 2023, https://www.eenewseurope.com/en/intel-tsmc-both-set-to-report-on-stacked-cfets-at-iedm/
88. M. Lapedus, New transistor structures at 3/2 nm: gate-all-around FETs will replace finFETs, but the transition will be costly and difficult, 25 Jan 2021
89. Y. Park et al., Ultralow dielectric constant SiCOH films by plasma enhanced chemical vapor deposition of decamethylcyclopentasiloxane and tetrakis (trimethylsilyloxy) silane precursors, in *Thin Solid Films*, vol. 727 (Elsevier, Amsterdam, The Netherlands, 2021)
90. International Roadmap for Devices and Systems (IRDS™) 2022 Update: More Moore, https://irds.ieee.org/images/files/pdf/2022/2022IRDS_MM.pdf
91. X. Kang et al., Cu/Airgap integration on 90 nm Cu BEOL process platform, in *2012 IEEE 11th International Conference on Solid-State and Integrated Circuit Technology (ICSICT)*
92. D. James, IEDM 2014– Monday was FinFET Day, 18 Dec 2014, http://chipworksrealchips.blogspot.com/2014/12/iedm-monday-was-finfet-day.html
93. D. Kanter, Intel 4 Process Scales Logic with Design, Materials, and EUV, Realworldtechnologies. https://www.realworldtech.com/intel-4/
94. S. Jones, Intel 4 Deep Dive, SemiWiki, 13 June 2022, https://semiwiki.com/semiconductor-manufacturers/intel/314047-intel-4-presented-at-vlsi/
95. M.H. van der Veen et al., Barrier/liner stacks for scaling the Cu interconnect metallization, in *2016 IEEE International Interconnect Technology Conference/Advanced Metallization Conference (IITC/AMC)*
96. W. Steinhogl et al., Size-dependent resistivity of metallic wires in mesoscopic range. Phys. Rev. B **66**, 075414 (2002)
97. H.-W. Kim, Recent trends in copper metallization. Electronics (2022), https://www.mdpi.com/2079-9292/11/18/2914

CMOS Circuit, Layout and Library Design

4

4.1 Introduction

Although it was already invented in the sixties, it took until the mid-eighties before CMOS became the leading technology for VLSI circuits. Prior to that time, only a few circuits were designed in CMOS. These early designs were generally limited to analogue circuits and digital circuits that dissipated little power. Examples include chips for calculators, watches and remote controls. CMOS offers both n-type and p-type MOS transistors. Initially, this meant that CMOS circuits were costlier than their nMOS equivalents.

The majority carriers in pMOS and nMOS transistors are holes and electrons, respectively. The zero-field mobility of holes is about three times less than electron mobility. This makes pMOS circuits significantly slower than nMOS circuits of equal chip area. The continuous drive for increased integrated circuit performance therefore led to the early disappearance of pMOS technologies. The demand for higher packing densities and performance led to an increase in the complexity of nMOS processes.

In particular, the quest for a lower τD product (power delay product) necessitated the availability of several different transistor threshold voltages in a single nMOS process. These included a few enhancement threshold voltages ($V_T > 0$) and different depletion threshold voltages ($V_T < 0$). Even threshold voltages of zero volts had to be available. These threshold voltages were provided at the cost of additional masks and extra processing steps, which rapidly elevated the complexity of nMOS processes to about the level of CMOS processes. A few advantages afforded by CMOS processes therefore led to their domination of the MOS IC world.

Modern manufacturing processes make it possible to integrate increasingly complex circuits and even complete systems on a single chip. The resulting number of transistors per logic chip may run into hundreds of millions up to a hundred billion. The associated power dissipation can easily exceed the critical 1 W

H. Veendrick, *Nanometer CMOS ICs*,
https://doi.org/10.1007/978-3-031-64249-4_4

maximum limit for cheap small plastic IC packages. Circuits that are manufactured in CMOS processes generally consume less than one-tenth of the power dissipated by an nMOS equivalent. Moreover, CMOS circuits have better noise margins. These advantages have led to the use of CMOS for the integration of most modern VLSI circuits. These include memories, digital signal processors, microprocessors, speech synthesisers, chips for consumer, data communication, automotive, medical and aviation applications, as well as complete Systems-On-Chip (SOC).

The various CMOS processes and their characteristic properties are extensively treated in Sect. 3.9. This chapter starts with a discussion on basic nMOS circuits to be able to understand CMOS circuit properties more easily. Basic design principles and problems associated with CMOS are subjects of this chapter. Several different types of both static and dynamic CMOS circuits are discussed. Related reliability issues, such as latch-up, are discussed in Chap. 9, together with other topics that improve the IC's robustness. The chapter ends with a section on CMOS layout design. A layout design method is illustrated by means of an example.

Finally, it should be noted that many examples are based on an n-well CMOS process. Initially, this process was chosen because of its compatibility with the conventional nMOS process. In addition, many dynamic CMOS circuits are 'nMOS-mostly'. Currently, most processes are twin-well CMOS processes, in which the nMOS and pMOS transistors can both be realised with optimum performance.

4.2 The Basic nMOS Inverter

4.2.1 Introduction

Generally, the electrical properties of a static nMOS circuit are completely determined by its DC behaviour and transient response. These will be explained with the aid of one of the most elementary MOS circuits, i.e., the inverter. In the following we treat the nMOS-only circuits as if they were designed in a 45 nm technology node, along with their supply and threshold voltages.

Figure 4.1 shows schematics of an inverter and its different types of *'load elements'*.

The inverter's DC behaviour and transient response are discussed for its different types of load elements. The discussions are based on formulae (1.18) which express the current in a transistor as follows:

$$\text{Linear region:} \quad I_{ds} = \beta(V_{gs} - V_T - V_{ds}/2)V_{ds} \quad (V_{ds} < V_{gs} - V_T)$$

$$\text{Saturation region:} \quad I_{ds} = \beta/2(V_{gs} - V_T)^2 \quad (V_{ds} \geq V_{gs} - V_T)$$

$$\text{Where:} \quad V_T = V_x + k\sqrt{V_{sb} + 2\phi_f}$$

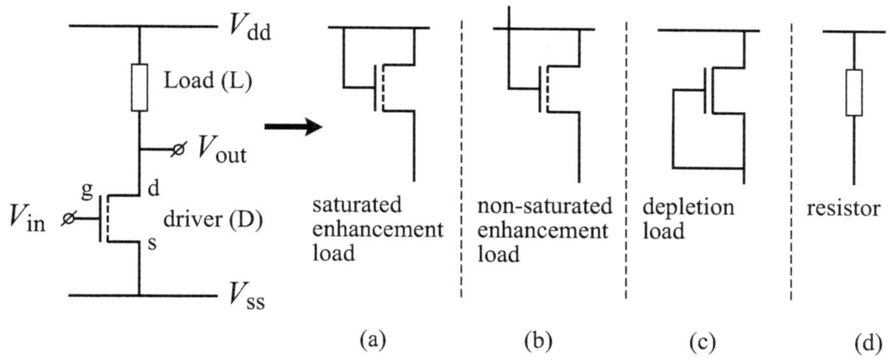

Fig. 4.1 An inverter and its different types of load elements

Two criteria are important when determining the dimensions of transistors in MOS logic gates:

- The location of the operating points. These are the output voltages V_L and V_H, which correspond to the logic values '0' and '1', respectively. Output voltage V_L, for example, must be a *'noise margin'* less than the threshold voltage V_{T_D} of the n-type enhancement driver transistor. The noise margin ensures that subsequent logic gates always interpret V_L correctly. V_{T_D} is about 0.3 V and a noise margin of about 0.15 V is normally used. This implies that $V_L \leq 0.15$ V in nMOS circuit design.
- The transient response. This implicitly refers to the rise and fall times associated with changes in the output's logic levels.

In the next sections, these criteria are discussed for the four types of inverters shown in Fig. 4.1.

4.2.2 The DC Behaviour

The DC behaviours of inverters with different types of load elements are explained separately below with the aid of Fig. 4.2. This figure shows the *'driver transistor'* characteristic $I_{ds} = f(V_{ds})|_{V_{gs}} = V_H$ together with the *'load lines'* of the different load elements in Fig. 4.1. The shapes of the load lines are characteristic of the respective load elements.

The output voltage of an inverter is 'low' ($V_{out} = V_L$) if its input voltage is 'high' ($V_{in} = V_H$) and vice versa. The output low level values corresponding to the different load elements are determined by the intersection of the driver characteristic and the relevant load line. These values are indicated by V_{L_a}, V_{L_b}, etc. in Fig. 4.2. The indicated positions are chosen for clarity and are not typical for the various load elements. The point of intersection between a load line and the driver characteristic

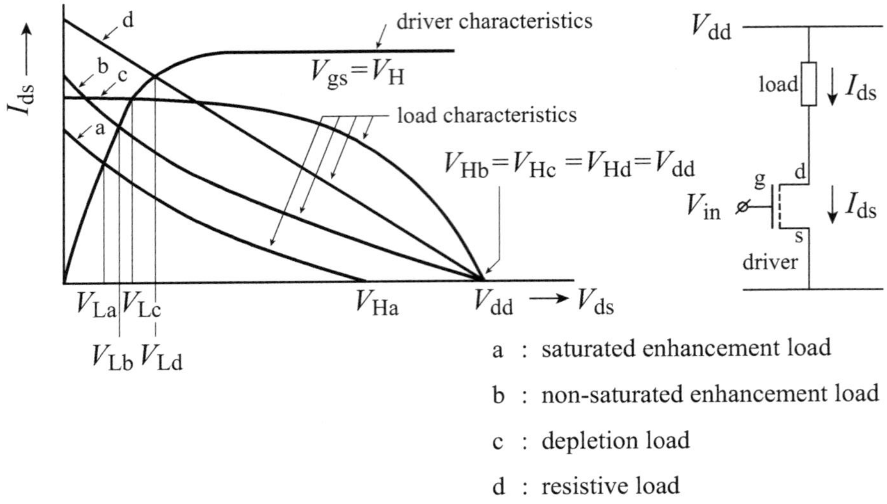

Fig. 4.2 Inverter characteristics for different types of load elements

is in fact chosen by the designer. For inverters that use transistors as load elements, this point is determined by the *'aspect ratio'* A, which is expressed as follows:

$$A = \frac{(\frac{W}{L})_D}{(\frac{W}{L})_L}$$

Achieving a correct 'low' level in static nMOS logic clearly requires a minimum ratio between the driver and load transistor sizes. This type of circuit is therefore called *ratioed logic*.

Saturated Enhancement Load Transistor

The DC behaviour of an inverter with a *saturated enhancement load* transistor is explained with the aid of Fig. 4.3, which shows a schematic diagram of the inverter. The load line and four driver characteristics, for different values of V_{in}, are also shown.

The minimum drain-source voltage of the load transistor is a threshold voltage, i.e., as $V_{ds} = V_{gs}$, $V_{ds_L} > V_{gs_L} - V_{T_L}$ always applies. Therefore, the load transistor always operates in the saturation region. The application of formulae (1.18) yields the following expression for the current in the load transistor:

$$I_{ds_L} = \frac{\beta_L}{2}(V_{ds_L} - V_{T_L})^2$$

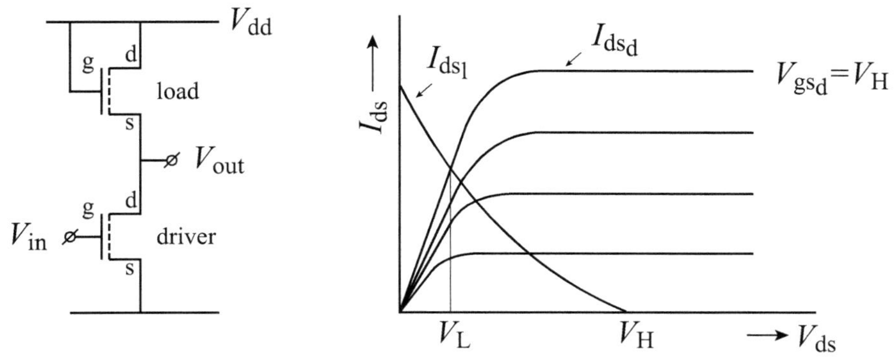

Fig. 4.3 An inverter with a saturated enhancement load transistor

The DC operation of an inverter with a saturated enhancement load transistor is described as follows:

- If $V_{in} = V_L < V_{T_D}$, then the driver transistor is 'off' and $I_{ds_D} = I_{ds_L} = 0$.
  According to the above expression for I_{ds_L}, the output voltage is then: $V_{out} = V_H = V_{dd} - -V_{T_L}$.
- If $V_{in} = V_H \gg V_{T_D}$ then $V_{out} = V_L$. The driver current I_{ds_D} and the load transistor current I_{ds_L} will then be equal:

$$\mathrm{I}_{\mathrm{ds}_\mathrm{D}} = \mathrm{I}_{\mathrm{ds}_\mathrm{L}}$$

$$\Rightarrow \underbrace{\beta_D \cdot \left(\mathrm{V_H} - \mathrm{V}_{\mathrm{T_D}} - \frac{\mathrm{V_L}}{2}\right) \cdot \mathrm{V_L}}_{\text{driver transistor in linear region}} = \underbrace{\frac{\beta_L}{2}\left((\mathrm{V_{dd}} - \mathrm{V_L}) - \mathrm{V}_{\mathrm{T_L}}\right)^2}_{\text{load transistor always saturated}}$$

Assuming $V_L \ll V_{dd}$ and $V_L/2 \ll V_H - V_{T_D}$ yields:

$$\left(\frac{W}{L}\right)_D \cdot (V_H - V_{T_D})V_L = \left(\frac{W}{L}\right)_L \cdot \frac{1}{2} \cdot (V_{dd} - V_{T_L})^2$$

With $V_{dd} - V_{T_L} = V_H$, this reduces to the following expression for the aspect ratio A of this inverter:

$$A = \frac{\left(\frac{W}{L}\right)_D}{\left(\frac{W}{L}\right)_L} \geq \frac{V_H^2}{2(V_H - V_{T_D})V_L} \tag{4.1}$$

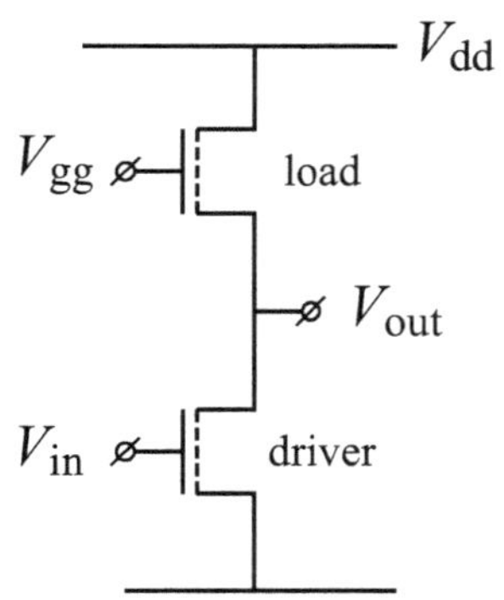

Fig. 4.4 An inverter with a non-saturated enhancement load transistor

The use of a saturated enhancement load transistor is disadvantaged by the associated *'threshold loss'*, which produces a high level V_H, and this is only $V_{dd} - V_{T_L}$ rather than V_{dd}. The corresponding relatively low input voltage applied to a subsequent logic gate results in a lower speed. The use of a non-saturated enhancement or depletion load transistor overcomes this problem and produces a V_H equal to V_{dd}.

The Non-saturated Enhancement Load Transistor

An inverter with a *non-saturated enhancement load* transistor is illustrated in Fig. 4.4.

The gate of the load transistor is connected to an extra supply voltage V_{gg} instead of the supply voltage V_{dd}. The extra supply voltage is large enough to ensure that the load transistor always operates in the non-saturated region, i.e., $V_{gg} > V_{dd} + V_{T_L}$.

The DC operation of the above inverter is described as follows:

- $V_{in} = V_L < V_{T_D} \Rightarrow I_{ds_L} = 0\,\text{mA}$ and $V_{out} = V_H = V_{dd}$.
- $V_{in} = V_H \gg V_{T_D} \Rightarrow V_{out} = V_L$.

The driver now operates in the linear region. The driver and load transistor currents are equal:

$$
\begin{aligned}
I_{ds_D} &= \left(\frac{W}{L}\right)_D \cdot \beta_\square \left(V_H - V_{T_D} - \frac{V_L}{2}\right) V_L \\
&= \left(\frac{W}{L}\right)_L \cdot \beta_\square \left(V_{gg} - V_L - V_{T_L} - \frac{V_{dd} - V_L}{2}\right)(V_{dd} - V_L) \\
&= I_{ds_L}
\end{aligned}
$$

Assuming $V_L \ll V_{dd}$, $\frac{V_L}{2} \ll V_H - V_{T_D}$ and $V_{gg} - V_{T_L} \gg V_L$ yields the following expression for the inverter's aspect ratio A:

$$A = \frac{\left(\frac{W}{L}\right)_D}{\left(\frac{W}{L}\right)_L} = \frac{\left(V_{gg} - V_{T_L} - \frac{V_{dd}}{2}\right) \cdot V_{dd}}{(V_H - V_{T_D}) \cdot V_L}$$

Since $V_H - V_{T_D} < V_{dd}$, the aspect ratio A is expressed as follows:

$$A = \frac{\left(\frac{W}{L}\right)_D}{\left(\frac{W}{L}\right)_L} \geq \frac{V_{gg} - V_{T_L} - \frac{V_{dd}}{2}}{V_L} \tag{4.2}$$

The use of a non-saturated enhancement transistor as load element has the following advantages:

- High $V_H (= V_{dd})$;
- Large noise margin;
- Fast logic.

The most significant disadvantage is the extra supply voltage required V_{gg} ($V_{gg} \geq V_{dd} + V_{T_L}$), which may necessitate an extra pin on the chip package. Alternatively, V_{gg} can be electronically generated on the chip. This results in a *'bootstrapped load'* element, as shown in Fig. 4.5.

The DC operation of the bootstrapped inverter is explained as follows:

- If $V_{in} = V_H$, then $V_{out} = V_L$ and $V_1 = V_{dd} - V_{T_A}$. The MOS *'bootstrap' capacitance* C therefore charges.

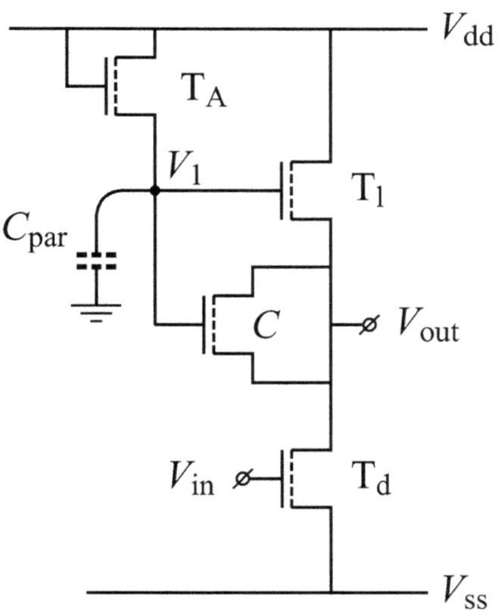

Fig. 4.5 The bootstrapped inverter

- When V_{in} switches from V_H to V_L, then V_{out} increases by ΔV_{out} and V_1 increases by ΔV_1. The magnitude of ΔV_1 is determined by the values of the bootstrap capacitance C and the parasitic capacitance C_{par} such that:

$$\Delta V_1 = \frac{C}{C + C_{par}} \cdot \Delta V_{out}$$

This means that V_1 immediately passes the $V_{dd}-V_{T_A}$ level and transistor T_A therefore no longer conducts. The voltage V_1 can then further increase to a voltage greater than V_{dd}. The maximum value of V_1 is determined by the capacitance ratio:

$$a = \frac{C}{C + C_{par}}$$

The value of a required to produce a 'high' output voltage is: $V_H = V_{dd}$ and is derived as follows:

$V_H = V_{dd} \text{when} V_1 \geq V_{dd} + V_{T_L}$.
$\Delta V_1 = a \cdot \Delta V_{out}$ and $V_1 = V_{dd} - V_{T_A} + a \cdot \Delta V_{out}$.

The load transistor T_L must remain in the linear operating region. The following equation therefore applies:

$V_1 - V_{T_L} > V_{dd}$
$\Rightarrow V_{dd} - V_{T_A} - V_{T_L} + a \cdot \Delta V_{out} > V_{dd}$
$\Rightarrow \Delta V_{out} > \frac{V_{T_A}+V_{T_L}}{a}$

The output high level must be equal to the supply voltage, i.e., $V_{out} = V_H = V_{dd}$. Therefore, $\Delta V_{out} = V_{dd} - V_L$. Assuming $V_{T_A} \approx V_{T_L}$ yields the following expression for a:

$$a > \frac{2V_{T_L}}{V_{dd} - V_L} \tag{4.3}$$

- If $V_{in} = V_H$, then $V_{out} = V_L$ and the gate voltage of the load transistor T_L is $V_{dd} - V_{T_A} \approx V_{dd} - V_{T_L}$. Load transistor T_L therefore operates in the saturation region when $V_{out} = V_L$. The aspect ratio A of the bootstrapped inverter is therefore identical to that given in equation (4.1) for the inverter with a saturated enhancement load transistor.

The bootstrapped inverter has the following advantages:

1. There is no threshold loss when the bootstrap capacitance C is correctly dimensioned.

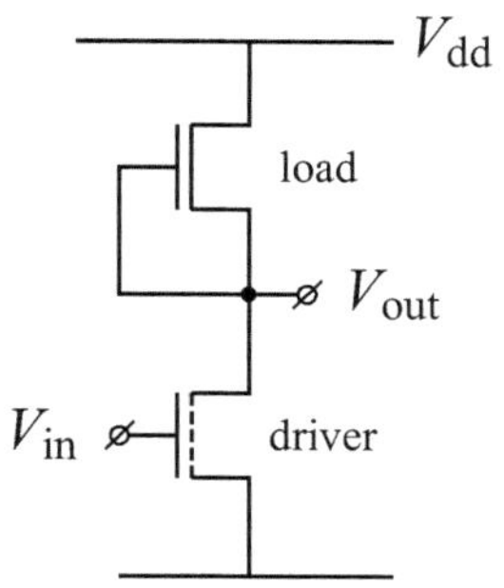

Fig. 4.6 An inverter with a depletion load transistor

2. There is no extra supply voltage required, because the voltage V_l is pumped to more than a threshold voltage above V_{dd}.
3. This basic bootstrap mechanism is also called a *charge-pump*, which is used in many E(E)PROMs and flash memories to generate the much higher programming and/or erasing voltages. To achieve such high voltages (≥ 10 V), several of these charge pumps are put in series.

The Depletion Load Transistor

The manufacture of depletion transistors requires an extra mask (DI) and additional processing steps. There are, however, considerable advantages associated with the use of a depletion transistor as load element. These include the following:

- The output high level equals V_{dd}, i.e., $V_H = V_{dd}$;
- There is no extra supply voltage required;
- Circuit complexity is minimal and bootstrapping is unnecessary;
- Noise margins are high.

For these reasons, before the move to CMOS, most nMOS processes were '*E/D technologies*' and contain both enhancement and depletion transistors. Some manufacturers, today, even include depletion transistors in their CMOS technologies.

Figure 4.6 shows an inverter with a *depletion load* transistor.

The DC operation of the inverter with a depletion load transistor is described as follows:

- The depletion load transistor has a negative threshold voltage which was usually between -1 and -3 V. Therefore, $V_{out} = V_H = V_{dd}$ when $V_{in} = V_L < V_{T_D}$.
- When $V_{in} = V_H > V_{T_D}$, then $V_{out} = V_L$ and $V_{gs_L} (= 0\,\text{V}) < V_{ds_L} + V_{T_L}$. In this case, the load transistor operates in the saturation region while the driver transistor operates in the triode region. Equating the currents in the load and driver transistors yields:

$$I_{ds_D} = I_{ds_L}$$

$$\Rightarrow \left(\frac{W}{L}\right)_{\rm D} \cdot \beta_\Box \cdot \left(V_{\rm H} - V_{\rm T_D} - \frac{V_{\rm L}}{2}\right) \cdot V_{\rm L} = \left(\frac{W}{L}\right)_{\rm L} \cdot \frac{\beta_\Box}{2} \cdot V_{\rm T_L}^2$$

If $\frac{V_{\rm L}}{2} \ll V_{\rm H} - V_{\rm T_D}$, then the aspect ratio A of the depletion-load inverter can be expressed as follows:

$$A = \frac{\left(\frac{W}{L}\right)_{\rm D}}{\left(\frac{W}{L}\right)_{\rm L}} \geq \frac{V_{\rm T_L}^2}{2V_{\rm L} \cdot (V_{\rm H} - V_{\rm T_D})} \tag{4.4}$$

The Resistive Load

VLSI circuits may consist of tens of millions to more than a billion logic gates which may dissipate between 0.1 and 20 nW each. A supply voltage of 1.2 V therefore requires a *resistive load* of several tens of MΩ per logic gate. Both diffusion and polysilicon have a *sheet resistance* of about 500 $\Omega/\Box$ in a 40 nm CMOS process. Realisation of a 10 MΩ resistance with a 50 nm wide polysilicon track therefore requires a length of 1 mm. At the cost of extra processing complexity; however, large resistances can be realised on small chip areas. For random-access memories (RAMs), the disadvantages of complex processing were justified by very large production quantities. The addition of a second polysilicon layer with very high resistivity in static RAM processes facilitated the realisation of memory cells that were considerably smaller than the *full-CMOS* cells. The use of resistive load elements (Fig. 4.7) was therefore mainly limited to application in static memories and was not normally encountered in VLSI circuits. Because this circuit's operation resembles that of an inverter with a depletion load transistor, we will no longer focus on this type of load.

4.2.3 Comparison of the Different nMOS Inverters

NMOS inverters with different load elements are now compared.

Adopting a 1 pF load capacitance, a *circuit analysis* program was used to simulate the charging and discharging characteristics that correspond to these load

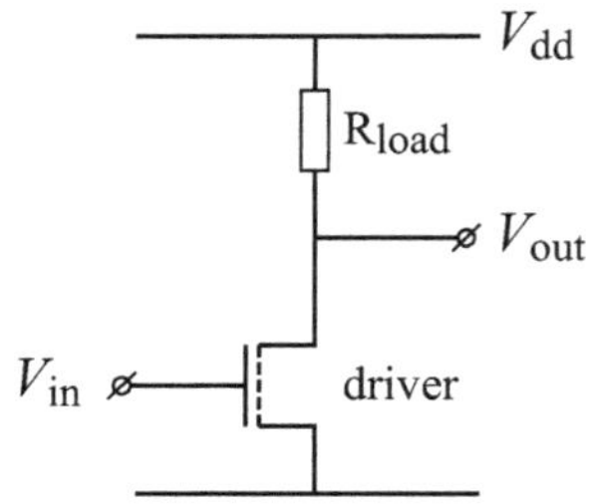

Fig. 4.7 An inverter with a resistive load

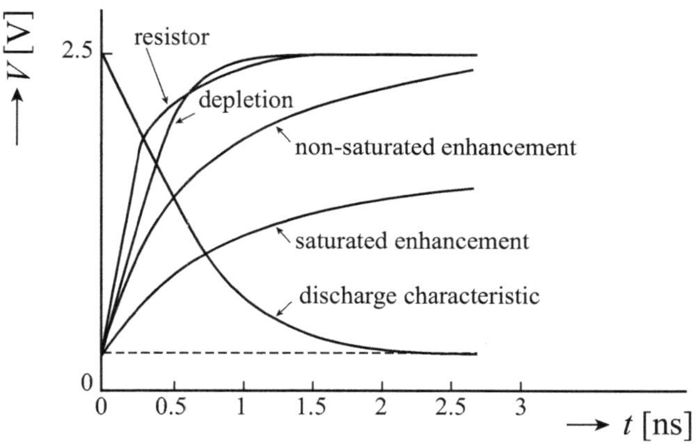

Fig. 4.8 Charging characteristics of nMOS inverters with different types of load, identical load capacitances and the same initial current I_0

and driver transistors, respectively. The charging characteristic associated with the load resistance was also simulated. The results are shown in Fig. 4.8.

The performance of logic gates, built with the previously presented different types of load elements, was quite different. Particularly the saturated enhancement load, because of the relatively large voltage drop at high level, made this logic much slower than logic with a non-saturated enhancement load or with the depletion load (Fig. 4.8). Logic with the non-saturated enhancement load required either an additional supply voltage to keep the load always in its saturation region, or it required bootstrapping techniques to do so. Both solutions were not so well appreciated and have made E/D nMOS logic (so, logic with the depletion load) the most popular technique to realise VLSI chips, until the mid-eighties when CMOS became the most dominant IC technology. For this reason, the next paragraph uses the depletion load for further evaluation of nMOS logic gates.

4.2.4 Transforming a Logic Function Into an nMOS Transistor Circuit

An inverter is transformed into a logic gate by replacing the driver transistor by a combination of MOS transistors. The combination may comprise series and/or parallel transistors. Each transistor gate is controlled by a logic signal. A complex logic function can therefore be implemented in a single logic gate with an associated propagation delay. The following transformation rules apply:

1. An *AND* function is realised by a *series connection* of transistors.
2. An *OR* function is realised by a *parallel connection* of transistors.

Because logic gates are an adaptation of the basic inverter, the output signal is always the inverse of the function that is derived when the transistors in the driver section are interpreted according to the above rules. In fact, implementations always comprise NAND, NOR or AND-OR-NOT functions.

Example A 'full adder' is described by the following logic functions (see also Sect. 7.3.5):

$$S = x\,\bar{y}\,\bar{z} + \bar{x}\,\bar{y}\,z + \bar{x}\,y\,\bar{z} + x\,y\,z$$
$$C_o = x\,y + x\,z + y\,z$$

Symbols x and y represent two bits which must be added. Symbol z represents the 'carry-in'. S represents the binary sum of x, y and z while C_o represents the 'carry-out'.

The logic function S can also be written as:

$$S = x\,(y\,z + \bar{y}\,\bar{z}) + \bar{x}\,(y\,\bar{z} + \bar{y}\,z)$$

This function corresponds to the implementation in Fig. 4.9, which realises the inverse ($\overline{S}$) of the sum function.

Figure 4.10 shows a realisation of the inverse of the carry function.

An nMOS transistor's gain factor β equals $\beta_\square \cdot \frac{W}{L}$. The gain factor β_{total} of n transistors connected in series is expressed as follows:

$$\beta_{\text{total}} = \left(\frac{1}{\beta_1} + \frac{1}{\beta_2} + \ldots \frac{1}{\beta_n}\right)^{-1}$$

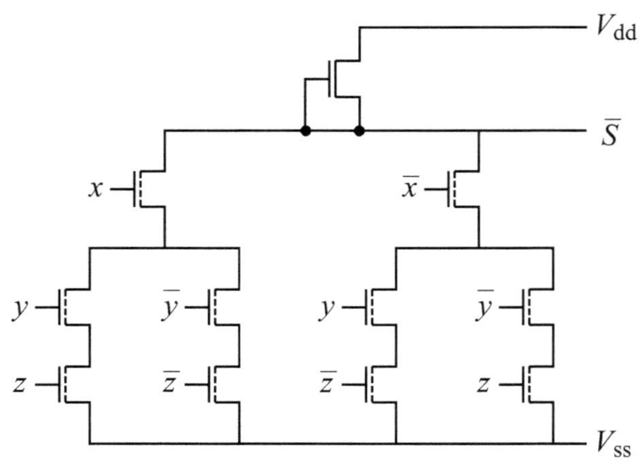

Fig. 4.9 An implementation of the function $\overline{S}$

Fig. 4.10 Implementation of the full adder inverse 'carry-out' function

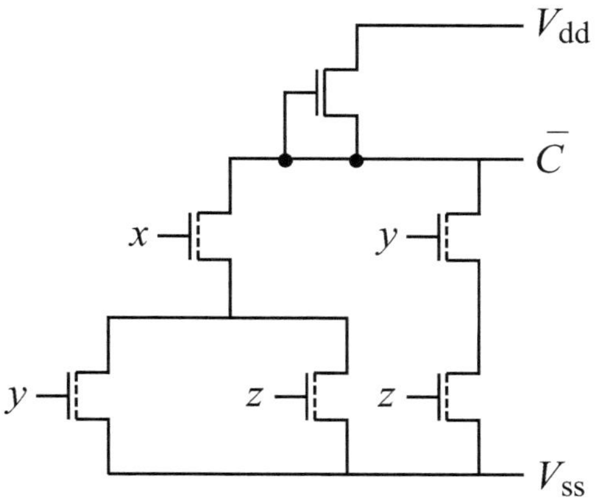

Fig. 4.11 NAND gate with n inputs and thus n transistors in series

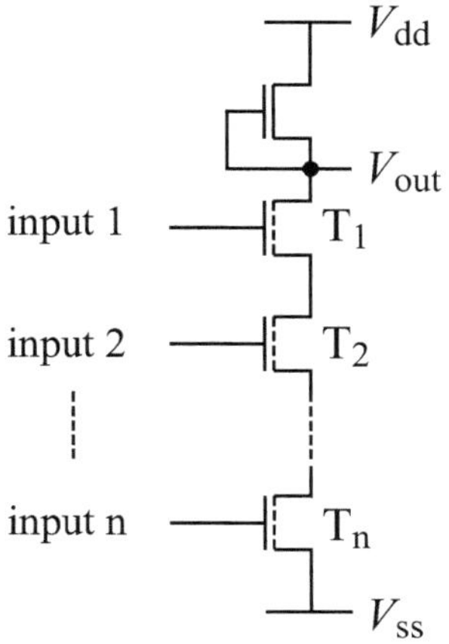

If all the transistors have equal dimensions, then:

$$\beta_{\text{total}} = \beta/n$$

The discharge time constant associated with these n transistors is then directly proportional to n. In fact, the speed of a logic gate is largely determined by the number of transistors that are connected in series in the driver section. It is thus generally advisable to keep this number to a minimum. Figure 4.11, for example, shows a NAND gate with n driver transistors in series. The effective $\left(\frac{W}{L}\right)$ ratio of these n transistors is expressed as follows:

$$\left(\frac{W}{L}\right)_{\text{total}} = \frac{1}{\left(\frac{W}{L}\right)_1^{-1} + \left(\frac{W}{L}\right)_2^{-1} + \ldots + \left(\frac{W}{L}\right)_n^{-1}} \tag{4.5}$$

The $\left(\frac{W}{L}\right)$ aspect ratio of the driver transistor in an inverter can be calculated using the formulae in Sect. 4.2.2. For a NAND gate with n inputs, the inverter's driver transistor (D) must be replaced by n transistors in series. The NAND gate will be as fast as the inverter if its transistors each have an aspect ratio $n \cdot (\frac{W_i}{L_i})$, where W_i and L_i are the width and length, respectively, of the inverter's driver transistor.

The number of parallel sections in a logic gate is also a critical implementation factor. The circuit area and hence the parasitic capacitances associated with the logic gate increase with the number of parallel sections. This causes an increase in the gate's propagation delay.

This section presented an overview on the electrical design of nMOS circuits and the creation of basic nMOS logic gates. A major disadvantage of nMOS logic is the associated power consumption. Each logic gate with a low level at its output consumes DC power. Therefore, even when a large logic nMOS chip has no signal transitions, there is a large DC power consumption. CMOS circuits, which require more complex technologies than nMOS circuits, do not consume DC power when there is no activity. This is the most important reason for the domination of CMOS circuits in the integrated circuit market.

4.3 Electrical Design of CMOS Circuits

4.3.1 Introduction

The acronym CMOS stands for *C*omplementary *M*etal *O*xide *S*emiconductor'. The word 'complementary' indicates that transistors of different types can be manufactured in CMOS processes. The types are n-channel and p-channel, or 'nMOS' and 'pMOS'. The nMOS transistor and its operation have been extensively treated before. The pMOS transistor has been briefly mentioned. Where necessary, additional details about its operation are provided in this chapter. The nMOS and pMOS transistors used in CMOS processes are both of the enhancement type. Section 1.7 reveals that the threshold voltage of the nMOS transistor is therefore positive while that of the pMOS transistor is negative. This is shown in Fig. 4.12.

The formulae discussed in Sect. 1.5, which describe the back-bias effect on the threshold voltages of nMOS and pMOS transistors, are as follows:

$$V_{T_n} = V_{X_n} + K_n\sqrt{V_{sb} + 2\phi_f} \qquad \text{(enhancement type: } V_{X_n} > 0,\; K_n > 0)$$

$$V_{T_p} = V_{X_p} + K_p\sqrt{V_{ws} + 2\,|\phi_f|} \qquad \text{(enhancement type: } V_{X_p} < 0,\; K_p < 0)$$

In the CMOS process that is considered in this section, the pMOS transistor is integrated in an n-well. Voltage V_{ws} in the above expression for the threshold voltage V_{T_p} of a pMOS transistor represents the voltage between the n-well and the source of the transistor.

The above expressions and figure show that the operation of the pMOS transistor is the exact complement of the nMOS transistor's operation. The electrical operation

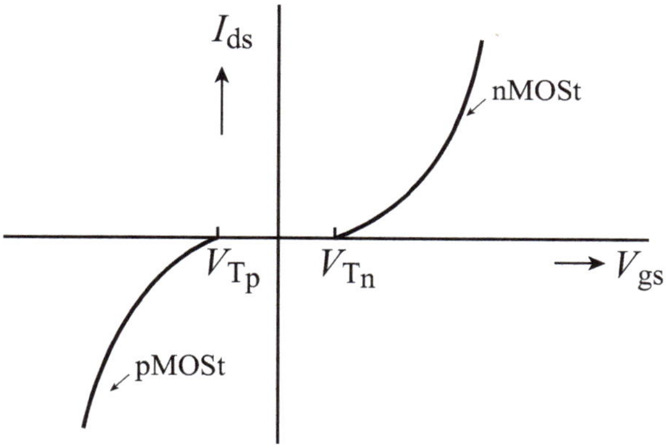

Fig. 4.12 The $I_{ds} = f(V_{gs})$ characteristics of nMOS ($V_{T_n} > 0$) and pMOS ($V_{T_p} < 0$) enhancement transistors

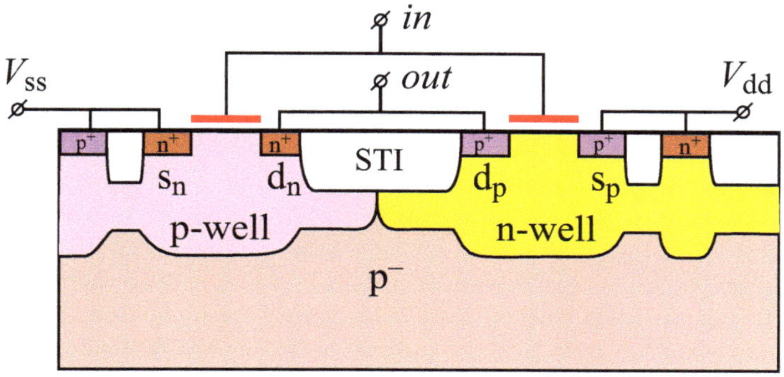

Fig. 4.13 Transistor connections for a CMOS inverter

of the nMOS and pMOS transistors can be summarised as follows: the pMOS transistor's behaviour with respect to the supply voltage is identical to the nMOS transistor's behaviour with respect to ground and vice versa.

4.3.2 The CMOS Inverter

A basic *CMOS inverter* consists of an nMOS transistor and a pMOS transistor connected as shown in Fig. 4.13. The n-well serves as a substrate for the pMOS transistor. It is formed by the diffusion or ion implantation techniques discussed in Chap. 3.

Figure 4.14 shows the circuit diagram of a CMOS inverter.

The influence of substrate voltage on the threshold voltage of a transistor is discussed in Sect. 1.5. This back-gate effect is proportional to the square root of the channel dope of the transistor and is represented by the K-factor. The K-factor of the pMOS transistors in a retrograde twin well process can be of the same order as that of the nMOS transistors.

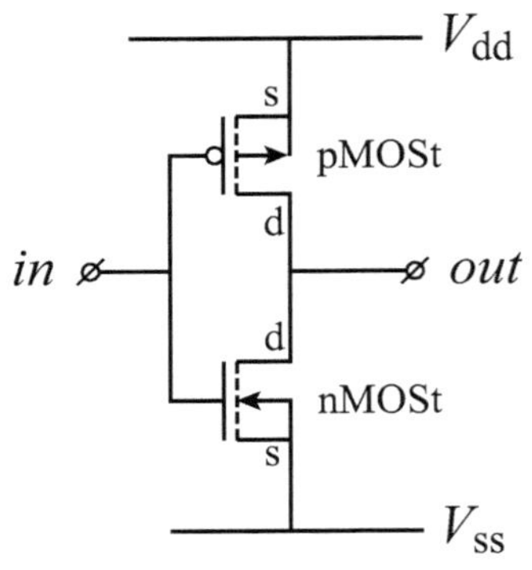

Fig. 4.14 Circuit diagram of a CMOS inverter

In technologies with channel lengths above 100 nm, the performance of the pMOS transistor is hampered by the mobility of holes, which is approximately two to four times lower than the mobility of electrons, depending on the amount of dopants in the channel. This leads to the following relationship between the effective $\beta_\square$ factors of nMOS and pMOS transistors (including second order effects):

$$\beta_{\square_n} \approx 2.5 \cdot \beta_{\square_p}$$

For equal absolute threshold voltage values, the pMOS transistor in the layout of an inverter with symmetrical behaviour will therefore be about 2.5 times the size of the nMOS transistor. This size ratio is expressed in the 'aspect ratio' A of the CMOS inverter as follows:

$$A = \frac{(\frac{W}{L})_p}{(\frac{W}{L})_n} = \frac{A_{T_p}}{A_{T_n}} = \frac{\beta_{\square_n}}{\beta_{\square_p}} \tag{4.6}$$

In many processes, all polysilicon areas and the sources and drains of nMOS transistors in an n-well CMOS process are n^+ areas. The sources and drains of the pMOS transistors are p^+ areas. It should be clear from Fig. 4.13 that p^+ and n^+ areas may *never* be directly connected to each other, not even in a stick diagram. Such an interconnection would produce a pn diode which only conducts in one direction. Connections between n^+ and p^+ areas must therefore always be made in metal. Many CMOS processes currently include *double-flavoured polysilicon*, or *dual-dope polysilicon*: n^+ polysilicon gate for the nMOS transistor and p^+ polysilicon for the pMOS transistor.

In advanced nanometer CMOS technologies, the difference between the effective $\beta_\square$ factors of nMOS and pMOS transistors is reducing to between $A = 1$ and $A = 2$, due to mobility reduction effects, such as discussed in Chap. 2, and mobility enhancement techniques, such as discussed in Chap. 3. In this textbook, we will use the value $A = 1.5$ in examples and exercises, unless otherwise stated.

The Electrical Behaviour of the CMOS Inverter

An nMOS inverter comprises a driver and a load transistor. However, the pMOS and nMOS transistors in a CMOS inverter are both driver transistors. Figure 4.15 shows a CMOS inverter and its transfer characteristic $V_{out} = f(V_{in})$. The gates of the pMOS (T_p) and nMOS (T_n) transistors are connected to form the inverter input. It is important to remember that $V_{T_p} < 0$ and $V_{T_n} > 0$.

The transfer characteristic is explained as follows:

T_n is 'off' and T_p is 'on' for $V_{in} < V_{T_n}$. The output voltage V_{out} then equals V_{dd}.
T_p is 'off' and T_n is 'on' for $V_{in} > V_{dd} + V_{T_p}$. V_{out} then equals V_{ss}.

In both of the above stable situations, one transistor is always 'off' and no DC current can flow from supply to ground. The current characteristic $I = f(V_{in})$ in Fig. 4.15b reflects this fact. The absence of DC current in the two stable situations is the most important advantage of CMOS when compared with nMOS. A current (often referred to as *short-circuit current*) only flows from supply to ground during an input voltage transition, for which the following conditions apply:

$$V_{T_n} < V_{in} < V_{dd} + V_{T_p}$$

Figure 4.15b shows the trajectory of the transient current associated with the input voltage transition from V_{ss} to V_{dd}. The areas where T_n and T_p operate in their respective saturation and triode regions are indicated in Fig. 4.15a.

The saturation areas are described for the two transistors as follows:

T_n : $V_{ds}|_{T_n} > V_{gs} - V_{T_n}$ i.e., $V_{out} > V_{in} - V_{T_n}$. This is the area above the dotted line $V_{out} = V_{in} - V_{T_n}$ in the transfer characteristic.
T_p: $V_{ds}|_{T_p} < V_{gs} - V_{T_p}$ i.e., $V_{out} - V_{dd} < V_{in} - V_{dd} - V_{T_p}$. This is the area below the dotted line $V_{out} = V_{in} - V_{T_p}$ in the transfer characteristic.

Figure 4.15 shows that the transistors in an inverter are both saturated during transitions between logic levels. Theoretically, their output impedances are then infinite. Application of Ohm's Law reveals that a finite current should then cause an infinitely large change in the output voltage. In practice, the output impedances are always finite and the maximum voltage change is limited. However, the transfer characteristic of the CMOS inverter is still very steep.

It must be noted that Fig. 4.15 is drawn on the basis of the assumptions that $V_{T_n} = -V_{T_p}$ and $V_{dd} > V_{T_n} + |V_{T_p}|$. The reader should verify that the transfer characteristic of the inverter displays hysteresis when $V_{T_n} + |V_{T_p}| > V_{dd}$.

The charging and discharging behaviour of a CMOS inverter can also be described by means of the static characteristic $I = f(V_{out})$ shown in Fig. 4.16. This characteristic is obtained when a pulse V_{in} with rise and fall times of 0 ns is applied at the inverter input. Capacitance C is the load capacitance present at the transistor's output. The currents through the pMOS and nMOS transistors are I_p and I_n, respectively.

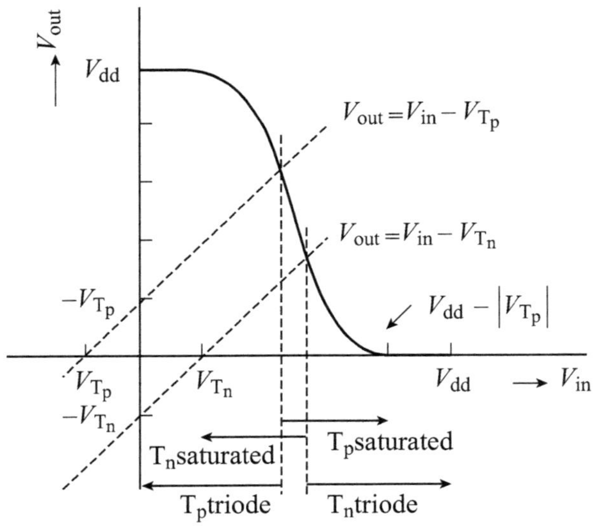

(a)

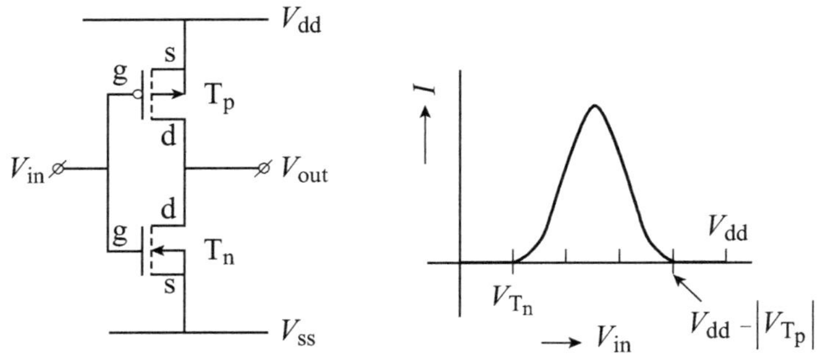

(b)

Fig. 4.15 Transfer characteristic (**a**) and current characteristic (**b**) of a MOS inverter

The curves in Fig. 4.16 are explained as follows:

Trajectory I : I_p rises from 0 to $I_{p_{max}}$ when V_{in} falls from V_{dd} to V_{ss}.
Trajectory II : C charges to V_{dd} and I_p decreases to 0.
Trajectory III : I_n rises from 0 to $I_{n_{max}}$ when V_{in} rises from V_{ss} to V_{dd}.
Trajectory IV : C discharges to V_{ss} and I_n decreases to 0.

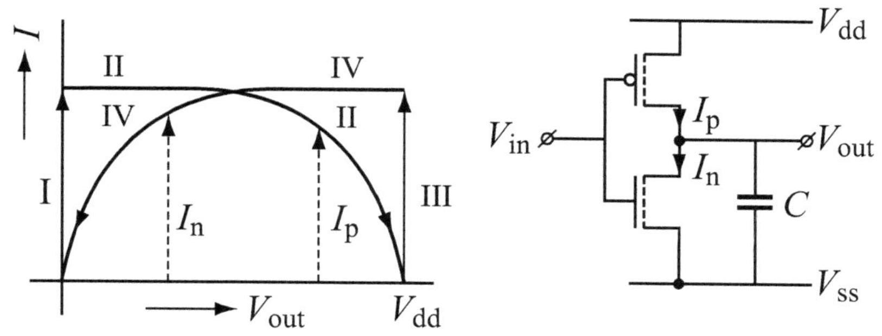

Fig. 4.16 Static CMOS-inverter characteristic

Fig. 4.17 Discharging a load capacitance through an nMOS transistor

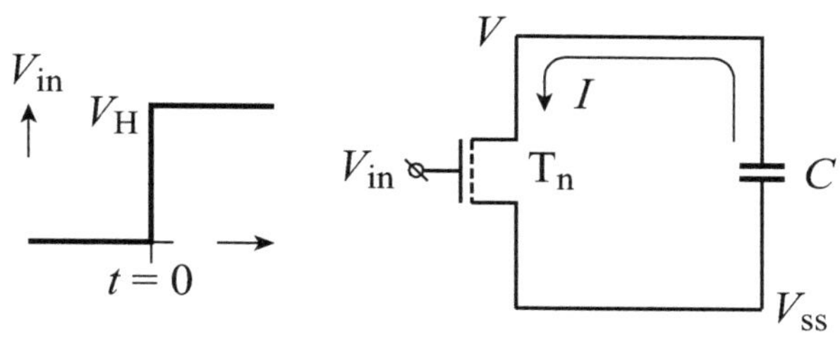

In Fig. 4.16, it is assumed that the βs and the V_Ts of the nMOS and pMOS transistors are equal. The current characteristics are therefore symmetrical with respect to $V_{out} = \frac{1}{2}V_{dd}$.

Designing a CMOS Inverter

A true CMOS logic gate contains a pMOS transistor for every nMOS transistor. A *pseudo-nMOS* version, however, uses just one active pull-up pMOS transistor with its gate connected to ground. Here, a DC current flows from supply to ground when the output is 'low'. The complementary behaviour of the transistors in true CMOS circuits ensures the absence of DC current at both the low and high stable operating points. This type of CMOS logic is therefore 'ratioless', and the voltages V_H and V_L associated with the respective 'high' and 'low' output levels are independent of the transistor sizes. In fact, V_H equals the supply voltage V_{dd} while V_L equals 0 V. The dynamic discharge characteristic of a CMOS inverter is obtained when a step voltage (which rises from 0 V to V_{dd} in 0 ns) is applied to its input. This is illustrated in Fig. 4.17. As shown in Fig. 4.18, the dynamic charge characteristic is obtained when the input step voltage falls from V_{dd} to 0 V in 0 ns.

A simple expression, derived from equalising the current expressions for the transistor and the capacitor, defines the gain factor β_n of an nMOS transistor which will discharge a capacitance C from V_{dd} to V in time t when a step voltage with amplitude V_{dd} is applied to its gate at $t = 0$:

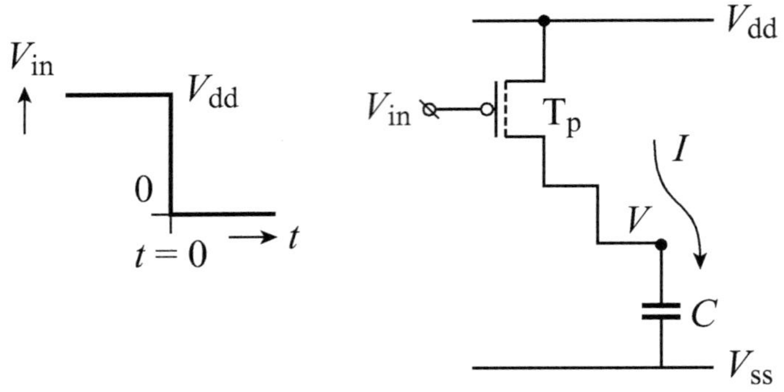

Fig. 4.18 Charging a load capacitance through a pMOS transistor

$$\beta_n = \frac{4 \cdot C}{V_{dd} \cdot t} \tag{4.7}$$

The required dimensions of the nMOS transistor are obtained by equating the gain factor β_n to $\beta_{\square_{n,eff}} \cdot A_{T_n}$, where A_{T_n} is the aspect ratio of the transistor and equals $(W/L)_{T_n}$. $\beta_{\square_{n,eff}}$ includes the second-order effects described in Chap. 2, which reduces the effective mobility, and as a consequence, it reduces $\beta_{\square_n}$ to $\beta_{\square_{n,eff}}$.

Example Given: A 28 nm CMOS process with $\beta_{\square_{n,eff}} = 100\,\mu A/V^2$ and $V_{dd} = 1\,V$.

Problem Determine the aspect ratio A_{T_n} of an nMOS transistor T_n which will discharge a load capacitance $C = 20\,fF$ from V_{dd} to $0.1 \cdot V_{dd}$ in 100 ps when a voltage V_{dd} is applied to its gate.

Solution Substituting in (4.7) yields:

$$\beta_n = \frac{4 \cdot 20 \cdot 10^{-15}}{1 \cdot 10^{-10}} = 792\,\mu A/V^2$$

Equating β_n to $\beta_{\square_{n,eff}} \cdot A_{T_n}$ and substituting $\beta_{\square_{n,eff}} = 100\,\mu A/V^2$ yields:

$$A_{T_n} = \left(\frac{W}{L}\right)_{T_n} \approx 7.9$$

In this example, the mobility-reduction effects (Chap. 2), are included in the simple basic current equations and are represented by the reduced value of $\beta_{\square_{n,eff}}$, which is about four to five times less than the $\beta_{\square_n}$ calculated from the zero field mobility.

The charging of a load capacitance through a pMOS transistor illustrated in Fig. 4.18 is analogous to discharging through an nMOS transistor. The expression for the gain factor β_p of a pMOS transistor, which will charge a capacitance C from 0 V to a voltage $V = 0.9 \cdot V_{dd}$ in time t when its gate voltage falls from V_{dd} to 0 V in 0 ns is therefore simply obtained by the same equation (4.7).

Example Given: The information in the previous example plus $\beta_{\square_{p,eff}} = 67\,\mu A/V^2$.

Problem 1 Determine the aspect ratio A_{T_p} of a pMOS transistor T_p which will charge the load capacitance C from 0 V to $0.9 \cdot V_{dd}$ in 100 ps when 0 V is applied to its gate.

Solution 1 This problem is the complement of the previous example.

Therefore, the following expression applies (see equation (4.6)):

$$A_{T_p} = A_{T_n} \cdot A = 7.9 \cdot 1.5 = 11.8$$

The rise and fall times of *buffer circuits* must be equal. These circuits must therefore use the previously mentioned value of about 1.5–2 for the aspect ratio A expressed in formula (4.6). Even for CMOS logic in conventional technology nodes with $A > 2$, the ratio between the channel width of the pMOSt (W_p) and the channel width of the nMOSt (W_n) is limited to roughly 1.5. Larger values yield larger pMOS transistors and thus increase the load capacitance presented to previous logic gates. For CMOS circuits other than inverters, factors $\left(\frac{W}{L}\right)_p$ and $\left(\frac{W}{L}\right)_n$ in formula (4.6) are the effective values which apply to the transistors in the p and n sections, respectively. The dimensions of these transistors must be selected so that the value for A is optimal. For technologies beyond 65 nm, different stress techniques and device orientations are used to improve the mobility of nMOS and pMOS transistors. The pMOSt mobility, however, will benefit most from these techniques, such that it is expected that in the near future, pMOS and nMOS transistors will exhibit almost equal performance at the same W/L ratio.

Dissipation of a CMOS Inverter

During the last two decades, CMOS technology has become the most dominant technology for VLSI circuits. The most important reason for this is its low static power consumption. This is because of the absence of DC currents during periods when no signal transients occur in static CMOS circuits. However, a *short-circuit current* flows from supply to ground when a change in a logic circuit's input voltage causes the output voltage V_{out} to change. This short-circuit current leads to additional power dissipation [19].

The *power dissipation* of a basic CMOS inverter is explained with the aid of Fig. 4.19.

Only the nMOS transistor conducts when the input voltage V_{in} of this static CMOS inverter is 'high' (V_{dd}). Similarly, only the pMOS transistor conducts when

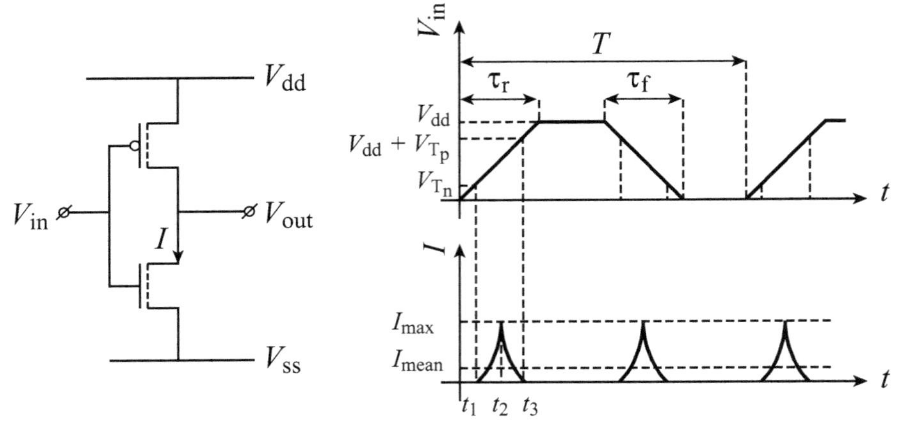

Fig. 4.19 Current through an unloaded inverter

the input voltage V_{in} is 'low' (V_{ss}). Therefore, the inverter does not dissipate power when the input is in either of the above stable states. However, during a transient at the input, there is a period when both the nMOS and pMOS transistors conduct. A short-circuit current then flows from supply to ground while the input voltage is between V_{T_n} and $V_{dd} - |V_{T_p}|$. This current I is shown in Fig. 4.19. If a load capacitance C_L is connected to the inverter output, then the dissipation consists of two components:

1. *Dynamic power dissipation*:

$$P_1 = C_L \cdot V^2 \cdot f \tag{4.8}$$

2. *Short-circuit power dissipation*:

$$P_2 = I_{mean} \cdot V \tag{4.9}$$

In the above equations, f $(= 1/T)$ is the frequency at which the voltage change V occurs on C_L and I_{mean} is the average short-circuit current. Clearly, the dynamic component P_1 is independent of transistor dimensions when parasitic capacitances at the output, such as pn-junction capacitances, are neglected. It is expressed in equation (4.8) and is explained with the aid of Fig. 4.20.

Capacitance C is charged and discharged via resistors with values R_1 and R_2, respectively. During charging, the energy dissipated in R_1 equals:

$$E_{R_1} = \int_0^\infty i^2(t) \cdot R_1 \cdot dt \qquad \text{with } i(t) = \frac{V}{R_1} \cdot e^{-t/(R_1 C)}$$

The solution to this integral is as follows:

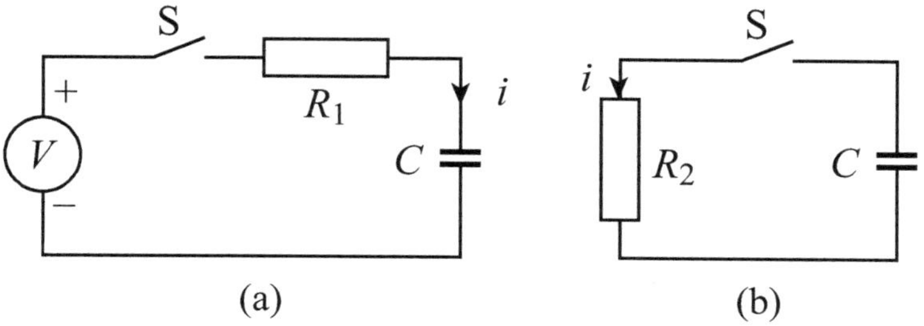

Fig. 4.20 Charging and discharging a capacitance

$$E_{R_1} = \frac{1}{2} \cdot C \cdot V^2$$

E_{R_1} is thus independent of R_1. Similarly, the energy E_{R_2} dissipated during discharging is independent of the value of R_2 and also equals $C \cdot V^2/2$.

The total energy E supplied by the voltage source V during a complete charge-discharge cycle is the sum of E_{R_1} and E_{R_2}, i.e., $E = C \cdot V^2$. For f cycles per second the total power dissipation is:

$$P = C \cdot V^2 \cdot f$$

This dynamic power dissipation appears in all types of logic, including static MOS circuits, bipolar circuits, TTL circuits, etc.

The short-circuit component P_2, however, is proportional to transistor dimensions; it also depends on the size of the load capacitance. An expression for I_{mean} in formula (4.9) is derived on the assumption that the inverter's load capacitance is zero [19]. Although an asymmetric inverter is not fundamentally different, the inverter is also assumed to be symmetric. In this case, the following equations apply:

$$\beta_n = \beta_p = \beta \text{ and } V_{T_n} = -V_{T_p} = V_T$$

During the period t_1 to t_2 in Fig. 4.19, the short-circuit current I increases from 0 to I_{max}. Throughout this period, the output voltage V_{out} is more than a threshold voltage V_{T_n} larger than the input voltage V_{in}. The nMOS transistor is therefore saturated and application of the simple MOS formulae (1.15) yields the following expression for I during this period of time:

$$I = \frac{\beta}{2}(V_{in} - V_{T_n})^2 \text{ for } 0 \leq I \leq I_{max}$$

The symmetry of the inverter produces a maximum value for I when V_{in} equals $V_{dd}/2$. In addition, the current transient during the period t_1–t_3 is symmetrical with respect to the time t_2. The mean current I_{mean} (i.e., the effective current which flows during one cycle period T of the input signal) can therefore be expressed as follows:

$$I_{\text{mean}} = 2 \cdot \frac{2}{T} \int_{t_1}^{t_2} I(t)\mathrm{d}t = \frac{4}{T} \int_{t_1}^{t_2} \frac{\beta}{2} (V_{\text{in}}(t) - V_{\text{T}})^2 \mathrm{d}t \tag{4.10}$$

The input voltage V_{in} is assumed to have a symmetrical shape and linear edges, with rise and fall times equal to τ. The value of V_{in} as a function of time t during an edge is therefore expressed as follows:

$$V_{\text{in}}(t) = \frac{V_{\text{dd}}}{\tau} \cdot t$$

The following expressions for t_1 and t_2 can be derived from Fig. 4.19:

$$t_1 = \frac{V_{\text{T}}}{V_{\text{dd}}} \cdot \tau \quad \text{and} \quad t_2 = \frac{\tau}{2}$$

Substituting these expressions for $V_{\text{in}}(t)$, t_1 and t_2 in equation (4.10) yields:

$$I_{\text{mean}} = \frac{2\beta}{T} \cdot \int_{\frac{V_{\text{T}}}{V_{\text{dd}}} \cdot \tau}^{\tau/2} \left(\frac{V_{\text{dd}}}{\tau} \cdot t - V_{\text{T}} \right)^2 \mathrm{d}t$$

The solution to this equation is:

$$I_{\text{mean}} = \frac{1}{12} \cdot \frac{\beta}{V_{\text{dd}}} \cdot (V_{\text{dd}} - 2V_{\text{T}})^3 \cdot \frac{\tau}{T}$$

Substituting this expression for I_{mean} into formula (4.9) yields the following expression for the short-circuit dissipation of a CMOS inverter with no load capacitance:

$$P_2 = \frac{\beta}{12} \cdot (V_{\text{dd}} - 2V_{\text{T}})^3 \cdot \frac{\tau}{T} \tag{4.11}$$

Formula (4.11) clearly illustrates that the short-circuit dissipation is proportional to the frequency $f = 1/T$ at which the input changes. Voltages V_{dd} and V_{T} are determined by the application and the process. Therefore, the only design parameters that affect P_2 are β and the rise and fall times (τ) of the inverter's input signal. For an inverter with a capacitive load, the transistor β values are determined by the required output rise and fall times. In this case, the short-circuit dissipation only depends on the input signal's rise and fall times, i.e., τ_{r} and τ_{f}, respectively. This is particularly true for buffer circuits which have transistors with large β values.

In the chapter on low-power design (Chap. 8), the CMOS power contributions are discussed extensively. However, the design of large buffer circuits is discussed in this section on basic CMOS circuit design.

CMOS Buffer Design

Large capacitances associated with integrated circuits include those presented by bus lines and 'off-chip' circuits. These capacitances must often be driven at high frequencies. The required 'buffer' driving circuits dissipate a relatively large part of the total power consumed by a chip. Optimising these buffers therefore requires considerably more effort than the approach adopted for CMOS logic. Formula (4.11) shows that short-circuit dissipation is directly proportional to the rise and fall times (τ) of an input signal. The input signals of buffers which drive bus lines connected to large numbers of different sub-circuits on a chip must therefore have particularly short rise and fall times.

Suppose the signal on a bus line with capacitance C_L must follow a signal at the output node A of a logic gate which is capable of charging and discharging a capacitance C_0 in τ ns. An inverter chain such as illustrated in Fig. 4.21 can be used as a buffer circuit between node A and the bus line.

From formula (4.11), it is clear that the rise and fall times on each input of the inverters in the above chain should be short. Moreover, it has been shown in literature [19] that minimum dissipation can be achieved when the rise and fall times on each of these inputs are equal to the rise and fall times at the buffer output. The inverter chain must therefore be designed so that the rise and fall times on the inputs of each of its component inverters are τ ns. According to literature [5], a minimum propagation delay time for the buffer is obtained when the *tapering factor* r between the βs of successive inverters is e, the base of natural logarithm. In terms of dissipation and silicon area, however, this will not lead to an optimum design. Design optimisation for minimum dissipation and silicon area requires a different approach [19].

When a signal is produced by a logic gate and must be buffered to drive a relatively large capacitive load $C_L = C_N = 10\,\text{pF}$, then the design of this buffer is

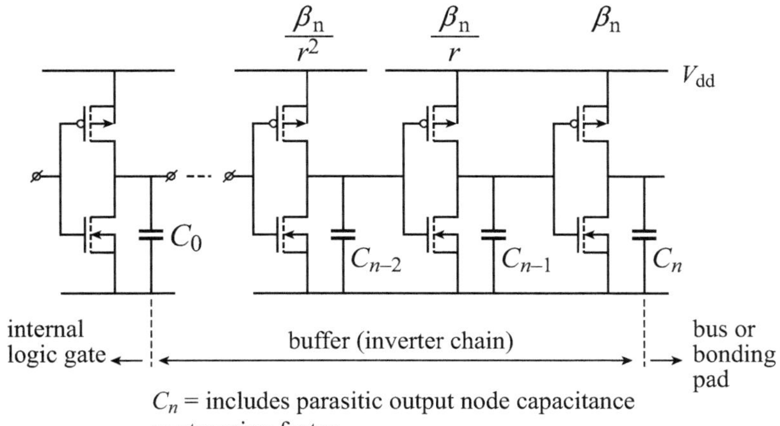

Fig. 4.21 A buffer circuit comprising an inverter chain

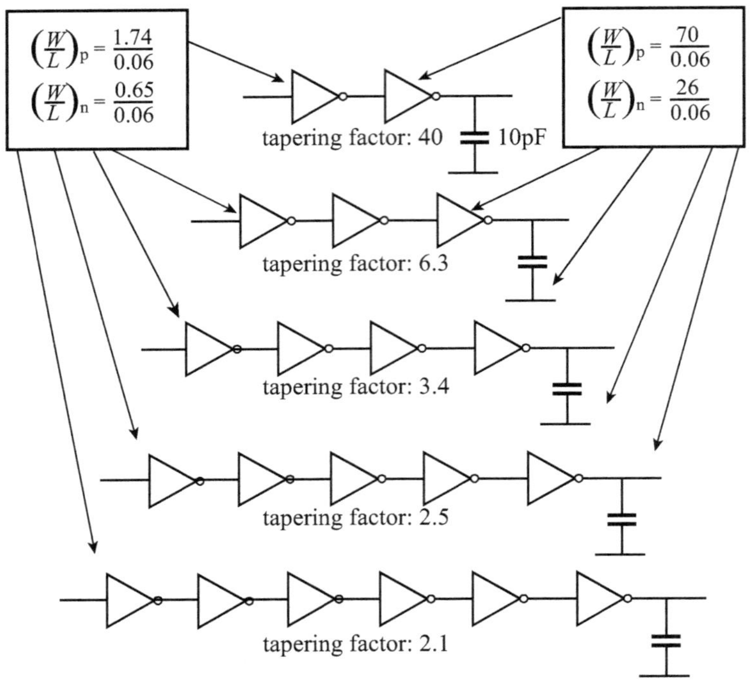

Fig. 4.22 Inverter chains with different tapering factors, all driving the same load

not defined by the power consumption only. Also area and switching-current peaks (di/dt) are important parameters that have to be seriously dealt with.

The size of the final driver stage n, is defined by the timing specifications of (dis)charging the output load C_n. The driving strength of the $n-1$ stage is defined by the required tapering factor. In conventional CMOS technologies, with around 1 µm channel lengths, the optimum tapering factor was close to 10, in order to achieve identical rise and fall times on the input and output of the individual driver stages. A larger tapering factor would lead to smaller pre-driver stages and so to more delay and larger short-circuit power consumption. A smaller tapering factor would lead to larger pre-driver stages, less delay, but to increased switching (di/dt) noise. When porting the driver to the next technology node, assuming a scaling factor $s \approx 0.7$ and maintaining the same W/L ratio, the transistor sizes W and L of the final driver stage N also scale with an average factor of 0.7 ($= s$). Because the gate-oxide thickness scales with the same factor, the total fan-in capacitance of the final stage becomes 1.4 times smaller. In order to achieve the same rise and fall times on the input of the final stage, this allows an increase of the tapering factor with the same amount. In other words: the tapering factor r is dependent on the technology node. Figure 4.22 shows six different drivers (inverter chains) with different tapering factors designed in a 1.2 V 60 nm CMOS technology.

Table 4.1 Characteristics of various 10 pF inverter chains with different tapering factors

Number of inverters	2	3	4	5	6	Unit
Tapering factor	40	6.3	3.4	2.5	2.1	
Relative area	1	1.15	1.35	1.61	1.85	
Total power	738	742	746	750	754	μW
Dynamic power	721	722	723	723	723	μW
Short-circuit power	12	12	11.5	11.5	11.5	μW
Max. di/dt (rel.)	1	5	10	13	14	
Output delay	896	657	590	580	575	ps

The input driving stage, which mimics the logic gate drive strength, and the output driving stage, driving a 10 pF load, are the same for all six drivers. The circuit simulations are done for a clock frequency of 50 MHz. Table 4.1 shows the major characteristics of these six drivers.

Because supply noise (see Chap. 9), which has a linear relation with the di/dt, is a real issue in nanometer CMOS ICs, it is a dominant factor in choosing the right driver (tapering factor). The first column refers to the first driver of Fig. 4.22. It shows 36% more delay than the second inverter chain. This second inverter chain, however, only shows 10% more delay than number 3, etc.

The short-circuit power consumption in this table only represents the short-circuit power consumed by the final inverter stage, because the pre-driver's short-circuit power is negligible. The table shows that the short-circuit power consumption is only a fraction of the dynamic power consumption. Generally, if a tapering factor equal to or larger than 10 is chosen, then, in nanometer CMOS ICs, the short-circuit power maybe completely neglected in the power discussions. Also the di/dt value is relatively low for this tapering factor. These considerations hold for on-chip clock drivers and bus drivers, as they usually consume more power than an average logic gate.

Because many output drivers still use 3.3 V, 2.5 V or 1.8 V supply voltages, these definitely will show different optimum tapering factors, which reduce with increasing output voltages. An example of an output driver circuit is given in Fig. 4.22. In such output drivers also, a tapering factor of close to 10 or larger would be the best choice: it guarantees a relatively short delay, a small short-circuit power consumption and a relatively low switching noise (di/dt). For advanced process nodes the signal rise/fall times are so short that the short-circuit power consumption can generally be neglected.

Noise Margins

The maximum amplitude of a noise signal that can be superimposed on all nodes of a long inverter chain without causing the output logic level to change is called *noise margin*. Figure 4.23 shows the transfer characteristic of a CMOS inverter for three different gain factor ratios. The noise margins for both high and low levels are very large because of the almost rectangular shape of these transfer characteristics. For the symmetrical inverter, with $\beta_n = \beta_p$ and $V_{T_n} = -V_{T_p}$, the noise margins are

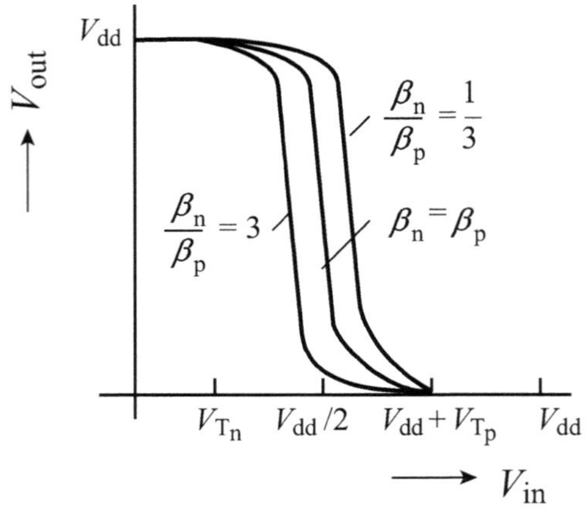

Fig. 4.23 CMOS inverter transfer characteristics for different aspect ratios

equal for both levels. Of course, not every inverter is symmetrical. In such cases, the noise margin is different for the two levels. However, the difference is only significant for highly asymmetrical inverters.

Generally, the operation of CMOS logic circuits is very robust. Even when the supply voltage is reduced to below the lowest of the threshold voltages (V_{T_n} and V_{T_p}), it will still deliver a correct logical output. However, this so-called *subthreshold logic* will then operate at relatively low frequencies.

4.4 Digital CMOS Circuits

4.4.1 Introduction

CMOS circuits can be implemented in static or dynamic versions. The choice is mainly determined by the type of circuit and its application. Two important factors which influence this choice are chip area and power dissipation. The differences between these factors for the two types of implementation are treated in this section.

4.4.2 Static CMOS Circuits

A logic function in static CMOS must be implemented in both nMOS and pMOS transistors. An nMOS version only requires implementation in nMOS transistors. A single load transistor is then used to charge the output. This load transistor also conducts when the output is 'low'. A current therefore flows from supply to ground and causes DC dissipation while the output of an nMOS logic gate is 'low'.

In a CMOS logic gate, a current only flows between supply and ground during output transitions. Figure 4.24 shows some static CMOS logic gates. Back-bias connections for both the nMOS and the pMOS transistors are indicated in the inverter in Fig. 4.24a. The respective back-bias voltages, V_{sb} and V_{ws}, are both 0 V. The back-bias connections are no longer shown in Fig. 4.24b, c and all subsequent

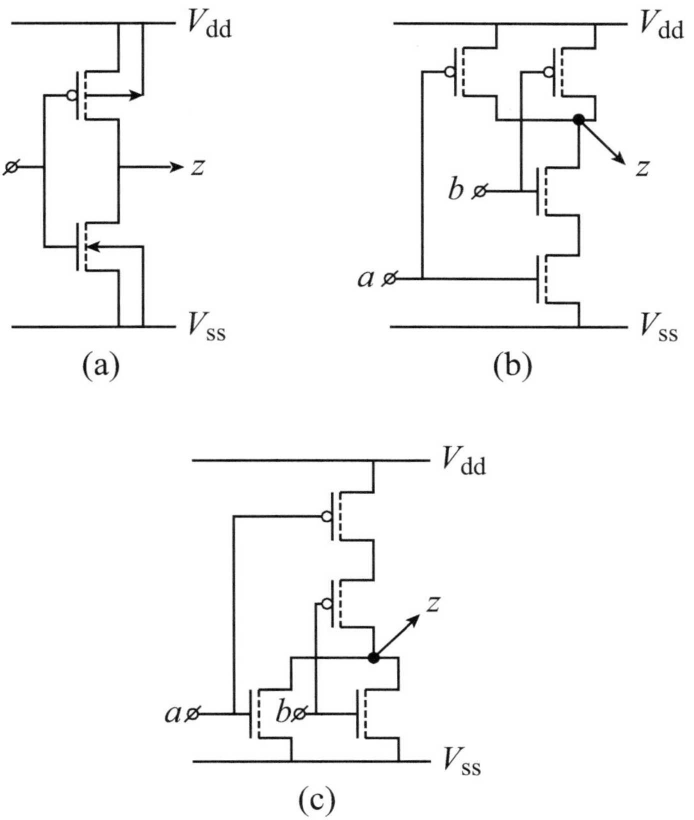

Fig. 4.24 Examples of static CMOS logic gates. (**a**) INVERT: $z = \bar{a}$. (**b**) NAND: $z = \bar{a} \cdot \bar{b}$. (**c**) NOR: $z = \overline{a + b}$

figures. Unless otherwise stated, the substrate voltages are assumed to be V_{ss} for the nMOS transistors and V_{dd} for the pMOS transistors. Figure 4.24b and c show nMOS and pMOS transistors, respectively, connected in series. The sources of some of these transistors are not connected to V_{ss} or V_{dd}. The back-bias effect has a considerable influence on nMOS and pMOS transistors whose sources are not connected to V_{ss} and V_{dd}, respectively. This is particularly true when the source is loaded.

In general, a series connection of transistors in the nMOS section of a CMOS logic gate will reflect a parallel connection of transistors in the pMOS section and vice versa. This is illustrated in Fig. 4.25, which shows an example of a static CMOS implementation of a complex logic function and its equivalent logic gate diagram.

The performance of a pMOS transistor is less than that of an nMOS transistor. The number of pMOS transistors in series in a CMOS logic gate should therefore be minimised. If this number becomes very large then, only in exceptional cases, can a pseudo-nMOS implementation be used.

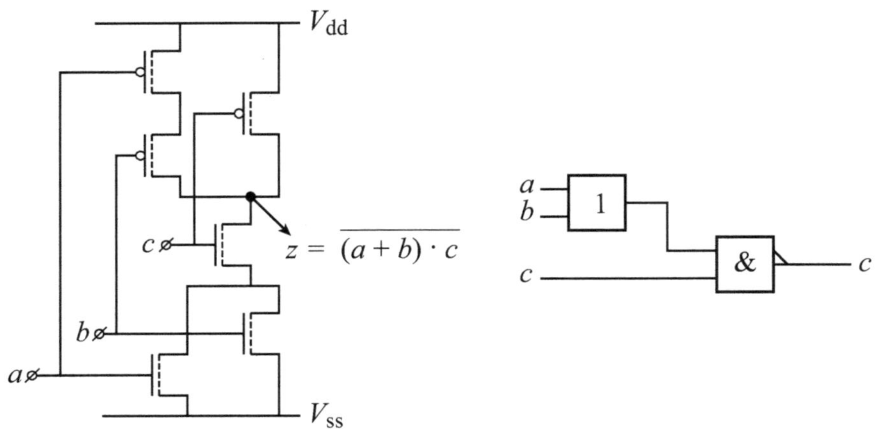

Fig. 4.25 An example of a more complex static CMOS logic gate

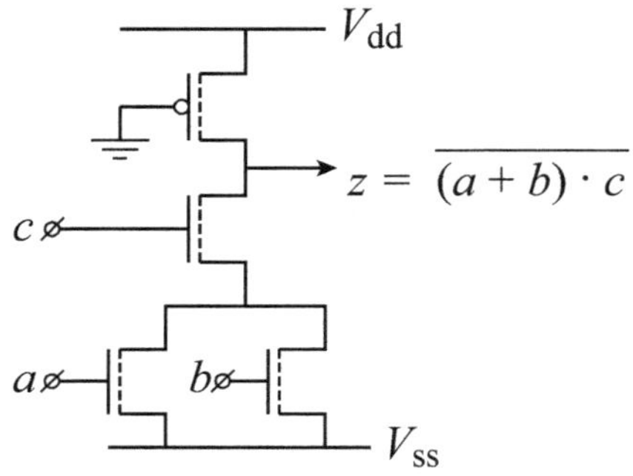

Fig. 4.26 A pseudo-nMOS logic gate

Figure 4.26 is an example of a pseudo-nMOS implementation of the CMOS equivalent in Fig. 4.25. The pseudo-nMOS version is identical to its nMOS counterpart except that the nMOS load element is replaced by a pMOS transistor with its gate connected to V_{SS}. Both nMOS and pseudo-nMOS logic gates have the advantage of the same low input capacitance. The output rise time of a pseudo-nMOS logic gate is determined by only one pMOS transistor and should therefore be short. A disadvantage of such a gate is the static power dissipation when the output is 'low'. The output low level and noise margins are determined by the ratio of the widths of the nMOS and pMOS transistors. Pseudo-nMOS logic is therefore also a form of ratioed logic, as discussed in Sect. 4.2.2. This type of logic is only very sparingly used and certainly not in today's low-power designs.

The CMOS Transmission Gate (Pass Transistor)

Figure 4.27 shows a *transmission gate* comprising a complementary pair of transistors. This is an important component in both static and dynamic circuits. It is used to control the transfer of logic levels from one node to another when its control

Fig. 4.27 CMOS transmission gate

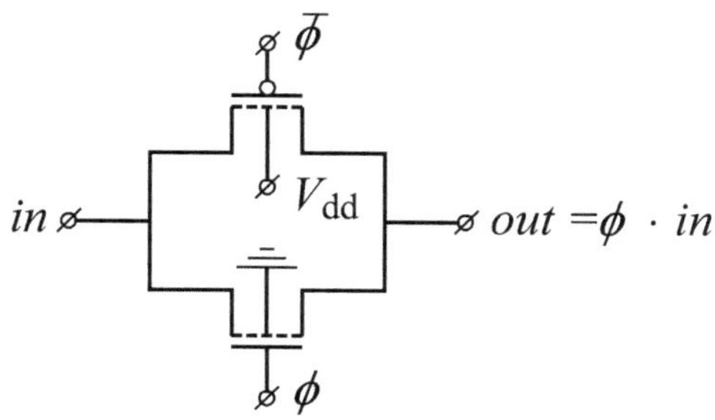

signals are activated. A single nMOS enhancement transistor can also be used to implement a transmission gate. Such an implementation has only one control signal but is disadvantaged by threshold loss. The threshold voltage of the transistor may be relatively high because of the body effect and the maximum high output level equals a threshold voltage below the control voltage. For this reason, the CMOS implementation is preferred.

If the gate of the nMOS transistor in the CMOS transmission gate is controlled by a signal ϕ, the gate of the pMOS transistor must be controlled by the complementary signal $\overline{\phi}$. When the input voltage is 0 V and ϕ is 'high', the output will be discharged to 0 V through the nMOS transistor. The complementary behaviour of the pMOS transistor ensures that the output voltage equals V_{dd} when the input voltage is at V_{dd} level and $\overline{\phi}$ is 'low'.

Figure 4.28 shows the contributions of both MOS transistors to the charge and discharge characteristics of a CMOS transmission gate. The pMOS and nMOS transistors prevent threshold loss on the output 'low' and 'high' levels, respectively.

Pass-Transistor Logic

In static CMOS circuits, transmission gates are used in latches, flip-flops, 'pass-transistor logic' and in static random-access memories. Examples of pass-transistor logic are exclusive OR (EXOR) logic gates and multiplexers. Figure 4.29 shows pass-transistor logic implementations of an EXOR gate. The nMOS transmission gate implementation in Fig. 4.29a is disadvantaged by high threshold loss resulting from body effect. The complementary implementation in Fig. 4.29b yields shorter gate delays at the expense of larger chip area. When connecting the outputs of these gates to a latch circuit (e.g., two cross-coupled pMOS loads), a static CMOS logic family is created (Fig. 8.17). The threshold voltage loss over the nMOS pass gates is compensated by the level restoring capability of the latch.

A general disadvantage of pass-transistor logic as presented in Fig. 4.29 is the series resistance between the inputs a and $\overline{a}$ and the output z. The charging and discharging of a load at the output through the pass transistor cause additional delay. Other disadvantages include the need for complementary control signals. The potentials of pass-transistor logic challenge the creativity of the designers. Several

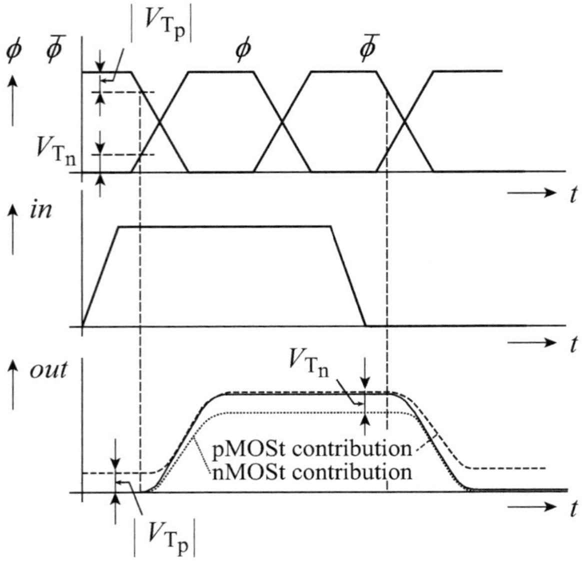

Fig. 4.28 CMOS transmission gate behaviour and the individual contributions of the nMOS and pMOS transistors to the charge and discharge characteristics

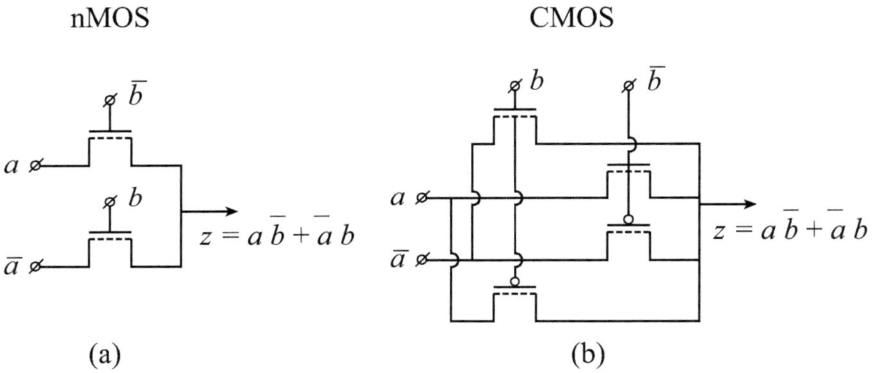

Fig. 4.29 Pass-transistor logic implementations of an EXOR logic gate with (**a**) nMOS pass transistors (**b**) CMOS pass-transistor gates

alternatives have been published. These are discussed in detail in the low-power Chap. 8, together with their advantages and disadvantages.

Finally, circuit designs implemented with pass-transistor logic must be simulated to prevent unexpected performance degradation or even erroneous behaviour caused by effects such as charge sharing (Sect. 4.4.4). With decreasing voltages in current and future processes, the performance of pass-transistor logic tends to drop with respect to standard static CMOS logic. Therefore, the importance and existence of pass-transistor logic are expected to decrease in the coming years. The forms of CMOS logic discussed above can be used in both asynchronous circuits and synchronous, or ‘clocked’, circuits. The latter type of circuits are the subject of the next section.

4.4.3 Clocked Static CMOS Circuits

Signals which flow through different paths in a complex logic circuit will ripple through the circuit asynchronously if no measures are taken. It is then impossible to know which signal can be expected at a given node and time. Controlling the data flow inside a circuit therefore requires synchronisation of the signals. Usually, this is done by splitting all the different paths into sub-paths with a uniform delay. The chosen delay is the worst case delay of the longest data ripple. In synchronous static CMOS circuits, the sub-paths are separated by means of ‘latches’ and/or ‘flip-flops’ which are controlled by means of periodic clock signals. Dynamic circuits may also use latches and flip-flops. Alternatively, data flow in dynamic circuits may be controlled by including the *clock signals* in every logic gate.

Static Latches and Flip-Flops

Latches and flip-flops are used for temporary storage of signals. Figure 4.30 shows an example of a static *CMOS latch* and an extra transmission gate. The transmission gate on the left-hand side is an integral part of the latch, which also comprises two cross-coupled inverters. Complementary logic values can be written into this latch via the transmission gates when the clock signal is high, i.e., when $\phi = 1$ and $\overline{\phi} = 0$. Feedback in the latch ensures that these values are held when $\phi = 0$ and $\overline{\phi} = 1$. This basic principle is used in static full-CMOS memory cells and flip-flops.

A flip-flop can temporarily store data and is controlled by one or more clock signals. The maximum clock frequency of a clocked static CMOS circuit is determined by the worst case delay path between two flip-flops. This path has the longest propagation delay as a result of a combination of logic gates and/or long signal tracks with large capacitances. There are several implementations of static

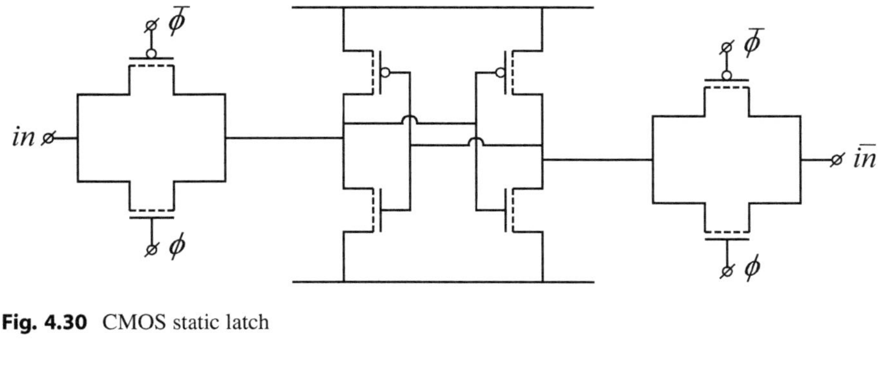

Fig. 4.30 CMOS static latch

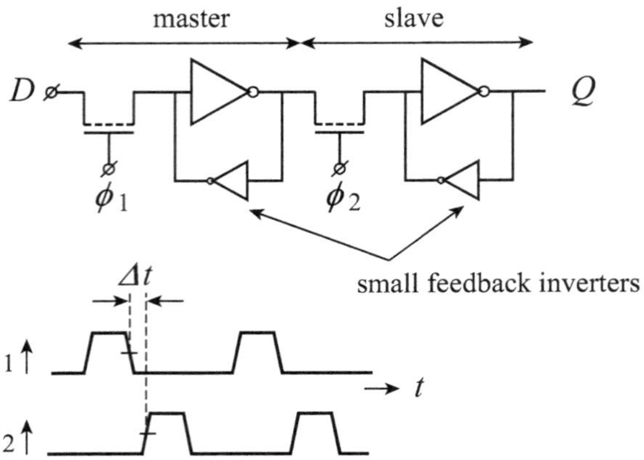

Fig. 4.31 D-type flip-flop with nMOS transmission gates and its 2-phase non-overlapping clock signals

CMOS flip-flops. The discussions below are limited to different forms of D-type flip-flops.

A *D-type flip-flop* can be built by connecting two latches in series, as shown in Fig. 4.31. The latches in this example use nMOS transmission gates. When the clock ϕ_1 goes 'high', data at the D input are latched into the 'master' latch of the flip-flop while the 'slave' latch maintains the previous input data. The D-input has to compete with the latch's feedback inverter via the nMOS transmission gate. The W/L aspect ratios of the transistors in the feedback inverter are therefore very small. The threshold voltage loss of the nMOS transmission gate produces a 'poor' high level at the input of the large inverter.

The aspect ratio, as expressed in equation (4.6), used for the large inverter must ensure that its output is 'low' when the poor high level is present at its input. The high level is then regenerated by the small feedback inverter. Static dissipation therefore does not occur. In practice, the aspect ratio of the large inverter must be close to 1. This ensures that the inverter's switching point is lower than half the supply voltage.

The flip-flop in Fig. 4.31 is therefore not very robust in its operation. Its operation is much more reliable, when it is implemented with complementary transmission gates. In this case, however, the nMOS and pMOS transistors in the first transmission gate are controlled by ϕ and $\overline{\phi}$, respectively. The nMOS and pMOS transistors in the second transmission gate are controlled by $\overline{\phi}$ and ϕ, respectively.

Another implementation of the D-type flip-flop is shown in Fig. 4.32. The additional transmission gates in the feedback loops of each latch interrupt these loops when data are being written into the latch. This reduces the driving requirements of the input circuit and the master, which makes it easier to change the state of the flip-flop.

Two clocks must be routed in chips with flip-flops which require complementary clocks, such as ϕ_1 and ϕ_2 in Fig. 4.31 or ϕ and $\overline{\phi}$ in Fig. 4.32. If the routing area is critical, a single clock flip-flop must be used. Such a flip-flop must then include an inverter to locally generate the inverse of the routed clock. However, there is then an

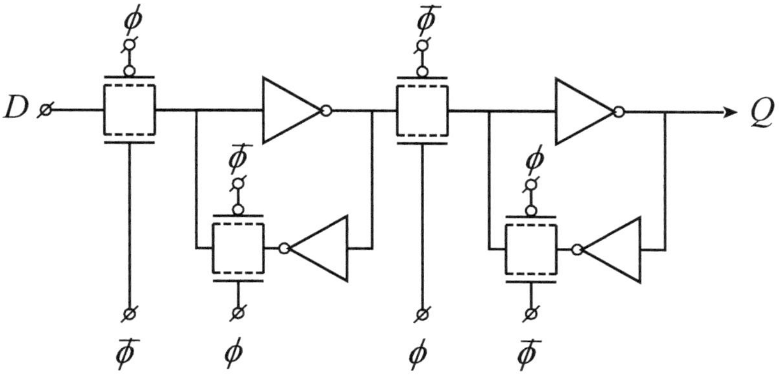

Fig. 4.32 Another implementation of a D-type flip-flop with complementary transmission gates

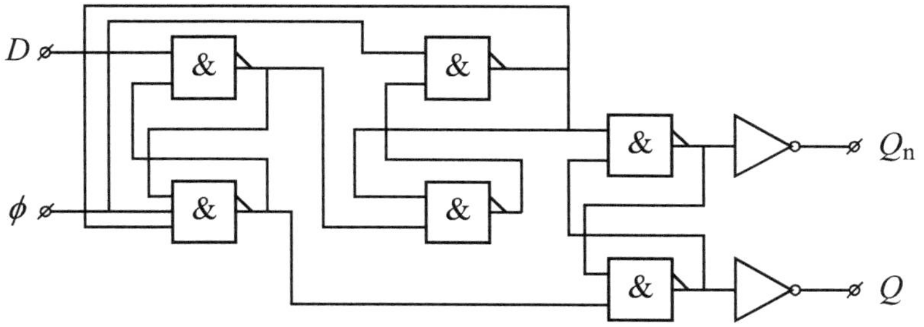

Fig. 4.33 A D-type flip-flop comprising NAND logic gates

increased risk of 'transparency'. This occurs when the *'clock skew'* causes a flip-flop's transmission gate to simultaneously conduct for a short period of time. This causes the flip-flop to be briefly transparent and data can *'race'* directly from the input to the output. This effect occurs when the flip-flop's complementary clocks arrive via different delay paths. If the clock ϕ_1 in Fig. 4.31, for instance, is delayed by more than a time period Δt with respect to clock ϕ_2, the flip-flop would be briefly transparent.

Clocks ϕ_1 and ϕ_2 in Fig. 4.31 are *non-overlapping*, i.e., ϕ_1 is 'low' before ϕ_2 goes 'high' and vice versa. The use of non-overlapping clocks is a good means of preventing transparency in flip-flops.

A discussion of the many more types and variants of static D-type flip-flops is beyond the scope of this book. However, the D-type flip-flop presented in Fig. 4.33 is particularly interesting. This flip-flop is primarily implemented with NAND logic gates. It requires only a single clock and is very robust. Unfortunately, it consists of 15 nMOS and 15 pMOS transistors and therefore requires considerably more chip area than the 10-transistor flip-flop in Fig. 4.31. A 'high-density gate array' layout of the flip-flop in Fig. 4.33 is shown in Fig. 7.38.

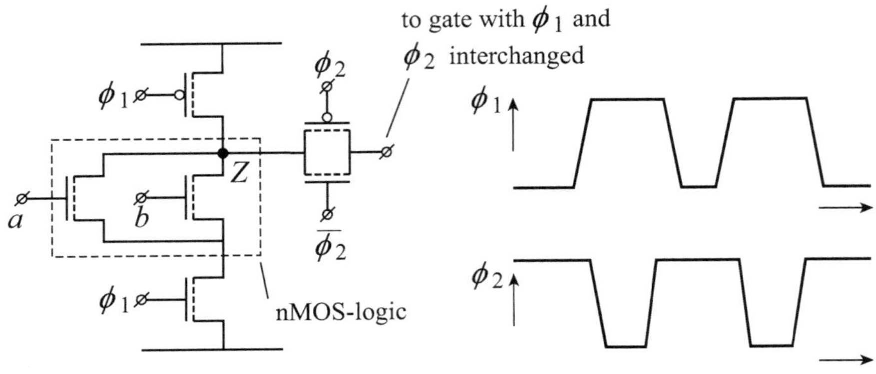

Fig. 4.34 A dynamic CMOS implementation of $Z = \overline{a + b}$

4.4.4 Dynamic CMOS Circuits

The main advantage associated with dynamic CMOS circuits is the small chip area that they require. The explanation lies in the fact that logic functions are only implemented in nMOS transistors. Only one pMOS transistor is used per logic gate to charge its output node. Dynamic CMOS circuits are therefore *'nMOS-mostly'* and can occupy significantly less chip area than their static CMOS equivalents. This is particularly true for complex gates.

Figure 4.34 shows a *dynamic CMOS* implementation of a NOR gate. A dynamic CMOS gate of this type requires four different clocks for proper operation, i.e., ϕ_1, $\overline{\phi_1}$, ϕ_2 and $\overline{\phi_2}$. Inputs a and b must be generated by a gate in which ϕ_1 and ϕ_2 are interchanged. The output may also only serve as an input for a gate with ϕ_1 and ϕ_2 interchanged.

The operation of the NOR gate is described as follows:

- Node Z is precharged to V_{dd} when clock ϕ_1 is 'low'.
- When ϕ_1 goes 'high', Z will be discharged if either a or b is 'high'.
- Clock ϕ_2 is then 'low' and the transfer gate passes the value on Z to the input of another logic gate.

There is a wide variety of dynamic CMOS logic implementation forms. These include the race-free, pipelined CMOS logic from the Catholic University of Leuven and Bell Labs' DOMINO-CMOS. In contrast to the form of dynamic CMOS shown in Fig. 4.34, all logic gates in a DOMINO-CMOS circuit are simultaneously precharged during the same part of the clock period. The logic gates sample their inputs when the precharge period ends. In keeping with the domino principle, however, each logic gate can only switch state after its preceding gate has switched. Figure 4.35 shows an example of a DOMINO-CMOS logic gate. The output Y of the dynamic gate is precharged when the clock ϕ is 'low'. The output Z of the

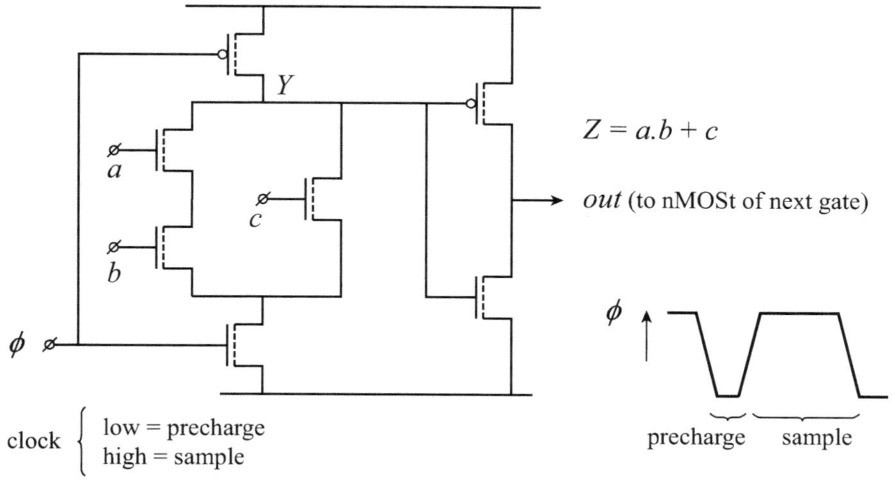

Fig. 4.35 An example of a DOMINO-CMOS logic gate

static inverter is then 'low'. In fact, the inverter output nodes of all logic gates are 'low' during precharge. These outputs can therefore either stay 'low' or switch to 'high' when ϕ is 'high'. Clearly, each node can only make one transition during this sample period. A node stays in its new state until the next precharge period begins. The data must obviously be given enough time to ripple through the worst case delay path during a sample period. The sample period will therefore be much longer than the precharge period. An important disadvantage of DOMINO-CMOS logic is that all gates are non-inverting. Circuit adaptations are therefore required to implement logic functions with inverse inputs, e.g., an EXOR gate.

Another disadvantage is the need to buffer each logic gate with an inverter; this requires extra silicon area. Today, DOMINO-CMOS logic is often used in high-performance processors. Particularly the most delay-critical circuits, like multipliers and adders are implemented in some style of DOMINO-CMOS [12]. Such high-performance processors require logic with large drive capability. This is particular an advantage of DOMINO logic, because only the inverter stage in a logic gate needs to be upscaled instead of every transistor within the logic function. With respect to power dissipation, several remarks on dynamic circuits are made in Chap. 8.

Dynamic CMOS Latches, Shift Registers and Flip-Flops

There are many variations of dynamic CMOS shift registers. However, most of them (like their static CMOS counterparts) basically consist of inverters and transfer gates. A *shift register* is in fact a series connection of flip-flops. Dynamic versions of latches and flip-flops therefore also exist. A dynamic *flip-flop* is also referred to as a dynamic shift register cell because it dynamically shifts data from its input to its output during a single clock cycle.

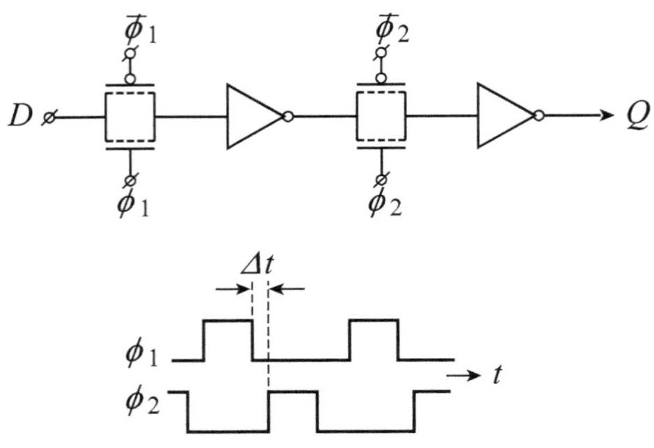

Fig. 4.36 Dynamic D-type flip-flop with non-overlapping clock signals

A minimum clock frequency is required to maintain information in circuits that use dynamic storage elements. This minimum frequency is usually several hundred Hertz and is determined by the subthreshold leakage current and the leakage current of the reverse-biased diffusion to substrate pn-junctions in both nMOS and pMOS transistors. There are many different types of dynamic CMOS storage elements.

By deleting the feedback inverters in Fig. 4.31, we get the dynamic D-type flip-flop shown in Fig. 4.36. Of course, this flip-flop comprises two dynamic latches.

The input data D in the above flip-flop are dynamically stored on the input capacitance of the first inverter when ϕ_1 is 'high'. When ϕ_2 is 'high', the output level of the first inverter is dynamically stored on the input capacitance of the second inverter. The *non-overlapping clocks* are intended to prevent the latch from becoming transparent and allowing data to race through the cell during a clock transition. Just as in the static flip-flop, however, this flip-flop will become transparent if the clock skew exceeds Δt. A shift register operates incorrectly when transparency occurs in its flip-flops.

Figure 4.37 presents another type of dynamic CMOS shift register cell. An advantage of this implementation is the reduced layout area resulting from the absence of complementary transfer gates. The clocks in the first section could also be switched and used in the second section. The resulting risk of transparency requires considerable attention.

Critical Phenomena in Dynamic Circuits

The operation of dynamic MOS circuits relies on the parasitic capacitances that store the logic levels. During a certain period of the clock cycle, several nodes in a dynamic circuit become floating, which makes them very susceptible to such effects as charge sharing and cross-talk.

- **Charge sharing**

  A typical example of *charge sharing* is shown in Fig. 4.38.

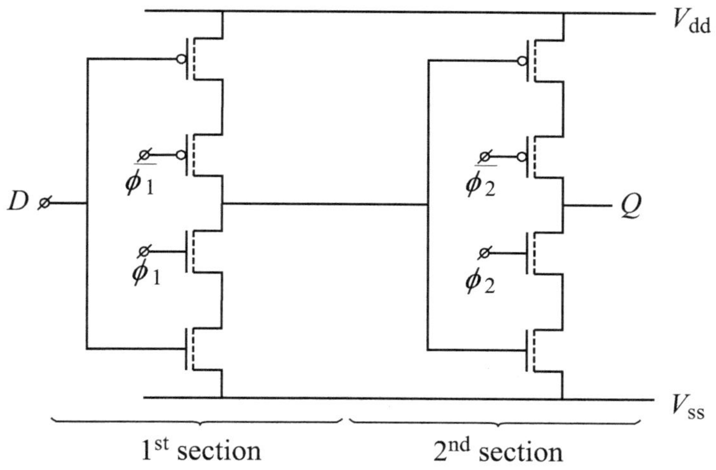

Fig. 4.37 Another dynamic CMOS shift register cell

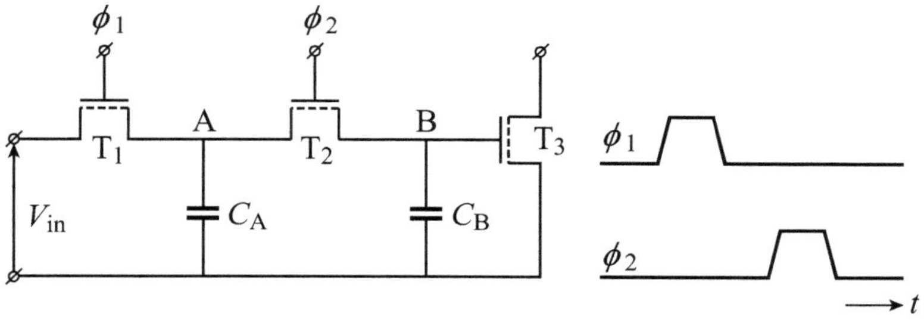

Fig. 4.38 An example of charge sharing

The high levels of clocks ϕ_1 and ϕ_2 are assumed to cause no threshold loss in transistors T_1 and T_2. When ϕ_1 goes 'high', C_A is charged to the voltage V_{in} and remains at this level when ϕ_1 goes low again. During the period when ϕ_2 is 'high', the charge on C_A is shared between C_A and C_B. The voltages at nodes A and B are then described by:

$$V_A = V_B = \frac{C_A}{C_A + C_B} \cdot V_{in} \tag{4.12}$$

As long as $C_B \ll C_A$, then $V_A \approx V_{in}$. However, if C_B is relatively large, then a 'high' level will be significantly degraded when charge is shared between C_A and C_B. Charge sharing circuits must therefore be used with caution and, if possible, should be avoided.

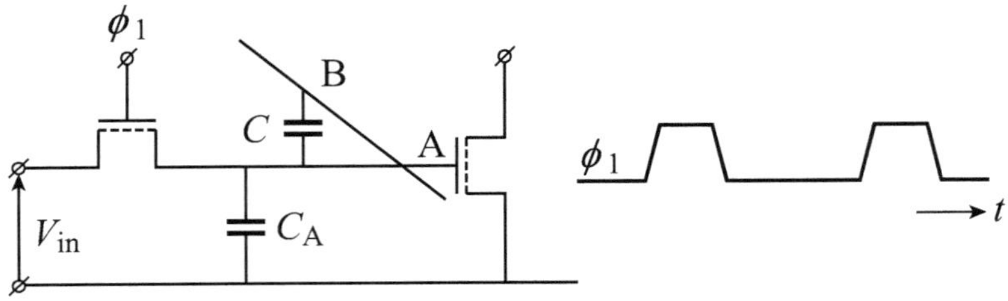

Fig. 4.39 A potential cross-talk situation

- **Cross-talk**
  Figure 4.39 shows a schematic of a situation in which *cross-talk* can occur. A capacitance C exists between node A and a signal track B which crosses it. When ϕ_1 goes from '1' to '0', capacitance C_A is supposed to act as temporary storage for the logic signal that was at A when ϕ_1 was '1'. However, node A has a very high impedance when ϕ_1 is '0', and a voltage change ΔV_B on the signal track B results in the following voltage change at node A:

$$\Delta V_A = \frac{C}{C_A + C} \cdot \Delta V_B$$

The value of the 'cross-over' capacitance C is proportional to the area of the overlap between node A and track B. A large value for C can lead to a disturbance of the logic levels at node A. The area and the number of potentially dangerous crossings must therefore be kept to a minimum during the layout phase of dynamic circuits. Each dynamic node in the finished layout must be checked to ensure that cross-talk noise remains within acceptable margins.

The properties of dynamic MOS circuits can be summarised as follows:

- dynamic MOS circuits have less fan-in capacitance and consume less chip area than static equivalents.
- phenomena such as charge sharing and cross-talk make the electrical design and layout of dynamic nMOS circuits considerably more difficult than for static circuits.

As static CMOS circuits are very robust and operate with large noise margins, almost all CMOS libraries are based on static CMOS.

4.4.5 Other Types of CMOS Circuits

The most important characteristics of different CMOS circuits have been presented. These include the small chip area associated with dynamic implementations of logic gates, the low power dissipation associated with static implementations, large logic swings and large noise margins, etc. The advantages and disadvantages associated with an implementation choice can therefore be weighed up. Power dissipation, for instance, can be sacrificed for speed, or speed can be achieved when lower noise margins are accepted.

In the past, several articles have appeared on specialised forms of CMOS, including Cascode Voltage Swing Logic (CVSL) [9]. A CVSL logic gate is obtained by replacing the pMOS transistors in a conventional static CMOS logic circuit by nMOS transistors, which require inverse input signals. The reduction in chip area (at the expense of speed) is particularly noticeable when complex logic gates are implemented in static or dynamic CVSL. A modified form of CVSL called Differential Split Level (DSL) Logic uses a reduced logic swing. It therefore operates about two to three times faster but dissipates more power than CVSL. These type of logic families were more used in conventional CMOS technologies with higher supply voltages.

Some advice which may simplify the task of selecting the right logic implementation is given in the next section.

4.4.6 Choosing a CMOS Implementation

An important decision at the start of a new CMOS design is the choice of logic implementation. The choice of a static or dynamic form is determined by a number of factors. The most dominant ones are power dissipation, speed, chip area and noise immunity. These factors are examined below.

Power Dissipation

As previously shown, static CMOS circuits do not dissipate power when the circuit is stable. Except for the subthreshold leakage power, power is only dissipated in gates that change state. In clocked static CMOS circuits, most power dissipation occurs during and immediately after clock transitions. In clocked dynamic CMOS, however, each gate output is precharged every clock cycle.

Consider the dynamic inverter as an example. If the input remains 'high' during successive clock periods, then the output should be 'low'. However, the output is precharged during every clock period. This repeated charging and discharging of the output leads to high power consumption. A static CMOS inverter in the same situation would not change state and would therefore consume no power. Circuits for low-power or battery-operated applications and many memory circuits are therefore implemented in static CMOS. Chapter 8 presents extensive discussions on low-power issues.

Speed and Area

Dynamic CMOS logic circuits are generally faster than their static CMOS counterparts. The nMOS-mostly nature of dynamic CMOS logic means that pMOS transistors are largely reserved for precharge and/or transfer functions while logic functions are only implemented in nMOS transistors. The input capacitance of a dynamic logic gate is therefore lower than a static equivalent. In addition, complex logic gates implemented in static CMOS may contain many pMOS transistors in series in the 'pull-up' path. A dynamic CMOS implementation offers increased speed and a smaller area because it uses only one pMOS transistor as an active pull-up.

Noise Immunity

In a static CMOS logic circuit, there is always a conduction path from a logic gate's output to ground or to the supply. Therefore, static CMOS circuits never show floating input and output nodes. Noise-induced voltage deviations on their logic levels are automatically compensated by current flows which restore levels. Dynamic circuits suffer from charge sharing and cross-talk effects, as already mentioned. There is also always a minimum clock frequency required because of the leakage of charge from floating nodes. As a result, static circuits are more robust. For this reason, most semi-custom design libraries are implemented in static CMOS.

4.4.7 Clocking Strategies

Advantages and disadvantages of several implementations of single-phase and multi-phase *clocking strategies* have been described in the previous discussions of static and dynamic CMOS circuits. Single-phase circuits are the most efficient in terms of routing area. However, they may require more transistors than multi-phase alternatives. Today's flip-flops include two inverters to generate the intended ϕ and $\overline{\phi}$. The many transistors required for a NAND gate implementation of a flip-flop should also be remembered. In addition, the timing behaviour of *single-phase* circuits is critical and requires many circuit simulations to ensure equivalent functionality for best and worst cases, i.e., when delays are shortest and longest, respectively. *2-phase* circuits that use non-overlapping clocks have less critical timing behaviour.

Clock skew is always present in clocked circuits. Chapter 9 describes clocking strategies and alternatives and also extensively discusses potential timing problems involved in designs with relatively large clock skew(s).

4.5 CMOS Input and Output (I/O) Circuits

The electrical 'interfaces' between a CMOS IC and its external environment must ensure that data are received and transmitted correctly. These input and output interfaces must be able to withstand dangers that they may be reasonably expected to encounter. CMOS input and output circuits and the associated protection circuits are discussed below.

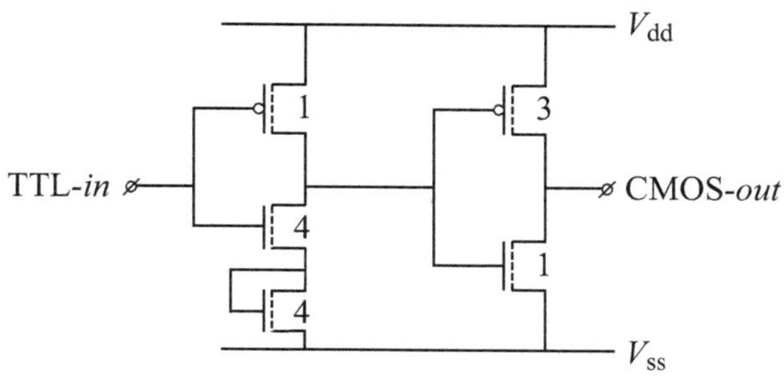

Fig. 4.40 TTL-CMOS input buffer

4.5.1 CMOS Input Circuits

MOS ICs often have to communicate with several other types of logic, such as ECL and TTL. All CMOS Inputs therefore require level conversion through a so-called *level shifter*. A *TTL-compatible* input buffer must interpret an input voltage below 0.8 V as 'low' while voltages above 2 V must be interpreted as 'high'. The switching point of a TTL-compatible CMOS inverter must therefore be about 1.5 V. However, the switching point of a symmetric CMOS inverter (i.e., an inverter with equal transconductances for the nMOS an pMOS transistors) is half the supply voltage. The effects of asymmetry on the switching point of an inverter are shown in the transfer characteristic in Fig. 4.23. This figure clearly illustrates that a TTL-compatible CMOS inverter must be asymmetric.

Figure 4.40 shows a TTL-CMOS input buffer with the approximated transistor aspect ratios. The first inverter converts the TTL input signal to a CMOS level. Today's I/O voltages support 2.5 V or 1.8 V with threshold voltages close to 0.5 V. For this reason, the nMOS input transistor is in series with an nMOS diode so that is off when the input is the maximum TTL low level of 0.8 V.

An input buffer is usually located quite a distance from the logic gates that it drives. The required routing then forms a considerable load capacitance. A clock signal's input buffer is even more heavily loaded. The size of the load capacitance determines the required widths of the nMOS and pMOS transistors in an input buffer's second inverter. To achieve equal rise and fall times, the ratio of these widths must be approximately as shown.

The widths and lengths of manufactured transistors may vary independently as a result of processing variations. The effects of these variations are particularly evident for smaller dimensions. Minimum allowed dimensions should therefore not be used to achieve the required accuracy for the switching point of about 1.5 V for the first inverter in Fig. 4.40. In a 2.5 V CMOS input buffer, with an nMOS threshold voltage of 0.5 V for instance, a minimum channel length of 0.25 μm, and a minimum channel width of 0.3 μm, the first transistor could be dimensioned as follows:

$$\left(\frac{W}{L}\right)_\mathrm{p} = \frac{0.5}{0.4}\,\mu\mathrm{m} \text{ and } \left(\frac{W}{L}\right)_\mathrm{n} = \frac{1.6}{0.4}\,\mu\mathrm{m}$$

Not using the minimum transistor sizes makes them less sensitive to process variations.

4.5.2 CMOS Output Buffers (Drivers)

There are many different output buffer designs. They usually contain a tapered chain of inverters, as discussed in Sect. 4.3.2. Transistor sizes in the output buffer are determined by the specifications of the output load and the clock frequency. Output load capacitances usually range from 10 to 30 pF, and I/O clock frequencies range from several MHz to 1 GHz.

Several problems arise when many outputs switch simultaneously at a high frequency. The resulting peak currents through metal tracks may exceed the allowed maxima. These currents also cause large *voltage peaks* across the intrinsic inductances in the bond wires between a chip's package and its bond pads. The accumulation of peak currents in power and ground lines leads to relatively large noise signals on the chip. These problems (which are also discussed in Chap. 9) must be taken into account when designing output buffers.

The very large transistors required in output drivers could result in unacceptably large *short-circuit currents* between supply and ground if the charge and discharge transistors were allowed to conduct simultaneously. Figure 4.41 shows an example of a short-circuit free 2.5 V output buffer. This tri-state buffer is combined with an output flip-flop and can drive a 10 pF load at 250 MHz. Signals 1, 2 and

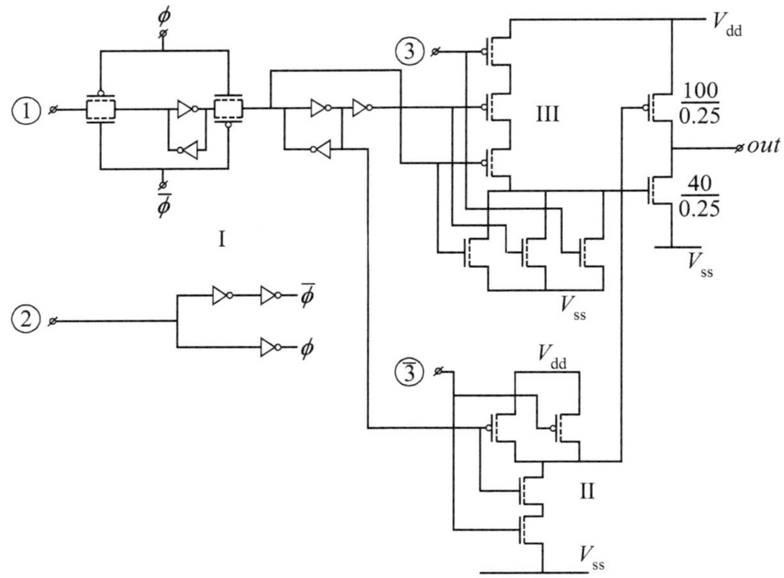

Fig. 4.41 Short-circuit free tri-state CMOS 2.5 V output buffer

3 represent the input data, the clock and the tri-state control, respectively. The logic circuits II and III control the gates of the nMOS and pMOS output driver transistors, respectively. These circuits ensure that the driver transistors never conduct simultaneously. The pre-driver logic gates must be designed such that they fulfil the tapering factor requirements as described in Sect. 4.3.2. This output driver is just one example of the many existing driver types. Many alternatives are available through publications and through the internet, each with its own schematic which is targeted at the specific application area.

Many I/O libraries also contain bi-directional pads, which contain both an input and output driver cell. An enable signal is then used to select the function (input or output). All I/O cells include a lot of additional circuits to protect the chip for latch-up, ESD and other reliability phenomena. These are treated in Chap. 9.

With today's system-on-chip integration, the number of I/O's is rapidly increasing. Consider a 64-bits bus interface. When all outputs switch simultaneously, it will introduce a large supply noise on either of the supply lines. Therefore, libraries contain output driver cells of which only a limited number may be connected to the same output supply line. An *SSO*-6 driver is a *simultaneously switching output* driver of which only six maybe supplied by the same output supply and ground pad. Chapter 9 also discusses the design strategies related to supply noise.

4.6 The Layout Process

4.6.1 Introduction

In this section, we present a simple set of basic design rules for a *CMOS process* containing a single polysilicon and a single metal layer. Normally foundries provide a *design-rule manual (DRM)* which may contain several hundred pages of design rules. Traditionally, the design rules consist of a set of geometric constraints within or between the various mask layers from which the chip is built. These constraints are defined such that a chip should fully match the specified requirements whenever the mask misalignments, lithography imperfections and process parameter variations are within their specified limits, independent of what circuits we create with them. These design rules also include special rules for the bonding process, such as bond pad sizes and spacings, but also support reliability topics as seal rings, guard rings and scribe lanes. Today, a DRM also covers *design for manufacturability (DfM)* rules, OPC rules, antennae design rules, pattern density rules (the use of tiles), reliability rules (NBTI, PBTI, electromigration, hot-carrier injection, latch-up and ESD and several stress sources) and scribe line rules. They also contain specific design rules and recommendations for analogue circuits and special components, such as resistors and capacitors. To include all the above aspects, the number of design rules has grown from around 1000 in the 60 nm node to close to 6000 in the 10 nm node. This has also increased the design complexity and design costs, which are discussed in more detail in Sect. 11.5.

This book includes a simple one-page design rule manual, which clearly shows the layout process. It gives a very good impression of the structure of a CMOS layout, as well as of the physical and geometrical shapes of the transistors and their connections. The resulting layout is rather close to a realistic layout. These layout *design rules* represent a virtual 28 nm CMOS process. Although such a process usually incorporates about six to ten metal layers, only one metal layer will be used in this layout design process. This is because many of the libraries only use the first metal layer for the local interconnections inside each *library cell*. After a description of each individual mask, the creation of a stick diagram and the layout process are demonstrated with an example. Finally, a process cross section shows the real silicon implementation.

4.6.2 Layout Design Rules

The process *masks* of the chosen technology are listed below in the order of the process sequence. Many of these masks are described in Sect. 3.9.

ACTIVE (layout colour: green)
This mask defines the *active areas* inside which the transistors will be created. Outside the active areas, thick oxide will be formed with STI (or LOCOS in the past). The width of an ACTIVE pattern determines the transistor channel width.

NWELL (layout colour: yellow)
This mask defines the areas where the pMOS transistors will be located. The n-well actually serves as a substrate for the pMOS transistors. As the CMOS process offers complementary transistors, the creation of a p-type substrate (p-well) for nMOS transistors is also required. This is usually automatically generated from the NWELL mask: a p-well will be created everywhere where no n-well pattern is defined. In most CMOS processes, the p-well areas are not fully complementary to the n-well areas, as wells may be used, for example, to implement resistors in analogue circuits, as well as for guard rings to prevent latch-up or ESD, or to isolate a sensitive analogue circuit better from substrate noise caused by the digital. So, normally there is a PWELL and an NWELL mask, however, for this simple set of design rules, it is assumed that the n-well and p-well are fully complementary, and therefore, the PWELL mask need not be drawn in the stick diagram and layout.

POLY (layout colour: red)
This mask defines the polysilicon pattern. A transistor channel is formed where POLY crosses an ACTIVE region. On top of thin gate oxide, polysilicon acts as a MOS transistor gate. Outside the active areas, polysilicon is used as a local interconnection only over small distances inside the library cells. The minimum width of the polysilicon determines the transistor channel length.

NPLUS (layout colour: orange)

	The sources and drains of nMOS transistors need n^+ implants. The NPLUS mask defines the areas in which n^+ is implanted. During the n^+ implantation, the STI (thick oxide regions) and the polysilicon gate act as barriers, e.g., we get self-aligned n^+ regions (sources and drains) everywhere within ACTIVE which is surrounded by NPLUS and not covered by POLY.
PPLUS	(layout colour: purple) Complementary to the NPLUS mask, sources and drains of the pMOS transistor are p-type doped by means of the PPLUS mask.
CONTACT	(layout colour: black) This mask defines contact holes in the dielectric layer below the first metal layer (METAL). Through these contact holes, the metal layer can contact polysilicon (POLY) and source or drain regions (ACTIVE).
METAL	(layout colour: blue) This defines the pattern in the first metal layer, which can be aluminium and tungsten in the 180 nm CMOS node and above, and copper in the 120 nm CMOS node and beyond. A track in this layer can be used for both short and long interconnections because its sheet resistance is relatively low.

Note Only a one-metal-layer process is used in the layout here. Current CMOS technologies use six to ten metal layers. However, for educational purposes, a layout with one metal layer gives a very good insight into the layout process. More metal layers only means: 'more of the same'.

Design Rules for a Virtual 28 nm CMOS Process

The following set of design rules of a virtual 28 nanometer CMOS process will be used in an example of a layout and in several exercises at the end of this chapter. Figure 4.42 serves as an illustration of each of these design rules.

ACTIVE		
a.	Track width	40
b.	Track spacing	60
NWELL		
c.	Track width	200
d.	Track spacing	200
e.	Extension NWELL over ACTIVE	52
POLY		
f.	Track width	24
g.	Track spacing	64
h.	Extension POLY over ACTIVE (gate extension)	64
i.	Extension ACTIVE over POLY (source/drain width)	60
j.	Spacing between ACTIVE and POLY	25

NPLUS		
k.	Track width	125
l.	Track spacing	125
m.	Extension NPLUS over ACTIVE (n^+ ACTIVE)	50
m1.	Spacing between n^+ ACTIVE and POLY	60
n.	Spacing between n^+ ACTIVE and NWELL	50
PPLUS		
o.	Track width	125
p.	Track spacing	125
q.	Extension PPLUS over ACTIVE (p^+ ACTIVE)	50
q1.	Spacing between p^+ ACTIVE and POLY	50
CONTACT		
r.	Minimum and maximum dimensions	32×32
s.	Spacing between contacts	55
t.	Extension ACTIVE over CONTACT	12
u.	Extension POLY over CONTACT	12
v.	Extension METAL over CONTACT	12
w.	Spacing CONTACT and POLY gate	24
x.	CONTACT on gate regions not allowed	!
METAL		
y.	Track width	40
z.	Spacing between tracks	40

The minimum width and spacing in a certain mask pattern are defined by the different processing steps involved. For instance, the ACTIVE is defined by the STI formation process, while a METAL pattern is the result of deposition and etching techniques. Minimum overlaps or separations between patterns in different masks are defined by alignment tolerances with respect to a common reference location and by the different processing steps involved. The minimum width of the POLY mask pattern determines the channel length of the transistors and is usually referred to in the process node name, e.g., a 40 nm CMOS process means that the minimum POLY width is close to 40 nm. However, below the 28 nm node, the name of the process node does no longer correspond with the transistor channel length or any other size (see the final conclusions section in the last chapter of this book).

The previous set of design rules is reflecting a virtual 28 nm CMOS process, although the channel length (= poly width) = 24 nm. For educational purposes, these rules have been simplified and rounded. For example, rules t, u and v assume an extension of 12 nm in all directions. However, in many layout design rule Manuals, these values could be less in one direction, under the condition that it is more in the other direction.

Usually, when a complex layout has to be developed, a stick diagram is first drawn to explore the different possibilities of layout interconnections. The use of a stick diagram is discussed first.

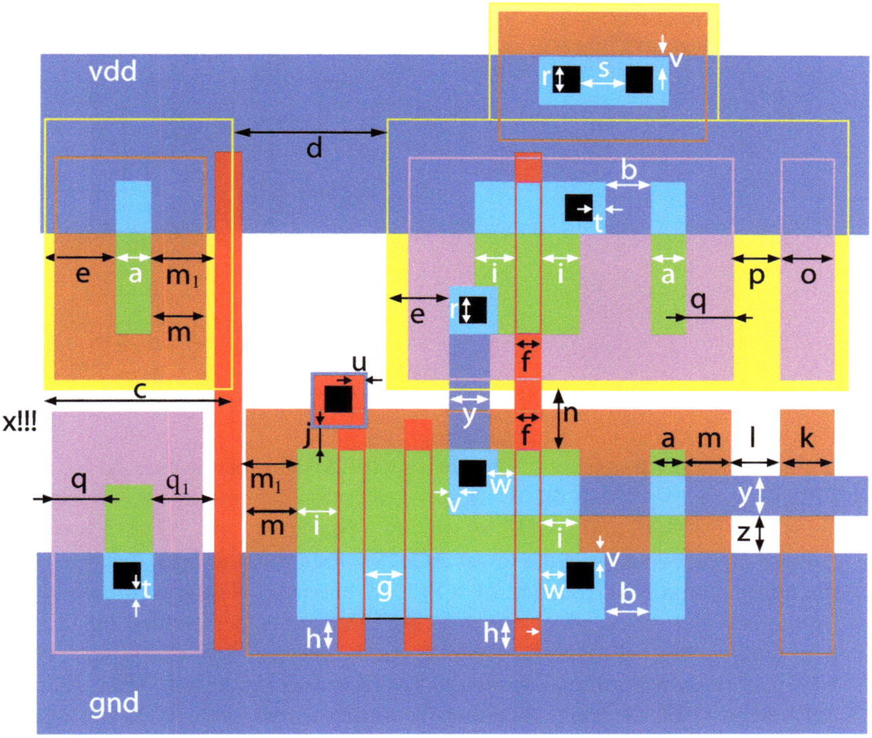

Fig. 4.42 Illustration of each of the design rules of the previous page

4.6.3 Stick Diagram

A *stick diagram* is used as an intermediate representation between circuit diagram and layout. This topological representation of the circuit is drawn in colours which correspond to those used in the layout. Only the connections of the different mask patterns are depicted, without paying attention to the sizes. The EXNOR circuit of Fig. 4.43 serves as an example for the development of a stick diagram. This EXNOR circuit represents the Boolean function: $Z = \overline{(a+b)\overline{ab}} = a\,b + \bar{a}\,\bar{b}$

Figure 4.44 illustrates the procedure for the generation of the stick diagram for the EXNOR logic gate.

The creation of this topological view is divided into three phases, represented by (a), (b) and (c) in the figure. These phases are explained as follows:

(a) Two horizontal parallel thin oxide (ACTIVE) regions are drawn. The lower ACTIVE region is usually reserved for nMOS transistors while the upper region

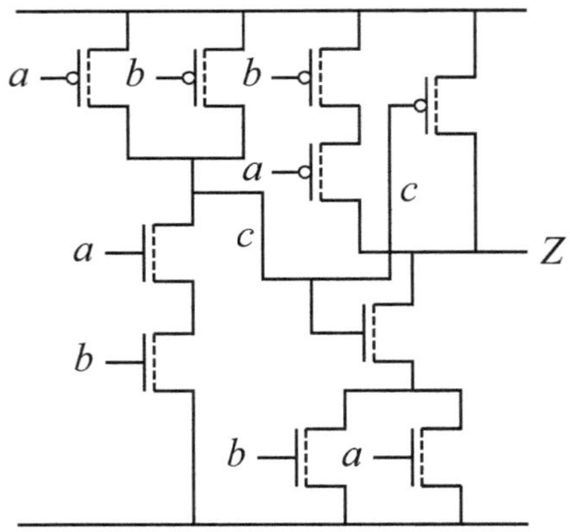

Fig. 4.43 Circuit diagram of a CMOS EXNOR logic gate

is for the pMOS transistors. The envisaged CMOS process uses NPLUS and PPLUS masks to define the n^+ and p^+ diffusion regions of the source/drain areas of the nMOS and pMOS transistors, respectively. An NPLUS boundary is therefore drawn around the lower ACTIVE region in the stick diagram while the upper region is surrounded by a PPLUS boundary. The n-well is indicated by the NWELL area, which overlaps ACTIVE areas surrounded by PPLUS. It is not required to draw the PWELL mask, because it is the inverse of the NWELL mask; everything outside the NWELL area becomes PWELL. Parallel polysilicon (POLY) gates are drawn vertically across both ACTIVE regions. Metal (METAL) supply and ground lines are drawn horizontally over the PPLUS and NPLUS regions, respectively.

(b) Additional METAL and POLY lines indicate transistor connections according to the function to be implemented. The source/drain diffusion areas of neighbouring transistors are merged and black crosses represent contacts. These transistor connections are implemented from left to right. The two nMOS transistors on the left of the stick diagram, for example, correspond to the nMOS transistors of the NAND gate on the left of the circuit diagram in Fig. 4.43. The drains of two pMOS transistors and one nMOS transistor are connected with METAL to form the NAND gate output. This connection is represented by a metal interconnection of n^+ and p^+ diffusion areas. A direct diffusion connection between an n^+ and p^+ area is not possible as it would form a diode. Connections between n^+ and p^+ areas therefore always occur via metal. The NAND gate output is connected to the gate of the most right nMOS and pMOS transistors.

(c) The third nMOS source/drain area from the left in Fig. 4.44b, is connected to ground and to another node. This is clearly not according to the required functionality, and such diffusion areas are therefore split into separate diffusion areas in Fig. 4.44c. Finally, we have to enable connections to the inputs and the outputs. Because first metal is already used for supply lines and internal cell connections, we have to enable connections to the second metal layer. This is done through adding vias to the input and output terminals of the cell. These vias are represented by the black circles. On top of these vias, small second-metal

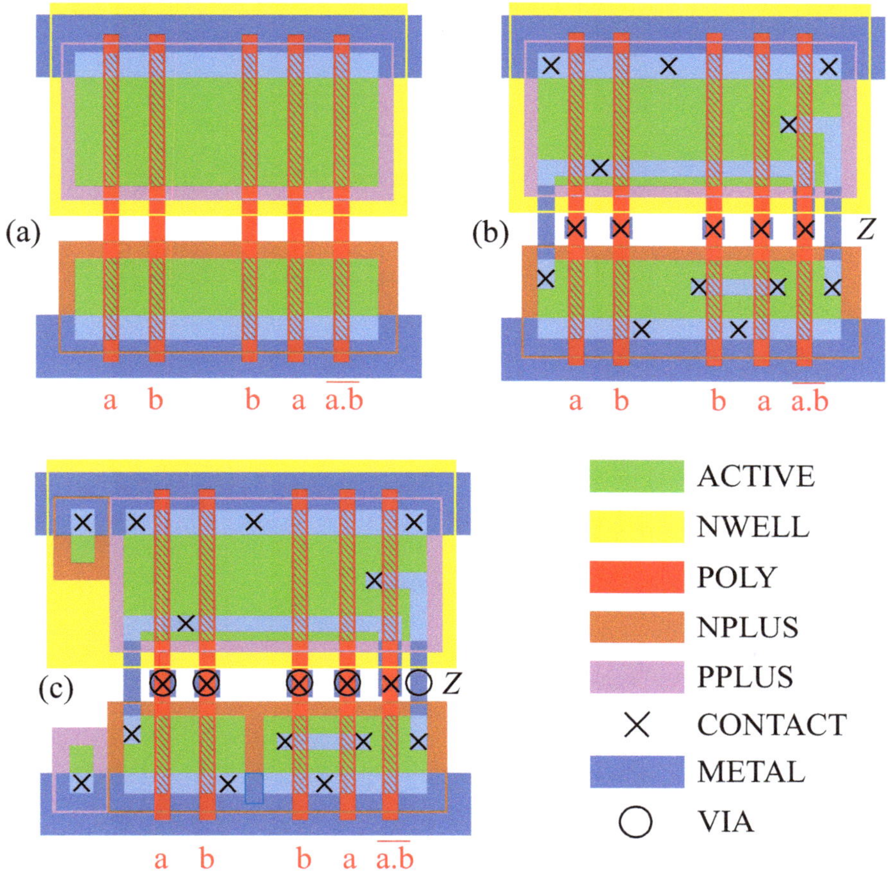

Fig. 4.44 Various steps in the design of a stick diagram

areas must be positioned to form the real terminals. However, for educational purposes, this is not drawn in this stick diagram, as it would make the figure less clear.

No back-bias voltage is used in the chosen process. The p-type substrate is therefore connected to ground, and the n-well is connected to the supply. These substrates and n-well connections are indicated at the left side in the figure.

There should be enough connections from PWELL to ground and from NWELL to V_{dd} to keep latch-up sensitivity to a low level. (latch-up is discussed in Sect. 9.5.5) These contacts reduce the values of R_1 and R_2, respectively, in Fig. 9.34. In current advanced CMOS libraries, these PWELL and NWELL contacts are included in a separate standard cell, which can be placed according to the needs of the specific design, e.g. 30–40 μm apart. This subject is further addressed in the layout discussion below.

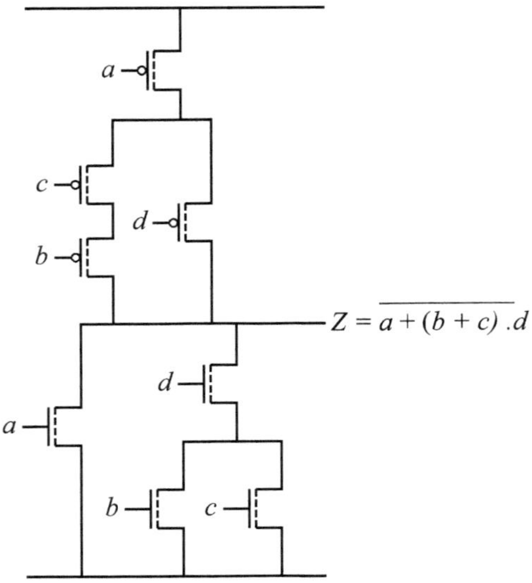

Fig. 4.45 Circuit diagram implementing $Z = \overline{a + (b + c) \cdot d}$

4.6.4 Example of the Layout Procedure

The following example shows the complete layout process from a basic Boolean function, through Boolean optimisation, circuit diagram and stick diagram to a layout. Consider the following Boolean function:

$$Z = \bar{a}\,\bar{b}\,\bar{c} + \bar{a}\,\bar{c}\,\bar{d} + \bar{a}\,c\,\bar{d} + \bar{a}\,\bar{b}\,c\,\bar{d}$$

To optimise this function for implementation in CMOS, an inverse Boolean expression in the format $Z = \bar{f}$ must always be found, because every single CMOS logic gate implements an inverted expression:

$$\begin{aligned} Z &= \bar{a}\,\bar{b}\,\bar{c} + \bar{a}\,\bar{c}\,\bar{d} + \bar{a}\,c\,\bar{d} + \bar{a}\,\bar{b}\,c\,\bar{d} \\ &= \bar{a}\,(\bar{b}\,\bar{c} + \bar{c}\,\bar{d} + c\,\bar{d} + \bar{b}\,c\,\bar{d}) \\ &= \bar{a}\,(\bar{b}\,\bar{c} + (\bar{c} + c + \bar{b}\,c)\,\bar{d}) \\ &= \bar{a}\,(\bar{b}\,\bar{c} + \bar{d}) \\ &= \overline{\overline{\bar{a}\,(\bar{b}\,\bar{c} + \bar{d})}} = \overline{a + \overline{(\bar{b}\,\bar{c} + \bar{d})}} = \overline{a + (b + c)d} \end{aligned}$$

Therefore, the optimised function for implementation as a single CMOS logic gate is: $Z = \overline{a + (b + c)d}$. The circuit diagram for this logic function is shown in Fig. 4.45.

The corresponding CMOS layout can be found in Fig. 4.46 (top). Figure 4.46 (bottom) shows a cross section through the line A-A' in the layout.

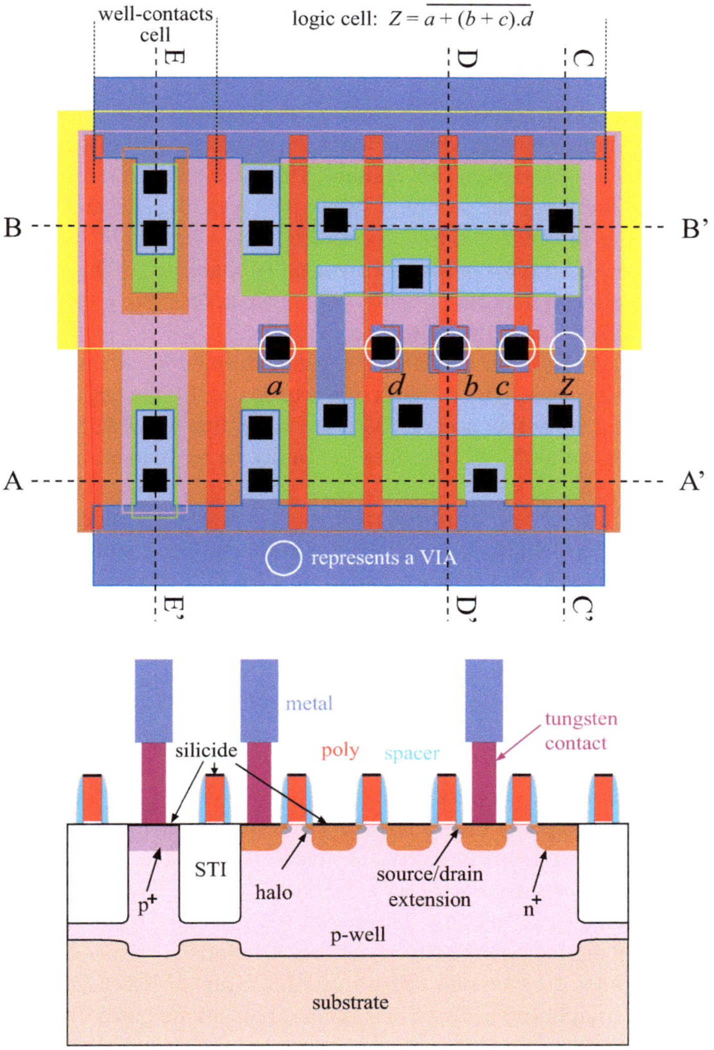

Fig. 4.46 Layout (top) and cross section (bottom) of the sample logic gate along the line A–A'

The layout contains one substrate (p-well) and one n-well contact. The use of extra n-well and p-well contacts reduces latch-up sensitivity but may lead to an increased layout area. In a 0.25 μm 2.5 V CMOS technologies and above, a practical compromise was to place at least one substrate and n-well contact per five nMOS and pMOS transistors, respectively. Because the supply voltage of current advanced CMOS technologies is close to 1 V, the probability of latch-up has reduced significantly and requires much less PWELL and NWELL contacts. Current libraries in 65 nm technologies and beyond can have cells with fixed NWELL and PWELL contacts in the cells (*tapped cells*) as well as cells that don't (*tap-less cells*). Libraries with tap-less cells also have special library cells which only contain an n-well and a p-well contact. For latch-up prevention, these *well-tap* cells (or shortly: tap cells) must be placed in between the normal logic cells, at the required interval as defined in the design rule manual. The large transistors in driver and I/O circuits which usually operate at higher supply voltages (3.3, 2.5 and 1.8 V) require considerably more substrate and n-well contacts.

The n-wells in a CMOS circuit layout are usually connected to the supply voltage. Generally, different neighbouring n-wells (which are connected to the same voltage) should be extended to form one large well.

The output node of a static CMOS logic gate is formed by an interconnection of n^+ and p^+ diffusion areas. The p^+ diffusion area is usually the larger. The parasitic capacitance of such an output node is therefore larger than its nMOS counterpart. In addition, the width of a pMOS transistor is usually larger than an nMOS transistor width.

As a result of silicided p^+ diffusion regions, the series resistances of sources and drains are low, and usually only one contact is sufficient per connected node. These resistances are only several ohms per square in CMOS technologies with silicided source and drain regions. Minimum source and drain areas can then be used to keep parasitic capacitances small.

The process cross section in Fig. 4.46 is made along the line A-A'. The cross section includes n^+ source, drain and gate areas, STI oxide isolation areas, a p-well area, a p-well contact and a source contact. A detailed study of the relationship between the cross section and the layout should enable the reader to draw a cross section at a line anywhere in the layout.

Circuit density and performance are often improved by using several polysilicon layers (memories) and seven to ten metal layers (VLSI). The area reduction must compensate for the costs associated with the additional masks and processing steps. However, with the ever-increasing current density, more and more metal layers are required to distribute the power properly across the chip.

In CMOS technologies beyond 100 nm, all nMOS and pMOS transistors are only allowed to have straight (no L-shape or Z-shape) polysilicon gates. In 65 nm CMOS and beyond, the regularity of the transistor patterns is increasing to support *litho-friendly design*. In these technologies, an increasing fixed-pitch approach is adopted to ease the lithographic process and to improve reproduction and yield. Therefore, the layout in Fig. 4.46 includes *dummy polysilicon* tracks at the left and right side of each library cell to increase the regularity of the polysilicon pattern.

4.6.5 Guidelines for Layout Design

Designing a correct layout involves more than just a translation of the circuit diagram into a layout that meets the relevant design rules. Attention must be paid to several key issues:

- Minimise layout area.
  A minimum layout area will especially reduce the overall silicon costs with the development of a new library that is to be used for the design of numerous chips. Moreover, when ICs become smaller, they generally show a higher performance, consume less power and are cheaper.
- Pay attention to parasitic elements.
  Each design, whether a library cell or a large logic block, must be optimised with respect to parasitic capacitances (source and drain junctions, metal interconnects) and resistances (mainly of long interconnections). This is necessary to achieve better performance and again reduces the power consumption.
- Pay attention to parasitic effects.
  Effects such as cross-talk, charge sharing and voltage drop across supply lines particularly greatly reduce the performance as well as the signal integrity. Such effects are extensively discussed in Chap. 9.

Table 4.2 shows some typical values of the capacitances and resistances of different components and materials used in a virtual 28 nm CMOS technology with a gate oxide thickness $t_{ox} = 12\,\text{Å}\quad(1.2\,\text{nm})$.

Note $*$ on thin oxide

It is clear that polysilicon and n^+/p^+ junctions can only be used for very short connections inside library cells as a result of the relatively high sheet resistance values.

Current nanometer CMOS logic processes include five to more than ten layers of metal. Most processes support one or more thick top metal layers, as explained in Chap. 3, to create a structured and adequate overall power distribution network.

The above discussions on CMOS layout implementation conclude this paragraph. More information on the design of CMOS circuits and layouts can be found in the reference list.

Table 4.2 Parasitic capacitances and resistance values in a virtual 28 nm process with $t_{ox} = 1.2\,\text{nm}$

Material	Capacitances	Resistances
Polysilicon	Gate cap: 17 fF/μm^2$*$	Poly 0.7–1 kΩ/□
(POLY)	Edge cap: 0.22 fF/μm $*$	Polycide 8 Ω/□
Copper (Cu)	Average track cap: 0.19 fF/μm	$\frac{5\cdot10^{-8}}{H}$ Ω/□
Metal layer My		
Source/Drain	Track cap: 1 fF/μm^2	$n^+ \approx$ 200–500 Ω/□
implants	Thick oxide edge cap: 0.3 fF/μm	$p^+ \approx$ 300–700 Ω/□
(ACTIVE)	Cap to POLY edge: 0.3 fF/μm	Silicided $n^+ \approx 8$ Ω/□
		Silicided $p^+ \approx 8$ Ω/□

4.7 Libraries and Library Design

In the previous section, we focussed on the layout design of individual (library) cells, with a focus on the layout design principle and layout design rules. The strategy is the same for all cells in a library. A *library* is a collection of pre-characterised *standard cells*, which introduces a certain design consistency to create, simulate and test large functional blocks (cores). The key aspect in these libraries is that all library cells have the same fixed height, which enables them to be placed in rows, to support the design automation process. A standard cell is a collection of transistors and wires which together implement a logic function (e.g., INVERT, AND, OR, NAND, NOR, EXOR, or combinations thereof e.g., full-adders, comparators and multiplexers) or storage functions (e.g., flip-flop or latch). It requires a huge design effort to develop a library from scratch. One strives to achieve the smallest possible area, while also fulfilling the electrical performance requirements, such as optimum power and speed. Because we have to deal today with an extremely broad spectrum of diverging applications, they can't all be served by just one library. Mobile (smartphones, tablets, wearables, etc.), automotive, healthcare, identification, security, industrial, Internet of Things (IoT), PCs and server stations, they all require their own specific process and library performance. Therefore, foundries support different libraries for the same technology node but targeted at different application domains: high-performance (HP), low-operating power (LoP) and low-standby power (LSTP). These are also the CMOS processes as categorised by the ITRS roadmap. Often the high-speed ICs used in PC and server stations, for example, are made in an *HP-CMOS* process, while ICs that are always active and need not run at the highest speed (e.g. an MP3 decoder), will most probably be manufactured in a *LOP-CMOS* process. Most ICs that are used in mobile applications with relatively long periods of standby time, are manufactured in *LSTP-CMOS* processes. It will be clear, now, that each foundry develops several different libraries for the same technology node to support the various application domains. The development of these libraries from scratch for every new technology node is very time consuming. Therefore, to create a library for a new process node, it has become common practice to reuse as much as possible from the library IP created for the previous process node. However, layout and performance requirements may change due to different lithography and fabrication tools and changing application domains, respectively. This may limit the similarity between libraries in successive process nodes and prevents a fast library migration. It may require more manual layout changes or the development of specific scripts to (semi-) automate these manual changes.

Standard-cell libraries from foundries or library vendors come with fully characterised cells for the complete range of specified temperatures, voltages and process parameters. Today's advanced designs require more than just process corner data to allow accurate modelling of cores that may run at different voltage levels and temperatures, depending on their use in the application. With adaptive voltage and frequency scaling, the library must be characterised over the full range over

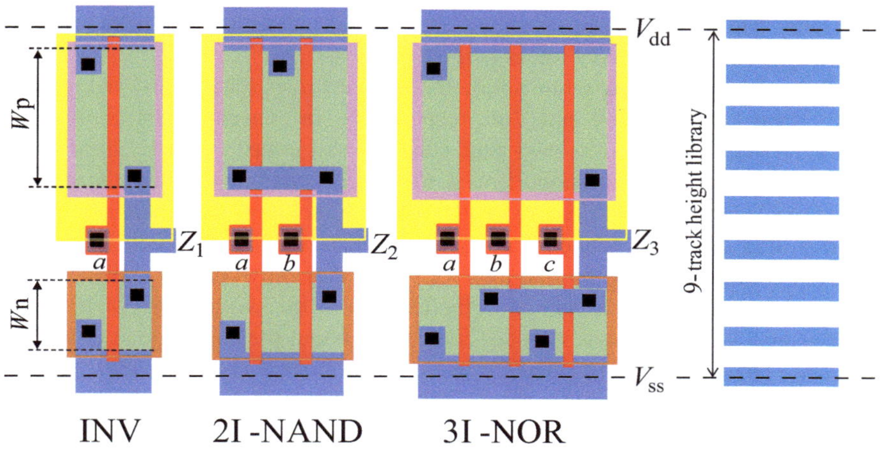

Fig. 4.47 Example of three basic standard library cells

potentially applied voltages and frequencies to guarantee first-time-right products. Also designers can no longer over-design to include all random process variations around specified process corners. This would lead to (area) overkill and requires more application relevant process corners, based on statistical timing analysis.

Today's libraries may contain several thousand to sometimes even more than ten thousand, due to a large variety of drive strengths, different threshold voltages (HV_T, SV_T and LV_T, respectively referring to high, standard and low-V_T) and different channel lengths. This enables the synthesis tools to better adapt a design to the aforementioned performance categories: high speed, low operating power or low standby power. Figure 4.47 shows a few different standard cells, as a first step to explain library architecture.

All cells have their V_{dd} and V_{ss} lines at the same height. These cells are placed in rows, such that their V_{dd} and V_{ss} lines form long continuous supply and ground lines along each row of cells through abutment. Most place and route tools (Chap. 7), require that the width and height of the library cells are equal to an integer number of first metal routing tracks that fit across the cells. The cells in the figure correspond to a 9-track *library cell height*. The widths of the nMOS and pMOS transistors, W_n and W_p, respectively, depend on the required *drive strength* as defined by the application. Most cells come with different drive strengths. For example, a 2-input NAND cell (2-I NAND) can have drive strength of 1/2 x drive, 1 x drive, 2 x drive, 3 x drive, 4 x drive, depending on the required speed, power and area requirements of the application. Usually the 1/2 x drive is used for low-power applications, while the larger drive cell support higher-speed applications, of course, at the cost of additional area. Subthreshold leakage currents are a real burden for applications that require low standby power. These LSTP applications are supported by the libraries in that they also offer so-called high-V_T (HV_T) cells. High-speed circuits, on the contrary, require faster transistors and benefit from low-V_T (LV_T) library cells.

This latter category of applications are usually realised with larger cells (e.g., 14 track height cells instead of 9 tracks height).

All typical CMOS logic circuits require their substrates (or p-wells) and n-wells to be connected to ground and Vdd, respectively. In conventional technologies, each library cell contained its own fixed substrate and n-well contact. Libraries for 60 nm CMOS and beyond support, next to *tapped cells*, which have these fixed well contacts included, also *tap-less cells*, which do not have these contacts. These libraries also contain so-called *well-tap cells*. These are library cells that have exactly the same height as normal logic cells, but they only contain a substrate (p-well) and n-well contact, as shown in the left part of the layout in Fig. 4.46. Most libraries also include similar well-tap cells, but without the connections to ground and Vdd. These cells are targeted for use in logic cores that require their p-wells and n-wells to be connected to separate back-bias voltages to increase the threshold voltages in standby mode. These well-tap cells have terminals connected to their respective wells and can automatically be placed and connected to the corresponding separate back-bias voltage supply lines by the place and route tools at distances as prescribe by the corresponding design rule manual (DRM).

Next to the previously discussed options for digital library cells, several semiconductor vendors and foundries also offer libraries built from transistors with longer than minimum channel length. These long-channel devices, with a channel length that is approximately 10% larger than nominal, exhibit a larger V_T and consume about a three times lower subthreshold leakage [13]. Intel makes extensive use of these long-channel devices in their Nehalem-EX processor, on paths with timing slacks, which increases V_T in non-critical paths.

All logic cells need to be fully characterised, meaning that key cell parameters, such as rise- and fall times, propagation delay and power consumption are captured by extensive simulations under worst-case, typical and best case environment and process conditions (*PVT* variations). These simulations are performed on fully extracted netlists, which also contain all parasitic capacitances and resistances and inductances. These netlists are generated by a so-called *parasitic extraction* tool, which creates a very accurate model of the circuit. Usually the physical layout of an IC is represented in an ASCII format called Design Exchange Format (*DEF*). It includes a netlist, placement of the cells and routing information, but has no information about the cells. The Library Exchange Format (*LEF*) is used to describe the library cells or macros and includes an abstract layout view, dimensions and cell area, design rules (technology data), routing blockages, pin positions and propagation delays from input(s) to output(s). All standard cells are checked against the design rules (see Sect. 4.6.2 for an example set of design rules). Although LEF and DEF formats are developed by Cadence, they have become industry-wide standards for design and layout information exchange.

Library efficiency depends on the ability to quickly create large variety of different cores on the chip. Therefore, most libraries also include memory generators for embedded SRAM and flash memory blocks, microprocessor cores and controllers, FIFOs and a lot of analog IP. In Chap. 7 we will continue the design trajectory with VLSI design and explain the tools that use the library cells to create larger logic core (blocks).

4.8 FinFET Layout

Section 4.6 presented detailed discussions on traditional CMOS layout design. This section discusses a layout approach for FinFET technologies, which is very similar to the traditional layout style. Next to the fact that FinFETs transistors carry their current along the edges of the 3D fins, FinFET processes are also a little different in the lowest transistor interconnection layers. Figure 4.48 shows a layout example of a planar MOS transistor (left) and of a FinFET (right).

In fact, there is not so much of a difference. Of a lateral MOS transistor, we could vary both the transistor channel width and length, as required by the application. In Sect. 4.6, we have shown that advanced CMOS processes already only use a single fixed polysilicon pitch (fixed width and fixed spacing) to support litho-friendly design. In fact, a FinFET layout consists of a row of fins in the horizontal direction covered with a row of metal (e.g. W for nMOS and TiAlN for pMOS) gates in the vertical direction. The metal gates are created by replacement gate process steps similar to a planar metal-gate process shown in Fig. 3.58. Wherever the fins are within the active area, they act as transistor source, drain or channel. The fins in the source and drain areas need to be connected to each other in order to parallelise these transistors to create a wider transistor. This is often done by an additional *local interconnect layer (LIL)*, in the above figure called Lilfin (local interconnect layer fin). Some vendors also call this the Metal0 layer. This layer is directly deposited onto the fins. The commonly used number of fins available in a library cell for both nMOS and pMOS FinFETs may vary from five to ten. The number of parallel connected fins depends on the required drive strength. So, the FinFET width is quantised and in the example layout, the total transistor width W equals that of five parallel fins. As is shown in Chap. 3, the transistor width W is than equal to: W = 5x(2x fin height + fin width). Possible physical dimensions in an example

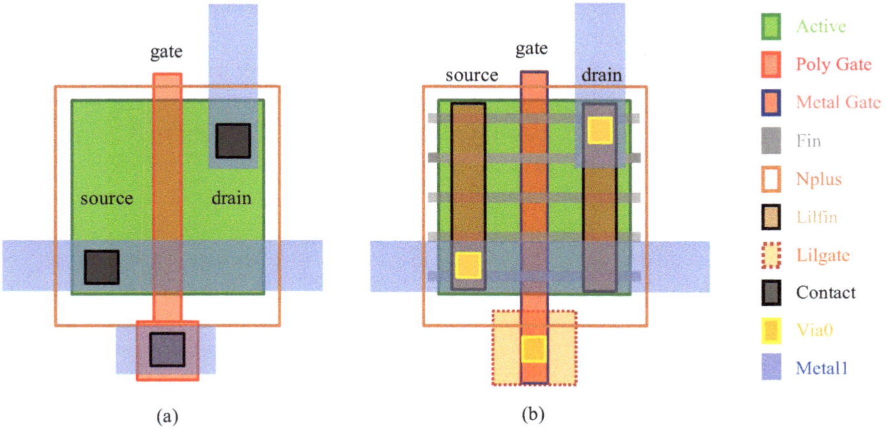

Fig. 4.48 Example layout of a planar nMOS transistor and a FinFET

16 nm FinFET process are: 28 nm for the gate length, 70 nm for the gate pitch, 8 nm for the fin width, 40 nm for the fin pitch and 40 nm for the fin height. In this example of 14 nm FinFET process, the fin height is even larger than in a 22 nm FinFET process to increase the current drive. This means that the *fin aspect ratio* (height/width) is 5. With these values, the total five-fin wide transistor width $W = 5 \times (2 \times 40 + 8) = 440$ nm (a lateral transistor that fits within five fin pitches would have a width of 200 nm). Because of this, FinFETs may show a two times higher drive current compared to a planar transistor with the same footprint. Advanced FinFET processes show aspect ratios from 5 toward 8, to further increase the drive current, which will allow a reduction of the number of fins to reduce library cell area.

Some vendors also use a second local interconnect layer, which enables direct connection between metal gate lines. In our example this mask is called Lilgate. Although the metal gates in our example are not connected to each other, we still need to cover these metal gate lines with a small Lilgate area, since we need to make the connections to these gates. The contact between the Lilfin and Lilgate layer on the one hand, and metal interconnect on the other, is often called Via0 while in the planar layout, the contact enables a metal connection to the source, drain and polysilicon regions.

An alternative is to use only a single LIL layer, which combines the patterns of the Lilfin and Lilgate in one layer, but then the vertical lines may have minimum feature size width, while the horizontal lines are somewhat wider with relaxed size control.

The next two-input NAND gate layout (Fig. 4.49) is an example of a FinFET library cell, where five fins have been used to create the nMOS and pMOS FinFETs. A FinFET library also includes low and high drive cells. A low-drive cell, for example, may only have two fins for the nMOSt and pMOSt connected in parallel. A high drive cell may even have two five-fin cells in parallel.

A real litho-friendly FinFET layout of a logic core consists of a large grid of horizontal fins and vertical metal gate lines. To separate the metal gate lines of the cells in one row from those in the row above and below, the metal gate is removed in the areas covered by the Gate Cut mask shapes in the upper and lower parts of the cell. In the above example layout cell, the fins start under the dummy gate at the left side and end under the dummy gate at the right side. Fins are fabricated by a double (SADP) or quadruple (SAQP) patterning technique, and with a Fin Cut mask (not drawn in the figure) etching step, they can be separated into individual (cell) fins.

An alternative is that all fins run from left to right across the complete logic core. The dummy gates of the nMOSt and pMOSt are then isolated from each other by also removing (etching) the gate in the extra Gate Cut shapes in the centre of the cell. Next, the lower dummy gates are then connected to ground (or a negative voltage), while the higher dummy gates are connected to V_{dd}(or a higher positive voltage), to switch these dummy transistors off. These dummy 'switched-off' transistors serve as isolation between the logic cells, which allow a higher cell density, but at the cost of (subthreshold) leakage current of these switched-off cell-isolation transistors. In FinFET libraries, the number of routing tracks across the cells has a strong influence

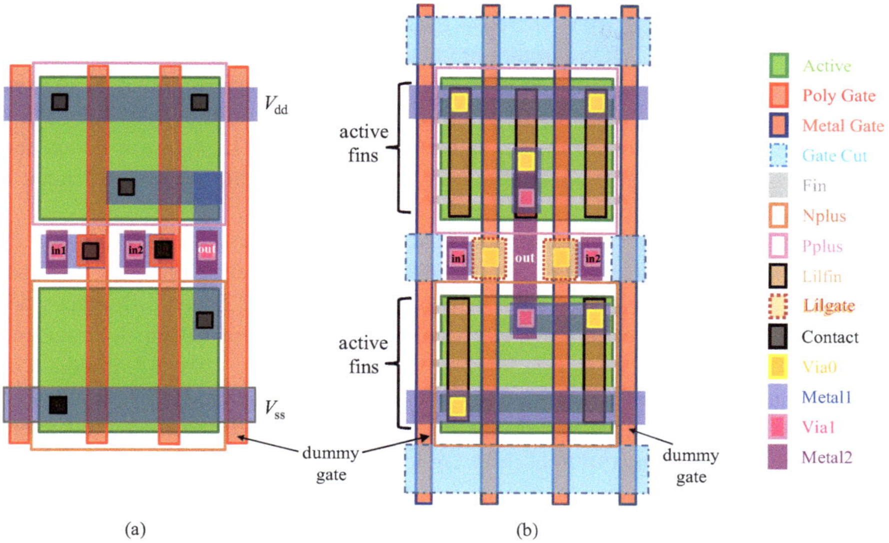

Fig. 4.49 Example of a two input NAND gate in planar CMOS and in FinFET technology in comparable technology nodes

on the height of the library cells. Libraries with a reduced cell height will cause more routing congestion problems during the place and route phase. There is also a focus on reducing the pitch in the lower metal layers.

With this summary on FinFET layout, we will close the discussions in this chapter. More information on the design of CMOS circuits and layouts can be found in the reference list, but also in the internet [14–16].

4.9 Conclusions

CMOS has become the major technology for the manufacture of VLSI circuits and now accounts for more than 90% of the total IC market. The main advantage of CMOS is its low power dissipation. This is an important requirement in current VLSI circuits, which may contain hundreds of millions to more than a billion of transistors.

Static CMOS circuits are characterised by high input and parasitic capacitances and relatively large logic gate structures. The silicon area occupied by a static CMOS logic circuit is about twice that of an nMOS counterpart. Dynamic CMOS circuits are nMOS-mostly and are therefore generally smaller than their CMOS counterparts. The use of a static rather than a dynamic implementation must therefore be justified by a sufficient reduction in power dissipation. Generally, static CMOS shows the lowest τD product and is thus the most power efficient implementation for VLSI. Moreover, its robustness is very important in current nanometer ICs as these show increasing noise, caused by cross-talk and supply voltage drops. *Static CMOS circuits* are also characterised by the fact that during operation, all inputs and outputs of every CMOS logic gate is either at ground or V_{dd} level and never floating. Therefore, all CMOS libraries used for digital circuits to be made in logic fabs are static CMOS libraries. Low-power issues and maintaining signal integrity at a sufficiently high level are the subjects of Chaps. 8 and 9, respectively.

Basic technologies for the manufacture of MOS devices are explained in Chap. 3. Various nMOS circuit principles are introduced. This chapter emphasises the most important differences between CMOS and nMOS circuits. These differences are evident in the areas of technology, electrical design and layout design. A structured CMOS layout design style is presented in this chapter while using a limited set of representative design rules. The combination of the CMOS and nMOS circuit design and layout principles discussed in this chapter should afford the reader sufficient insight into the basic operation of different CMOS circuits.

4.10 Exercises

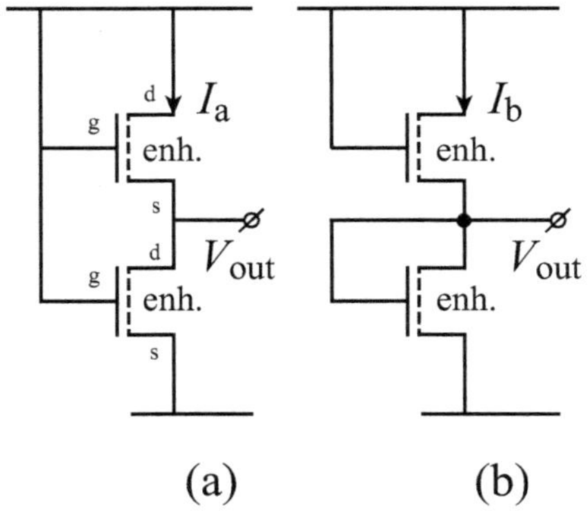

1. The following values apply for the above circuits:
   $V_{dd} = 1\,V$
   $K = 0\,V^{1/2}$
   $|V_x| = 0.25\,V$
   All transistors are of the same size; assume ideal transistors (no leakage currents).
   (a) What can be said about V_{out} in circuits a and b: $V_{out} <>= V_{dd}/2$? Explain.
   (b) Which of the currents I_a and I_b is larger and why?

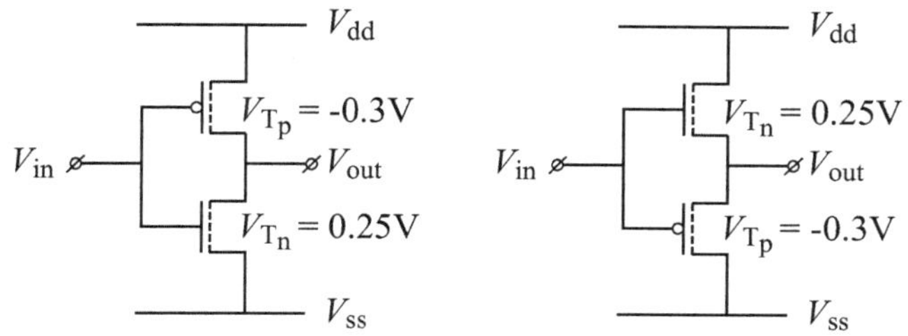

2. (a) If $V_{dd} = 0.4\,\text{V}$ in the above figure, explain what would happen at the output of the left circuit when V_{in} switches from 0 V to V_{dd} and back. Draw this in the inverter characteristic: $V_{out} = f(V_{in})$.
   (b) Repeat (a) for $V_{dd} = 1\,\text{V}$.
   (c) If $V_{dd} = 1\,\text{V}$ in the right circuit, and V_{in} switches from 0 V to V_{dd} and back, draw $V_{in} = f(t)$ and $V_{out} = f(t)$ in the same diagram (assume $K = 0\,\text{V}^{1/2}$).

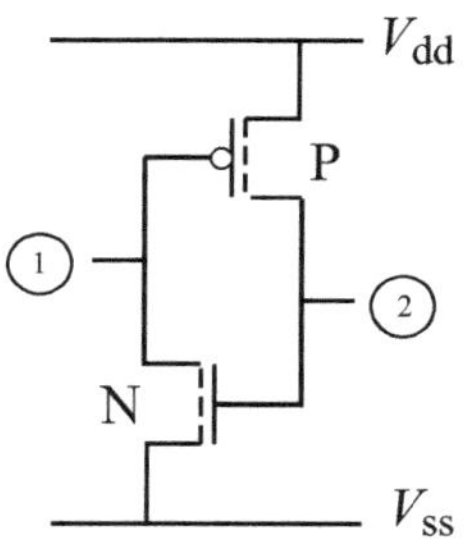

3. The above circuit is called a tie-off cell.
   (a) Describe its behaviour during power up.
   (b) What will the voltages at nodes 1 and 2 be after the power-up situation has been stabilised?
   (c) What are the major differences between this circuit and a CMOS inverter?
   (d) Assume that the transistor sizes in this circuit are identical to the transistor sizes of a CMOS inverter. Assume that the input of the inverter is connected to V_{dd}. Which of the circuits, tie-off cell or inverter, would represent more capacitance between the V_{dd} and V_{ss} terminals and why?
4. (a) Explain in no more than 10 lines the cause of short-circuit dissipation.
   (b) What is the reason that the short-circuit power consumption in large drivers (bus- and output drivers) in modern CMOS processes ($L < 90\,\text{nm}$) has become negligible.
5. Draw a process cross section along the line indicated by B–B' in the layout in Fig. 4.46.
6. The following function must be implemented in a CMOS logic circuit: $z = c(\overline{a}\,b + a\,\overline{b})$

(a) Draw a circuit diagram of a static CMOS implementation of the required logic circuit. The required inverse signals must also be generated in this circuit.

(b) Adopt the approach presented in this chapter and draw the CMOS stick diagram and layout of the logic circuit
Assume $\left(\frac{W}{L}\right)_n = \frac{200\,\text{nm}}{50\,\text{nm}}$ and $\left(\frac{W}{L}\right)_p = \frac{300\,\text{nm}}{50\,\text{nm}}$
(scale: 5 nm≡ 1 mm).

7. Consider the following logic function: $z = c + ab + \overline{a}\overline{b}$
   (a) Rewrite this function such that it is optimised for implementation in MOS.
   (b) Draw a circuit diagram of a static CMOS implementation.
   (c) Adopt the approach presented in this chapter and draw the CMOS stick diagram and layout of the logic circuit
   Assume $(\frac{W}{L})_n = \frac{200\,\text{nm}}{50\,\text{nm}}$ and $(\frac{W}{L})_p = \frac{300\,\text{nm}}{50\,\text{nm}}$
   (scale: 5 nm≡ 1 mm).
8. A static CMOS inverter has been implemented in a 28 nm CMOS technology for 1 V operation with $V_{T_n} = |V_{T_p}| = 0.25$ V.
   (a) For which of the logic gates, NOR or NAND, would you limit the number of inputs and why?
   (b) Explain what will happen if you would run this inverter at only 0.2 V?
   (c) Which of the library cells would you consider to be the most critical one in reducing the supply voltage to subthreshold voltage levels?
9. The following values are given for the parameters in the circuit below:
   $V_{X_n} = 0.25$V
   $V_{X_p} = -0.3$ V
   $K_p = K_n = 0\,\text{V}^{1/2}$
   $V_{bb} = -1$ V

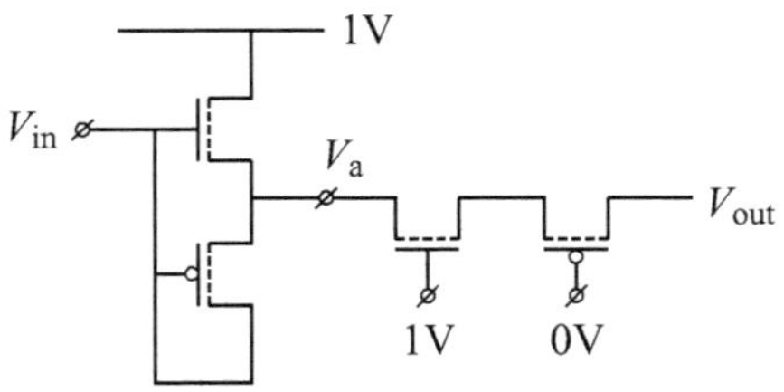

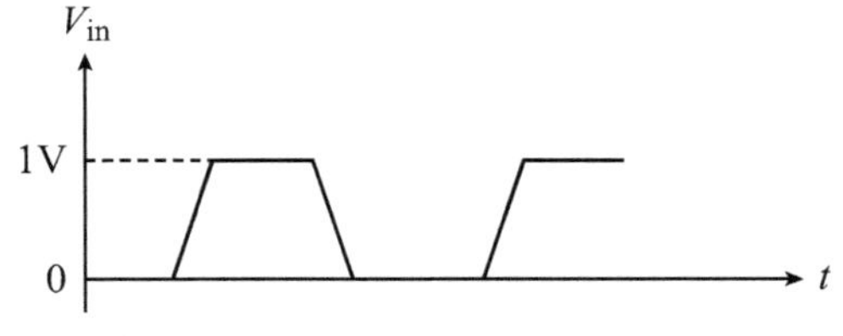

Explain what happens to voltages V_a and V_{out} when V_{in} switches as shown. Draw V_{in}, V_a and V_{out} in one diagram.

References and Further Reading

CMOS physics and technology (see also Chap. 3)

1. R.C. Jaeger, *Introduction to Microelectronic Fabrication.* Modular Series on Solid-State Devices, vol. 5 (Prentice Hall, New Jersey, 1988)
2. Y. Sakai et al., *Advanced Hi-Cmos Device Technology* (IEEE IEDM, Washington, DC, 1981), pp. 534–537
3. S.M. Sze, *Modern Semiconductor Device Physics* (Wiley, 1997)
4. S. Wolf, R.N. Tauber, *Silicon Processing for the VLSI Era*, vol. 1 (Process Technology, Lattice Press, 1986)

CMOS Design Principles (General)

5. C. Mead, L. Conway, *Introduction to VLSI Systems* (Addison-Wesley, 1980)
6. N. Weste, K. Eshraghian, *Principles of CMOS VLSI Design, a Systems Perspective* (Addison-Wesley, 1993)
7. L.A. Glasser, D.W. Dobberpuhl, *The Design and Analysis of VLSI Circuits* (Addison-Wesley, 1985)
8. M. Annaratone, *Digital CMOS Circuit Design* (Kluwer Academic Publishers, 1986)
9. L.G. Heller et al., *Cascode Voltage Switch Logic*, IEEE Digest of technical papers of the ISSCC, 1984
10. J.M. Rabaey, *Digital Integrated Circuits: A Design Perspective* (Prentice Hall, 1995)
11. K. Bernstein et al., *High Speed CMOS Design Styles* (Kluwer Academic Publishers, 1999)
12. International Solid-State Circuits Conference Digest of Technical papers, February 2000, pp. 90–11, pp. 176–177, pp. 412–413, pp. 422–423
13. S. Rusu, IEEE Distinguished Lecture, in *Microprocessor Design in the Nanoscale Era*, IEEE Penang Joint Chapter, July 29, 2013
14. T. Cui et al., 7nm FinFET standard cell layout characterization and power density prediction in near- and super-threshold voltage regimes, in *Green Computing Conference (IGCC), 2014 International*, 3–5 Nov 2014
15. P. McLellan, FinFET Custom Design, SemiWiki.com, The open Forum for Semiconductor Professionals, 04 Feb 2014
16. K. Vaidyanathan et al., Exploiting sub-20-nm complementary metal-oxide semiconductor technology challenges to design affordable systems-on-chip. J. Micro/Nanolith. MEMS MOEMS **14**(1), 011007 (2014)
17. K. Bhanushali, et al., FreePDK15: An Open-Source Predictive Process Design Kit for 15nm FinFET Technology, Conference: Proceedings of the 2015 International Symposium on Physical Design, Monterey, USA, March 2015
18. Hao Chen, et al., AutoCRAFT: Layout Automation for Custom Circuits in Advanced FinFET Technologies, Session 13: Advances in Analog and Full Custom Design Automation, ISPD 2022, Virtual Event, Canada, March 27–30, 2022

Power Dissipation in CMOS

19. H.J.M. Veendrick, Short-circuit dissipation of static CMOS circuitry and its impact on the design of buffer circuits. IEEE J. Solid State Circuits **SC-19**(4), 468–473 (1984)

For Further Reading

20. IEEE Journal of Solid-State Circuits
21. ISSCC and ESSCIRC conferences, VLSI and ISLPED symposia, digests of technical papers

Special Circuits, Devices and Technologies 5

5.1 Introduction

This chapter discusses a number of special circuits, devices and technologies. These circuits and devices can be used in digital, analogue and mixed analogue/digital applications. They are realised in various MOS technologies or their derivatives, which include the BICMOS technologies discussed in this chapter.

The chapter begins with an explanation of circuits that operate as image sensors. We distinguish *charge-coupled devices (CCDs)* as well as *CMOS image sensors*. Their ability to capture images finds its usage in all kinds of cameras. Their operation is based upon the conversion of light into electrons.

The second category of special devices covered in this chapter are MOS transistors capable of delivering high power. These power MOS field-effect transistors, or *power MOSFETs*, are feasible as a result of improvements in technology, which enable the manufacture of transistors capable of withstanding high voltages as well as large current densities. Power MOSFETs obviously operate according to the same field-effect principle as ordinary MOS transistors. This principle is discussed in Chap. 1.

Finally, devices based on mixed bipolar and CMOS technologies, such as IGBT transistyors, are also discussed. They were/are particularly used to enhance the performance of both digital and mixed analogue/digital circuits.

5.2 CCD and CMOS Image Sensors

5.2.1 Introduction

Charged-coupled devices (CCDs) are almost exclusively used as image sensors. They basically operate by transferring charge from below one transistor gate to below another one in a ‘channel’. CCD implementations include *surface-channel*

H. Veendrick, *Nanometer CMOS ICs*,
https://doi.org/10.1007/978-3-031-64249-4_5

(SCCD) and *buried-channel* (BCCD) devices. Also, for analogue applications, there must be a relationship between the size of the packet and the signal which it represents. The packet size must therefore be maintained during transfer. An alternative to CCD imaging is CMOS imaging, which is currently used in large volumes in cheap consumer and communication products, such as digital cameras and mobile phones. However, recent advancements in CMOS imaging resulted in CMOS image sensor quality approaching that of CCD sensors. This makes them increasingly attractive for use in a broad range of applications.

5.2.2 Basic CCD Operation

CCD sensors are composed of an array of semiconductor capacitors capable of holding an electrical charge. When light hits the sensor, these capacitors, acting as individual photosites, absorb the light and convert it into an electrical charge packet. Each charge packet is then transferred via successive neighbouring photosites, like a serial chain, through the array until it reaches the corner of the array where it is read. The charge packets that are created in each photosite are thus transported through the array by serial shift registers. Such *CCD shift registers* can be realised with 2-phase, 3-phase and other multi-phase clock systems. The operation of a CCD is explained below with the aid of the 2-phase SCCD structure shown in Fig. 5.1. A diagram of the 2-phase clocks ϕ_1 and ϕ_2 is also shown in this figure. The gates indicated by bold lines are polysilicon *'storage gates'*, under which charge is stored. The remaining gates are *'transfer gates'* created in a second polysilicon or metal layer. They lie on a thicker oxide than the storage gates and therefore have a much higher threshold voltage ($V_T \approx 1$ V). These transfer gates serve as a barrier between the storage gates. Operation of the 2-phase SCCD is explained on the basis of the surface potential distributions under the gates.

Suppose the first and third storage gates contain a full and an empty charge packet, representing the logic levels '1' and '0', respectively.

The charge packet corresponding to the first storage gate is then full of electrons. This is represented by a full *'charge bucket'* under the gate in Fig. 5.1. The charge bucket corresponding to the third storage gate, however, is almost empty, i.e., it is practically devoid of electrons. At time point 1, both ϕ_1 and ϕ_2 are 'low' and the storage gates are separated from each other. At time point 2, ϕ_1 has switched from a low to a high level and the charge is transferred from the ϕ_2 storage gates to the ϕ_1 storage gates. At time point 3, both ϕ_1 and ϕ_2 are 'low' again and the charge is now stored under the ϕ_1 storage gates. The description of the shift behaviour at time points 4 and 5 is obtained by replacing ϕ_1 by ϕ_2 in the above descriptions for time points 1 and 2, respectively.

A comparison of the time points 1 and 5 in Fig. 5.1 shows that the charge has been transferred from the first to the third bucket in one complete clock period. In fact, the charge is transferred from one CCD *'cell'* to another during a single clock period. Each cell clearly requires two storage elements which each comprise a bucket, a *transfer gate* and a *storage gate*. The two storage elements in a CCD

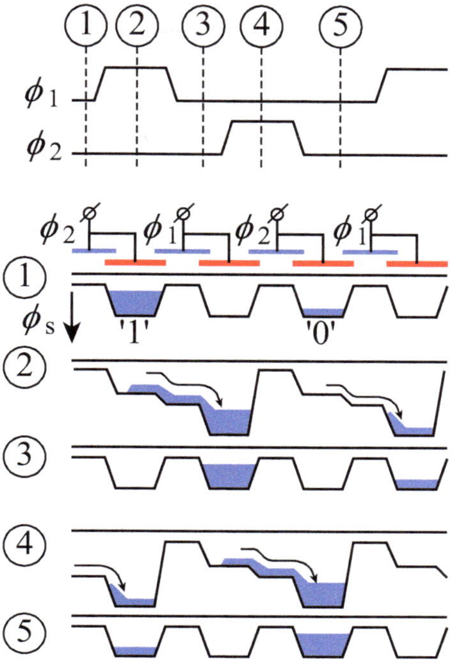

Fig. 5.1 The shift operation in a basic 2-phase SCCD

cell are analogous to the master and slave latches in a *D-type flip-flop*. Clearly, the implementation of a 2-phase CCD register comprising 1 Mb, for example, requires 2 million storage elements on a chip. In practice, a better ratio between the number of storage elements and the number of register cells is obtained by using another type of clocking strategy.

The discussion of charge transfer in Fig. 5.1 is based on the assumption that one bucket was full with electrons and another one was empty. The operation of an SCCD clearly relies on the filling of these buckets. Figure 5.2a shows a simplified SCCD comprising some sensor cells and an output section.

In an image sensor, photons reach the silicon surface through a lens. The silicon then converts the photons into electrons locally. A complete image is then captured in an array, which is read out by shifting (transferring) its contents to the CCD array output. The *charge transfer* in an SCCD occurs right at the silicon surface under the gates.

Unfortunately, the surface is inhomogeneous and therefore plagued by *surface states*. These surface states have a certain energy and can trap electrons which have higher energy. During charge transfer, the associated change in surface potential profile causes the surface states to release the trapped electrons. If this occurs before the transfer is complete, then the released electrons will simply rejoin the rest of the electrons in the packet and *'transfer efficiency'* is maintained. However, if an electron is released from a surface state when the transfer is complete, then it cannot

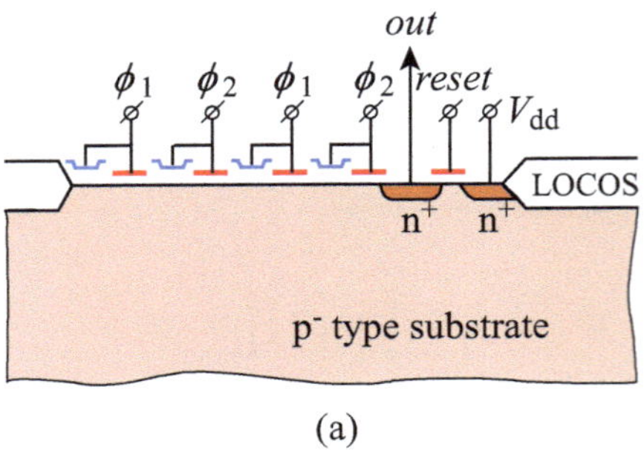

Fig. 5.2 (**a**) Basic structure of an SCCD and (**b**) a BCCD

rejoin its charge packet. This reduces transfer efficiency. The surface states continue to release the trapped electrons until a new charge packet arrives. The new packet will not be degraded by surface states that are still full when the packet arrives. The empty surface states will, however, be filled by the new packet and the process will repeat itself.

Clearly, transfer efficiency depends on the number of surface states. In previous generations of CCDs, transfer efficiency was increased by using a small charge to represent a '0'. This *'fat zero'* ensures that surface states remain filled. Transfer efficiency is also reduced by incomplete transfer of charge packets at high clock frequencies.

Leakage current accounts for another problem related to CCDs and, of course, to other dynamic memories as well. This *'dark current'* is caused by thermal generation of minority carriers and slowly fills the buckets of a CCD. The result is a *'maximum storage time'*, during which the data in a CCD will remain correct. In addition, dark current causes a fixed noise pattern on the data that is read from a CCD.

Both transfer efficiency and dark current largely determine the operating limits of a CCD. These factors therefore require considerable attention during CCD design.

The previous section clearly indicates that surface states form an important limiting factor for the performance of SCCDs. These surface states are unavoidable. Therefore, the only way to improve performance is to realise a CCD in which storage and transfer of charge occur in a channel which is 'buried' a short distance below the silicon surface. A buried n-channel can be realised by creating a thin n-type layer on top of a p-type substrate. Compare the SCCD and BCCD structures in Fig. 5.2a and b, respectively.

The operation of an SCCD is closely related to the characteristics of a MOS capacitor with a thick depletion layer. However, the operation of a BCCD is related to the characteristics of a MOS capacitor with a fully depleted layer. Therefore, the first requirement for the successful operation of a BCCD is that the thin n-type layer is fully depleted of electrons. This is achieved by using clock and control signals with an offset voltage. This voltage produces a potential maximum a short distance below the silicon surface. Electrons (representing data bits) injected into the device are stored at this potential maximum. The depleted n-type layer prevents the charge carriers from reaching the surface states and a high transfer efficiency is therefore achieved. The operation of a BCCD is otherwise identical to that of an SCCD.

Buried-channel CCDs were developed for two important reasons. The first is their *immunity* to surfaces states. The second is the increased *operating frequency* which they afford compared to surface-channel CCDs. The increase is caused by the fact that charge is transferred at a speed determined by the bulk mobility instead of the surface mobility. The maximum clock frequency of a BCCD is therefore twice that of an SCCD of equivalent dimensions. However, the definition of the buried channel in a BCCD requires an extra mask. BCCDs are also subject to many problems when their dimensions are reduced. In addition, it is inherently difficult to control the charge in a BCCD because it is stored at a distance from the gate which is longer than for an SCCD. Currently, all image sensor CCDs are implemented as BCCDs. Because of the large number of pixels, a lot of transfers are required. The immunity to surface states then outweighs the disadvantages of BCCDs.

The charge-coupled device principle can be used in both analogue and digital applications. As stated, the bulk part of the applications is in image sensors. Professional cameras now use sensors with over 100 Megapixels. In video camera applications, conventional CCD image ICs consisted of separate sensor and memory parts. Currently, the chip only contains a sensor (Fig. 5.3), which captures the image when the shutter is open and it temporarily acts as a storage device when the shutter is closed.

Fig. 5.3 A 24 milion pixel image sensor 36 x 24 mm^2 for digital photography in LCC package for Leica Camera (Courtesy of CMOSIS)

The main advantage of CCD image sensors over the CMOS imaging devices (discussed in the next section), is the outstanding image quality of the CCDs. Their fabrication technology is optimised with one main goal: imaging performance. For that reason, CCDs are still very popular in established markets such as digital still photography, camcorders, but also in high-end markets such as broadcast and astronomy.

5.2.3 CMOS Image Sensors

MOS image sensors already exist since the late 1960s. Due to problems with noise, sensitivity, scalability and speed, CCD sensors became much more popular. In the early 1990s, however, CMOS image sensors regained their popularity. The efforts were driven by low-cost, single-chip imaging systems solutions. During the last 15 years, the developments in and applications of CMOS imaging have intensified so much that still complete sessions at the major IC conferences, like IEDM and ISSCC, are devoted to them [1].

Another driving factor for an increased activity in CMOS image sensors is the continuous improvement in CMOS technology. Scaling of the sensor pixel size is limited by both optical physics and costs [2] and occurs at a lower pace than the scaling of the CMOS feature size, see Fig. 5.4a. This allows to combine the CMOS image sensor with image processing on a single chip at relatively lower costs.

The ability to capture low-light images depends on the efficiency to convert incoming photons into electrons, which subsequently discharge the pixel capacitor. We distinguish between both passive and active pixels. An *Active Pixel Sensor (APS)* includes an active amplifier in every pixel. Figure 5.4 shows three different pixels. When the pass transistor (TX) in Fig. 5.4b is accessed, the photodiode is connected to a bit line. Its charge is converted into a voltage by the readout circuit (amplifier) located at the bottom of a bit line. Due to the small pass gate, this single transistor pixel allows the smallest pixel size and consequently, the highest *fill factor* (ratio of sensor area to total area of sensor plus support electronics). The performance of a pixel was improved by adding active amplifier circuitry to the cell, see Fig. 5.4c, resulting in average fill factors between 20% and 30%. Transistors RST and RS are for resetting and row selection, respectively.

The photogate APS in Fig. 5.4d integrates charge under the gate. Its readout architecture looks simular as in CCDs [2]. The addition of an extra junction at the right side in the substrate forms a *pinned photodiode (PPD)*, which is connected to the pixel read-out circuit through an additional pass transistor TX, which separates the photodiode from the read-out node. Almost all products on the market these days make use of this pixel architecture, and it is the PPD that really boosted the introduction of CMOS image sensors into commercial products [3].

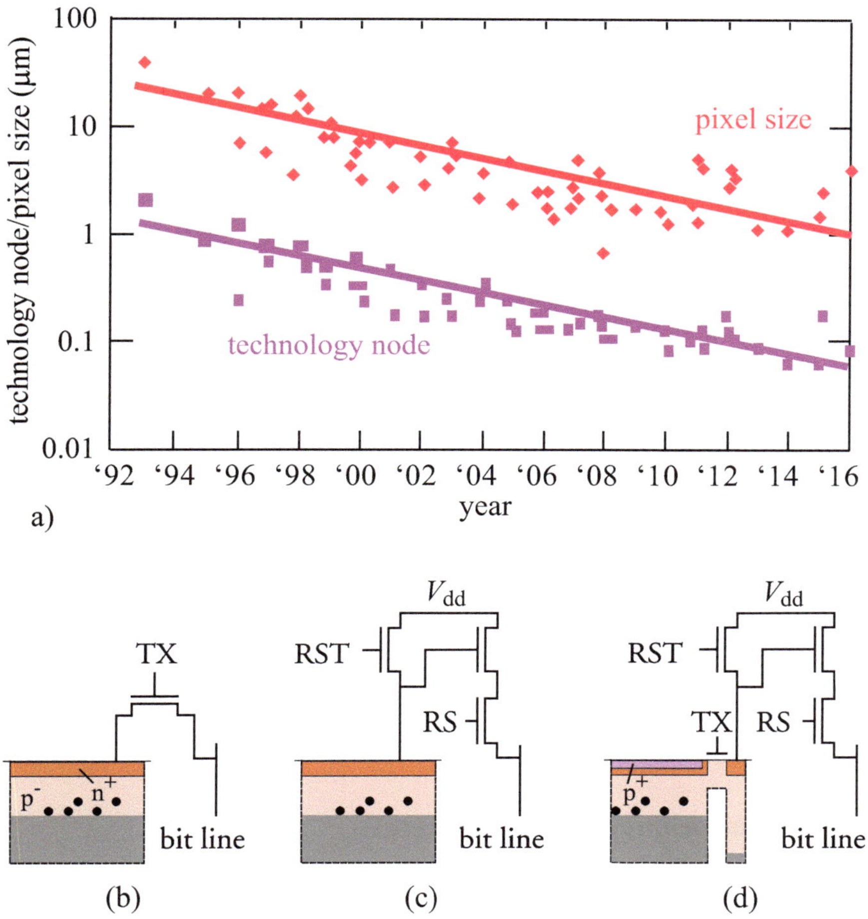

Fig. 5.4 (**a**) Scaling of MOS pixel and technology node, (**b**) passive pixel, (**c**) photodiode active pixel sensor (APS) and (**d**) photogate APS pixel (Source: [3])

The low power consumption, the limited number of external connections and supplies, high level of integration and low production costs allow CMOS sensors to be used in a variety of applications: multimedia, low-cost cameras [4], PC camera, security and machine monitoring, video phone, fax, etc.. It is expected that further development of CMOS image sensor technology will create completely new imaging markets that were not accessible in the past with CCDs. Moreover, application of CMOS imaging technology is no longer restricted to consumer applications. Examples are the automotive industry and the ambient intelligence applications. More sophisticated and high-resolution imaging applications will become available as CCD and CMOS imagers continue to improve [1]. However, the development of very powerful signal processing chips enables CMOS imaging systems with high image quality. Today's image processing cores can correct for single pixel defects and defect columns without any noticeable effect for the customer.

An important contributor to the noise in image sensors is the photon shot noise, which is caused by statistical variations in the number of photons captured by a pixel during exposure. This shot noise component has a lot of influence on the signal-to-noise behaviour of the imaging system [3]. The requirements to achieve a sufficiently high signal-to-noise ratio ($\geq$ 40 dB) can be translated into an interesting rule of thumb: to make decent images for consumer applications, one needs 10,000 electrons in every pixel. This number can be reduced with extensive use of image processing and noise suppression. So, scaling of the pixel sizes is no longer limited by the minimum feature sizes of the CMOS process, but by the required minimum number of electrons that can be stored in each pixel.

To overcome the problems in sharing the pixel area by a sensor and an active amplifier, advanced CMOS image sensors are currently more and more built from two or three-layer solutions. The two-layer transistor pixel *stacked CMOS image sensor* technology separates photodiodes and pixel transistors on different substrate layers [5]. This new structure approximately doubles the saturation signal level relative to conventional image sensors and widens the dynamic range and reduces noise. Another example of a stacked CMOS image sensor is presented in [6]. This sensor consists of an upper (90 nmMOS) chip, which contains a back-illuminated pixel array, and a lower (40 nm CMOS) chip containing the logic.

5.2.4 Conclusions on Image Sensors

Due to the presence of multiple transistors near each photosite on a CMOS sensor, its light sensitivity tends to be lower, as some photons land on the transistors rather than reaching the photocell. CCD sensors yield high-quality images with minimal noise, while CMOS sensors are typically more sensitive to noise. Because CMOS sensors can be produced in standard silicon wafer fabs, they are cheaper than CCD sensors. The power consumption of CCD sensors is often two orders of magnitude larger than that of a comparable CMOS sensor. In general, the continuous growth in performance in combination with a significantly cheaper production process makes CMOS sensors increasingly attractive for the broadening imaging market. The use of stacked CMOS image technology enables a relatively larger sense area per pixel, leading to increased attractiveness of this type of sensor. However, certain high-performance industrial and scientific applications, e.g., space imaging, may still necessitate the use of CCD sensors [7].

5.3 BICMOS Circuits

5.3.1 Introduction

Since the mid-eighties, a growing interest in BICMOS technologis has resulted in a lot of commercially available ICs. The *BICMOS technology* facilitates a combination of both bipolar and CMOS devices on a single IC and enables the simultaneous exploitation of the advantages of both device types.

The penalty of more complex processing restricted the use of BICMOS technologies to fairly specialised applications. It is estimated that a BICMOS wafer after full processing will cost 20% to 30% more than a baseline CMOS wafer. In several applications, this price increase will be offset by the performance enhancement. Performance characteristics of BICMOS devices and their technology are explained further. Future expectations and market trends are also discussed.

5.3.2 BICMOS Technology

There are several ways of obtaining a BICMOS process. It could, for instance, be based on an existing bipolar process or a completely new BICMOS process could be developed. The conventional approach, however, was to start from a CMOS process. An associated advantage was that existing CMOS design and CAD tools could then be used for BICMOS designs. A BICMOS process based on an n-well CMOS process is considered here. This is a logical choice because of the considerable similarities between this BICMOS process and the n-well CMOS process discussed in Chap. 3.

The development of the BICMOS process from a baseline n-well CMOS process is explained with the aid of the cross sections in Fig. 5.5. The source and drain implants are typically less than a tenth of a micron deep. The depth of the n-well is less than a micron. The realisation of an npn transistor requires an additional p-type implant in the n-well. This implant forms the base of the npn transistor and is shown in Fig. 5.5. The npn transistor exhibits a relatively high collector series resistance. This is also the base resistance of the pnp transistor in the *parasitic thyristor*, as discussed in Chap. 9 (latch-up). This 'simple' structure is therefore rather susceptible to *latch-up*. The above disadvantages are largely overcome when the structure shown in Fig. 5.5b is used. In the associated process, n^+ implants are created in the p-type substrate prior to the growth of a p^- epitaxial layer. The resulting 'buried layer' n^+ areas subsequently become part of the n-wells. The npn transistor obtained in this process is basically isolated and latch-up via the parasitic pnp transistor is largely prevented by the n^+ buried layer. The creation of the buried collector areas and the base implant requires two more masks than in a standard n-well CMOS process.

In the late eighties and early nineties, BICMOS was also used for digital circuit categories that needed to operate beyond the performance limits of CMOS. The temporary increase in market volumes around 1996 was caused by falling MOS

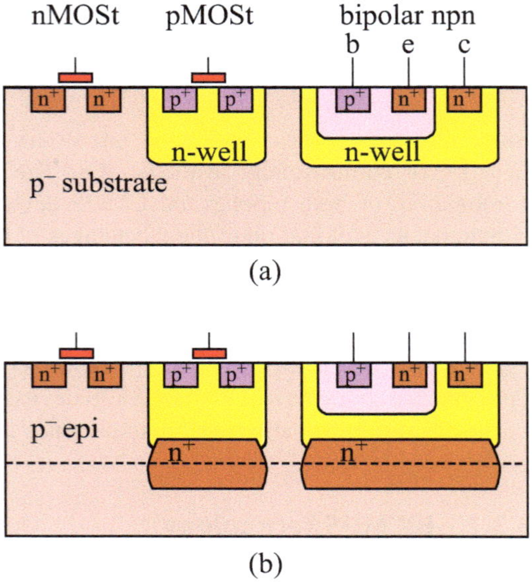

Fig. 5.5 Cross section of a basic BICMOS process (**a**) and BICMOS process cross section with additional n^+ buried well for improved reliability (**b**)

memory prices (of DRAMs in particular) and to the growing high-end market for microprocessors (high demand for BICMOS-based Pentium™ chips), ASICs and SRAMs. At voltages below 2.5 V, the performance gap between bipolar and CMOS gradually narrowed, in favour of full-CMOS products. Over the years, as a result of intensive R&D, more and more analog and RF functions have been realised in CMOS, because of cost considerations. This had reduced BICMOS usage in the nineties to only mixed-signal and RF-circuits. Growing demands for multi-Gb data communication and wide-bandwidth radio communication systems caused renewed interest in BICMOS technology. This has put stringent requirements on the technology for providing sufficient performance at affordable cost.

A combination of SiGe heterojunction bipolar transistors (*HBT*) and CMOS was the most promising technology solution to meet these requirements. It needed full optimisation of the device architecture of the bipolar transistor: a shallow-base, a dedicated Ge profile in combination with low-power concepts and low-cost solutions [8]. Figure 5.6 shows an example cross section of a SiGe HBT for excellent RF performance [9, 10]. The Ge profile plays a dominant role in the characteristics of the SiGe HBT device. It is obvious that a SiGe BICMOS technology would allow to combine the performance and density requirements for the afore-mentioned communication systems [11, 12].

Some vendors offer a BICMOS process which has built-in options that allows a more optimal integration of passives (resistors, capacitors and inductors) for RF system-in package products [13].

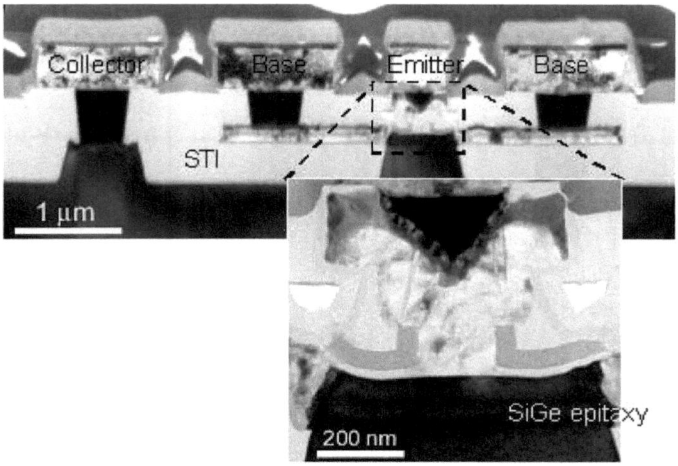

Fig. 5.6 Transmission Electron Microscope (TEM) picture of a SiGe HBT device cross section, showing the Emitter, Base and Collector metal connections and the shallow trench isolation (STI). The inset shows a magnification of the actual intrinsic device, with an epitaxially grown SiGe base layer as the engine for excellent RF performance. (Photo: NXP Semiconductors)

5.3.3 BICMOS Characteristics

Its higher *gain factor* and lower *noise* generally renders bipolar technology more suitable than CMOS for analogue applications. However, CMOS is more attractive for digital control, storage and signal processing circuits because of its low quiescent power, reasonable performance and high packing density. The mixture of the two technologies offers unique possibilities in both analogue, digital and mixed analogue/digital applications.

BICMOS was first introduced in digital I/O circuits, where it provided increased output driving capability. It was subsequently applied in the peripheral circuits of SRAMs to shorten the access times. These circuits included sense amplifiers, word line and bit line drivers.

Low-voltage bipolar transistors and high-voltage lateral DMOS transistors, incorporating both n-type and p-type channels, are combined in some BICMOS processes. These processes allow the integration of truly analogue high-voltage circuits, such as the video output amplifiers mentioned in Sect. 5.4.3.

The previously mentioned applications of BICMOS technologies illustrate their potential benefits. However, in addition to the increase in costs compared to an average CMOS technology, there are other drawbacks associated with BICMOS. For instance, the CMOS digital parts of a BICMOS chip may generate considerable transient noise on the supply and ground lines. This 'bounce' is discussed in Chap. 9. Considerable efforts are required to prevent it from entering analogue parts of the chip. Moreover, the reduced density of BICMOS logic limits its usage to critical functions on a VLSI chip. This reduces the potential performance advantage. The

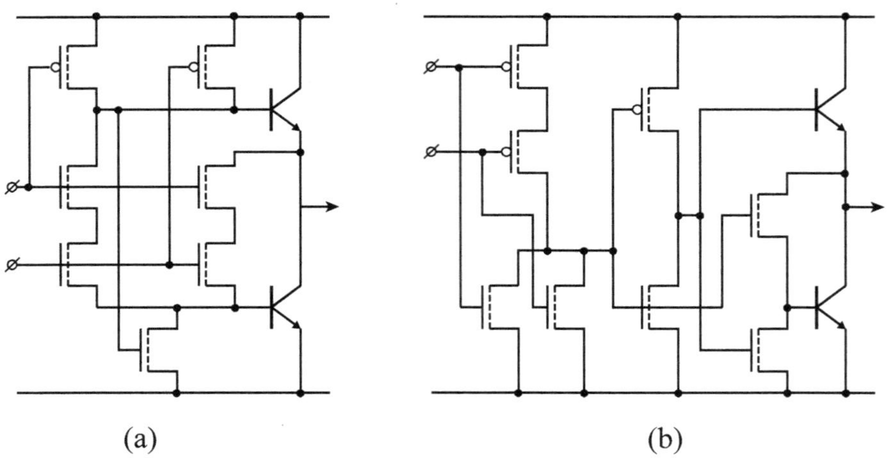

Fig. 5.7 Typical BICMOS implementations of a NAND gate

commercial use of BICMOS technology for digital ICs is therefore only justified when the additional costs are compensated by increased performance.

5.3.4 BICMOS Circuit Performance

BICMOS logic gates usually employ CMOS transistors to perform the logic function and bipolar transistors to drive the output loads. The two typical BICMOS implementations of a *NAND gate* shown in Fig. 5.7 illustrate this two-stage structure.

The two-stage structure of a BICMOS logic gate leads to a larger propagation delay for an unloaded BICMOS gate than for its CMOS counterpart. The performance advantage of a BICMOS implementation over a CMOS implementation therefore only applies in the case of gates with larger fan-outs. Figure 5.8 shows a frequently published comparison of the propagation delay as a function of fan-out for typical CMOS and BICMOS NAND gates. The comparison was made for nMOS and pMOS transistor widths of 4 μm and 7 μm, respectively, in a process with a 0.35 μm gate length. The cross-over point lies between a fan-out of two and three. For higher fan-outs, the performance of a BICMOS circuit is better.

Figure 5.8 suggests that BICMOS is preferable to CMOS even for relatively low fan-outs. For large capacitive loads, the figure shows that the propagation delay can be reduced by a factor of 2.5 when BICMOS is used. However, the presented comparison does not account for the extra area required by the driver stage in the BICMOS implementation. A more representative comparison is obtained when the CMOS logic gate is also equipped with a CMOS output driver. The resulting comparison is shown in Fig. 5.9 for BICMOS and CMOS NAND gates implemented as NOR gates followed by bipolar and CMOS drivers,

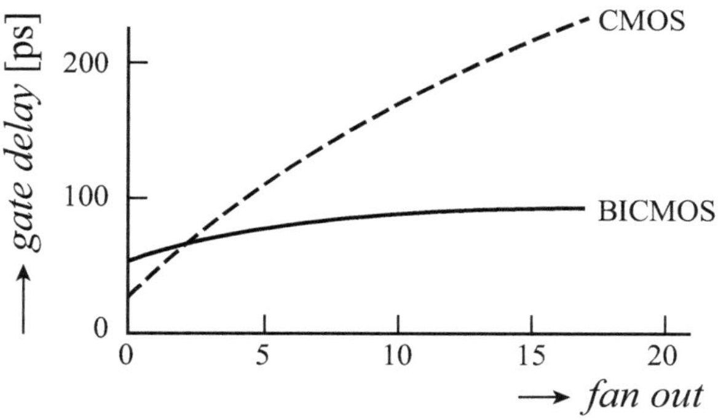

Fig. 5.8 Gate delays of typical CMOS and BICMOS NAND gates

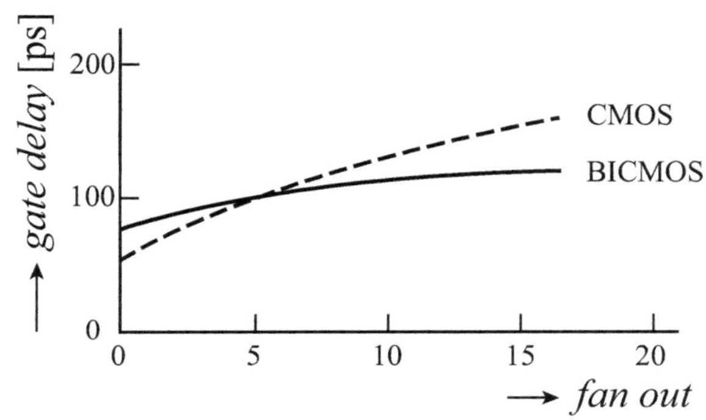

Fig. 5.9 Propagation delays of CMOS and BICMOS NAND gates implemented as NOR gates with CMOS and bipolar drivers, respectively

respectively. Such a comparison shows a dramatic reduction in speed advantage and reveals that BICMOS only affords a small performance improvement for gates with a high fan-out. In practice, this means that implementation of logic gates in BICMOS is not cost-effective for low- to medium-speed applications. Its usage in VLSI circuits and *Application-Specific ICs (ASICs)* is therefore limited to circuits that have to drive large capacitances, e.g., driver and I/O circuits. BICMOS is also used in ICs that have to operate beyond the performance limits of CMOS.

Supply voltage dependence, temperature dependence and process parameter dependence are also important factors that must be included in a comparison of the performance of BICMOS and CMOS circuits. These factors are compared further.

CMOS current levels are quadratically reduced when the supply voltage is reduced. This results in a reduction of the speed of both CMOS and BICMOS circuits. Bipolar circuits, however, are also hampered by inefficient operation at lower supply voltages. Manufacturers of BICMOS ICs will therefore face a dilemma when supply voltage standards below 1.8 V become more accepted as minimum feature size decreases to below 180 nm. Innovative design may reduce this dilemma.

The influence of temperature on the performance of CMOS and BICMOS circuits is closely related to the different origins of transistor currents. In bipolar transistors, the current is caused by diffusion. This current is less affected by temperature than the MOS transistor drift current discussed in Sect. 2.3. As a consequence, the switching speed of BICMOS is less dependent on temperature than that of CMOS.

It has been empirically found that variations in CMOS parameters caused by processing spread have a greater influence on circuit performance than variations in bipolar process parameters.

Finally, it should be noted that a BICMOS driver implementation shows a reasonable power dissipation advantage over a CMOS driver.

It is clear that the application of BICMOS technology is not trivial. This explains its limited application in semiconductor products.

5.3.5 Future Expectations and Market Trends

From a performance point of view, the future for BICMOS technologies originally looked promising. However, a fair comparison of BICMOS and CMOS circuit performance reveals that the advantages afforded by BICMOS are really only significant in mixed analog/RF/digital circuits.

The major issue with BICMOS, as compared to a combination of separated bipolar and CMOS ICs in one package, is cost. But also such aspects as performance, power consumption and integration density are essential to determine the overall system benefits. During the last decade, SiGe BICMOS has become very popular in (ultra)high-speed applications [14–16]. Intensive research on performance improvements has led to highly complex HBT device architectures with f_T and f_{max} values of 505 GHz and 720 GHz, respectively [12, 17]. As long as BICMOS enables the integration of different functions onto a single die in a cost-effective way, it will remain as an effective platform for those systems that require performance beyond the limits of CMOS.

5.4 Power MOSFETs

5.4.1 Introduction

The invention of the bipolar junction transistor in 1947 provided the foundation for modern integrated circuits and power electronic circuits. The first power devices based on semiconductor technology were demonstrated by Hall in 1952. He used germanium stacked junctions to achieve a continuous forward current of 35 A and a punch-through voltage of 200V. Since about 1955, silicon has been preferred for power devices. By 1960, such junctions allowed the implementation of 500 V rectifiers. Currently, silicon rectifiers are available with continuous current ratings of 5000 A and reverse voltages of 6000 V. The application of MOS technology in *power transistors* has been a major focus point for the industry since the late seventies.

The prospects of high-speed and high-input impedance in many low-voltage applications are particularly attractive. *Double-diffused MOS transistors* were originally introduced during the mid-seventies. The *DMOS transistor* allowed increased performance without reducing the source–drain distance, whilst excessive electric fields were avoided. Originally, the introduction of DMOS power FETs was seen as a major threat to the bipolar power transistor. However, their advantages only render *power MOSFETs* suitable for a limited part of the power electronics application area.

Improvements in technology and yield have resulted in better performance for MOS power transistors. Power MOSFETs can be implemented as discrete devices or can be integrated with other devices on a single chip. Usually the integrated power MOS devices deal with lower voltages and less power consumption than the discrete ones. Breakdown voltages over 1000 V are now possible with discrete devices. The *breakdown voltage* V_B of a power MOSFET is related to its typical resistance $(R_{on} \cdot Area)$. Typical corresponding values might be $(R_{on} \cdot Area) = 0.1\Omega{\cdot}\text{mm}^2$ at $V_B = 100\,\text{V}$ for a discrete power n-type MOSFET and $(R_{on}{\cdot}Area) = 0.15\Omega{\cdot}\text{mm}^2$ at $V_B = 100\,\text{V}$ for an integrated nMOS device. In practice, power dissipation is limited by the maximum power rating of the power MOSFET's package. Figures between 100 W and 350 W have been realised for packaged discrete power MOSFETs. Discrete power MOSFETs with die sizes of 200 mm^2 have been reported in the literature. Large-area low-voltage devices are designed for use as synchronous rectifiers, replacing diodes in power supplies (e.g., in PCs and laptops). When the current levels of power devices exceed about 1 A at operating voltages in excess of 150 V, monolithic integration of the power devices with the rest of the circuit is no longer cost-effective.

5.4.2 Technology and Operation

All high-voltage devices use a so-called drain extension (as discussed in Chap. 3), which is used as drift zone to distribute the voltage across. The resistivity of this drift zone strongly depends on the current and gate voltage. Most discrete power MOSFETs use vertical drain extensions, while integrated power MOSFETS usually apply lateral drain extensions.

This section will first focus on the technology aspects of discrete power devices and then continue with an example technology for the integration of power MOSFETs with analog and digital circuits onto one chip.

The high-voltage *vertical double-diffused MOS (VDMOS)* transistor shown in Fig. 5.10 is an example of a discrete power MOSFET.

The threshold voltage of the above VDMOST is generally 2–3 Volts. When the gate voltage is increased from 0 V to about 12 V, the p-well area near the surface is inverted (see Sect. 1.3). A channel then exists between the n^+ source and the n^- epitaxial layer. The charge carriers will flow *vertically* from the source to the drain when a high voltage is applied to the latter. The drain voltage can vary from 50 V to 1000 V, depending on the application. VDMOS transistors are usually n-type rather than p-type because of their higher channel mobility. Because of the scaling of the

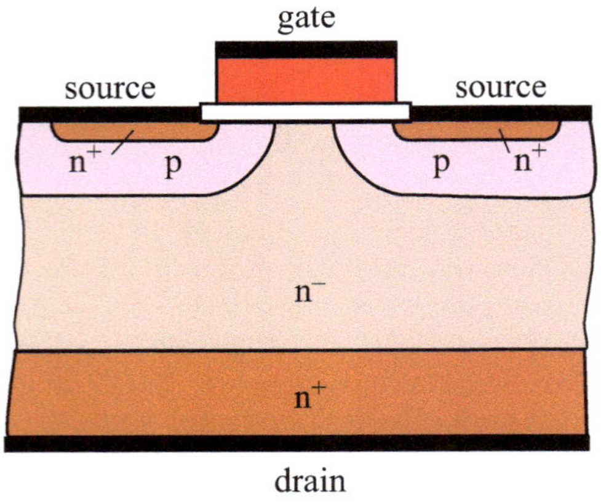

Fig. 5.10 Cross section of a VDMOS discrete power transistor

gate oxide thickness, devices with a gate voltage below 5 V and a threshold voltage below 1 V have become available as well.

A power MOSFET is a regular array of n transistors connected in parallel, with an equivalent resistance (RDS_{on}) equal to $1/n$ of the individual transistor's RDS_{on}. So, a larger die results in a lower on-resistance, a larger parasitic capacitance and so in a reduced switching speed. Therefore, there is a continuous drive to minimise the size of the individual transistor in order to reduce the on-resistance while keeping the capacitances constant. The use of the trench technology in power MOSFETs has resulted in significant improvements of on-resistance for low voltage devices. Instead of being planar, the gate structure is now built in a trench, which may reduce the transistor area by about 50%, depending on the operating voltage range.

Figure 5.11 shows a schematic and an SEM photograph of a cross section of a trench MOSFET [18] for applications with a voltage range up to 100 V. For these applications, transistors are used in which the gate is incorporated in a trench. The way to reduce the on-state resistance of these devices is by increasing the cell density. This is done by using innovative integration technologies. Transistor densities as high as 640 million cells per square inch are available in recent power MOSFET trench technologies.

The drive for smaller form factors and reduced system costs will force the power device products towards low-cost *high-voltage CMOS (HV-CMOS)* technologies, which begin to offer functionality comparable to BCD technologies but with much less process complexity. 40 V LCD driver products are already produced in high volumes in a HV-CMOS process. It is expected that other applications, such as printer head drivers and bus transceivers, will soon follow [19]. In R&D new process/substrate combinations are examined to further increase the voltage levels. An example is the design and fabrication of level shifters with more than 900 V operating voltage used in three-phase power applications [20].

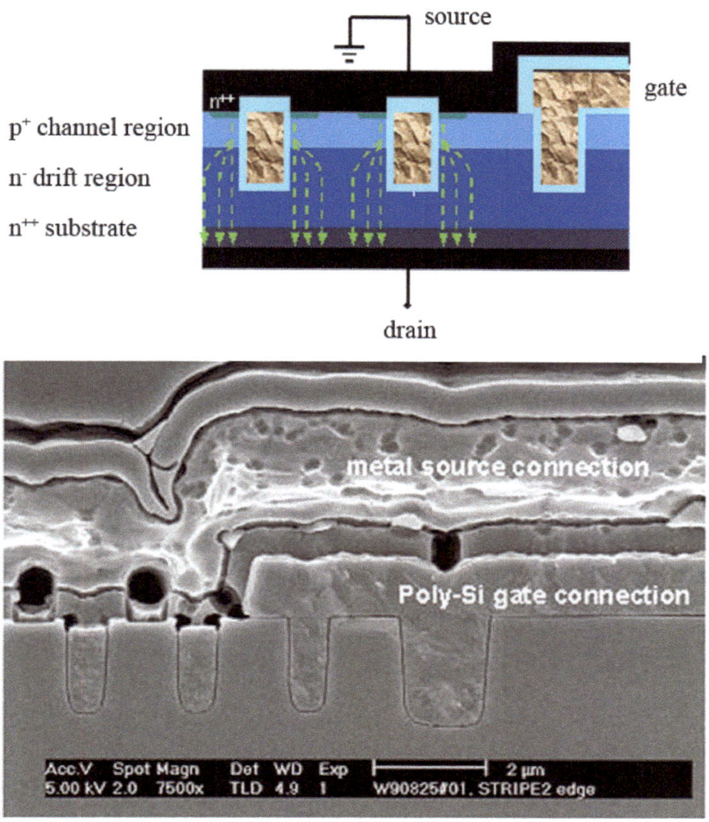

Fig. 5.11 Schematic cross section, showing the electron flow in the on-state (positive voltage on the gate), and an SEM photograph of a trench MOSFET (Source: NXP Semiconductors)

5.4.3 Applications

Power MOSFETs have a relatively long history as discrete power switches in fluorescent lamp ballasts, switch-mode power supplies and automotive switching applications. In electrical shavers, they are used both in the form of discrete devices and as part of larger integrated circuits, e.g., automatic supply voltage adaptors and battery chargers. Their high current capability makes power MOSFETs suitable for use in driver circuits, e.g., for stepper motors. On the other hand, plasma display drivers, for example, only require relatively small driving currents.

Power MOSFETs are easily integrated in bipolar and BICMOS circuits (see next Sect. 5.5) because they do not inject minority carriers. The combination of low-voltage bipolar transistors and high-voltage lateral DMOS transistors of both n and p types facilitates production of analogue high-voltage circuits [21]. Examples include video output amplifiers [22] and [23].

In the world of today, electronic devices are getting more and more mobile. Obvious examples are smart phones, laptops, tablets, GPS, cameras, etc. The absence of direct connection to the mains, as well as the ever-increasing demands on environmental friendliness, put strong demands on the power consumption of the circuitry in these devices. Large operating times and low energy consumption when charging the battery both require the use of efficient power converters. Power MOSFETs are also key components in these converters. Improving these components implies the search for the best trade-off between the off-state breakdown voltage and the power losses. For a given breakdown voltage, which is determined by the application, these power losses should be as low as possible, in order to make the power converters as efficient as possible.

From the above it may be clear that power MOSFETs will be used in an increasing number of applications, such as industrial systems, consumer electronics and electric vehicle. The global power MOSFET market was US$ 6.55 billion but is expected to almost double by the year 2030 [24]. In summary, we can conclude that there are several important factors in the development of a power MOSFET technology: device architecture, device density, current capability, on-resistance, break-down voltage, etc. The order of priority, however, depends on the application area.

5.5 Bipolar-CMOS-DMOS (BCD) Processes

(C)MOS technologies are best suited to efficiently implement digital circuits and memories. However, with the inclusion of special process options, they also enable the integration of various analogue/mixed signal products in so-called *mixed signal CMOS* processes. Figure 5.12 shows a variety of circuits which can be integrated in such a process.

RF CMOS circuits are often used to transmit or receive wireless signals, such as used in satellite (GPS), bleutooth, WIFI, NFC, 3G/4G, terrestrial broadcast and automotive radar applications, to name a few. Transmitters and receivers use high-quality oscillators to tune their transmit (broadcast) and receive frequencies, which require high-accuracy passive components: resistors, capacitors and inductors. CMOS is most commonly used to implement analogue/digital and low-mid performance RF circuits, which require several additional process options (add-ons). High-frequency wireless transmitter and power amplifier products even require different materials, such as gallium-arsenide (GaAS), which dominates these product markets with frequencies in excess of 250 MHz and with less noise.

As said before: MOSFETs are best suited to implement high-density logic and (SRAM and ROM) memory circuits and they accommodate low-power products. Bipolar transistors are known for their high amplification, low output resistance and high current drive capability. They are therefore better suited for high-frequency analog amplifiers. From the previous paragraph it is clear that power MOSFETs allow high-voltage switching and enable power stages. Figure 5.13 shows the individual process cross sections of these three basic devices.

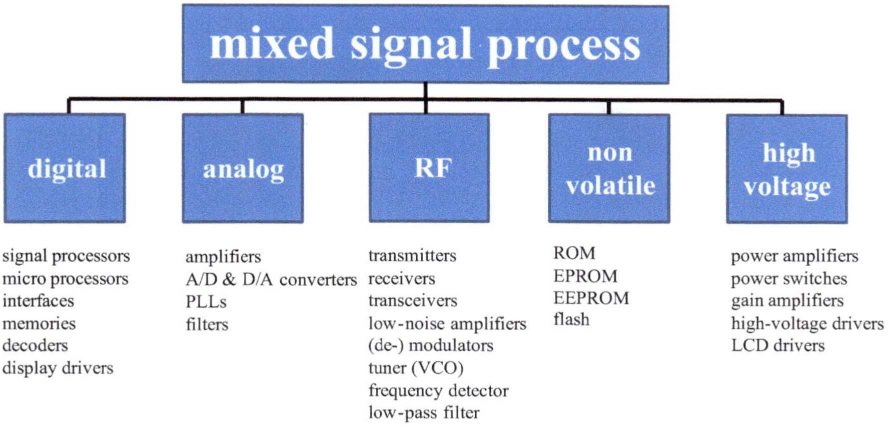

Fig. 5.12 A selection of potential circuits that can all be integrated with a mixed-signal CMOS process

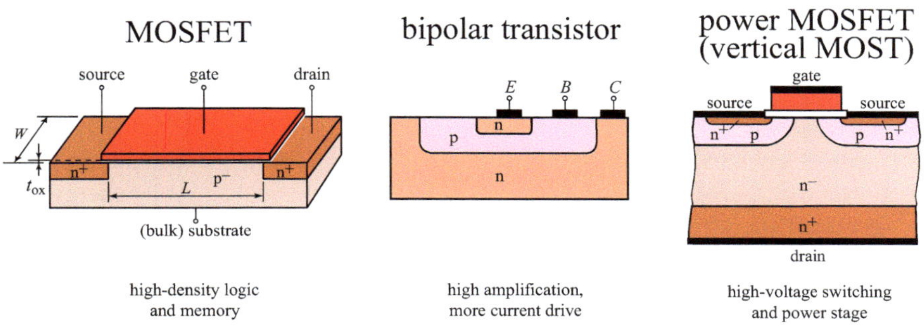

Fig. 5.13 Cross sections of a MOS transistor, a bipolar transistor and a power MOSFET

There are many products that simultaneously require the best of these three components. A process that integrates all of these devices is called *Bipolar-CMOS-DMOS process*, or shortly: *BCD process*. Its voltage often ranges from 5 V to 700 V. The high-voltage devices often use SOI substrates which enable even higher voltages in the range of 900–1000 V. A primary driver of BCD technology growth was in the power management of many systems. Only in this field of application we can distinguish a variety of applications:

- power management IC (PMIC) in mobile: smart phones, tablets, laptops, GPS, portable medical device, MP-3 players, Bluetooth, headsets, USB chargers
- battery management in general
- consumer display power and back lighting
- power management in computer systems: PCs and servers
- wireless chargers

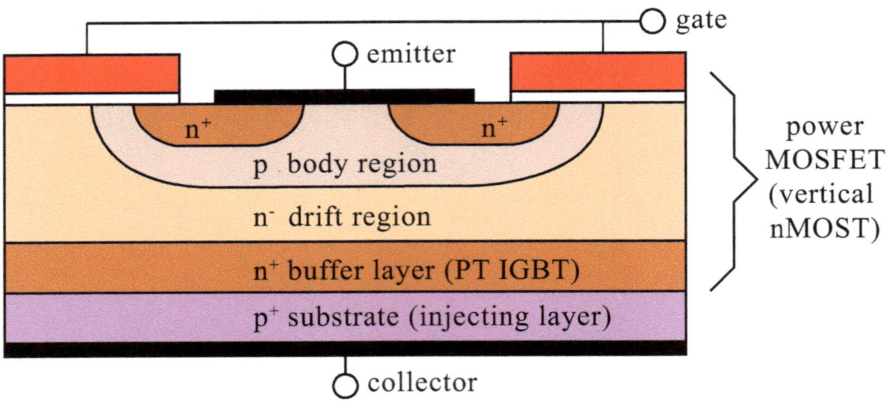

Fig. 5.14 Example cross section of an IGBT transistor

- light management: LED control, power switches and TL light switching
- solar array: voltage conversion, inverters
- industrial: stepper precision and servo motor control
- automotive: electronic fuel injection systems, window lifts, wipers, airbags, steering wheel
- medical: cochlear implant, implantable management for neuro stimulation
- power management products that convert vibration, thermal and solar energy into electrical energy for energy harvesting

Other examples of BCD products are: audio amplifiers in audio applications, (dimmable) LED drivers and power switches in lighting applications, lighting, motor control (mirrors, wipers, seat positioning) in automotive applications and spindle speed control in hard-disc drives. Some applications require embedded non-volatile memory (EEPROM of eFlash). There are also companies that use discrete *insulated gate bipolar transistors (IGBTs)*. Figure 5.14 shows the architecture of an IGBT transistor.

When the gate voltage exceeds the threshold voltage, the p-type body regions below the gates become inverted, thereby creating an n-channel in the p-body region between the n^+-emitter regions and the n^--drift region. With a positive voltage at the collector, the n^+/p^+ diode opens causing a large current flow from collector to emitter. The conductivity is increased by electrons injected from the n^+ region and holes injected from the p^+ region. The lightly doped n^--drift region makes it suitable for high-voltage power switching. IGBTs combine the best properties of MOS and bipolar devices. It has the high input impedance and switching speed of a MOS transistor and the large current capability of a bipolar device. It has a much smaller on-state resistance than a MOS transistor of the same size. On-currents may exceed 1 kA, while voltage ratings may exceed 2 kV and the device exhibits medium switching speeds. A disadvantage is its higher fabrication cost. After the power

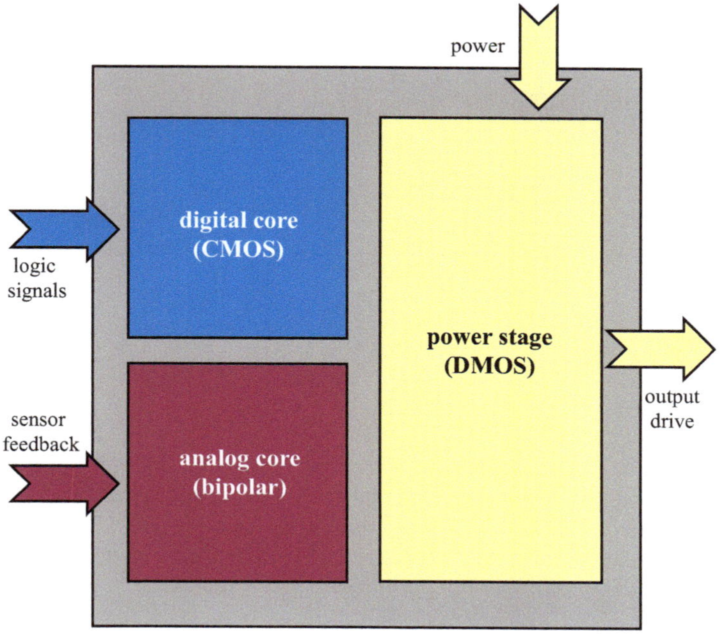

Fig. 5.15 Generic representation of a BCD chip

MOSFET, the IGBT is the most widely used power device. Generally, for most applications, it holds that the IGBT is the better choice for power conditioning at voltages over 200 V, currents above 500 A and frequencies below 100 kHz, while power MOSFETs operate better at voltages below 200 V, currents below 200 A and frequencies over 100 kHz. IGBTs are often found in applications as motor drives, kitchen induction appliances (microwave ovens, cookers), welding machines and in switch-mode power supplies. The latter ones convert the mains AC voltage input down to a low-voltage DC output. AC-DC switched-mode converters are widely used in external power supplies for laptops, tablets, mobile phones, etc.

Most of the above applications require a digital controller, an analog block that processes the analog inputs and a power stage that drives the high currents. Figure 5.15 shows a simplified block diagram that represents a generic BCD chip and needs no further explanation.

Finally, Fig. 5.16 shows a cross section of a BCD process, which integrates the three devices of Fig. 5.13.

As many of the BCD products relatively use a small digital content, they are often fabricated in a conventional process node. Many producers still use an ‘artificial 140 nm CMOS process’, which is a linear shrink of their basic 180 nm CMOS process, in which many of the add-ons are available for the required BCD products. Today, more and more add-ons are/become available in the 90 nm, 60 nm and 40 nm nodes.

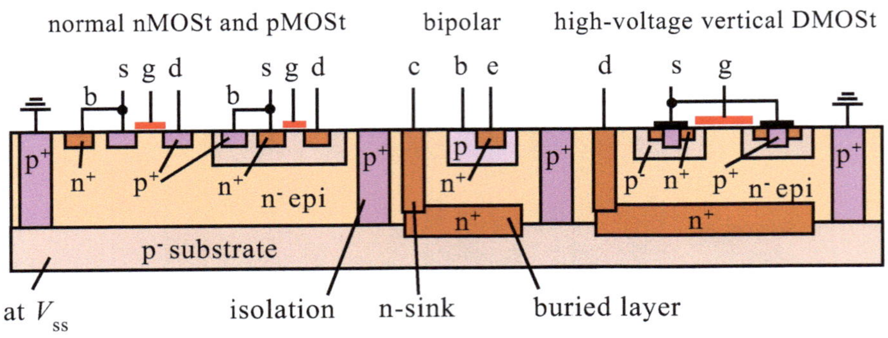

Fig. 5.16 Example of a BCD process integrating MOSFETs, bipolar transistors and high-voltage MOSFETs

Because the amount of electronics in cars is rapidly increasing, the automotive market is increasingly using BCD technologies with many applications using SOI based power devices [25]. Some companies claim the name *ABCD process*, where the *A* represents either analogue or advanced, depending on the company that uses it.

5.6 Conclusions

A number of devices and technologies that can be used in both purely digital and mixed analogue/digital ICs are discussed in this chapter. Because this is the only link between the presented topics, no general conclusions are presented here. The reader is therefore referred to the application sections associated with the CCD and MOS power transistor topics and the section on future expectations and market trends associated with the BICMOS topic.

5.7 Exercises

1. A dynamic shift register can be implemented as discussed in the chapter on CMOS circuits. It can also be implemented as a charge-coupled device (CCD). What are the main differences between the former implementations and the CCD implementation? State advantages and disadvantages to support your answer.
2. Assume that the transfer of a logic '1' through an SCCD is represented by a full charge packet. Explain what happens if the temperature increases when a series of data bits consisting of a hundred '1's, one '0' and again a hundred '1's, i.e., 111...1111011111...111, is transferred through the device.
3. Explain the main differences between a low-voltage MOS transistor which operates at 1.2 V and a power MOSFET.

4. Explain why BICMOS circuits exhibit a longer propagation delay than their CMOS counterparts for small capacitive loads and a shorter propagation delay for large capacitive loads.
5. Explain why BICMOS circuit performance relatively reduces with technology scaling, compared to CMOS circuit performance.

References

1. Image Sensor, Session at the ISSCC conferences: ISSCC Digest of Technical Papers, 2015–2024
2. E.R. Fossum, CMOS image sensors: electronic camera-on-a-chip. IEEE Trans. Electron Dev. **44**, (1997)
3. A.J.P. Theuwissen, Better pictures through physics; the state of the art of CMOS image sensors. IEEE Solid State Circuits Mag. (2010)
4. K.-B. Cho, et al., A 1/2.5 inch 8.1 Mpixel CMOS Image Sensor for Digital Cameras. ISSCC Digest of Technical Papers, 2007, pp. 508–509
5. 2-Layer Transistor Pixel, Sony, Technology, Image Sensor for Mobile (2022). https://www.sony-semicon.com/en/technology/mobile/2-layer-pixel.html
6. C. Okada, et al., A high-speed back-illuminated stacked CMOS image sensor with column-parallel kT/C-cancelling S&H and Delta-Sigma ADC, in *2021 IEEE International Solid-State Circuits Conference* (pp. 166–118)
7. CCDs will Continue to Provide a Crucial Imaging Capability for Science, Teledyneimaging.com, January 22, 2020
8. K. Washio, et al., SiGe HBT and BiCMOS Technologies. IEDM, Digest of Technical Papers, Session 5.1.3., 2003
9. J. Donkers, et al., *Vertical Profile Optimisation of a Self-Aligned SiGeC HBT Process with an n-Cap Emitter* (IEEE/BCTM, 2003)
10. J. Donkers, et al., Metal Emitter SiGe:C HBTs, IEDM, Digest of technical papers, 2004
11. L.J. Choi, et al., A Novel Isolation Scheme featuring Cavities in the Collector for a High-Speed 0.13 μm SiGe:C BiCMOS Technology, SiRF, 2007
12. H. Rücker, et al., High-performance SiGe HBTs for next generation BiCMOS technology, Semiconductor Science and Technology, IOPScience. Semicond. Sci. Technol. **33** 114003 (2018)
13. P. Deixler, et al., QUBiC4plus: a cost-effective BiCMOS manufacturing technology with elite passive enhancements optimized for 'silicon-based' RF-system-in-package environment, in *Bipolar/BiCMOS Circuits and Technology Meeting*, 2005 (pp. 272–275)
14. Y. Ruan, et al., A Monolithic SiGe BiCMOS Power Amplifier For UHF RFID Application, in *11th International Conference on Wireless Communications, Networking and Mobile Computing (WiCOM 2015)*, 2015 (p. 3)
15. E. Öztürk, et al., An ultra-wideband SiGe BiCMOS LNA for W-band applications. Microwave Opt. Technol. Lett. **57**(6), 1274–1278 (2015)
16. V. Valenta, et al., SiGe: C BiCMOS components and integrated solutions for F-band radar frontends. https://hal.archives-ouvertes.fr/hal-01131093/document, march 2015
17. T. Zimmer, et al., SiGe HBTs and BiCMOS technology for present and future millimeter-wave system. IEEE J. Microwaves **1**(1), 288–298 (2021)
18. F. Udrea, et al., *SOI-based devices and technologies for high voltage ICs* (BCTM 2007), pp. 74–79
19. H. Gensinger, High-Voltage CMOS Technologies for Robust System-on-Chip Design. HVCMOS_FSA Forum, June 2006
20. R. Lerner et al., Comparison of different device concepts to increase the operating voltage of a trench isolated SOI technology to above 900 V", in *Facta Universitatis Series: Electronics and Energetics*, vol. 28(4), pp. 645–656 (2015)

21. A. Ludikhuize, A versatile 250/300V IC process for analog and switching applications. IEEE Trans. Electron. Dev. **ED-33**, 2008–2015 (1986)
22. P. Blanken, P. van der Zee, An integrated 8 MHz video output amplifier. IEEE Trans. Consumer Electron. **CE-31**, 109 (1985)
23. P. Blanken, J. Verdaasdonk, An integrated 150 V_{pp}, 12kV/μs class AB CRT-driving amplifier. ISSCC, Digest of Technical Papers, 1989, New York
24. Power MOSFET market, https://straitsresearch.com/report/power-mosfet-market, 2022
25. P. Flatresse et al., FD-SOI, the technology shaping the future of automotive radars. SemiWiki, October 16 2023

Memories

6

6.1 Introduction

Memories are circuits designed for the storage of digital values. In a computer system, memories are used in a large variety of storage applications, depending on memory capacity, cost and speed. Figure 6.1 shows the use of memory storage at different hierarchy levels of a computer system.

The high-level memories directly communicate with the computer's processor. They must deal with its high data communication bandwidth and therefore need high performance, but they are expensive. As we move down the hierarchy, both the memory capacity and the access time increase, resulting in a reduction of the cost per bit. A cache memory acts as an intermediate storage between the CPU and the main memory and stores the most-frequently and/or most-recently used data and instructions for fast access.

A memory may constitute a single IC or be part of a larger IC. These types are referred to as *stand-alone* and *embedded* memories, respectively.

The digital values in a memory are each stored in a *'cell'*. The cells are arranged in a *matrix* or *array*, which affords an optimum layout.

Memories that lose their data when power is removed are referred to as *volatile*. Memories that retain their data in the absence of power are called *non-volatile* memories. The *'data retention time'* of a memory is the period for which it keeps its data when the supply voltage is removed. A finer division of memories yields the following four types:

- Serial memory;
- Content-addressable memory (CAM);
- Random-access memory (RAM);
- Read-only memory (ROM).

H. Veendrick, *Nanometer CMOS ICs*,
https://doi.org/10.1007/978-3-031-64249-4_6

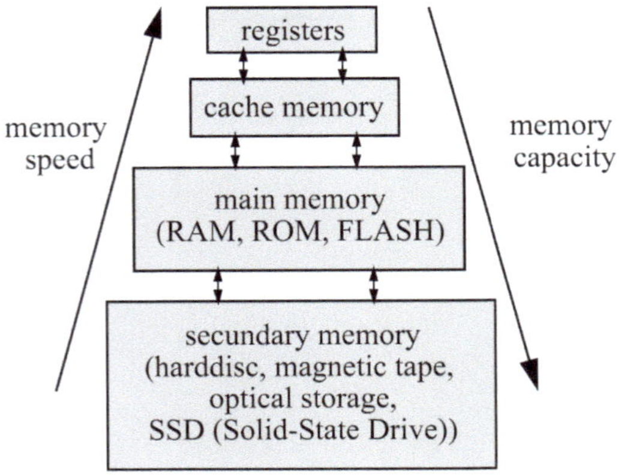

Fig. 6.1 Memory hierarchy in a computer system

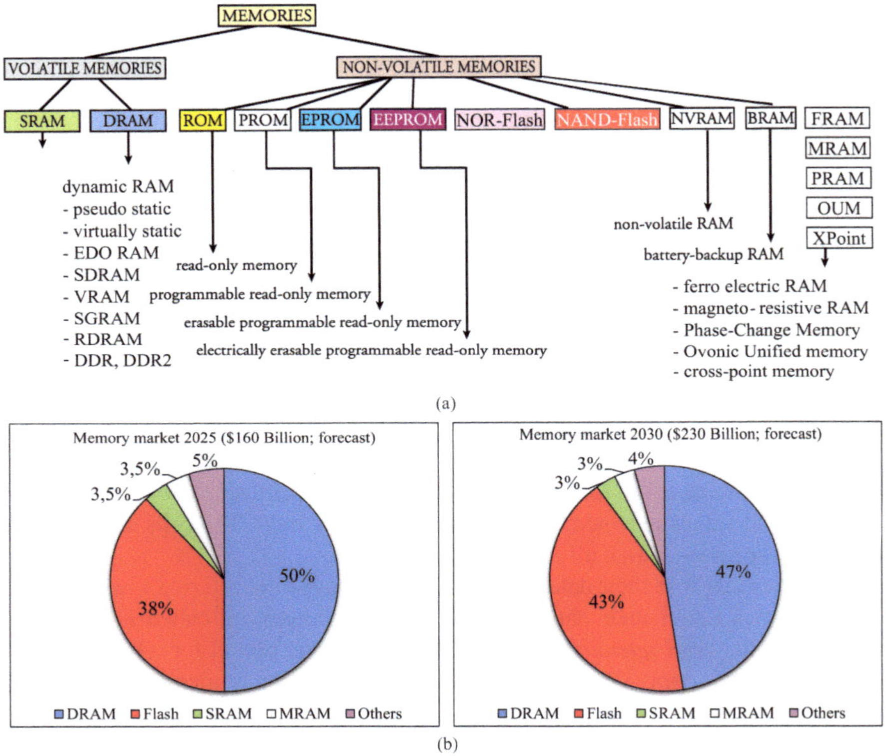

Fig. 6.2 (**a**) Overview of different types of memories (**b**) Relative memory market shares in 2025 and expected market shares 2030 (Source: Mordor Intelligence/Grand View Research)

Figure 6.2 presents an overview of the various implementation possibilities for memories. This figure also shows the respective market shares in 2025 and the expected market shares in 2030. The increased market share gained by the DRAMs is mainly the result of the rise of new high-speed architectures, which make them particularly suited for the growing high memory bandwidth applications such as games, video and graphics applications and printers.

Volatile memories include *'static'* and *'dynamic'* RAMs. Electrical feedback in the memory cell of a *static RAM (SRAM)* ensures that voltage levels are maintained and data are retained as long as the power supply remains. The data in a *dynamic RAM (DRAM)* memory cell are stored as a charge on a capacitor. Gradual leakage necessitates periodic refreshing of the stored charge. A dynamic RAM that internally refreshes its own data automatically is called a *pseudo-static* or *virtually static* RAM.

The cells in serial memories form one or more shift registers, which can each store a 1-bit data stream. The 'first in, first out' *(FIFO)* operation of shift registers ensures that data enter and leave a serial memory in the same sequence. Examples of their use include delay lines in video applications.

The cells in a content-addressable memory may contain an SRAM cell plus additional comparison circuitry, because an access is not based on offering a dedicated address, but on the comparison of input data bits with stored data bits. When a match occurs, the corresponding output data bits are returned.

The cells in a RAM or ROM array must have individual unique *'addresses'*. Alternatively, they may be connected in parallel groups. In this case, each group or *'word'* has a specific address. The capacity of a RAM or ROM that is divided into words is specified by the number of words and the number of bits per word. Examples are 1 Gb×4, 512 Mb×8 and 256 Mb×16. These three specifications all refer to a 4 Gb memory, which can store over 32,000 newspaper pages or 9 hours of MP3 music. The quantification of bits in this chapter is according to international conventions: 1 Gb equals 1 gigabit (1 Mb=1 megabit, etc.), 1 GB equals 1 gigabyte (1 MB=1 megabyte, etc.). The word *byte* is a short notation for *by eight* and so it contains 8 bits.

The data in a ROM can only be read, whereas the data in a RAM can be written and read. The sequence in which data are read from a ROM or RAM is unrestricted. Therefore, *access* is in fact *random* in both cases. The term RAM, however, is generally only used to refer to memories that allow reasonably high-frequency read and write operations at random locations.

A RAM requires both data and address inputs and data outputs. Figure 6.3 is a general schematic representation of an addressable memory.

The memory shown is obviously a RAM. The read (R) and write (W) inputs are often combined in one single input which controls the mode of operation. A ROM requires no data inputs but does require address inputs and data outputs. The schematic of a ROM is therefore obtained if the data (I) and W inputs in Fig. 6.3 are removed. The schematic of a serial memory is obtained if the address inputs are removed. Flash memories have either random access (NOR flash) or serial access (NAND flash) (see Sect. 6.5.4).

The *'access time'* of a memory is the time interval between the initial rising clock edge in a read operation and the moment at which the data are available at the output terminals. The *'cycle time'* of a memory is the minimum possible time

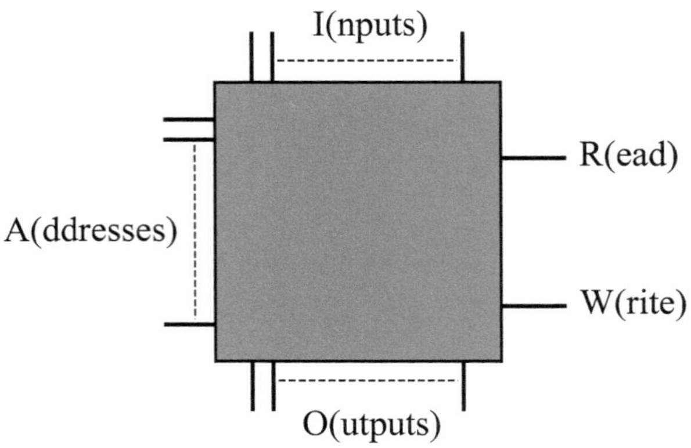

Fig. 6.3 General representation of a memory

between two successive accesses. The cycle time of an SRAM may be greater than, smaller than or equal to its access time, while the typical cycle time of a DRAM is about twice the access time. This is because the accessed cells in a DRAM must be refreshed after each read and write operation. Although access times are often used for comparison of the different memories available from different manufacturers, cycle time comparison would be better for benchmarking purposes. There are many techniques that improve the access and cycle times of DRAMs. These are discussed in Sect. 6.4.4.

Various types of memories are discussed in this chapter. Their operation and properties are explained and possible applications are given. A brief discussion of the structure of a simple 4 kb SRAM provides considerable insight into memory operation.

6.2 Serial Memories

Serial memories are usually encountered in the form of static or dynamic shift registers. Modern *video memories* are an important exception. These memories are serial by nature, and random access is therefore often not required. However, they are implemented as DRAMs, in which the cells are serially accessed. Many of these memories include a buffer memory, such as a *FIFO* (first-in first-out) or *LIFO* (last-in first-out) to change the sequence of the data bits. Serial memories are used in video and graphics applications. Such a memory is sometimes called a *video RAM* or *VRAM* (see Sect. 6.4.4).

Small serial memories may be implemented using the CMOS shift register cells presented in Chap. 4. The discussion on shift registers in Chap. 4 makes further elaboration on serial memories unnecessary.

6.3 Content-Addressable Memories (CAM)

In a *content-addressable memory (CAM)*, also called *associative memory*, each cell has its own comparison circuit to detect a direct match between the search bit and the stored bit. Instead of supplying a memory address, in a CAM, an input word is supplied, and a search operation is executed through the whole memory within one clock cycle. The response to a search is a list of one or more addresses, depending on a single or multiple match of the stored data in the memory. Some CAM memories directly return the full corresponding data contents on those addresses. A more detailed description of a CAM is beyond the scope of this book. A rather complete tutorial on CAM can be found in [1].

6.4 Random-Access Memories (RAM)

6.4.1 Introduction

Random-access memories can be subdivided into the two following classes:

- Static RAM (SRAM)
- Dynamic RAM (DRAM)

These two types of RAM are discussed separately further. The basic operation of a RAM is explained with the aid of a 4 kb SRAM. A subsequent discussion of the major differences between SRAMs and DRAMs illustrates the considerable difference in their operation.

6.4.2 Static RAMs (SRAM)

A true static memory is characterised by the time between a change on its address inputs and the presence of valid bits at its data outputs. Dynamic memories often require a considerably more complex pulse pattern with very stringent timing requirements.

SRAM Block Diagram

For most stand-alone SRAMs, every possible combination of address inputs can be decoded. A memory with n address inputs therefore contains 2^n addresses. An SRAM with 12 address inputs, for example, therefore, has at least 4096 memory words. Figure 6.4 shows the block diagram of such a 4 kb SRAM. This example represents a so-called by-1 memory (4k by-1), meaning that at each address selection only one cell (one bit) is accessed. This has been done for educational purposes. In most memories, a complete word, which may contain 4, 8, 16, 32 or even 64 bits, is stored at one address.

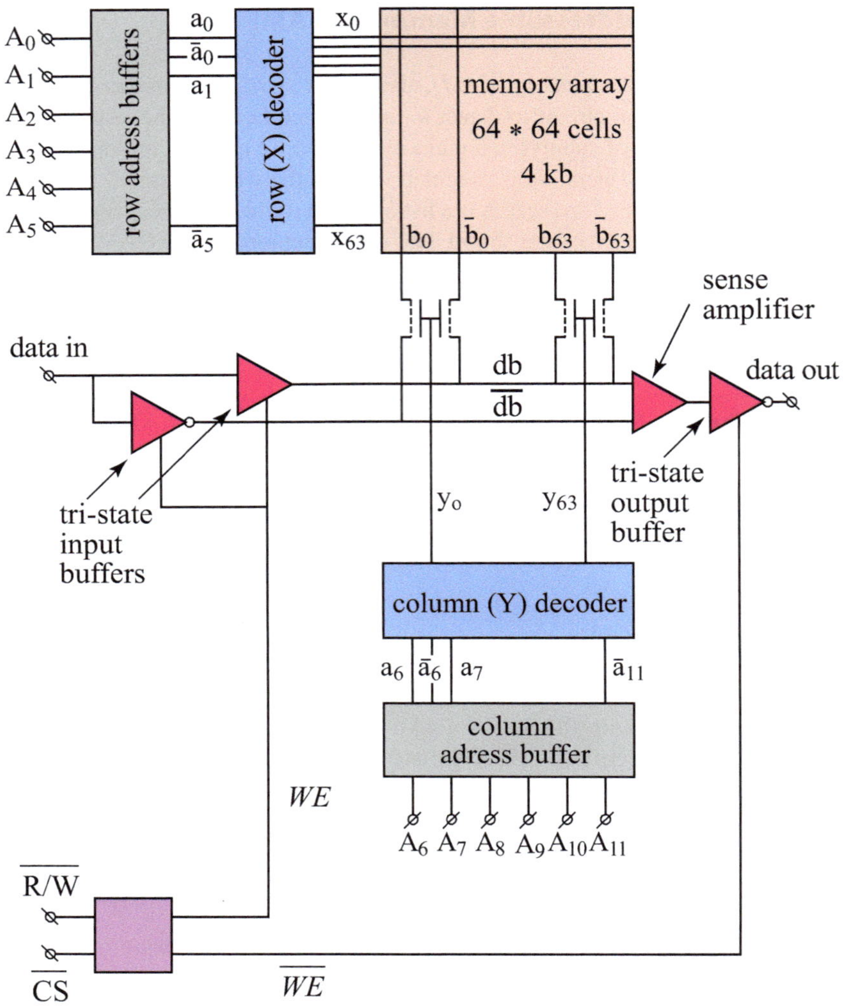

Fig. 6.4 Block diagram of a 4 kb SRAM

Its 4096 memory cells are organised in an array of 64 rows and 64 columns. Each row and column can therefore be addressed by 6 address inputs. In addition to an array of memory cells, an SRAM requires control logic circuits. These circuits will now be described.

- A *row decoder* selects the *'word line'* x_i of the row in which the addressed word (or cell) is located. The row decoder is also known as an *x-decoder*.
- A *column decoder* selects the *'bit line select'* line y_j of the column in which the addressed word (or cell) is located. The column decoder is also known as a *y-*

decoder. The addressed cell is located at the point of intersection of the selected row and column and is referred to as cell x_i,y_j. The y_j signal selects the *bit lines* b_j and $\overline{b_j}$ of the addressed cell.

- *Address buffers* connected to the address inputs drive the row and column decoders. The output lines of the row and column address buffers traverse the length and width, respectively, of the array. They therefore form large capacitive loads for the address buffers.
- The tri-state *data input buffers* drive data buses db and $\overline{db}$ when the memory is being written. These buffers drive the large capacitive load of the data bus line and the selected bit line. They must also be capable of forcing the memory cell into another logic state. Current memories have separated data bus drivers and bit line drivers.
- A *sense amplifier* detects the contents of the selected cell via the complementary bit lines b_j and $\overline{b_j}$ and data bus lines db and $\overline{db}$. The detection must occur as fast as possible so that the access time is reduced to a minimum. The sensitivity of the sense amplifier may be as low as 70–100 mV. Current sensing and differential voltage sensing are alternative techniques for optimised memory performance. Because of the reducing voltage headroom, current sensing becomes less popular, because it consists of more analogue circuitry. The 4 kb SRAM in this example only includes one sense amplifier. Preferably one sense amplifier per column should be used, but it does no longer fit in the memory cell pitch. Therefore, today's memories may include one sense amplifier for every four columns combined with a multiplexer circuit for selection.
- The tri-state *data output buffer* transfers the data from the sense amplifier to the SRAM output when the memory is being read. Because memories are often used in a memory bank, it must be possible to enable one single memory, while others are disconnected from the bus. This requires a high-ohmic third state of the output (tri-state).

The SRAM Control Signals

The control signals required in an SRAM are described further. For the sake of simplicity, the commonly used *output enable (OE)* signal is omitted.

- The *write enable* ($\overline{WE}$) signal determines whether data are written to the selected cell or read from it. During writing, the bit line signals are derived from the input. Depending on the data to be stored, only one of the two bit lines is pulled to ground, while the other is kept at $V_{\rm dd}$ level by the precharge transistors. During reading, the cell pulls one of the bit lines low and the bit line signals are then transferred to the output.
- The *chip select* ($\overline{CS}$) signal facilitates selection of a single SRAM when many are combined to form a large memory system. Such a system consists of one or more *memory banks*. The memories in such a system may be connected to common address and data buses. Although more than one memory (or even a complete bank) can be selected at the same time, only one at a time can put data on the data bus. The $\overline{CS}$ signal of the relevant memory is activated by decoder

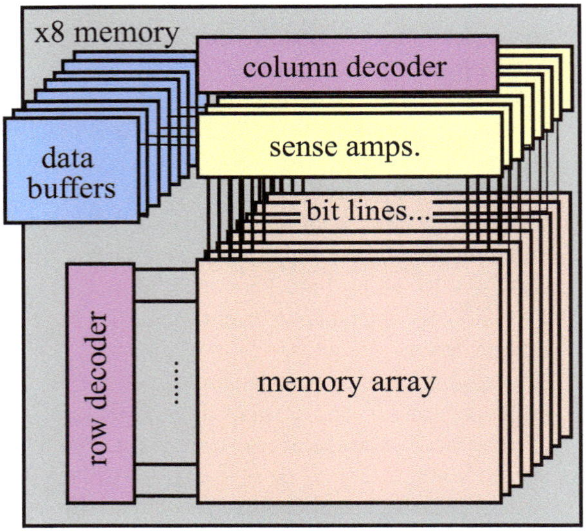

Fig. 6.5 Physical representation of a x8 RAM architecture

logic in the memory bank. This logic produces 'high' logic levels on the $\overline{CS}$ inputs of the remaining memories. Their output buffers are therefore placed in the high-impedance mode, and the memories are isolated from the data bus.

Normal memory architectures are by 2, by 4, by 8, etc., meaning that 2, 4, respectively 8 memory arrays can be accessed simultaneously. Figure 6.5 shows the physical representation of a x8 RAM architecture.

The more parallel accesses a memory allows, the higher the communication bandwidth to interfacing CPUs.

The SRAM Read Operation

The read operation performed in an SRAM is explained with the aid of an example in which data are read from the cell x_{35},y_{40}, see Fig. 6.4. The steps required to achieve this are as follows:

- The word line x_{35} is activated by placing the decimal value 35 on address inputs A_5 to A_0: $A_5A_4A_3A_2A_1A_0 = 100011$.
- The bit line select signal y_{40} is activated by placing the decimal value 40 on the address inputs A_{11} to A_6:
  $A_{11}A_{10}A_9A_8A_7A_6 = 101000$.
- The $\overline{CS}$ signal is driven 'low' to select the memory.
- The $\overline{WE}$ signal is driven 'high' so that the information in the selected cell can be read via the selected bit lines, the sense amplifier and output buffer. The logic '1' on the $\overline{WE}$ signal activates the output buffer and places the tri-state input buffers in the high-impedance state. At the beginning of each read cycle, all bit lines b_i and $\overline{b_i}$ are precharged through clocked transistors (not drawn in Fig. 6.4) to the high level (other precharge levels, such as half-V_{dd} or low (V_{ss}) levels are

also also used). If the value '0' is subsequently read from the selected cell, then bit line $\overline{b_{40}}$ remains 'high' while bit line b_{40} discharges slightly via the cell. The bit line voltage levels are transferred to the respective $\overline{db}$ and db data buses. The sense amplifier quickly translates the resulting voltage difference to a logic '0', which is then transferred to the output via the buffer. A similar explanation applies when the value '1' is read from the selected cell, but then b_{40} remains high and $\overline{b_{40}}$ will discharge.

The SRAM Write Operation

The write operation performed in an SRAM is explained with the aid of an example in which data are written to the cell x_{17},y_{15}, see Fig. 6.4. The steps required to achieve this are as follows:

- The word line x_{17} is activated by placing the decimal value 17 on the address inputs A_5 to A_0: $A_5A_4A_3A_2A_1A_0 = 010001$.
- The bit line select signal y_{15} is activated by placing the decimal value 15 on the address inputs A_{11} to A_6:
  $A_{11}A_{10}A_9A_8A_7A_6 = 001111$.
- The $\overline{CS}$ signal is driven 'low' to select the memory.
- The $\overline{WE}$ signal is driven 'low' so that the information on the data input can be written to the selected cell via the data input buffers and the selected bit lines. The value on the db data bus is then equal to the value on the data input while the $\overline{db}$ data bus has its inverse value. The logic '0' on the $\overline{WE}$ signal activates the input buffers and places the tri-state output buffer in the high-impedance state.

SRAMs are designed in a variety of synchronous and asynchronous architectures and speeds. An asynchronous SRAM is activated when an address change is detected. As a result, a clock signal is generated and stored data are accessed. However, this type of SRAM is limited in its speed. Therefore, the fastest SRAMs are generally synchronous. Controlled by one or more clocks, synchronous SRAMs show reduced access and cycle times, boosting their clock frequencies to the same level as those of the high-performance RISC processors and PCs. Improved performance can be achieved when several words are selected simultaneously by a single address. In *burst mode* operation, the address is incremented by an on-chip counter and the parallel read words are serialised to form a large sequence of high-speed data bits. Several burst addressing sequences can be supported, including those used in Pentium^{TM} and PowerPC^{TM} processors. Nowadays, burst mode is also supported by relatively cheap microcontrollers.

Static RAM Cells

Access time is an important RAM specification and is mainly determined by the signal propagation time from the memory cell to the output. A satisfactory *access time* requires an optimum design of the memory cell, selection circuits, bit lines, sense amplifiers and output buffers. Possible *memory cell implementations* for SRAMs are discussed in detail further.

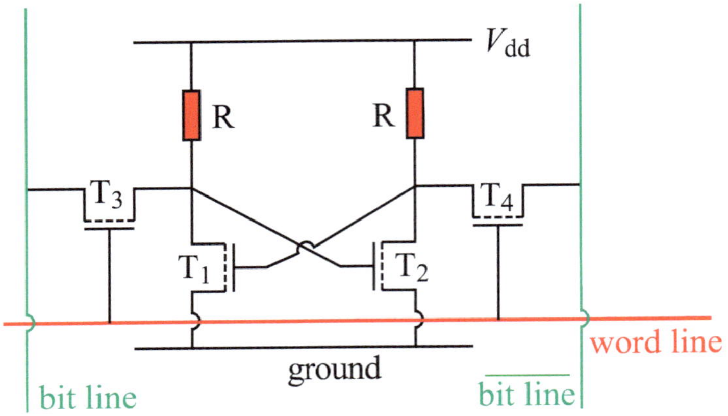

Fig. 6.6 Four-transistor static RAM cell

1. *Four-transistor/R-load SRAM cell*
   Figure 6.6 shows a memory cell consisting of four transistors. This cell contains two cross-coupled inverters with resistive loads. These types of inverters are discussed in Sect. 4.2, and they lead to continuous static power dissipation in the memory cell. This dissipation is kept as low as possible by forming the resistors in an extra high-ohmic polysilicon layer. Typical values are 10 GΩ or more. This polysilicon layer necessitates a more complex manufacturing process than for the full-CMOS cell. An advantage of the four-transistor cell, however, is its reduced cell area, because the resistors are implemented in a second polysilicon layer and folded over the transistors. These memories are hardly or no longer used today, mainly because of their reduced operating margins, but also because of their relatively large power consumption, in both active and standby modes.
2. *Four-transistor loadless SRAM cell*
   The introduction of a loadless four-transistor cell [2, 3] allows a 35% area reduction using the same design rule. Comparing Fig. 6.6, in the loadless cell, the resistors R and the V_{dd} connection are completely removed and transistors T_3 and T_4 are replaced by pMOS transistors. This allows the cell nodes to store full-swing signals after writing. In the standby mode, bit lines are precharged to V_{dd}, and the data are maintained in the cell when the leakage current of the pMOS transistors is more than an order of magnitude larger than that of the nMOS transistors. However, because its operation partly depends on the value of the leakage currents, it will be very difficult to realise large memories with it, since leakage currents are not very well controlled. This is particularly due to the large intra-die variations in V_T in current CMOS technologies.
3. *Six-transistor/full-CMOS SRAM cell*
   The two afore-discussed memory cells were used in the 1970s and 1980s. Today, almost all SRAM memories use the cell shown in Fig. 6.7. Transistors T_1 to T_4 comprise two cross-coupled inverters which function as a latch. Pass transistors

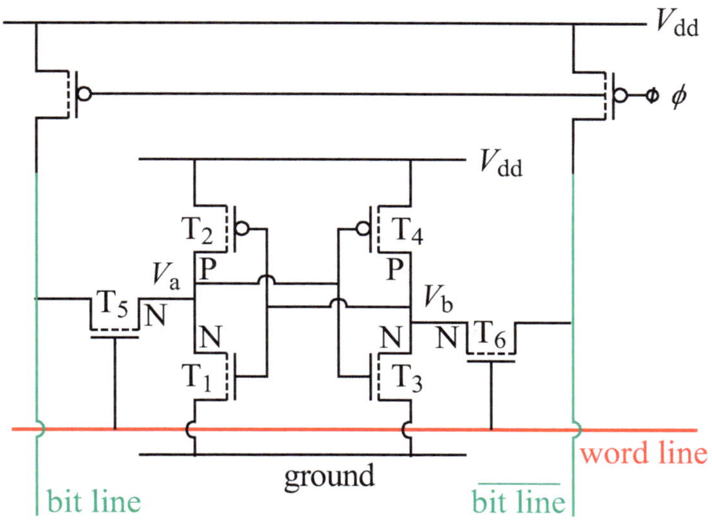

Fig. 6.7 Six-transistor static RAM cell

T_5 and T_6 provide access to the latch. During a write operation, the write data are transferred to the bit lines, the word line goes 'high' and the data on the bit lines are transferred to the latch through pass transistors T_5 and T_6.

During a read operation, first both bit lines are precharged to V_{dd}, by switching signal ϕ only shortly to zero. Then the word line goes 'high' and the contents of the cell cause a slight discharge on one of the precharged bit lines. The discharge takes place through the relevant pass transistor, T_5 or T_6, and inverter nMOS transistor, T_1 or T_3.

A small voltage difference of about 50–100 mV between the two bit lines is sufficient for an SRAM sense amplifier to determine the logic level in the memory cell. This logic level is then transferred to the output pin via the output buffer. The 4 kb SRAM in Fig. 6.4 has only one sense amplifier close to the output driver of the memory. Because the number of rows in one memory block (and thus the number of memory cells in one column) has increased dramatically, every column needs an amplifier at the bottom, to be able to still sense the small voltage differences between the bitlines during a read operation.

The small subthreshold and gate leakage currents are the only currents that flow in the six-transistor cell when it is simply retaining data. Memories containing full-CMOS cells are therefore suitable for low-power applications. However, the relatively large distance required between nMOS and pMOS transistors requires quite a large chip area for this memory cell. Over the last four decades, this full-CMOS 6T cell has been used for almost all embedded random access memories on all logic, microprocessor and mixed analog digital ICs.

The word lines in both the six-transistor memory cells are implemented in a stack of different metal layers. The considerable parasitic capacitance and resistance of long word lines cause the cells furthest from a row decoder in an SRAM to exhibit a greater RC-delay than those closest to the decoder. This situation is often addressed by dividing the arrays of large memories into several smaller sections with separate row decoders between them. The resulting word lines have lower parasitic capacitance and resistance, and their RC-delays are at least a factor 4 lower than for a single array. The silicides mentioned in Chap. 3 are also used to reduce resistance of polysilicon word lines.

Previous discussions made it clear that the 6-transistor SRAM cell is most commonly used in stand-alone and embedded SRAMs. Due to the continuous scaling of threshold and supply voltages, the noise margins of SRAM memories *(SNM = static-noise margin; WM = write margin)* have reduced dramatically, basically due to the increase in transistor parameter spread, combined with a reduction of the supply voltages. Particularly the threshold-voltage variation in the SRAM narrow-width transistors due to random doping fluctuations has a great influence on the SRAM robustness of operation. Due to the strict area requirements, an SRAM cell may easily loose its state when variations in transistor sizes (W and L), in supply voltage (V_{dd}), in threshold voltage (V_T), in temperature and STI stress, to name a few, become too large. The SNM of an SRAM cell describes how well it can cope with these variations. Figure 6.8 shows a diagram representing the stability of the SRAM cell of Fig. 6.7.

The curve I in the diagram shows the response of V_a on the stimulus of V_b, while the response of V_b on the stimulus of V_a is represented by curves II. For each mode of memory operation (hold or read mode), the SNM is represented by the largest square that can be drawn between the related curves I and II. A large size of the square represents a large SNM. Traditional worst-case values for SNM were in the order of 120–200 mV. Because of the reducing supply voltages and increasing process parameter spread, current SRAMs show SNMs of only a few tens of millivolts.

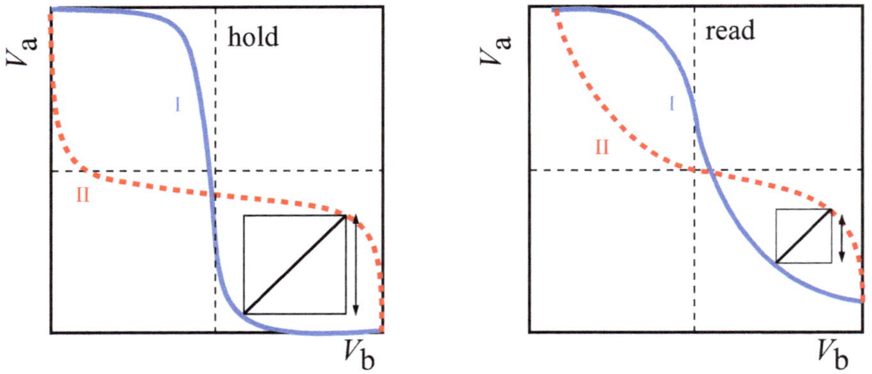

Fig. 6.8 Stability diagram of the SRAM cell of Fig. 6.7 and its SNM during hold and read mode

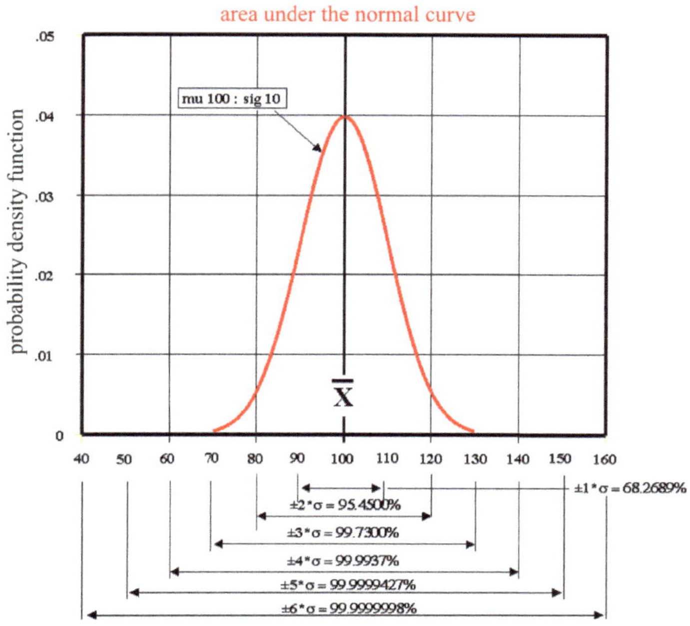

Fig. 6.9 Probability density function versus the spread σ around the mean value μ of a certain parameter

Usually a design parameter is targeted at a certain mean value, with 3-σ margins ($\approx$0.13% of the parts fail) at each side of the mean value (Fig. 6.9), assuming a normal distribution of the probability density function of a parameter with a mean value μ and a standard deviation σ.

By including a 3-σ margin at both sides of the mean value, the probability of a failure per cell is $\approx$0.13% at each side. For the early memories, a 3-σ parameter spread was considered sufficient to design the relatively small SRAM memories with several kilobits capacity. *Worst-case (WC)* and *best-case (BC) corner simulations* do not cover all possible combinations already for a long time, due to the increasing occurrence of transistor mismatch, both in value and in number. In other words, due to mismatch, not all memory cell transistors are in the same point in the same corner at the same time, like in the *WC* and *BC simulations*. For most of the embedded memories with capacities of up to 10 Mb a 6-σ parameter spread is taken into account, due to the reduced voltage margins and increased number of memory cells. According to Fig. 6.10, which shows this *parametric yield* loss as a function of the read or write margin, the yield of that memory is close to 99% (equivalent to 1% yield loss). This yield loss is derived from the probability density function (Fig. 6.9), which shows that in 0.0000002% of the trials, the 6-σ margins would be exceeded. In many cases, we only need to take one side of this probability density function, e.g., when a parameter exceeds the +6-σ value (at the right side of

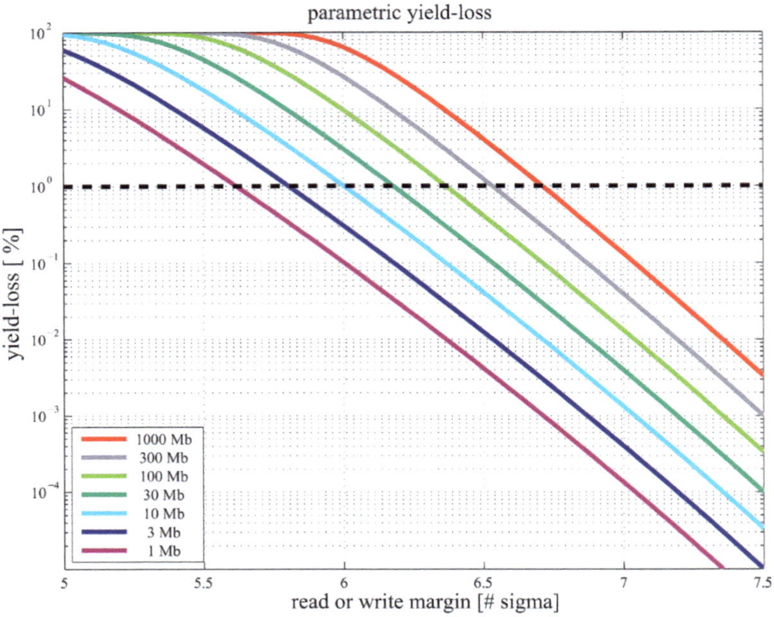

Fig. 6.10 Parametric yield loss as a function of the read or write margin

the diagram), leading to a failure, then in case the same parameter would exceed the -6-σ value (at the left side of the diagram), it would usually not lead to a failure.

Sometimes even a 6.4-σ spread is taken for this size of memory to achieve 99.9% yield without redundancy (see Sect. 6.8), which means that only 1 in 10 billion cells fails. A 7-σ spread, which is also under discussion for sub-10 nm SRAM design, refers to 99.99% yield in a 100 Mb memory without redundancy.

The previously mentioned yield numbers can be achieved without redundancy. The diagram in Fig. 6.11 shows the number of redundant cells that are required for various memory capacities to achieve 90% yield. It shows that we can exchange redundancy with read or write margin in the design. The decision, which solution will be supported, is most commonly based on the amount of area overhead.

Even in advanced CMOS process nodes, it is very hard to run the memories at voltages below 1 V. This is a combined result of the increasing number of bits and the different and almost conflicting transistor sizing requirements from the read and the write operation. Another problem is the increasing leakage currents, particularly when many memory cells share the same bit line.

In the example in Fig. 6.12, we assume that the bit lines are connected to 1 k memory cells and that the lower cell is to be read. If the total leakage current of 1023 cells (I_{leak}) is about equal to or higher than the cell read current (I_{cell}), the sense amplifier may read the wrong state, because also bit line is discharged through 1023 (subthreshold-)leaking pass transistors.

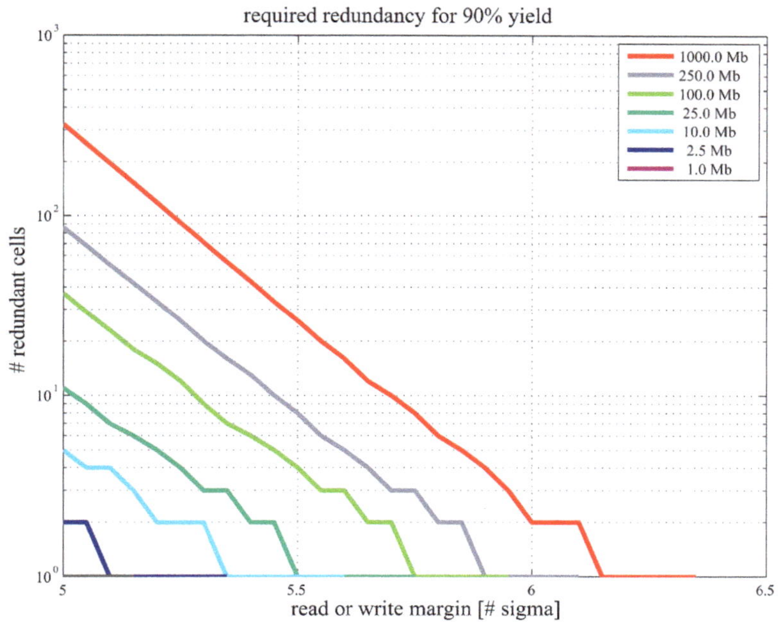

Fig. 6.11 Assuming a yield of 90%, then this figure shows the amount of redundancy required for different memory capacities

Fig. 6.12 Representation of a disturbed read operation

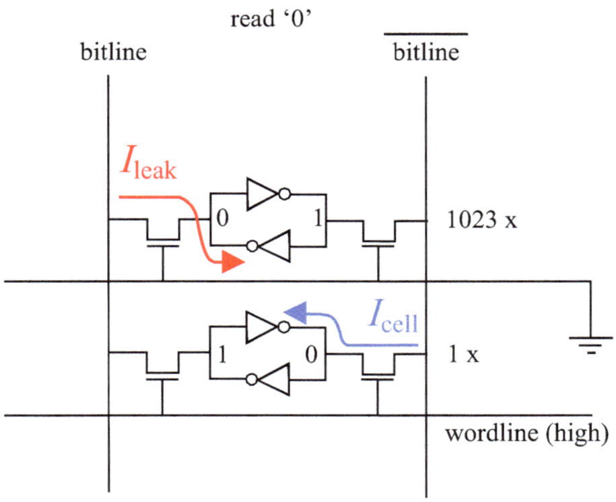

SRAM margins still get worse and do not allow the cell area to be scaled each next technology generation with a factor of two anymore. As a consequence, the memory density increase will no longer double each new generation, but is expected to increase by only 50%. Common stand-alone SRAM cell areas are between 80 and $150F^2$/bit, compared to $6F^2$/bit for a stacked DRAM cell (see section on

DRAMs) and only $2F^2$/bit for a multi-level NAND-flash cell. For *embedded SRAM (eSRAM)* memories, it holds that it will face the same problems as the stand-alone SRAMs, but only one or a couple of technology generations later. Massive simulations are required to completely validate SRAM designs. One solution to maintain future SRAM operation robustness is to use larger transistor sizes in the cell, since $\sigma \propto 1/\sqrt{WL}$. Alternatives are to use a more complex cell (7–10 transistors/bit) [4, 5] or to use a more complex technology, such as the 3D approach by Samsung [6], where the pMOS load and nMOS pass transistor are stacked on top of the planar pull down or to use error detection and correction techniques to increase the robustness of operation [7]. An increasing amount of SRAMs are implemented in FinFET technology. The quantised device width and limited process knobs have greatly affected *FinFET SRAM* design. The transistors in the 6T cell in Fig. 6.7 may be built with a different number of parallel fins. For high-density FinFET SRAM cells, the number of fins for the pull-down, the pass-gates and the pull-up transistors may be 1, 1 and 1, respectively. For low-voltage SRAM cells, the respective number of fins (= drive strength) are: 2, 1 and 1, while this may be 3, 2 and 1 for high-speed SRAM cells. There is also an alternative to create the different drive strengths while using only one single fin for all three transistors. For instance, for the low-voltage FinFET SRAM cell, the drive strength of 2 for the pull-down transistors can also be achieved by adjusting the channel strain and channel dope of these transistors. It also may require the introduction of *assist circuits* to enable operation at reduced supply voltages [8, 9]. These assist circuits are located in the periphery of the memory and create different voltage boosting levels at the word lines and bit lines to improve write and read ability and increase static noise margins. These assist circuits are also used in today's 6T SRAMs in planar CMOS technologies. Still a lot of R&D is being carried out to find alternative SRAM cell architectures for nanometer process nodes [10].

SRAM Applications

Static Random Access Memories (SRAMs) are widely used as memory elements in various electronic devices and systems. A very important application is cache memory in computer systems, including processors, GPUs and even in-memory databases, where SRAMs act as interface between DRAM and CPU. The fast read and write access times of these caches help improve performance by storing frequently accessed data closer to the processing units. Next to this SRAMs are also extensively used in embedded systems, such as microcontrollers and system-on-chips (SoCs). These systems require fast on-chip memory for storing critical code and data. Because of their fast and reliable access and low latency, SRAMs are also used in numerous other applications such as networking and telecommunications, automotive, consumer electronics, industrial and medical equipment. Last but not least, SRAMs are increasingly used in the so-called in-memory computing applications as discussed in Sect. 6.7. The scaling of SRAM memory cells was not an issue for a long period of time, not even in FinFET process nodes between 16 and 7 nm. However in the 5 and 3 nm nodes, the SRAM cell sizes remain nearly identical [11]. This means that in the smaller process nodes, SRAM would take a larger portion of the total chip area.

As mentioned before, scaling of the supply voltage required increasingly complex SRAM architectures or the use of assist circuits. Because this boosts the cost of embedded memories, more and more emphasis is put on embedded alternatives, such as embedded DRAM (*eDRAM*) with 1T, 3T or 4T cells. Some vendors even use embedded ZRAM (Sect. 6.4.4) on their microprocessors.

6.4.3 Dynamic RAMs (DRAM)

The basic block diagram of a DRAM is quite similar to that of an SRAM. The main difference between an SRAM and a DRAM is the way in which information is stored in the respective memory cells. All stand-alone DRAMs consist of n-type cells because of the high-performance requirements. DRAMs may use back-bias voltages to have a better control on the threshold voltage to limit leakage for improving refresh characteristics and to reduce junction capacitances. When DRAMs are embedded in a logic chip, p-type cells were often chosen, because the n-well in which the DRAM is located can then be separately connected to an additional positive back-bias to achieve the previous advantages. In triple-well technologies, this is no longer necessary, because each individual n- and p-well are electrically isolated.

Figure 6.13 shows the basic circuit diagram and a water model of a DRAM cell realised with one transistor and one capacitor, which is called a *1T-1C cell*. Single-transistor capacitor-less DRAM cells (i.e. real *1T cells*) are in a research phase [12]. Note: the standard one-transistor one-capacitor cell is often referred to as 1T DRAM, especially in comparison to earlier 3T and 4T DRAM cells. A logic '1' is written into the cell by placing a high level on the bit line while the word line is active. The capacitor in the cell is then charged to a high level. This is also applicable with reverse polarities for p-type cells. The data in a cell are thus determined by the presence or absence of a charge on its capacitor. Parasitic junction leakage and

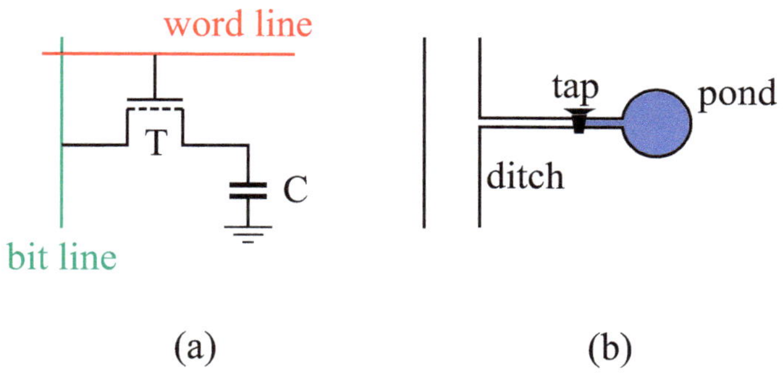

Fig. 6.13 (**a**) Circuit diagram of a DRAM (**b**) Water model of a DRAM cell

transistor subthreshold leakage cause this charge to leak away, just like the water in the pond evaporates as time progresses. The information in the cell must therefore be frequently refreshed.

In addition to leakage, the information in a DRAM memory cell is also destroyed when it is read. This so-called *destructive read-out (DRO)* is caused by the cell capacitance being much smaller than the bit line capacitance. The cell contents must therefore be restored immediately after each read operation. For this reason, each bit line is equipped with a *refresh amplifier*, which consists of a sense amplifier and some restore circuits. This sense amplifier detects the bit line level and writes its amplified value back into the cell. The operation is called a *row refresh* because it is done simultaneously for all cells that are addressed by an active word line.

In practice, the *refresh operation* for each cell must be performed every 2–256 ms, depending on the cell size and the technology. In many large memories, today, the static refresh period is 64 ms. A 1 Gb DRAM, e.g., a 1 Gb DDR SDRAM, where two internal rows are refreshed in parallel, 16,368 row addresses require a periodic refresh interval time of (64 ms/16,368)/2 = 7.81 μs [13]. During the refresh cycle, the internal control keeps track of which rows have been refreshed and which have not.

130 nm DRAMs also used a negative bias (e.g. −1.5 V, generated on chip) to reduce subthreshold leakage in standby mode. However, this additional field between gate and drain increases the GIDL leakage. 90 nm DRAM process generations and beyond therefore used a special *recessed-channel array transistor (RCAT)* as access transistor in the DRAM cell to reduce GIDL. This transistor is therefore only used in the memory array [14]. It has a much longer channel length (in the vertical direction, so that the lateral cell sizes can still scale) to reduce the subthreshold leakage current. Gate leakage is reduced by creating a stack of different dielectrics, fabricated by using atomic layer deposition (see Chap. 3) resulting in an overall high-ϵ equivalent dielectric in this transistor. Both leakage reduction techniques are driven by the need to enhance the capacitors' data retaining properties and minimise the refresh frequency in order to reduce the power consumption in both active and standby modes. A DRAM can therefore not be accessed for a certain percentage of the time. This percentage is typically between 1 and 5% and is one of the reasons why DRAMs are more difficult to use than SRAMs. The requirements for the DRAM leakage currents are much tighter than for logic circuits. The total sum of all leakage components may not be more than about 1 fA per cell [15].

The read operation in a DRAM requires a reasonable signal level on the bit line. For a long time, smaller cell sizes came hand in hand with smaller storage capacitance values. However, this value determines the sensing voltage, the data retention times, the sensing speed and the soft-error rate. Particularly the increasing sensitivity to soft errors (Chap. 9) has put a lower limit to this capacitance value. Advanced memory generations (4 Gb or more), include memory cells with a capacitor value close to 5 fF. Due to the continuous scaling of the parasitic node capacitances in an SRAM, these memories have become more susceptible to soft errors than DRAMs.

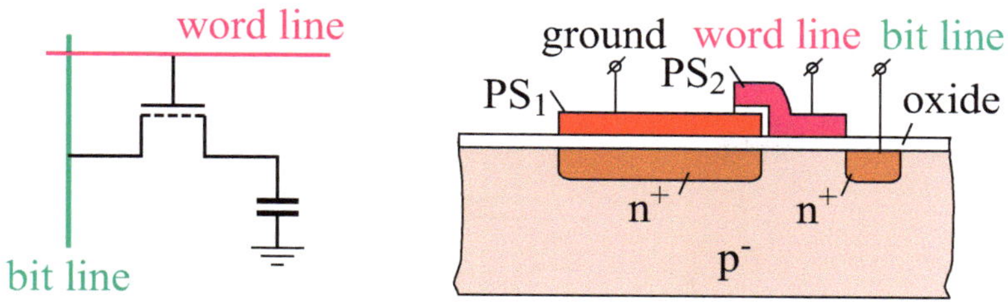

Fig. 6.14 The planar DRAM cell

The use of planar (C)MOS processes for the implementation of DRAM memories was limited to capacities of up to 1 Mb. A typical example of the *planar DRAM cell* used in these DRAMs is shown in Fig. 6.14.

The relatively small capacitance in planar cells makes them unsuitable for advanced DRAMs. Let's go back to the analogy with the water in the pond in Fig. 6.13. Assuming the pond is just 10 cm deep and the water evaporates 1 cm a day at 30 °C and assuming a minimum water level of 5 cm to keep the fish alive, we need to refresh the water every 5 days. However, if we make the pond deeper, e.g., 1 m, than the 'refresh rate' reduces to only once every 95 days. This is exactly what the DRAM vendors do: use the third dimension by building trench capacitor or stacked capacitor cells. These three-dimensional cells which afford increased storage capacitance in a reduced planar surface area are (or were) therefore used for large DRAMs. These include the *stacked capacitance cell (STC)* and the *trench capacitance cell* shown in Fig. 6.15.

These manufacturing processes are much more complex. In a stacked-capacitor technology, the capacitor is commonly fabricated after the transistors with less thermal processing (Fig. 6.17a). The dielectric can be sandwiched between various electrode materials that allow the use of high-ϵ materials. Until 0.15 µm DRAMs, an *oxide-nitride-oxide (ONO)* dielectric has been used, while Ta_2O_5 (Tantalum Pentoxide, with $\epsilon_r \approx 25$) has been used in the 0.12 µm generation. These high-ϵ dielectrics cannot be used as normal transistor gate oxide that needs to contact polysilicon. Increase of the capacitance value is achieved by reducing the dielectric thickness. The minimum thickness, however, is typically limited by the afore-mentioned leakage current of 1 fA per cell. The ONO sandwiches achieve a thickness of about 3.5 nm. Another way to increase the capacitance in a stacked-capacitor cell is to fabricate an uneven surface on the bottom electrode of the capacitor (Fig. 6.16), by using a so-called *hemispherical grain (HSG)* process flow.

This increases the surface area as well as the total capacitance value. Compared to a cell with an even surface, the HSG cell shows a capacitance increase of more than a factor of 2.

In a trench-capacitor technology (Fig. 6.17b) the capacitor is fabricated before the selection transistor.

As the transistor fabrication needs a high-temperature step, (future) high-ϵ capacitor dielectrics need to survive this harsh treatment, which is an important

Fig. 6.15 (**a**) Stacked capacitance and (**b**) trench capacitance three-dimensional DRAM cells

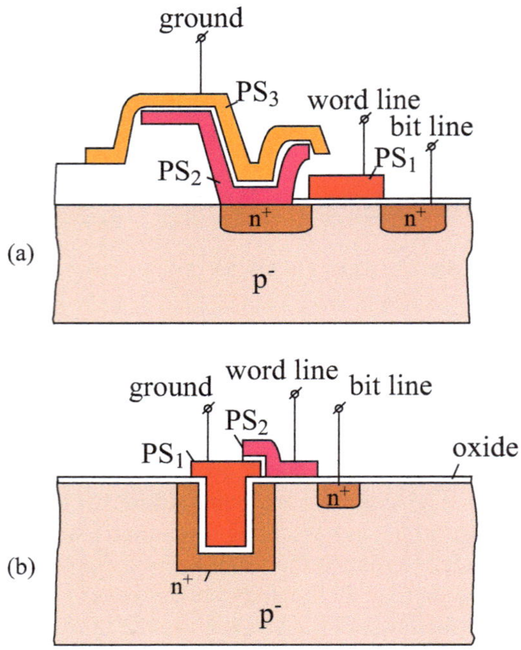

Fig. 6.16 Cross section of a cylindrical stacked-capacitor cell using a hemispherical process flow

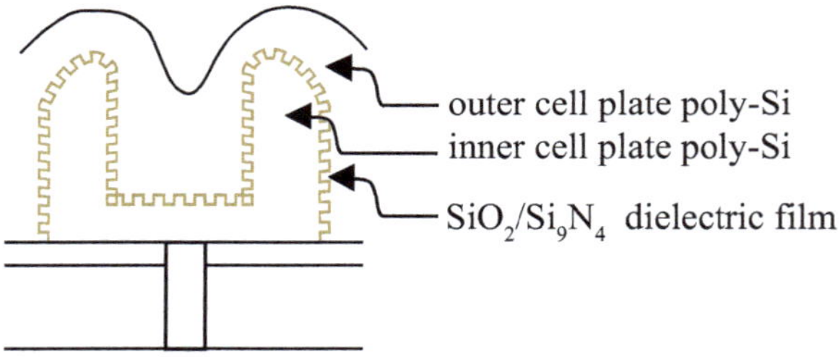

materials selection criterion. The afore-mentioned Ta_2O_5 is suitable for stacked capacitor applications but cannot be used in trench cells, since these then need to survive the temperature cycles needed for transistor fabrication.

Some companies used Al_2O_3 *silicon-insulator-silicon (SIS)* trench capacitors [16]. *BST* ($Ba_{1-x}Sr_xTiO_3$) materials showed dielectric constants, which were even higher, i.e., about 250–600. However, the processes for producing BST are not yet fully developed.

Comparing stacked and trench capacitor DRAM cell trends leads to the following observations:

- Both types of memories use(d) a high-ϵ dielectrics which could not have been fabricated without the so-called *atomic layer deposition (ALD)* step (see Chap. 3).

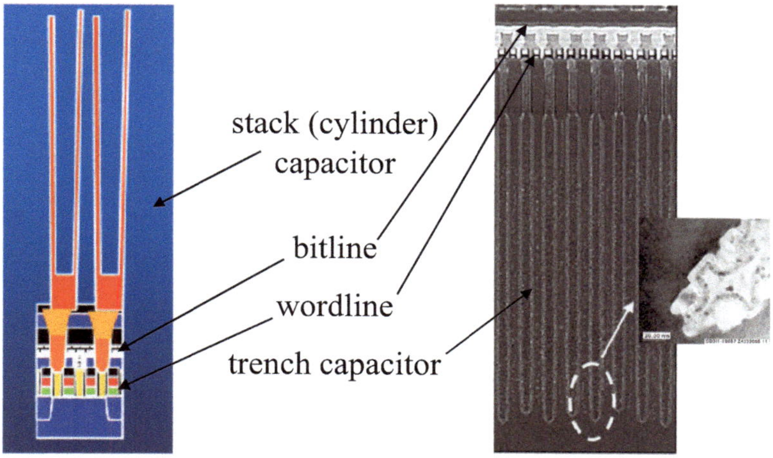

Fig. 6.17 Example of a stacked capacitor cell architecture (**a**) and a trench-capacitor cell in 70 nm CMOS (**b**) (drawing + photo: Qimonda)

- Due to the high aspect ratio (= height/width ratio $\approx$ 60–80) of the storage capacitor for both trench- and stacked-capacitor cells, their series resistance can be a bottleneck in further scaling. This is due to the fact that the planar trench-cell area scales quadratically with the feature size, while its depth almost remains constant.
- While trench capacitors were buried beneath the silicon surface, stacked capacitors create significant topographies across these designs and put stringent demands on both lithography and mechanical stability.
- One of the most important life-time reliability tests is the *burn-in test* (see Chap. 9), to identify devices which are subject to infant mortality or excessive parameter drift. During a standard burn-in test, the device is operated for 160 hours (or 240 hours) at an elevated temperature, usually 125 °C. This test activates very substantial wafer stress mechanisms, which dramatically affect the dielectric layer homogeneity in the deep-trench capacitor cells of a DRAM. Particularly in nanometer CMOS processes (65 nm and beyond), the dielectric layer may be damaged, causing unacceptably high cell leakage currents.
- Trench-capacitor storage nodes were more sensitive to soft errors and to substrate noise.
- Stacked-capacitor DRAMs are built from twin cells, meaning that two cells share the same bit line contacts. Due to a different topology, trench capacitor cells showed etching problems which do not allow to share the bit line contact between two neighbouring cells. Trench-capacitor areas ($8F^2$/bit) were therefore usually larger in area than stacked-capacitor cells ($6F^2$/bit), with F being the half pitch (in 65 nm CMOS process: $F = 65$ nm). Beyond 50 nm only stacked-capacitor memories are manufactured. Because of the use of *vertical pillar capacitors*; see Fig. 6.19), a single memory cell area is expected to reduce to $4F^2$/bit.

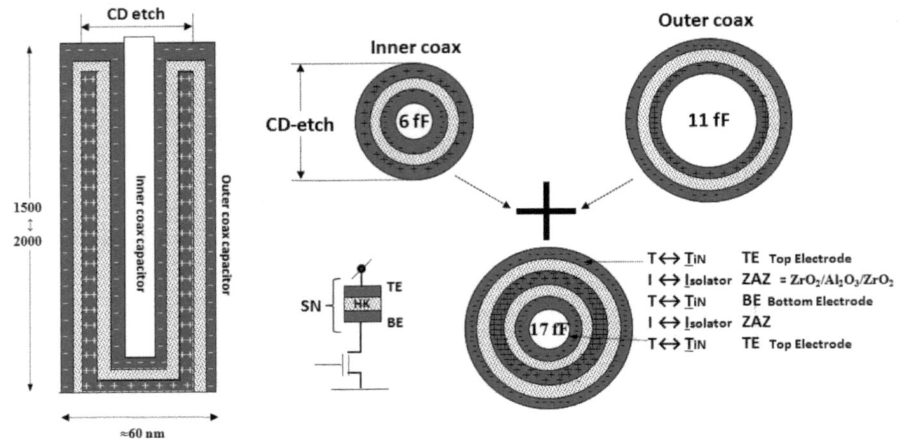

Fig. 6.18 Double-sided charge capacitor (crown capacitor) for DRAM cells (Source: Ewoud Vreugdenhil; ASML)

- Due to the high-aspect ratio (60 or more) of the trench capacitor, the trench edge required an etching angle of approximately $89^{c}irc$ in order to still have sufficient width in the bottom of the trench. Beyond 90 nm technologies, it became much more difficult to fulfill this requirement. A trench-capacitor DRAM wafer was full of holes, causing severe strain and stress in the substrate and a lot of highly leaking capacitors (yield loss) after burn-in.

From the above it is clear that the scaling of trench-capacitor cells incorporated much more problems than those of stacked-capacitor cells. Therefore, all original trench-DRAM vendors have either moved to the stacked-DRAM camp or have completely withdrawn from the DRAM scene. These stacked DRAMs can basically be fabricated with either of the two different memory-cell architectures, dependent on the processing sequence of the capacitor and the bit line: *capacitor-over-bit line (COB)* and *capacitor-under-bit line (CUB)* architecture. Because of the so-called *overlay problem* (alignment problem of the contact hole connecting the bit line, along two storage capacitors down to the common source of the access transistors of the twin cells) in the CUB architecture, all stacked DRAMs used the COB cell architecture (until the 90 nm node).

To reduce the aspect ratio of the cylindrical capacitor, the double-sided charge capacitor (*crown capacitor*) has been developed. This cell was already introduced around 2005. Figure 6.18 shows a cross section of this dual-cilinder cell in a 25 nm DRAM technology.

This capacitor represents a capacitance value of 17 fF, which is built from two cylinders. The inner and outer cylinders represent 6 and 11 fF of storage capacitance, respectively. The thickness t_d of tri-layer high-ϵ dielectric ZAZ (Fig. 6.19, left: zirconium oxide—aluminium oxide—zirconium oxide) is around 9 nm. The dielectric constant for the aluminium oxide (Al_2O_3) layer is around 8. Because of its

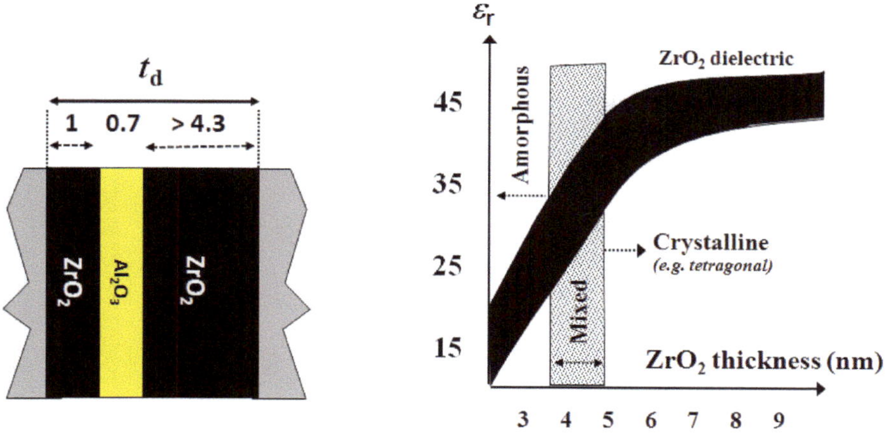

Fig. 6.19 Detailed cross section of the dielectric tri layer (left) and the relation between the dielectric constant of zirconium oxide and its thickness (Source: Ewoud Vreugdenhil; ASML)

small thickness, this layer need not be scaled for the next generations. The dielectric constant of zirconium oxide (ZrO_2) is around 45. However, below a thickness of 5 nm its value dramatically reduces with the thickness. The diagram in Fig. 6.19 (right) shows this relationship. As a result, the thickness of the tri-layer dielectric cannot be scaled below 6 nm. This crown capacitor cell has been used until and including the 20 nm node.

Further scaling can only be performed using the cylindrical stacked-DRAM (coaxial) capacitor cell (*pillar cell*) with a titanium nitride (TiN metal) inner electrode. An example of such a memory in a 22 nm technology is shown in Fig. 6.20. Current DRAM die areas are close to 30 mm^2, which allows about 30 dies in one exposure field (reticle).

To achieve the minimum required charge-storage capacity of 17 fF per cell, the aspect ratio H/D of the capacitor is large ($\approx$80–100). Today, the charge-storage capacity of advanced DRAMs is only 5 fF, while the bitline capacitance is about a factor of 5 larger. Also the contacts and vias have relatively large aspect ratios (Fig. 6.21), which will put increasing pressure on the accuracy, homogeneity and reliability of the lithographic, deposition and etching steps.

An alternative to *piller DRAM cell* is the tubular cell (Fig. 6.22). In the pillar cell, the sizes of the central storage electrode (SE) and circular dielectric layer are determined by a combination of the lithographic quality and hole etching homogeneity. The tubular cell is much more robust in terms of process variability. First, the contact hole is etched. Next, the hole is filled with successive ALD layers (ground plate, dielectric layer, storage electrode), whose thickness is extremely well controlled. Finally, the interior of the hole is filled with dummy material. Litho variations between different holes and etching across the depth of the holes are now being absorbed by the internal dummy fill and no longer by an essential part of the

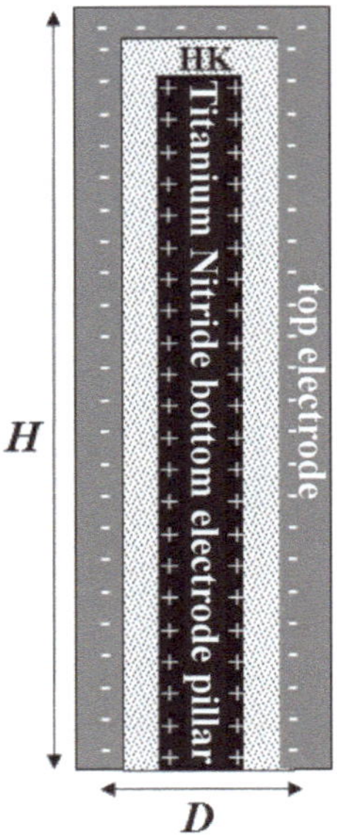

Fig. 6.20 Example of a single-sided charge capacitor cell (pillar cell) in 22 nm DRAM technology (Source: Ewoud Vreugdenhil; ASML)

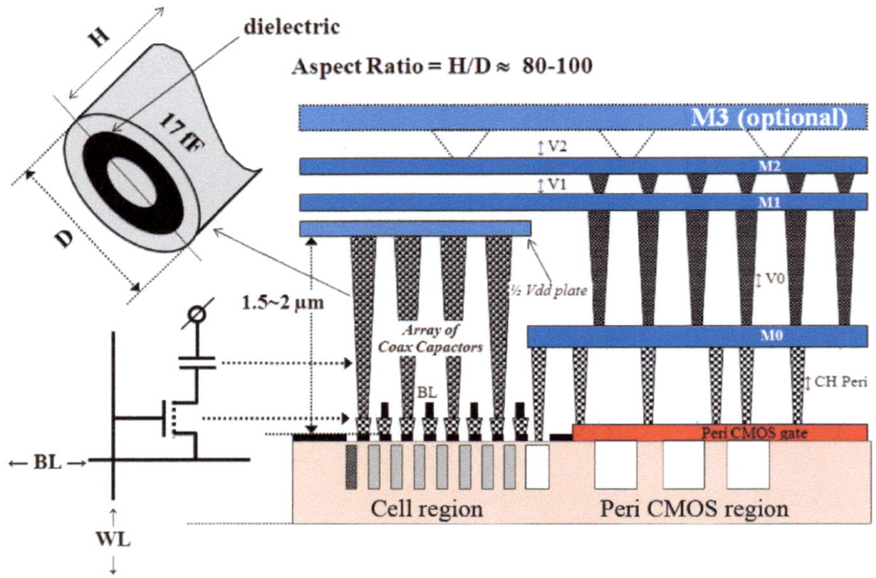

Fig. 6.21 3D cylindrical stacked memory cell in 22 nm DRAM technology (Source: Ewoud Vreugdenhil; ASML)

capacitor. Disadvantage of the cell is its larger capacitor area, but its aspect ratio is 3.5 times smaller, allowing faster DRAM operation.

The average charge leaking time of a 17 fF DRAM cell is roughly 1 s. With a *refresh rate* of 64 ms, this means that the worst-case (weakest) cell can have a 16

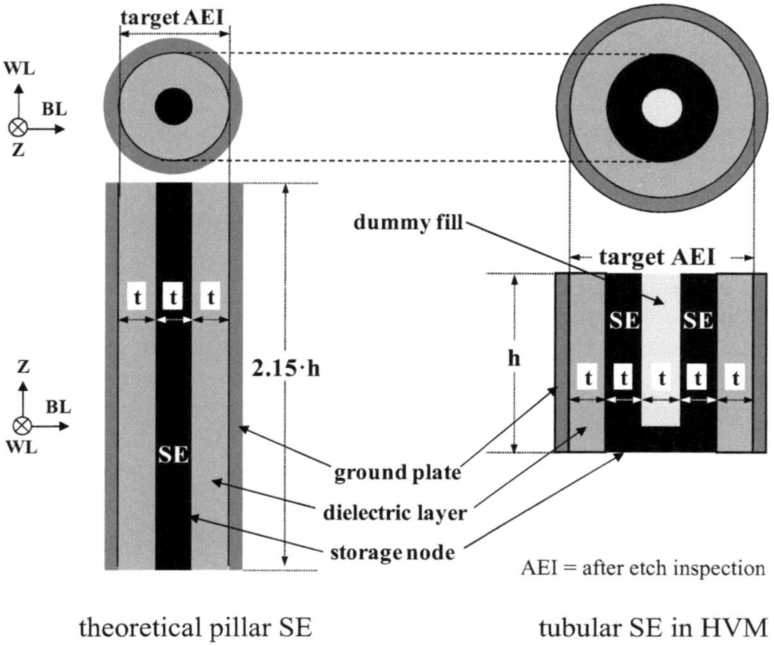

Fig. 6.22 Comparison of DRAM pillar cell and tubular cell

times higher leakage than the average cell. Certain DRAMs apply adaptive refresh: the refresh that is needed per block, because some memory blocks may require a much lower refresh rate. In that case, ECC may be applied for the (weak) cells that might still fail.

Today, the storage cell capacity has reduced to only 5 fF with the introduction of error detection and correction techniques. Redundancy accounts for a few percent of additional cells and, as such, the memories need not be designed for the weakest cells, as these memory blocks can easily be disconnected by fuses.

In the past, DRAM memory density increased with a factor of 4 every new technology node. A factor of 2 (s^{-2}) could be achieved from conventional technology scaling, assuming a size-scaling factor of $s \approx 0.7$. A factor of 1.4 was achieved by shrinking the lateral cell area with more than s^2. Another factor of 1.4 in density was achieved by increasing the chip size with this factor. Today, for reasons of process complexity and yield, the chip sizes can no longer be increased. Since the third dimension (stacked or trench cells) is fully exploited today, also the lateral cell area cannot be scaled more than with s^2. Therefore, today, current DRAM capacity only doubles every new technology node. The continuous scaling of the DRAM cells also causes a relative increase in *random doping fluctuations (RDF)* leading to a larger spread in the drive strength (threshold voltage) of the cell's access transistor, thereby reducing the operating margins of the DRAM memories in general. This forces the

DRAM vendors to introduce metal-gate transistors into the memory cells, where the Vt is more determined by the work function, rather than by the channel dope. There are couple of trends to be distinguished today:

- Advanced process steps.
  Different *high-ϵ metal-gate (HKMG)* materials are being used in DRAM cell design. This combination of materials increases the cell capacitance, which allows lower refresh rates and improves transistor switching speed. It also increases the robustness of memory operation, while reducing power consumption and leakage currents.
- DRAM process node naming.
  In the last decade, memory vendors no longer used precise numbers in their *DRAM process node names*. Although 22 nm process technology was still referred to as the 22 nm node, for process generations below that node, vendors began using such terms as: $1x$, $1y$ and $1z$. Specifically for DRAM, these node names typically refer to the cell's active area 'half-pitch'. In the half-pitch range from 19 nm down to 10 nm, they started by calling the first generation D$1x$, the next D$1y$ and then D$1z$. Then they reached the end of the Roman alphabet and therefore continued by using D1α, D1β, D1γ, etc. The D1α process generation is comparable to a 14 nm DRAM process node [17].
- 3D DRAM.
  Over the last couple of decades, DRAM scaling could still be achieved by a continuous shrinking of the thickness of dielectric layers between the metal lines in the memory array. As a result of that, bit line signals start to interfere with one another. This also holds for the word lines and results in increased crosstalk, higher power consumption and heat, less performance and a reliability threat. Regarding the transistors in the memory periphery, it is also expected that the polysilicon-gate transistors will be/are replaced by high-ϵ metal-gate (HKMG) transistors to enable the required speed increase for the new DDR families. Finally, there is a limitation in scaling of the cell capacitor footprint, as it becomes extremely hard to create its very high aspect ratio by the etch and deposition processes [18]. Although not envisioned soon, but *3D DRAM* technology may be required to enable further DRAM scaling into the next decade. The above reference gives an insight in its possibilities.
- 5th and 6th generation of DDR: DDR5 (Double Data Rate 5).
  DDR5 is currently the latest and fastest generation of DRAM technology. It offers higher data transfer rates, increased memory capacity and lower power consumption compared to its predecessor, DDR4. DDR5 memory modules are capable of reaching speeds of more than 8000 MT/s (mega transfers per second) and have larger capacities ranging from 8 GB to 1 TB per module. Section 6.4.4 represents more details on DDR memory generations.
- Advanced Memory Architectures.
  Various advances in memory architectures aim to improve performance, efficiency and reliability. One example is the introduction of *High Bandwidth Memory (HBM)*, which utilises vertically stacked memory chips, which increases

memory capacity without significantly increasing the module's physical footprint. It provides high-speed data transfer and reduced power consumption (see Sect. 6.4.4). This technology is commonly used in high-performance graphics cards and specialised computing systems.

- Persistent Memory.
  Persistent Memory, also known as *Storage Class Memory (SCM)* or *Non-Volatile DRAM (NVDRAM)* is a hybrid memory technology that seeks to bridge the gap between traditional DRAM and storage devices, such as SSD and hard disks [19]. It retains data even during power loss, combining the speed and low latency of DRAM with that of the non-volatility of storage. Persistent memory has the potential to greatly enhance system performance because it is more durable than DRAM, and it can read and write data as much as 10 times faster than NAND drives. It enables new applications in areas such as database management, AI, big data analytics and in server stations. Early *non-volatile RAMs (NVRAM)* combined SRAM and non-volatile memory on the same chip. These expensive chips were used in applications, such as medical systems and aviation, where loss of data was not allowed.

Despite associated processing and operational disadvantages, the DRAM has achieved a dominant market position. This is mainly because of the relatively low area per bit, which is generally 15–20 times smaller than those of SRAMs. This leads to cost advantages of a factor of 5–10.

General Remarks on DRAM Architectures

There are important differences between the basic DRAM and SRAM operation. Both SRAMs and DRAMs have similar secondary and sometimes even tertiary amplifiers in the I/O path.

The access time of a DRAM was approximately two to four times longer than that of an SRAM. This is mainly because most SRAMs were designed for speed, while DRAM designers concentrate on cost reduction.

DRAMs usually consist of many memory arrays. A typical array may consist of 700 rows by 1000 columns. The 700-cell high columns yield shorter bitlines and faster memories. Die sizes are often in the range of 25–40 μm^2. DRAMs are generally produced in high volumes. Minimising the pin count of DRAMs by row and column address multiplexing makes DRAM operation slower but cheaper as a result of the smaller chip size. Because of the optimisation of DRAM cells for a small area, the higher DRAM processing costs can be regained by the larger number of dies on the wafer. Moreover, DRAM technologies only used two metal layers up to the 65 nm node. In 45 nm and beyond, DRAM memories use three or four metal layers to provide the optimum balance between density (area) and process cost. This low number of metal layers is possible due to the very simple and regular structure of bit lines, word lines and supply rails.

In addition to minimising cell area, other techniques are also used to reduce the total area of memories. One such technique reduces the number of bond pads on

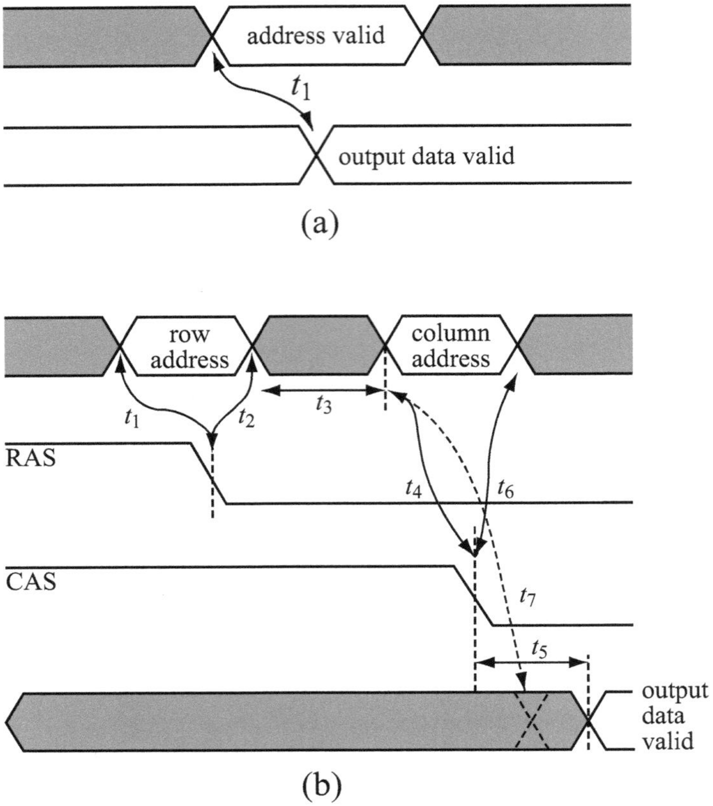

Fig. 6.23 Access times of (**a**) an SRAM: access time = t_1 and (**b**) a DRAM: access time = $t_1 + t_2 + t_3 + t_4 + t_5$ or improved access time = $t_1 + t_2 + t_3 + t_7$

stand-alone DRAMs by multiplexing the row and column addresses through the same bond pads.

Stand-alone SRAMs use separate bonding pads for the row and column addresses to achieve fast access times. The access time of a stand-alone SRAM is therefore considerably shorter than that of an equivalent stand-alone DRAM. This is illustrated in Fig. 6.23a. The *RAS* and *CAS* signals represent the *row-address signal* and *column-address signal*, respectively. This figure compares the access times of a stand-alone SRAM and a stand-alone DRAM, which uses row and column address multiplexing. The access time of the SRAM is only determined by the time interval t_1 whereas the total access time of the DRAM is determined by the sum of several set-up, hold and delay times. The improved DRAM access time in Fig. 6.23b is achieved by omitting the column address latches and implementing a *static column access*.

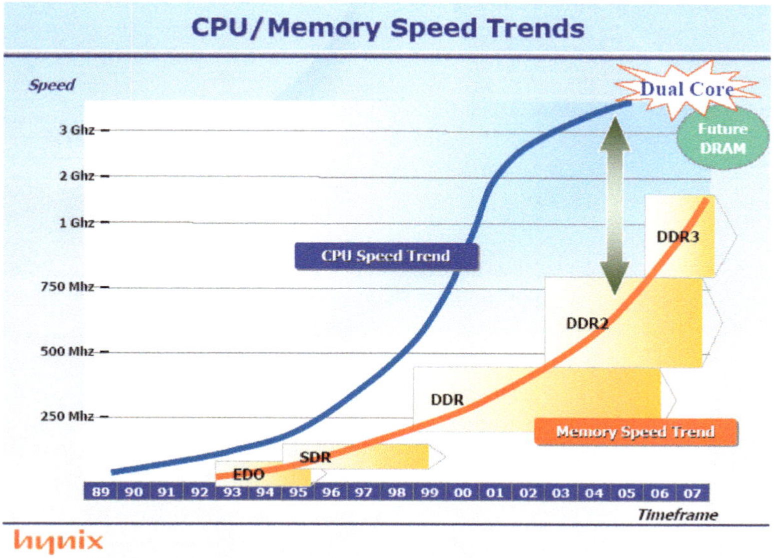

Fig. 6.24 Memory versus CPU speed trends in the past (Source: Hynix)

The data rate of a RAM is determined by the cycle time. This has already been defined as the minimum possible time between two successive accesses to a memory. The cycle time of an SRAM can be equal to its access time. In a DRAM, however, the cycle time is the sum of the access time, the precharge time of the bit lines and the refresh time. This holds for full random access. In page mode (or EDO), precharge and refresh times do not add to the (page mode) cycle time. Therefore, page mode cycle times are about two to three times shorter than full random-access cycle times.

One of the biggest problems over the last decades is the fact that microprocessors showed much larger speed improvements than the DRAMs (see Fig. 6.24). This gap is the primary drive for DRAM manufacturers to dramatically increase random access and data rates and offer Gb-bandwidth interfaces.

6.4.4 High-Performance DRAMs

The increased overall system performance required the DRAM performance to increase at the same pace. Several solutions have been developed to improve DRAM performance during reading. These relatively new generation DRAMs include *Fast Page Mode (FPM)*, *Extended Data Out (EDO)* Mode, burst data using *synchronous DRAMs (SDRAM)* and *Rambus DRAM (RDRAM)*.

All four approaches are based on the ability to access complete pages without requiring the start of a new memory cycle. A *page*, which represents all the memory

Table 6.1 Overview of different DRAM types

Interface	Type of DRAM
Asynchronous	FPM, EDO
Synchronous	SDRAM, DDR, DDR2, DDR3, RLDDR2, RLDDR3, DDR4
Graphics	EDO, VRAM, SGRAM, SDRAM, DDR, DDR2, DDR3, DDR5, GDDR2, GDDR3, GDDR4, GDDR5
Rambus	RDRAM, XDR, XDR2

cells that share a common row address, can have a length of as many as several kilobit. The drawback of page mode is the segmentation of the data, increasing the chance that the required data will not be on the accessed page. Particularly graphics applications benefit from page mode access.

Another advantage of page mode architectures is their reduced power consumption, because there are no sense and refresh currents during page mode access. Most DRAMs are asynchronous; these include conventional DRAMs, FPM and EDO RAMs. A memory operation is then initiated on the arrival of input signals.

The differences between a synchronous and an asynchronous DRAM involve more than just the presence or absence of a clock signal. With SDRAMs, for instance, a precharge cycle is independent of a RAS, to allow multiple accesses on the same row. Internally, a refresh cycle is identical to a read cycle. No column addresses are needed during refresh, since no output data are required. FPM DRAM, EDO DRAM, SDRAM and RDRAM are all based on the same core memory. Therefore, their internal timing diagrams look very similar. The differences are mainly determined by how they communicate with the outside world. These differences include the speed at which address and control signals can propagate through the DRAM and the speed at which data propagate from the DRAM to the memory controller [20]. Table 6.1 shows an overview off the different DRAM types.

In the following, a brief overview of the different high-speed DRAM architectures is presented.

Fast Page Mode DRAM

An *FPM DRAM* offers faster access to data located within the same row because the row command doesn't need to be repeated. This means that only one RAS signal needs to be given, followed by four CAS signals, because the four words all come from the same row. The column address set-up starts as soon as the column address is valid, so that the column address can be latched at the falling edge of CAS. This is different from conventional page modes in which a column address access was initiated by the falling edge of the CAS signal. It was, therefore, required to wait with the column address set-up until the falling edge of CAS. In this way, a reduced page cycle can be achieved in comparison to conventional page mode DRAMs.

Extended Data Out DRAM

The *EDO DRAM* architecture looks very similar to the FPM DRAM. However, it contains an additional register that holds the output data. This allows the start of the next cycle before the previous one is finished. The possibility to "overlap" output data with input data of a next cycle results in a 30% speed improvement over comparable page mode DRAMs. Most EDO DRAMs contain a single bank architecture and must therefore process memory operations serially. A memory operation cannot start before the previous one is completed.

Synchronous DRAMs

When the transfer of address, data and control signals to a DRAM is synchronised by the system clock, such a DRAM is called a synchronous DRAM. Both *SDRAMs* and *RDRAMs* have synchronous architectures and interfaces. Different synchronous DRAM architectures are presented here.

1. SDRAM architectures

   In an SDRAM, in addition to a given external starting address, the next column addresses during a burst are generated by an on-chip counter, while an asynchronous DRAM requires the memory controller to generate a new column address for each access. SDRAMs and RDRAMs are generally built with multiple memory banks (two, four...). Each bank is a memory of its own [21], allowing individual and parallel operation for maximum performance. SDRAM architectures use burst features to accommodate fast external transfer at increasing burst rates. Synchronous DRAMs (SDRAM, SGRAM and RDRAM) use the system clock as their clock input. Therefore, they are targeted to meet the speed requirements of commonly-used PC systems. A trend in increasing the memory's bandwidth is the use of *Double Data Rate (DDR)* I/Os, which are already available since 1998. Figure 6.24 shows how the memory interface speed tries to keep pace with the increasing CPU speed [22]. The *memory controller* plays a dominant role in the interface between memory and CPU. It is therefore required that these controllers also exhibit sufficient performance. In the design of very high-speed controllers, maintaining signal integrity even requires impedance matching techniques between signal wires and their loads to prevent signal reflections.

   In the DDR mode, both the falling and rising edges of the clock are used to double the data throughput, with data sample frequencies from 200 to 400 MHz at 2.5 V. New versions of the DDR concept drive the data rates rapidly further, but also at reduced voltages: DDR2 with 400–800 MHz bus rates at 1.8 V and DDR3 with bus rates to above 1 GHz (800–1600 MHz) at 1.5 V. Although the bandwidth of DDR3 is doubled compared to DDR2, the transition from DDR2 to DDR3 did not require a major speed change of the DRAM core (the memory cell array). To achieve this double bandwidth, DDR3 uses a prefetch of eight words instead of a four-word prefetch for DDR2. This means that for every single read or write operation, a total of eight words are accessed in parallel in the DRAM core to enable the high data rate at the interface. A major difference between DDR4 and DDR3 is that the DDR4 interface between the CPU and the memory is no longer based on a bus architecture but on point-to-point connections or channels. This simplification in the connections increases the

Table 6.2 Summary of DDR memory speed parameters

Type	V_{dd} [V]	Clk freq. [MHz]	Data rate [Mb/s/pin]
SDRAM	3.3	100	100
DDR1	2.5	100–200	200–400
DDR2	1.8	200–400	400–800
DDR3	1.5	400–800	800–1600
DDR4	1.2	800–1600	1600–3200
DDR5	1.1	2000–4000	4000–8000

complexity of the memory controller, because it has to manage these memory channels in a reliable way (source: JEDEC DDR4 standard). DDR4 will not completely replace DDR3. Low-power versions are created to support this rapidly increasing market segment, like LPDDR4, which is developed for the wireless market. In the mean time, several new DDR generations and derivatives have been developed: DDR5, DDR6. The DDR6 standard is still in development, with the first production tests in 2024. Speeds of about 24 GB/s are expected [23]. *GDDR*s are specially developed for graphic cards and also these come in various generations: GDDR3, GDDR5, and GDDR6. Even the specification for the GDDR7 memory standard has recently (2024) been published by JEDEC. It is impossible to include all details of all DDR standards in this book. Therefore, the reader is advised to search the internet for further information and updates. There are also speculations that the next memory interface will be a serial interface [24]. Table 6.2 summarises some DDR memory speed parameters.

Many SDRAMs can also operate in a random-access mode, in which they show similar timing as FPM or EDO DRAMs. Advanced wireless and wired networking applications require DRAM memories with fast random access (or a very low latency) and a low power consumption. The relatively new *Reduced Latency DRAM (RLDRAM)* architecture offers low latency through smaller array blocks (i.e. more banks) compared to DDR3. For example, the latency of RLDDR3 memories is <10 ns, while typical latencies of standard DDR3 memories are in the range of 40–50 ns. Reduced power consumption is realised by operating the I/O circuits at a lower voltage than the memory core (Source: Micron). SDRAMs may have 64-bit or even 128-bit wide I/O formats. Besides commodity DRAM applications, this allows them to also serve in applications with extremely high memory bandwidths. For this purpose, an SDRAM architecture includes: burst feature, more than one memory bank for parallel memory operation and a clocked or synchronous interface. Particularly graphics applications (which are characterised by high-speed and wide I/O buses) require extremely high bandwidths. Video RAMs (VRAMs) and Synchronous Graphics RAMs (SGRAMs) are specially designed for graphics applications.

2. Video RAM architectures

As the pixels on a computer terminal or a TV are refreshed serially, the first *Video RAMs (VRAM)* provided continuous streams of serial data for refreshing the video screen. It acts as a buffer between the (video) processor and the display. Most

VRAMs were a dual-port version of a DRAM meaning that the display is reading its image from the video RAM, while the processor is writing a new image into it. The standard DRAM had to be extended with a small serial access memory and a serial I/O port to support the storage of video pictures [25]. However, all VRAMs still have the original standard random-access DRAM port also available. During a serial read, the DRAM array is accessible via the DRAM port for a separate read or write operation. Special features, such as block write and flash write are supported by additional circuits. However, the rapid rise of special SDRAM architectures, such as SGRAMs became so cheap and dense that they made the original VRAMs obsolete.

3. SGRAM architectures

   Synchronous Graphics RAM (SGRAM) architectures are similar to SDRAMs but optimised for graphics applications. They contain similar additional hardware, such as registers and mask registers to support block write and write-per-bit functions. This results in faster and more efficient read and write operations. These features are supported by special registers and control pins. Colour registers are mainly used to store the colour data associated with large areas of a single colour, such as a filled polygon [25]. The data in these colour registers can be written in consecutive column locations during block-write operation. Write-per-bit allows the masking of certain inputs during write operations; it determines which memory locations are written. Most SGRAMs are specially designed for use in video cards in PCs.

   They lag by about a factor 2 in memory capacity behind commodity DRAMs. 32–256-bit wide SGRAMs are being developed. A major difference with a VRAM is the additional synchronous interface of the SGRAM. Current SGRAMs have I/O data rates of several GBps (gigabytes per second). DDR interfaces can push the SGRAM's graphics peak bandwidth even further. The popularity of SGRAMs has increased such that they are currently used in many graphics systems. Another DRAM version, called the *Rambus*^{TM} *DRAM (RDRAM)*, is gaining popularity as well, particularly in graphics applications.

4. RDRAM architectures

   The *RDRAM* (particularly the *Direct RDRAM*) provides high bandwidth for fast data transfer between the memory and the programming parts in a system. The Rambus^{TM} interface is licensed to many DRAM manufacturers and, at certain engineering fees, they can get customised interfaces to their existing products. Because of the high bus clock rates (600–800 MHz) and the use of DDR, RDRAMs claim extremely high bandwidths, competing with that of SDRAMs and require fewer data lines than the wide-word DRAM. The Direct RDRAM had only little overhead on a standard DRAM architecture and offers several modes from power-down (only self-refresh) to selective powered-down memory blocks [26]. RDRAM memory was popular until the 2000s. An alternative to the Direct RDRAM is the *XDR RDRAM (extreme data rate RDRAM)* which offers extremely high bandwidth and low latency. It can offer several times the bandwidth of a DDR3 memory. However, as DDR SDRAM is an open public standard, RDRAM and its derivates lost attraction and by 2004, Intel and others moved to high-bandwidth DDR SDRAM modules.

5. High Bandwidth memory (HBM)
To make optimal use of the speed and/or bandwidth of a particular component is to directly integrate (embed) it onto the die. Many microprocessors contain static random-access memories (SRAMs), which are often used as internal cache memories. *Cache memory* is designed to temporarily store frequently used data and instructions in order to speed up the data retrieval process. When the CPU needs data that is stored in memory, it checks the cache memory first. If the data are found in cache, it can be retrieved more quickly than if it had to be obtained from main memory. Although these embedded caches show extremely high-bandwidth, they also occupy a large amount of die area, as they are implemented as six transistors/bit (see Fig. 6.7). These memories show capacities up to several hundred megabytes, rather than gigabytes. *High Bandwidth Memory (HBM)* DRAM is a type of main memory used in computer systems for high-performance applications, such as gaming, artificial intelligence, and scientific computing. It provides both increased bandwidth and reduced power consumption compared to traditional DDR memory. It is a stacked memory solution that consists of multiple memory dies (*stacked dies*) that are interconnected vertically, rather than being placed side by side on a circuit board. Figure 6.25 shows an example comparison of a graphics DDR5 (GDDR5) (left) and a HBM memory. Today, the number of stacked DRAM dies has increased to 16. This vertical stack design results in a smaller physical footprint and shorter signal paths, which improves efficiency and speed. The major advantage of HBM DRAM is its high bandwidth, which is achieved by using wide data buses and high-speed signaling techniques. The 32-bit GDDR5 offers 28 GB/s per chip, while the 1024-bit HBM offers more than 100 GB/s. It also has a lower power consumption than traditional memory solutions, because of its shorter signal paths and lower voltage requirements. Additionally, its compact size allows for improved thermal performance and a better form factor in smaller devices. This makes it ideal for data-intensive workloads that require large amounts of memory bandwidth, such as in server

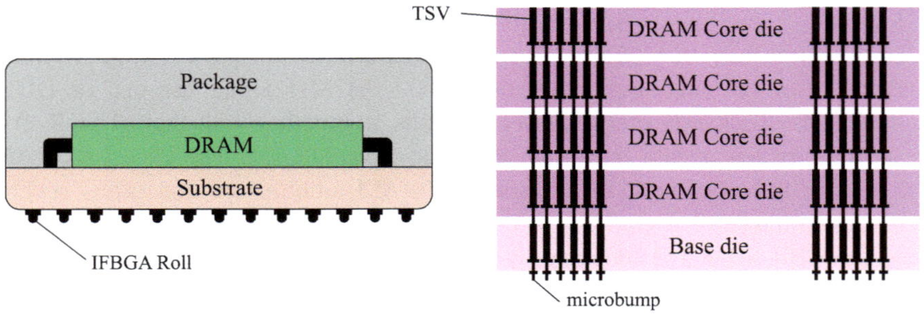

Fig. 6.25 Example comparison of GDDR5 (left) and HBM memory (right) (Source: AMD)

stations, neural network computing, lithography simulation, mask simulation, image processing, AI training, weather forecasting and real-time rendering. The afore-mentioned HBM with 16 memory dies [27] stacked on top of each other is used in the Nvidia's Grace-Blackwell GB200 Superchip developed for high-speed AI and computer operations (see also Fig. 8 in the Preface section).

Currently, DRAMs have passed the gigabit level, with production versions available of 8 Gb and 16 Gb. The first 32 Gb DRAM chip has just been announced [28]. As the application area increases, the hunger for increased densities and higher speeds will drive the complexity and performance of SDRAMs and DRAMs to incredibly high levels. DRAM memory speed is also increased by using more blocks with smaller block sizes: e.g. a 4 Gb Samsung DRAM may have 1024 bit lines and 768 word lines. Even a *Deca-Data Rate SDRAM* with an I/O error detection scheme for high-end servers and network applications has already been developed [29]. In many applications, there is also a pressure on the DRAM standby power consumption. Another example of a DRAM shows an extended data-retention sleep mode, with longer refresh cycles to reduce standby power consumption. This, of course, led to more bit failures, which, on their turn, were then corrected by using *error-correction code (ECC)* combined with the conventional redundancy [30]. These last two examples show in which direction memory vendors are thinking when optimising their memories for high speed or low power.

The move of high-speed microprocessors from bulk CMOS to an SOI technology has also initiated the exploration of SOI for memories. This has resulted in the presentation of the *Zero Capacitor DRAM*, which is a capacitor-less, single transistor DRAM, also called *Z-RAM*. In the operation of MOS transistors in an SOI technology, the floating-body effect (Sect. 3.2.4), was seen as a parasitic phenomenon. In the Z-RAM, the floating-body charge is even enhanced and used to store ones and zeros (Fig. 6.26).

If the body of an nMOS device is positively charged, its threshold voltage is reduced and the device carries a 'large' current, representing a logic '1'. A negatively charged body has the opposite effect and represents a logic '0'. Because it uses no specific capacitor device, its scalability potentials are expected to be better than that of SRAM and DRAM [31].

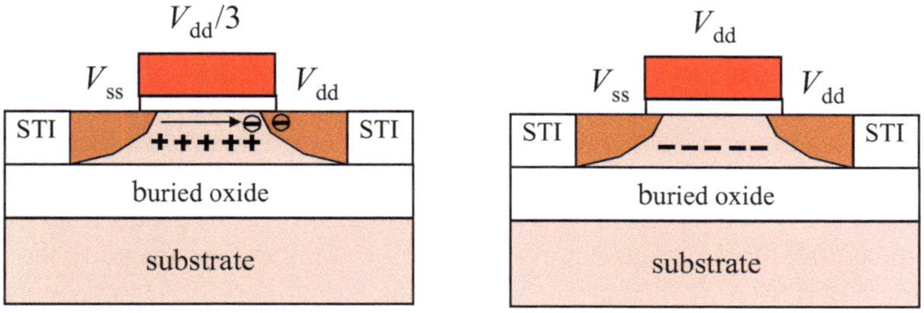

Fig. 6.26 Storage of a logic 1 and logic 0 in a Z-RAM cell

Another realisation of a Zero-Capacitor DRAM architecture is the capacitor-less (indium-gallium-zinc-oxide) *IGZO-based DRAM* cell architecture. This 2-transistor-0-capacitor (2T0C) DRAM cell architecture offers a solution to the challenges faced by the traditional 1-transistor-1-capacitor (1T-1C) DRAM in terms of density scaling. These challenges include the high off-current of Si transistors at small cell sizes and the large space taken up by the storage capacitor. With the 2T0C IGZO-TFT (thin-film transistor) DRAM cell, the need for a storage capacitor is eliminated as the parasitic capacitance of the gate of the read transistor serves as the storage element. Additionally, IGZO-TFTs are known for their exceptionally low off-current, improving memory cell retention. Moreover, the ability to process IGZO-TFTs in the back-end-of-line (BEOL) stage enables the reduction of the DRAM memory's footprint by placing the memory cell under the memory array and the stacking of individual DRAM cells. These advancements pave the way for high-density 3D DRAM, which is crucial in meeting the demands of data-intensive applications like data centres, internet-of-things, artificial intelligence and cloud computing [32]. A recent study has resulted in creating cell on-currents of 40 μA/μm at a positive threshold voltage of 1.78 V. This enables the creation of a non-volatile DRAM with 10 ns write access times and data-retention times close to 25 hours [33]. As this type of memory is still in the R&D phase and far from high-volume processing and applications, more details go beyond the scope of this book. The reader is advised to check the internet for publications regarding further trends and improvements made in this type of memory.

6.4.5 Single- and Dual Port Memories

Most modern memories are *single-port memories*. These memories cannot perform a read and write operation at the same time, because these operations share the same port. Several applications, e.g., communications, video and graphics processing, will benefit from an architecture in which both operations can be performed simultaneously. A *dual-port memory* (or *multi-port memory*) supports writing on one memory position through one port while reading from another position through the other port, which increases the communication bandwidth. As discussed before, video (D)RAMs are often available as dual-port memory. Also many SRAMs are available as dual-port memory. Compared to the single-port SRAM cell of Fig. 6.7, in a dual-port cell, all transistors and connections are doubled, except for the two feed-back inverters. This leads to an area increase of approximately 70–80%. Moreover, because of the increased cell capacitances, an individual access takes more time, leading to a bandwidth improvement of much less than a factor of 2.

The increased complexity and layout area of a multi-port memory, compared to a single-port memory, come at a cost. The relatively small production volumes drive the costs of these memories even higher. In summary: compared to the use of two single-port SRAMs, for certain applications, a dual-port SRAM may only offer minor advantages such that the choice between applying a dual-port SRAM or two single-port SRAMs becomes difficult.

6.4.6 Error Sensitivity

The logic value in a RAM cell may change as a result of radiation caused by α-*particles* and *cosmic particles*. The α-particles may come from impurities in the metal layer (e.g., aluminium), from the package, or from other materials in the close vicinity of the chip. The particle radiation generates a relatively large number of electron–hole pairs, which may randomly change the data in memory cells. This random loss of stored information occurs in both DRAM and SRAM cells. SRAMs are particularly prone to the resulting *'soft errors'*, which become more influential as densities increase and stored charges decrease. DRAMs based on CMOS technology have reduced susceptibility to α-particles, because the storage capacitance per cell remains constant.

Memories can also be covered with a *polymide layer* to protect them against external α-particle radiation. This reduces soft-error rates by a factor of 1000 or more. This does not apply to the cosmic particles, which can even pass through half a metre of concrete. This is one of the reasons why the cell charge is not decreased every new DRAM generation. Chapter 9 presents more details on soft errors.

6.5 Non-volatile Memories

6.5.1 Introduction

Since their introduction in the early 70's, non-volatile memories have become key components in many electronic systems. Until the explosion in the growth of flash memories, the market was relatively small and mainly driven by mobile applications. Today, the flash memory market has become the second largest after the DRAM market.

As discussed in Sect. 6.1, a non-volatile memory keeps its stored data when the memory is disconnected from the supply. Non-volatile memories include ROM, PROM, EPROM, EEPROM, flash (E)EPROM, FRAM, MRAM and PCM. In the following paragraphs, their basic operation is discussed in some detail, including their fundamental properties.

6.5.2 Read-Only Memories (ROM)

A *ROM*, also known as *mask ROM* or *mask-programmable ROM*, is in fact a random-access memory which is written during the manufacturing process. The information is therefore lasting and non-volatile. It can be read but it can never be altered. With the exception of the write facility, the architecture of a ROM is similar to that of a RAM. Subsequent discussions are therefore restricted to the different techniques for writing the information during the manufacturing process. The ROM memory cells required by each technique are examined separately.

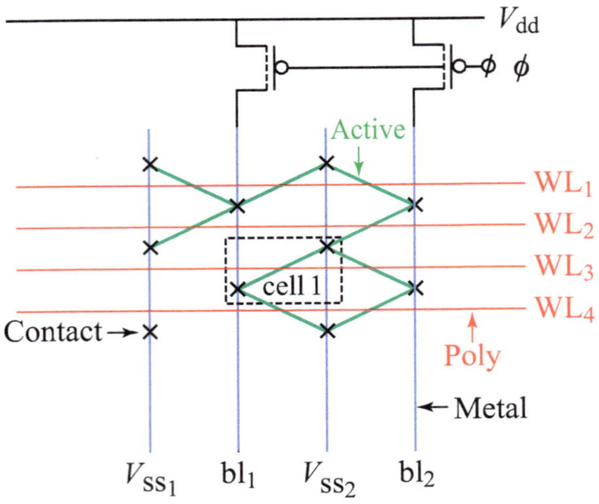

Fig. 6.27 ROM with information in the ACTIVE mask

Different processing layers could, in principle, be used to store information in a ROM. Two choices, made for educational purposes, are the diffusion and contact layers. ROM cells and structures based on the corresponding ACTIVE and CONTACT masks are discussed further.

ROM Cell with the Information in the ACTIVE Mask

Figure 6.27 shows the structure of a ROM which is programmed by means of the ACTIVE mask, see Sect. 4.6. The ROM cell is enclosed by a dashed line in the figure. An example of the layout of such a cell is given in Fig. 6.28.

All bit lines in the ROM in Fig. 6.27 are precharged when ϕ is 'low'. The V_{SS_1} line is switched to ground when ϕ goes 'high'. The cell enclosed by a dashed line is read when the corresponding word line WL_3 goes 'high'. Bit line bl_1 will then be discharged if ACTIVE is present in the cell. Otherwise, bl_1 will remain charged. The information in the ROM is therefore stored in the ACTIVE mask, corresponding to the presence or absence of a memory transistor at the selected cell position.

Figure 6.29 shows a photograph of a ROM array based on the cell of Fig. 6.28.

ROM Cell with the Information in the CONTACT Mask

Figure 6.30 shows the structure of a ROM which is programmed by means of the CONTACT mask. All bit lines in this ROM are precharged through the pMOS transistor when ϕ is 'low'. A word line is activated when ϕ goes 'high'. The bit lines of cells connected to the selected word line and containing a CONTACT hole are then discharged. The CONTACT hole in the cell locally connects the aluminium (METAL) bit line to the drain of a transistor, which has its source connected to a grounded diffusion (ACTIVE) track. The series resistance of the ACTIVE tracks is reduced by means of an extra aluminium ground line which is implemented every 8–10 bit lines.

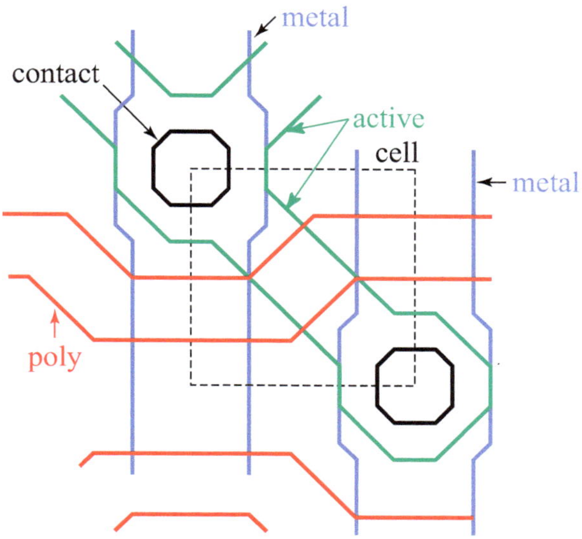

Fig. 6.28 Layout of an ACTIVE-mask programmed ROM memory cell

Fig. 6.29 Photograph of an array of ROM cells with 45° mask shapes (Source: NXP Semiconductors)

Comparison of the ACTIVE-Mask and CONTACT-Mask Programmed ROM Cells

A fair comparison of the chip area of the ACTIVE-mask and CONTACT-mask programmable ROM memory cells requires the inclusion of a suitable fraction for the area of the extra ground line in the latter cell. This gives the following values for a 28 nm CMOS process:

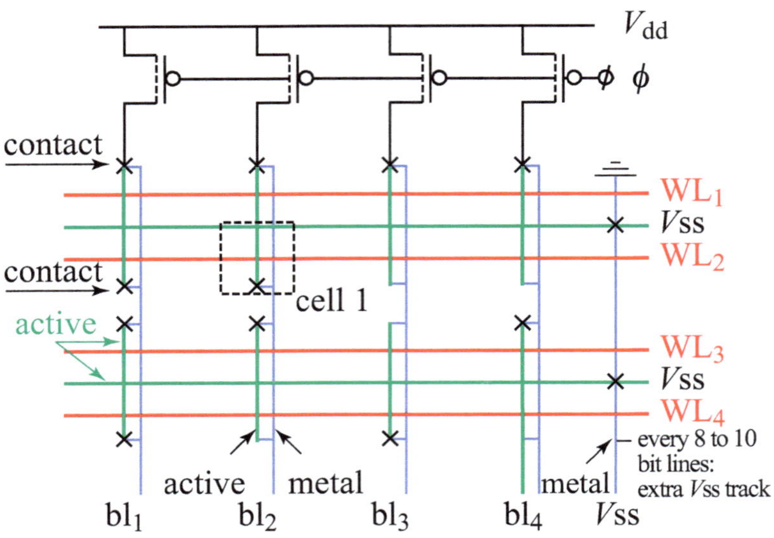

Fig. 6.30 ROM with information in the CONTACT mask

ROM cell in Fig. 6.27 : 0.03 μm^2 $\leftrightarrow$ROM cell in Fig. 6.30 : 0.05 μm^2

Although the second cell is the larger of the two, it has the advantage that its information is programmed in the CONTACT mask. This mask is used in one of the last steps in the manufacturing process. Therefore, ROMs which store information in the CONTACT or one of the METAL or VIA masks can be largely prefabricated. Now, only a small number of manufacturing steps are required to realise a ROM with specific contents. In contrast, the ACTIVE mask is usually one of the first in the manufacturing process. The *turn-around time* between order and delivery is therefore much shorter for a ROM with its information in the CONTACT or METAL or VIA masks than for a ROM with information in the ACTIVE mask. Therefore, in current CMOS processes, the programming is increasingly done in one of the BEOL mask layers, e.g., the 2nd or 3rd metal mask or a via mask.

There are some other types of ROMs as well. In a *serial ROM*, a NAND type of structure is used to discharge the bit line. In such a ROM, a V_T-implant is used for program storage (enhancement or depletion type of memory transistor). The series connection of the cells allows a much smaller number of contacts. This results in a small area, but also in a relatively low speed of operation.

In certain applications, the contents of a ROM can be directly copied into a part of a RAM, from which it can be accessed much faster. This (part of a) RAM is then called *shadow RAM*. The BIOS-code in a PC was usually stored in a ROM; however, in most PCs, it was directly copied into a shadow RAM during booting.

In general, the maximum ROM memory capacity is lagging behind the flash memory (four to eight times) and in many of its original applications, ROM is replaced by flash memory, which has experienced an incredibly high growth in both capacity and market volume. A ROM is used in high-volume applications, where it is absolutely certain that the contents need not to be changed.

6.5.3 Programmable Read-Only Memories

Introduction

The three different types of programmable Read-Only Memory are PROM, EPROM and EEPROM. Respectively, these ROMs are *programmable*, *erasable-programmable* and *electrically erasable programmable*. They are programmed by users rather than during manufacturing. Although they are programmed by users, these memories are still called read-only memories because the number of programming/erasing cycles is rather limited in normal usage.

PROMs (Programmable Read-Only Memories)

A *PROM* is a one-time programmable read-only memory which can be programmed only once by the user. Each cell contains a fuse link which is electrically blown when the PROM is programmed. Traditional PROMs were usually manufactured in a bipolar technology. The fuses were then implemented in a nickel-chromium (NiCr) alloy. The resulting cell is relatively large and is about four times the size of a ROM cell.

Today, poly-fuse cells are used in standard CMOS processes, which can also be electrically blown. These fuses are then silicided, so that larger programming currents are generated at the same programming voltages. In high-ϵ metal-gate processes, the poly fuses have been replaced by metal fuses. Poly and metal fuses are smaller than the traditional NiCr cells.

Currently, PROMs move towards the 3D dimension, where four or eight layers of memory arrays are stacked on top of each other. Memory cells are located between two successive layers of metal and positioned at the crossroads of the metal wires in each of these layers, which run in perpendicular directions. Each cell consists of a diode in series with an anti-fuse. By applying a large electrical field across the anti-fuse (by selecting the corresponding metal tracks in two successive layers, between which the cell is located), its physical state changes, causing a dramatic reduction of its resistance. The cells that are not programmed maintain their high-resistive state. Since the cells are only fabricated between layers above the silicon, the silicon can be used to implement all selection, programming and reading circuitry. Each cell is only $4F^2$ in size, but because of the four or eight layers, their effective area is only $1F^2$ or $0.5F^2$ respectively. These 3D *one-time-programmable (OTP) memories* exhibit a relatively large *area efficiency (AE)*, which may be larger than 85%. The first 3D OTP memory chips were introduced on the market in 2003 by Matrix Semiconductor, mostly for gaming applications. In the 45 nm node, this memory has a potential bit capacity as high as 64 Gb. Still, this memory did not become a commercial success because it could not compete with the low price per bit of NAND flash memories. Figure 6.31 shows an example of an advanced 3D OTP

Fig. 6.31 3D OTP chip with 4 layers of memory arrays (Source: Matrix Semiconductor)

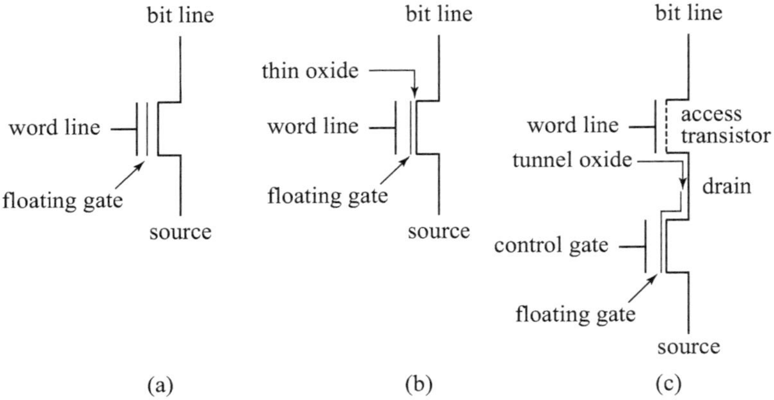

Fig. 6.32 Schematic representation of (**a**) an EPROM cell, (**b**) a flash (E)EPROM cell and (**c**) a full-featured EEPROM cell

chip. The wish for rewritability in many applications has increased the demand for erasable architectures. These are discussed in the following sections.

EPROMs

Figure 6.32a shows a schematic representation of an *EPROM* memory cell.

The data in this cell, as in an EEPROM cell, is represented by the presence or absence of charge on the *'floating gate'* of the memory transistor. The floating gate is charged by means of a large electric field between the transistor's source and drain. This accelerates electrons in the channel to very high energy levels. Some of the resulting *'hot electrons'* (see Chap. 9) penetrate through the gate oxide to the floating gate. This type of programming is called *channel hot electron injection (CHEI)*. Sufficient charge is collected on the floating gate when high drain-to-source voltages of over 3.2 V (in a 65 nm process) and gate-source voltages of about 8–9 V are applied. This causes currents of the order of 0.3 mA in the cell. The number of programming/erasing cycles in an EPROM is limited (10.000–100.000 cycles). Currently, the higher voltages are often generated on-chip by means of charge pumps. Alternatively, an EPROM can be removed from the system and programmed in a special PROM programmer. It then uses a second power supply of around 8–9 V, depending on the technology node.

EPROMs are erased by exposing the cells to ultraviolet (UV) light. This is done through the transparent (quartz) windows in EPROM packages. In many applications, EPROMs are only programmed once. They are therefore also available as *one-time-programmable (OTP)* devices in cheap standard plastic DIL packages with no transparent windows. As a result of its complex reprogramming operation (non field-programmable UV erase), the use of EPROMs, today, is very limited.

6.5.4 EEPROMs and Flash Memories

Floating-gate PROM structures, which allow electrical erasing and programming, were developed at the end of the seventies. The two resulting categories are electrically erasable PROM (EEPROM) and Flash memories.

EEPROM

Unlike with EPROM and Flash memory, EEPROM data can be changed on a bit-by-bit basis. This is also called a *full-featured EEPROM* or *double EPROM*, whose basic cell architecture is shown in Fig. 6.32c. Because of the separate access transistor in the cell, EEPROMs feature relatively low bit densities compared to EPROM and Flash memories. This transistor allows selective erasure of cells. Erasure is often done per byte.

Figure 6.33 shows a cross section of the storage transistor of a full-featured EEPROM cell.

Data storage and erasure are achieved by moving electrons through a small thin oxide tunnel region between the floating gate and drain. This is done by applying a high electric field of about 10 MV/cm across the tunnel oxide, which induces so-called *Fowler–Nordheim (FN) tunnelling*. The cell is programmed by applying a high voltage to the drain with respect to the top gate, which causes the electrons to flow back to the drain. The cell is erased when a voltage of about 12–15 V is applied between the gate and drain (substrate or source, depending on the technology). Now, electrons tunnel through the thin oxide and produce a negative charge on the floating gate. This increases the threshold voltage of the memory transistor. Therefore, the memory transistor in an erased cell has a high threshold voltage. The small currents

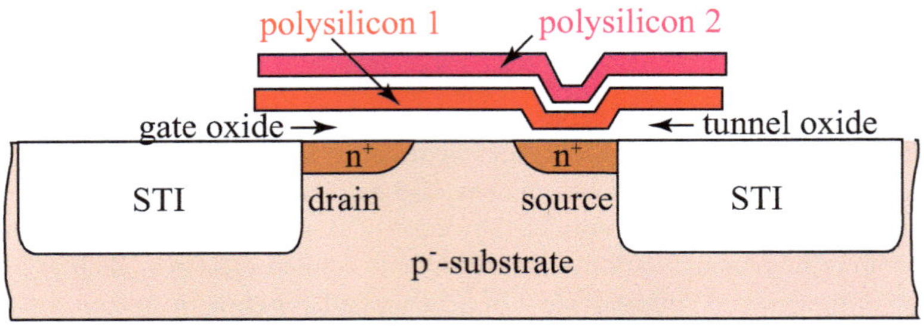

Fig. 6.33 Example of floating-gate EEPROM cell

Fig. 6.34 Endurance characteristic of a full-featured EEPROM cell

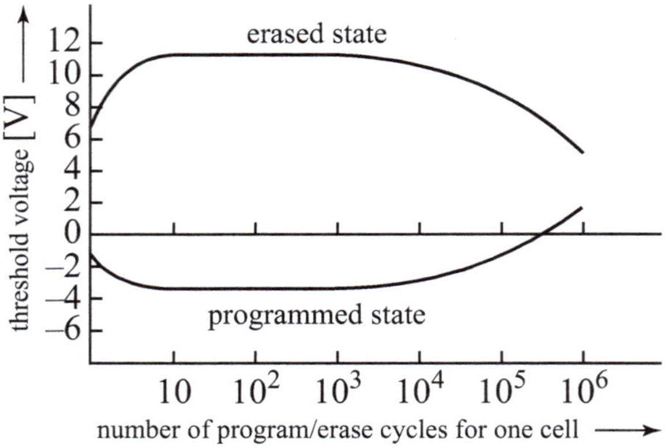

involved in the tunnelling mechanism used in full-featured EEPROMs facilitate on-chip generation of the 12–15 V for programming and erasing the memory.

An important characteristic of a full-featured EEPROM is the variation in memory transistor threshold voltage associated with successive programming/erasing cycles. Eventually, the threshold-voltage difference between a programmed and an erased cell becomes too small due to charge trapping in the oxide. This imposes a limit on the number of times that a cell can be erased and programmed. The maximum number of program/erase cycles that can be applied to the smallest programming unit (cell, page or block, depending on the type of memory) of a E(E)PROMs and flash memories is called *endurance*. The plot of the threshold-voltage variation is called the *endurance characteristic*, see Fig. 6.34 for an example. The threshold difference enabled a few hundred thousands to more than a million programming/erasing cycles for the individual cells.

The *data retention time* of all EEPROMs is more than 10 years. The various applications [34] of EEPROMs: conventional consumer applications, universal remote controls, cordless telephones, garage door openers, cameras, automotive, home audio and video and smart cards. Attention is also focused on cellular telephones and pagers. Innovative features have been added to EEPROMs, such as Block Lock which allows users to combine alterable data with secured data.

Although EEPROM technology offers more flexibility, it is facing increased competition from flash memory, which allows much higher densities, as a result of the absence of the bit-by-bit change feature of an EEPROM, see Fig. 6.32.

Flash Memories

The name *flash memory* originated from the first EPROMs, which used (a flash of) UV light through the transparent window in the EPROM package (see Sect. 6.5.3) during erasure. Current flash memories are memories in which the complete memory or complete sectors (blocks) of the memory can be erased simultaneously. Today, the most important flash-memory categories are: *NAND-flash* and *NOR-flash*.

All flash memories are arranged into blocks. The typical block capacity is 64–512 kB for a NOR-flash and 16 kB to several MB for a NAND-flash, respectively. Erasure cannot be done individually, but is done by complete blocks in 'one flash'. The lifetime of a flash chip is determined by the maximum number of program/erase cycles per block, which was typically specified around 100,000 cycles about three decades ago. However, as we will learn in this chapter, due to scaling, the operating margins of advanced NAND-flash are dramatically reduced, so that in some applications, that number has reduced to even below 1000. It is therefore crucial that the erase cycles are evenly distributed over the blocks. Today's flash memories include control circuitry which distributes the number of program/erase cycles evenly over the total memory.

Each cell in a NOR-flash (Fig. 6.35) contains a bit line contact, which makes it relatively large, but gives it a relatively fast random access. A cell in a NAND-flash is part of a serial chain. It is therefore small (cell area is only $4F^2$/bit, compared to $6F^2$/bit for a DRAM with stacked-capacitor cell and 80–100F^2/bit for a 6T SRAM cell), has a slow random access (typically 20 µs) but allows fast sequential

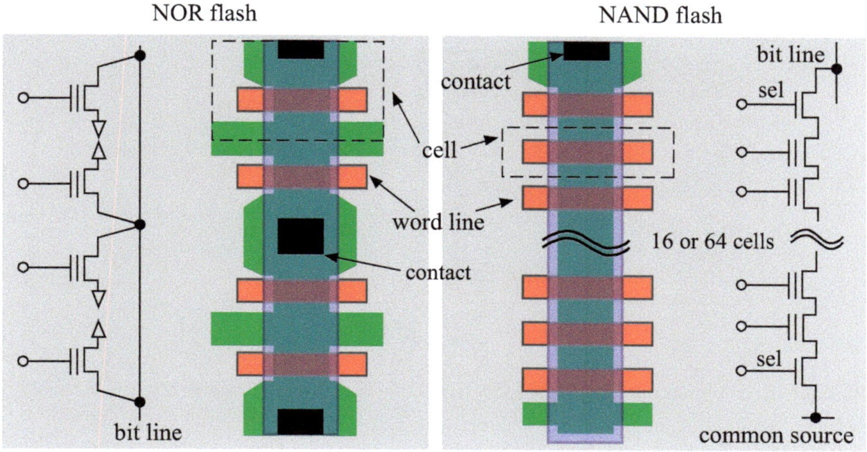

Fig. 6.35 Basic difference between NOR and NAND-flash architecture (Source: Samsung)

access (typically 20 ns). Programming a NOR-flash is done by using *channel hot-electron injection (CHEI)* , for example by connecting the source to GND, the drain to +5 V and the top gate to +12 V, respectively, while FN tunnelling is used for its erasure, by connecting the gate to GND and the source to +12 V. In a NAND-flash, FN tunnelling is used for both programming and erasure. Programming (typically 200 μs per word line) can be done by connecting the gate to +20 V and the substrate to GND and erasure (typically 2 ms for a 128 kB block) can be done by reversing these voltages.

45 nm NAND-flashes used 16–32 transistors in series (Fig. 6.35). The number of cells on a word line is typically between 2048 and 16,348. This is also called a *page*. During a read operation, first the bit lines (Fig. 6.35) are precharged to V_{dd}. Then the selection transistors (sel) are turned on and all unselected word lines are set to such a high positive voltage that all cells (programmed or not) function as pass transistors. If there is only one bit stored per cell, the word line of the selected cells is set to GND, assuming that the erased cells have a negative V_T (normally-on) and will conduct and discharge the corresponding bit line. The programmed cells, with a positive V_T will not conduct and cannot discharge the bit line.

Figure 6.36 shows an example architecture of a 2 Gb NAND-flash. Data are loaded from a page in the array into the data register. Like with SRAM and DRAM, also NAND-flash architectures are being optimised for enhanced throughput. Some NAND-flash architectures therefore have an additional *cache register*. In this case, a copy of the data register is added in between the memory array and the drawn data register. This top data register is then called the cache register. This allows loading the next sequential access from the array into the data register, while reading the previously accessed data from the cache register. This increases the data throughput by about one-third. There are only three basic operations in a NAND Flash: read a page, program a page and erase a block.

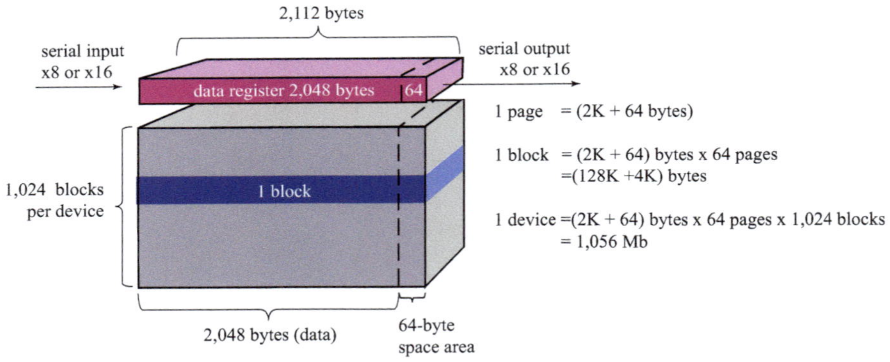

Fig. 6.36 Example architecture of a 2 Gb NAND-flash memory (Source: Micron Technology, Inc.)

Because of its random access, NOR-flash has traditionally been used to store relatively small amounts of executable code in applications such as mobile phones and organisers.

Because of its serial architecture, the NAND flash has a longer read access. However, the need for low-cost high-density memory drove the NAND-flash into the newer mobile devices with increasing performance, such as mobile phones, MP3 players, cameras and streaming media. Today's feature phones combine all of these applications in one device, which are commonly supported by NAND-flash. NAND-flash is also replacing NOR-flash in code-storage applications. The slow read action in a NAND flash, compared to a NOR flash, is compensated by internal parallelism. NAND-flash architecture looks similar to hard-disk drive, in that it is also sector-based and may also have bad blocks. It therefore requires error-correcting code to guarantee correct data retention. The capacity of NAND flash chips, today, is a result of the combined advances in memory capacity increase of the individual dies and the increase of the number of dies that can be stacked on top of each other in a single MCM module. Several NAND dies are currently packaged into a memory stick, creating a total memory capacity of 512 Gb, or more, particularly suited for memory hungry consumer electronics devices.

In all (E)EPROM and flash memories, the minimum thickness (most commonly 7–8 nm) of the dielectric layers above and below the floating gate is determined by the accumulated dielectric leakage charge over the specified data-retention time (usually $>$10 years). This has limited the scaling of the vertical dimensions and voltages in these memories, which also has a negative impact on the lateral scaling. The first generations of flash memories contained only 1 b/cell (*single-level cell (SLC)*). The relatively large signal-to-noise ratio in these types of memories allows to store more levels in one cell to further reduce the cost per bit. A *multi-level cell (MLC)* can store 2b/cell represented by four different amounts of charge on the floating gate. During a read cycle, the control gate is set to a high level and the current through the cell is inversely proportional to the charge on the floating gate. Current sensing requires three differential sense amplifiers, which each compare the cell current with that from one of three reference cells. The outputs of these sense amplifiers directly represent the stored cell data. Multilevel storage has been known for quite some time. However, reduced noise margins and increased design complexities created a lack of commercial interest. The first *multilevel-storage* memory has been delivered since 1991. In a serial-EEPROM technology, analogue data samples were directly stored at a resolution of 256 levels in each cell, without the need for conversion of the analogue sample to binary words. The first commercial multi-level flash memory products were announced at the end of 1996. Both Micron Technology and SK Hynix recently started production of their 232 and 300 layer TLC NAND flash memory, respectively [35, 36]. The architecture of the Micron chip is shown in Fig. 9 in this book's Preface.

Four-bit memory cells are also in development. In a multi-level memory, the distance between adjacent threshold voltage charge distributions on the floating gate is becoming very small and may lead to a decrease in reliability with respect to read operation and data retention. Therefore, a multi-level flash memory may allow only a few thousand program/erase cycles per physical sector, while a single-level flash memory is capable of a hundred thousand of these cycles. The use of on-chip error correction coding (ECC) and *wear-levelling* algorithms may alleviate these problems. Wear-levelling algorithms equalise the number of erase/write cycles evenly over the total memory to significantly increase the lifetime of the chip. In many NAND flash memories, the blocks with only 10% of the maximum specified number of program/erase cycles (maximum endurance) must retain their data for at least 10 years. The blocks with 100% endurance cycles only need a data retention time of one year! Another important parameter that reduces the lifetime of a flash memory is its *Write Amplification Factor (WAF)*. This factor reflects the total number of physical memory bits that need to be written compared to the total number of data bits that are required to change. This is because flash memories are written in full blocks, although only a fraction of their memory cells need new data. For example, if the minimum block size is 2 MB and we need to replace only 1 MB of data, WAF is equal to 2.

The flash memory has penetrated many markets which were previously dominated by magnetic and optical discs, ROMs, EPROMs and EEPROMs. Many of the NAND flash memories are used in *solid-state drives (SSDs)*, also called *electronic disks*. SSDs replace hard disks in several applications. SSDs are smaller, lighter, consume less power, operate faster and do not make noise. However, generally they have a smaller total memory capacity and are more expensive. They are therefore particularly suited for mobile phones, tablets and mini laptops. Small SSDs are also used in normal laptops to store the BIOS data used during start-up. Today, particularly the high-capacity 3D flash memories are also used for cloud storage, where the number of program-erase cycles is very limited. Being able to continuously increase the density of flash memories would speed this process up even more. Next to using more electrical levels in the memory cell to increase the density of flash memories, also multiple layers of stacked memory cells are introduced. After the explanation of some basic charge storage techniques, we will come back to the scaling trends and 3D stacking techniques required to continue the flash roadmap for the next 10 years.

Competing Flash Technologies

One of the problems related to the scaling of floating-gate devices is the relatively large minimum thickness of the dielectric isolation layers above and below the floating gate. This is required to limit charge leakage from the floating gate to guarantee a sufficiently long data-retention time. An alternative to the floating-gate device is the *charge trap flash (CTF)* device. Its operation is similar to the earlier SONOS cell, which will be discussed first. Figure 6.37 shows a cross-sectional view of a SONOS memory cell.

Basically, the cell is a conventional nMOS transistor, in which an oxide-nitride-oxide stack is embedded between the polysilicon gate and the channel. This stack is built from a gate dielectric consisting of an ALD oxide layer of approximately 2 nm thickness, a silicon nitride layer of about 5 nm and a second oxide layer with a thickness between 5 and 10 nm. Programming of the cell is done as follows. When

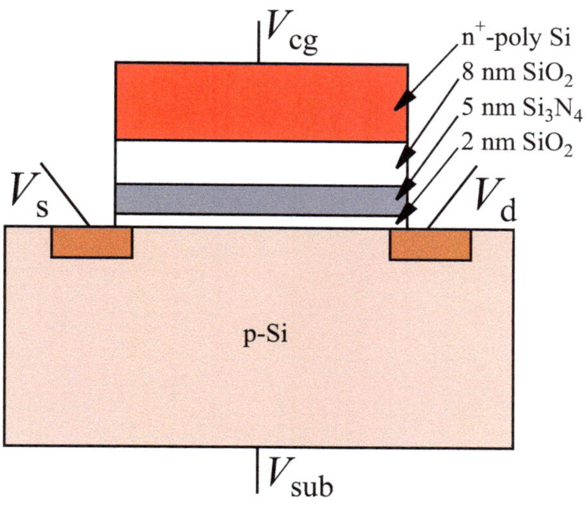

Fig. 6.37 Cross section of a SONOS memory cell

the gate is connected to a sufficiently large positive voltage, electrons will tunnel from the substrate through the ultra-thin oxide layer to the nitride layer. Because silicon nitride has the intrinsic property to trap electrons, it will trap this negative charge, which causes the threshold voltage of the transistor to increase. Likewise, during erasure, the threshold voltage can be decreased with a negative voltage on the gate, by which holes are injected through the same tunnel oxide.

This nitride layer acts about the same as a floating gate, but shows some different properties. Compared to conventional floating-gate devices, a SONOS device offers a dramatic improvement of the *radiation hardness*. This is due to the fact that the charge is stored in discrete isolated deep traps in the nitride dielectric layer, which is therefore not a conductive layer like the polysilicon floating gate. A high density of traps in the nitride layer, which is somewhat dependent on the nitride film thickness and growth technique, allows sufficient charge storage and memory operation of SONOS devices. Like SONOS devices, charge trap flash (CTF) devices also use a charge trapping nitride layer between two isolation layers. Figure 6.38 compares the CTF cell with the floating-gate cell (FG cell).

This isolated-charge storage effect makes CTF devices highly immune to tunnel oxide defects, while even a single (point) defect in the tunnel oxide of conventional floating-gate devices can completely destroy the stored data. This offers a strong reliability advantage, particularly after many program and erase cycles and improves the data retention time. This property can also be exchanged with thinner oxide layers.

A large dose of cosmic charge particles, originated from solar or galactic radiation, may completely discharge floating-gate devices and limit their radiation hardness. In CTF devices, such a high-dose radiation will only discharge the nitride locally, making them particularly suited for high robustness and reliability demanding applications, e.g., medical, aviation and military.

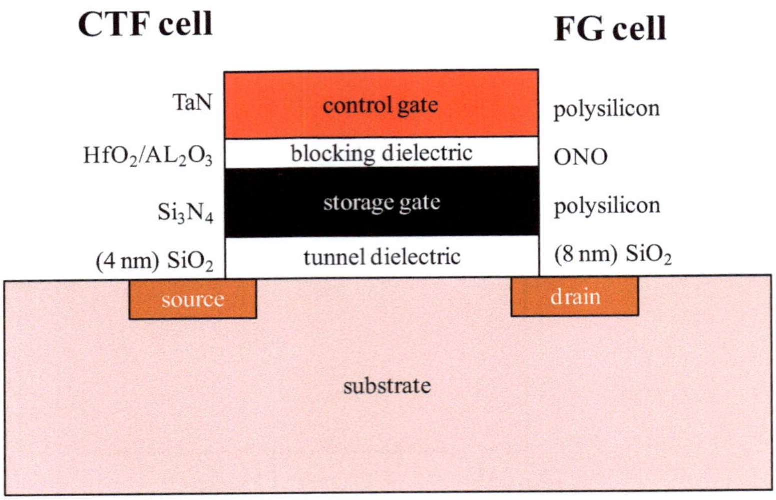

Fig. 6.38 Comparison between the charge trap cell and floating gate cell architecture

Such applications require large operating margins, large endurance levels and long data retention times (>10 years). These can best be implemented with single-level cells (SLC) with a relatively large discrepancy between the two amounts of electrons that represent the two logic states. MLC memories may have very small endurance levels and short data retention times. Most modern planar NAND flash memories use the floating-gate (FG) cell. In the example of a 14 nm planar three-bit-per-cell *(triple-level cell (TLC))* FG cell NAND flash memory, the three bits in each cell are represented by eight different charge levels. These charge levels correspond to eight different threshold voltages (V_{T_i}). Figure 6.39 shows the various target threshold voltages and the Poisson distributions of the number of cells around these target values.

The lower part of the figure shows the average number of charge carriers that correspond with these target threshold voltage values. It also shows that the average discrepancy between two charge levels is only 10 electrons. The maximum charge leakage of a memory cell in such a NAND flash is therefore specified as less or equal to one electron per year! It is clear that to distinguish between the very small amounts of charges both the program and read operations required extremely high accuracy. The programming of such memories is done by a slow and well-controlled Fowler–Nordheim tunnelling process, called *write and verify programming* also called *ISPP* (incremental step pulse programming ([37]; pp. 21–28)). The lowest threshold voltage (V_{T_0}) corresponds with the erased state. The distribution around the 111 (erase) level is wider, since an erase operation is different from the program write and verify cycle, needed for the other charge levels. Moreover, programming is always done at page (# cells/word line) level (see Fig. 6.36), while erasure is always done at block level. The definition of assigning the three bits to a charge level (and

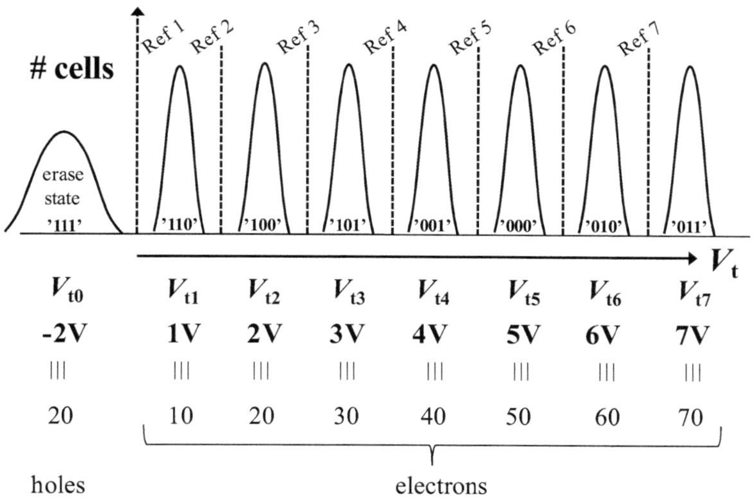

Fig. 6.39 Representation of the eight charge levels in a 3 bit/cell planar NAND flash memory, in combination with the threshold voltage and the number of carriers that each level represent (Source: Ewoud Vreugdenhil; ASML)

to a specific V_T) e.g., V_T of -2 V as 111, the V_T of 1 V as 110, etc., lies in the fact that this is a Gray-code sequence, which allows less complex error detection and correction schemes.

It was common practice to set the read reference voltages at fixed values at design time. However, over time, due to program/erase cycling and wear out, charge loss may occur such that individual distributions shift and their distribution tail may overlap a neighbouring read reference, leading to a read error. Some flash memories, today, include a *read-retry* mechanism, in which changes in the distributions are tracked and the related read reference value is dynamically adjusted to reduce the number of read errors, or to fully eliminate them [38].

Due to the afore-described reduced operating margins, all advanced NAND flash memories incorporate *wear-levelling compensation* and increasingly complex *error detection and correction codes (ECC)* to maintain reliable flash memory operation. A NAND flash memory therefore requires a memory controller that contains the circuits that enable these compensation, detection and correction operations.

Because the wear out of flash memories heavily depends on the number of program/erase cycles *(endurance)* it is required that over the memory life time all cells experience an equal number of program/erase cycles. This administration is performed by the wear-levelling algorithm executed by the memory controller. Memory cells that store 3 bits/cell (3bc) can also be used to store 2 bits (2bc) or 1 bit per cell (1bc). Over the lifetime of a flash memory, the controller can also perform another wear-level compensation in that it can redefine its cell mode: $(n + 2)$bc $\Leftrightarrow$ $(n + 1)$bc $\Leftrightarrow$ nbc. In this way, for example, a 3bc NAND flash memory can also be used as a 1bc memory in applications that require high reliability standards. Despite all these problems with reducing margins, endurance, retention times and memory life time, memory vendors are still optimising/exploring 4bc and even 5bc NAND flash memories.

For many decades, the retention time of all non-volatile memories was specified as at least 10 years, when operated within the specified endurance and within the minimum and maximum specified operating and storage temperatures. Today, there is a close relationship between data retention time and endurance. *Endurance* is the number of program/erase (P/E) cycles that can be applied to a block of *flash* memory before the storage media becomes unreliable. According to the JEDEC standard JESD47G.01 [39], all NAND flash blocks that have experienced 10% of the specified maximum cycling endurance must retain their data for at least 10 years. Blocks that have experienced 100% of this maximum have to retain data for at least one more year.

For the past four decades, flash memories were built with planar structures using floating gates. Some companies continued to do so, even for their sub-20 nm nodes, as shown by the above 14 nm NAND flash example. However, further planar floating-gate scaling comes to an end due to the increasing interference between neighbouring cells and the extremely high electrical field (due to required 25 V programming voltage) across the dielectric between these cells, which both reduce reliable operation of NAND flash products. This has resulted in the development of so-called *vertical NAND flash (V-NAND flash)* structures, which require a totally different vertical interconnect scheme. Most V-NAND flashes use CTF cells to build the vertical string. Since the charge trapping nitride layer is non-conductive, meaning that it intrinsically isolates all trapped charges, its isolation and patterning accuracy requirements are less stringent than for floating gate devices. V-NAND flashes therefore do not require the most advanced lithography tools and fabricated using relatively relaxed design rules. The V-NAND cell consists of a basic Gate-All-Around (GAA) transistor, as discussed in Sect. 3.9.4, but with an additional circular *charge-transfer gate (CTG)* . Figure 6.40 shows the architecture and electric field distribution in this 3D GAA NAND flash cell and a stack of three cells to show the V-NAND structure.

There are several important advantages of the 3D GAA CTG NAND flash memory compared to the planar FG NAND flash memory:

- The channel length L.
  In planar CMOS technologies, the minimum values of the transistor channel W and L are primarily determined by the lithography. In most GAA flash cells L is determined by the layer thickness (see also Fig. 6.42), which is close to 32 nm and, with an average $W/L \approx 6$, it means that the width is close to 200 nm. This means that the CTG gate in a GAA cell is much larger than the FG in a planar cell in 14 nm, which only has a $W \approx L \approx 14$ nm enabling much more charge storage on the CTG of the GAA cell. While the eight different charge levels in the 14 nm planar FG cell are only represented by 20 holes (representing 111) to 70 electrons (representing 011) as shown in Fig. 6.39, the charge levels in the

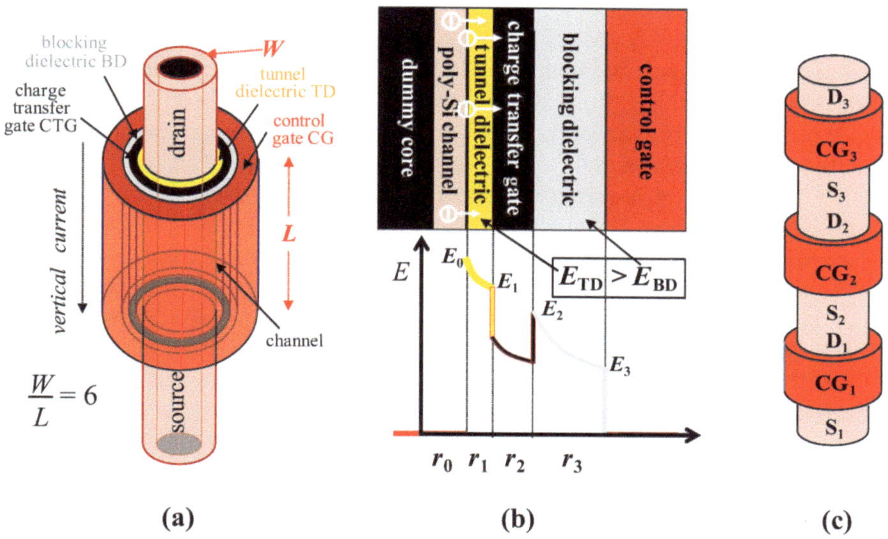

Fig. 6.40 (**a**) Architecture of the GAA charge-transfer gate (CTG) cell (**b**) Electric field distribution through the cross section of the cell (**c**) Stack of three of such cells as used in a 3D Vertical NAND Flash memory (Source: Ewoud Vreugdenhil; ASML)

CTF cell are much larger and vary from 250 holes (111) to 1750 electrons (011). This makes the 3D CTG GAA cell much more robust requiring much less ECC coding and wear-out compensation algorithms. This would even enable 3D GAA 4bits/cell NAND flash memories.

- Junction-less transistors.
  These GAA transistors do not have doped source (S) and drain (D) regions. In fact, when selected, the voltage on the control gate (CG) not only creates a circular channel below itself, as shown in Fig. 6.40, also its fringing field creates a channel in the source and drain regions between the control gates, so that they become conducting as well. This is due to the fact that the layers below the CG are so thin that the CG sidewalls (in fact, the word-line layers (WL in Fig. 6.42) represent the CGs in Fig. 6.40), as they are in one plane) are very close to the S/D regions and can create conducting channels in them.
- The curvature effect.
  The cross section in Fig. 6.40b shows the electric field across the various parts that together form the total radius of the cell. For attracting the charge to the CTG during programming, a large electric field between the channel and CG needs to be created. In a planar cell, there is still a high field between the control gate and the storage gate (Fig. 6.38) during programming and the blocking dielectric must prevent charge carriers to pass from the storage gate to the control gate. Due to the circular GAA cell, the field in the tunnel dielectric is (much) larger than that in the blocking dielectric. This so-called curvature effect enhances the electrical

field in the tunnel dielectric (pp. 280 in [37]), while the field in the blocking dielectric is decreased. The smaller the diameter of the polysilicon channel, the higher the field enhancement would be. This effect allows a reduction of the programming voltage, from 20–25 V for the planar cell to 15–18 V for the GAA cell, which results in a two times power reduction during programming.

- The fabrication process.
  The fabrication of a *vertical NAND (V-NAND)* GAA flash memory is quite different from that of a planar NAND flash memory. Today's V-NAND flash memories contain already several hundred stacked layers of memory cells and that number is still expected to increase to maybe 800 in 2030. However, we will describe here the architecture and process overview of a 48-layer Samsung gate-replacement process, which is still more or less representative for state-of-the-art NAND flash today. To create a V-NAND of 48 effective memory layers, the wafer needs to be deposited with a large stack of 48 NONO (nitride oxide nitride oxide) layers plus a few additional layers at the bottom and at the top for selection and redundancy purposes. So, to create 48 effective memory layers on top of each other, there are roughly 55 (or more) combined oxide/nitride layers deposited on the wafer. After this deposition, 100 nm holes are etched completely through the stack from top to bottom. This is a very advanced etching step, since the stack is not a mono layer, but consists of different materials: oxide and nitride. This one-shot etch etches through these different layers with only one gas compound. Figure 6.42 represents this V-NAND architecture, showing the 100 nm wide holes. Then this hole is used to successively deposit the various memory cell layers with atomic layer deposition (ALD) techniques in which the successive materials are deposited in gas phase on the cylindrical surface inside the holes. The sequence of process steps will be discussed on the basis of Fig. 6.41a.

  This process starts with the deposition of a ZrO_2 etch stop layer (the control gate will be created later), followed by the blocking dielectric (12–15 nm), the CTG dielectric ($\approx$5 nm), the tunnel dielectric ($\approx$4 nm), the polysilicon (channel: 5 nm) and dummy ($\approx$40 nm). So, all circular layers have now been deposited, except for the control gate. Note that the source, drain and channel are created on a lightly doped hollow polysilicon pipe, rather than by a massive central polysilicon cylinder. The next step is to etch deep high-aspect ratio (HAR) trenches from the top to the bottom. First of all, these trenches are used for an isotropic (wet) etch step, in which all (!!!) nitride layers (coloured in Fig. 6.42) Fig. 6.42 are completely removed. Then these hollow layers as well as the trenches are deposited with tungsten, which now also forms the control gates completing the CTG cell. In fact, all cells between two HAR trenches in one layer are connected to the same (tungsten) word line. Next, with an isotropic etching step, the tungsten in the deep trenches is completely removed. ($\approx$6 nm) deposition step along the edges of the empty trenches. Finally, the remaining of the trenches is completely filled with tungsten again, which now serves as a low-resistance ground. A very important advantage of the above described sequence of ALD deposition of the GAA CTG cell stack layers from the outer edge of the 100 nm holes to the dummy core (see core (see Fig. 6.40) is that all cells have

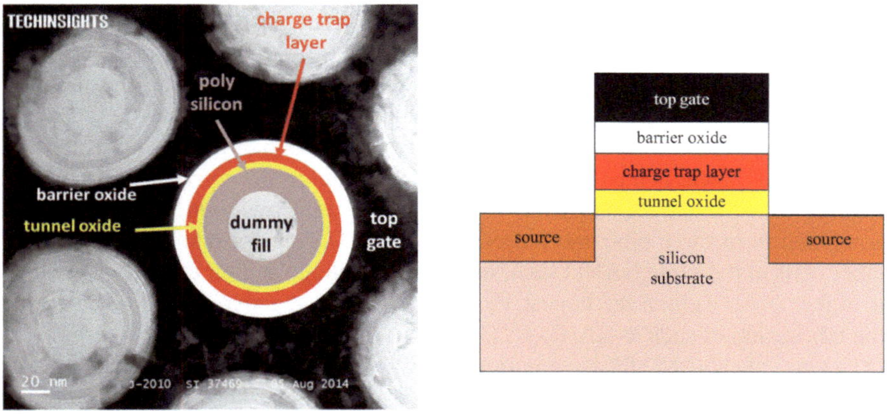

(a) (b)

Fig. 6.41 (**a**) TEM cross section through a word line layer (see Fig. 6.42), showing the sequence of ALD-deposited layers the V-NAND flash memory holes, which correspond to the stack of layers in a planar memory cell (**b**) (Source: TechInsights)

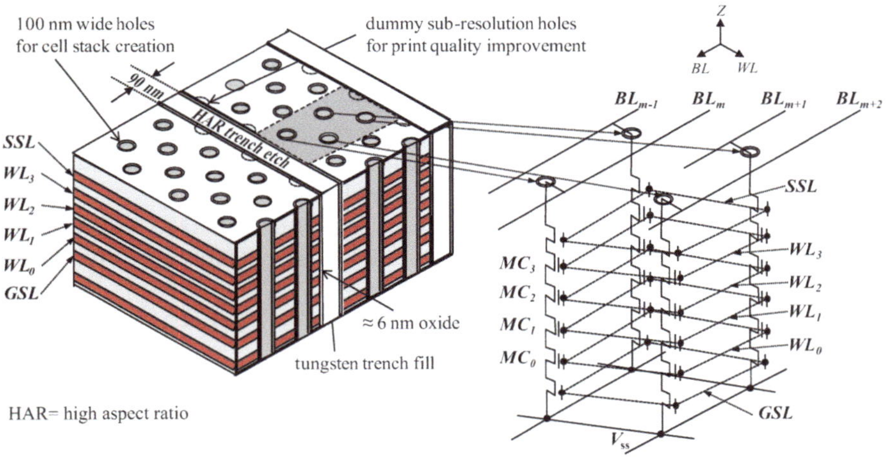

Fig. 6.42 3D representation of a GAA NAND flash memory array (Source: Ewoud Vreugdenhil; ASML)

the same thickness of the circular cell layers: BD, CTG and TD, respectively, independent of the variations in the diameter (target: 100 nm) of the holes. In other words: if the hole diameter varies over its total depth, or due to process variations, the thickness of the circular cell layers is still very constant, but the variations are then transferred to variations in the diameter of the dummy core, which does not affect the cell behaviour.

Reading of the data stored in a NAND flash cell requires a complex combination of voltages on the transistors in the selected stack. Because we can store three bits in one cell, represented by eight different charge levels (V_T s), reading a cell is done by slowly ramping the voltage at the control gate of the selected cell through the different reference voltages (Ref 1 to Ref 7 in Fig. 6.39). At every reference voltage, the current is measured. As soon as a current in the corresponding bit line is detected, at a certain reference voltage, then it is known which V_T (and what combination of three bits) was stored in the cell. During the reading of a cell, all other cells in the same NAND stack must have a voltage at their control gate which is higher ($\approx$10 V) than the largest V_T ($\approx$7 V). From the above, it will be clear that the fabrication of such V-NAND CTG GAA flash memories is completely different from planar FG NAND flash memories. The sizes of and pitches between the various cells in the various stacked memory layers are rather relaxed and do not require the most advanced (EUV) lithography tools. Scaling of this type of memories is no longer dependant on scaling of the physical sizes, but much more on the ability to stack more memory layers on top of each other. It means that scaling problems in planar NAND flash memories related to lateral size reduction, such as lithographic and physical process steps as well as with increasing cell interference, have now turned into problems with vertical process steps, such as layer deposition, etching of holes with extremely high aspect ratios, cylindrical ALD of various dielectric layers, etc. [40]. The thickness between the successive memory cell layers is around 50 nm, but further scaling is very limited, since the nitride storage electrode of the individual cells is in fact a long thin nitride cylinder from the top layer to the bottom. As the nitride does not conduct, the individual storage cell charge cannot flow from one cell to the cells in the layers directly above or below that cell.

Figure 6.43 shows a vertical SEM cross section of a 48L V-NAND flash memory array. In fact, it shows 55 layers: 48 memory cell layers, 4 dummy cell layers and 3 column selection transistor layers [41]

This V-NAND array is used in the 256 Gb V-NAND CTG GAA flash memory, which was announced already in 2016 [42].

The above discussions were related to a Samsung approach, where the 100 nm holes (Fig. 6.42) are created in one etching step through a stack of 56 combined oxide/nitride layers and where the nitride layers are later being replaced by wet-etching and deposition steps by tungsten. You cannot create these holes through a stack of oxide/tungsten layers in one etching step. Therefore, Intel-Micron started with a stack of polysilicon/oxide layers through which they created the holes with only one etch step. The polysilicon then acts as top gate of the gate-all-around transistors [43]. This saved a lot of process steps compared to the above explained replacement (nitride by tungsten) technique. During this decade, the number of stacked NAND flash memory layers has increased to several hundred. The larger the number of stacked layers, the more difficult it becomes to etch through them from top to bottom in just one single etching step. Therefore, industry has found a solution by creating a limited number of these layers in a deck and then stack several of these decks on top of each other. Figure 6.44a shows an example of a V-NAND flash architecture consisting of 3 decks [44]: an upper, a middle and a

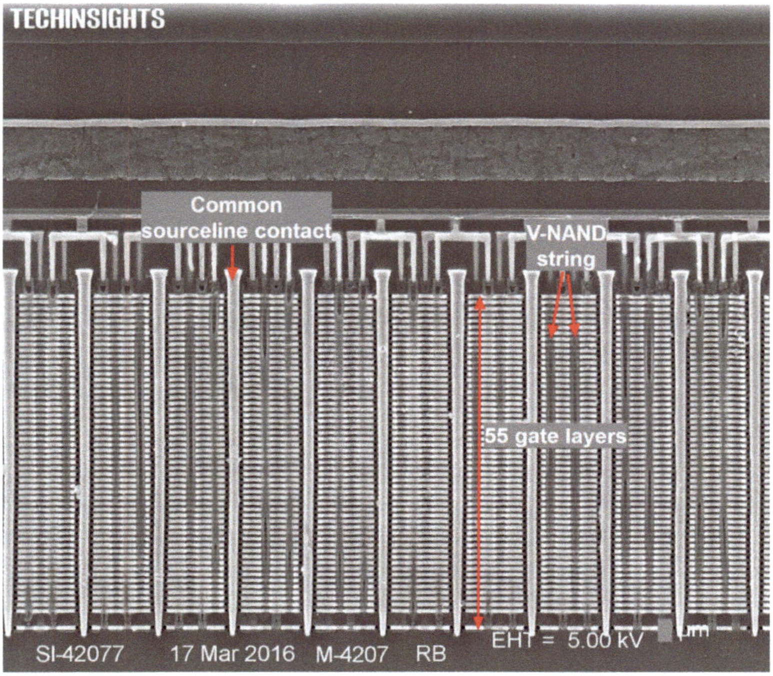

Fig. 6.43 Vertical SEM cross section of a 48L V-NAND flash memory array. (Source: TechInsights)

lower deck. Assuming that each deck will contain 48 memory cell layers, then the process starts with the deposition of roughly 112 layers (48 memory cell layers + 8 dummy word line layers and 56 isolation layers), creating the first deck. Next holes are etched through these layers analogous to those in Fig. 6.42. Then another deck of 112 layers is deposited on the first deck and again holes are etched through these layers. Next, this process is repeated for the third deck. The holes in the middle and upper deck need to be accurately aligned with those of the lower deck, because of the successive deposition of atomic layers in these holes, analogous to the ones shown in Fig. 6.41a, through all three decks.

This solution, however, is time-consuming. Therefore, several device manufacturers equipment vendors are developing processes to increase the layer count by etching deeper vertical holes. Tokyo Electron, an etching tools manufacturer, is developing a method for rapidly drilling more than 10-micron (10 μm) vertical channels for 400-layer 3D NAND nodes [45]. According to a publication, its High-Aspect-Ratio (HAR) dielectric etching technology employs a cryogenic ($-273\,°C \leq T \leq -150\,°C$) wafer stage and new gas chemistry to create 10-micron-high channels with an 'excellent' etching profile in just 33 minutes and with an 84% reduced carbon footprint. Analogous to the *triple-level cell (TLC)* (3b/cell) NAND flash memory example in Fig. 6.39, also most 3D NAND flash memories use an increasing number of bits per cell. Many of them also use a *quad-level cell (QLC)* (4b/cell), to increase the memory capacity. As discussed below Fig. 6.39, also these 3D NAND flash memories require wear-leveling compensation and increasingly complex error-detection and correction codes (ECC) to maintain reliable memory operation. Even 5b/cell (*penta-level cell (PLC)* [46]; 32 different charge levels!!) NAND flash memories are being engineered to reduce SSD costs. They will need even more complex error-correction/detection requiring more horse power from the memory controller. High-volume production and intensive use of PLC NAND flash in SSDs may be expected by the end of this decade.

The larger number of stacked layers results in reduced performance and reduced operating margins. Therefore, memory vendors increase the partitioning of the 3D NAND devices into more planes, currently up to 16, to achieve higher degrees of parallelism for improved performance.

Because every word-line layer needs an individual connection for selection, each successive layer must be a little larger than the one above it, forming a staircase at the periphery. The continuous increase of memory layers causes a relatively large periphery area for such a staircase (Fig. 6.44b). In case of a 128-layer architecture, the staircase width would be 80 μm wide. Patterning the staircase would require as many masks as there are layers in the stack, leading to hundreds of masks for state-of-the-art 3D NAND flash memories. Today, a large fraction of the staircase pattern is created by deposition and etching techniques. Figure 6.45a shows the sequence of process steps to create a four-step staircase. It starts with the deposition of a thick hard-mask layer and then pattern this layer with a lithographic step. Next, a so-called pull-back wet-etch step is performed on this layer, which etches both the

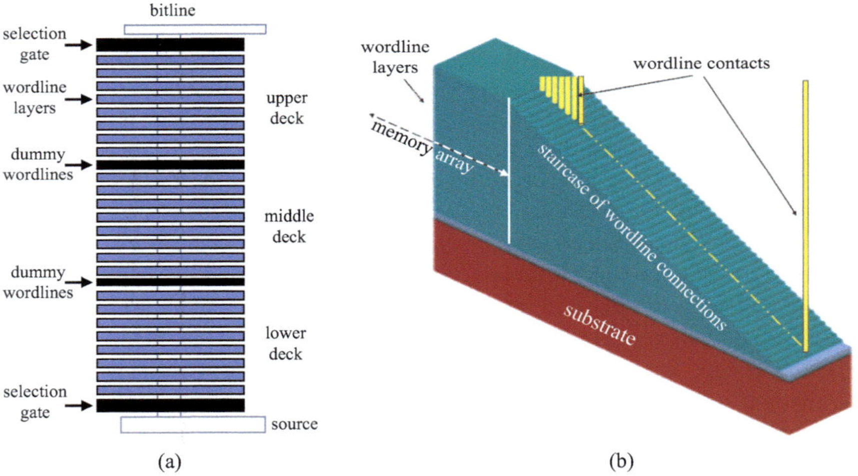

Fig. 6.44 (**a**) V-NAND architecture consisting of three decks [44] and (**b**) the periphery of a V-NAND memory array showing a unidirectional staircase for connection of the word-line layers (Courtesy of Applied Materials)

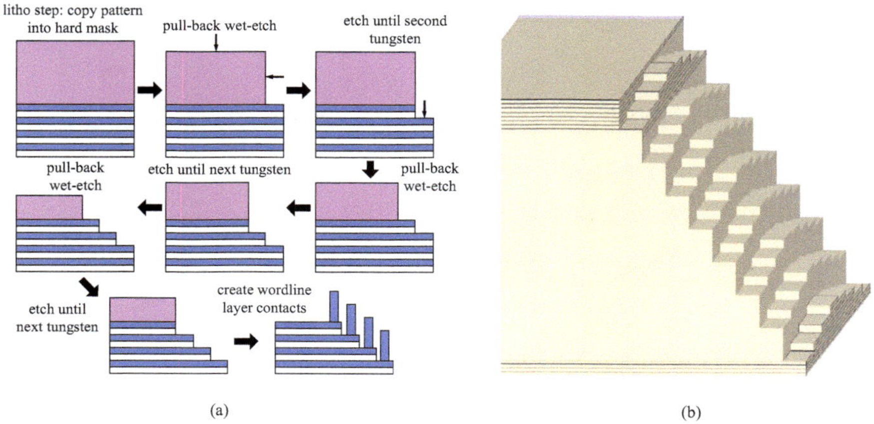

Fig. 6.45 (**a**) Creation of part of the staircase pattern by a combination of deposition and etching techniques (**b**) bidirectional staircase, i.e. they ascend or descend from two directions reducing the die area consumed by the staircase (Courtesy of Objective Analysis; printed with permission)

top and the sidewall of this hard-mask layer. This is then followed by etching steps to remove the top oxide and tungsten layer. Next, again, a pull-back wet-etch step on the hard-mask layer is performed, followed by etching steps to remove the next oxide and tungsten layer. This process is repeated another two times to create a 4-step staircase. As the deposition and etching techniques progress rapidly over time, 16- and 32-step staircases also exist already.

As discussed earlier, a unidirectional staircase would consume a lot of silicon area. A solution to this problem is to make a bidirectional staircase, where the stairs ascend or descend in two directions rather than one (Fig. 6.45b). Also here, pull-back wet-etch steps are used to reduce the number of lithographic steps [47].

Not only the NAND flash memory cells use the third dimension, also 3D packaging techniques are used to create huge memory capacity on a relatively small footprint. Figure 6.46 shows a package cross section with 16 dies stacked one on top of the other (*stacked dies*) and connected using conventional wire bonding technology. The dies are an outstandingly 40 μm thin, which might be close to the thinnest that can be achieved with 300 mm diameter wafers without a carrier wafer.

As the demand for high-capacity memory solutions increases, manufacturers are developing 3D NAND memories with even higher storage densities. As explained before, this is achieved by stacking more memory cells vertically, which allows for more storage capacity in a smaller space. The latest 3D NAND memories can store between 1 Tb and 2 Tb of data on a single chip. This has been achieved by increasing the number of stacked memory layers to 232 or more [35, 36] and/or increase the number of bits per cell: respectively 2-bits-per-cell *(MLC)*, 3-bits-per-cell *(TLC)*, 4-bits-per-cell *(QLC)*. Even a 5-bits-per-cell *(PLC)* NAND flash memory has been published [46]. It will be clear that with an increasing number of bits per cell, the performance, the operating margins, the data-retention time and memory-endurance

Fig. 6.46 Package cross section with 16 dies connected by conventional wire bonding (Source: TechInsights [41])

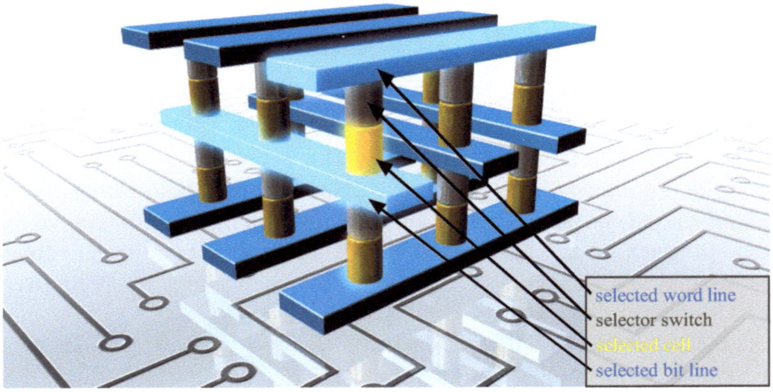

Fig. 6.47 Cross section of Intel/Micron 3D XPoint™ memory (Source: Micron Technology)

reduce. Even when using error-detection and correction and wear-levelling schemes, the reliability of these memories becomes a real challenge, particularly for the QLC and PLC NAND memories. There is still no physical limitation for improving 3D NAND flash memories, and so it is expected that by the year 2030, the number of memory layers in a 3D NAND may exceed one thousand!

An alternative non-volatile memory technology, first presented in 2015, was the *3D Xpoint™memory* [48]. It was co-developed by Intel and Micron Technology and built from a transistor-less architecture in which the cells are created between two successive metal layers at the intersection of word lines and bit lines. As such, each cell can be individually written (programmed and erased) and read at much faster rates than flash memories, as these only enable block-level erasure and programming. 3D XPoint™ is not seen as a replacement technology for either NAND flash or DRAM. The endurance is expected to be a few million program/erase cycles, which is roughly a thousand times higher than that for NAND flash. Figure 6.47 shows a cross section of this cross point memory. The cell consists of two parts: an ovonic switch to select the cell in series with a material-property change storage element, which changes the particular property over the full bulk of its material. It may use chalcogenide materials (although not fully confirmed by the inventors) for both selector and storage parts. This is claimed to increase scalability, stability, endurance, robustness and performance. It can have a low-resistance or a high-resistance state, depending whether there was a logic 'one' or logic 'zero' stored (programmed) in the cell. So far, this technology has not brought what was promised, due to some serious production problems. The most important one was the jump from the lab to the fab. This was the reason that Micron has quit the 3D XPoint business and sold the Lehi fab [49]. In 2021, Intel discontinued its consumer line of Optane products based on XPoint technology, and in July 2022, Intel closed the Optane division and discontinued the development of 3D XPoint.

Also Samsung worked on higher performance memory technologies which they call Z-NAND memory [50]. Initially, Z-NAND was a high-performance derivative of their V-NAND 3D NAND flash memory based on a SLC (one bit per cell) and a four times smaller page size to support higher speeds. The company announced that they were also working on a second generation which will be an MLC version of Z-NAND. However, it is very difficult to find additional publications on further improvements in this respect. If it remains SLC-based, which means low density, than it would be too expensive to compete with the TLC and QLC NAND flash versions.

6.5.5 Non-volatile RAM (NVRAM)

A *non-volatile RAM* combines SRAM and EEPROM technologies. This kind of memory is sometimes called a *shadow RAM*. Read and write actions can be performed at the speed of an SRAM during normal operation. However, the RAM contents are automatically copied to the EEPROM part when an on-chip circuit detects a dip in power. This operation is reversed when power returns. An NVRAM therefore combines the retention time of an EEPROM with the high performance of an SRAM. Recently, modules that combine NV memory and DRAM memories are introduced on the market. Such a module contains a NAND flash memory, a high-speed (DDR3 or higher) DRAM and a super-capacitor in a DIMM sub-system package. When a power dip or system failure occurs, the data are transferred from the DRAM to the flash memory. NVRAMs were targeted at applications in which it is not allowed to loose any data during power failures, such as medical and aviation and aerospace.

6.5.6 BRAM (Battery RAM)

A *BRAM* comprises an SRAM and a battery which provides sufficient power to retain the data when the memory is not accessed, i.e., when the memory is in the *standby mode*. The battery is used when power is absent. An SRAM is chosen because of its low standby power consumption. The battery is included in the BRAM package and the data retention time is close to 10 years. It was used to store configuration data in many digital products, such as bios data in PCs.

6.5.7 FRAM, MRAM, PRAM (PCM) and RRAM

A lot of research effort is devoted to develop the Holy Grail: a universal memory that could someday replace SRAM, DRAM and flash. Several alternatives are currently in development and/or small-volume production.

Ferroelectric RAM technology has "been available" for quite some time. The basic concepts of *FRAM* operation have been known since the 1950s. However, with the focus on the costs and the quality of silicon memories, progress in FRAM technology is at a much lower pace than that in SRAM and DRAM technology.

The first FRAM realised on silicon was unveiled in 1988 [51]. It contained 256 bits, which were built up from a six-transistor, two-capacitor array per cell. Compared to DRAM technology, this FRAM consumed a lot of area. Using a two-transistor, two-capacitor cell from 1992, currently, densities up to 16 Mb are being commercialised, with many different standard interfaces, in nanometer technologies with one-transistor, one capacitor per bit. This basic cell looks very similar to a basic DRAM memory cell, see Fig. 6.48.

The operation of a DRAM cell is discussed in Sect. 6.4.3. The operation of an FRAM cell is based on the polarisation state of its ferroelectric capacitor. The dielectric material used in this capacitor belongs to a certain class of dipole materials, which are called Perovskite crystals. By applying an electric field across this dielectric, these crystals polarise. This polarised state is maintained after the electric field is eliminated. The dielectric is depolarised when an electric field of the opposite direction is applied.

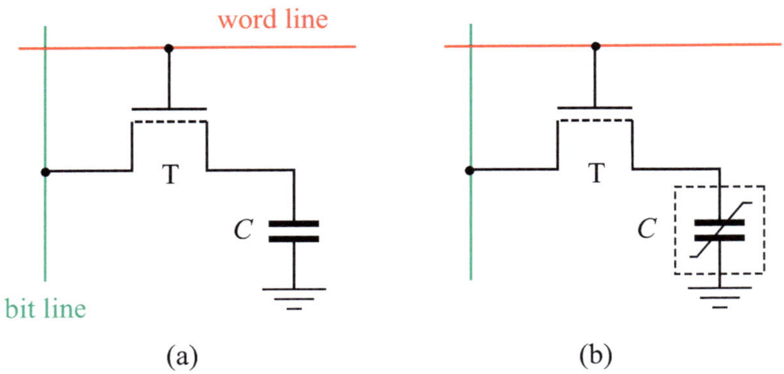

Fig. 6.48 (**a**) Basic DRAM cell, (**b**) basic 1T, 1C FRAM cell

During a read operation, an electric field is applied across the capacitor. Similar to a DRAM, the current is detected by a sense amplifier. When the dipoles switch state, the sense current is higher. Again similar to a DRAM, the data in a FRAM cell are destroyed during reading (*Destructive Read-Out (DRO)*). The cell contents must therefore be rewritten (refreshed) immediately after each read operation. A complete read cycle includes a precharge period, a read operation and a rewrite operation. Because of higher dielectric constants, an FRAM's cell charge density is higher than that of DRAM cells, allowing smaller cell sizes.

Advances in FRAM technology have resulted in trench capacitor and stacked capacitor architectures, analogous to DRAM technology evolution. Currently, Fujitsu and Infineon (Cypress) are offering 8 and 16 Mb FRAM memory, respectively. Basically, an FRAM operation depends on voltages rather than currents. This makes FRAMS particularly suited for low-power applications. FRAMs are therefore considered as the ideal memory for emerging low-power applications, such as smart cards and RF identification [52]. Potential applications include smart phones and tablets. Compared to EEPROM and flash memories, the number of read/write operations (endurance cycle) for FRAMs is several orders of magnitude higher (up to 10^{10}–10^{12}); however, several wearout/fatigue problems of the ferro-electric material have not really been solved yet, and as such it has limited compatibility with (advanced) CMOS processes and it faces severe scaling problems. The use of high-ϵ dielectrics, such as HfO_2 is well established in most state-of-the-art CMOS logic and memory processes. Utilising the ferroelectric phase of HfO_2 (FE-HfO_2) enables a much simpler approach to create manufacturable and scalable, CMOS compatible 1T and 1T/1C ferroelectric memories, based on *ferroelectric FETs (FeFETs)* [53, 54]. This technology can be used both in *ferroelectric DRAM (FeDRAM)* [55] and *ferroelectric NAND (FeNAND)* flash memory [56]. The used ferroelectric (FE) material creates non-volatility by retaining its polarised state in the absence of an electric field. It can be programmed by a relatively low programming voltage in the order of 3–5 V, as compared to the NAND flash programming voltage of 15–20 V. Explanation of this technology is beyond the scope of this book. Lots of it is contained in the above related references.

Next to FRAM technology, there are a few other alternative memory technologies in development. *Magneto-resistive RAM (MRAM)* was one of the emerging memory technologies. An MRAM acts as the magnetic counterpart of an FRAM. An MRAM cell consists of a selection transistor and a *magnetic tunnel junction (MTJ)* stack for data storage. This stack is built from a sandwich of two ferro-magnetic layers separated by a thin dielectric barrier layer. One of the magnetic layers has a fixed polarisation direction, while the polarisation direction of the other one can be controlled by the direction of the current in the bit line. The MTJ stack resistance is increased in the case of anti-parallel magnetisation orientation. During a read operation, a current tunnels from one ferro-magnetic layer to the other through the dielectric layer and the resistance state is detected.

This state is compared with the electrical resistance of a reference cell, which is always in the low resistance state.

The MTJ cell can be integrated above the selection transistor to achieve a small cell size and a cost-effective memory solution. It allows a virtually unlimited number of fast read and write cycles, comparable to DRAM and SRAM. The first prototype 16 Mb MRAM was already announced in 2004 [57]. Volume production of the first commercial 4 Mb stand-alone MRAM has been announced in 2007. An MRAM requires a relatively high current (several milli-amps) to program one bit. This high program current, combined with the reducing quality (robustness and reliability) of its magneto-resistive operation, forms severe roadblocks for scaling MRAM-based devices beyond the 65 nm node.

The more recently developed technique *spin transfer torque (STT)* uses spin-aligned electrons for the polarisation of the MTJ stack. The STT-MRAM requires a much lower programming current than the traditional MRAM. The lower current enables scaling beyond 65 nm. In 2021, Everspin presented the world's first 1 Gb STT-MRAM product in a 28 nm process. Currently, NETSOL has launched 1–32 Mb serial and 1–64 Mb parallel STT-MRAM products which are fabricated in the Samsung 28 nm FDSOI process. It plans to move to the 14 nm FinFET process to increase memory density and performance for a broader range of applications, such as IoT, AI, industrial, wearable, medical, consumer and automotive applications. MRAMs become increasingly popular in many embedded applications. In collaboration with TSMC, NXP Semiconductors will offer automotive embedded MRAM (*eMRAM*) in TSMC's 16 nm FinFET technology. Also Samsung and Renesas are developing eMRAM.

Another interesting non-volatile RAM alternative is the so-called *Phase-Change Memory (PCM)* , also known as *PRAM* or phase-change RAM *PCRAM* and *Ovonic Unified Memory (OUM)*. Its basic operation uses a unique property of polycrystalline *chalcogenide alloy*. This so-called phase-change property is also used for recording and erasing in optical media (re-writable CD and DVD). In these media, the required heat for programming is generated through exposure to an intense laser beam.

Figure 6.49 shows a cross section of a basic PRAM storage cell.

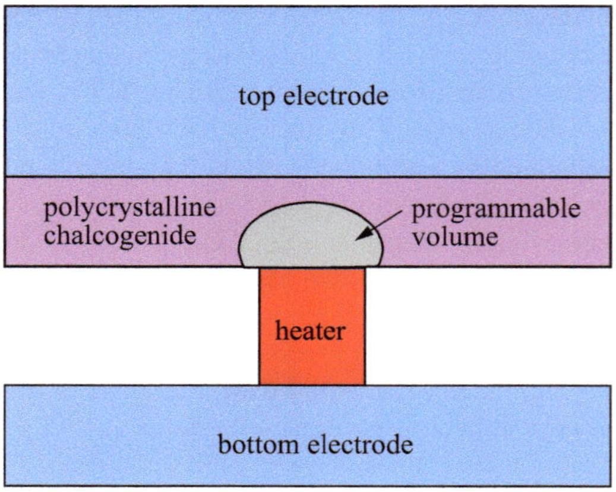

Fig. 6.49 Basic cross section of a Phase-Change Memory

Under the influence of heat generated by Joule heating, the polycrystalline state can be changed into an amorphous state and back. Each of these states corresponds to a different value of electrical resistivity representing a logic 1 and logic 0. When heated above melting point ($\approx$650°C), the chalcogenide alloy totally loses its crystalline structure. In the cell in Fig. 6.49 only a small programmable volume of the material is locally heated to above its melting point. This heat can be supplied by a current pulse through the heater. When rapidly cooled, the chalcogenide material is locked into its amorphous high-impedance state. By heating the material above its crystallisation, but below its melting temperature, the cell switches back to its low-impedance crystalline state. The difference in impedance between the two states is between one and two orders of magnitude. During a read operation, a voltage is applied to the cell, and the read current is measured against the read current of an identical reference cell with a fixed logic state. Similar to all memories that use a reference cell during reading, this cell must be placed well within the memory array to reduce array edge effects or must be surrounded by dummy cells. In sub 50 nm technologies, the use of reference cells is becoming more and more difficult due to variability. Instead of using reference cells, a dedicated circuit for creating the reference current is becoming more and more common. This reference current generator provides the optimum reference current for fast and reliable read operations, including aging effects.

Today's PRAM complexity is still far behind that of DRAM and SRAM, but due to the extending application areas with limited power budgets, particularly in handheld devices, there is more pressure to develop a real non-volatile RAM. Volume production of the first commercial PRAM was announced for the second half of 2007. Examples of PRAM designs can be found in [58, 59] and [60]. In the 2000s, PRAMs were seen as the most promising successor of NOR-flash [61, 62]. In 2008 the first multi-level 256 Mb PRAM prototypes were presented. While the '0' and the '1' were represented by the fully amorphous and fully crystalline state, the two additional states were represented by different levels of partial crystallisation [63]. PRAM operates at much faster speeds than NAND flash memory and is close to the performance of DRAM. However, the number of write operations per cell is limited, and so, bits cannot be changed when this limit is reached, resulting in a limited lifetime. A lot of techniques are currently explored to increase this lifetime [64].

Many other non-volatile techniques are currently in basic R&D phase. Two look promising. The first one is the *resistive RAM (RRAM or ReRAM)* , a memory, whose operation is also based on resistance change. It consists of a metal/perovskite-oxide/metal sandwich structure, in which a reversible resistance switching behaviour can be triggered by the application of short voltage pulses [65, 66]. It has been shown that RRAM can swich fast (<0.3 ns) [67] and can be used in low-power applications [68]. A 32 Gb demonstrator in 24 nm technology has been presented in 2013 [69]. RRAM has potential to break the performance bottleneck of traditional architectures in logic computing and can be used in compute-in-memory (CiM) applications, such as AI, neural networks and brain-like computing [70]. Finally, the second one is the *conductive bridging memory (CBRAM)*, in which the cell consists of a thin electrolyte layer, sandwiched between two electrodes. The logic state of a cell is defined by deposition or removal of metal, e.g., Ag, within this electrolyte layer. The mechanism is promising because of its scalability, its multi-level capability and low-power potentials. A 2 Mb demonstrator has recently been published [71] and commercial 1 Mb products are available on the market as a replacement for EEPROM. Recently, a study has been published for the use of CBRAM in memory and neuromorphic computing applications [72].

All memory types discussed in this section form a kind of 'exclusive memory' category, in that they are accredited certain exclusive properties beyond those of the mainstream SRAM, DRAM, ROM, EEPROM and flash memories. Some of them exist already for more than 30 years, and they seem to regain popularity every now and then. All in all, because of their exclusive properties, they are either complex and expensive to make, or they face severe scaling challenges. Moreover, since their potential applications also change rapidly over process node and over time, the reader is advised to search the internet for more detailed information about this category of memories as further details are beyond the scope of this book.

6.6 Embedded Memories

The integration of complete systems-on-a-chip (SoC) include the combination of logic circuits (logic cores) with memories (memory cores). There are several reasons to put memories and logic on the same chip. In many cases, this is (and will be) done to:

- offer higher bandwidth
- reduce pincount
- reduce system size
- offer a more reliable system
- reduce system power

Also the low cost of interconnect at chip level may be a good reason to embed memories or other cores. The diagram [73] in Fig. 6.50 shows the relative cost of interconnect as a function of the distance from the centre of the chip. It clearly shows that the chip level interconnect is by far the cheapest one.

Another reason to embed memories is to fill the design productivity gap. Figure 6.51 shows this gap with respect to the growth in IC complexity according to the ITRS roadmap [74]. The solid line represents the number of logic transistors per chip. The dotted line shows the design productivity. Many of the transistors made available by the technology, but unused by the design, may be used to increase the amount of embedded memory.

Basically, there are three different approaches to implement an SOC. The first one is to embed memories in a logic chip, integrated in a logic-based process (*embedded memory*). In essence, all common types of memories can be embedded: SRAM, DRAM, ROM, E(E)PROM and flash memories. Except for the SRAM and ROM, they all require several additional masks and processing steps to efficiently embed

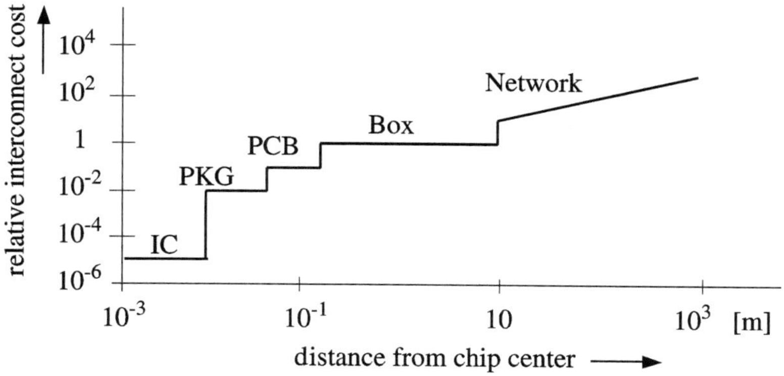

Fig. 6.50 Relative cost of interconnect

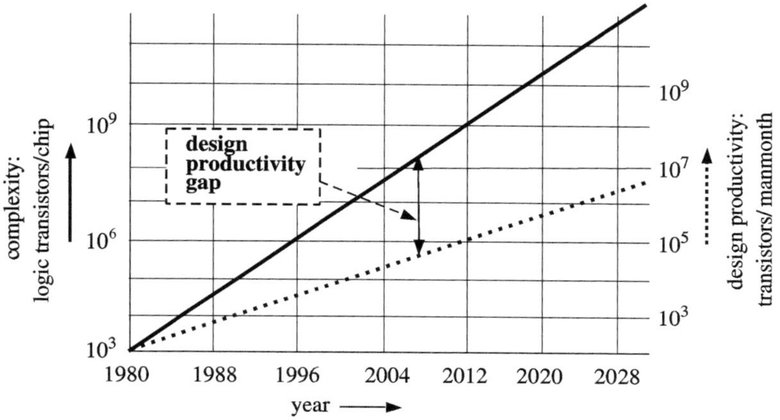

Fig. 6.51 The design productivity gap with respect to the growth in IC complexity

these memories. Traditionally, the advantage of stand-alone SRAMs was twofold: they offered a higher speed than DRAMs and consumed less power in standby mode. Both advantages, however, are losing ground. The speed of the SRAMs has become so high that their usage is more and more hampered by the limited bandwidth of the interface between ICs. As already discussed in Sect. 6.4.2, SRAM leakage has reached so high levels, that, in combination with increasing variability, a read operation can cause a parasitic write, which may flip the cell and results in different and almost conflicting transistor sizing requirements for the read and the write operation. The increasing leakage is not only a limiting factor in achieving low-standby power levels, it also contributes to a reduced static noise margin of an SRAM cell.

Compared to stand-alone memories, embedded memories show a much lower bit density. For embedded SRAMs, this difference may be in the order of a factor of 2. For embedded DRAMs, this factor can be much more (from three to eight), depending on the amount of extra masks and processing steps that are added to the standard CMOS logic process.

Embedded SRAM can be found in practically every application, as discussed in Sect. 6.4.2 Also the usage of embedded flash spans a wide range of applications: micro controllers, industrial, office automation, networking, consumer, smart cards and RFID tags. Today, the increasing requirements in number of features and bandwidth have led to a growth of both the capacity and number of distributed embedded memory instances on a single chip. Complex microprocessors may incorporate level-one, level-two and sometimes even level-three caches and contain several billion transistors, of which most (>80%) are located in these embedded memories.

The following trends are observed in embedded memories. As discussed in Sect. 6.4.2, to enable low-voltage operation and to improve noise margins, many SRAM designs use assist circuits to enable operation at reduced supply voltages [8, 9]. Also 7T, 8T and even 9T SRAM cell alternatives are explored for this reason. For embedded PROM, traditional poly-fuses are being replaced by anti-fuses based on a transistor with a thin oxide region. A high voltage is used to break-down the thin oxide to create a conducting channel between source and drain.

Certain applications demand even higher memory capacities and require the highest possible density of the memory blocks. Therefore, the second approach was to embed logic (processors etc.) in a memory (mostly DRAM) process (*embedded logic*). A DRAM in a logic-based process will not be as compact as in a DRAM-based process, because this process has been optimised for it. Analogous to this, logic in a DRAM-based process will not be as compact as in a logic-based process, partly because DRAM processes use fewer metal layers than current logic processes. Next to a higher memory density, embedded logic potentially offers much higher memory bandwidth due to the fact that bus width can be much larger (>1024b) since these are not pin limited. There are several examples of embedded logic processes: flash-based embedded logic process [75] and DRAM-based embedded logic process [76].

However, the emerging graphics market requires very high speed DRAMs at limited power consumption, which drives the need for *high bandwidth memories (HBM)*, in which the memory dies are made in a DRAM process and stacked on a logic die that contains all address and control logic and sense electronics. This solution is discussed in Sect. 6.4.4.

Non-volatile memories are also very popular to embed on ICs in a large variety of applications. As discussed before, the program of a ROM is easily implemented in the second (or third) metal layer and is fully compatible with a normal CMOS process. All programmable *embedded non-volatile memories* use a floating gate to store the data. This requires several extra masks and processing steps added to a base line CMOS process, thereby increasing the fabrication cost of the products. Full embedded non-volatile capability, in which the cell contains a stack of the tunnel oxide, storage gate, blocking oxide and control gate, as shown in Fig. 6.38, for example, requires so many additional process steps, that it may increase overall

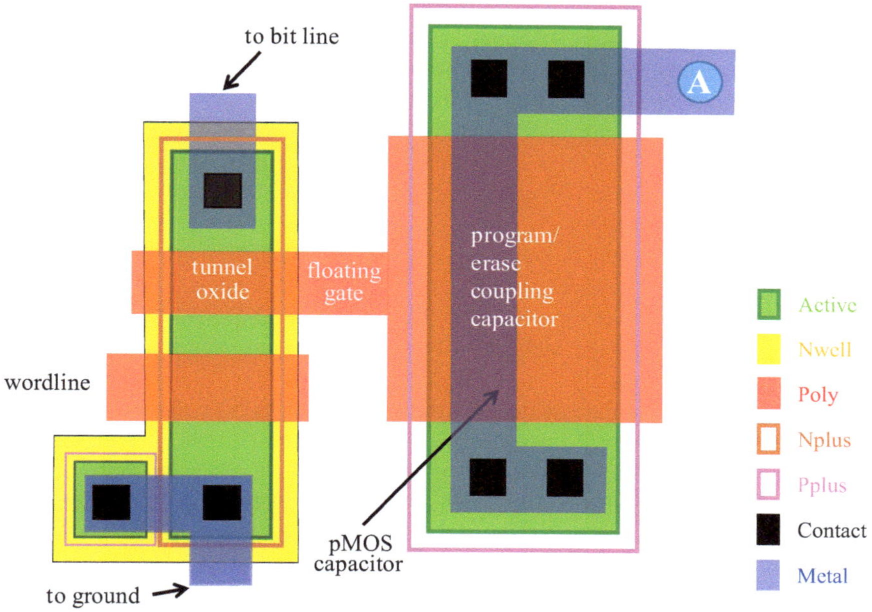

Fig. 6.52 Example architecture of an embedded non-volatile MTP memory cell with a minimum of additional process steps for small embedded memory capacity

production cost by 30% or more. Some applications, however, only demand that each die has its own ID like lot and wafer number, location on the wafer, test results and pass flag. Others require some parameter trimming (e.g., adjustment of reference voltage or current), storage of failing rows or columns (in memories with redundancy), configuring of I/O pad functionality or small code updates in the application (some codes are country-dependent). They therefore require only a small amount of embedded non-volatile memory in the order of a few tens of bits up to 1 kb, for which full embedded non-volatile capability is not economically viable. These products may use a folded floating-gate cell, which can be seen as a lateral version of a stacked non-volatile cell. They are therefore larger in size. Figure 6.52 shows an example architecture of such a cell.

Because the memory is made in a baseline process, the programming voltages must be limited, since there are no real thick (7 nm) oxides available. Therefore, the coupling capacitor must have a high coupling efficiency. To program the cell, the metal connections left (to bit line and to ground) are connected to ground, while node A is pulled high (e.g., 10 V). The floating gate is then also pulled high through the program/erase coupling capacitor. This creates a high voltage across the tunnel oxide in the cell causing Fowler–Nordheim tunnelling of electrons from the channel to the floating gate. This generates a negative charge on the gate of this storage transistor, such that it is always off, representing a logic ‘1’, for instance. To erase the cell, node A is pulled to ground, while the bit line and ground connection (left) are pulled to the high voltage. Since the nwell is then also at the high voltage, it forces the reverse tunnelling of electrons back from the floating gate, such that this storage transistor is always on, representing a logic '0'. Using high program and erase voltages in the cell requires the use of thicker gate oxide, both for the transistors and the coupling capacitance. In many cases, the thicker oxide available from the I/O circuits (2.5 V or 3.3 V) is used. During the erase cycle, the word line voltage is pulled to around 5 V; otherwise, there will be 10 V across its gate oxide, then.

This device is often called a *multi-time programmable (MTP)* memory, because it only has a limited endurance (number of program/erase cycles) and a relatively short data retention time compared to EEPROMs and NAND flash memories. This is often compensated by the inclusion of error-correction codes and parity bits to guarantee low ppm levels.

Testing is a problem that arises with the merging of huge memory blocks with logic on a single chip. In a large-complexity, stand-alone memory, true memory performance can be measured because of the accessibility of the memory through the I/O pads. When such complex memories are embedded, direct accessibility through the pads is obviously less, because this is often done by multiplexing to I/O pads. BIST techniques are required to minimise testing costs and wafer handling (see Chap. 10).

Finally, as mentioned in Sect. 6.5.7, MRAMs become increasingly popular in many embedded applications. In collaboration with TSMC, NXP Semiconductors will offer auto-motive embedded MRAM (eMRAM) in TSMC's 16 nm FinFET technology. Also Samsung and Renesas are developing eMRAM.

6.7 In-memory Computing

In traditional computing systems, data are typically stored on disk drives and/or SSDs or in other forms of persistent storage. When a computation or query needs to be performed on this data, it is read from the disk into the computer's memory, where it can be processed by the CPU. However, this I/O process can be a bottleneck as the disk access is much slower compared to semiconductor-memory access. A very interesting brain-mimicking approach, which overcomes this limitation by keeping data in fast, volatile memory (RAM), allowing for near-instantaneous access, is *in-memory computing (IMC)* [77, 78], also called *compute-in-memory (CiM)*. One of the main advantages of in-memory computation is its ability to handle large volumes of data with low latency. As modern datasets continue to grow exponentially, traditional disk-based/SSD systems struggle to keep up with the demand for quick data processing. By storing data in RAM and processing it in parallel, CiM enables complex calculations and analyses on vast datasets without experiencing performance degradation. This allows for real-time processing and analysis of complex queries, enabling businesses to make faster, data-driven decisions. Another advantage is the ability to support real-time analytics and interactive data exploration. There are several applications for CiM:

- Real-time analysis of large volumes of data
  - in fraud detection to detect potential fraud patterns in vast amounts of data
  - in supply chain management for improved inventory management, demand forecasting and logistics optimisation
  - in Internet of Things (IoT) applications to facilitate real-time monitoring, data analysis, and decision-making in smart city initiatives, industrial IoT and connected vehicles
  - in high-performance computing tasks like scientific simulations, weather forecasting, computational fluid dynamics, lithography simulation, mask simulation, image processing and genome (subset of genes/DNA) sequencing
  - in *deep neural networks (DNN)* AI applications, such when training deep learning models which requires frequently updating model parameters.

These are just a few examples of the diverse applications for CiM. The technology can be applied across various industries and use cases where real-time data processing and analysis are critical. However, it is important to note that CiM also has some considerations. Firstly, the cost of memory can be considerably higher than disk/SSD storage, particularly when dealing with large datasets. Proper capacity planning and optimisation techniques may be required to ensure cost-effectiveness. Secondly, the volatile nature of memory means that data is lost when the system is powered off or experiences a failure. Therefore, appropriate data backup and recovery strategies should be implemented to mitigate the risk of data loss. CiM techniques can be used at different levels of the memory hierarchy and based on different memory technologies, including SRAM-based on-chip memories, or off-chip DRAMs and nonvolatile memories [79]. In conclusion, CiM is a powerful approach that leverages the speed and efficiency of main memory to process large volumes of data quickly. It allows one to store data in RAM, across a cluster of computers, and process it in parallel [80].

6.8 Redundancy in Memories

Stand-alone memories are sold in very high volumes and must therefore be very cheap to produce. Methods to achieve a low price include yield-improvement techniques which may, for example, result in a yield in excess of 70% for areas greater than 100 mm^2. However, many stand-alone memories have one or more cells that do not function properly. For this reason, most stand-alone memories include several redundant memory rows and/or columns which can be used to replace defective cells. The faulty cells are detected by means of memory testers and a laser beam is used to isolate their corresponding rows or columns. This so-called *laser-fusing* technique is also used to exploit spare rows and columns and re-address the faulty ones to the redundant (spare) ones. Currently, the *poly fuse* has replaced the traditional laser fuse, because it does not require the special laser equipment. Poly fuses are smaller and can be programmed by a tester by providing normal signals and voltages. Figure 6.53 shows an example of row-redundancy implementation. In

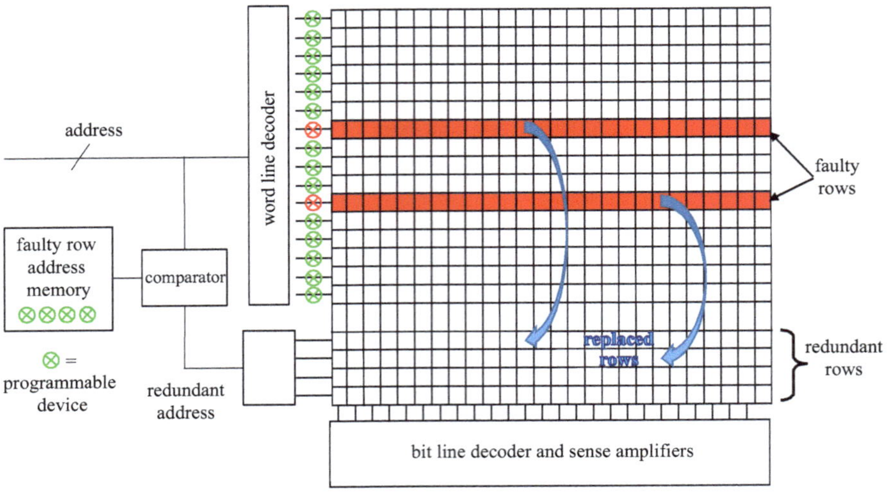

Fig. 6.53 Example of row redundancy implementation

this example, the memory array is extended with four additional dummy (redundant) rows, which can be addressed by the redundant address decoder. Let's assume that during the memory test, two failing bits in two different rows have been detected. The addresses of the corresponding faulty rows are then stored in the faulty-row address memory. This memory is built from programmable devices, which can be poly silicon fuses that are blown by a laser during the test. In ICs that already contain embedded non-volatile (EEPROM or flash) memory, the faulty row address memory may be built from the same non-volatile cells. Everytime that a faulty row (word line) would be addressed, the comparator finds a match between this address and the stored faulty-row address and produces a redundant address which activates the corresponding redundant row. In this way, the two faulty rows are replaced by two redundant rows. This type of redundancy can be used in both embedded SRAMs and in stand-alone DRAMs. Today, with increasing performance requirements and high-speed DDR DRAM interfaces, this solution has a significant drawback in that the total word line access latency increases because the row address comparison and the redundant address generation cost additional time and increase the cycle time of the memory. Many solutions to this problem can be found on the internet, but are beyond the scope of this book.

Redundancy techniques may be used to improve the yield by a factor of as much as 20–50 during the initial development phase of new memory process generations. During memory production ramp up, the memory may include more redundancy (e.g. close to 5% additional bits) than during high-volume production in mature processes (e.g., less than 5% additional bits). While redundancy allows hardware correction of failing bits to increase yield, many memories (DRAM, SRAM, ROM and Flash) may also include a form of error correction of (weak) bits by using

ECC (error-correction code). Because of scaling, process variations increase while memory noise margins reduce and the 'electrical distance' between a 1 and a 0 reduces. This latter effect may also be the result of wear-out of memory cells on a NAND-flash memory chip during use. ECC may also be used to support reduced refresh rates in DRAMs or lower-voltage operation. All of this results in a steadily increasing bit error rate. In most memory applications that use ECC, the ECC operation is performed on the memory controller chip. The ECC code, however, may be stored, either on a separated memory chip, or inside the memory itself [81].

Certain applications, e.g., medical and automotive sytems, require extremely high reliability standards. These may require so-called *built-in repair analysis (BIRA)*, which is commonly used to repair memories post-manufacturing [82]. For autonomous driving, for instance, a crucial requirement is detecting latent faults. This requires periodical tests and, possibly, repairs during automotive operation. The reader is advised to search the internet for more information on this rapidly changing world.

6.9 Classification of the Various Memories

Table 6.4 provides an overview of the different types of memories with respect to some important parameters that characterise them. The numbers in Table 6.4 are orders of magnitudes and may vary between different memory vendors. The characteristic values of these parameters render each type of memory suitable for application areas. These areas are summarised in Table 6.3.

6.10 Conclusions

The total stand-alone MOS memory market, which is for almost 90% dominated by DRAM and flash memories, currently represents about 25% of the total IC market turnover. This indicates the importance of their use in various applications. Most applications have different requirements on parameters such as memory capacity, power dissipation, access time, data retention time, endurance and reprogrammability, etc. Modern integrated circuit technology facilitates the manufacture of a wide range of memories that are each optimised for one or more application domains. The continuous drive for larger memory performance and capacity is leading to ever-increasing bit densities and the limits are not yet in sight. The DRAM and flash markets show the largest volumes and, not surprisingly, the highest demand for new technologies (Fig. 6.2b).

This decade, the focus in the DRAM sessions at the ISSCC conferences is on high-bandwidth DRAM, for graphics and high-performance computing applications based on various high-speed interfaces and on reducing power consumption. Regarding flash memories, further lateral scaling is limited, due to increasing electrical interference between neighbouring cells as well as due to increasing lithographic problems. Therefore, most innovations in flash memories are related to

3D solutions, where the production of memory chips with over 300 stacked memory layers has recently started. It was expected that by the year 2020, the cost per bit of non-volatile semiconductor memories has reached the level of hard disks. However, even today, SSDs are almost always much more expensive than hard disks with the same storage capacity. This is often a factor of two or more, depending on the speed of the SSD. In many smaller memory capacity applications, such as smart-phones, tablets and mini-laptops, the harddisks have already been replaced by these solid-state drives (SSDs).

Table 6.3 Application areas for the various memory types

Memory type	Application areas
SRAM	Mostly as embedded memory on processors used in Super-fast systems, low-power systems, cache memories in PCs (hard disks, DVD R/W caches), workstations, telecommunication, multimedia computers, networking applications, mobile phones, supercomputers, mainframes, servers, embedded memories
DRAM	Medium to high speed, main memory in computer systems, desktop, server, low-cost systems, networking, large volumes, PC, hard disk drives, graphics boards, printer applications, mobile phones
FRAM	Low-power, non-volatile applications, smart cards, RF Identification, replacement of non-volatile RAM and potentially high-density SRAM
ROM	Large volumes, video games, character generators, laser printer fonts, dictionary data in word processors, sound source data in electronic musical instruments embedded memories
EPROM	CD-ROM drives, modems, code storage, embedded memories
EEPROM	Military applications, flight controllers, consumer applications, portable consumer pagers, modems, cellular and cordless telephones, disk drives, printers, air bags, anti-lock braking systems, car radios, smart card, set-top boxes, embedded memories
FLASH	Portable systems, communication systems, code storage, digital TV, set-top boxes, memory PC cards, BIOS storage, digital cameras, ATA controllers, flash cards, palm tops, battery powered applications, smart phones embedded memories, MP3 players, e-books, tablets
NVRAM BRAM	Systems where power dips are not allowed, medical systems, space crafts, etc., which require fast read and write access
Persistent memory SCM; NVDRAM	Database management, AI, big data analytics and in server stations

Table 6.4 Characteristics of different memory types

DEVICES		SRAM	DRAM	ROM	NOR-flash	NAND-flash
Physical cell size		150–200F^2	4–8F^2	4F^2	8–10F^2	4–5F^2
Capacity [bit]		<256k	<64G	<1G	<16G	<2T
Max. number of programming cycles		∞	∞	1	10^4–10^5	10^3–10^4
Programming time (write)		5–40 ns	20–100 ns		5–10 µs	100–300 µs
Accesstime (read)		5–20 ns	10–70 ns	5–20 ns	Random: 80–150 ns serial 80–12 ns	Random: 10–20 µs serial 5–50 ns
Retention time	No power supply	0	0	∞	>10 year	>10 year
	Power supply	∞	2–64 ms			

This chapter has given insights into the basic operating principles of the most popular range of embedded and stand-alone memory types. Their application areas are summarised in Table 6.3, while their characteristic parameters are compared in Table 6.4.

Note that the memory capacity in the table above is specified per chip. Multi-chip modules and so-called DIMM modules that are used in computers contain several memory chips. The physical size of the cells is expressed in F^2, where F represents the minimum feature size of the process (e.g., 22 nm in the 22 nm process node), which is typically equal to half the minimum word line or bit line pitch in memory processes.

6.11 Exercises

1. Assume that the column decoder in Fig. 6.4 is implemented in CMOS as shown in the adjacent figure and the column address is
   $a_{11}a_{10}a_9a_8a_7a_6 = 010110$.

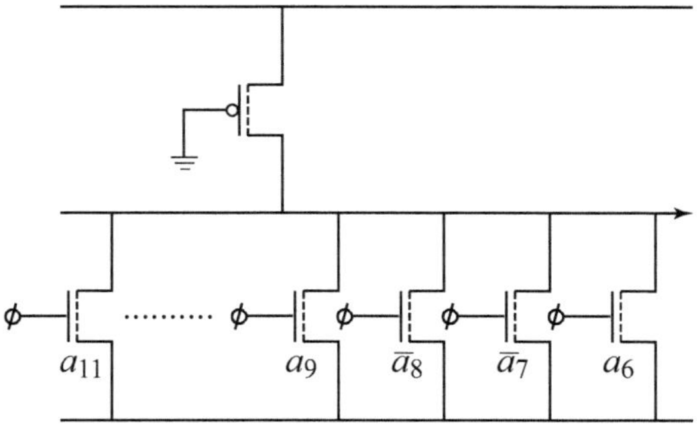

(a) Describe the data flow in Fig. 6.4 during a read operation when word line x_{20} is also selected.
(b) What is the major disadvantage of such a decoder?
(c) What would be the problem if this decoder were implemented in static CMOS?
2. Describe the major differences between the ROM realisations of Figs. 6.27 and 6.30. Explain their relative advantages and disadvantages.
3. Why does a stand-alone flash EPROM sometimes require one more power supply than a full-featured EEPROM?
4. Table 6.4 gives a summary of some important memory parameters.
   (a) Explain the difference in chip area between a non-volatile RAM and an SRAM.
   (b) Explain the difference in access times between an SRAM and a DRAM.
5. The adjacent figure shows a dynamic memory cell which consists of three transistors. This is a so-called *3 T-cell.*

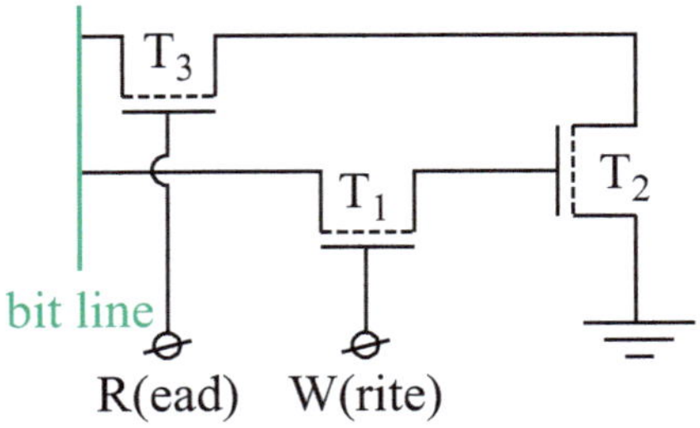

   (a) Explain the operation of the 3 T-cell.
   (b) What can be said about the read-out data after one write and read cycle?
   (c) Comment on the size of the storage nodes in the 3 T-cell and the 1 T-cell?
6. What is a multilevel flash memory?
   What is the reason of its existence?
   What would be their main problem for future process generations?
7. What are the major differences between a polysilicon floating gate NAND flash memory and a charge trap flash memory?
8. Explain the difference between an embedded memory and embedded logic.

References

Information about memories is usually confidential and is often proprietary. Many of the relatively few books available on the subject are therefore outdated. This reference list therefore only contains a few published books and the titles of interesting journals and digests on relevant conferences. In this edition, it is extended with many references on state-of-the-art material from conferences, publications, and internet sites.

1. K. Pagiamtzis et al., Content-addressable memory (CAM) circuits and architectures: a tutorial and survey. IEEE J. Solid-State Circuits **41**(3), 712–727 (2006)
2. K. Noda et al., A 1.9 μm^2 loadless CMOS four transistor SRAM cell in a 0.18 μm logic technology, in *IEDM Digest of Technical Papers* (1998), pp 643–646
3. K. Takeda et al., 'A 16 Mb 400 MHz loadless CMOS 4-transistor SRAM Macro, in *ISSCC Digest of Technical Papers* (2000)
4. L. Chang et al., Stable SRAM cell design for the 32 nm node and beyond, in *2005 Symposium on VLSI Technology, Digest of Technical Papers*, pp. 128–129
5. J. Keane et al., 5.6Mb/mm^2 1R1W 8T SRAM arrays operating down to 560 mV utiliing small-signal sensing with charge-shared bit line and asymmetric sense amplifier in 14nm FinFET CMOS technology charge-shared bitline and asymmetric sense amplifier in 14nm FinFET CMOS technology, in *ISSCC 2016, Digest of Technical Papers* (2016), pp. 308–309
6. S.M. Jung et al., Highly area efficient and cost effective double stacked S (stacked single-crystal Si) peripheral CMOS SSTFT and SRAM cell technology for 512 Mb SRAM, in *IEDM 2004, Digest of Technical Papers* (2004), pp. 265–268
7. M. Khayatzadeh, M. M., J. Wang, M. Alioto, D. Blaauw, D. Sylvester, A reconfigurable dual-port memory with error detection and correction in 28nm FDSOI, in *ISSCC 2016, Digest of Technical Papers* (2016), pp. 310–311
8. E. Karl et al., The impact of assist-circuit design for 22 nm SRAM and beyond, in *IEDM Technical Digest* (2012), pp. 561–564
9. T. Song et al., A 10nm FinFET 128Mb SRAM with assist adjustment system for power, performance and area optimization, in *ISSCC 2016, Digest of Technical Papers* (2016), pp. 306–307
10. S. Ahmad et al., A comprehensive review of design challenges and techniques for nanoscale SRAM: a cell perspective, TechRxiv (2022)
11. A. Klotz, CPU and GPU SRAM caches are not shrinking, which could increase chip cost or reduce performance. TECHSPOT (2022). https://www.techspot.com/news/97000-cpu-gpu-sram-caches-not-shrinking-which-could.html
12. J. Wan et al., Z^2-FET used as 1-transistor high-speed DRAM, in *ESSDERC, Digest of Technical Papers* (2012)
13. Designing for 1 GB DDR SDRAM. Micron Technology, Technical Note (2003)
14. C. Cho et al., A 6F^2 DRAM technology in 60 nm era for gigabit densities, in *2005 Symposium on VLSI Technology, Digest of Technical Papers* (2005), pp. 36–37
15. J.A. Mandelman et al., Challenges for future directions for the scaling of DRAM. IBM J. Res. Dev. **46**(2/3), 187–212 (2002)
16. H. Seidl et al., A fully integrated Al_2O_3 trench capacitor DRAM for sub-100 nm technology. IEDM (2002)
17. Micron D1α, '14 nm'! The Most Advanced Node Ever on DRAM!', TechInsights, DRAM Memory Technology (2021)
18. B. Vincent, 3D DRAM Is Coming. Here's a Possible Way to Build It', Newsroom Lam Research (2023)
19. J. Gerend et al., Understand and deploy persistent memory. https://learn.microsoft.com/en-us/azure-stack/hci/concepts/deploy-persistent-memory, Article 18 April 2023
20. C. Hampel, High-speed DRAMs keep pace with high-speed systems, in *EDN* (1997), pp. 141–148
21. C. Green, Analyzing and implementing SDRAM and SGRAM controllers, *EDN* (1998), pp. 155–166
22. R. Faramarzi, High speed trends in memory market, in *Keynote Adress, Jedex Conference*, Shanghai (2006). http://www.jedexchina.org/program.htm
23. P. Hardware, DDR6 RAM: The Next Big Thing in Memory Technology, TechAmaze (2023). https://tecamaze.com/ddr6-ram/
24. T. Schmitz, The Rise of Serial Memory and the Future of DDR, Xilinx, WP456 (v1.1) (2015)
25. Understanding Video (VRAM) and SGRAM Operation. http://www.chips.ibm.com/products/memory

26. D. Bursky, Graphics-optimized DRAMs deliver top-notch performance. Electr. Design (1998), 89–100
27. D. Chiang, SK Hynix to reveal 16-layer HBM3E to main the lead, DIGITIMES Asia (2024)
28. A. Shilov, Samsung Unveils Industry's First 32Gbit DDR5 Memory Die: 1TB Modules Incoming, AnandTech (2023)
29. K.-h. Kim et al., An 8 Gb/s/pin 9.6 ns Row-Cycle 288 Mb Deca-Data Rate SDRAM with an I/O error-detection Scheme, in *ISSCC Digest of Technical Papers* (2006), pp.154–155
30. T. Nagai, A 65 nm low-power embedded DRAM with extended data-retention sleep mode, in *ISSCC Digest of Technical Papers* (2006), pp.164–165
31. P. Fazan, Z-RAM zero capacitor embedded memory technology addresses dual requirements of die size and scalability, Innovative Silicon Inc (2005). http://clients.concept-web.ch/is/en/technology_white_paper.php#
32. G. Sankar Kar, Capacitorless DRAM Cell on IGZO Base Shows Promising Values, EETimes Europe (2021)
33. Q. Hu, et al., True nonvolatile high-speed DRAM cells using tailored ultrathin IGZO, in *Advanced Materials, Wiley Online Library* (2023)
34. B. Dipert, EEPROM, survival of the fittest, in *EDN* (1998), pp 77–90
35. T. Cibean, Micron's 232-layer NAND starts shipping with up to 2TB per package, TECHSPOT (2022). https://www.techspot.com/news/95431-micron-232-layer-nand-starts-shipping-up-2tb.html
36. B. Kim, A high-performance 1Tb 3b/Cell 3D-NAND flash with a 194MB/s write throughput on over 300 layers, in *ISSCC Digest of Technical Papers* (2023), pp. 402–403
37. S. Aritome, *NAND Flash Memory Technologies*, (Wiley, Hoboken, 2015)
38. Y. Cai et al., Threshold voltage distribution in MLC NAND flash memory: Characterization, analysis, and modeling, in *Proceedings of the Conference on Design, Automation and Test* (2013), pp. 1285–1290
39. JEDEC Solid State Technology Association, Stress-Test-Driven Qualification of Integrated Circuits, JESD47G.01 (2010). http://www.jedec.org/
40. E. Korczynski, 3D-NAND Deposition and Etch Integration, Semiconductor Manufacturing & Design Community (2016). http://semimd.com/blog/tag/3d-nand/
41. K. Gibb, First Look at Samsungś 48L 3D V-NAND flash, EETIMES (2016)
42. D. Kang et al., 256Gb 3b/Cell V-NAND flash memory with 48 stacked WL layers, in *ISSCC Digest of Technical Papers* (2016) , pp. 130–131
43. B. Moyer, How It's Built: Micron/Intel 3D NAND, EEJournal (2016). https://www.eejournal.com/article/20160201-micron/
44. A. Khakifirooz et al., A 1Tb 4b/Cell 144-Tier floating-gate 3D-NAND flash memory with 40MB/s program throughput and 13.8Gb/mm 2 bit density, in *2021 IEEE International Solid-State Circuits Conference (ISSCC)* (2021)
45. A. Shilov, Kioxia and WD to Present Details on 3D NAND With 300+ Layers, Tom's Hardware (tomshardware.com) https://www.tomshardware.com/news/kioxia-and-wd-present-details-on-3d-nand-with-300-layers. 04 May 2023
46. A. Khakifirooz et al., A 1.67Tb, 5b/cell flash memory fabricated in 192-layer floating gate 3D-NAND technology and featuring a 23.3Gb/mm2 bit density, in *ISSCC Digest of Technical Papers* (2023), pp. 399–400
47. J. Handy, How Samsung Will Improve 3D NAND Costs, The Memory Guy Blog. https://thememoryguy.com/tag/v-nand/. August 2017
48. R. Merritt, 3D XPoint Steps Into the Light, EETIMES (2016)
49. Timothy Prickett Morgan, 3D XPoint Memory At The Crossroads, The NextPlatform (2021)
50. B. Tallis, Samsung at Flash Memory Summit: 64-layer V-NAND, Bigger SSDs, Z-SSD. http://www.anandtech.com/show/10560/ 11 August 2016
51. B. Dipert, FRAM: ready to ditch niche?, in *EDN* (1997), pp. 93–107
52. FRAM RFID LSI for UHF TagSolutions and Embedded RF Solutions, Fujitsu Application Notes
53. J. Muller et al., Ferroelectric hafnium oxide based materials and devices: assessment of current status and future prospects. ECS J. Solid State Sci. Technol. **4**(5), N30–N35 2015

54. J. Müller et al., Ferroelectric hafnium oxide: a CMOS-compatible and highly scalable approach to future ferroelectric memories, in *International Electronic Device Meeting (IEDM)* (2013), pp. 280–283
55. T.P. Ma, FEDRAM: A capacitor-less DRAM based on ferroelectric-gated field-effect transistor, in *2014 IEEE 6th International Memory Workshop (IMW)* (2014)
56. Z. Xizhen, FeFET logic circuits for operating a 64 kb FeNAND flash memory array. Integr. Ferroelectr. **132**(1), 114–121 (2016)
57. R. Wilson, MRAM Steps to 16Mbit, Electronics Weekly (2004)
58. W.Y. Cho et al., A 0.18 μm 3.0 V 64 Mb nonvolatile phase-transistion random access memory (PRAM). IEEE J. Solid-State Circuits **40**, 293–300 (2005)
59. S. Kang et al., A 0.1 μm 1.8 V 256 Mb 66 MHz synchronous burst PRAM, in *ISSCC Digest of Technical Papers* (2006), pp.140–141
60. M. LaPedus, Samsung to ship MCP with phase-change, EE|Times 28 April 2010
61. G.W. Burr et al., Phase change memory technology. J. Vacuum Sci. Technol. B **28**(2), 223–262 (2010)
62. Kurzweil AI, Will phase-change memory replace flash memory? (2013)
63. X. Zhou et al., Phase transition characteristics of Al-Sb phase change materials for phase change memory application. Appl. Phys. Lett. **103** (2013)
64. M. Mohseni et al., A survey on techniques for improving Phase Change Memory (PCM) lifetime, Elsevier. J. Syst. Archit. **144**, 103008 (2023)
65. X. Chen et al., Buffer-enhanced electrical-pulse-induced-resistive memory effect in thin film perovskites. Jpn. J. Appl. Phys. **45**(3A), Part 1, 1602–1606 (2006)
66. P. Clarke, Resistive RAM sets chip companies racing, EETimes, 24 April 2006
67. H.Y. Lee et al., Evidence and solution of over-RESET problem for HfO_x based resistive memory with sub-ns switching speed and high endurance, in *Procedings of the IEDM* (2010), pp.19.7.1–19.7.4
68. L. Goux et al., Ultralow sub-500nA operating current high-performance TiN\Al2O3\HfO2\Hf\TiN bipolar RRAM achieved through understanding-based stack-engineering, in *Digest of Technical Papers., Symposium on VLSI Circuits* (2012), pp. 159
69. T.-Y. Liu, T. Hong Yan et al., A 130.7 mm^2 two-layer 32-Gbit ReRAM memory device in 24-nm technology, in *2013 IEEE International Solid-State Circuits Conference Digest of Technical Papers 12.1* (2013)
70. C. Wang et al., Correction: Research progress in architecture and application of RRAM with computing-in-memory, Royal Society of Chemistry, Nanoscale Advances, Issue 7 (2023)
71. H. Hönigschmid et al., A non-volatile 2 Mbit CBRAM memory core featuring advanced read and program control, in *Proceedings of the 2006 Symposium on VLSI Circuits* (2006), pp. 138–139
72. H. Abbas et al., Conductive bridge random access memory (CBRAM): challenges and opportunities for memory and neuromorphic computing applications. MicroMachines **13**, 725 (2022)
73. W.J. Donath, Placement and average interconnections lengths of computer logic. IEEE Trans. Circ. Syst. **26**(4), 272 (1979)
74. Semiconductors Industrial Associations, ITRS Roadmap, yearly update. http://www.itrs.net
75. A. Fazio et al., ETOXTM Flash Memory Technology: Scaling and Integration Challenges. http://developer.intel.com/technology/itj/2002/volume06issue02/art03_flashmemory/vol6iss2_art03.pdf. 16 May 2002
76. Linley Gwennap, Day dawns for eDRAM, EETimes. http://www.eetimes.com/op/showArticle.jhtml?articleID=16500906. 14 April 2003
77. What is in-memory computation? https://hazelcast.com/glossary/in-memory-computation/
78. In-memory processing, https://en.wikipedia.org/wiki/In-memory_processing
79. M. Ali et al., Compute-in-memory technologies and architectures for deep learning workloads. IEEE Trans. Very Large Scale Integr. Syst. **30**(11), 1615–1629 (2022)

80. A. Sebastian, How the future of computing might look like nothing before. https://research.ibm.com/blog/abu-sebastian-ieee-in-memory-computing. 11 Jan 2023
81. B. Moyer, More Errors, More Correction In Memories; New technologies increase the cost of accuracy as density increases Semiconductor Engineering (2021)
82. V. Sontakke et al., Memory built-in self-repair and correction for improving yield: a review. Int. J. Electr. Comput. Eng. **14**(1), 140–156 (2024)

Further Reading

83. B. Prince, *Semiconductor Memories: A Handbook of Design, Manufacture and Application* (Wiley, New York, 1996)
84. W.J. McClean, Status 1999, A report on the IC industry, ICE Corporation, Scottsdale, Arizona (1999)
85. B. Prince, *High Performance Memories* (Wiley, New York, 1996)
86. IEEE digest of technical papers of the International Solid State Circuit Conference. The ISSCC is held every year in February in San Francisco
87. IEEE Journal of Solid-State Circuits. This is a general reference to a monthly magazine on Integrated circuits
88. IEDM Digest of Technical Papers. The IEDM is a conference held every year on different locations in the world

Very-Large-Scale Integration (VLSI) and ASICs 7

7.1 Introduction

The continuing development of IC technology during the last couple of decades has led to a considerable increase in the number of devices per unit chip area. The resulting feasible IC complexity currently allows the integration of a complete *system on a chip (SOC)*, which may comprise hundreds of millions to even more than a hundred billion transistors.

Consequently, the design of such chips no longer simply consists of the assembly of a large number of logic gates. This poses a problem at a high level of design: how to manage the design complexity. Besides this, the growing influence of parasitic and scaling effects (see Chaps. 2, 9 and 11), which may reduce chip performance dramatically, requires a lot of additional design resources to take and implement adequate measures.

Such ICs combine signal processing capacity with microprocessor or microcontroller cores and memories. The dedicated signal processing parts take care of the computing power (workhorse), while the microprocessor or controller serves to control the process and possibly performs some low-performance computation as well. The memories may store programme code and data samples. Finally, since the world is analog, most ICs also contain on-chip analog interface and pre- and post-processing circuits as well as an increasing number of wireless interfaces. The development of such heterogeneous systems on one or more ICs, for instance, may require tens to even hundreds of man-years, depending on their complexity. Microprocessors for standard PCs and servers, usually referred to as mainstream *MPUs* (Intel and AMD processors), may even require several thousand man-years of development time.

A significant amount of the total IC turnover is generated in the "low-end market". This market consists of low-complexity ICs and was originally controlled by the large IC vendors. During the 1980s and 1990s, however, a change took place,

H. Veendrick, *Nanometer CMOS ICs*,
https://doi.org/10.1007/978-3-031-64249-4_7

and the low-end market is now dominated by *application-specific integrated circuits (ASICs)*. These are ICs which are realised for a single end-user and dedicated to a particular application. ASICs therefore implement customer-specified functions, and there are various possibilities for the associated *customisation*. This can be an integral part of an IC's design or production process, or it can be accomplished by programming special devices.

ASICs do not include ICs whose functionality is solely determined by IC vendors. Examples of these *"application-specific standard products" (ASSPs)* include a standalone USB interface chip but also Nvidia Grace Blackwell 200 (GB200) "superchip" for AI applications. These ASSPs are so-called vendor-driven ICs, of which the vendor wants to sell as many as possible to every customer he can find. ASICs are customer-driven ICs, which are only tailored to the specific requirements of one single customer. Actually, *user-specific integrated circuits (USICs)* would be a more appropriate name for ASICs. The use of USICs would clearly be preferable because it emphasises the fact that the IC function is determined by the customer's specification and not simply by the application area.

The *turnaround time* of an ASIC is the period which elapses between the moment a customer supplies the database and the moment the vendor supplies the first samples. The turnaround time associated with an ASIC depends on the chosen implementation type. A short turnaround time facilitates rapid prototyping and is important to company marketing strategies. In addition, ASICs are essential for the development of many real-time systems, where designs can only be verified when they are implemented in hardware. There exist many different market segments for which we can distinguish different ASIC products:

- Automotive: networking, infotainment, GPS, tire pressure monitor, body electronics
- Mobile communications: mobile/smart phones (GSM, UMTS), tablets, modems, wireless local loop (WLL), GPS
- Medical: patient monitoring, diagnostics, ultrasound
- Display: LCD TV, flat panel, projection TV
- Digital consumer: CD/DVD, MP3, audio, TV, media box, set-top box, encoders/decoders
- Connectivity: WLAN, Bluetooth, USB, NFC, FireWire
- Identification: smart cards, electronic car keys, e-passports and RF-ID tags, such as animal tags and product tags
- Industrial: robotics, motor/servo control
- Military: image, radar and sonar processing, navigation
- AI (artificial intelligence): pattern recognition, chatbots, face recognition, machine learning, neural networks

Suitable *computer-aided design (CAD)* tools are therefore essential for the realisation of this rapidly expanding group of modern ICs. Growing design complexity combined with shorter product market windows requires the development of an efficient and effective design infrastructure, based on a (application-) domain-specific *SoC design platform*. In this respect, a platform is an integrated design environment, consisting of standard-cell libraries, IPs and application-mapping tools, which is aimed at providing a short and reliable route from high-level specification to correct silicon. The convergence of consumer, computing and communications domains accelerates the introduction of new features on a single chip, requiring a broader range of standards and functions for an increasing market diversity. This makes a design more heterogeneous, with a large variety of domain-specific, general-purpose IP and memory cores. Next to this, there is a tremendous growth in the complexity of embedded software, which may take more than 50% of the total SoC development costs, particularly in multiprocessor design.

This puts very high demands on the flexibility and reusability of a platform across a wide range of application derivatives, requiring a large diversity of fast-compiling IPs in combination with efficient verification, debug and analysis tools. Such a platform needs to be scalable and must also enable the addition of new IP cores without the need for changing the rest of the system.

The design process is discussed on the basis of an ASIC design flow. The various implementation possibilities for digital VLSI and ASICs are discussed, and factors that affect a customer's implementation choice are examined. These implementations include standard-cell, gate array, field-programmable gate array (FPGA) and programmable logic devices (PLD). Market trends and technological advances in the major ASIC sectors are also explained.

7.2 Digital ICs

Digital ICs can be subdivided into different categories, as shown in Fig. 7.1. ASICs can be classified according to the processing or programming techniques used for their realisation. A clear definition of the types and characteristics of available digital ICs and ASICs is a prerequisite for the subsequent discussion of the trends in the various ASIC products. Figure 7.1 presents quite a broad overview of digital ICs but excludes details such as the use of *direct slice writing (DSW)* or masks for IC production. Several terms used in this figure and throughout this chapter are explained on the next pages.

Definitions

ASSP: *Application-specific standard products* are ICs that are suitable for only one application, but their availability is not restricted to a single customer. Examples include video ICs for teletext decoding and ICs for D/A conversion in DVD players.

Core: Predesigned industry (or company) widely used standard building block: RAM, ROM, microprocessor (e.g. ARM, MIPS and SPARC), graphics processing unit (GPU), interfaces (Bluetooth, USB and NFC), etc.

Custom: A *custom IC* is an IC in which all masks are unique for a customer's application. The term *full-custom IC* is often referring to an IC in which many subcircuits are new handcrafted designs. In this book, full-custom ICs fall under

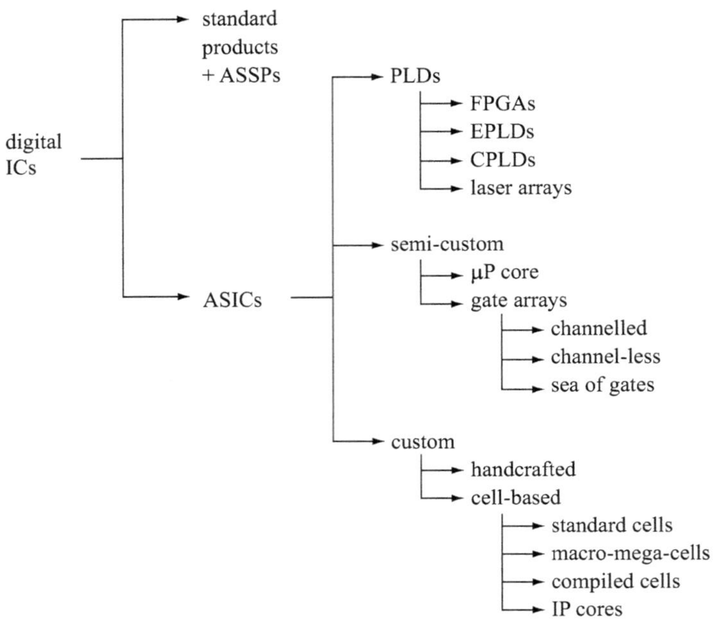

Fig. 7.1 An overview of digital ICs

the category of custom ICs. Cell-based custom IC designs are based on standard cells, *macro cells*, *mega cells* and possibly *compiled cells*. Macro and mega cells or cores are large library cells like multipliers, RAMs, ROMs and even complete microprocessor and signal processor cores. Compiled cells are automatically generated by modern software libraries. These cells are used for dedicated applications and are generated as a function of user-supplied parameters.
The customisation of PLD-based ASICs takes place after IC manufacture. Customisation of custom and semi-custom ASICs, however, is an integral part of IC manufacture. The *turnaround time* of ASICs from database ready to first silicon varies enormously and depends on circuit complexity and the customisation technique. This time can range from a few hours for a PLD to between 6 and 12 weeks for a custom design.

FPGA: A *field-programmable gate array* is an IC that has the ability to change its functionality after the manufacture. It contains programmable logic and programmable routing channels. It belongs to the group of ICs that is usually referred to as programmable logic devices (PLDs).

HDL: *Hardware description language*. This language is used for formal description of the structure and behaviour of electronic circuits. It provides the circuit designer to describe (model) a circuit before it is physically implemented. Verilog and VHDL have become the two most popular HDLs for coding the design of integrated circuits. Synthesis tools are able to read these HDL codes, extract logic operation, and transfer these into a netlist of logic gates.

IP: *Intellectual Property*. With the complexity of ICs reaching a billion or more transistors, the traditional way of designing can no longer be continued. Therefore, the concept of *virtual component* has been introduced in 1996 by the *Virtual Socket Interface Alliance (VSI* Alliance: www.vsi.org), which was an international forum trying to standardise reusable cores, concepts, interfaces, test concepts, support, etc. Licensing and royalty issues of IP were also addressed. Due to the low efficiency of the alliance to create standards for the development of IP cores, VSIA was dissolved in 2008. However, this standardisation is a prerequisite to fully exploit the potentials of design reuse. The cores (or IP) can be represented in three forms.

A *soft core* is delivered in the form of synthesizable HDL and has the advantage of being more flexible and the disadvantage of not being as predictable in terms of performance (timing, area, power). Soft cores typically have increased intellectual property protection risks because RTL source code is required by the integrator.

Firm cores have been optimised in structure and in topology for performance and area through floor planning and placement, possibly using a generic technology library. The level of detail ranges from region placement of RTL sub-blocks to relatively placed data paths to parameterised generators to a fully placed netlist. Often, a combination of these approaches is used to meet the design goals. Protection risk is equivalent to that of soft cores if RTL is included and is less if it is not included.

Finally, *hard cores* have been optimised for power, size or performance and mapped to a specific technology. Examples include netlists fully placed, routed and optimised for a specific technology library, a custom physical layout or the combination of the two. Hard cores are process- or vendor-specific and generally expressed in the GDSII format. They have the advantage of being much more predictable but are consequently less flexible and portable because of process dependencies. The ability to legally protect hard cores is much better because of copyright protections and there is no requirement for RTL.

Figure 7.2 is a graphical representation of a design flow view and summarises the high-level differences between soft, firm and hard cores.

Due to the convergence of digital communications, consumer and computer, there are an increasing number of real-time signals to be processed: voice, professional audio, video, telephony, data streams, Internet of Things (IoT), etc. This processing is usually performed by high-performance analog and digital signal processors.

Today's integrated circuits are complex heterogeneous systems: they consist of many different types of processing, storage, control and interface elements. Many of these elements are available as a kind of (standard) IP. Examples of IP are:

- Microprocessors (CPU): use software to control the rest of the system
  - Intel Xeon & Itanium, Oracle SPARC, IBM Power10, ARM, MIPS, 80C51, TI TMS9900, etc.
- Digital signal processors (DSP): manipulate audio, video and data streams

  - Omaps, TMS320 and DaVinci (TI), DSP56000 series (Freescale), DSP16000 series (Agere), EPICS and Trimedia and EPICS (NXP), Oak, TeakLite, ARC VPX (Synopsys).
  - Most DSPs are for wireless products.
- (F)PGA-based accelerators: decoders, encoders, error correction, encryption, graphics or other intensive tasks
- Memories
  - Synopsys, Artisan, embedded memories and caches
  - Memory controllers (Denali): controlling off-chip memories
- Interfaces: external connections
  - USB, FireWire, Ethernet, UART, Bluetooth, NFC, keyboard, display or monitor
- Analog
  - A/D, D/A, PLL (e.g. for use in clock generation), oscillator, operational amplifier, differential amplifier, bandgap reference, SerDes, PHYs

	Design flow	Representation	Libraries	Technology	Portabillity
Soft not predictable very flexible	system design RTL design	behavioural RTL	N/A	technology independent	unlimited
Firm flexible predictable	floor planning synthesis placement	RTL & blocks netlist	reference library • footprint • timing model • wiring model	technology generic	library mapping
Hard not flexible very predictable	routing verification	polygon data	process-specific library & design rules • characterised cells • process rules	technology fixed	process mapping

Fig. 7.2 Graphical representation of soft, firm and hard cores (Source: VSIA)

PLD: The first *programmable logic devices* were customised by fuses or antifuses. Modern PLDs are programmed by on-chip memory cells. Most PLDs can be customised by end users themselves in the field of application, i.e. they are *field-programmable* devices (FPGA). The customisation techniques used are classified as *reversible* and *irreversible*. PLDs include *erasable* and *electrically erasable* types, which are known as *EPLDs* and *EEPLD*, respectively. The former are programmed using EPROM techniques, while the EEPROM programming technique is used for the latter devices. These programming techniques are explained in Sects. 6.5.3 and 6.5.4, respectively. Complex PLDs (CPLDs) are often based on the combination of PAL™and PLA architectures.

Reuse: Future design efficiency will increasingly depend on the availability of a variety of predesigned building blocks (IP cores; see IP definition). This *reuse* requires easy portability of these cores not only between different ICs but also between different companies and between different process nodes. Standardisation is one important issue, here (see IP definition). Another important issue concerning reuse is the quality of the (IP) cores. Similar to the Known Good Die (KGD) principle when using different ICs in an MCM, we face a Known Good Core (KGC) principle when using different cores in one design. The design robustness of such cores must be so high that their correctness of operation will always be independent of the design in which it is embedded.

RTL: Register transfer level. See Sect. 7.3.4.

Semi-custom: These are ICs in which one or more but not all masks are unique for a customer's application. Many semi-custom ICs are based on "off-the-shelf" ICs which have been processed up to the final contact and metal layers. Customisation of these ICs therefore only requires processing of these final contacts and metal layers. This results in short turnaround times. A gate array is an example in this semi-custom category.

Standard product: Standard products, also called *standard commodities*, include microprocessors, memories and standard logic ICs, e.g. NAND, NOR, QUAD TWO-INPUT NAND. These ICs are produced in large volumes and available from different vendors. Their availability is unrestricted, and they can be used in a wide variety of applications. They are often put into a product catalogue.

Usable gates: The number of gates in a PLD or (mask-programmable) gate array that can actually be interconnected in an average design. This number is always less than the total number of available gates.

Utilisation factor: The ratio between that part of a logic block area which is actually occupied by functional logic cells and the total block area (gate array and cell-based designs).

7.3 Abstraction Levels for VLSI

7.3.1 Introduction

Most of today's complex VLSI designs and ASICs are synchronous designs, in which one or more clock signals control the data flow to, on and from the chip. On the chip, the data is synchronised through flip-flops, which are controlled by a clock ϕ (Fig. 7.3). Flip-flops temporarily store the data and let it go on clock demand. At any time, the positions and values of all data samples are known (by simulations).

The logic gates in between the flip-flops perform the functionality of the logic block from which they are part. So, in a synchronous chip, the signal propagates through the logic path from one flip-flop to the next. The logic path with the longest propagation delay (usually one with many complex gates and/or large wire delays) is called the *worst-case delay path*. This path determines the maximum allowed clock frequency. Next to many different functional logic blocks, most systems also contain memory, interface and peripheral blocks.

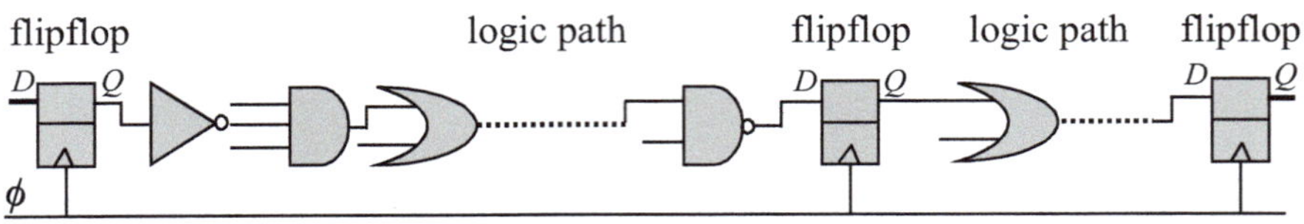

Fig. 7.3 Representation of a logic path in a synchronous design

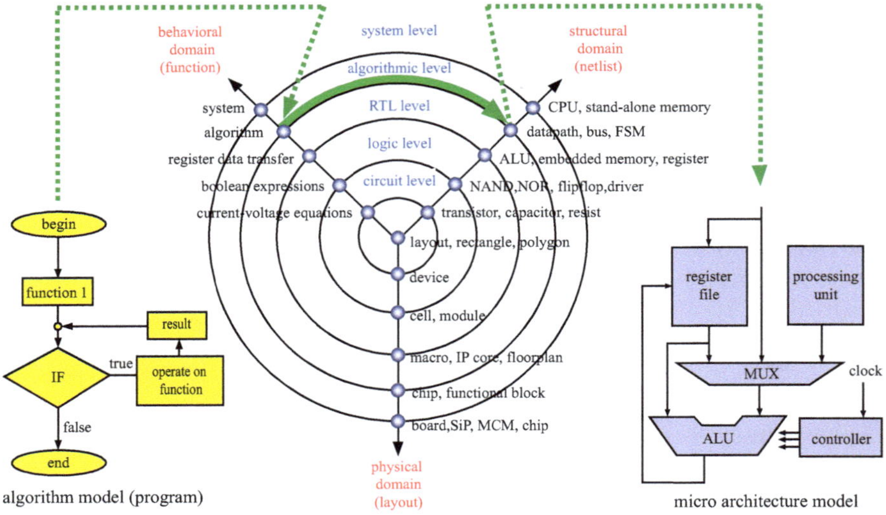

Fig. 7.4 Gajski-Kuhn VLSI design abstraction-level chart

The implementation of a complete system on one or more ICs starts with an abstract system-level specification. This specification is then analysed and transformed into a set of algorithms or operations. Next, an optimum architecture that efficiently performs these operations must be chosen. A model that represents the different abstraction levels is the *Gajski-Kuhn chart*, which is named after the two developers in 1983 (Fig. 7.4).

It distinguishes three domains of VLSI design representation: a behavioural, a structural and a geometrical domain. At the design start, a behavioural description is provided. Due to its high abstraction level, it does not contain any information on the design structure, whether it is synchronous/asynchronous, and no timing constraints are considered. Let's take the example of an elevator function: the elevator (Z) goes up when its door (c) is closed and when somebody (a) in the elevator or somebody (b) on another floor has pushed a button. Then its function in the behavioural domain could be described as follows: El (Z) goes up when door (c) is closed AND (button push (a) in elevator OR button push (b) on other floor).

Its structural and physical representations are shown in Fig. 7.5. A structural description describes the system as a collection of components and their

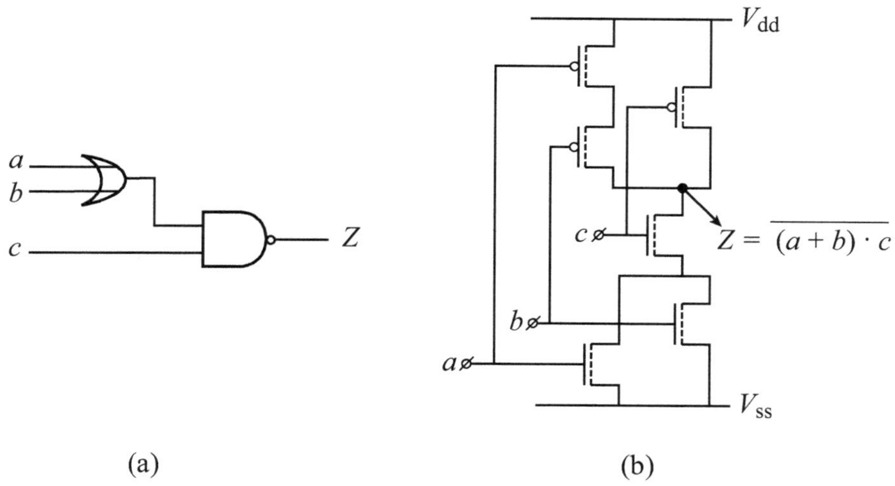

Fig. 7.5 Structural (**a**) and physical (**b**) representation of the elevator function example

interconnections, while the physical description relates to the basic devices and interconnections.

Each of the domains in Fig. 7.4 is divided into five levels of abstraction, represented by concentric rings. Starting with system level, at the outer ring, the design details refine as we go more and more towards the centre point of the diagram, ending at the layout level.

At the *system level*, the basic specification of an electronic system is determined. Usually, the system at this level is represented by one or more block diagrams.

The *algorithmic level* specifies how the data in the system is manipulated (processed and stored) so that the system does what it has to do.

At *register-transfer level*, the behaviour is described in more details as communication between registers. Figure 7.6 shows an example representation of a function at algorithmic level and a microarchitecture model of the same at RTL level.

It is clear that the microarchitecture is much closer to the real on-chip implementation.

We will use a signal processor as an example function to describe and explain the various abstract levels in the following (sub) sections. The chosen processor must perform an adaptive FIR filter. As a consequence, this processor must repeatedly fetch numbers from a memory, multiply or add them and then write the result back into the memory. Such a chip may contain several ROM and/or RAM memory units, a multiplier, an adder or accumulator, data and control buses and some other functional modules.

The design of an IC comprises the transformation of a specification into a layout. The layout must contain all pattern shapes in every mask layer needed to fabricate

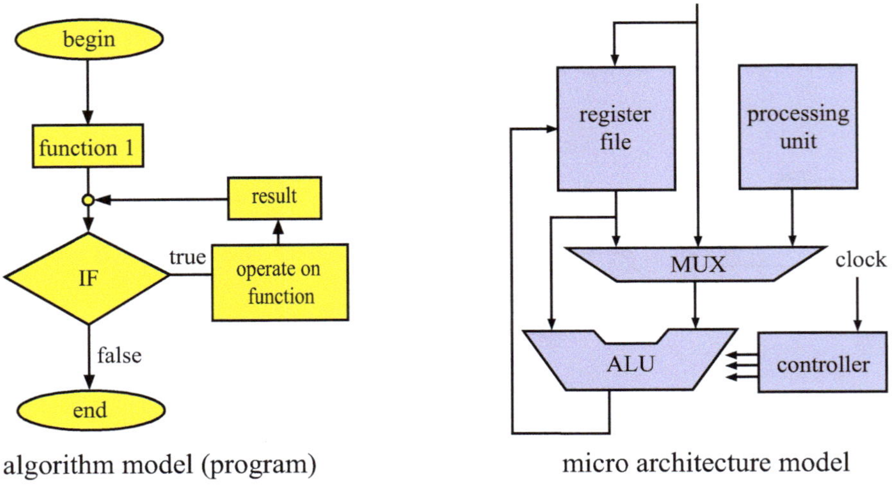

Fig. 7.6 Representation of a function at algorithmic and RTL level

Fig. 7.7 Abstraction levels in the design and implementation/verification paths of VLSI circuits

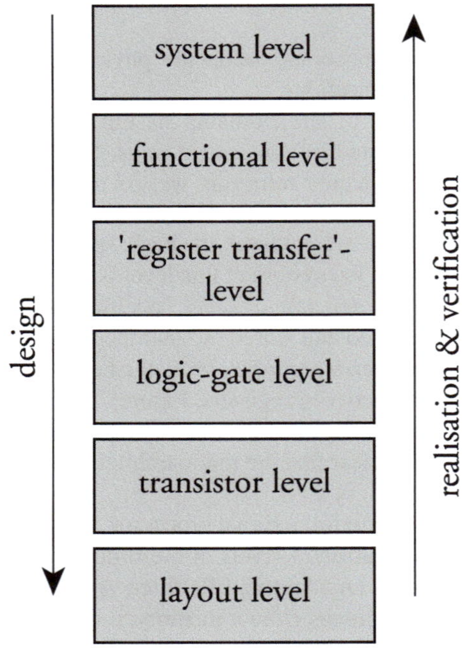

the chip. Clearly, the *design path* starts at the top (or system) level and ends at the bottom (or silicon) level. This *top-down* process is illustrated in Fig. 7.7.

The various design phases are accompanied by several different *abstraction levels*, which limit the complexity of the relevant design description. The top-down

Table 7.1 Design complexity at different levels of abstraction

Level	Example	Number of elements	
System	Heterogeneous system	10^8–10^{10}	transistors
Functional	Signal processor	10^5–10^9	transistors
Register	Digital mixer	10^3–10^5	transistors
Logic gate	Library cell (NAND, full adder)	2–30	transistors
Transistor	nMOSt, pMOSt	1	transistor
Layout	Total SoC	10^8–10^{11}	rectangles

design path allows one to make decisions across abstraction levels and gives high-level feedback on specifications. The *bottom-up* path demonstrates the feasibility of the implementation of (critical) blocks. This process begins at the layout level of a single part and finishes with the verification of the entire IC layout. The abstraction levels that are used in the design path are described on the following pages. Table 7.1 shows the design complexity at these levels of abstraction.

The number of transistors in the right column refers to a state-of-the-art complex system built from various complex ICs.

7.3.2 System Level

A system is defined by the specification of its required behaviour. Such a system could be a multiprocessor system and/or a *heterogeneous system*, consisting of different types of processing elements: microprocessor, DSP, analog, control, peripheral and memory cores. Advanced heterogeneous architectures, today, also include the integration of graphics processing units (GPU) to increase graphics processing speed by one or two orders of magnitude, compared to running it on a CPU. Figure 7.8 shows a heterogeneous system, containing a signal processor, a microprocessor (IP core), embedded software, some glue logic (some additional overall control logic), local buses, a global bus and the clock network. The transformation of a system into one or more ICs is subject to many constraints on timing, power and area, for example.

While a heterogeneous system consists of several different types of processing and storage elements, there is today also an increased focus on architectures with multiprocessor cores and even architectures built from only a limited number of different cores, such as in AI applications. In the ultimate case, an architecture can be built from a multiple of identical cores (tiles) to create a *homogeneous sytem*. Figure 7.9 (top) shows a layout of a massively parallel processor for video scene analysis implemented as a homogeneous design [1], as opposed to the heterogeneous chip (bottom).

System decisions taken at the highest level have the most impact on the area and performance parameters. Decisions regarding functions that are to be implemented in hardware or software are made at the system level. Filter sections, for example,

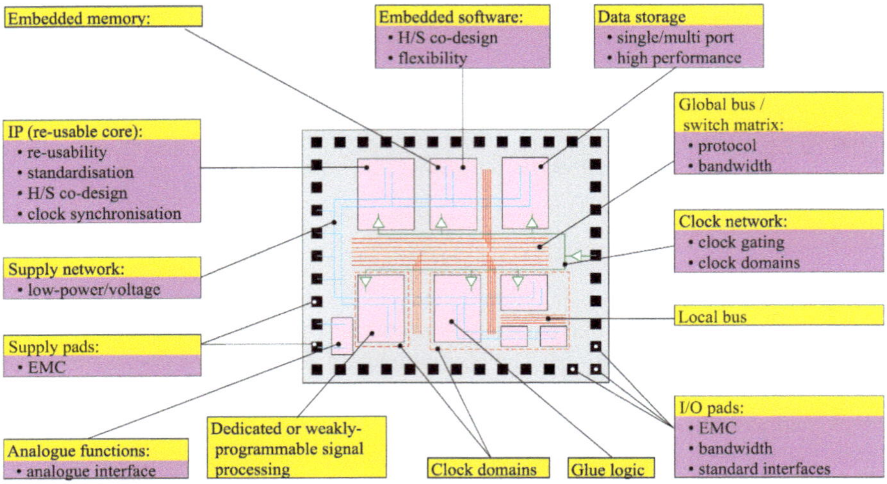

Fig. 7.8 Systems on a chip; an example of a heterogeneous system

are frequently programmed in software. A system-level study should also determine the number of chips required for the integration of the chosen hardware. It is generally desirable to subdivide each chip into several sub-blocks. For this purpose, *data paths* and *control paths* are often distinguished. The former is for data storage and data manipulation, while the latter controls the information flow in the data path and to and from the outside world. Each block in the data path may possess its own *microcontrol unit*. This usually consists of a decoder which recognises a certain control signal and converts it into a set of instructions.

The block diagram shown in Fig. 7.10 represents a description of the weakly programmable signal processor of Fig. 7.8 at the system abstraction level. The double-bus structure in this example allows parallel data processing. This is typically used when very high throughputs are required. For example, data can be loaded into the arithmetic logic unit (ALU) simultaneously from the ROM and the RAM. In this type of architecture, the data path and control path are completely separated. The control path is formed by the program ROM, which may include a programme counter, control bus and the individual microcontrol units located in each data path element.

Other system implementations may not show such a clear separation of data and control paths.

7.3.3 Functional Level

A description at this level of abstraction comprises the behaviour of the different processing elements and other cores of the system. In case of the signal processor

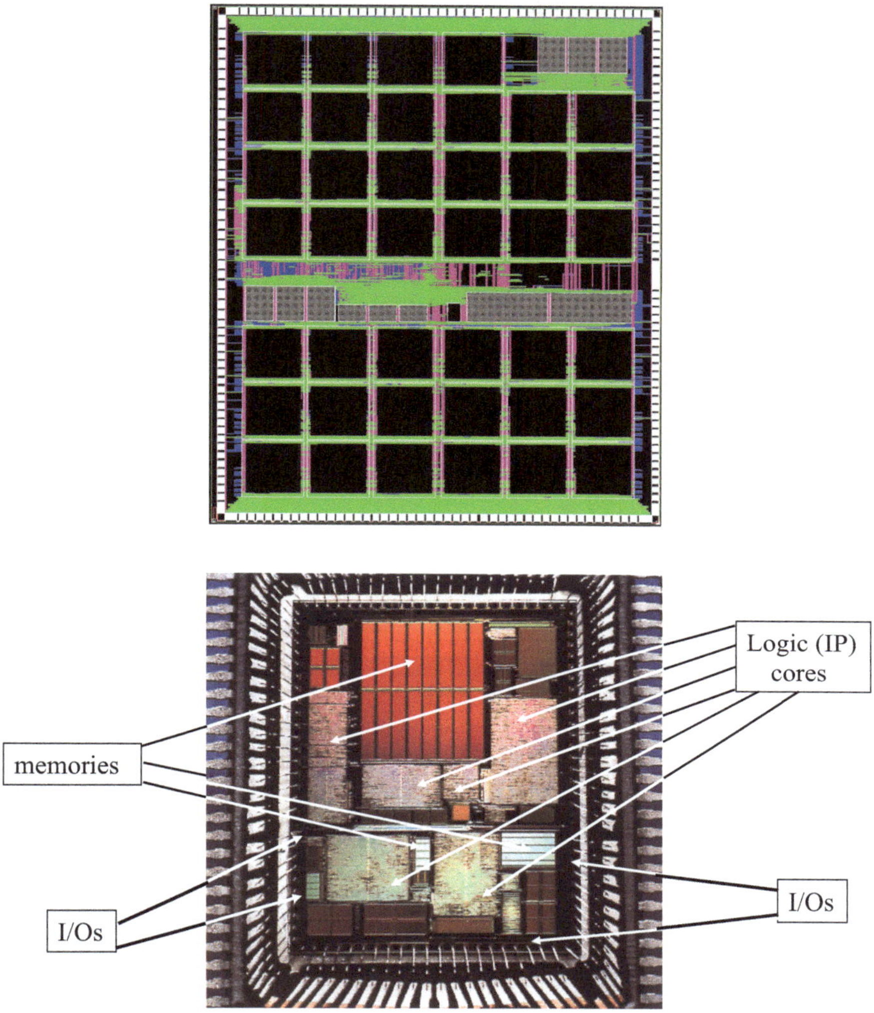

Fig. 7.9 Example of a homogeneous design, consisting of a multiple of identical cores (tiles) and a heterogeneous chip consisting of various different cores (Source: NXP Semiconductors)

of Fig. 7.10, we distinguish an ALU, a digital mixer, a RAM, an ROM and the I/O element.

RAMs, ROMs and I/O elements are usually not very complex in their behaviour. As a result of the simplicity of their behaviour, however, they are mostly described in the next, lower level of abstraction, the RTL level.

Let us take the digital mixer as an example. Also, this one, because of its simple architecture, will be described at the lower RTL level.

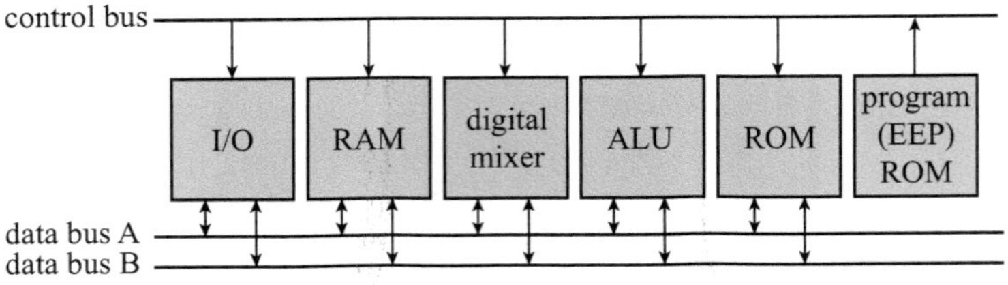

Fig. 7.10 Block diagram of a signal processor

There are some tools, e.g. MATLAB, Simulink and SystemC, that allow a description of complex blocks at functional level. They allow high-level system evaluation and verification in different use cases across different hierarchy levels and exploration of alternative solutions for certain functions.

The chosen mixer, at this hierarchy level (RTL level), consists of different arithmetic units (adder, multiplier, subtractor), which are functions as well, so the RTL level and functional level show some overlaps (see also Fig. 7.17).

7.3.4 RTL Level

RTL is an abbreviation for *register-transfer language*. This notation originates from the fact that most systems can be considered as collections of registers that store binary data, which is operated upon by logic circuits between these registers. The operations can be described in an RTL and may include complex arithmetic manipulations. The *RTL description* is not necessarily related to the final realisation.

To describe a function at this level is a difficult task. A small sentence in the spec, e.g. *performs MPEG4 encoding*, will take many lines of RTL code, and its verification is extremely difficult. Logic simulation and/or even emulation may help during the verification process but cannot guarantee full functionality, since it is simply impossible to fully cover all possible cases and situations. Let us return to our digital mixer example. The behaviour of this mixer can be described as:

$$Z = k \cdot A + (1 - k) \cdot B$$

The value of k determines the level of mixing of the signals A and B. When $k = 0$, Z will be equal to B, and when $k = 1$, Z will be equal to A. When k=$\frac{1}{2}$, then: $Z = \frac{1}{2} \cdot A + \frac{1}{2} \cdot B$.

The description does not yet give any information about the number of bits in which A, B and k will be realised. This is one thing that must be chosen at this level. The other choice to be made here is what kind of multiplier must perform the required multiplications. There are several alternatives for multiplier implementation, of which some are discussed as examples.

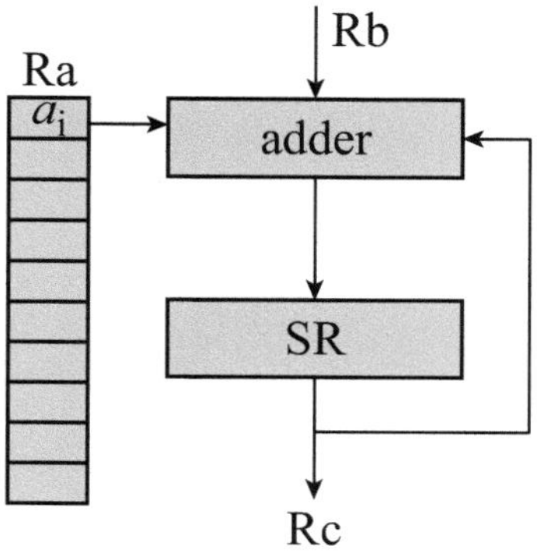

Fig. 7.11 Example of a bit-serial iterative multiplier

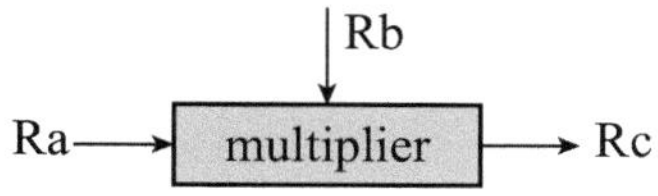

Fig. 7.12 A parallel multiplier

- *Serial-parallel multiplier:* Input Ra input is bit-serial, and the Rb input is bit-parallel; see Fig. 7.11.
  During the execution of a multiplication, the *partial product* is present on the multiplier's parallel output bits (Rc). These are initially zero.
  If a_i=1, for instance, then the Rb bits must be added to the existing partial product and then shifted one position to the left. This is a *shift-and-add operation*. When a_i=0, the Rb bits only have to be shifted one place to the left in a shift operation and a zero LSB added to it.
- *Parallel multiplier:* The bits of both inputs Ra and Rb are supplied and processed simultaneously. This *bit-parallel operation* requires a different hardware realisation of the multiplier. Options include the *array or parallel multiplier*, schematically presented in Fig. 7.12.
  The array multiplier necessitates the choice of a structure for the addition of the partial products. The possibilities include the following:
- *Wallace tree*: Here, bits with equal weights are added together in a tree-like structure; see Fig. 7.13. An advantage of the architecture is that the two input signals for each single adder always arrive at the same time, since they have propagated through identical delay paths. This will reduce the number of glitches at the outputs of the individual adder circuits, which may occur when there is too much discrepancy between the arrival times of the input signals.
- *Carry-save array*: Fig. 7.14 illustrates the structure of this array, which consists of AND gates that produce all the individual $x_i \cdot y_j$ product bits and an array of full adders which produce the total addition of all product bits.

As an example, at this level, we choose the array multiplier (parallel multiplier) with carry-save array. This would lead to a different behaviour from the serial multiplier and thus to a different RTL description.

An example of RTL-VHDL description for the mixer is given in Fig. 7.20.

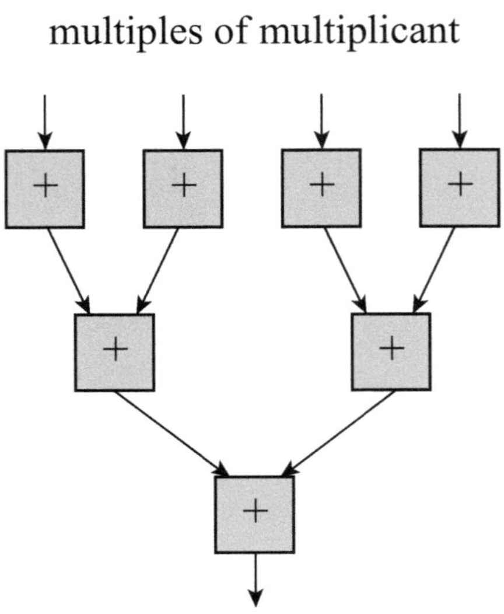

Fig. 7.13 Wallace tree addition

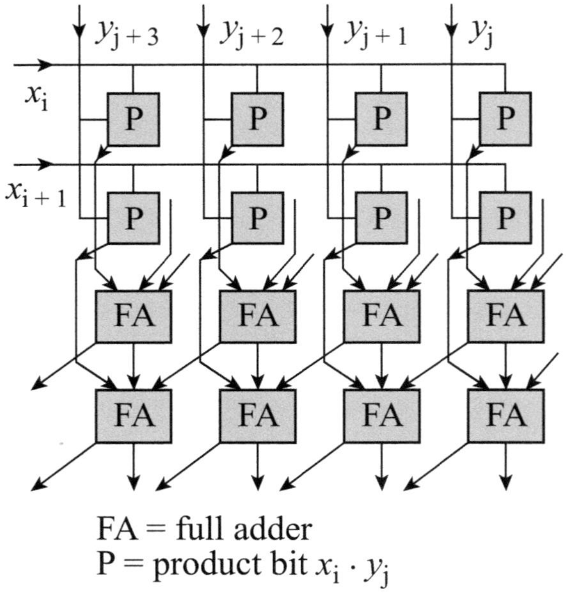

Fig. 7.14 Array multiplier (parallel multiplier) with carry-save array

7.3.5 Logic-Gate Level

As stated in Sect. 7.4, the RTL description is often specified through *hardware description languages (HDL)*, such as VHDL and Verilog. It is then mapped onto a library of cells (logic gates). This is done by a logic synthesis tool, which transforms a VHDL code into a netlist (see example in Fig. 7.27). A *netlist* contains a list of the library cells used and how they are connected to each other. Examples of such library cells (logic gates) are AND, NAND, flip-flop, full adder, etc. As an example of the decisions that need to be taken at this logic level, we choose the full adder, from which we will build the array multiplier of Fig. 7.14. A full adder performs the binary addition of three input bits (x, y and z) and produces sum (S) and carry (C) outputs. Boolean functions that describe the operation of a *full adder* include the following:

(a) Generation of S and C directly from x, y and z:

$$C = xy + xz + yz$$
$$S = x\overline{yz} + \overline{xy}z + \overline{x}y\overline{z} + xyz$$

(b) Generation of S from C:

$$C = xy + xz + yz$$
$$S = \overline{C}(x + y + z) + xyz$$

(c) Generation of S and C with exclusive OR gates (EXORs)

The choice of either one of these implementations depends on what is required in terms of speed, area and power. Implementation (b) will contain fewer transistors than (a) but will be slower because the carry must first be generated before the sum can evaluate. The implementation in (c) is just to show another alternative. Supposing our signal processor is used in a consumer video application where area is the most dominant criterion, then, at this hierarchy level, it is obvious that we choose implementation (b) to realise our full adder. A logic-gate implementation is shown in Fig. 7.15.

7.3.6 Transistor Level

At this level, the chosen full adder must be mapped onto a number of transistors. In some design environments, the logic-gate level is not explicitly present, and the higher-level code is directly synthesized and mapped onto a "sea of transistors". These are discussed in Sect. 7.6.6. The transistor-level description depends on the chosen technology and the chosen logic style, such as dynamic or static CMOS. For the realisation of our full adder, we choose a static CMOS implementation, as shown in Fig. 7.16.

As this full adder consists of a relatively low number of transistors (30), it is efficient, both in terms of area and power dissipation, compared to the one realised with AND, OR and INVERT gates in Fig. 7.15. Note that both sum S and carry C circuits are symmetrical with respect to their nMOS and pMOS transistor

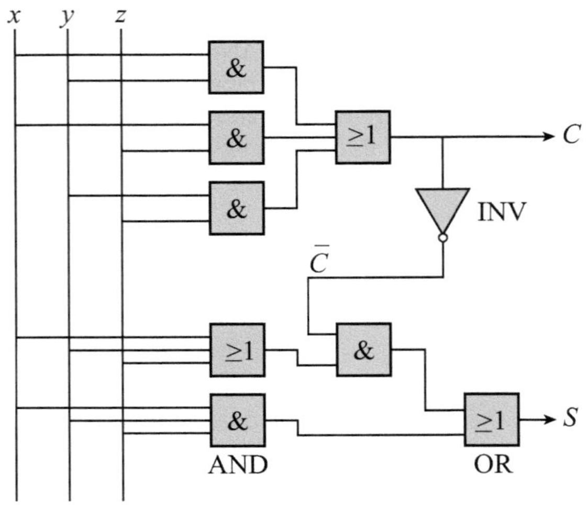

Fig. 7.15 Basic logic-gate implementation of a full adder

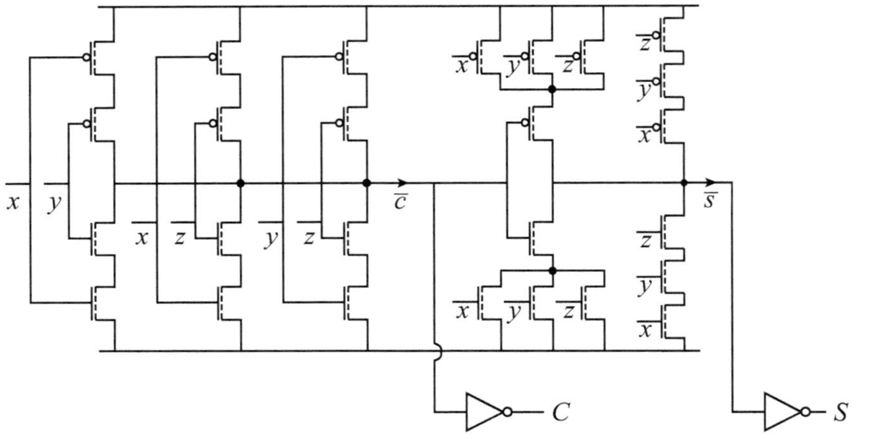

Fig. 7.16 Static CMOS realisation of the chosen full-adder cell

schematics, because the full adder is one of the few symmetrical logic functions, next to the half adder and the multiplexer.

Thus, the transistor-level implementation of the logic gate is determined by either speed, area or power demands, as is actually every IC implementation. In this example, we choose the implementation of Fig. 7.16 for our full adder.

7.3.7 Layout Level

The chosen transistor implementation must be translated into a layout-level description at the lowest abstraction level of a design. Most of the time, these layouts are made by specialists, who develop a complete library of different cells in a certain technology. To support high-performance, low-power and low-leakage applications, today, a library may consist of 1500 different cells. There may be different cell versions of the same logic function but with a different drive strength, a different threshold voltage and/or a different gate oxide thickness. However, special requirements on high speed or low power may create the need for custom design to optimise (part of) the chip for that requirement. In Chap. 4, the layout process is explained in detail.

7.3.8 Conclusions

As shown in the signal processor example before, in the top-down design path, decisions about different possible implementations have to be made at each level. In this way, a *decision tree* arises. Figure 7.17 shows an example of a decision tree for the previously discussed signal processor system.

The decision tree starts at the highest level, i.e. the system level. Every time we move one level down in the tree, we focus on a smaller part of the design, which allows us to add sufficient details to take the right decision at this level and

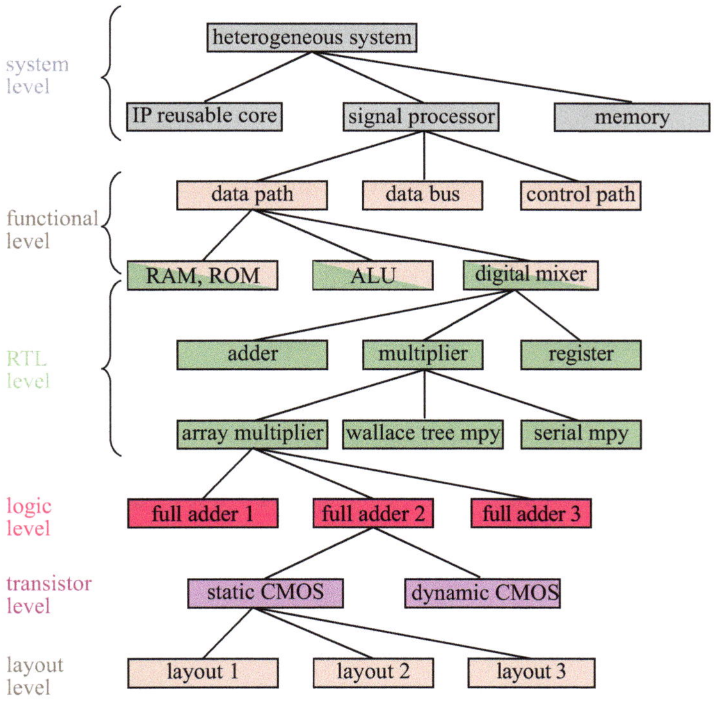

Fig. 7.17 Decision tree for a complex system on a chip

then move to the next level. However, the decisions at each level can be strongly dependent on the possibilities available at a lower or at the lowest level. System designers who wish to achieve efficient area implementations therefore require a reasonable knowledge about the consequences of their decision at implementation level. For instance, the decision to implement a double data bus structure (Fig. 7.10) requires twice as many interconnections as a single-bus implementation. As a result, the implementation of a double bus will not only take twice the "bus" area, but it also doubles the noise contribution since it doubles the level of the simultaneously switching current.

Decision trees and abstraction levels basically reduce the complexity of design tasks to acceptable levels. However, the abstraction levels are also accompanied by verification problems. More levels can clearly increase verification difficulties. Requirements at a certain level of abstraction depend on details at a lower level. Details such as propagation delays, for example, can influence higher-level timing behaviour.

For example, the final layout implementation of a full adder clearly influences its electrical behaviour. Delay times are also determined by factors such as parasitic wiring capacitances.

The bottom-up implementation and verification process begins at the layout level. Cell layouts are assembled to form modules, and these are combined to form the larger units that are indicated in the *floor-plan* of the IC. The floor-plan is a product of the top-down and bottom-up design process and is an accurate diagram which shows the relative sizes and positions of the included logic, analog, memory and interface cores. Cores that are identified as *critical* during the design path are usually implemented first. These are cores which are expected to present problems for power dissipation, area or operating frequency. Verification of their layouts reveals whether they are adequate or whether an alternative must be sought. This may have far-reaching consequences for the chosen architecture.

The interdependence of various abstraction levels and implementations clearly prevents a purely top-down design followed by purely bottom-up implementation and verification. In practice, the design process generally consists of various iterations between top-down and bottom-up paths.

Abstraction-level descriptions which contain sufficient information about lower-level implementations can limit the need for iterations in the design path and prevent wasted design effort. The maximum operating frequency, for example, of a module is determined by the longest delay path between two flip-flops. This *worst-case delay path* can be determined from suitable abstraction-level descriptions and used to rapidly determine architecture feasibility. As an example, the multiplier in the previously discussed signal processor is assumed to contain the worst-case delay path.

The dimensions of logic cells in a layout library, for example, could be used to generate floor-plan information such as interconnection lengths. These lengths, combined with specified delays for the library cells (e.g. full adder, multiplexer, etc.), allow accurate prediction of performance. The worst-case delay path can eventually be extracted from the final multiplier layout and simulated to verify that performance specifications are met.

The aim of modern IC design environments is to minimise the number of iterations required in the design, implementation and verification paths. This should ensure the efficient integration of systems on silicon. Beyond the 30 nm node, designers face a continuously increasing design complexity caused by additional lithography, process and variability issues on top of the area, timing, power, leakage and noise issues that already exist since the 100 nm node. System integration and verification of multibillion transistor designs, with multi-clock and power domains (Chaps. 8 and 9), require smooth integration of reusable externally and/or in-house developed IP with newly designed IP. For many process nodes already, design verification has become the costlier part of the design cycle. For the above-described complex ICs, it may take even more than 50% of the total design costs, particularly when they also include a variety of analog IP.

7.4 Digital VLSI Design

7.4.1 Introduction

The need for CAD tools in the design and verification paths grows with increasing chip complexity. The different abstraction levels, as discussed in the previous subsection, were created to be able to manage the design complexity at each level.

7.4.2 The Design Trajectory and Flow

The continuous growth in the number of transistors on a chip is a drive for a greater integration of synthesis- and system-level design. The increasing complexity of the system-level behaviour, combined with an increasing dominance of physical effects of devices (e.g. variability), supply lines (e.g. voltage drop and supply noise) and interconnections (e.g. propagation delay and cross-talk), is a drive for a greater integration of synthesis and physical design.

Figure 7.8 shows a *heterogeneous system on a chip (SOC)*. First, the entire design must be described in a complete specification. For several existing ICs, such a specification consists of several hundreds of textual pages. This design specification must be translated into a high-level behavioural description, which must be executable and/or emulatable.

In many cases, software simulation is too slow and inaccurate to completely verify current complex ICs. Also, the interaction with other system components is not modelled. Logic *emulation* is a way to let designers look before they really act. Emulation allows the creation of a hardware model of a chip. Here, proprietary emulation software is used, which is able to map a design on reprogrammable logic and which mimics the functional behaviour of the chip. Emulation is usually done in an early stage of the design process and allows more effective *hardware/software codesign*. The validation/verification problem has also led to the introduction of hybrid simulator tools [2], which claim to speed up simulation by 10–100 times for a single-chip or multi-chip system. Once the high-level behavioural description is verified by simulation or emulation, all subsequent levels of design description must be verified against this top-level description. The introduction of standard verification methods, such as OVM (Open Verification Methodology) and UVM (Universal Verification Methodology), is another attempt to deal with the verification complexity. These standards are supported by the major CAD vendors. Figure 7.18 shows a general representation of a design flow.

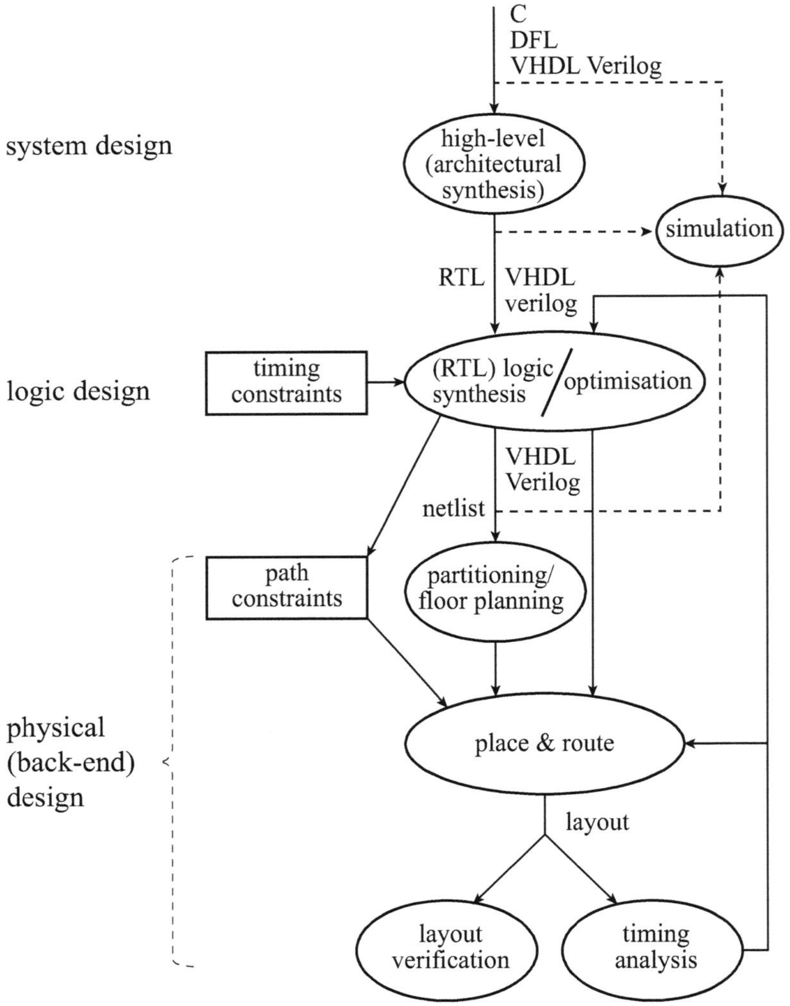

Fig. 7.18 General representation of a design flow

Synthesis tools automatically translate a description at a higher hierarchy level into a lower level one. These tools are available at several levels of abstraction. As systems continuously demand more performance improvements than available from the limited intrinsic performance improvements by scaling to the next technology node, the focus is currently more towards improved and more efficient algorithms. These algorithms require a higher level of design, MATLAB, C/C++, SystemC or similar platforms, rather than RTL. High-level synthesis transforms a behavioural description into a sequence of possible parallel operations which must be performed on an IC. The derivation of ordering operations in time is called *scheduling*.

The *allocation* (or *mapping*) process selects the required data path components. These high-level components include complete signal processor and microprocessor cores, as well as coprocessors, ALUs, RAMs, I/O blocks, etc. With some exceptions, *high-level synthesis (HLS)* tools are focussed at specific application domains, such as DSP and data path designs, which are driven by regular streams of data samples. The design workflow requires knowledge of both software to write C applications and hardware to parallelize tasks and resolve timing and memory management issues [3,4]. For telecom and audio processor ICs, there are tools which are different from those that are created and used for the development of video signal processors. *Behavioural synthesis* tools, also called *high-level synthesis (HLS)* tools, generate RTL hardware descriptions in *VHDL* or *Verilog* from the system specification. The *RTL code* of a logic block describes its functionality in detail; in fact, it describes the behaviour of every bit in that block at every clock cycle. Although research on high-level synthesis (HLS) started already in the late 1980s, industrial adoption has taken of slowly because of the long learning curve one has to master and it was difficult to formally prove the equivalence between the high-level description and the synthesised RTL models. So far, it has been successfully applied in video and signal processing environments but has found only limited use in other areas [5].

Current and future *systems on silicon* (Fig. 7.8) are, and will be, designed by using a wide variety of predesigned building blocks. This design *reuse* requires that these *intellectual property (IP)* parts, such as microcontrollers, microprocessors, graphics processors, memories and interfaces, can easily be ported from one chip design to another. Such a reuse must be supported by tools. Design reuse will be fuelled by the sharing of cores among companies. In many cases, a reduced instruction set computer (RISC) microprocessor core (ARM, MIPS, SPARC) is used. If we include the application (programme) in an on-chip ROM or other types of memory, this is called *embedded software*.

Two surveys, respectively conducted in 2013 and 2021, each with several hundred responses from design and verification engineers (Fig. 7.19) show that, on average, two thirds of their ASIC design content is reused IP, of which more than 60% is internally (in-house) developed IP [6]. As the diagram shows, there has not been much change in IP reuse over a period of 8 years in the recent past. We may assume that that will not change much in this decade either. The development of new design content is often done using an IP-based design approach, in which the design is partitioned into IP modules.

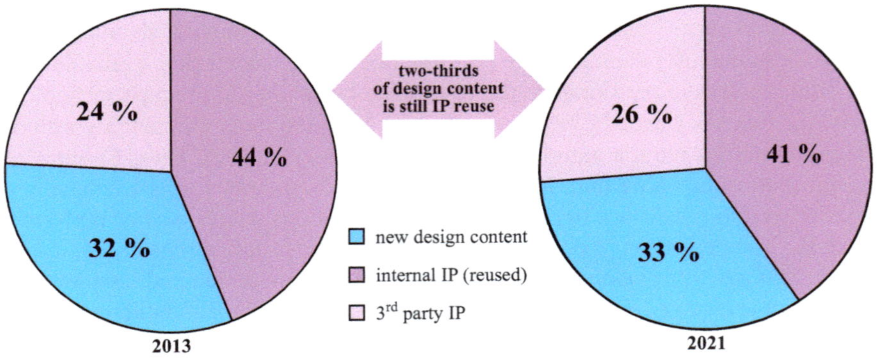

Fig. 7.19 Non-memory SoC and IC content in 2013 and 2021, respectively (Source: IC Manage) [6]

An overall ASIC design style thus requires several engineering teams working in parallel on managing and executing various design tasks: new IP creation, integration of new and reused IP, chip assembly (floor-planning) and verification.

Synthesis tools must play a key role in integrating such predesigned building blocks with synthesised glue logic onto one single chip. The most used type of synthesis is from the RTL level to a netlist of standard cells. Each *system on a chip* can be considered to consist of many registers which store binary data. Data is operated upon by logic circuits between these registers. The operations can be described in a *register-transfer language (RTL)*. Before the VHDL code (or Verilog) is synthesised at this level, the code must be verified by simulation.

At higher functional levels, software (VHDL) simulators are often sufficiently fast. However, in many cases, RTL-level simulation is a bottleneck in the design flow. Besides an increase in the complexity of ICs, longer frame times (as in MPEG video and DAB) must also be simulated. Such simulations may run for several days, resulting in too long iteration times and allowing only limited functional validation of an RTL design.

A *hardware accelerator*, with accompanying software, is a VHDL simulator platform in which the hardware is often realised with reconfigurable logic, e.g. with field-programmable gate arrays (FPGAs) or with a large multiprocessor system, which is connected to the network or a host system. Gate-level descriptions as well as memory modules can be downloaded into a hardware accelerator. However, most non-gate-level parts (RTL and test bench) are kept in software. The accelerator hardware speeds up the execution of certain processes (i.e. gates and memory) and the corresponding events. In fact, the accelerator is an integral part of the simulator and uses the same type of interface. Generally, the raw performance of a hardware accelerator is less than with *emulation*.

When the RTL description is simulated and proven to be correct, RTL synthesis is used to transform the code (mostly VHDL or Verilog) into an optimised netlist.

Actually, the described function or operation at RTL level is *mapped* onto a library of (standard) cells. Synthesis at this level is more mature than high-level synthesis and is widely used. The synthesis of the functional blocks and the composition of the complete IC are the work of the physical or back-end designer. Next to the logic synthesis, *back-end design* tasks include the place and route of the logic cells in the generated netlist and the *floor-planning*, which assigns the individual logic blocks, memories and I/O pins to regions in the chip. It also includes tasks that maintain signal integrity (crosstalk, supply noise, voltage drop, etc.), variability (parameter spread, transistor matching, etc.), reliability (electromigration, antenna rules, etc.) and *design for manufacturability (DfM)* (via doubling, metal widening or spreading, dummy metals, etc.). This back-end design is no longer a straightforward process, but it requires many iterations to cover all of the above design objectives simultaneously. This shows that the back-end design has become a very complex task, which needs to be supported by appropriate tools, smoothly integrated in the design flow.

Finally, the *design verification* is also a growing part of both the front-end and back-end design trajectory. CAD tools are also used for the validation in the IC design verification path. Simulation is the most commonly used *design verification* method. *Behavioural simulation* is usually done on an IP block basis at a high abstraction level (algorithm/architecture). It runs quickly because it only includes the details of the behaviour and not of the implementation. *Logic simulation* is performed at RTL or netlist level and relates to the digital (or Boolean) behaviour in terms of logic 1s and 0s. *Circuit simulation* is the transistor-level simulation of the behaviour of a schematic or extracted layout. It usually includes all device and circuit parasitics and results in a very accurate and detailed analog behaviour of the circuit. Due to the rapid increase in the IC's complexity, it is impossible to completely simulate a system on a chip and verify that it will operate correctly under all conditions. Moreover, it is very difficult to envision and simulate all potential event candidates that may lead to problems. Achieving 100% verification coverage would require huge time-consuming simulations with an unlimited number of input stimuli combinations.

Luckily, there are other verification methods that complement the simulation. *Formal verification* is a mathematical method to verify whether an implementation is a correct model for the specification. It is based on reasoning and not on simulation. This verification may include the comparison of design descriptions at different levels of abstraction. Examples of this so-called *equivalence checking* are the comparison between behavioural description and RTL description, which checks whether the synthesis output is still equivalent to the source description, and the comparison between the RTL description and the synthesized netlist to prove equal functional behaviour. It does not prove that the design will work.

Timing verification is done at a lower hierarchy level. During a (deterministic) *static timing analysis (STA)* , each logic gate is represented by its worst-case propagation delay. Then, the worst-case path delay is simply the sum of the worst-case delays of the individual gates in that path. Due to the increasing process-induced parameter spread in devices and interconnect structures, these worst-case numbers are often so high that this type of static timing analysis leads to design overkill, to less performance than in the previous technology node or to incorrect critical paths. This has led to the introduction of a *statistical static timing analysis (SSTA)* tool, which uses probability distributions of random process variations and tries to find the probability density function of the signal arrival times at each internal node and primary output. This type of analysis is considered necessary, particularly for complex high-performance ICs [7–9]. However, probability density functions are difficult to compute, and the method needs to be simplified to make it a standard component of the verification process.

As a result of the growing number of transistors on one chip and with the inclusion of analog circuits or even sensors on the same chip, verification and analysis have become serious bottlenecks in achieving a reasonable design turnaround time. Extensive verification is required at each level in the design flow, and as discussed before, there is a strong need for cross verification between the different levels. Verification often consumes 40–60% of the total design time. With increasing clock speed and performance, packaging can be a limiting factor in the overall system performance. Direct attachment of chip on board, flip-chip techniques and chiplets stacking (see Sect. 10.3.3) continues to expand to support system performance improvements. Verification tools are therefore needed across the chip boundaries and must also include the total interconnect paths between chips.

Finally, a set of data and script files, called the *process design kit (PDK)*, is used to enable the use of various *EDA (electronic design automation)* tools to support the full-custom design flow of the IC, from schematic entry to verified layout. In fact, it acts as an interface between design and foundry. It is developed by the foundry. A PDK mainly consists of symbols; device models (transistors, capacitors, inductors and resistors); technology files (including process parameter spread); parameterised cells (Pcells); verification decks; design rule manual containing layout, electrical and reliability design rules; etc.

Since PDKs are foundry-specific and include a complex set of files, a detailed description of their contents, their languages and formats is beyond the scope of this book but can be found on the internet.

7.4.3 Example of Synthesis from VHDL Description to Layout

This paragraph discusses the design steps of the digital mixer (see Sect. 7.3.4), starting at the RTL description level (in VHDL) and ending in a standard-cell layout. Figure 7.20 shows the RTL-VHDL description of this mixer.

Figure 7.21a shows a high abstraction-level symbol of this mixer, while a behavioural-level representation is shown in Fig. 7.21b.

After synthesis, without constraints, our mixer looks as shown in Fig. 7.22.

Figure 7.23 shows the multiplier and adder symbolic views after synthesis.

Figure 7.24 shows the schematics of the adder, after synthesis with no constraints.

Figure 7.25 shows the schematics of the adder, after synthesis with a timing constraint for the worst-case delay path.

```
LIBRARY IEEE;

USE IEEE.std_logic_1164.ALL;
USE IEEE.std_logic_arith.ALL;
USE IEEE.std_logic_unsigned.ALL;

ENTITY potmeter IS

   GENERIC (par_width: natural := 4;
            operand_width : natural := 12);

   PORT (A, B: IN std_logic_vector(operand_width-1 DOWNTO 0);
         K: IN std_logic_vector(par_width-1 DOWNTO 0);
         Z: OUT std_logic_vector(par_width+operand_width-1 DOWNTO 0));

END potmeter;

ARCHITECTURE behaviour OF potmeter IS

BEGIN

   PROCESS (A, B, K)

      CONSTANT K_max: integer := 2**par_width-1;
      VARIABLE K_int: integer;

   BEGIN

      K_int := conv_integer(K);
      Z <= K*A + conv_std_logic_vector(K_max-K_int, par_width) * B;

   END PROCESS;

END behaviour;
```

Fig. 7.20 RTL-VHDL description of mixer

The additional hardware in Fig. 7.25 compared to that of Fig. 7.24 is used to speed up the carry ripple by means of carry lookahead techniques. Figure 7.26 shows the relation between the delay and the area. The figure clearly shows that reducing the delay by timing-constraint synthesis can only be achieved with relatively much additional hardware (area).

Figure 7.27 shows a part of the *netlist* of logic (library) cells onto which the mixer function has been mapped. In fact, it shows only a fraction of the gates of the circuit schematic in Fig. 7.25. A netlist may contain instances, modules, pins and nets. An instance is the materialisation of a library cell or a module. A module, itself, is built from several instances and their connections. Pins, also called ports or terminals, represent the connection points to an instance or module, and finally, a net represents a connection between pins. The figure shows the different library cells and the nodes to which their inputs and outputs are connected.

The next step is to create the layout of this block. The *place-and-route (P&R)* tool places the netlist cells in rows and also creates the interconnections between the pins of the cells (or modules). Due to the growing complexity of IP cores in combination with the need to accommodate higher chip performance, the physical design of these

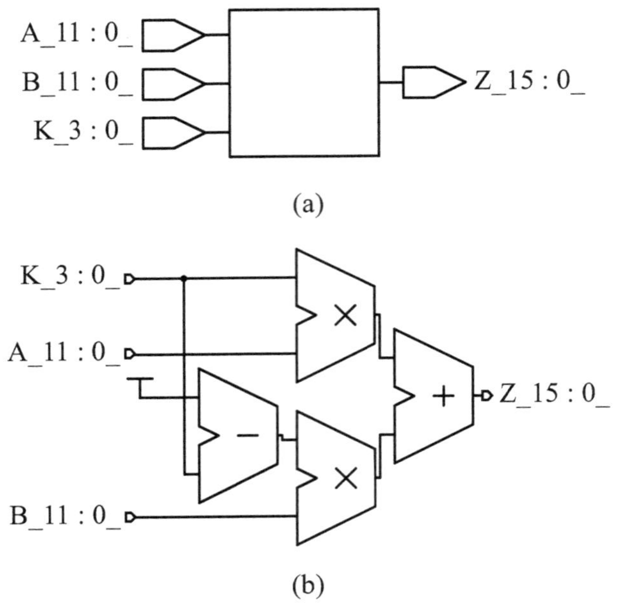

Fig. 7.21 (**a**) Abstraction-level symbol and (**b**) behavioural-level representation of the mixer

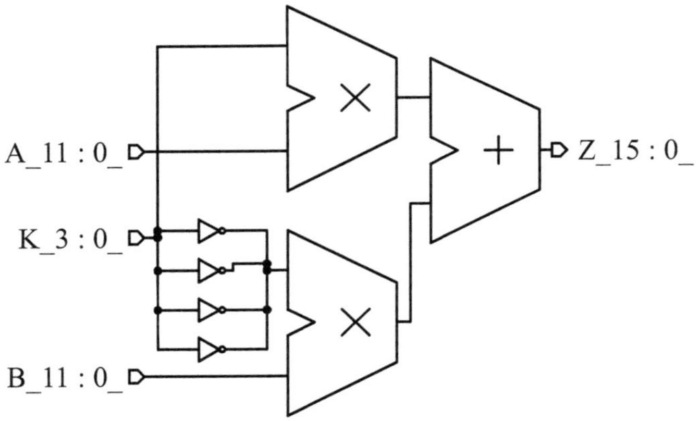

Fig. 7.22 Mixer schematic after synthesis with no constraints

cores becomes a real challenge. To achieve timing closure on such complex blocks with very tight area, timing and power constraints are a difficult task. Cell placement is a critical part of the back-end design flow, as it has severe impact on core area, wire length, timing and power requirements. P&R tools, today, allow area-driven, wire length-driven, timing-driven as well as power-driven placement [10] and thus allow placement optimisation for various different application domains. A timing-

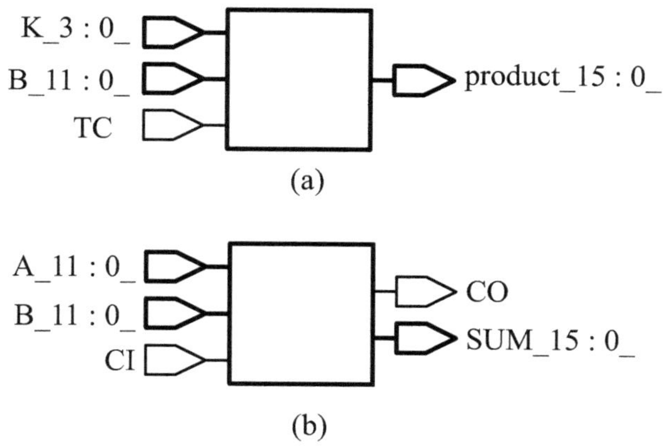

Fig. 7.23 Multiplier (**a**) and adder (**b**) symbolic views

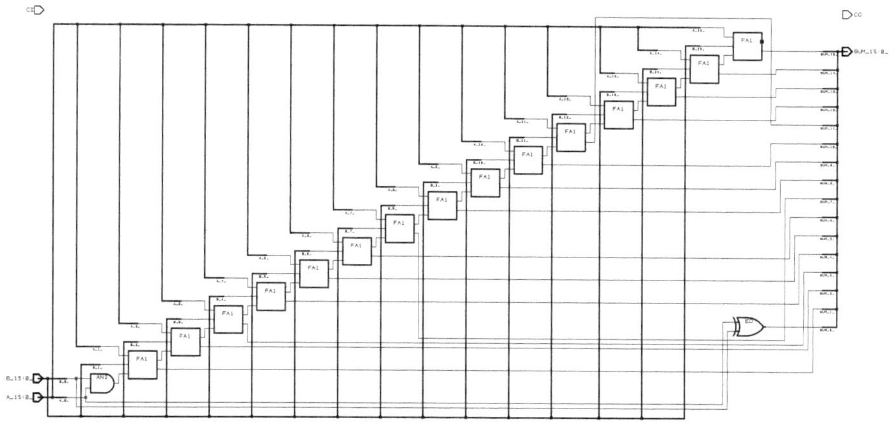

Fig. 7.24 Adder schematics after synthesis with no constraints

driven placement, for example, can assign higher weights to critical nets to reduce their wire length as well as select faster cells, e.g. with higher drive capability and/or reduced threshold voltage, to reduce the critical path delay (see also Sect. 4.7).

After the use of place-and-route tools, a standard-cell design of the mixer is created. Figure 7.28 shows the result. The figure shows a standard-cell block composed of various rows of standard cells that are automatically connected according to the schematics in Fig. 7.25. The metal layers in the standard-cell block are not shown because they would shield the visibility of the cells. The netlist and layout are the result of the chosen description of the mixer's functionality according to:

$$Z = k \cdot A + (1 - k) \cdot B$$

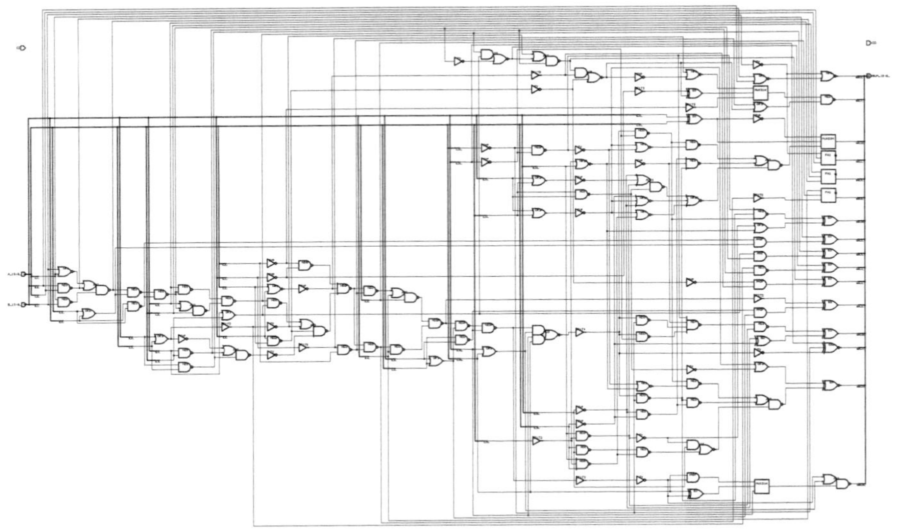

Fig. 7.25 Adder schematics after timing-constraint synthesis

Fig. 7.26 Relation between maximum delay and the amount of hardware (area)

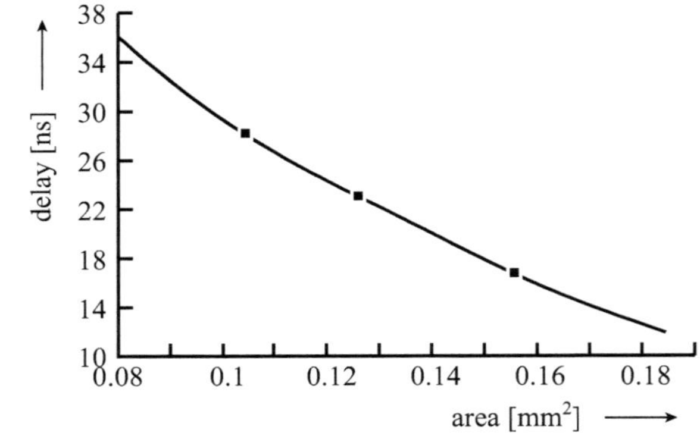

This implementation requires two adders and two multipliers. However, an obvious optimisation of the same function may lead to a more efficient implementation. The following description

$$Z = k \cdot (A - B) + B$$

requires only two adders and one multiplier. This example shows that the decision taken at one hierarchy level can have severe consequences for the efficiency of the final silicon realisation in terms of area, speed and power consumption.

Although the synthesis process uses tools which automatically generate a next level of description, this process is controlled by the designer. An excellent design

```
module potmeter_DW01_add_14_1 ( A, B, CI, SUM, CO );
input  [13:0] A;
input  [13:0] B;
output [13:0] SUM;
input  CI;
output CO;
    wire n52, n53, n54, n55, n56, n57, n58, n59, n60, n61, n62, n63, n64, n65,
        n66, n67, n68, n69, n70, n71, n72, n73, n74, n75, n76, n77, n78, n79,
        n80, n81, n82, n83, n84, n85, n86, n87, n88, n89, n90, n91, n92, n93,
        n94, n95, n96, n97, n98, n99, n100, n101, n102, n103, n104, n105;
    BF1T1 U5 ( .Z(SUM[2]), .A(A[2]) );
    BF1T1 U6 ( .Z(SUM[0]), .A(A[0]) );
    BF1T1 U7 ( .Z(SUM[1]), .A(A[1]) );
    AO6 U8 ( .Z(n52), .A(n53), .B(n54), .C(n55) );
    AO6 U9 ( .Z(SUM[3]), .A(n56), .B(n57), .C(n58) );
    AO32 U10 ( .Z(n59), .A(n60), .B(n61), .C(n62), .D(n63) );
    AO32 U11 ( .Z(n64), .A(n59), .B(n65), .C(n54), .D(n55) );
    NR2 U12 ( .Z(n66), .A(n67), .B(n68) );
    AO6 U13 ( .Z(n69), .A(A[7]), .B(B[7]), .C(n70) );
    NR2 U14 ( .Z(n71), .A(n65), .B(n63) );
    NR2 U15 ( .Z(n72), .A(n73), .B(n74) );
    AN2 U16 ( .Z(n75), .A(n76), .B(n77) );
    EO U17 ( .Z(SUM[9]), .A(n66), .B(n78) );
    EO U18 ( .Z(SUM[8]), .A(n79), .B(n80) );
    EO U19 ( .Z(SUM[6]), .A(n81), .B(n82) );
    EO U20 ( .Z(SUM[5]), .A(n71), .B(n83) );
    MUX21N U21 ( .Z(SUM[13]), .A(B[13]), .B(n84), .S(n85) );
    IV U69 ( .Z(n84), .A(B[13]) );
    IV U70 ( .Z(n105), .A(A[10]) );
    IV U71 ( .Z(n96), .A(B[7]) );
    IV U72 ( .Z(n79), .A(n95) );
endmodule
```

Fig. 7.27 Part of the mixer netlist after synthesis with 14 ns timing constraints

is the result of the combination of an excellent tool and a designer with excellent skills in both control of the tools and knowledge of IC design.

7.4.4 Floor-Planning

When all required external, in-house and newly synthesised IP cores are available, these have to be integrated to create a compact chip, as shown in Fig. 3.14. *Floor-planning* is an important part of the chip design cycle. The major modules are manually placed. Next, the blocks that have very intensive direct communication with each other must be positioned in each other's close vicinity in order to limit power consumption and/or propagation delay across their signal interconnect wires.

Floor-planning is also supported by the P&R tools in that they can change the aspect ratio of the synthesised standard-cell cores. The shape of such a *chiplet* is fully adjusted to the area requirements as defined by the floor-plan. Other tools support further placement of the cores, based on their aspect ratios and pin positions. Some of these tools can also create and implement multi-voltage domains (see Chap. 8) to support on-chip power management [11]. DSPs, graphics processors, microprocessors and DDR memory interfaces are critical floor-plan elements as they are often critical in timing and relatively take a large share of the total power consumption. Therefore, in certain applications, these blocks must be distributed

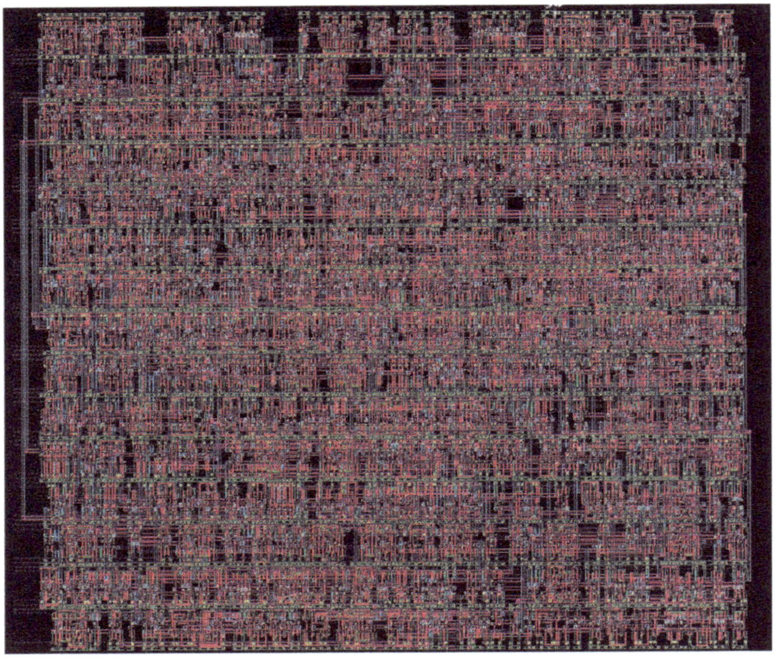

Fig. 7.28 Standard-cell implementation of mixer

over the chip to prevent local overheating. Important other floor-planning issues are:

- Chip-level signal wiring and wire estimation
- Insertion of feed-throughs
- Distribution of power nets
- Clock distribution

Figure 7.29 shows an example of a floor-plan. For educational purposes, this example chip only contains a limited number of cores.

With the continuous growth of chip complexity, even state-of-the-art SoCs for mobile and consumer applications may contain more than 100 different cores, which may be distributed over different voltage and/or different clock domains and surrounded by 1000–2000 I/O and power supply pads. This is one of the reasons why interfaces have moved from parallel to serial architectures.

It is clear that floor-planning has become one of the most crucial, critical and time-consuming tasks in an SoC design. It is not just to create the smallest chip area with the given area and pin constraints of the individual cores; it is also extremely important to position them such that all chip-level timing and power requirements are achieved.

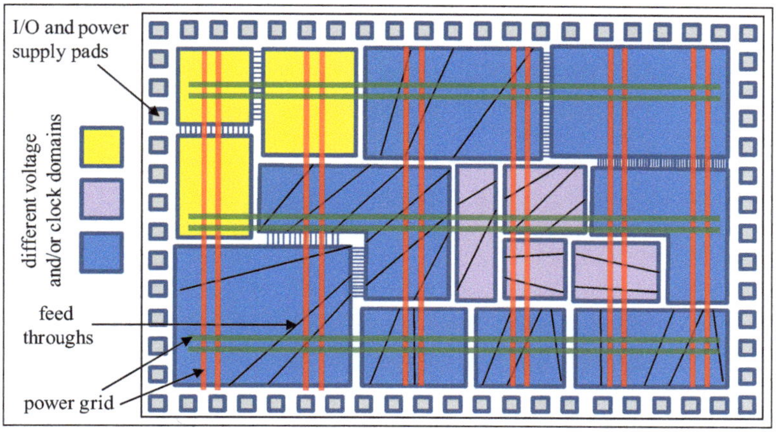

Fig. 7.29 Example floor-plan of a chip

7.5 The Use of ASICs

The growth in the ASIC business is primarily the result of the increasing number of application areas and of the general increase in the use of ICs. *ASICs* often provide the only solution to problems attributed to speed and/or space requirements. Another incentive for the use of ASICs is the degree of concealment which they afford. This concealment poses extra difficulties to competitors interested in design duplication.

ASICs make it reasonably easy to add new functionality to an existing system without an extensive system redesign. In addition, the increased integration of system parts associated with the use of ASICs has the following advantages:

- Reduced physical size of the system
- Reduced system maintenance costs
- Reduced manufacturing costs
- Improved system reliability
- Increased system functionality
- Reduced power consumption

The advantages afforded by ASICs can have a positive influence on the functionality/price ratio of products and have led to the replacement of standard ICs in many application areas. However, there are also disadvantages associated with the use of ASICs. These include the following:

- The costs of realising an ASIC are quite substantial and less predictable than those associated with standard ICs.
- Unlike standard products, ASICs are not readily available from a diverse number of suppliers. Inaccurate specifications or errors in the design process may

cause delays in ASIC turnaround time and result in additional *non-recurring engineering (NRE)* costs. These are costs incurred prior to production. Typical NRE costs include the cost of:
- Training and use of design facilities
- Support during simulation
- Placement and routing tools
- Mask manufacturing (where applicable)
- Test development
- The delivery of samples

Furthermore, standard products are always well characterised and meet guaranteed quality levels. Moreover, small adjustments to a system comprising standard products can be implemented quickly and cheaply.

The advantages and disadvantages associated with the use of ASICs depend on the application area and on the required ASIC type and quantities. Improved design methods and production techniques combined with better relationships between ASIC customers and manufacturers have a considerable influence on the transition from the use of standard products to ASICs.

An ASIC solution in the above discussions does not necessarily imply a single chip or *system-on-a-chip (SoC)* solution, but it might also refer to a *system-in-a-package (SiP)* solution. For a discussion on SoC versus SiP system solutions, the reader is kindly requested to read the appropriate subsection in Chap. 10.

7.6 Silicon Realisation of VLSI and ASICs

7.6.1 Introduction

In addition to the need for computer programmes for the synthesis and verification of complex ICs, CAD tools are also required for the automatic or semi-automatic generation of layouts. The development of Intel's Pentium and Xeon processors, for example, took several thousands of man-years. The same holds for the IBM PowerPC. Figure 7.30 shows a photograph of an Intel Xeon processor. This 12-core Haswell-E/EP i7 core processor in the Xeon family combines 8 processor cores with a 768 kB L1 cache, a 1.56 MB L2 cache and a 20 MB L3 cache memory, resulting in a 356 mm^2 chip containing 2.6 billion transistors. It uses a core voltage of 0.735 V and consumes 140 W when running at its maximum clock frequency of 3 GHz.

In fact, the increased use of CAD tools in recent years has very often merely facilitated the integration of increasingly complex systems without contributing to a reduction in design time. This situation is only acceptable for very complex high-performance ICs such as a new generation of microprocessors. Less complex ICs, such as ASICs, require fast and effective design and layout tools. Clearly, the need for a fast design and layout process increases as the lifetimes of new ICs become shorter. The lifetime of a new generation of ICs for many mobile gadgets, for instance, is close to 1 year. This means that the design process may take only

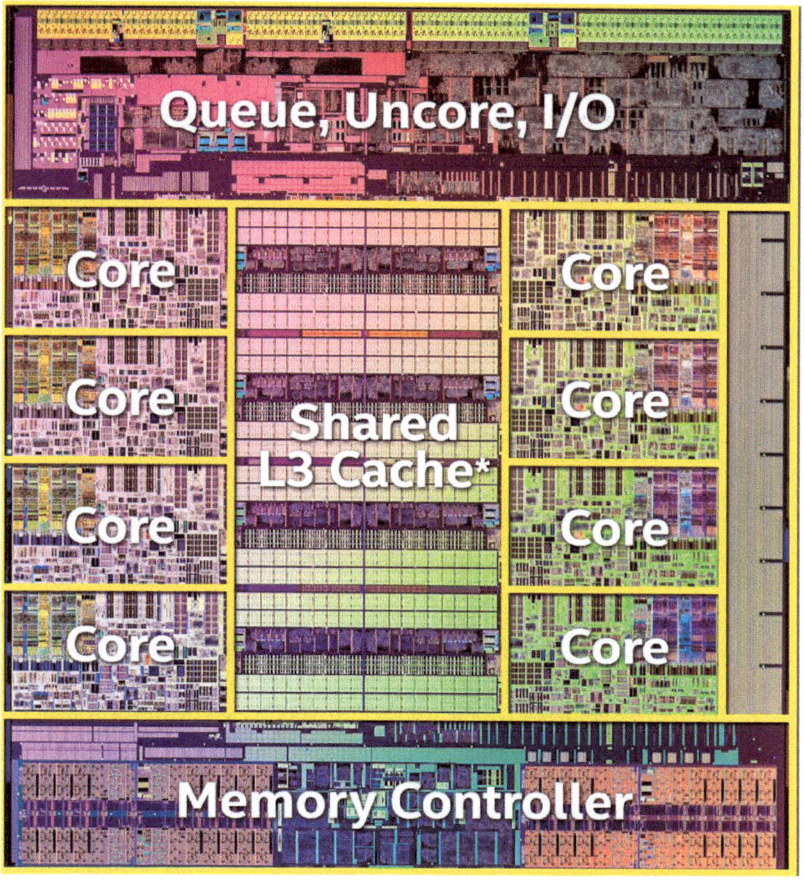

Fig. 7.30 The Intel Haswell-E/EP 8-core processor of the Xeon family (courtesy of Intel)

a couple of months. Each layout design must be preceded by a thorough floor-plan study. This must ensure that the envisaged layout will not prove too large for a single-chip implementation in the final design phase. As discussed before, a floor-plan study can take considerable time and only leads to a definite floor-plan after an iterative trial-and-error process. Layouts of some parts of the chip may be required during the floor-plan study. Although we distinguish between the different ASIC categories of custom ICs, semi-custom ICs and PLDs in this book, the differences are rapidly diminishing as a result of the pace at which improvements in IC technologies are realised. PLDs are moving towards gate arrays, gate arrays are moving towards cell-based designs, and cell-based designs may use sea-of-gates structures such as embedded arrays to implement the glue logic as well as for mapping of cores onto such arrays. Each category uses the best features of the others.

Table 7.2 Comparison of performance of different layout implementation forms

	Performance	
Implementation form	Speed	Area
Handcrafted layout	+ + + + +	+ + + + +
Bit-slice layout	− + + + +	− + + + +
Cell-based design	− − + + +	− − + + +
Structured array ASIC	− − − + +	− − − + +
(Sea-of-gates) gate array	− − − − +	− − − − +
PLD (FPGAs and CPLDs)	− − − − +	− − − − +

The choice of implementation is determined by the required development time, production volume and performance. Table 7.2 summarises the performance of various *layout implementation forms*. This table is only valid in general terms.

The different layout implementation forms are discussed separately in the next subsections.

7.6.2 Handcrafted Layout Implementation

A *handcrafted layout* is characterised by a manual definition of the logic and wiring. This definition must account for all relevant layout design rules for the envisaged technology. The design rules of modern technologies are far more numerous and complex than those used in the simple initial nMOS process. However, various CAD tools have emerged which ease the task of creating a handcrafted layout. These include interactive computer graphic editors (or *polygon pushers*), compactors and *design rule check (DRC)* programmes.

An example of a handcrafted layout is illustrated in Fig. 7.31. Such an implementation yields considerable local optimisation. However, the required intensive design effort is only justified in MSI circuits and limited parts of VLSI circuits. The use of handcrafted layout is generally restricted to the design of basic logic and analog cells. These may subsequently be used in standard-cell libraries, module generators, bit-slice layouts, etc. Very high-speed designs may still require handcrafted design techniques, whether or not supported by simple CAD tools, but this is limited to those circuits that cannot achieve the performance targets using synthesis tools.

7.6.3 Bit-Slice Layout Implementation

A *bit-slice layout* is an assembly of parallel single-bit data paths. The implementation of a bit-slice layout of a signal processor, for example, requires the design of a circuit layout for just one bit. This bit slice is subsequently duplicated as many times as required by the word length of the processor. Each bit slice may comprise one or more vertically arranged cells. The interconnection wires in a bit slice run over the cells with control lines perpendicular to data lines. CAD tools facilitate the efficient

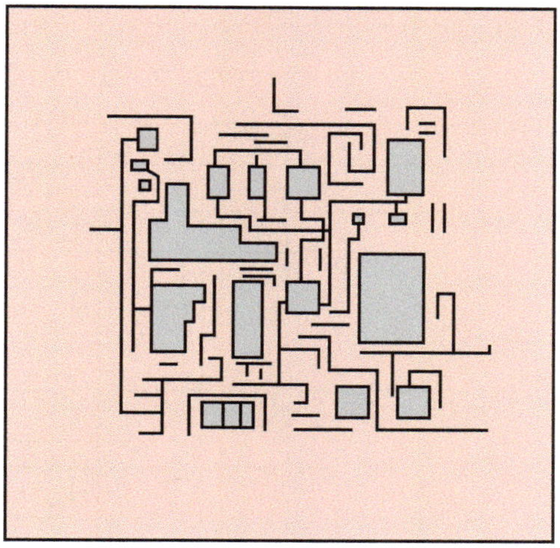

Fig. 7.31 Typical contours of a handcrafted layout

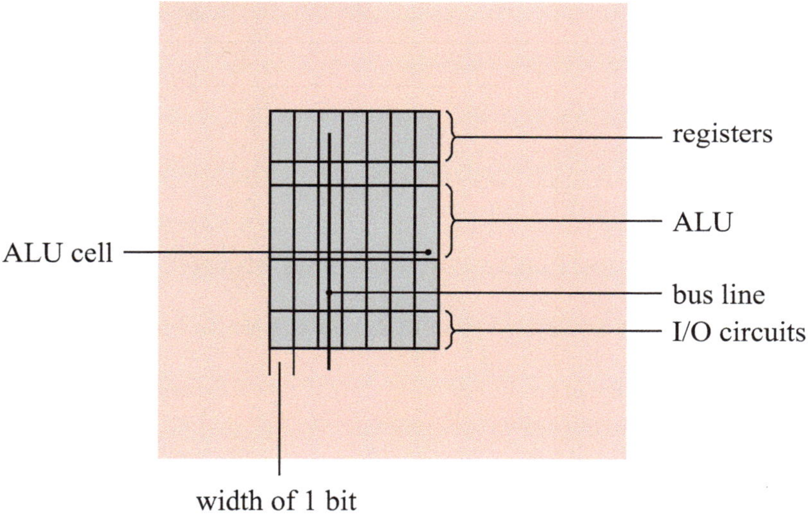

Fig. 7.32 Basic bit-slice layout

assembly of bit-slice layout architectures. The bit-slice design style is characterised by an array-like structure which yields a reasonable packing density. Figure 7.32 illustrates an example of a bit-slice layout architecture. A bit-slice section is also indicated in the chip photograph in Fig. 7.50.

The AMD Am2901 is an example of a bit-slice architecture. Today, this layout style has become less popular, because it requires a lot of manual design effort compared to the availability of a fully synthesizable alternative with the standard-cell approach, discussed in Sect. 7.6.5

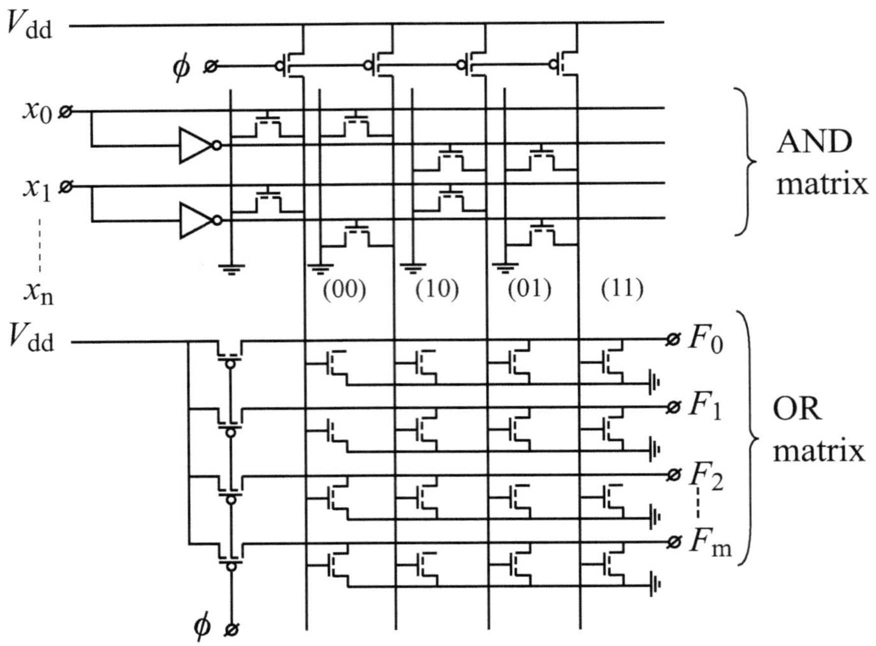

Fig. 7.33 Logic functions realised with an ROM

7.6.4 ROM, PAL and PLA Layout Implementations

In addition to serving as a memory, an ROM can also be used to implement logic functions. An example is shown in Fig. 7.33.

Only one vertical line in this ROM will be "high" for each combination of address inputs $x_n \cdots x_0$. This vertical line drives the gates of $m + 1$ transistors in the OR matrix. The outputs F_j, that are connected to the drains of these transistors, will be "low". If, for example, the address inputs are given by $x_0 x_1 = 10$, then the second column line will be "high". A "low" will then be present on outputs F_1 and F_2. The information stored in the ROM in Fig. 7.33 is thus determined by the presence or absence of connections between MOS transistor drains and the output lines. In this way, the structure of an ROM can easily be used to realise logic functions. Table 7.3 shows a possible truth table, which could be implemented with the ROM in Fig. 7.33.

Clearly, the set of logic functions that can be realised in an ROM is merely limited by the number of output and address bits. The regular array structure of an ROM leads to a larger transistor density per unit of chip area than for random logic. A large number of logic functions could, however, require an excessively large ROM, while the use of an ROM could prove inefficient for a small number of logic functions. In general, an ROM implementation is usually only cheaper than random logic when large volumes are involved.

Table 7.3 Example of a truth table implemented with the ROM in Fig. 7.28

x_n	-	-	-	x_1	x_0	F_m	-	-	-	F_1	F_0
0	-	-	-	0	0	0	-	-	-	1	1
0	-	-	-	0	1	1	-	-	-	0	1
0	-	-	-	1	0	0	-	-	-	0	0
0	-	-	-	1	1	0	-	-	-	0	0

Unfortunately, there are no easy systematic design procedures for the implementation of logic functions in ROM. Other disadvantages are as follows:

- Lower operating frequency for the circuit.
- The information in an ROM can only be stored during manufacturing.
- Increasing the number of input signals by one causes the width of the ROM to double.
- A high transistor density does not necessarily imply an efficient use of the transistors.

It is clear from Fig. 7.33 that the vertical column lines in an ROM represent the *product terms* formed by the address inputs x_i. These product terms comprise all of the logic AND combinations of the address inputs and their inverses. Only the OR matrix of an ROM can be programmed.

Figure 7.34 illustrates the basic structure of a *programmable logic array (PLA)* . Its structure is similar to that of an ROM and consists of an *AND matrix* and an *OR matrix*. In a PLA, however, both matrices can be programmed, and only the required product terms in the logic functions are implemented. It is therefore more efficient in terms of area than an ROM. Area requirements are usually further reduced by minimising the number of product terms before generating the PLA layout pattern.

The logic functions implemented in the PLA in Fig. 7.34 are determined as follows: a_0 is "high" when x and $\overline{z}$ are low, i.e. $a_0 = \overline{x}z$. Similarly, $a_1 = x\overline{yz}$ and $a_2 = \overline{x}yz$.

The outputs are therefore expressed as follows:

$F_0 = \overline{a_1} = \overline{x\overline{yz}}$

$F_1 = \overline{a_0 + a_2} = \overline{\overline{x}z + \overline{x}yz}$

$F_2 = \overline{a_0 + a_1} = \overline{\overline{x}z + x\overline{yz}}$

A PLA can be used to implement any combinatorial network comprising AND gates and OR gates. In general, the complexity of a PLA is characterised by $(A + C) \times B$, where A is the number of inputs; B is the total number of product terms, i.e. the number of inputs for each OR gate; and C is the number of outputs, i.e. the number of available logic functions.

Sequential networks can also be implemented with PLAs. This, of course, requires the addition of memory elements. A PLA can be a stand-alone chip or an integral part of another chip such as a microprocessor or a signal processor. PLAs are frequently used to realise the logic to decode *microcode instructions*

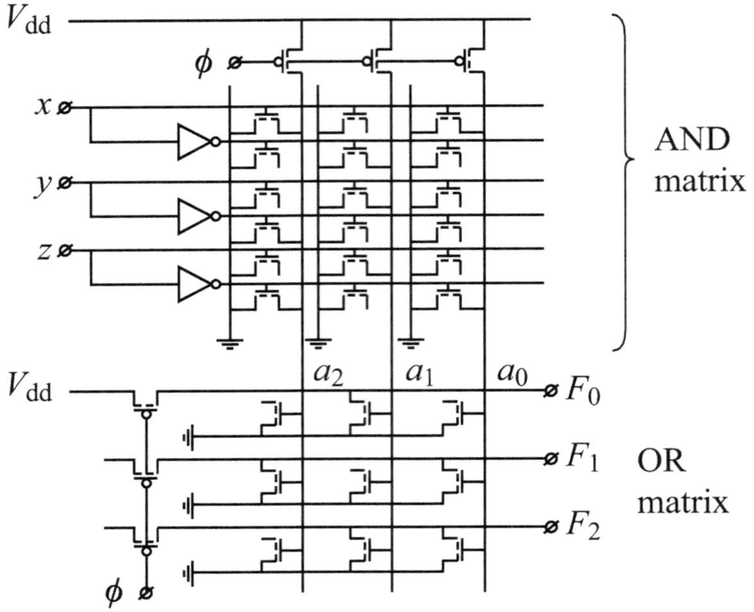

Fig. 7.34 Basic PLA structure

Table 7.4 Programmability of AND and OR planes in ROM, PAL or PLA devices

	Programmable	
Device	AND-plane	OR-plane
ROM	No	Yes
PAL	Yes	No
PLA	Yes	Yes

for functional blocks such as memories, multipliers, registers and ALUs. Several available CAD tools enable a fast mapping of logic functions onto PLAs. As a result of the improvements in cell-based designs, ROM and PLA implementations are becoming less and less popular in VLSI designs. Another realisation form is the *programmable array logic (PAL)*. In this concept, only the AND plane is programmable and the OR plane is fixed. Figure 7.43 shows an example of a PAL architecture.

Table 7.4 summarises the programmability of planes (AND, OR) in the ROM, PAL and PLA devices. Programmable techniques include fuses (early and smaller devices), floating-gate transistors ((E)EPROM) and flash devices. In some cases, an ROM (PLA) block is still used in a custom design; the programming is done by a mask. These are then called mask-programmable ROMs (PLAs). Most of the former ROM, PAL, PLA applications are currently implemented by the more flexible field-programmable gate arrays (FPGA) and complex PLDs (CPLDs), which are discussed in Sect. 7.6.7.

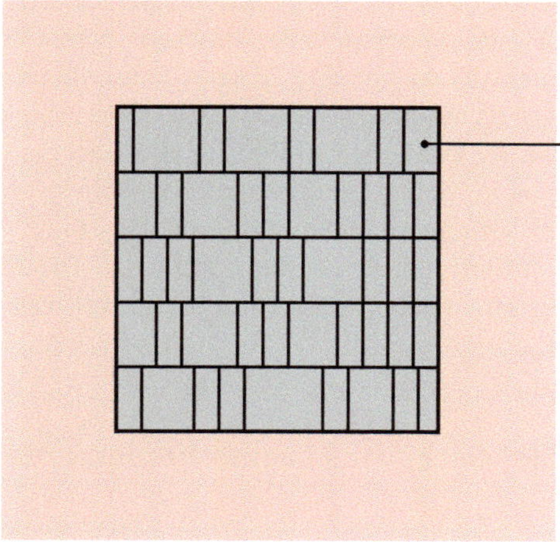

Fig. 7.35 Basic standard-cell layout

7.6.5 Cell-Based Layout Implementation

Figure 7.35 shows a basic layout diagram of a chip realised with *standard cells*.

In this design style, an RTL description of the circuit is synthesized and mapped onto a number of standard cells which are available in a library; see Sect. 4.7. The resulting netlist normally contains no hierarchy. The standard-cell library usually consists of a large number of different types of logic gates, which are all of equal height (Fig. 4.47).

Today's libraries may contain between 500 and 2000 cells, due to a large variety of drive strengths and different threshold voltages (HV_T, SV_T and LV_T, respectively referring to high, standard and low-V_T). This enables the synthesis tools to better adapt a design to such performance requirements as high speed, low power or low leakage, for example.

The standard-cell design flow is supported by mature synthesis and P&R tools (Sect. 7.4). Routing is done at a fixed grid across the logic gates. The supply lines are specially structured to create a supply network with minimum resistance and are usually an integral part of the standard-cell design approach. The clock network is usually generated by a clock-tree synthesis tool, which creates balanced clock trees to reduce intrinsic clock skew and also deals with timing constraints. However, many clock synthesis tools often balance different clock paths by compensating interconnect RC delay in one path with buffer delays in another, leading to a different path sensitivity to PVT variations. High-speed processors use relatively large clock grids leading to less clock skew and less sensitivity to PVT variations but at increased power levels. In addition, they require a detailed analysis of all parasitic resistive, capacitive and inductive effects, including the modelling and simulation of the current return paths. Modern standard-cell design environments facilitate the inclusion of larger user-defined cells in the library. These *blocks*, *macros* or *cores* may include multipliers, RAMs, signal processor cores, microprocessor cores, etc.

During the late 1980s, extra attention was paid to advanced circuit test methods. These include *scan test* and *self-test* techniques; see Sect. 10.2.1. The scan technique uses a sequential chain of intrinsically available flip-flops to allow access to a large number of locations on an IC or on a printed circuit board. The self-test technique requires the addition of dedicated logic to an existing design. This logic generates the stimuli required to test the design and checks the responses. The result is a logic circuit or a memory which is effectively capable of testing itself. Details of IC testing are discussed in Chap. 10.

7.6.6 (Mask-Programmable) Gate Array Layout Implementation

A gate array is also called a *mask-programmable gate array (MPGA)*. A conventional *gate array* contained thousands of logic gates, located at fixed positions. The layout could, for example, contain 10,000 three-input NAND gates. The implementation of a desired function on a gate array is called *customisation* and comprises the interconnection of the logic gates. The interconnections were located in dedicated *routing channels*, which were situated between rows of logic gates. In these conventional channelled gate arrays, the routing was often implemented in two metal layers.

This type of gate array is depicted in Fig. 7.36a. The channels are essential for interconnecting the cells when production processes with one or even two metal layers are involved.

In a conventional gate array, the ratio between the available cell and routing channel areas was fixed. Obviously, the actual ratio between the areas used was dependent on the type of circuit. In practice, the available area is rarely optimally used. This feature is especially important for larger circuits. Furthermore, larger circuits require more complex interconnections, and this increases the density in routing channels. The *channel-less gate array* architecture was therefore introduced. Other names encountered in literature for this architecture include *high-density gate array (HDGA)*, *channel-free gate array*, *sea of gates*, *sea of transistors* and *gate forest*.

Figure 7.36b shows the floor-plan for a channel-less gate array. It consists of an array of transistors or cells. It does not contain any specially reserved routing channels. In the 1990s, more advanced gate arrays comprised an array of *master cells*, which consist of between four and ten transistors. In some cases, the master cells are designed to accommodate optimum implementations of static RAMs, ROMs or other special circuits. A given memory or logic function is implemented by creating suitable contact and interconnection patterns in three or more metal layers. The master cells in a gate array can be separated by *field oxide isolation*, which is created by using the STI technique described in Chap. 3. An example of such a gate array master-cell structure is shown in Fig. 7.37, which also shows an example of a gate array floor-plan.

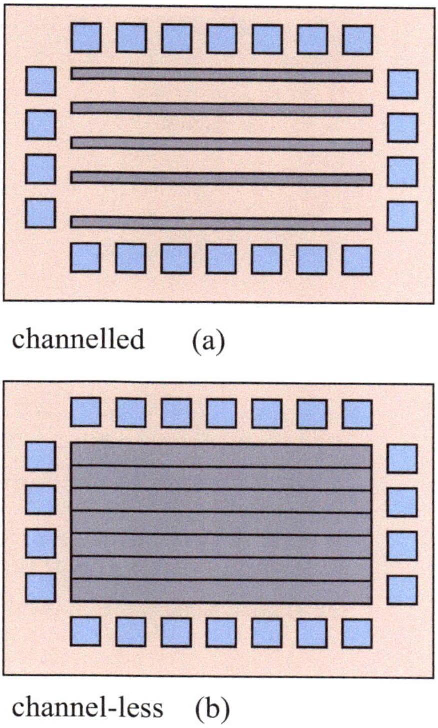

Fig. 7.36 Floor-plan for (**a**) conventional and (**b**) channel-less gate arrays

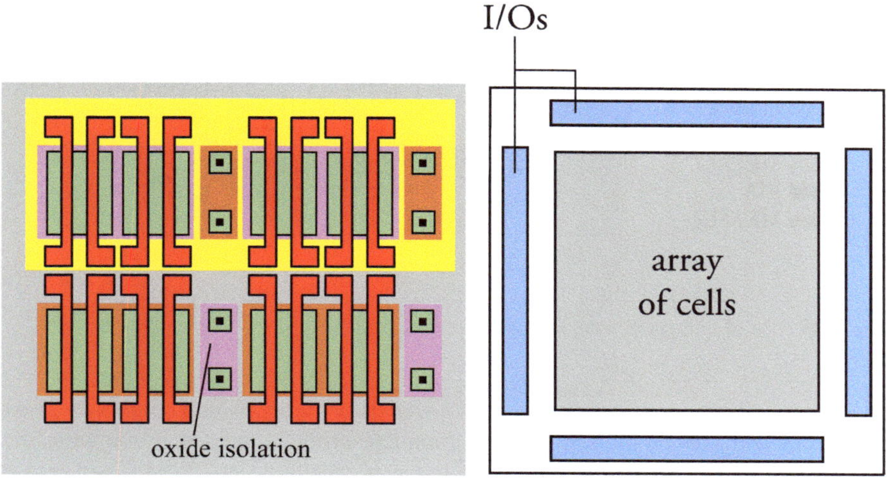

Fig. 7.37 An example of a gate array master-cell structure and floor-plan

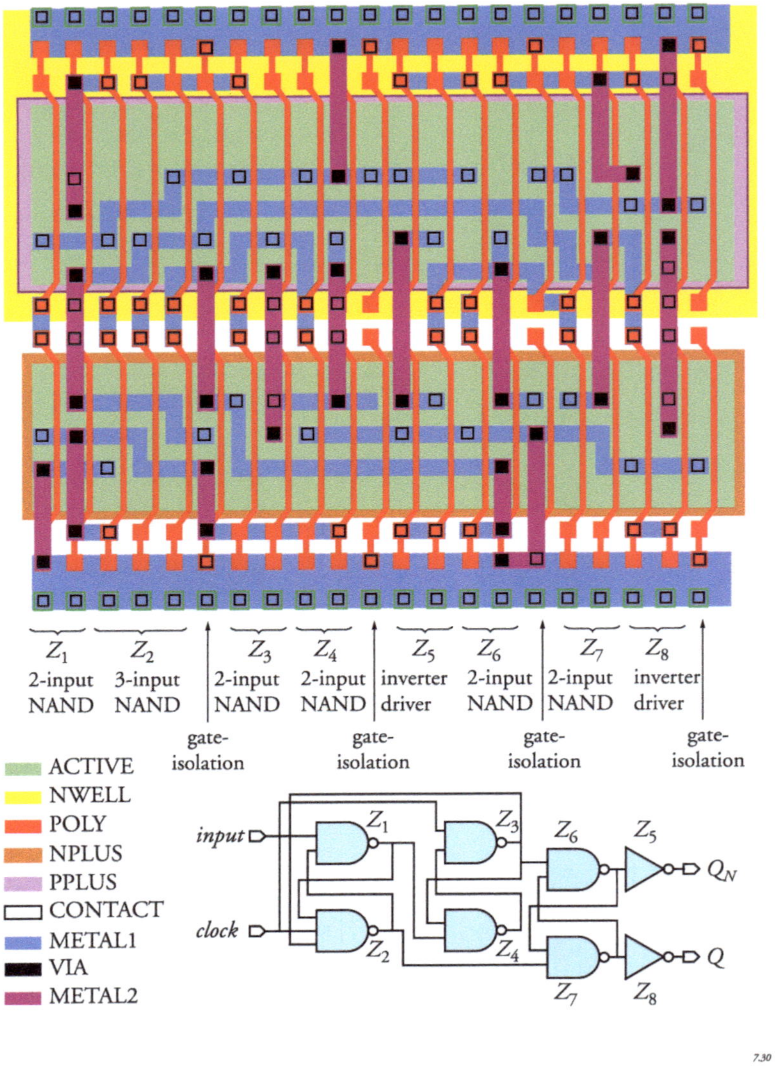

Fig. 7.38 Sea-of-transistors array with gate isolation

Figure 7.38 shows a section of a sea-of-transistors array, which comprises a row of pMOS and nMOS transistors. The complete array is created by copying the section several times in the horizontal and vertical directions. These gate arrays are also often called *continuous arrays* or *uncommitted arrays*. The rows are not separated by routing channels, and the floor-plan is therefore the same as shown in Fig. 7.36b. These gate array architectures facilitate the implementation of large VLSI circuits on a single gate array using a large number of metal layers. The logic and memory functions are again realised through the interconnection and contact hole patterns.

The various logic gates and memory cells in a sea-of-transistors architecture are separated by using the *gate-isolation technique* illustrated in Fig. 7.38.

The layout in the figure is a D-type flip-flop, based on the logic diagram shown. The gate-isolation technique uses pMOS and nMOS isolation transistors, which are permanently switched off by connecting them to supply and ground, respectively. This technique obviously requires both an nMOS and a pMOS isolation transistor between neighbouring logic gates [12].

The NRE costs of these devices depended on circuit complexity and were in the order of 100 k\$-1 M\$. Small transistors placed in parallel with larger transistors facilitate the integration of logic cells with RAMs, ROMs and PLAs in some of these HDGA architectures [13].

The design methods used for gate arrays are becoming increasingly similar to those used for cell-based design. This trend facilitates the integration of scan test techniques in gate array design. As a result of the increasing number of available cells, the software for gate array programming resembles that of cell-based designs. Also, the availability of complete cores that allow reuse (IP) is becoming available to gate array implementation.

Off-the-shelf families of gate arrays are available and include the full transistor manufacture with source and drain implants. Customisation therefore only requires the processing of several contact and metal masks. This facilitates a short *turnaround time* in processing and renders gate arrays suitable for fast prototyping.

Gate array publications include advanced low-power schemes and technologies (SOI). For high-speed gate arrays, gate delays (three-input NOR with a fan-out of two) below 50 ps have been reported. The complexity of advanced gate arrays has exceeded several tens of millions of gates. The popularity of these (mask-programmable) gate arrays reached a maximum during the 1990s. The last decade showed a dramatic reduction in new gate array design starts, mainly due to the rapid cost reduction and gate complexity increase of the field-programmable gate arrays. These FPGAs have now completely taken the MPGA market. The reason why the subject is still in the book is twofold: First reason is that an MPGA shows that a digital circuit just consists of a bunch of identical transistors, whose functionality is only determined by the way they are interconnected. Their architecture is very similar to today's litho-friendly library cells, which have reached the regularity of the mask-programmable gate array architecture in that they also show fixed-pitch transistor gates in technology nodes at and beyond 60 nm. Second reason is that MPGA approaches are still being used in structured ASICs. An example is the Fit Fast Structured Arrays (FFSA™) of Toshiba, which can be configured by customizing only a few metal layers reducing turnaround time to as little as 5 weeks from RTL hand-off to sample delivery (see the product list at Toshiba website). FPGAs are the subject of the next paragraph.

7.6.7 Programmable Logic Devices (PLDs)

A *PLD* is a *programmable logic device*, which can be programmed by fuses, antifuses or memory-based circuits. Another name currently also used for a certain category of these devices is *field-programmable device (FPD)*. The first user-programmable device that could implement logic was the programmable read-only memory (PROM), in which address lines serve as logic inputs and data lines as output (see also Sects. 6.5.3 and 7.6.4). PLD technology has moved from purely conventional bipolar technology, with a simple fuse-blowing mechanism, to complex architectures using antifuse, (E)EPROM, flash or SRAM programmability.

As a result of the continuous drive for increased density and performance, simple PLDs are losing their market share in favour of the high-density flexible PLD architectures. In this way, PLDs are moving closer and closer towards a gate array or cell-based design and are a real option for implementing *systems on silicon*. Another piece of evidence for this trend is the fact that several vendors are offering libraries of embedded cores and megacells. In the following, several architectures are presented to show the trend in PLDs.

Field-Programmable Gate Arrays (FPGAs)

FPGAs combine the initial PLD architecture with the flexibility of an *in-system programmability (ISP)* feature. Many vendors currently offer very high-density *FPGA* architectures to facilitate *system-level integration (SLI)*. Current FPGAs are mostly SRAM-based and combine memory and *lookup tables (LUTs)* to implement the logic blocks. Vendors offering LUT-based FPGAs include Xilinx (Spartan for low-power footprint, extreme cost sensitivity and high-volume, Artix for cost-sensitive high-volume markets, Kintex midrange family, Zynq for high-end embedded systems), and ALTERA (Stratix for high-end applications, Arria midrange family, Cyclone for low-power cost-sensitive markets and Max10 with non-volatile capability for cost-sensitive markets).

Initially, FPGAs were used to integrate the glue logic in a system. However, the rapid increase in their complexity and flexibility makes them potential candidates for the integration of high-performance, high-density (sub)systems, previously implemented in gate arrays [14]. The potentials of an FPGA will be discussed on the basis of a generic FPGA architecture (Fig. 7.39).

Today, these architectures consist of a large array of hundreds of thousands of programmable (re)configurable logic blocks and configurable switch matrix blocks. A logic block generally offers both combinatorial and sequential logic. Figure 7.40 shows an example of a configurable block.

In many FPGA architectures, the configurable block includes one or more *lookup tables (LUTs)* , one or more flip-flops and multiplexers. Some also contain carry chains to support adder functions. The combinatorial logic is realised by the LUTs, which each may contain three to eight inputs. Figure 7.41 shows an example of a four-input LUT.

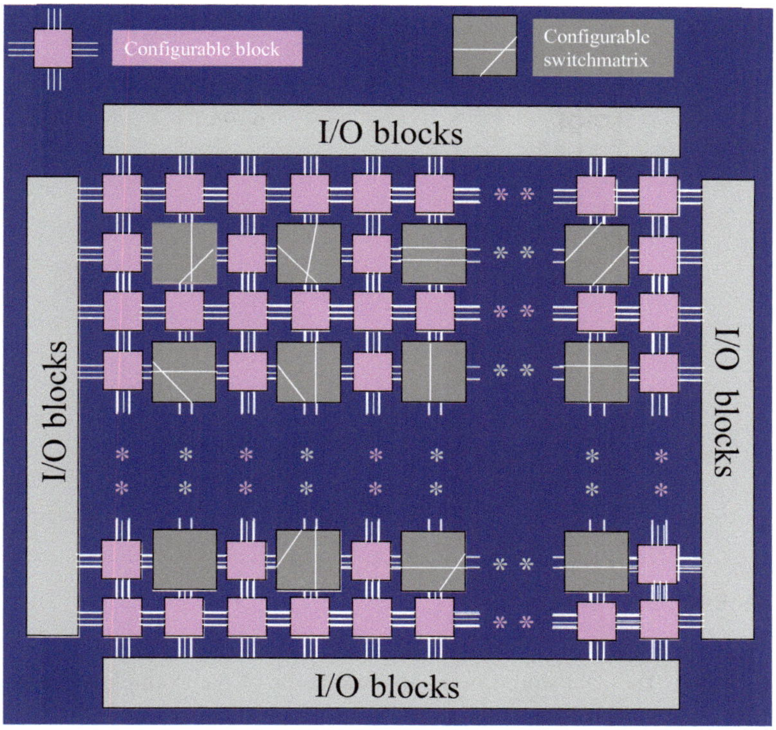

Fig. 7.39 General representation of an FPGA architecture

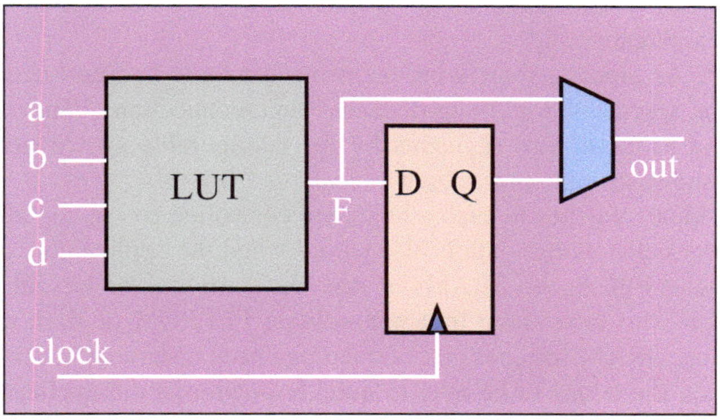

Fig. 7.40 Example of a four-input configurable block

It is basically a small memory consisting of 16 memory cells and a couple of multiplexers. By changing the values in these memory cells (when the application is loaded into the FPGA), any logic function (F) of the four inputs (a, b, c and d)

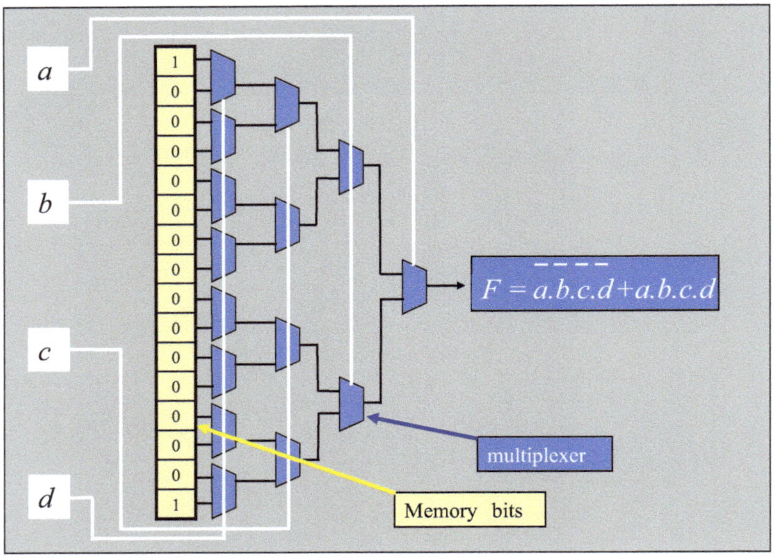

Fig. 7.41 Example of a four-input LUT

can be created. The data stored in the memory cells of the example represents the following logic function:

$F = \bar{a} \cdot \bar{b} \cdot \bar{c} \cdot \bar{d} + a \cdot b \cdot c \cdot d$

The LUT, however, can also serve as a distributed memory in the form of synchronous or asynchronous, single- or dual-port SRAM or ROM, depending on needs of the application.

Many FPGAs contain short wire segments for local interconnections as well as long wire segments for "long-distance" interconnections. The logic blocks are connected to these wire segments by the configurable switch matrix blocks. Figure 7.42 shows an example of such a block.

The individual switches in such a block are controlled by the so-called configuration memory cells, whose data is also stored when the application is loaded into the FPGA. Most FPGAs use SRAMs to store the configuration bits, although there are also a few who store them in a non-volatile EEPROM or flash memory. All FPGAs that use SRAM for configuration storage need a shadow non-volatile back-up memory on the board to be able to quickly download the application into the on-chip configuration memory. Downloading from a software programme would lead to relatively large configuration times, whenever the application is started again after a power down.

Next to the configurable logic and switch matrix blocks, many FPGA architectures include dedicated IP cores, digital signal processors (DSPs), microprocessors such as ARM and PowerPC, single- and/or dual-port SRAMS, flash memories and multipliers.

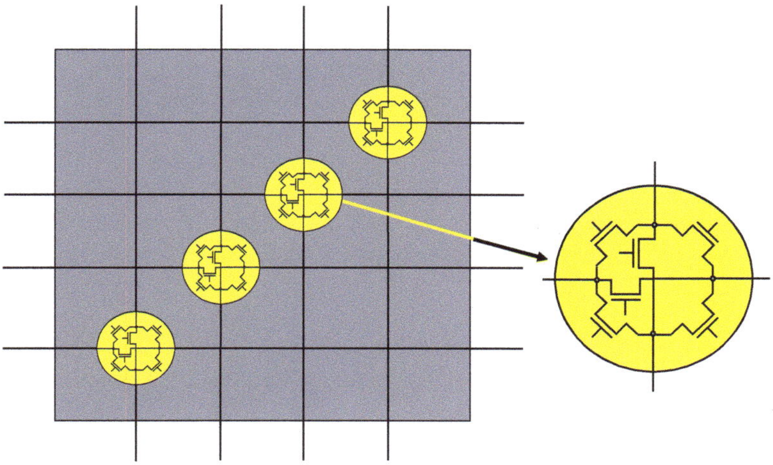

Fig. 7.42 Example of a configurable switch matrix

Finally, most I/O blocks support a variety of standard and high-speed interfaces. Examples of single-ended interfaces are LVTTL, LVCMOS PCI, PCI-X, I^2C, UART, GPIO, USB, GTL and GTLP, HSTL and SSTL. Examples of differential I/O standards are LVDS, extended LVDS (2.5 V only), BLVDS (bus LVDS) and ULVDS, HyperTransport™, differential HSTL and SSTL.

Of course also, several dedicated memory interfaces, such as DDR, DDR-2, DDR-3 and SDRAM and memory controllers, are supported.

Among the state-of-the-art FPGAs are the Xilinx Virtex™-7 family and Altera Stratix 10 FPGA. To get a flavour of the potentials of these FPGAs, some of the characteristic parameters of both families are presented. The Virtex UltraScale family includes about 5.5 million logic cells, 2,880 DSP slices and 88.6 Mb of block RAM and a maximum of 1,456 I/O pins. The Altera Stratix 10 FPGA contains 5.5 million logic elements, an integrated quad-core 64 bit ARM Cortex-A53, a floating-point DSP, 1,980 DSP blocks, 166 Mb embedded memory and a maximum of 1,680 I/O pins. Both FPGAs are fabricated in 14–16 nm FinFET technologies.

The design flow to develop an FPGA application has similarities with the previously discussed standard-cell design flow. An RTL-level VHDL or Verilog description is simulated to validate the system requirements. Next, a synthesis tool maps the design to a netlist, which is then translated into a gate-level description. At this level, the design is simulated again to verify its consistence with the original RTL-level simulation. Finally, this gate-level description is realised by the FPGA logic and sequential resources, while timing data is added. A final simulation, including these timing details, must then confirm whether the system requirements are met. Due to their increasing performance and versatility, FPGAs can be used in a broad spectrum of applications, such as automobiles, high-performance computing, communications and IoT. They can easily adopt new standards and quickly modify hardware for the specific application requirement even during use in the application. According to [15], the global FPGA business almost reached US$ 7 billion in 2021, and it is expected to reach about US$ 12 billion by 2027.

Further details of state-of-the art FPGAs can easily be found on the internet and are beyond the scope of this book.

This section is meant to present the basic architecture of an FPGA and a flavour of the potentials of current state-of-the-art FPGAs. As explained before, most FPGAs reconfigurability (logic as well as interconnect) is controlled by on-chip configuration SRAM memory bits and require additional non-volatile configuration back-up memory on the board.

Complex Programmable Logic Devices (CPLDs)

The structure of a PLD has evolved from the original PAL™devices, which implement sum of products (min terms), where the AND-array is programmable and the OR-array is fixed (see Sect. 7.6.4). Compared to PLAs, PALs lack flexibility but show shorter propagation delays and require less complex software. Figure 7.43 shows an example of the basic PAL architecture, which implements three logic functions.

The connections in the AND-array of the CPLD are commonly realised by programming non-volatile memory cells, which are floating-gate transistors (Fig. 7.44), which means that it can be (p)reprogrammed using in-system programmability and it will securely retain its programme, even when it is powered off.

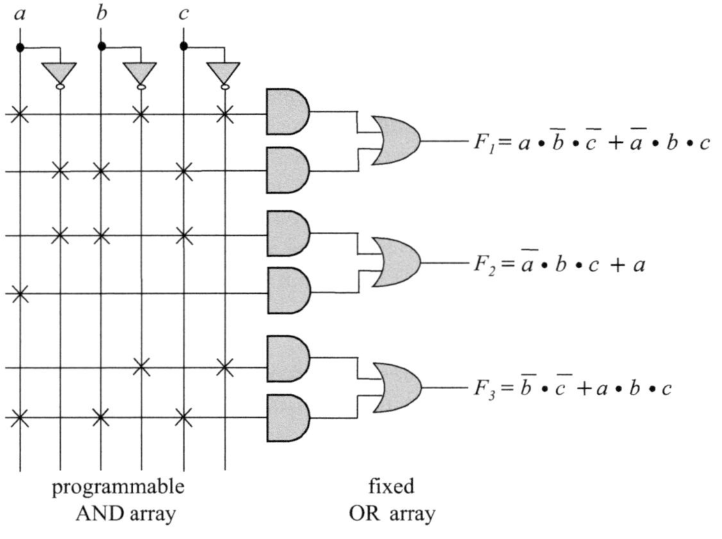

Fig. 7.43 Example of a basic PAL architecture implementing three different logic functions of three inputs

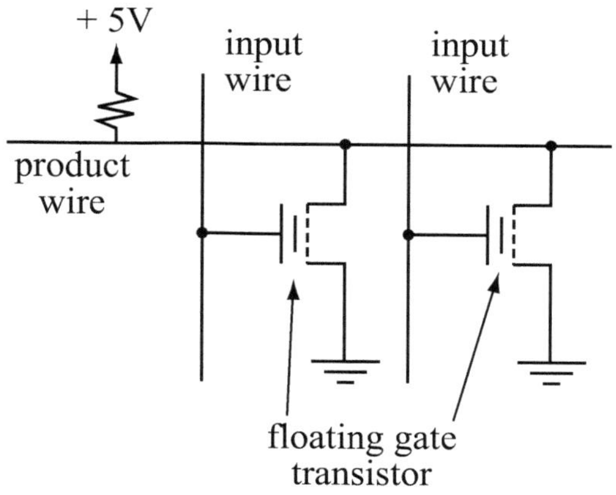

Fig. 7.44 Use of floating-gate transistors to realise AND-array connections in CPLD (Source: IEEE Design and Test of Computers)

There is no technical reason why the previously discussed FPGAs use SRAM or antifuse programming techniques instead of non-volatile, except that the fabrication process will be cheaper.

The original simple PLDs only implemented some tens of logic functions. A large design had to be mapped onto a couple of PLDs, which became a barrier for PLD usage. As a result, ASIC vendors started developing PLDs with much larger arrays, and the *complex PLD* or *CPLD* was born. CPLDs are offered by a large number of vendors, including Altera (MAX II and MAX V families), Xilinx (Coolrunner™II and XC9500XL™series), Lattice Semiconductors (MACHX03 family), Atmel (CPLD ATF15xx family), etc. Most CPLD architectures look very similar and are based on the previously discussed PAL and/or PLA architectures. Since the logic depth of these arrays is relatively short, even wide-input PLD functions offer short pin-to-pin propagation delays. Many of them also include registers, but their total complexity in terms of equivalent logic gates and flip-flops is usually relatively low, compared to FPGAs.

An example of a CPLD architecture is shown in Fig. 7.45.

As stated before, the total complexity of most CPLDs in terms of equivalent logic gates and flip-flops is relatively low, compared to FPGAs. They are therefore often fabricated in relatively conventional process nodes and used in small systems to implement complex finite-state machines, fast and wide decoders or high-performance control logic. Because the functionality is stored in a non-volatile way, most CPLDs are also suited for use in applications where they can be completely switched off during idle times, without losing their functionality as an SRAM-based FPGA would. The high-end (high-complexity) CPLD applications show some overlap with the low-end FPGAs. Because of the large number of flip-flops and their dynamic reconfigurability, FPGAs are much more flexible in use, compared to CPLD.

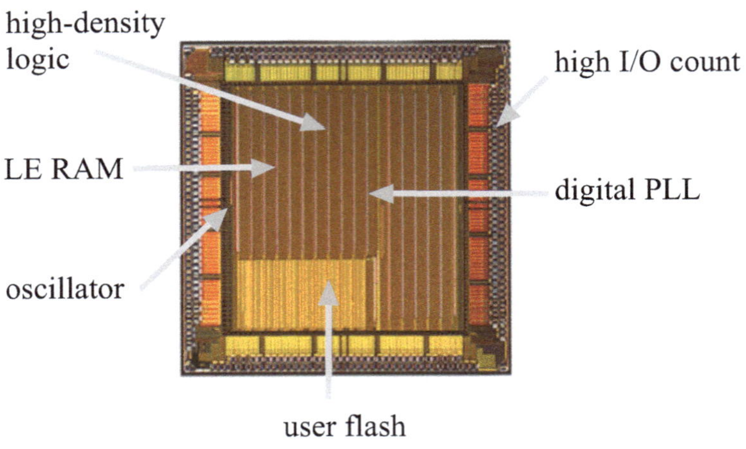

Fig. 7.45 Altera's MAX V CPLD architecture (courtesy of Altera)

Programmability of FPGAs and CPLDs

The most important switch-programming techniques currently applied in FPGAs are SRAM, antifuse and non-volatile memory cells. Figure 7.42 shows an example of a configurable switch matrix to configure the routing of signals through available interconnect patterns. SRAM cells or flip-flops are also used in a lookup table to configure logic functions (Fig. 7.41).

In the majority of current commercially available CPLDs, the switches are implemented as floating-gate devices, like those in (E)EPROM and flash technologies (Fig. 7.44) [16]. However, CPLDs with SRAM programmability appear on the market. Here, the switches are used to programme the AND- and OR-array of the PAL; see Fig. 7.43. In 90% of the CPLDs, the connections are made through programmable multiplexers or full cross-point switches. If an input is not used in a product term (minterm) in an AND plane on a CPLD, the corresponding EPROM gate transistor is programmed to be in the off-state. Similar architectures can be built with EEPROM transistors.

7.6.8 Embedded Arrays, Structured ASICs and Platform ASICs

The previously discussed cell-based designs (Sect. 7.6.5) may include standard cells, macro cells, embedded memory blocks, IP cores, etc. A different approach to cell-based designs is the inclusion of *embedded arrays*. In most cell-based designs that include an embedded array, all masks are customised, as in the cell-based designs. Embedded arrays combine a gate array-like structure and large cells such as microprocessor cores, memories and I/O functions. Cores either can be mapped onto the sea-of-gates array (see Sect. 7.6.6) or can be implemented as a separate block. Figure 7.46 shows the architecture of an embedded array ASIC.

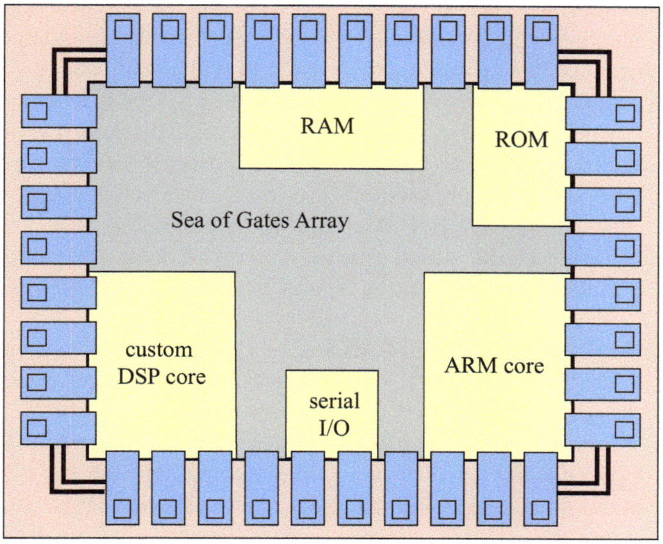

Fig. 7.46 Architecture of an embedded array ASIC (Source: ICE)

The idea behind such an ASIC is to reduce the total turnaround time from spec definition to first silicon. During the first 20% of the spec development time, almost 80% of the system is defined. So, at that time, the engineers know already which memory type (SRAM, DRAM, flash, etc.) and how much memory is needed, what type of IP cores (CPU, DSP, ARM, analog IP, etc.) are needed and also what type of I/Os the application requires. Also, a rough estimation of the required number of logic gates can be made at that time. These are then implemented as a kind of mask-programmable (sea-of-gates) array. The chip is then sent to the fab and is being processed up to the final back-end masks (metal layers and vias), in parallel to the design team defining the remaining 80% of the spec to come to the final spec definition. After completing the spec, only the final metal and via masks need to be defined and processed, thereby reducing the turnaround time and more specifically the time to market. Even last-minute design (spec) changes are allowed. Due to the very small lifetimes of today's products in many consumer and communication markets, it has become very important to have the ability to put prototype products quickly to the market, perform a fast customer product review and transfer it, if necessary, into a high-volume standard-cell design. Toshiba uses this embedded array concept in their UniversalArray ASIC architecture, where the customer can define his own ASIC, with a selection of various available IPs and I/Os and with the logic implemented on a (sea-of-gates) gate array, available in their Fit Fast Structured Array (FFSA) series [17]. In normal standard-cell blocks, the empty areas are filled with *filler cells*, which do not contain any transistor, but are only used to extend the supply lines and n-wells and p-wells and allow routing in most metal layers. Due to the sea-of-gates approach in the universal array architecture, the "empty areas", here, also contain unused transistors and offer additional flexibility for creating small design changes. The first product needs to undergo all mask and processing steps, but redesigns, or derivatives with small changes in the logic content, can be quickly realised by changing only the final metal and via masks and performing only the back-end processing. You need to do the design yourself, using the vendor's technology and design kit. The NRE costs for the first run may be in the order of a few hundred thousand US$ for a 120 nm CMOS design with a few million gates and a few Mb of embedded SRAM up to a million US$ or more for a 60 nm design. It includes the mask costs and delivery of about 100 samples. A new run, with only minor metal mask changes, may cost several hundred thousand US$.

Structured ASICs and Platform ASICs

(Mask-programmable) gate arrays have suffered from a declined popularity over the last decade. This has increased the gap between the cell-based design ASICs and FPGAs. A *structured ASIC* or *platform ASIC* is a combination of the cell-based and FPGA design concepts, which targets prototyping applications and relatively low-volume markets (10k–100k). It offers a large selection of IP cores, which can be customised through a limited number of masks. Basically, personalisation can be done by customising all metal and via masks, by customising only a subset of the metal and via masks or by customising only one via mask. NRE costs are relatively low (from 50k$ to several 100k$), but the price per individual chip can be four to six times the cell-based design version. In the following, a structured array ASIC example is presented to show some capabilities of this category of ASIC products.

eASIC's Nextreme structured array ASIC family This structured (array) ASIC is an example of customisation through only one top level via mask. The Nextreme family (see http://www.easic.com/products) consists of three members, each with different sub-members, offering from 350k to 13 million gates and 56 Mb of embedded dedicated block memory. The most advanced eASIC is processed in 28 nm CMOS [18]. Customisation is done, only through the VIA-6 mask, allowing very short production turnaround times. Figure 7.47 shows an example of eASIC architecture.

It combines various processor and memory cores with peripherals and interfaces. eASIC claims 2–6 weeks design time, followed by 4 weeks of manufacturing. It allows rapid software changes using Diamond processors.

Configurable PLLs and DLLs are embedded for clock generation and clock-phase shifting purposes. Next to a variety of interfaces and I/O standards, also SERDES (serialiser-deserialiser), differential and DDR interfaces are supported through a library of input, output and bidirectional I/Os, which can be configured into a large variety of options and drive strengths.

For prototyping and other low-volume applications, a direct-write e-beam machine is used to perform this VIA-6 customisation to avoid the costly mask production. For high volumes, the custom VIA-6 mask is generated from the same design database.

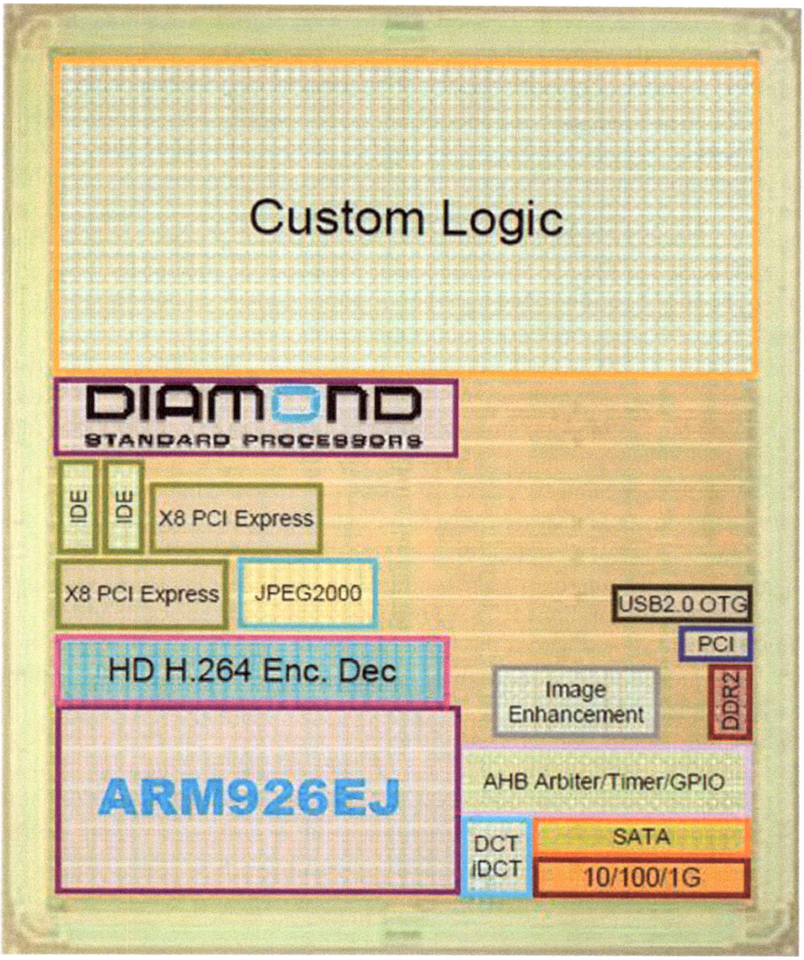

Fig. 7.47 Example architecture of Nextreme structured ASIC (courtesy of eASIC)

Structured ASICs attack the low end of the ASIC market. Although there has already been a "structured arrays ASICs vendor" shake-out, there are more vendors than the ones referred to in this section. The selection that has been made here presents a good flavour of the potentials of available products of this ASIC category.

7.6.9 Hierarchical Design Approach

The *hierarchical layout* design style is characterised by a modular structure (as shown in the heterogeneous chip in Fig. 7.9). The different modules are identified during the design path. With a complex system on chip, for example, the various required functional modules emerge from the specification. These modules may include microprocessor core, ROM, RAM, peripherals and interfaces, etc.

A top-down design strategy generally leads to a satisfactory implementation of a hierarchical layout. The hierarchical division allows various designers or design teams to simultaneously produce layouts of the identified modules. Reasonable gate or bit densities are combined with a reasonable speed. The afforded performance renders the hierarchical layout design style suitable for most VLSI and ASIC designs. The design time for hierarchical layouts can be drastically reduced with good CAD tools. Available libraries contain parameterised *module generators*.

These (mostly) software descriptions are synthesised to produce netlists, which can be used to create layouts of required modules. Assembly of the resulting instances and bond pads leads to the creation of a complete chip layout. Even the assembly and interconnection are automated in *placement and routing* programmes (using P&R and floor-planning tools).

The hierarchical design style, can, of course, include modules which are created by using different layout design styles, e.g. standard-cell or handcrafted module layouts. The hierarchical style, for a conventional two-metal layer design, was disadvantaged by the relatively large routing areas that could be necessary. However, with the present availability of six to more than ten metal layers, interconnections and buses can be routed across the logic blocks. In some cases, however, the chip area may not be optimum as a result of the *Manhattan skyline* effect, which results from different block shapes.

Figure 7.48 shows the *meet-in-the-middle strategy* used in the hierarchical design approach. This strategy was already introduced by Hugo de Man in the early 1980s [19]. Here, the high-level system description is used to synthesise a design description comprising macro blocks at the implementation level. This

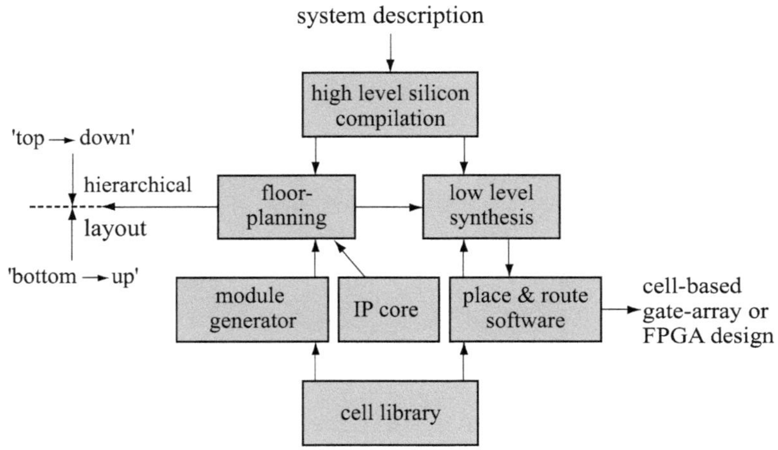

Fig. 7.48 Meet-in-the-middle strategy

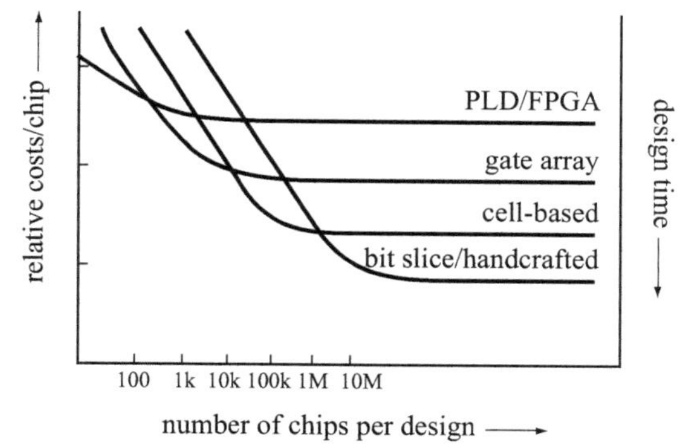

Fig. 7.49 Cost comparison of the different layout implementation forms

implementation level lies roughly in the middle of the top-down design path. The choice of implementation form is still open at this level, and possibilities may include a cell-based, gate array or FPGA design. It must be possible to generate these macros from existing design descriptions. Sometimes, module generators are also used to generate a core. The (re)use of IP cores allows a fast "plug-in" of different functional blocks, which are standardised to a certain extent. Clearly, the results of design and layout syntheses meet at the implementation level.

7.6.10 The Choice of a Layout Implementation Form

The unique characteristics of each form of layout implementation determine its applicability. The choice of implementation form is determined by chip performance requirements, initial design costs, required volumes and time-to-market requirements. Figure 7.49 shows a cost comparison of the different forms of layout implementation.

A single chip may combine different implementation forms. The previously discussed embedded array ASICs and structured ASICs are examples of this. Figure 7.50 shows a photograph of a conventional microprocessor in which handcrafted, bit-slice and memory layout styles are combined. Particularly ICs that require fast and complex data paths, which usually include a memory, one or more address counters and ALUs, may combine data path layout with standard-cell, memory and full-custom design.

An implementation technique that was popular in the 1980s and early 1990s and is still used in some cases today is the symbolic layout and compaction technique. A *symbolic layout* is a technology-independent design, which can be used for every layout implementation form. In a symbolic layout, transistors and contacts are represented by symbols whose exact dimensions are unspecified while wires are represented by lines whose widths are also unspecified. The abstract symbolic layout is transformed to an actual layout by a compaction programme, which accounts for all of the design rules of the envisaged manufacturing process.

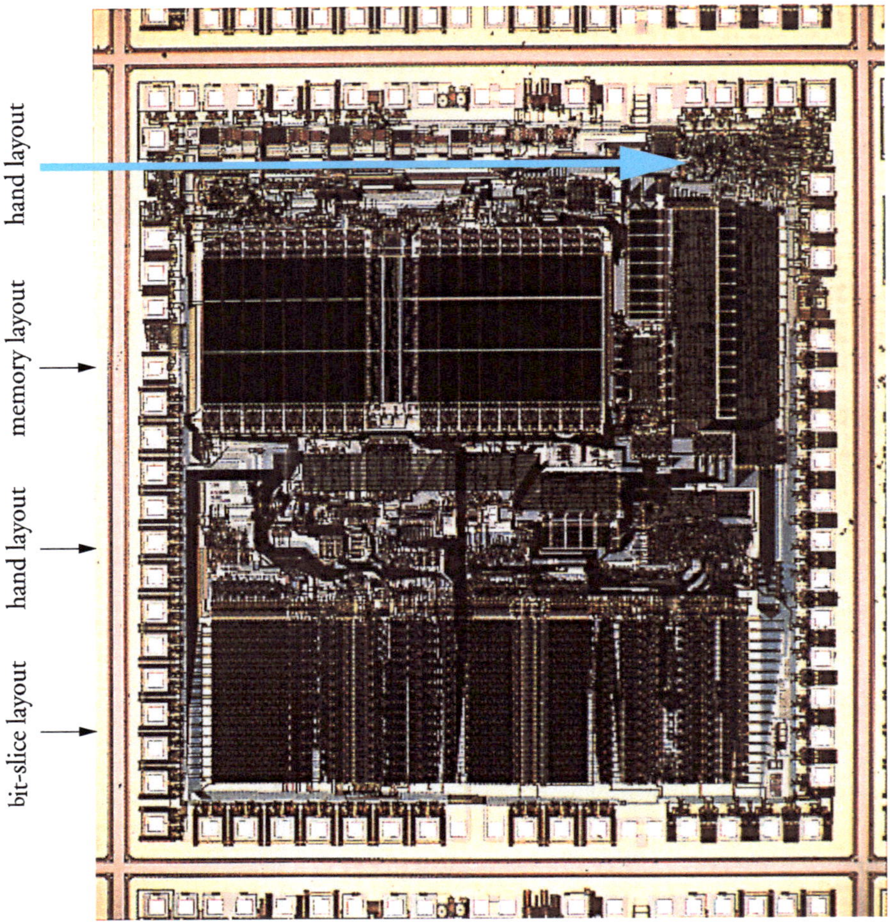

Fig. 7.50 A conventional microprocessor chip which combines different layout implementation forms (Source: NXP Semiconductors)

The symbolic layout technique allowed a short design time and relieves designers of the need to know specific layout and technology details. The technique is, however, disadvantaged by the associated relatively low gate density and low switching speed. These compare unfavourably with handcrafted layout results. Furthermore, the abstract nature of a symbolic layout only loosely reflects technological aspects. This may result in fatal design errors. While the concept of symbolic layout may not be as prevalent in modern chip design, the principles of design hierarchy, abstraction and verification are still fundamental in creating efficient and reliable semiconductor devices.

Finally, the dimensions of all circuit components and wiring in an IC layout are scaled versions of the actual on-chip dimensions. This *geometric layout representation* is generally described in a *geometric layout description language (GLDL)*. Such languages are common to many CAD tools and usually serve as the data-interchange format between IC design and manufacturing environments. A GLDL has the following typical features:

- It facilitates the declaration of important layout description parameters, e.g. masks, resolution and dimensions.
- It facilitates the definition of geometrical forms, e.g. rectangles and polygons.
- It facilitates the definition of macros, e.g. patterns or symbols.
- It enables transformations, e.g. mirroring and rotation.
- It contains statements for the creation of matrices.

Currently, *GDSII* (GDS = graphic design system) is the de facto standard for physical chip design exchange in the semiconductor industry.

7.7 Conclusions

This chapter introduces various VLSI design and layout realisations and their characteristic properties. A top-down design approach, combined with a bottom-up implementation and verification through a hierarchical layout style, appears suitable for most VLSI circuits. In practice, the design process consists of a number of iterations between top-down and bottom-up paths, the aim being to minimise the number of iterations.

The use of IP cores that are available from in-house resources and external vendors is fuelling the reuse of existing functionality, such as microprocessor and signal processing cores and memories, analog, interfaces, etc. This reuse increases the problems with timing and communication between cores from different origins. Chapter 9 discusses these problems in detail.

During the last decade, the design complexity of an ASIC has dramatically increased and caused the design costs to increase almost an order of magnitude (see Chap. 11). This has put a permanent pressure on the efficiency of the design process. Semiconductor companies have built application domain-specific platforms, which are key to a higher design productivity and improved product quality. Since IC production fabs are becoming extremely expensive, more companies will share the same production facility and production process and become *fab-lite* (outsourcing more and more of the manufacturing operations) or even *fabless*. So, semiconductor (design) houses can then only differentiate themselves to design better products faster and cheaper.

Various ASIC design and implementation styles have been presented. Standard-cell designs, mask-programmable and field-programmable gate arrays and structured ASICs all are different in the way they are designed, in the way they are fabricated and in the way they are used in an application. The choice of ASIC style largely depends on the required turnaround time and product volume. In the continuous search for increased design productivity, generative AI may open new ways to assist the design effort [20].

A good IC design must be accompanied by a good test and debug strategy. Testability and debug are discussed in Sects. 10.2 and 10.6, respectively, and require considerable attention during the design phase. The use of an extra 5% of chip area to support testability and debug might, for instance, lead to a 50% reduction in test costs.

7.8 Exercises

1. Why are abstraction levels used for complex IC designs?
2. What is meant by floor-planning?
3. Explain what is meant by logic synthesis.
4. What does the term "Manhattan skyline" describe in relation to a VLSI layout?
5. Assume that a standard-cell and a gate array library are designed in a CMOS technology. The libraries consist of logic cells with identical logic functions. Describe the main differences between the two libraries in terms of:
   (a) Cell design
   (b) Chip area
   (c) Production time and cost
   (d) Applications
6. Random logic functions can, for instance, be implemented using an ROM or a standard-cell realisation. Explain when each of these possibilities is preferred.
7. Draw a schematic diagram of a PLA which implements the following logic functions:
   $F_0 = \overline{\overline{xy} + xyz}$ $F_1 = \overline{x\overline{y} + \overline{x}y + x\overline{z}}$ $F_2 = \overline{xyz + \overline{xyz}}$
8. Explain what is meant by mixed-level simulation.
9. Explain in your own words what is meant by IP. What is the cause of its existence? How can it affect design efficiency, and what are the potential problems involved with it?
10. Explain the differences between an FPGA and a CPLD.
11. Explain the "meet-in-the-middle" strategy.
12. Explain why a cell-based design implementation is much smaller than a design implemented with an FPGA.

References

1. A. Abbo et al., 'XETAL-II: A 107 GOPS, 600 mW massively-parallel processor for video scene analysis, ISSCC Digest of Technical Papers, San Francisco (2007)
2. R. Goering, Startup Liga promises to rev simulation, EETIMES, 17 July 2006
3. P. Coussy et al., An introduction to high-level synthesis. IEEE Design Test Comput. **26**, 8–17 (2009)
4. E. Oruklu, et al., System-on-chip design using high-level synthesis tools. Circuits Syst. **3**, 1–9 (2012)

5. K. Karras et al., High-level synthesis case study: Implementation of a memcached server, in *1st International Workshop on FPGAs for Software Programmers (FSP 2014)*, Munich, Germany (2014)
6. D. Dako, 5 Elements of Future-Proof Multisite Design & IP Management (2022). https://besttechviews.com/design-ip-management-5-elements-of-future-proof-multisite-ddm/
7. C. Forzana, D. Pandini, Statistical static timing analysis: a survey. Integr. VLSI J. **42**, 409–435 (2009). Elsevier
8. S.J. Wan et al., Fast and accurate statistical static timing analysis, in *2014 IEEE International Symposium on Circuits and Systems (ISCAS)* (2014)
9. A. M. Baker, Y. Jiang, Modeling and architectural simulations of the statistical static timing analysis of the non-gaussian variation sources for VLSI circuits. Int. J. Sci. Res. Pub. **3**(1) (2013)
10. A. Malik, et al., VLSI: Techniques for Efficient Standard Cell Placement, ISSN: 2347-2200/V2-N1/PP-17-21, IJSE-ITS: Race-2014
11. A. Hassan, Fundamentals of floor planning a complex SoC. Electronic Design, 21 March, 2012
12. I. Okhura et al., A novel basic cell configuration for CMOS gate-array, in *CICC* (1982), pp 307–310
13. H.J.M. Veendrick et al., An efficient and flexible Architecture for High-Density Gate Arrays, ISSCC Digest of Technical Papers, San Francisco (1990)
14. Z. Or-Bach, FPGAs as ASIC Alternatives: Past & Future, EETIMES. 21 April 2014
15. V. Srinivas Durga Prasad, FPGa market trends with next-gen technology. Design & Reuse (2022)
16. See current CPLD architectures on the CPLD vendors websites: Altera, Xilinx, Lattice, Cypres, etc. (2016)
17. Structured Arrays/Gate Arrays; FFSA/Fit FAST Structured ARRAY (2014). http://toshiba.semicon-storage.com/eu/product/asic/structured-arrays.html
18. M. Maxfield, Is it an ASIC? Is it an FPGA? No, itś eASIC!, EETIMES 14 Sep 2015
19. H. de Man et al., An intelligent module generator environment, in *Proceedings of the 23rd Design Automation Conference* (1986), pp. 730–735
20. R. Merritt, S. Volley, Designers Tap Generative Ai for a Chip Assist (2023, October 30). https://blogs.nvidia.com/blog/llm-semiconductors-chip-nemo

Less Power: A Hot Topic in IC Design

8

8.1 Introduction

Although already used in the 1970s, it took until the mid-1980s before CMOS became the leading edge technology for VLSI circuits. Prior to that time, only a few designs were implemented in CMOS, particularly the ones that really required the low-power features of CMOS. Most examples, then, were battery-supplied applications, such as wristwatches (tens of millions per year), pocket calculators, portable medical devices (hearing aids and implantable heart controls) and remote controls.

From the 1970s until today, however, the number of transistors increased from only a few thousand to more than a hundred billion, while chip frequencies, particularly in the high-performance processor category, increased from several megahertz to several gigahertz. In that period, the power consumption of these ICs increased from less than 1 W to above 100 W, while the power consumption of the ASIC category of ICs has reached the level of one to several Watts, which is in the range of the maximum allowed power consumption of a cheap plastic package. This is one of the main driving forces for low-power CMOS. It was also the reason for switching from nMOS to CMOS technology in the early 1980s.

Currently, the requirement to also have access to powerful computation at any location is another driving force to reduce CMOS power dissipation.

The increasing number of portable applications is a third driving force for low-power CMOS. In the consumer market, we can find examples such as games, smart phones, tablets, photo and video cameras, GPS systems and flat screen TVs. In the PC market, an increasing percentage of computers is sold as notebook or laptop computers. Digital cellular telephone networks, which use complex speech and video compression algorithms, form a low-power CMOS application in the telecommunication field.

Finally, the emerging multimedia market will also show many new products in the near future. At the time of going to print, we see the portable full motion video and graphics as examples of such low-power applications. The personal digital

H. Veendrick, *Nanometer CMOS ICs*,
https://doi.org/10.1007/978-3-031-64249-4_8

assistant (PDA) available during the 1990s has fully been replaced by smart phones and tablets. The development of these portable and handheld devices has increased the drive for significant battery performance improvements. Therefore, the next section in this chapter will present a short summary on existing battery technologies.

Another important driving force for low power is the future system requirement. In a 2-nm CMOS technology (high volume expected around 2026), more than ten trillion transistors can be packed on a board of 20 by 20 cm with very high-density packaging techniques (multi-chip modules (MCM), system-in-a-package (SiP) and system-on-a-package (SoP), stacked chiplets, etc.). These cannot be realised with the power levels of current systems. Particularly in AI applications, the power consumption may be so high that liquid or even immersion cooling is required [1].

In general, less power also leads to simpler power distribution, less supply and ground bounce and a reduction of electromigration and electromagnetic radiation levels. A low-power design attitude should therefore be common in every IC design trajectory, because it is beneficial for power consumption, robustness and reliability of current and future ICs and systems.

8.2 Battery Technology Summary

A *battery* is usually built from more than one cell, which can chemically store energy for a certain period of time. Based on the difference in the chemical process, we can distinguish two different types of batteries. *Non-rechargeable batteries* use the so-called primary cells with a non-reversible chemical reaction and must be hand in as small chemical waste, when empty. These *primary battery cells* perform much better in terms of charge capacity, charge storage and charge leakage but are less cost-efficient in high-performance systems or systems that are always on. *Rechargeable batteries* use *secondary battery cells*, which deliver energy by transforming one chemical substance into another. This transformation is reversible in that it can be put back into its original chemical state during recharging. In battery-operated systems that need frequent replacement of the batteries, rechargeable batteries would be a more economically viable solution. But in applications which need a long battery lifetime, e.g. a year for an electronic clock, rechargeable batteries must be recharged at least every 3 months, while a non-rechargeable battery may "tick" for more than a year. Most implantable pacemaker batteries are lithium-based, draw only 10–20 μA and last between 5–10 years [2].

The growing diversity of battery-operated systems, combined with the increasing performance and longer battery lifetimes, requires an improved *battery energy* efficiency, while smaller weight and shrinking dimensions require a reduced number of stacked battery cells. The performance of cells in series is substantially worse than that of individual cells. A single-cell battery with both high cell voltage and high energy efficiency is a real need in many applications.

Advances in rechargeable battery technologies are aimed at improving the battery capacity per unit of volume. Nickel-cadmium batteries have dominated the battery

Table 8.1 Characteristics of rechargeable batteries

Battery type	Nominal cell voltage [V]	Energy/volume [Wh/l]	Energy/weight [Wh/kg]	Self-discharge rate [%/month]
Nickel-cadmium (NiCd)	1.2	200	100	20
Nickel-metal hydride (NiMH)	1.2	300	150	30
Lithium-ion/lithium polymer	3.7	400	200	5–10

market, but they suffer from low cell voltage and low energy efficiency (see Table 8.1).

The nickel-metal hydride batteries are rapidly replacing the nickel-cadmium ones because of the higher energy capability. Both the NiCd and NiMH types of batteries suffer from the so-called memory effect. In many applications, these batteries are recharged before they are sufficiently or completely empty. After many of such recharge operations, the battery starts suffering from a so-called voltage depression, also known as the *battery memory effect*, which reversibly degrades its energy storage capacity. They therefore need a periodic deep discharge to prevent this memory effect.

During the last decade, single-cell lithium-ion and lithium-polymer (Li-pol) batteries have emerged as the more favoured choice. In a Li-pol battery, the lithium electrolyte is a solid polymer medium as compared to the organic solution in a Li-ion battery. They both offer a higher cell voltage and a higher energy density (up to 400 Wh per litre). Because lithium is one of the lightest elements (third) of the periodic system, it helps to save weight, particularly in tiny handheld devices. Moreover, the *self-discharge rate* is only a few per cent, and they hardly exhibit the memory effect. A major disadvantage of the lithium batteries is their sensitivity to over (dis)charge or short circuit, because this can cause them to ignite or even explode. Li-ion and Li-pol battery packs therefore may contain internal protection circuits to monitor its voltage to prevent battery damage and its temperature to disconnect from the application, in case it gets too hot. These batteries should therefore not be used in applications in which they could be exposed to high temperatures.

As the world becomes more mobile, the demand for better battery technology will continue to increase. Most of these applications are in the range of 10 mW (jpeg encoding in a cell phone) to 10 W (peak power in a mobile device). However, the incremental improvements in battery technology do not keep pace with this increase in battery demand, and, as such, it puts an additional burden on the shoulders of the IC design community by requiring a more intensive use of less-power design methods. Recently, there is an increased R&D effort on the development of *lithium-air batteries*, also known as Li-O_2, which are expected to provide more than four times the energy of current lithium-ion batteries [3]. However, a lot of challenges lie ahead and must be addressed before they become available for high-volume applications [4]. Finally, compared to today's battery energy density, the move to other materials for the cathode and anode will result in only a limited energy density increase of about a factor of five.

More information on battery technologies can be found in [5] and [6] and by scanning the internet for further trends in battery technology.

8.3 Sources of CMOS Power Consumption

During the operation of CMOS circuits, there are four different sources that contribute to the total power consumption:

$$P_{total} = P_{dyn} + P_{stat} + P_{short} + P_{leak} \tag{8.1}$$

where P_{dyn} represents the dynamic dissipation.

This is the power dissipated as a result of charging and discharging (switching) of the nodes and can be represented by the following equation:

$$P_{dyn} = C \cdot V^2 \cdot a \cdot f \tag{8.2}$$

where C is the total capacitance,
V is the voltage swing,
f is the switching frequency and
a is the activity factor.

The activity factor represents the average fraction of gates that switch during one clock period. This number can be as low as 0.05 (low activity), for example, but it can also be as high as 2–4 (very high activity) because of hazards; see Sect. 8.5.3.

P_{stat} represents the static dissipation. This is the power dissipated as a result of static (temporary or continuous DC) current. In Sect. 8.5, the basic causes of the different contributions are explained in detail.

The contribution of the short-circuit dissipation is represented by P_{short}. This is the power dissipated in logic gates as a result of short-circuit currents between supply and ground during transients.

Finally, the last contribution to the total power dissipation is made by the leakage dissipation P_{leak}. This is the power dissipated as a result of subthreshold leakage currents, gate leakage currents and substrate leakage currents. Both technology and design can affect several of these power dissipation contributors; see Table 8.2.

Table 8.2 Power dissipation contributors

Contributor	Technology-dependent	Design-dependent
P_{dyn}	x	x
P_{stat}		x
P_{short}		x
P_{leak}	x	x

The following sections discuss the technological and design measures that can be taken to reduce the different power consumptions.

8.4 Technology Options for Low Power

As can be seen in Table 8.2, technology can affect both the dynamic power dissipation and the leakage power dissipation.

8.4.1 Reduction of P_{leak} by Technological Measures

As a result of scaling the channel length over generations of technologies, we arrived at a point (when channel lengths became less than 0.5 μm), where we also had to reduce the supply voltage to limit the electrical fields inside an MOS transistor; see Chap. 2. Between the 0.8 µm CMOS technology node and the 120 nm node, the supply voltage has been gradually reduced from 5 to 1.2 V. Reducing the supply voltage means that the circuits become relatively slower. Therefore, the threshold voltage also has to be reduced.

This has severe consequences for the leakage currents as well as for the noise margin within digital circuits. Because of the subthreshold (weak-inversion) and gate leakage currents, as discussed in Chap. 2, we will have a leakage current through an nMOS transistor when its gate voltage is at zero volt. The higher the threshold voltage, the less leakage current will flow at $V_{\text{gs}} = 0\,\text{V}$.

Let us define the subthreshold slope $s_{\text{subthr.}}$ to be the change in threshold voltage, causing a tenfold increase of the subthreshold current at $V_{\text{gs}} = 0\,\text{V}$. In current technologies, $S_{\text{subthr.}}$ is between:

$$63\,\text{mV/decade}(I) < s_{\text{subthr.}} < 80\,\text{mV/decade}(I) \tag{8.3}$$

This means that a reduction of the threshold voltage of 100 mV leads to an increase of leakage current (at $V_{\text{gs}} = 0\,\text{V}$) of a factor close to 18. It should be clear that, for power and speed reasons, an optimum has to be found for the threshold voltages of both nMOS and pMOS transistors.

Example Assume a reference transistor with an aspect ratio $W/L = 140\,\text{nm}/28\,\text{nm}$. If $V_{\text{T}} = 0.35\,\text{V}$, then its leakage current might be 80 pA. Suppose the threshold voltage shifts to 0.25 V, now the current will increase to approximately 2 nA. Present standby currents in large RAMs can vary from nano-amperes to milli-amperes, depending on their storage capacity and application environment.

With decreasing channel lengths, the threshold voltage also decreases as a result of the small channel effects (threshold voltage roll-off; Chap. 2). Consequently, the threshold voltage can be as low as 0.2 V for a minimum channel length transistor in a 32-nm CMOS technology. Also, for real low-voltage applications, the threshold voltage should be low to allow for a certain speed. However, at these low-threshold voltages, the circuits suffer from a relatively large loss of power caused by leakage currents , especially in the standby mode.

There are several solutions to this problem. One is to vary the threshold voltage by applying a back-bias voltage during *standby mode* [7]. Depending on the K factor in the equation for the threshold voltage (Eq. (1.16)), the threshold voltage can be increased by about 100 millivolt by applying a negative p-well bias, for an nMOS transistor, or a positive n-well bias, for a pMOS transistor. These additional *back-bias* voltages (both nMOS and pMOS need back-bias in the standby mode) can either be supplied by additional supply pads or generated on the chip. The back-bias voltage can be offered to the complete chip or only to distinguished cores (e.g. processors or memory cores) that need to be put in standby, while others remain active. To be able to apply a different bias voltage to a limited number of logic and/or memory cores, the p-well areas of these cores need to be isolated from p-well areas of the cores that remain active and don't need a p-well bias. This can be realised by a so-called *triple-well technology* that offers an additional third well [8], also called *deep n-well* or *buried well.*

Figure 8.1 shows a cross section of a triple-well device. In this technology, the p-well and n-well areas can respectively be connected to V_{ss} and V_{dd} or to separate p-well and n-well bias voltages. The nMOS transistors are isolated from the substrate. A triple-well technology has another important advantage. Because it physically separates all p-well areas from the p^{-}-substrate, it is much more difficult for the noise induced into the nMOS substrates (p-wells) to propagate through the triple well into the p^{-}-substrate. This is of particular importance in designs that combine analog and digital circuits on one IC. In a triple-well technology, the analog circuits are better isolated from the digital noise. Finally, it also offers reduction of the n-well resistance leading to a reduced latch-up sensitivity and thereby improving reliability.

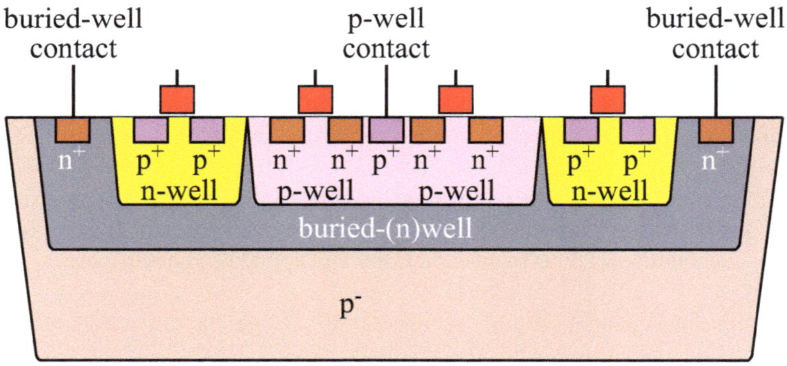

Fig. 8.1 Cross section of a triple-well device

The triple well itself also isolates the complete p-type wafer substrate from the transistor sources and drains. It prevents the electrons and holes generated in the substrate by an α-particle or neutron hit from reaching these source and drain areas. As such, using the triple well will also reduce the *soft error rate* (*SER*; Sect. 9.3.5).

Active Well Biasing for Leakage Power Reduction

The previous discussions are not only limited to the assignment of a fixed voltage to the substrate or n-well. Also, dynamic regulation of the well bias to vary the threshold voltage (V_T) for reduction of the *leakage power* is applied. However, the continuous scaling of the device feature sizes introduced short-channel effects (SCEs), in which the extended depletion layers around the source and drain junctions lead to the so-called V_T-roll off (Chap. 2). This has required the implantation of compensation dope, locally in the channel around the source and drain areas (halos). The negative side effect of this local dope is an increased junction leakage current, particularly when the junctions are reverse-biased, e.g. by using well biasing. Also in certain future technologies, the small gate oxide thickness causes the gate leakage to reach the same order of magnitude as the subthreshold leakage current. The diagram in Fig. 8.2 shows the leakage current in a 65-nm nMOS transistor as a function of the gate and well-bias voltages.

In this particular example, a negative well bias causes an increase rather than a reduction of the transistor off-current ($V_{gs} = 0$). Similar diagrams can be derived for the pMOS transistor. This effect is dependent on the process and the parameters (V_T) and physical dimensions of the transistors (channel length and gate oxide thickness) in that process. Devices in a general-purpose process show leakage properties different from those in a low-leakage process. A high-V_T device also behaves differently from a low-V_T device. So, the influence of a *well-bias* voltage on the *standby current* of a core is very much related to which device is used in which technology. The general trend is that this effect will only become stronger in smaller bulk-CMOS processes, since they require a continuous increase of the halo doping [9].

Consequently, for the 65-nm CMOS node and beyond, *well biasing* is an increasingly less effective instrument for reducing the leakage current during the standby mode. Active well biasing is also seen as a means to compensate for process parameter spread, because the circuit then only runs on subthreshold leakage currents. This can be done in an adaptive way. Both *adaptive body bias (ABB)* and *adaptive voltage scaling (AVS)* can be used to compensate for process variations [10, 11]. The design implications of these techniques are discussed in Sect. 8.4.2. The back-bias effect may be much reduced in case of FinFETs, because they usually have a relatively thin body which is fully depleted and has a relatively low channel dope. Also, since the gate fully surrounds the body on three sides, it has a much better control over the channel, and leakage paths far from the gate are reduced.

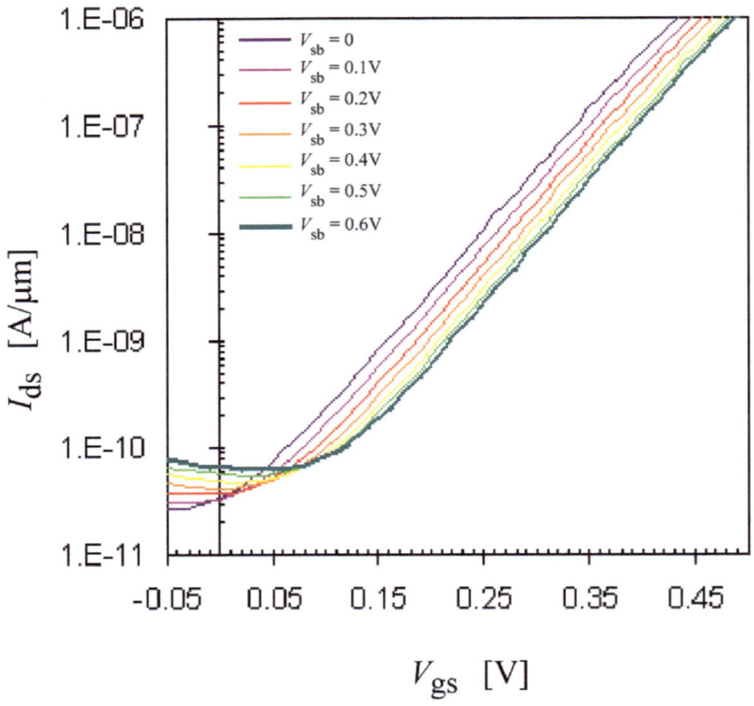

Fig. 8.2 Drain current versus gate voltage at different well-bias voltages

A second approach to reduce standby (leakage) currents is to use multiple thresholds [12]. Now, the power supply of the core (with low V_T circuits) is switched by a very large transistor with high V_T; see Fig. 8.3.

The width of this transistor is such that there is only a marginal voltage drop across it. However, all storage cells and memories in the core must be connected to the permanent power supply and have a high V_T, in order not to lose the cell data. Although the figure suggests the use of both an nMOS and a pMOS switch, usually only one *power switch* is used. An nMOS power switch has the advantage of occupying less area, due to their larger current capability. However, many CMOS circuits use the ground as a reference for signals and protection devices. An nMOS power switch would therefore need a change in related design concepts and introduces additional risks. Therefore, pMOS power switches are more commonly applied. The use of high-V_T power switches also introduces additional noise peaks into the power supply network. When a large core is switched of, its intrinsic decoupling capacitance will fully be discharged after a certain period of time. When the core is switched on again, large current peaks flow through the supply network to fully bring the core back to the supply voltage level. These peak currents can be reduced, by implementing the power switch as a combination of

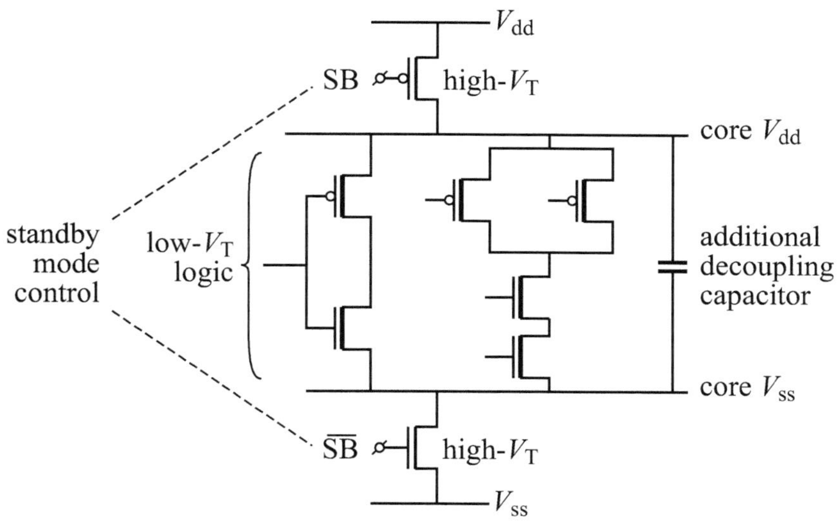

Fig. 8.3 Power supply switch in a multiple V_T environment

many smaller power switches in parallel and then successively switching them on one by one.

Another way of using a multi-V_T *(MTCMOS)* or dual-V_T technology is to design all library cells with low-V_T (LV_T), standard-V_T (SV_T) and high-V_T (HV_T) transistors. A smart synthesis tool can then implement most logic paths with HV_T cells and only use LV_T (or SV_T) cells in the critical paths. For many designs, this means that only about 10% of the logic is built from low-V_T cells, resulting in a leakage power reduction close to one order of magnitude.

In planar silicon-gate CMOS processes, different V_T is created by a different *threshold implant (V_T-implant)* . This changes the substrate dope concentration N_A in the related transistor channel (see expression (1.16) in Sect. 1.5). Every additional V_T-implant is at the cost of an additional mask and implantation step. A dual-V_T design therefore requires four V_T-implant masks and process steps: two for the nMOS and two for the pMOS transistors and four different implantation steps.

For *high-ϵmetal-gate transistors*, including FinFETs, a different V_T is created by the deposition of a different *work-function layer* in the transistor architecture, which changes the V_x parameter in the above-mentioned expression. Figure 3.56 shows such a work-function layer. A dual-V_T design based on these transistors, therefore, requires also four different masks and the deposition of at least four different work-function layers. This causes an increase in process complexity. In sub-20-nm process nodes, foundries may offer a large variety of threshold voltages to support all possible design specifications that customers may require for their specific application domain. As an example, in a 7-nm process, the following variants may be offered for the LV_T-transistor only: LV_T (low V_T), ULV_T (ultra-low

V_T), ELV_T (extreme-low V_T), LV_T LL (low-V_T low-leakage) and ULV_T LL (ultra-low-V_T low-leakage). Because not every foundry offers the same process options and these options may change rapidly over process node and over time, the reader is advised to search the internet for more related information.

An alternative way to reduce the subthreshold leakage currents is to use longer than minimum transistor channel lengths. Due to the V_T-roll off effect, as discussed in Chap. 2, the threshold voltage increases with the channel length. These *long-channel devices* exhibit both smaller leakage currents (I_{off}) and on currents (I_{on}). So, depending on the application area, the designer may decide to use library cells with non-minimum channel lengths, when available.

The other major component in the total leakage current of a transistor is the *gate oxide leakage current*, which is important for an oxide thickness below 2.5 nm (see Chap. 2). Particularly for an oxide thickness below 2 nm [13], this leakage component may become larger than the subthreshold leakage. Current low-leakage processes may offer, next to the dual or multi-V_T option, also a dual t_{ox} option. These options can only be fully exploited when they are supported by the libraries and tools to efficiently reduce the leakage power components in standby mode.

8.4.2 Reduction of P_{dyn} by Technology Measures

In the following formula for the dynamic dissipation, both capacitance C and voltage V are partly determined by the technology:

$$P_{dyn} = C \cdot V^2 \cdot a \cdot f$$

Generally, the load (capacitance) of a logic gate is formed by the interconnection capacitance, the gate capacitance (fan-in of the connected logic) and the parasitic junction capacitances in the driving logic gate itself.

A reduction of the gate capacitance means a thicker gate oxide, which also affects the β and thus the speed of a MOS transistor dramatically. So, this is no alternative to reduce the capacitance.

The reduction of the interconnect capacitances depends on the thickness and the dielectric constant of the oxide and on the track thickness; see Fig. 8.4.

As a result of resistive and electromigration effects, the thickness of the metal layers could only be reduced when other metals were used for routing. In this respect, copper was found to be a good candidate. The sheet resistance of copper is about 40% smaller than that of aluminium. However, copper diffuses through oxides and needs therefore to be fully encapsulated within a barrier (Chap. 3). This reduces the effective resistance improvement to only 25%. This advantage is used to reduce the copper track thickness with 25%, so that they exhibit about the same resistance as aluminium tracks, but at a much lower mutual capacitance. This maintains the signal propagation across the interconnect, while at the same time the cross-talk and power consumption are reduced.

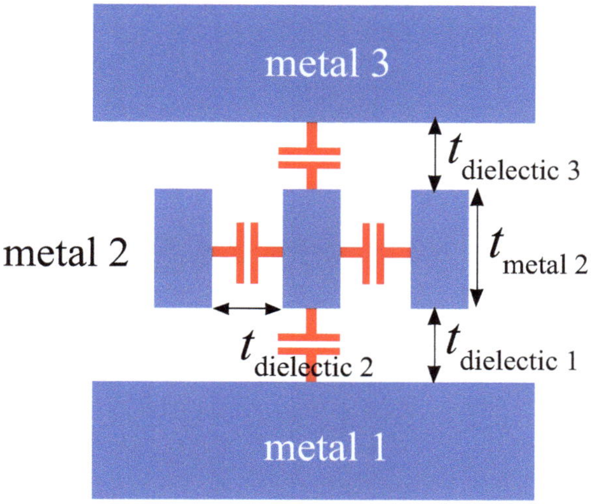

Fig. 8.4 Cross section of multilevel metal to show capacitance

Thicker oxides require more advanced planarisation steps. The space between two metal tracks in the same layer cannot be increased much, as it will increase the chip area as well. It thus hardly affects the power dissipation, because the metal lines would become longer then.

One way to decrease the dielectric capacitance is to find materials with a lower ε_r. The ε_r of SiO_2 is around 4; the ε_r of air is 1. Current values for ε_r are between 2.5 and 3. A value close to two may be achievable in the future. A lot of R&D has already been and is still being carried out on the creation of *air gaps* between closely spaced metal lines [14]. Air gaps influence the electromigration properties. It has been introduced in a product in 2016 but only in the lower metal layers, because they carry less current (see Chap. 3: Fig. 3.70 and reference [92] in Chap. 3).

Junction capacitances are formed by the depletion regions of the source and drain junctions of both nMOS and pMOS transistors. The thicknesses of the depletion regions and, therefore, the values of their capacitances are determined by the dope of the n^+ and p^+ regions. A reduction of the junction capacitances is not expected, because of the increasing dope of halo implants needed for the suppression of the short-channel effects (Chap. 2). An alternative to the mainstream current CMOS processes for low power might be a silicon on insulator CMOS process, which is discussed in Sect. 3.2.4.

8.4.3 Reduction of P_{dyn} by Reduced Voltage Processes

The decrease of the channel length over generations of technologies has increased the peak of the electrical field in the pinch-off region near the drain to unacceptable values. For a 0.7 μm technology, LDD structures (Sect. 9.5.3) brought a satisfactory reduction of this electrical field, primarily to reduce hot-carrier effects. However,

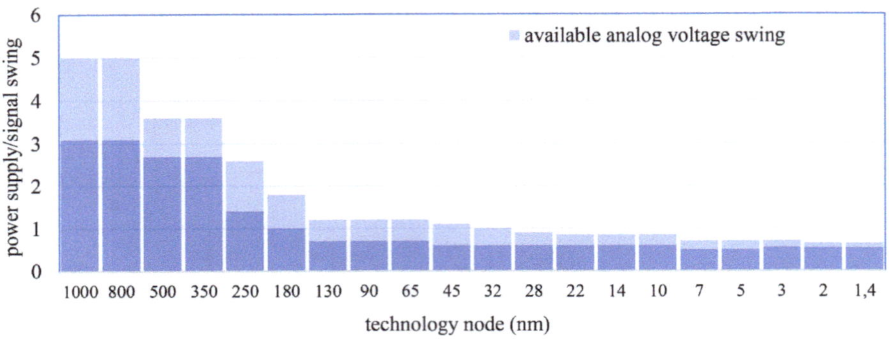

Fig. 8.5 Reduction of supply voltage as a function of the channel length (Source: IEEE IRDS and Bits on Chips)

from about 0.6 μm technologies and beyond, these LDD structures are no longer sufficient. The only way to reduce the peak electrical field is to lower the supply voltage. Figure 8.5 shows the trend in supply voltage reduction and available analog voltage swing as a function of the process node. Below the 14-nm node, the supply voltage reduction is very limited, due to the loss in operating margins, not just for the on-chip analog content but also for the embedded (SRAM) memories. This is why different circuits (e.g. memory, analog and digital cores and I/Os) on complex IC designs in advanced process nodes are often operated at different supply voltages (*multi-voltage domains*). It is also a reason why some companies move to *chiplet* implementation for complex designs.

Shorter channel lengths will require lower V_{dd} voltages. For performance reasons, the threshold voltage V_T had also to be reduced. However, this has led to an increase of the subthreshold (leakage) currents; see Sect. 8.4.1. The *power-delay product* (τD, product; τ, gate delay and D, dissipation) was the classic performance metric used in technologies above 120 nm, because the dynamic power consumption was the largest power component in those technology nodes:

$$\text{Energy} = \tau \cdot D = \text{delay} \cdot P_{\text{dyn}} = \frac{1}{f} \cdot CV^2\alpha f = CV^2\alpha \tag{8.4}$$

where α represents the average switching activity factor of the logic gates in a core. For traditional CMOS circuits, the following two expressions apply:

$$\text{power}(D) = C \cdot V^2 \cdot a \cdot f \tag{8.5}$$

$$\text{delay}(\tau) = \frac{2C \cdot V}{\beta(V - V_T)^2} \text{ from : } \left\{ \begin{array}{rl} Q = I \cdot t & = C \cdot V \\ I & = \frac{C \cdot V}{t} \\ \beta(V_{gs} - V_T)^2 & = \frac{2C \cdot V}{t} \end{array} \right\}$$

To reduce both the power and the delay, capacitance C must be reduced. From the previous two equations, the $\tau \cdot D$ product will be equal to:

$$\tau \cdot D = b \cdot \frac{V^3}{(V - V_{\mathrm{T}})^2} \tag{8.6}$$

where b is a constant.

The minimum will exist for $\frac{\delta \tau D}{\delta V} = 0$, which results in $V = 3V_{\mathrm{T}}$. Thus, when a ratio of three is used between the supply voltage and the threshold voltage, the process should allow for optimum performance.

The power-delay product assigns equal weight to the power and to the delay of a circuit. For circuits for which power has a higher priority than speed, we might give a higher weight to the power than to the delay, and the metric becomes:

$$\text{Power} \cdot \text{Energy} = P_{\mathrm{dyn}}^2 \cdot \tau \tag{8.7}$$

For high-speed circuits, we might give more weight to the delay instead of to the power. Then the metric becomes equal to the *energy-delay product*:

$$\text{Energy} \cdot \text{Delay} = P_{\mathrm{dyn}} \cdot \tau^2 \tag{8.8}$$

So, the chosen metric depends on the requirements of the application.

Most CMOS technology nodes (65 nm and below), today, support three different categories of ICs. The *HP (high-performance) process* is meant for those ICs that demand a high speed for their circuits. The *LSTP (low standby power) process* variant is targeted at application areas with reasonably long standby times of the logic and/or memory cores, because it offers both high-V_{T} nMOS and pMOS transistors. This may be combined with a thicker gate oxide for reduced gate leakage. Particularly large memory cores would benefit greatly from this low-leakage technology variant.

The third category of applications are those that are always on and therefore require *low-operating power (LOP)*. As an example, an MP3 player is either on, when selected, or completely switched off. For such applications or functions, the dynamic power consumption is the largest contribution to the overall power, which requires another power optimisation approach. For these applications, a reduction of both the threshold and the supply voltage would render them to run at the same speed, consuming less active power but more leakage power. This optimisation allows the exchange of operational power (dynamic power; active power) and leakage power, without sacrificing performance. The choice of V_{T} determines the amount of on-current (I_{on}) and leakage current (I_{off}). It turns out that the optimum power is achieved when the operational power (P_{dyn}) is about equal to the leakage power (P_{leak}) [15]:

$$P_{\mathrm{leak}} = I_{\mathrm{off}} \cdot V_{\mathrm{dd}} = C \cdot V_{\mathrm{dd}}^2 \cdot \alpha \cdot f = P_{\mathrm{dyn}} \tag{8.9}$$

with:

$$f = \frac{1}{T} = \frac{1}{L_d \cdot \tau} \text{and} C \cdot V_{\mathrm{dd}} = I_{\mathrm{on}} \cdot \tau \tag{8.10}$$

where L_d represents the average logic depth of the logic paths.

This results in:

$$I_{\mathrm{off}} \cdot V_{\mathrm{dd}} = I_{\mathrm{on}} \cdot V_{\mathrm{dd}} \cdot \tau \cdot \alpha \cdot \frac{1}{L_d \cdot \tau} = I_{\mathrm{on}} \cdot V_{\mathrm{dd}} \cdot \frac{\alpha}{L_d} \tag{8.11}$$

or:

$$\frac{I_{\mathrm{on}}}{I_{\mathrm{off}}} = \frac{L_d}{\alpha}. \tag{8.12}$$

The optimum $I_{\mathrm{on}}/I_{\mathrm{off}}$ ratio for a consumer IC, with a logic depth of about 40 gates and an activity factor of around 0.1, would be 400. For a high-performance video processor with a logic depth of 15 gates and an activity factor of 0.3, the optimum $I_{\mathrm{on}}/I_{\mathrm{off}}$ ratio would be equal to 50, requiring CMOS devices with a much lower V_{T}. To be able to use this optimisation concept in a given technology with fixed threshold voltages, synthesis tools have to be developed to match the L_d/α ratio of as many logic paths to the $I_{\mathrm{on}}/I_{\mathrm{off}}$ ratio of the transistors.

8.5 Design Options for Power Reduction

As shown in Table 8.2, we can also reduce the dynamic, the leakage, the static and the short-circuit dissipation by taking measures in the design. Because the measures for the latter two are clear and compact, we start with these two first. Leakage power reduction has already been discussed in Sect. 8.4.1.

8.5.1 Reduction of P_{short} by Design Measures

During an input transition at a CMOS logic gate, there may be a temporary current path from supply to ground. The resulting short-circuit power dissipation can be relatively high if no attention has been paid to this [16]. Consider the example of Fig. 8.6, which is currently still representative for output drivers.

With these numbers, the dynamic power dissipation becomes:

$$P_{\mathrm{dyn}} = C \cdot V^2 \cdot a \cdot f = 2\,\mu\mathrm{W},$$

and the short-circuit power dissipation becomes [16]:

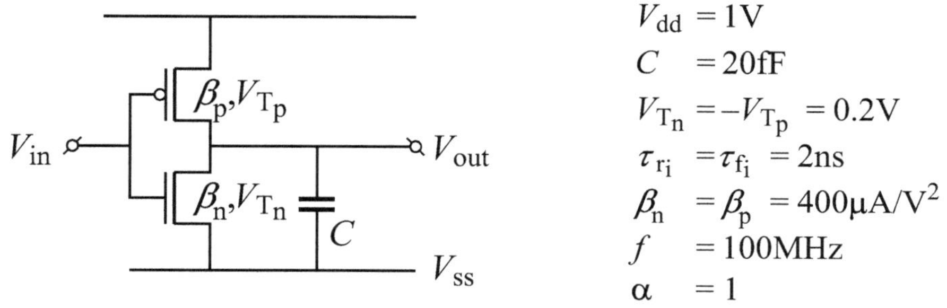

Fig. 8.6 Inverter example to illustrate the level of short-circuit power dissipation

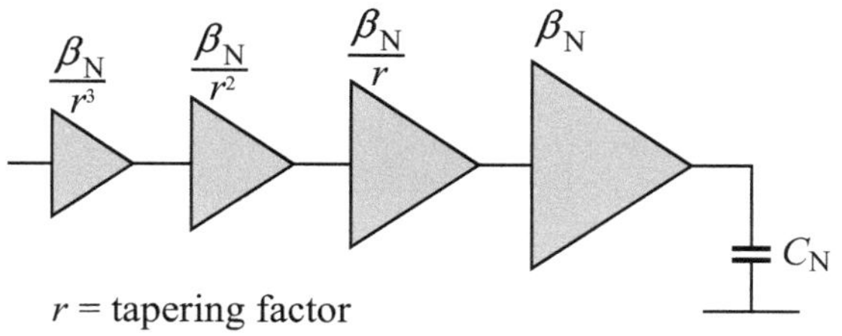

Fig. 8.7 Inverter chain with tapering factor

$$P_{\text{short}} = \frac{\beta}{12} \cdot (V_{\text{dd}} - 2V_{\text{T}})^3 \cdot \frac{\tau}{T} = 1.44\,\mu\text{W}.$$

In conclusion, either τ_f and τ_r on the inputs are much too large or the β of the pMOS and nMOS transistors must be reduced. In normal CMOS logic circuits, the rise and fall times on the transistor inputs are usually so small that the short-circuit power consumption inside the logic cores is negligible compared to the other power contributors. For CMOS drivers (internal, clock and output drivers), this short-circuit power can be minimised when τ_f and τ_r are equal on all nodes (Fig. 8.7). This requires tapering of the inverters in such a driver; see Chap. 4, Table 4.1.

In conventional CMOS technologies, a tapering factor between 8 and 16 usually resulted in a minimum short-circuit dissipation, which was less than 10% of the total dissipation [16]. In most CMOS processes, today (beyond 100 nm), the short-circuit power consumption is fully negligible (Sect. 4.3.2).

An important remark to be made here is that the pMOS and the nMOS transistors can never conduct simultaneously during a transient when $V_{\text{dd}} < V_{\text{T}_\text{n}} + |V_{\text{T}_\text{p}}|$, thereby eliminating the short-circuit dissipation completely.

8.5.2 Reduction/Elimination of P_{stat} by Design Measures

In complex logic gates which require many pMOS transistors in series (four or more input NOR gates, address decoder in memories, etc.), pseudo-nMOS solutions are

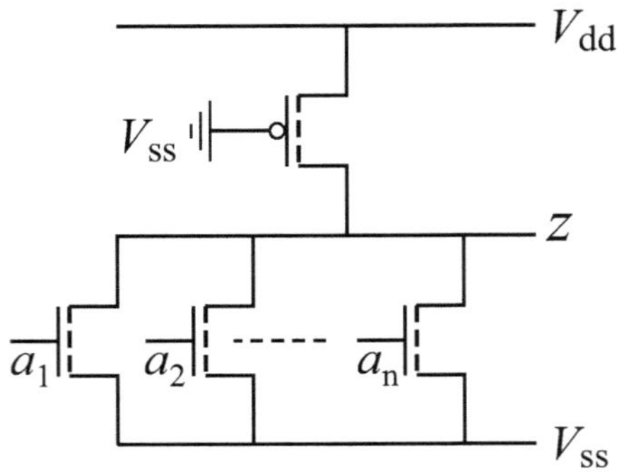

Fig. 8.8 n-Input pseudo-nMOS NOR gate

sometimes applied; see Fig. 8.8. When the output of such a gate is low, there is a continuous static current from V_{dd} to ground.

For low-power applications, this is not an efficient way of implementation. In this case, the power can be reduced by replacing the grounded pMOSt by a clocked pMOSt. This may reduce the power by a factor equal to the clock duty cycle. For a low-power design, this is not a good solution, because a *pseudo-nMOS logic* gate consumes about 10–20 times that of a full static CMOS realisation. Therefore, to eliminate static power consumption, no pseudo-nMOS should be used at all.

8.5.3 Reduction of P_{dyn} by Design Measures

The dynamic dissipation was expressed by:

$$P_{dyn} = C \cdot V^2 \cdot a \cdot f.$$

By means of design techniques, we are able to influence all parameters in this expression. We will therefore present several alternative measures for each parameter to reduce its contribution to the power consumption. Examples of these measures are given at various hierarchy levels of design: algorithm/architecture, logic and transistor level. They show that the decisions taken at the higher levels have much more impact on the power consumption than those taken at the lower levels.

Power Supply (V) Reduction

A lower voltage generally means less performance and less chance for latch-up. Let's assume we have the following circuit on a chip (see Fig. 8.9):

The total propagation delay time through the logic is equal to the sum of the propagation delays of the multiplier and accumulator. This total propagation delay determines the minimum duration T of the clock period. If we double this clock period, the propagation delay is allowed to be twice that of the original circuit. To achieve this doubling, we may reduce the supply voltage from 1.2 V to 0.95 V, for example, in a 1.2 V 65-nm CMOS technology. However, if the throughput is to

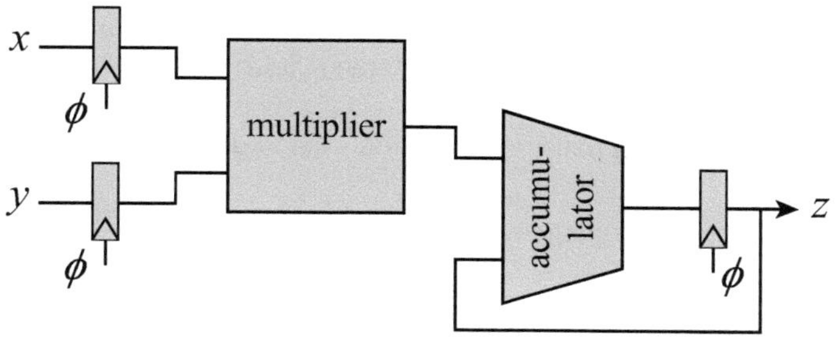

Fig. 8.9 A basic data path

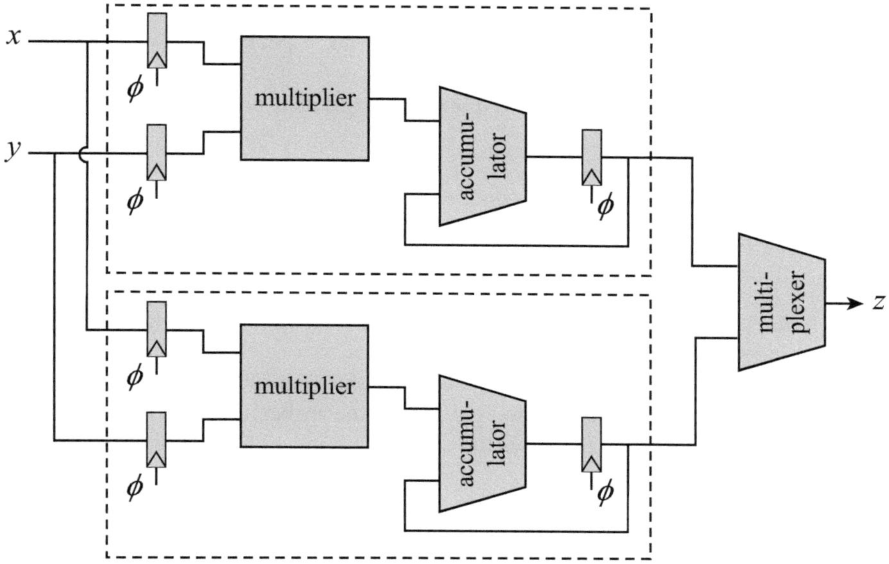

Fig. 8.10 Parallel implementation

be retained, two of these circuits can be connected in parallel and their inputs and outputs multiplexed (parallelism), or additional latches can be placed in between the logic functions to shorten the critical delay paths between two successive flip-flops (pipelining).

(A) *Parallelism*

Figure 8.10 shows a parallel implementation of the circuit. As a result of demultiplexing and multiplexing the signals, the same performance can be achieved as in the original circuit of Fig. 8.9, but now each block runs at only half the clock frequency.

When we include multiplexers and additional wiring, this parallel architecture results in an increase of the total capacitance to be switched by a factor of about 2.25. The power dissipation comparison for the circuits of Fig. 8.9 and the parallel implementation in Fig. 8.10 then yields:

$$P_{\text{dyn}}(\text{basic data path}) = C \cdot V^2 \cdot a \cdot f_{\text{ref}} = P_{\text{ref}}$$

$$P_{\text{dyn}}(\text{parallel data path}) = (2.25C) \cdot \left(\frac{0.95}{1.2}V\right)^2 \cdot a \cdot \frac{f_{\text{ref}}}{2} = 0.7 \cdot P_{\text{ref}}$$

where f_{ref} and P_{ref} represent the frequency and power consumption of the reference circuit of Fig. 8.9, respectively.
Thus, the parallel implementation of the data path results in a power reduction of a factor of about 1.42, however at the cost of area overhead of more than a factor of two. This is sometimes not allowed, especially in the cheap high-volume consumer markets.
Another way to maintain performance at a reduced power supply voltage is pipelining.

(B) *Pipelining*
In Fig. 8.9, the critical path is equal to:

$$T_{\text{crit}} = T_{\text{mpy}} + T_{\text{acc}} \Rightarrow f_{\text{ref}}$$

where T_{mpy} and T_{acc} represent the worst-case delay paths (critical paths) of the multiplier and the accumulator, respectively.
Let us assume that the propagation delays of the multiplier and the accumulator are about the same and that we put a pipeline in between the multiplier and accumulator. Figure 8.11 shows the circuit with the additional pipelines.
Now, the critical path is:

$$T_{\text{crit}} = \max[T_{\text{mpy}}, T_{\text{acc}}] \Rightarrow f_{\text{max}} > f_{\text{ref}}$$
$$\text{if} T_{\text{mpy}} \approx T_{\text{acc}} \Rightarrow f_{\text{max}} \approx 2 \cdot f_{\text{ref}}.$$

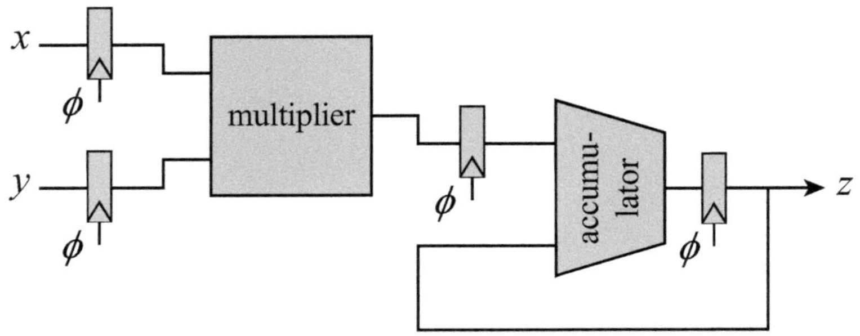

Fig. 8.11 Pipelined implementation

The additional pipeline allows a frequency which is about twice as high. Therefore, the voltage may reduce to about 0.95 V to maintain the same frequency again. As a result of the additional pipelines, the area increase will be about 20%. Comparing this pipelined architecture with the original one leads to the following result:

$$P_{\text{dyn}}(\text{basic data path}) = C \cdot V^2 \cdot a \cdot f_{\text{ref}} = P_{\text{ref}}$$

$$P_{\text{dyn}}(\text{pipelined data path}) = (1.2C) \cdot \left(\frac{0.95}{1.2}V\right)^2 \cdot a \cdot f_{\text{ref}} = 0.75 \cdot P_{\text{ref}}$$

Thus, with only an area penalty of 20%, we almost get the same result as with parallelism. An alternative is the combination of parallelism and pipelining.

(C) *Combination of parallelism and pipelining*
By using both parallelism and pipelining techniques, the critical path timing is relatively reduced by a factor of four. This also results in a reduction of the requirement on speed by a factor of four. To achieve this speed requirement, the voltage can be reduced to only $0.77 \cdot V_{\text{ref}}$. Comparing this technique with the original one leads to:

$$P_{\text{dyn}}(\text{basic data path}) = C \cdot V^2 \cdot a \cdot f = P_{\text{ref}}$$

$$P_{\text{dyn}}(\text{parallel/pipelined}) = (2.25 \cdot 1.2C) \cdot \left(\left(\frac{0.77}{1.2}\right)^2 \cdot V\right)^2 \cdot a \cdot \frac{f}{2} = 0.55 \cdot P_{\text{ref}}$$

Therefore, by using this combination of techniques, we can achieve an improvement (reduction) in power of a factor of 1.8, but this will lead to an increase in chip area by about a factor of 2.7. The choice between area and power is a matter of priority. However, a designer does not usually have the freedom to select the supply voltage level: he chooses a technology, and then the supply voltage is "fixed": for a 65-nm CMOS process, the supply voltage is generally fixed at 1.2 V, because the library cells are characterised for this voltage. Today, with nominal supply voltages below 1 V, the use of pipelining and parallelism will yield less power reduction but is still appropriate in such applications as hearing aids and some other medical applications that need to operate for a long period with only one battery charge.

(D) *Real low-voltage design*
A real low-voltage design may be connected to a supply voltage of, for example, 0.6 V. This gives a reduction in power dissipation according to:

$$P_{\text{dyn}}(1.2\,\text{V}) = C \cdot 1.44 \cdot a \cdot f$$

$$P_{\text{dyn}}(0.6\,\text{V}) = C \cdot 0.36 \cdot a \cdot f$$

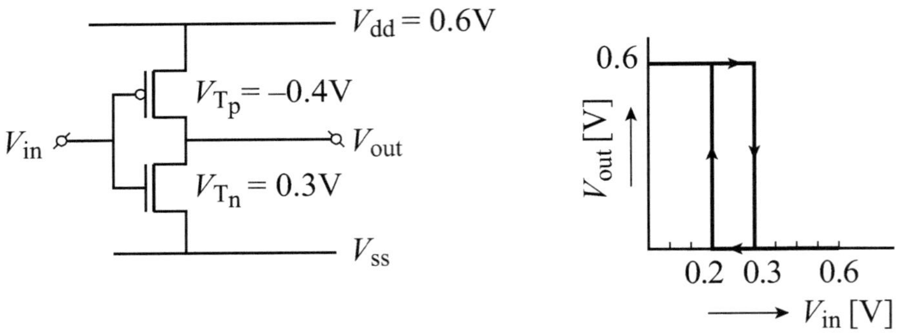

Fig. 8.12 CMOS inverter + characteristic, showing hysteresis when $V_{dd} < V_{T_n} + |V_{T_p}|$

This results in an improvement of more than a factor of 4. However, V_Ts are often between 0.2 and 0.5 V in absolute values. This means that the supply voltage, V_{dd}, can be less than:

$$V_{dd} < V_{T_n} + |V_{T_p}|.$$

This results in hysteresis in the inverter characteristic of a CMOS inverter; see the example in Fig. 8.12.
In this example, the following values are assumed:
$V_{dd} = 0.6$ V,
$V_{T_n} = 0.3$ V and
$V_{T_p} = -0.4$ V.

The operation of the inverter is as follows, when switching V_{in} from 0 to V_{dd} and back again:

- When $V_{in} \leq 0.2$ V, the pMOS transistor is on and the nMOS transistor is off; the output is at the high level (0.6 V).
- When 0.2 V< V_{in} < 0.3 V, both the pMOS and nMOS transistors are off, so the output remains (floating) at the high level.
- At $V_{in} = 0.3\ V$, the nMOS transistor switches on and the output V_{out} immediately goes to 0 V because the pMOS transistor remains off.
- When 0.3 V< V_{in} < 0.6 V, the nMOS transistor remains on, and the output remains at 0 V.
- When we switch V_{in} back to 0 V again, when 0.2 V< V_{in} < 0.3 V, both the pMOS and the nMOS transistors are off. This means that the output remains at 0 V but floating (high impedance state).
- When V_{in} becomes equal to 0.2 V, the pMOS transistor switches on, and the output switches to 0.6 V.

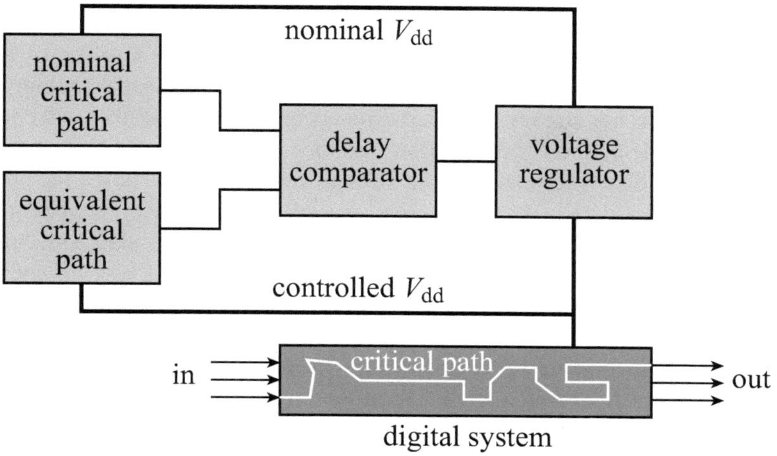

Fig. 8.13 Example of voltage regulator principle

- Finally, when $0\,V < V_{in} < 0.2\,V$, the pMOS transistor remains on, and the output remains at 0.6 V.

Although these kinds of circuits ($V_{dd} < V_{T_n} + |V_{T_p}|$) are relatively slow, they have been used for a long time in battery-operated products, e.g. watches. One advantage of these circuits is that a short-circuit current can never flow, because one transistor always switches off before the other one switches on. Therefore, there is no short-circuit dissipation at all. Not every library is suited for low-voltage operation. This means that a new low-voltage library must be developed and characterised, including a RAM, an ROM and other IP blocks. Moreover, because of the low-voltage supply, the threshold voltage (V_T) must be controlled very accurately, since the circuits are then much more sensitive to threshold voltage variations.

(E) *Voltage regulators*

Generally, ICs also contain low-performance parts which could actually run at lower supply voltages. These can be supplied externally or generated on chip by means of voltage regulators [17]; see Fig. 8.13.

If such a voltage regulator is used, attention should always be paid to its power efficiency. A better alternative is to run the chip at the lowest required voltage and perform a voltage up-conversion only for the high-performance circuit parts. Such voltage up-converters are used in single-cell hearing aids, for example. One can also use *DC-DC converters*. Here too, the power efficiency is an important factor in the use of such circuits. Currently, this efficiency is in the 90% range.

(F) *Reduced voltage swing*
Bus widths, both on chip and off chip, are tending to grow to 32, to 64 and even to 128 bits. This means that the number of simultaneously switching buses and/or outputs has already increased dramatically and this number will continue to increase. If the power dissipation becomes high with respect to other parts of the chip, then a lowering of the voltage swing on these buses (or outputs) has to be considered. As an example, we take the interface between a high-performance microprocessor and a 400-MHz DDR1 1-Gb SDRAM. Such an SDRAM may provide a 64-bit data path, while another 24 bits are needed to access all 64-bit words in the memory. In addition, several ECC bits (when available), clock and control signals are needed in this interface, which may lead to a total of about 96 interface pins. DDR1 supports an I/O voltage level of 2.5 V. If we assume that all bits, representing 15 pF of load capacitance each, would switch at the fastest rate, the total power consumption of this interface would be:

$$P_{\text{dyn}} = C \cdot V^2 \cdot f = 96 \cdot 15 \cdot 10^{-12} \cdot 6.25 \cdot 400 \cdot 10^6 = 3.5\,\text{W}$$

By replacing this 2.5 V DDR1 interface with a 1.8-V DDR2 interface, this power dissipation would reduce to about 1.8 W. Reduced voltage swing techniques are frequently used to reduce the power dissipation of large 32-bit or 64-bit processors. Over the last decades, also DDR3, DDR4 and DDR5 interfaces have been developed, as shown in Table 6.2 (Chap. 6). Also, special low-power DDR (LPDDR) versions (LPDDR1–LPDDR5) have been created for smart phones and tablets and other battery-operated products. Much more detailed info on these interfaces can be found in [18].

(G) *Dynamic voltage and frequency scaling (DVFS)*
In case the application does require a further reduction of the power by dynamically adjusting the voltage to the performance needs, there are two possibilities to do so. The first one is to adaptively control the voltage (*adaptive voltage scaling (AVS)*) by means of an analog feedback loop that continuously matches the required supply voltage to the performance needs of the running application. The second possibility is to have the voltage switched to one of the discrete (lower-)voltage levels supplied to the chip.
Continuous adaptation of the supply voltage to the performance demands of the system requires the integration of an on- or off-chip voltage regulator circuit into the system. On-chip voltage regulation can only be performed, by varying the resistance of a big transistor, which is positioned in between the core and the supply voltage (Fig. 8.14).
However, with this type of linear voltage regulators, also called *low-dropout (LDO) regulators*, the power reduces only linearly with the V_{dd}*core* instead of quadratically, since part of the power saving in the core is now consumed in the variable resistor. Until today, the main application of an LDO was to transform a relatively large battery voltage to the required operating voltage levels of an IC (e.g. 1 V in a 28 nm CMOS chip, derived from a 4.3–4.7-V lithium battery).

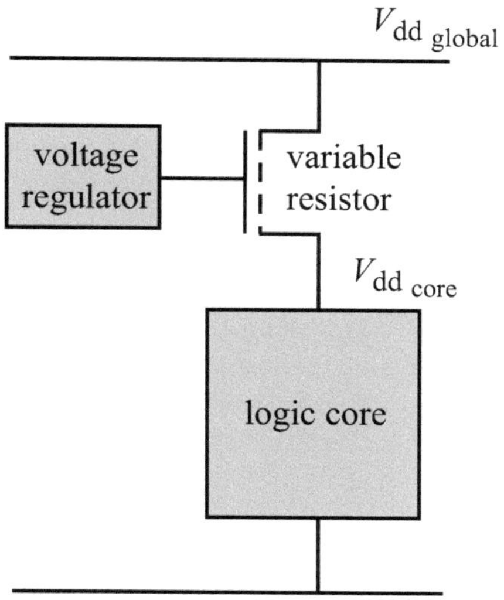

Fig. 8.14 On-chip voltage regulation

A more efficient power regulator uses a dynamically controllable off-chip DC-DC converter. Current DC-DC converters show power efficiencies close to 90%. Fully adaptive voltage scaling is more applicable to certain CPU cores (e.g. on a Pentium) or for a single MPU core (e.g. ARM core) on an ASIC, where relatively large design teams are assigned to the power management task and where most critical delay paths in that core are exactly known due to a huge series of extensive simulations, based on accurate circuit models. For a dedicated core, like an ARM core, a replica of its worst-case delay path could be used.

Applying adaptive voltage scaling generally in the cores of an ASIC is not easy because it becomes very difficult to find a *replicator* that mimics the voltage behaviour of all these cores correctly. These cores may then run at (unknown) supply voltages, which require full library characterisation across the full supply voltage range. Communication between cores in different voltage domains needs the use of *level shifters*. However, if the voltages of the different domains are automatically adjusted, these level shifters need to be very flexible in that on one IC it needs to transfer signals from a low-V_{dd} to a high-V_{dd} domain, while on another IC from a different batch of wafers it might need to do the opposite. Although the principle of using level shifters in itself looks easy, it can have far more design implications than seen at first sight. A level shifter introduces additional time delay in the communication path between two cores in different voltage domains. Particularly for high-performance communication, there is no time budget left to do level shifting at all. In all other applications, the level shifter has serious impact on the timing closure of the overall IC design.

Including the tolerances of the voltage regulators themselves, this makes synchronous communication between such cores very complex and sometimes even impossible. Moreover, time delay, complexity and risks are added to the design, as well as debug and diagnose time. Next to this, the scheduling of, and elapsed time required for, the new supply voltage to settle (often a hundred to a few hundreds of microseconds) is defined by the application.

The power management system needs direct knowledge about the current and future workload generated by (bursty) operations [19]. For non-periodic applications, this is not an easy task, and the information must be derived from monitoring the system load and by using appropriate models to predict future performance demands. Although the idea of dynamic voltage and/or frequency scaling exists for a long time, the technique is not widely used until now, mainly due to the problem of reliably predicting the performance demands.

Next to that, this type of dynamic power management must also be fully supported by the libraries and design flow, such that it becomes completely transparent for the designers.

To a lesser extent, the above considerations also hold for the case in which the supply is not fully adaptively regulated but where the power management system selects the appropriate voltage level from a few discrete voltages supplied to the chip.

It should be noted that switching the supply voltage to different levels, including a complete power on or off switch, may introduce large current peaks in the circuit or even in the board, affecting the signal integrity and/or EMC behaviour of other on-chip cores or on-board devices.

Finally, it has been shown [20] that although dynamic voltage scaling renders the lowest-energy dissipation for most microcontrollers, it is not always dramatically better than using a combination of dynamic frequency scaling and the built-in power-down modes, which is much less complex and less expensive to implement. A combination of simultaneous scaling of the voltage and frequency, dependent on the task (function and performance need) of the system, is *DVFS (dynamic voltage and frequency scaling)* [21]. While it may seriously reduce the active power consumption, it introduces a lot of challenges regarding system validation and verification.

For certain high-volume devices, the voltage assignment can be done after silicon realisation. In this so-called *power-binning concept*, every chip is measured and the assignment of the voltage to the different supply domains on the chip is based on the real silicon performance measurement. The supply connections are then made by using polysilicon fuses, analogous to the selection of redundant columns in a memory. Power binning allows manufacturers of power-critical devices to sell high-end and low-power versions at a premium price.

(H) *Subthreshold logic*

Most low-power applications require power levels in the milliwatt range. However, there are an increasing number of applications, e.g. RF-ID cards, sensor networks and some low-power processing in the idle mode in mobile applications, that require power consumption in the microwatt range. These *ultra-low-power* applications can be realised with various CMOS circuit architectures. The most simple one, which also allows the use of the standard CMOS design flow, is to operate CMOS logic close to, or beyond, the threshold voltage. This so-called *subthreshold logic* uses transistors that operate in the weak-inversion region, in which the current has an exponential relation with the voltage (Sect. 2.7.1). Normally, subthreshold logic is characterised by its limited supply voltage, which is then smaller than the sum of the threshold voltages of the nMOSt and pMOSt: $V_{\text{dd}} < V_{\text{T}_\text{n}} + V_{\text{T}_\text{p}}$. As already mentioned in Chap. 4, CMOS circuits exhibit large noise margins. In fact, a logic CMOS circuit in a 40-nm CMOS process node will still operate at supply voltages as low as 80 mV but then only at extremely low frequencies and with a huge spread in maximum achievable frequency. Workable voltage levels lie mostly between 0.2 and 0.4 V. At these supply voltage levels, CMOS logic is extremely sensitive to process parameter spread, because the circuit then only runs on subthreshold leakage currents. It is therefore a major R&D challenge to create subthreshold logic circuits with reduced sensitivity to this spread [22]. Also, additional electronic feedback circuitry may help to reduce the performance spread of subthreshold circuits and systems [23]. Some companies have successfully applied subthreshold logic in their products. Applications include pacemakers, hearing aids, wearables and contactless smart cards. Today, increasing interest is expected from applications in the automotive and *Internet of Things (IoT)* domains, particularly in (micro-) sensing and monitoring applications, which only allow extreme low-power budgets. As these application domains will rapidly grow in the next few years, the reader is advised to search the internet for recent developments in this respect.

Capacitance Reduction

The total capacitance to be switched on an IC can be reduced or limited at two levels: at system level and at chip level. The decisions taken at system level usually have more effect on the IC power than those taken at chip level. This is because a different architecture for an ALU/multiplier or for a filter structure can have more area consequences for the total hardware. This is shown in the following example:

(A) *System level*

Suppose we have to perform the following algorithm:

$$y(n) = \sum_{m=0}^{k-1} x(n-m). \tag{8.13}$$

A possible hardware implementation is shown in Fig. 8.15.

When k is large, many additions have to be performed. Here, the hardware configuration will contain a lot of full adders to generate the sum and carry functions. The data has to ripple through a large number of full adders, leading to long propagation times and a limited clock frequency. A high-performance implementation would even require additional pipelines and/or carry-look-

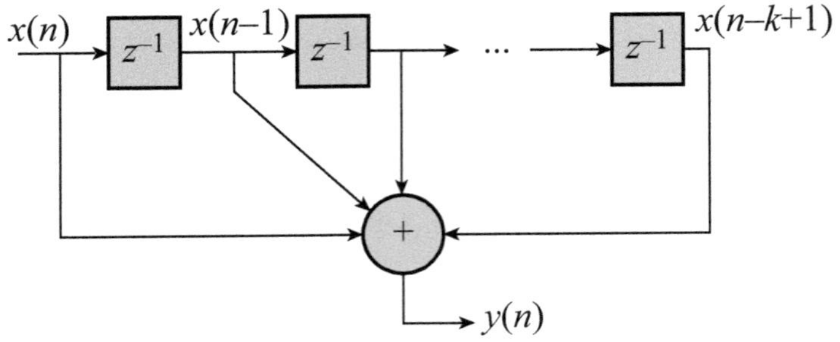

Fig. 8.15 Digital realisation of the running sum algorithm

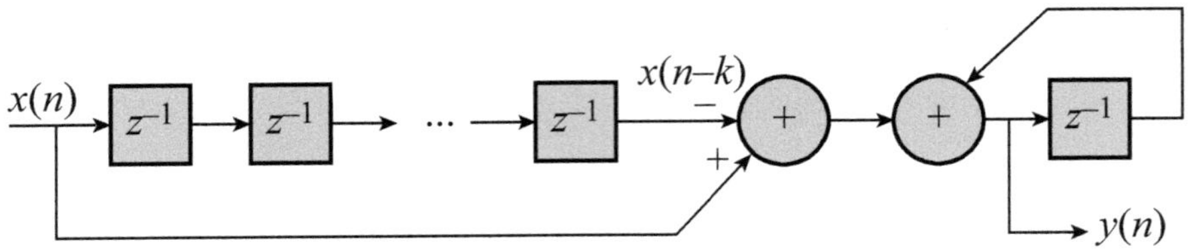

Fig. 8.16 Recursive realisation of the running sum algorithm

ahead techniques to improve speed. With regard to the power consumption, this implementation is very disadvantageous. Figure 8.16 shows an alternative recursive realisation:

$$y(n) = y(n-1) + x(n) - x(n-k). \tag{8.14}$$

Although it consists of two adders, each adder here has only two inputs, which means that much less hardware is involved.

From this example, we can conclude that the algorithm level is at least as important as the chip level for a low-power realisation of a certain function.

(B) *Chip level*

At chip level, there are many options for reducing the power consumption by capacitance reduction. This ranges from libraries, via tools and circuit techniques, to layout implementation:

- Optimised libraries.

  In many cases, low power implicates the use of a reduced voltage. This requires library cells with a low-gate complexity (maximum three inputs). These cells suffer from less body effect and show a better performance than cells with a higher complexity. Most libraries were designed for high performance. They contained relatively large transistors which consume power accordingly. Using these libraries for a low-power design is an overkill in both power and area. In a real low-power library, the transistor and cell sizes must be limited, such that the fan-in capacitance, the parasitic junction capacitances of source and drain regions and the total interconnect lengths after routing will also be reduced. Source and drain regions can be reduced by adapting a very regular layout style.

Flip-flops are probably the most frequently used cells of a library. In many synchronous chips, 10–50% of the total layout area is often occupied by flip-flops. They therefore play a dominant role in the performance, the area, the robustness and the power consumption of a chip. It is clear that the flip-flops should be designed for low power, not only for their internal power consumption but also for the clock driver power consumption. A low fan-in for the clock input combined with better clock skew tolerance (more output delay) allows smaller clock driver circuits, thereby reducing both power consumption and current peaks. Standard-cell libraries may be available with different cell heights. Usually this cell height is expressed in the number of metal grids (= minimum track width + minimum spacing). Cell heights of 12–14 grids are often used for common VLSI designs, while libraries with cell heights of 7–9 grids are often used for low-power designs. Current standard-cell libraries may contain between 1000 and 2000 cells, with different transistor drive strengths and threshold voltages to support both high-performance and low-power and low-leakage applications, as explained in Sect. 8.4.1.

- Pass-transistor logic (transfer gate, pass gate, transmission gate).

  This logic already existed in the nMOS era. The most efficient circuits to be implemented in *pass-transistor logic* are multiplexers and half-adder and full-adder cells. The basic difference between this logic and conventional static CMOS logic is that a pass-transistor logic gate also has inputs on the source/drain terminals of the transistors. A major disadvantage of nMOS pass-transistor logic is the threshold voltage loss ($V_{out} = V_{dd} - V_{T_n}$) at high output level. When such a signal is input to a CMOS inverter, a leakage current flows in this inverter when $V_{T_n} \geq |V_{T_p}|$.

  nMOS pass-transistor logic will thus not be an alternative for low-power design. For different reasons, it is usually not feasible to control the threshold voltages (i.e. $V_{T_n} \geq |V_{T_p}|$) at the technology level. To compensate for the threshold voltage loss and for other disadvantages of nMOS pass-transistor logic, several pass-transistor logic styles have been presented in literature. The most important ones will now briefly be discussed.

  Complementary Pass-Transistor Logic (CPL) [24]

  A CPL gate (Fig. 8.17) basically consists of two nMOS logic circuits, two small pMOS transistors for level restoration and two inverters for generating complementary outputs. Without the cross-coupled pMOS pull-up transistors, CPL would also show the same problems as the above-discussed nMOS pass-transistor logic.

  Because of the complementary logic circuits, the fan-in and the complexity of a CPL gate approach that of a conventional CMOS gate. Because of the availability and necessity of the complementary signals, much more routing

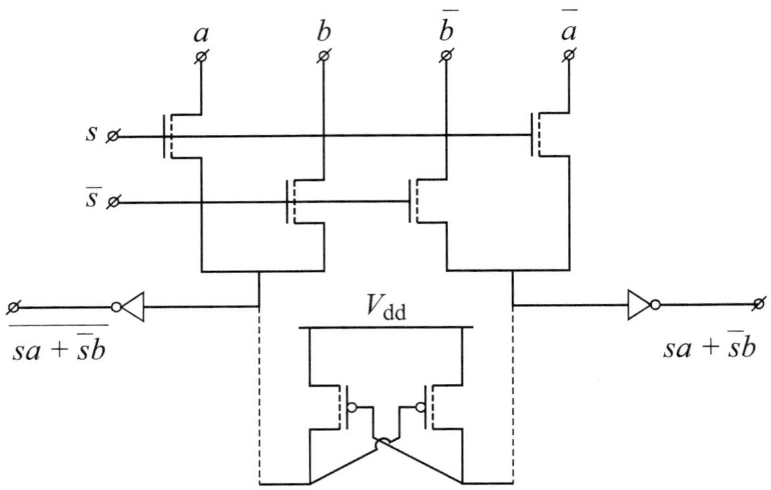

Fig. 8.17 Two-input multiplexer in CPL

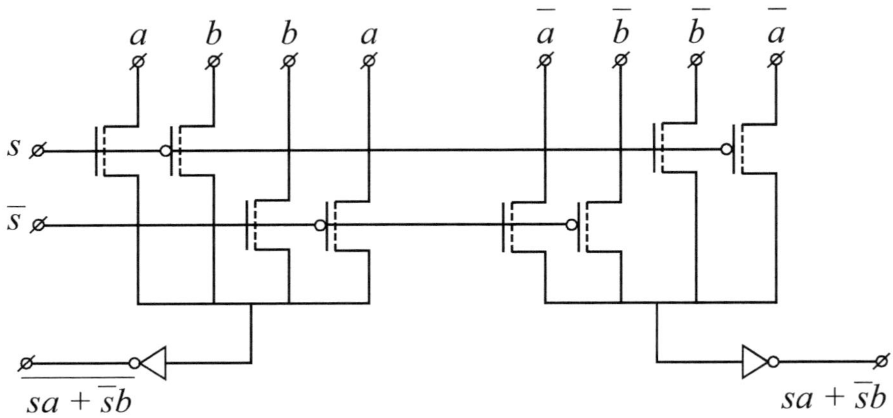

Fig. 8.18 Two-input multiplexer in DPL

area is required. Moreover, simple logic functions require a relatively high transistor count.

Double Pass-Transistor Logic (DPL) [25]

A DPL logic gate uses both nMOS and pMOS logic circuits in parallel, providing full swing at the outputs; see Fig. 8.18.

Because of the necessity of complementary signals, this logic style has the same routing complexity as CPL. Although it needs no swing restore circuit, it contains more transistors, particularly in complex gates, and has a higher

fan-in than CPL. This usually requires more switching power. It is therefore less attractive than other pass-transistor logic and standard CMOS logic.
Other Styles of Pass-Transistor Logic
There are several other styles of pass-transistor logic. One, *the Swing Restored Pass-Transistor Logic* (SRPL; [26]) is derived from CPL. Here, the output inverters are mutually cross-coupled (compare Fig. 8.17) and must be overwritten by the pass-transistor network. This makes this logic less robust for general implementation. In *Lean Integration with Pass-Transistor* (LEAP; [27]), single-rail pass-transistor logic is used. This basically contains an nMOS logic network and a level restore circuit, consisting of an inverter and a feedback pMOS pull-up transistor. This is slower than CPL. At supply voltages of $V_{dd} < 2V_{T_n}$, this pass-transistor style is no longer applicable because the output inverter can no longer be turned on.
Finally, new styles of pass-transistor logic are being invented continuously (DPTL [28], EEPL [29], PPL [30]). However, many are derived from previous styles with only minor changes. Whatever style of pass-transistor logic will be invented yet, they will all have the same disadvantages: either they will suffer from threshold voltage loss and need a level restore circuit or they will need the double-rail approach (complementary inputs and outputs).
Conclusions
Although different pass-transistor logic families are presented in literature, showing better performance in terms of power-delay products than conventional CMOS logic, the opposite is also published [31].
Initially, pass-transistor logic showed equal functionality with low transistor count. However, with reduced voltages, complex ICs and low-power focus, this advantage has been undone by the necessity of a level restore circuit and/or dual-rail implementation. Except for half- and full-adder functions, conventional CMOS circuits perform better than any pass-transistor style where both power and robustness are concerned. As a result of increasing process variations and extending application environments, the robustness will play an especially dominant role in the development of (standard) cell libraries.

- Synthesize logic functions into larger cells.
  Usually, logic functions are mapped onto library cells. This, however, is rather inefficient in terms of area and power. The full-adder function might serve as a good example, where S is the sum function and C represents the carry:

$$S = a\bar{b}\bar{c} + \bar{a}\bar{b}c + \bar{a}b\bar{c} + abc$$

$$C = ab + ac + bc$$

In a standard-cell library without a full-adder cell, the sum function would require four three-input AND functions and one four-input OR. With a dedicated full-adder library cell, the area will be roughly halved. Generally, a cell compiler, capable of optimising complex functions and creating logic gates, would be a good tool for optimising both area and speed. However, good characterisation tools must then also be available to generate accurate timing views of these compiled cells.

- Use optimised synthesis tools.
  Good tools are required for an optimum mapping of complex logic functions onto the library cells. These tools must include reasonably accurate timing models. Usually, the less hardware is used, the less power will be consumed.
- Use optimised place and route tools.
  Many current CAD tools for place and route are area- or performance-driven. Part(s) of the circuits can have different weights for high performance. These require priority in the place and route process. With a focus at low power, power-driven (activity/capacitance) place and route tools are required, resulting in minimum wire lengths.
- Use custom design, if necessary.
  Reduction of the interconnection lengths can be achieved by different layout styles. Especially *cell abutment* is a way to optimise data paths in very regular structures, such as bit slice layouts, multipliers, etc. Custom design must only be applied if the additional design time can be retrieved. Practically speaking, this only holds for high-volume chips or for chips with very tight power specifications, which cannot be achieved with other design styles.
- Make an optimum floor-plan.
  Although this sounds very commonplace, it is not self-evident. During floor-planning, the focus should be on wasting less area and on reducing bus and other global interconnections. The cores that have intensive communication with each other should be placed at minimum distance, to reduce the wire length of the communication buses.
- Optimise the total clock network.
  Clock signals run globally over the chip and usually switch at the highest frequency (clock frequency f, data frequency $<f/2$). As discussed, the number of flip-flops and their properties are a dominant factor in the total clock network. The flip-flops should be optimised for low fan-in and a better clock skew tolerance so that smaller clock drivers could be used. Section 9.2.2 presents a robust flip-flop, which is also very well suited for low-power designs.
- Use well-balanced clock trees.
  Balanced clock trees are those in which drivers and loads are tuned to one another, such that equal clock delays are obtained, anywhere in the chip. This reduces the clock skew, which allows for smaller clock drivers.
- Dynamic versus static CMOS.
  Chapter 4 presents implementations of static and dynamic CMOS logic gates. With respect to capacitance, a dynamic CMOS gate generally has less fan-in capacitance. This is because the function is usually only realised in an nMOS network, while the pMOSt only acts as a (switched) load. Because

every gate is clocked, we get very large clock loads. Moreover, as a result of the precharging mechanism, the average activity in a dynamic gate is higher than its static counterpart. A more detailed look into the activity of static and dynamic CMOS logic is presented in the following paragraph.

- Memory design.
  To reduce the total capacitance to be switched in a memory, the memory can be divided into blocks (block select), such that they can be selectively activated (precharge plus read/write). Divided word lines and divided bit lines mean that less capacitance is switched during each word line and bit line selection. Wider words (64 bits instead of 32 bits) reduce the addressing and selection circuit overhead per bit.
  The precharge operation can be optimised by selectively precharging the columns (only those to be read or written) instead of all simultaneously.

Reduction of Switching Activity

Most of the switching activity of a circuit is determined at the architectural and register transfer level (RTL). At the chip level, there are less alternatives for lowering the power consumption by reducing switching activity.

This paragraph presents several of these alternatives, starting at the architectural level.

(A) *Architectural level*
Choices made at the architectural and RTL level heavily influence the performance, the area and the power consumption of a circuit. This subsection summarises the effect that these choices have on the activity of the circuit:

- Optimum binary word length.
  The word length must be not only optimum in terms of capacitance but also in terms of activity, which means that only that number of bits is used that is really required to perform a certain function.
- Bit serial versus bit parallel.
  Figure 8.19 gives two alternative implementations for a 16 by 16 bit multiplier: a bit serial iterative multiplier and an array multiplier.
  The array multiplier only consists of logic that is really required for its function. In the bit serial approach, not only the required logic for multiplication is needed but also the logic for additional control and registers. For a fair comparison, a complete multiplication must be taken. For the parallel multiplier, we have power∗1 (period); for the bit serial one, we have power∗16 (periods). This means that for a full 16*16 bits multiplication, data has to go 16 times through the serial multiplier, while it only needs to go one time through the hardware of the parallel multiplier. From this example, we may conclude that a parallel implementation generally has less overhead than a bit serial one and will therefore consume less power.
- Optimise system power instead of chip power only.
  Complete systems use blocks such as DSP, A/D, D/A, memories, etc. As a result of the increasing communication bandwidth (data word length times

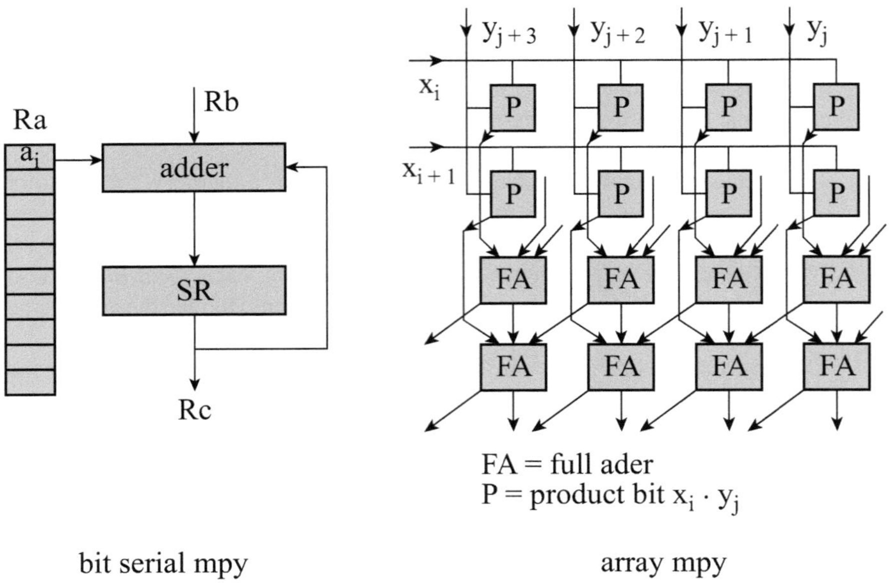

Fig. 8.19 Bit serial iterative and array multiplier

frequency) of signals between these blocks, a lot of power would be wasted in the I/O circuit if each block was a separate chip. If possible, all functions should be on one chip. This will increase the chip power, but it will reduce the system power. A concentration of high-performance system parts and low-performance system parts on different areas on one chip is attractive for power as well. The low-performance parts could then run at lower frequencies and reduced voltages to save power.

- Number representation.
  The choice of the number representation can also have an effect on the power consumption; see also Fig. 8.20. An example study of reducing the activity in 8x8 precision multipliers can be found in [32].
  It is clear that unsigned code is only favourable for positive numbers. The most significant bit is then also used for magnitude representation. The two's complement notation shows problems (discontinuity) at the transition from $7 \rightarrow -8$. The diagram shows two discontinuities for the sign-magnitude notation: at the transition from $7 \rightarrow -0$ and also at the transition from $0 \rightarrow -7$. It is therefore more difficult when used in counters.
  When small values are represented by many bits, the most significant bits in the two's complement notation adopt the value of the sign bit. If the signal is around zero, it will frequently switch from a positive to a negative value and vice versa. In the two's complement notation, a lot of bits will then toggle, while in the sign-magnitude notation only the sign bit will toggle,

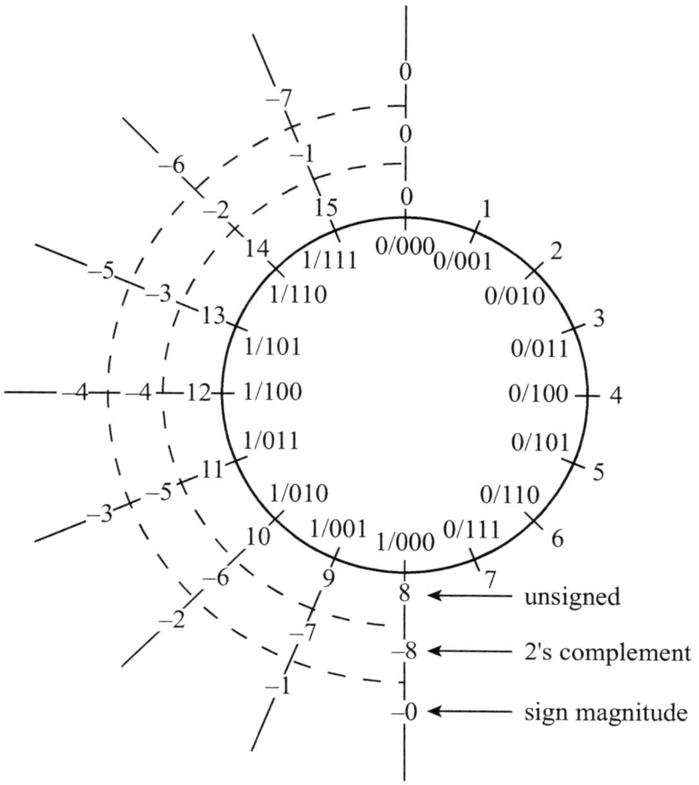

Fig. 8.20 Number representation diagram

Fig. 8.21 Representation of an 8-bit adder/subtractor

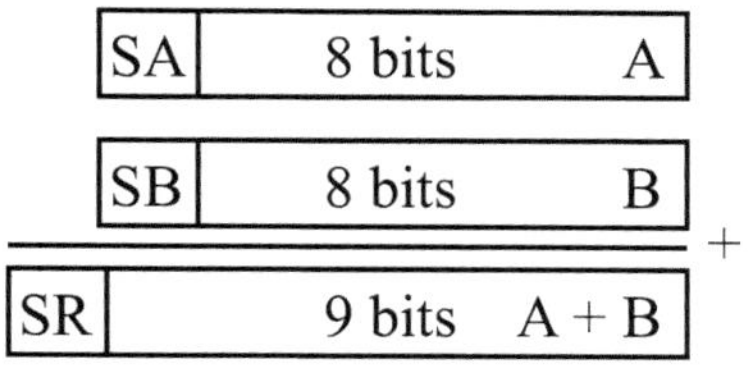

resulting in less power consumption. In the following example, the use of the two's complement notation and the sign-magnitude notation in adders and multipliers is compared.

Example 8-bit adder/subtractor. The representation is shown in Fig. 8.21:

In the two's complement notation, the addition/subtraction operation does not give any problems. However, in the sign-magnitude notation, additional comparators must be used for a subtraction:

$$\text{if} A < B \Rightarrow sum = B - A$$
$$\text{else} \Rightarrow sum = A - B$$

Implementation with synthesis and standard-cell place and route tools reveals a difference in silicon area of a factor of about three in favour of the two's complement notation.

Example Two's complement multiplication:

$$X = -X_{n-1} \cdot 2^{n-1} + \sum_{i=0}^{n-2} X_i \cdot 2^i \tag{8.15}$$

$$Y = \underbrace{-Y_{m-1} \cdot 2^{m-1}}_{\text{sign}} + \underbrace{\sum_{j=0}^{m-2} Y_i \cdot 2^j}_{\text{value}} \tag{8.16}$$

The result of multiplying X and Y is:

$$X \cdot Y = X_{n-1} \cdot Y_{m-1} \cdot 2^{n+m-2} + \sum_{0}^{n-2} \sum_{0}^{m-2} X_i Y_j 2^{i+j} \tag{8.17}$$
$$- \left(\sum_{j=0}^{m-2} X_{n-1} \cdot Y_j \cdot 2^{n-1+j} + \sum_{i=0}^{n=2} Y_{m-1} \cdot X_i \cdot 2^{m-1+i} \right).$$

The realisation in an array multiplier requires the last two product terms to be skipped. A nice alternative is the Booth multiplier, in which half the number of full adders is replaced by multiplexers and where these two product terms are automatically skipped.

Example Sign-magnitude multiplication:

$$X = -1^{X_{n-1}} \cdot \sum_{i=0}^{n-2} X_i \cdot 2^i \tag{8.18}$$

$$Y = -1^{Y_{m-1}} \cdot \sum_{j=0}^{m-2} Y_j \cdot 2^j \tag{8.19}$$

and the product:

$$X \cdot Y = \underbrace{-1^{X_{n-1} \oplus Y_{m-1}}}_{\text{sign}} \sum_{i=0}^{n-2} \sum_{j=0}^{m-2} \underbrace{X_i \cdot Y_j \cdot 2^{1+j}}_{\text{magnitude}} \tag{8.20}$$

In this notation, the sign bit of the product is just a simple EXOR of the individual sign bits, while the magnitude is just the product of only positive numbers.

Conclusions on Number Representation
Although the sign-magnitude notation is convenient for multiplier implementation, the Booth algorithm array multiplier is more popular. Such a multiplier requires relatively little hardware and is thus suited for low-power implementation.

The sign-magnitude notation is convenient for other applications. However, use is limited to representing absolute values in applications with peak detection, but even here it is still used more for number representation than for calculation. If only number representation is considered, the sign-magnitude notation shows less activity when the signal varies around zero. Note that, with compression techniques such as MPEG, a lot of zeros (000..00) are only represented by one bit. The use of compression techniques automatically reduces the power consumption.

- Optimum code.
  Even the code in which an operation is expressed can influence the power consumption. An example is shown in Table 8.3.
  Table 8.3 shows the switching activity of two 3-bit counters: a BCD counter and a Gray code counter. The table also shows that the BCD counter exhibits twice the switching activity of the Gray code counter.
- Alternative implementations for arithmetic multiplier and adder circuits.

Table 8.3 Comparison of switching activity in a BCD counter and a Gray code counter

Standard binary code (BCD)		Gray code	
	Number of changing bits		
000	3	1	000
001	1	1	001
010	2	1	011
011	1	1	010
100	3	1	110
101	1	1	111
110	2	1	101
111	1	1	100
	2	1	

↖↗
average/clock

Besides the previously discussed options (bit serial versus bit parallel and number representations), there are many other alternatives that can influence the power consumption of arithmetic logic. Alternatives for multiplier implementation include Booth multiplier, array multiplier, Wallace tree multiplier, etc. Alternatives for the addition process are carry select, carry ripple, carry save and carry lookahead techniques. With respect to power consumption, a general rule of thumb is "every implementation that speeds up an arithmetic process will require additional power". The choice of an arithmetic implementation depends on the priorities in an application with respect to speed, area and power consumption. Therefore, no fixed prescribed choice can be given for low power here.

- Microprocessor and microcontroller architecture.
  Many products use microprocessor cores: mobile phones, medical electronics, automotive and consumer products, watches and games. Maintaining or improving the performance while reducing the power consumption is a continuous challenge for the designers of new products in these fields. Generally, an instruction in a RISC architecture needs more execution cycles than in a CISC architecture. Pipelined RISC microprocessors use one or two cycles per instruction, while the CISC microprocessor often uses 10–20 cycles. However, complex algorithms mapped on a RISC machine generally require more instructions than a CISC machine. The CISC architecture may have too much hardware for only simple algorithms, which leads to a kind of overkill.
  In these cases, CISC power consumption may be more. From literature, it appears that each architecture (whether RISC or CISC) can in itself be optimised for low power. No real winner can be distinguished here because both architectures have many parameters to be adjusted for optimum low power.
- Limited I/O communication.
  In many applications, many I/O pins are used for communication between processor and memory and/or A/D or D/A converters. To reduce activity, these blocks have to be integrated on one single die. This may increase the chip power, but it certainly reduces the system power.
- Synchronous versus asynchronous.
  In synchronous circuits, the data transfer to, on and from the chip is usually controlled by a global clock signal. However, this clock signal does not contain any information. In contrast, asynchronous circuits proceed at their own speed. Here, the output of one circuit is immediately used as an input to the next. The relatively large difference in delay paths may lead to random operation and requires a special design style and test strategy. Actually, there are two kinds of asynchronous circuits: asynchronous subfunction(s) of synchronous designs and purely asynchronous designs (self-timed circuits).
- Asynchronous subfunction (of synchronous design).
  A synchronous chip is nothing more than a collection of asynchronous circuits which are separated by flip-flops (registers). Thus, asynchronous blocks are embedded between registers. A 4-bit counter may serve as an example.
  Figure 8.22 shows an asynchronous implementation and two synchronous alternatives of this counter. In the synchronous versions, each flip-flop is clocked at

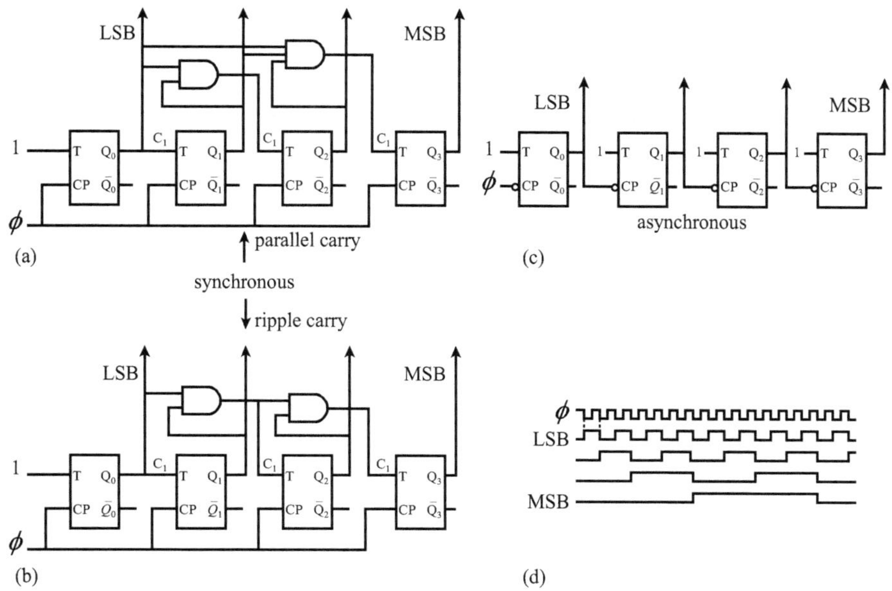

Fig. 8.22 Different versions of a 4-bit counter with timing diagram: (**a**) synchronous with parallel carry, (**b**) synchronous with ripple carry, (**c**) asynchronous and (**d**) timing diagram

Fig. 8.23 Self-timed logic cell

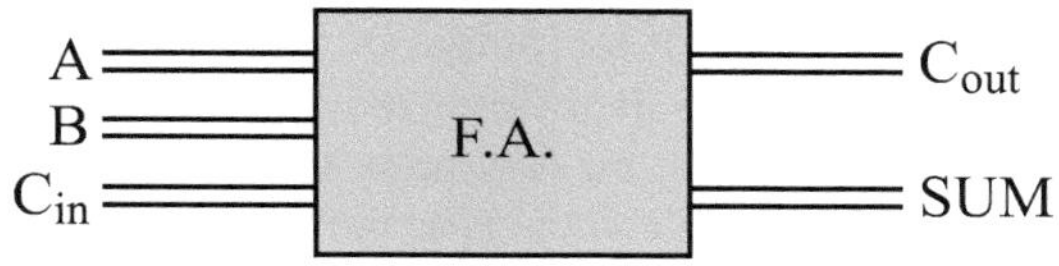

the highest frequency, which consumes a lot of power. The synchronous counter with parallel carry consumes the most power because it has more hardware than the ripple carry counter. In the asynchronous counter version, only the first flip-flop (LSB) runs at the highest frequency, whereas the others act as frequency dividers (divide by two). This version therefore requires much less power (about 1/3) than the best of the synchronous versions.

Asynchronous logic was already introduced in the early 1980s [33], but has not been used intensively since then, mainly because many design concepts and flow are different from the synchronous standard-cell design concepts and flow.

- Pure asynchronous designs *(self-timed circuits)*.

  A basic *asynchronous design* requires additional hardware to perform the necessary request (GO) and acknowledge (DONE) signals. Figure 8.23 shows a full-adder cell implemented as an asynchronous logic cell.

  In this technique, an enormous area overhead must be spent to implement and route the additional logic that is associated with each request and acknowledge signal. This overhead is at least a factor two. An advantage is that no glitches can

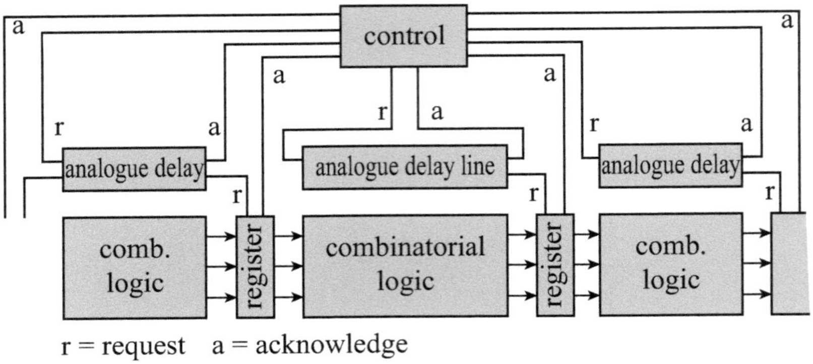

Fig. 8.24 Self-timed circuit by using analog delay that mimics combinatorial logic delay

occur (see next Subsection B). Another way of implementing self-timed circuits is to generate the request and acknowledge signals at a higher level of circuit hierarchy; see Fig. 8.24.

After the data is stored in the register, it generates an acknowledge signal a, which is transformed by the control block into a request signal r, which then propagates through the analog delay line at the same time that the logic signal propagates through the combinatorial logic block. Shortly after the data has become valid at the output of the logic block, the request signal arrives at the register, which then stores the data at the output of the logic block. If this data did not change, no new request data signal is generated. In this way, a higher component efficiency is achieved. A major disadvantage is that the dummy delay lines must be designed to be marginally slower than the slowest path through the combinatorial logic. This combinatorial logic also shows glitches as in synchronous logic; see next Subsection B.

The key to this form of self-timed logic is the ability to model the combinatorial logic delay with a very small analog delay line (inverter chain). Self-timed techniques are also used in synchronous systems, for instance, to generate the clocks needed in smaller parts of the chip. In RAMs, many self-timed clocks are generated on chip. A final discussion on power consumption of synchronous and asynchronous circuits leads to the following statement:

"Although asynchronous circuits are only active when necessary and thus operate at reduced power, these need not be *the* implementation for low-power circuits". Synchronous logic, optimised for low power, can achieve a power level that approaches that of asynchronous circuits. However, synchronous logic was mostly optimised for high speed (and, in some cases, for small area). Certain circuits are particularly suited for asynchronous implementation. But, for those that are not, the power consumed by the control circuit and the large test circuit can be greater than the advantage gained by having no clocks.

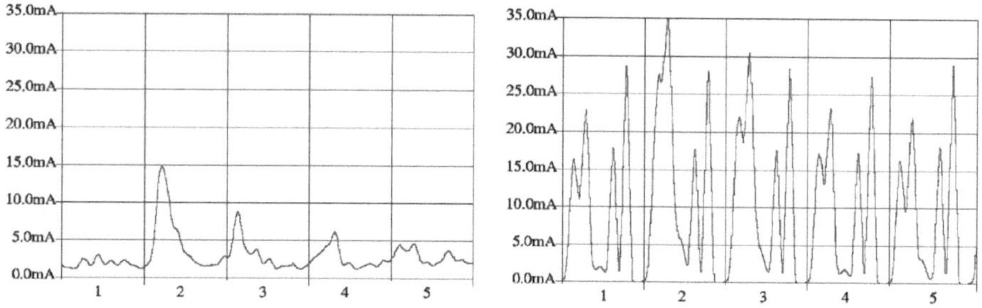

Fig. 8.25 Comparison of current peaks in an asynchronous (left) and synchronous ARM core (right) when executing the same benchmark operation

Several design houses are quietly replacing relatively small portions of their systems with asynchronous units. Already in the mid-1990s, Hewlett-Packard added an asynchronous floating-point multiplier to its 100-MHz RISC processor. These approaches are probably the wave of the future: asynchronous subunits residing in a synchronous framework [34], or vice versa. More recently, asynchronous designs are used in an increasing number of application domains, e.g. smart cards [35], automotive, internet routers (switches) [36] and wireless products (ARM cores [37]). Particularly in the smart card application, the reduction in power consumption directly leads to a performance increase, since these cards operate with a fixed power budget.

The increasing popularity of asynchronous design was caused after the successful combination of this design style with scan test, which resulted in stuck-at test coverage (see Chap. 10) better than 99%. This asynchronous test methodology is based on applying a synchronous full scan in asynchronous *handshake circuits* [38].

Another advantage of an asynchronous implementation compared to a synchronous one is the general reduction of interference and noise. Figure 8.25 shows the result of a typical standard Dhrystone benchmark instruction set running on an ARM11 core. Dhrystone compares the performance of the benchmarked processor core to that of a reference core, by measuring the average time the core takes to perform many iterations of a single loop containing a fixed sequence of the instructions of the benchmark.

The figure shows the current peaks in the supply lines of the asynchronous ARM core (left) during the execution of five Dhrystone loops within a period of 34 μs. Next, the clock period of the synchronous ARM core is adjusted such that the same five Dhrystone loops also last 34 μs, resulting in much larger current peaks (right), which also results in larger noise and emission levels. Although asynchronous designs exhibit a broader frequency spectrum of the generated supply noise, the amplitude is usually much less than that of their synchronous equivalents.

- Optimised memory design.

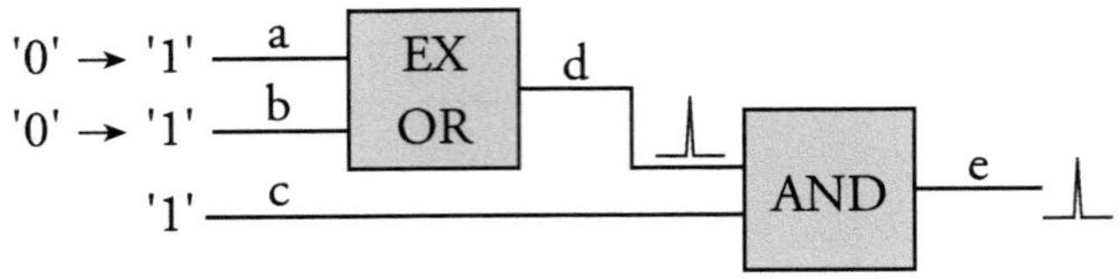

Fig. 8.26 Unnecessary transitions in a simple logic circuit

The previously discussed comparison can also be used in the realisation of memories. To reduce internal memory activities, self-timed techniques are used to generate a lot of different clocks or acknowledge signals which should be active according to some sequence. The alternative to performing one single operation (such as activate precharge, deactivate precharge, select word line, activate sense amplifier, select column, etc.) in one clock period means that a lot of clock periods are needed for only one read or write operation. This would be at the cost of increased power consumption.

(B) *Implementation level*

- Reduce glitching.

  Static CMOS circuits can exhibit *glitches* (also called dynamic hazards, critical races or *spurious transitions*) as a result of different propagation delays from one logic gate to the next. Consequently, a node can have multiple unnecessary transitions in a single clock cycle before it reaches its final state. Figure 8.26 gives an example.

  Suppose the (a, b) inputs of an EXOR gate switch from (0,0) to (1,1). In both situations, output d of the EXOR should remain low ("0"). However, because of a different delay in the switching of the input signals, the state diagram of the (a, b) inputs might follow the following sequence (0,0)→(0,1)→(1,1). Therefore, the (a, b) inputs are (0,1) for a very short period of time, resulting in a temporary "1" at output d. This glitch also propagates through the next AND gate.

  Such unnecessary transitions dissipate extra power. The magnitude of this problem is related to the kind of circuit to be realised. As a result of the occurrence of glitches, an 8-bit ripple carry adder with random input patterns consumes about 30% more power. For an 8*8-bit array multiplier, this number can be close to 100%; for a 16*16-bit array multiplier and for standard-cell implementation of a progressive scan conversion circuit, it can be as high as 200% ! Generally, the larger the logic depth, the larger the skew between the arrival times of input signals at a logic gate, and the higher the probability of a glitch at the output of that gate. Therefore, a large power saving could be achieved in such circuits if all delay paths were balanced.

  Different architectures can lead to a different percentage of unnecessary transients. A 16*16-bit Wallace tree multiplier has only 16% glitches,

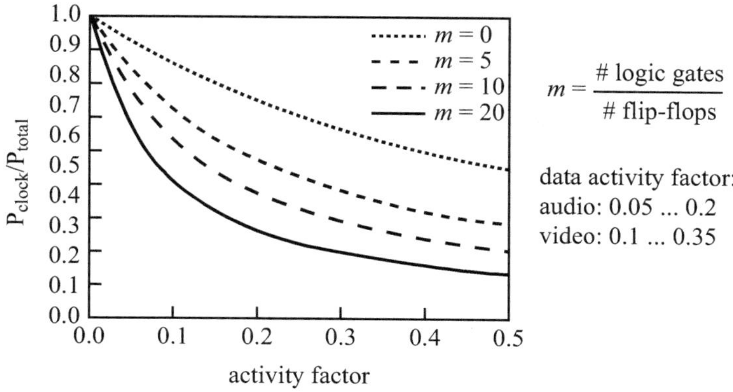

Fig. 8.27 Relative clock power consumption as a function of the activity

compared to the above 200% for a 16*16-bit array multiplier. The Wallace tree multiplier has far more balanced delay paths.
Finally, another way of reducing the number of glitches is to use retiming/pipelining to balance the delay paths.

- Optimise clock activity.
There are two reasons why clock signals are very important with respect to power dissipation. The first is that clock signals run all over the chip to control the complete data flow on the chip in a synchronised way. This means that clock capacitance caused by both very long tracks and a large number of flip-flops can be very large. In complex VLSI chips, the clock load can be as high as one to several picofarads. The second reason is that the clock signal has the highest frequency (the maximum switching frequency of data signals is only half the clock frequency). The total power consumed by the clock network depends heavily on the number of connected flip-flops and latches.
Figure 8.27 shows the relative clock power consumption as a function of the average activity on a chip. This is expressed as a fraction of the total power consumption.
Because the clock dissipation can be as high as 20–50% of the total chip dissipation, its activity should be reduced. This can be done because clock transitions carry no information. There are several ways to reduce clock activity. Including the use of *dual-edge Triggered (DET)* flip-flops. If a flip-flop could be triggered on both edges of the clock pulses instead of on only one edge, it would be possible to use a clock at half frequency for the same data rate, thereby reducing the power dissipation of the total clock network. A flip-flop that acts at both edges of the clock pulse is called a dual-edge triggered flip-flop, while the conventional positive- and negative-edge triggered flip-flops belong to the category of *single-edge triggered (SET)* flip-flops. However, the use of DET flip-flops has been limited up to now

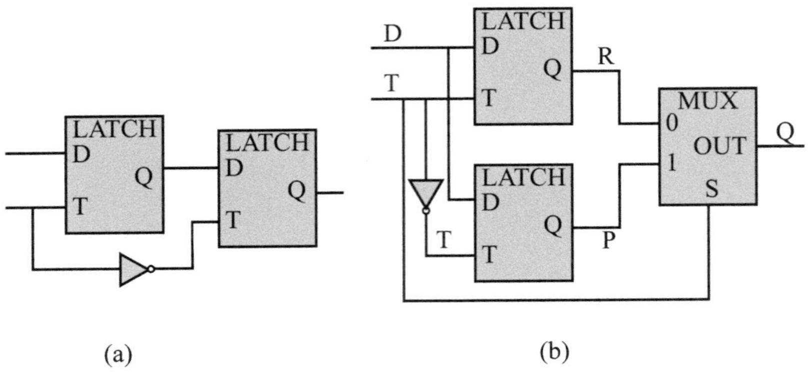

Fig. 8.28 Schematic diagram showing a SET flip-flop (**a**) and a DET flip-flop (**b**)

by the high overhead in complexity that these flip-flops require and because they are not fully compatible with the current design flow. Both the SET and DET flip-flops have two latches. Basically, in a DET flip-flop (see Fig. 8.28), the two latches are arranged in parallel, while in a SET flip-flop (see Fig. 8.28a) they are placed serially [39]. DET and SET flip-flops show comparable maximum data rates; however, DET flip-flops either require additional silicon area or are more difficult in use with respect to timing aspects [40, 41].

Since the clock contribution to the overall chip power consumption depends on the data activity, power savings of 10–20are possible, at the cost of some additional flip-flop area (10–20%), when using DET flip-flops.

In conventional synchronous designs, the complete system is clocked at the highest frequency, even when some portions of the chip could operate on fractions of this frequency. In some cases, clock dividers are used to generate these lower frequencies. From a low-power point of view, we should start from the opposite direction.

This means that we supply the chip with the lowest required frequency and generate higher clock rates locally, if needed. This can be achieved by PLL-like circuits. In this way, the globally distributed clock would run at the minimum clock frequency, and the higher clock frequencies would only be used where they are really needed. This might reduce the global clock activity drastically and also reduce the functional activity.

Another approach to reduce the total chip activity is to switch the clock off temporarily for certain functional blocks or even for the complete chip during moments that no useful operations are executed. In this respect, different names are used for the same issue: gated clocks, stop the clock, sleep mode, power-down mode, etc.

Table 8.4 Function table of a two-input NOR and an EXOR gate

Two-input NOR		EXOR	
a b	z	a b	z
0 0	1	0 0	0
0 1	0	0 1	1
1 0	0	1 0	1
1 1	0	1 1	0

A representative example is a coefficient ROM, whose power consumption can be relatively large. In many cases, such an ROM is often used for less than 1% of the time. Forcing this block to the power-down mode, e.g. by switching off its clock, saves 99% of its total power consumption.
When a signal processor enters the power-down mode, all its internal memory and register contents must be maintained to allow the operation to be continued unaltered when the power-down mode is terminated. Depending on the state of some control register(s), external devices can cause a wake-up of the DSP, e.g. when terminating an input operation. The processor enters the operating state again by reactivating the internal clock. The program or interrupted routine execution then continues.
A disadvantage of gated clocks (sleep modes, etc.) is that some logic operation has to be performed on the clock signal. This causes an additional delay for the internal gated clock, which may result in timing problems during data transfer between blocks that run at the main clock and those that run at a gated clock. Therefore, compensated delays must be used in those blocks that do not use a gated clock. Generally, gated clocks decrease the design robustness with respect to timing (see Chap. 9).

- Dynamic versus static CMOS.
  The decision to implement a circuit in dynamic or static CMOS logic not only depends on power considerations. Aspects of testability, reliability, ease of design and design robustness are also very important here. In the comparison of dynamic and static CMOS realisations, several differences show up with respect to power. As precharge and sample periods in dynamic CMOS circuits are separated in time, no short-circuit dissipation will occur. Also, the absence of spurious transitions (hazards) reduces the activity of dynamic CMOS. However, precharging each node every clock cycle leads to an increase of activities.

Example Let us assume that all input combinations in Table 8.4 are uniformly distributed.

Because each logic gate output in a dynamic CMOS chip is high during precharge, the output will be discharged in 75% of the input combinations of a two-input NOR $\Rightarrow$ activity factor 0.75 while for the EXOR activity factor 0.5. In static CMOS, power is only dissipated when the output goes high:

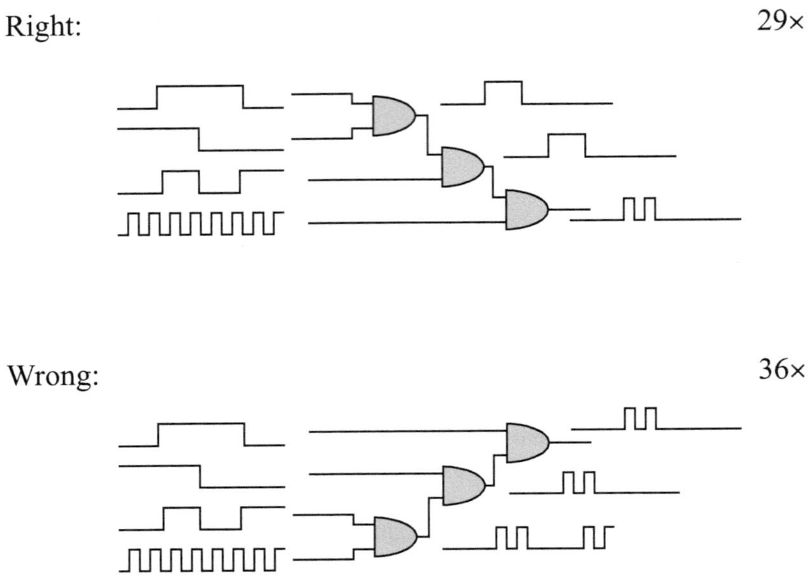

Fig. 8.29 Reduction of total activity by ordering signals

$$\text{NOR}: P_{0\to1} = P(0) \cdot P(1) = 3/4 \cdot 1/4 = 3/16$$
$$\text{EXOR}: P_{0\to1} = P(0) \cdot P(1) = 1/2 \cdot 1/2 = 1/4$$

Usually, the logic function in dynamic CMOS is realised with an nMOS pull-down network, while a pMOS transistor is used for precharge. This leads to small input capacitances, which makes dynamic logic attractive for high-speed applications. Besides the higher activity factor, the additional clock loads to control the precharge transistors also lead to much higher dissipation. The use of dynamic logic is not as straightforward and common as static logic. In terms of design robustness and ease of design, static CMOS is favourable as well. Finally, when power reduction techniques (such as power-down modes, in which the clock is stopped) are being implemented, dynamic CMOS is much more difficult to apply because of its charge leakage. Generally, it can be stated that dynamic logic is not a real candidate for low-power (low-voltage) realisation.

- Connect high-activity input signals close to the output of a logic gate.
  Figure 8.29 shows that connecting signals with high activity close to the output of the propagation chain will reduce the total switching activity and so the total power consumption of that chain.
- Exploit the characteristics of library cells.

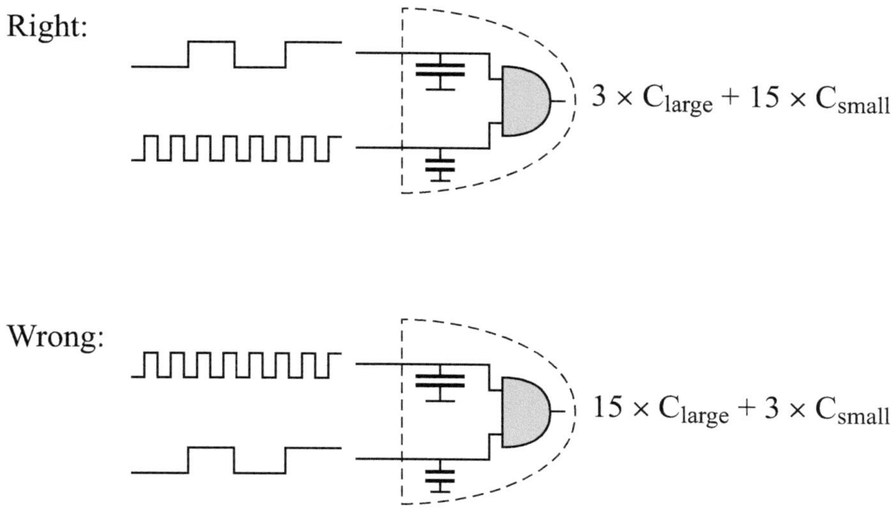

Fig. 8.30 Reduction of power dissipation by matching high-activity signals with low-capacitance inputs

Here again, when there are signals showing high activity, it is obvious that these will cause less power dissipation when they are connected to the low-capacitance inputs of logic gates. Figure 8.30 shows an example.

It should be clear that the power savings of these last two items can only be achieved by dedicated software programs, which perform some statistics on the signal activity inside a logic block.

8.6 Computing Power Versus Chip Power: A Scaling Perspective

The scaling process and its impact on the performance, reliability and signal integrity of MOS ICs are extensively discussed in Chap. 11. However, the scaling process with respect to the system performance of digital signal processors (DSPs) requires a different approach.

An important parameter reflecting this system performance is the computing power of a DSP. Generally, this computing power (U) is proportional to:

$$U = n \cdot f \tag{8.21}$$

where n represents the number of transistors and f is the frequency.

The power dissipation of the DSP is proportional to:

$$P = n \cdot f \cdot C \cdot V^2. \tag{8.22}$$

From these two equations, it can be derived that the computing power per Watt dissipation is equal to:

$$U/[\mathrm{W}] = \frac{1}{C \cdot V^2}. \tag{8.23}$$

In the following discussion on scaling, V_{T} effects and velocity saturation are neglected. If the scaling factor between two successive process generations is s (usually $s \approx 0.7$), then the number of transistors will increase to:

$$n_{\mathrm{s}} = n/s^2 \tag{8.24}$$

and the frequency to:

$$f_{\mathrm{s}} = p/s^2 \cdot f \tag{8.25}$$

where p equals the voltage scaling factor ($V_{\mathrm{p}} = p \cdot V$), as this factor may differ from s.

The capacitance C scales to:

$$C_{\mathrm{s}} = s \cdot C \tag{8.26}$$

Combining the previous equations results in the following expressions concerning computing power and chip power impact, the computing power scales to:

$$U_{\mathrm{s}} = n_{\mathrm{s}} \cdot f_{\mathrm{s}} = p/s^4 \cdot U \tag{8.27}$$

and the power dissipation per unit area with:

$$P_{\mathrm{s}} = n_{\mathrm{s}} \cdot f_{\mathrm{s}} \cdot C_{\mathrm{s}} \cdot {V_{\mathrm{p}}}^2 = p^3/s^3 \cdot P. \tag{8.28}$$

Therefore, the computing power per Watt after scaling increases to:

$$U_{\mathrm{s}}/[\mathrm{W}] = \frac{1}{s \cdot p^2} \cdot U/[\mathrm{W}]. \tag{8.29}$$

Remarkably, voltage scaling has more impact on the computing power per Watt than the process scaling. From the 0.8 µm CMOS to the 120-nm CMOS node, the voltage has continuously been scaled with a factor p that was about equal to the process scaling factor s. This means that, neglecting the second-order effects, the computing power per Watt for these generations of DSPs has increased according to:

$$U_{\mathrm{s}}/[\mathrm{W}] = \frac{1}{s^3} \cdot U/[\mathrm{W}]. \tag{8.30}$$

Because the subthreshold leakage current increases exponentially with a reduction in the threshold voltage, it has caused a slowdown in the reduction of the supply voltage, which has only marginally been reduced from the 120-nm CMOS node to the 28-nm CMOS node. Although this has limited the increase in standby (leakage) power consumption, it has reduced the power efficiency improvement when moving to the next technology node.

Each DSP generation will therefore still become more power-efficient but to a lesser extent. Second-order effects have a more negative impact on the transistor performance and thus on the DSP efficiency. However, even after such a reduction in efficiency improvement, a lot of new DSPs are still expected to enter the market with improved power efficiency.

8.7 Conclusions

With respect to conventional CMOS processes and design styles, large power savings could be achieved because they were optimised for speed and area, but not for power. Power can be reduced in different ways; however, the largest power savings can be achieved by reducing the supply voltage. In this respect, the scaling process from 0.8 μm CMOS technologies to the 28-nm node, in which the supply voltage gradually reduced from 5 to 0.9 V, has had a huge impact on the total power reduction. This was the basis for the integration of a large variety of functional features (camera, MP-3, phone, games, internet access, Bluetooth communication, audio, video, GSM, etc.) into one single mobile gadget: the smart phone (and of course also the tablet).

In CMOS process technology development, a few measures can be taken to reduce power: limit the leakage currents and limit the parasitic capacitances.

In the design, however, there are many options for reducing the supply voltage, the total capacitance and the activity on a chip. It has been shown that the decisions taken at the higher hierarchy levels have much more impact on the system power consumption than those taken at the lower levels (circuit and layout level) of design (Fig. 8.31). However, at these lower levels, there are more options to reduce power (library cell design, V_T choice, interlevel dielectric capacitance, back-bias, power switches, reduced voltage swing, etc.), such that together they can also contribute to a serious power reduction.

A complete and clear set of less-power design options/rules cannot be given, because the use of many of these depends on the application. This chapter is meant to present a rather complete summary of these options and to provide the designer with a less-power attitude. As most CMOS design teams currently outsource their advanced designs to a semiconductor foundry, the process is no longer a differentiator, since the competition does the same. In other words, you can only differentiate yourself from your competitor by creating the most efficient designs in terms of power consumption.

Finally, although several alternative low-power CMOS design styles have been presented at conferences and magazines, static CMOS logic is still favourable in

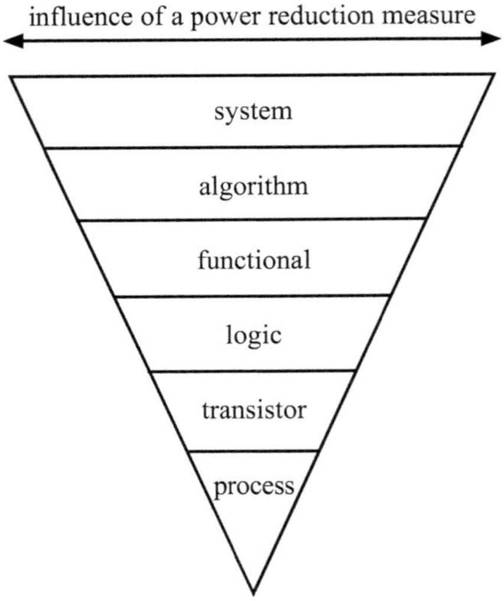

Fig. 8.31 Influence of a power reduction measure at different design hierarchy levels

many ways. It is very robust with respect to transistor scaling and supply voltage reduction. Besides this, design integrity is becoming a key issue in nanometer VLSI design, which also makes static (complementary) CMOS the best candidate for many process generations to come.

Both dynamic voltage and frequency scaling and the use of different power domains increase the challenge to completely verify all full-chip timing requirements and may lead to a reduced robustness of operation. The next chapter discusses most of the robustness-related chip aspects and presents a wide range of solutions to deal with them.

8.8 Exercises

1. Why must every designer always have a low-power attitude?
2. Explain all different power contributions. Which of these is larger and why?
3. How could the subthreshold leakage power dissipation be reduced?
4. In optimizing a complete library for low power, which of the library cells would you focus most of your attention to?
5. What is the greatest advantage of constant-field scaling with respect to power dissipation?
6. What would be the difference in activity factor between a static and dynamic CMOS realisation of the next Boolean function: $z = \overline{abc}$?
7. Repeat exercise 6 for $z = \overline{a + b + c}$.
8. What is the dynamic power consumption of a one-million gate CMOS logic core, running at 1 V with a clock frequency of 2 GHz, when we assume an average gate capacitance of 10 nF and an average activity factor of 0.2?

9. What is the difference between energy and power dissipation of a chip. Explain which is important for what device.
10. Mention several ways in which the process technology can be optimised to reduce (a) the dynamic power consumption and (b) the leakage power consumption.

References

1. H. Barry, AI, Liquid Cooling And The Data Center Of The Future, Forbes Technology Council (2023)
2. BU-106: Advantages of Primary Batteries. http://batteryuniversity.com/learn/article/primary_batteries. 15 April 2016
3. J.E. Harmon, New design for lithium-air battery could offer much longer driving range compared with the lithium-ion battery, Press Release by Argonne National Laboratory (2023)
4. P. Chen et al., Recent progresses and challenges in aqueous lithium–air batteries relating to the solid electrolyte separator: A mini-review, National Center for Biotechnology Information. Front. Chem. **10**, 1035691 (2022)
5. R. Van Noorden, The rechargeable revolution: A better battery. Nat. Int. Weekly J. Sci. **507**, 26–28 (2014)
6. http://batteryuniversity.com (2016)
7. K. Seta et al., 50% active-power saving without speed degradation using standby power reduction (SPR) circuit, in *IEEE Digest of Technical Papers* (1995), pp. 318–319
8. T. Kuroda et al., A 0.9 V, 150 MHz, 10 mW, 4 mm^2, 2D discrete cosine transform core processor with variable threshold voltage (V_T) scheme. IEEE J. Solid-State Circuits **31**, 1770–1779 (1996)
9. A. Montree et al., Limitations to adaptive back bias approach for standby power reduction in deep sub-micron CMOS ICs, in *Proceedings of the '99 European Solid-State Device Research Conference (ESSDERC)* (1999), pp. 580–583
10. T. Chen et al., Comparison of adaptive body bias (ABB) and adaptive supply voltage (ASV) for improving delay and leakage under the presence of process variation. IEEE Trans. Very Large Scale Integr. Syst. **11**(5), 888–899 (2003)
11. M. Meijer et al., Limits to performance spread tuning using adaptive voltage and body biasing, in *International Symposium on Circuits and Systems (ISCAS)* (2005), pp. 5–8
12. M. Izumikawa et al., A 0.25μm CMOS 0.9 V, 100 MHz, DSP core. IEEE J. Solid-State Circuits **32**, 52–61 (1997)
13. D. Lee et al., Gate oxide leakage current analysis and reduction for VLSI circuits. IEEE Trans. VLSI Syst. **12**(2), 155–166 (2004)
14. *International Technology Roadmap for Semiconductors*, 2011 Edition; Interconnect (2011), pp. 27–29
15. C. Piguet, Design Methods and Circuit Techniques to Reduce Leakage in Deep Submicron, *Faible Tension Faible Consommation, FTFC* (2003)
16. H.J.M. Veendrick, Short-circuit dissipation of static CMOS circuitry and its impact on the design of buffer circuits. IEEE J. Solid State Circuits **SC-19**(4), 468–473 (1984)
17. V. Kaenel et al., A voltage reduction technique for battery-operated systems. IEEE JSSC **25**, 1136–1140 (1990)
18. LPDDR description in Wikipedia, https://en.wikipedia.org/wiki/LPDDR, update June 2023
19. J. Pouwelse, Dynamic voltage scaling on a low-power microprocessor, in *Proceedings of the 7th Annual International Conference on Mobile Computing and Networking* (2001), pp. 251–259
20. R. Ghattas, et al., Energy management for commodity short-bit-width microcontrollers, in *Proceedings of the 2005 International Conference on Compilers, Architectures and Synthesis for Embedded Systems* (2005), pp. 32–42

21. V. Viswanath, The Power of Dynamic Voltage Frequency Scaling, EETIMES, 20 Aug 2015
22. N. Verma et al., Nanometer MOSFET variation in minimum energy subthreshold circuits. IEEE Trans. Electron Dev. **55**(1), 163–174 (2008)
23. M. Zangeneh et al., Designing tunable subthreshold logic circuits using adaptive feedback equalization. IEEE Trans. Very Large Scale Integr. Syst. **24**(3), 884–889 (2016)
24. K. Yano et al., A 3.8 ns CMOS 16x16-b multiplier using complementary pass-transistor logic. IEEE JSSC **25**, 388–393 (1990)
25. M. Suzuki et al., A 1.5 ns 32b CMOS ALU in double pass-transistor logic, in *Digest ISSCC* (1993), pp. 90–91
26. A. Parameswar et al., A swing restored pass-transistor logic-based multiply and accumulate circuit for multimedia applications. IEEE JSSC **31**, 805–809 (1996)
27. K. Jano et al., Top-down pass-transistor logic design. IEEE JSSC **31**, 792–803 (1996)
28. J.H. Pasternak, C. Salama, Differential pass-transistor logic. IEEE Circuits Dev. **9**, 23–28 (1993)
29. M. Song et al., Design methodology for high speed and low power digital circuits with energy economized pass-transistor logic (EEPL), in *Proceedings 22nd ESSCIRC Digest* (1996), pp 120–123
30. W.H. Paik et al., Push-pull pass-transistor logic family for low-voltage and low-power, in *Proceedings 22nd ESSCIRC Digest* (1996), pp. 116–119
31. R. Zimmermann, W. Fichtner, Low-power logic styles: CMOS versus pass-transistor logic. IEEE JSSC **32**, 1079–1090 (1997)
32. Vivekvati et al., A comparative study on reduction in power consumption by switching activity of 8 × 8 precision multipliers. Int. J. Adamas Techn. Rev. **2**(1), 1–6 (2015)
33. C. Mead, L. Conway, *Introduction to VLSI Systems*, Chapter 7 by C.Seitz (Addison-Wesley, Boston, 1980)
34. C. Maxfield, To be or not to be asynchronous that is the question, in *EDN* (1995), pp. 157–173
35. J. Kessels et al., A design experiment for a smart card application consuming low energy, Chapter 13 in *Principles of Asynchronous Circuit Design: A Systems Pespective* (Kluwer Academic Publishers, Dordrecht, 2001)
36. Ä. Lines, Asynchronous interconnect for synchronous (SOC) design. IEEE Micro J. **24**(1), 32–41 (2004)
37. A. Bink, ARM996HS, the first licensable, clockless 32-bit processor core. IEEE Micro J. **27**, 58–68 (2007)
38. F. te Beest et al., Synchronous full-scan for asynchronous handshake circuits. J. Electr. Test. Theory Appl. **19**, 397–406 (2003)
39. R. Hossain et al., Low power design using double edge triggered flip-flops. IEEE Trans. VLSI **2**(2), 261–265 (1994)
40. J. Yuang et al., New single-clock CMOS latches and flipflops with improved speed and power savings. IEEE J. Solid State Circuits **32**, 62–69 (1997)
41. A.G.M. Strollo et al., Low power double edge-triggered flip-flop using one latch. Electr. Lett. **35**, 187–188 (1999)

Robustness of Nanometer CMOS Designs: Signal Integrity, Variability and Reliability

9

9.1 Introduction

With shrinking feature sizes and increased chip sizes, the average load capacitance of a logic gate is now dominated by the interconnection (metal wires) rather than by the transistor itself. Most of the potential electrical problems, such as cross talk, critical timing, substrate bounce, clock skew, etc., are related to the signal propagation and/or high (peak) currents through these metal wires.

Currently, complex VLSI chips may contain hundreds of millions to more than a hundred billion transistors that realise complete (sub)systems on one single die. The design of these ICs requires a lot of different tools, as presented in Chap. 7. The sequence in which these tools are used, from the upper hierarchy levels down to the layout level, is called the "design flow".

IC design flows have been automated so much that "first-time-right silicon" is considered as natural. However, keeping control over all the tools used in the design flow (the high-level description language, the synthesis tools and the verification tools, to name a few) requires the complete attention of the designers. Thus, even when designers are familiar with the physical aspects of complex ICs, the potential physical and electrical problems do not get the attention that they require, particularly in nanometer technologies.

First silicon (especially of high-performance ICs) therefore shows first-time-right functionality but often at lower or higher supply voltages and/or at lower frequencies than required. Actually, at a time where designers are drifting away from the physical transistor level into abstract high-hierarchy levels of design, exactly the opposite would be required to get current and future VLSI chips operating electrically correctly. Many ICs are therefore no longer "*correct by design*" but are "designed by corrections".

This chapter deals with the robustness of digital circuits in relation with the continuous scaling process. It will cover most topics related to signal integrity (timing, cross talk, signal propagation, voltage drop, supply and substrate noise, soft

H. Veendrick, *Nanometer CMOS ICs*,
https://doi.org/10.1007/978-3-031-64249-4_9

errors, EMC, etc.), variability (systematic and random variability) as well as such reliability issues as electromigration, leakage, ESD, latch-up, hot-carrier injection and NBTI.

Because of the increasing clock frequencies and increasing chip complexity, e.g. multi-frequency and multi-voltage domains, timing closure has become one of the most time-consuming activities in the total design trajectory. Therefore, the next subsection will start with some of the related timing and clocking issues.

The reducing signal integrity is a result of two conflicting effects: the increase of noise and the reduction of the noise margins (V_{dd} and V_T). A relatively large section is therefore devoted to almost all aspects related to signal integrity and ways to maintain it at a sufficiently high level. A continuous reduction of the noise margins also has a severe impact on the quality of the IC test. The increasing discrepancy between chip operations during test and in the application will result in more customer returns and design (re)spins. The section will therefore also include some remarks on the effect of scaling on test coverage and complexity; however, the basics of IC testing are discussed in the next chapter.

As a result of the continuously reducing transistor sizes and voltages, the transistor behaviour is becoming much more sensitive to an increasing number of variability causes. Moreover, also, the diversity and level of variations increase. The fourth paragraph presents a flavour of the impact of these variations on both analog and digital circuit design.

The continuous scaling of the devices and interconnects also severely impact the reliability of the integrated circuit. In addition, a necessary move to new materials may even have dramatic consequences for the overall IC reliability. Section 9.5 is therefore devoted to reliability and ways to maintain it. A robust design not only refers to a robust electrical operation with respect to the specified performance; it also takes great discipline to set up a robust database and create a complete design documentation. This is needed to enable quick redesign (re)spins and support reuse. These aspects are discussed in the sixth paragraph. Paragraph seven presents some conclusive remarks. The negative impact of scaling on signal integrity and reliability depends on the way scaling is performed. The final chapter in this book discusses the effects of scaling in general but will also come back to the influence of constant voltage scaling, constant-field scaling and constant-size scaling on the robustness of operation of nanometer CMOS ICs.

9.2 Clock Generation, Clock Distribution and Critical Timing

9.2.1 Introduction

The majority of today's integrated circuits are synchronous designs in which the data transfers to the chip, across the chip and off chip are controlled by one or more clock signals. Clock frequencies may vary from a few tens of megahertz for extremely low-performance ASICs to a few gigahertz for high-performance microprocessors. The timing complexity of an IC is related not only to the frequency of its clock signals but also to the diversity, complexity and number of cores integrated on a single chip. The overall timing complexity is so high that full chip-level timing closure has become a real burden. Additionally, advanced power reduction techniques, such as clock disabling, different voltage and frequency domains, power switching and dynamic voltage and frequency scaling, in combination with increasing process parameter variations and circuit noise have made overall chip timing become a nightmare. Next to the fact that the corresponding design solutions need to be supported by the models, the tools and the design flow, also the back end designers must be well-educated so that they understand the issues and can manage the "global timing picture".

Nevertheless, this subsection will only discuss some of the basics of clock distribution and clock generation and is only meant to trigger and focus the attention of the design community to the huge challenges of timing closure.

9.2.2 Clock Distribution and Critical Timing Issues

Very complex designs may contain hundreds of millions to more than a hundred billion transistors on silicon die areas of one to several square centimetres. Most VLSI designs contain synchronous logic, which means that data transfer on the chip is controlled by means of one or more clock signals. These clock signals are fed to latches, flip-flops and registers, which temporarily store data during part of the clock period.

Current VLSI chips may contain millions of these latches or flip-flops, and the total wire length of the clock signals may exceed several hundred metres. To achieve high system performance, the clock frequency is often maximised. The combination of a large clock load and a relatively high clock frequency is the cause of many on-chip timing problems.

The following sections discuss potential timing problems, most of which are related to the clock signals:

Single-Phase Clocking

From Fig. 9.1, we can derive that the minimum clock cycle time is given by:

$$\tau_{\text{min}} = \tau_{\text{ff}} + \tau_{\text{logic}} + \tau_{\text{su}} + \tau_{\text{skew}} \tag{9.1}$$

where:

τ_{ff} is the flip-flop delay from clock to output.

τ_{logic} is the propagation delay through the logic.

τ_{su} is the set-up time of the data of flip-flop 2.

τ_{skew} is the maximum amount of time that the clock of flip-flop 2 can be earlier than that of flip-flop 1.

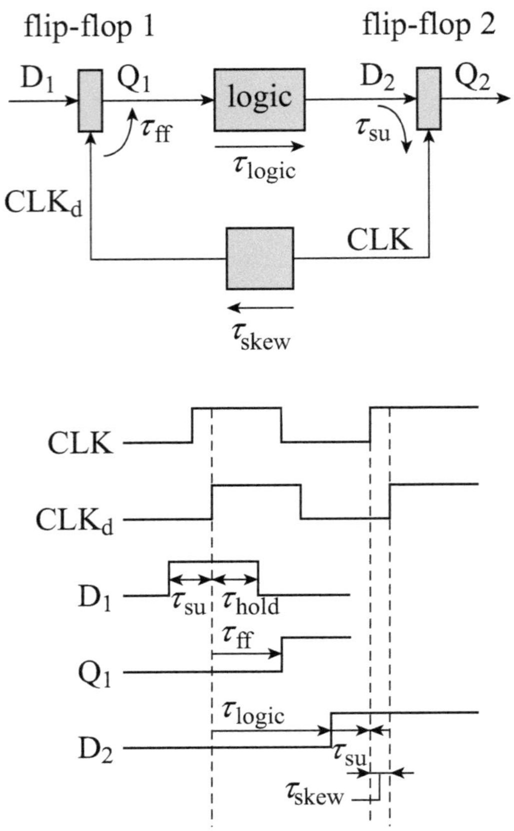

Fig. 9.1 Single-phase clock system and its timing diagram

Especially τ_{logic}, which is dominant in Eq. 9.1, must be carefully simulated to be sure that the required frequency (clock period) will be achieved. This "simulation" is usually performed by the static timing analysis tool, which adds the worst-case delay of each of the gates together to determine the total delay of the logic path. In combination with the synthesis tools, it should guarantee satisfactory timing results. As most standard-cell libraries include dual-V_T cells, e.g. standard-V_T (SV_T) and high-V_T (HV_T) cells, the synthesis and place and route tools may now select a combination of these cells in order to create power-efficient logic cores. To reduce subthreshold leakage, it uses as many HV_T cells as possible and only uses the SV_T cells in the critical (delay) paths. This adds additional complexity to the libraries, the cell characterisation tools and the overall timing verification of the logic cores.

Sometimes, the logic path between two flip-flops is absent. This is the case when pipeline and/or scan registers are implemented by using series connections of flip-flops (Fig. 9.2).

Especially in the scan mode during testing (see Chap. 10), the logic paths between two flip-flops are bypassed, and flip-flops are directly connected to other

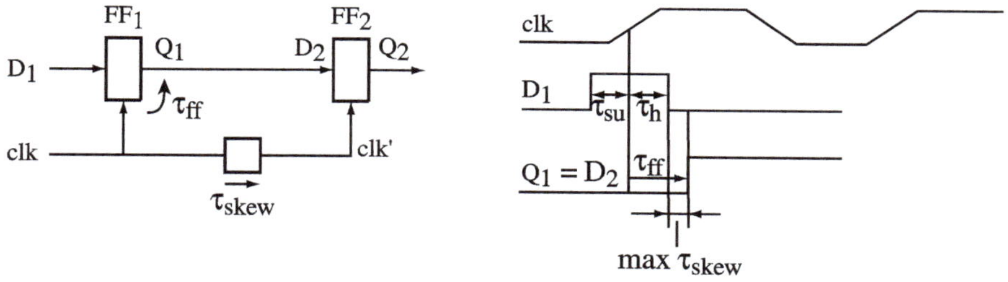

Fig. 9.2 Critical timing situation in case of direct communication between two flip-flops

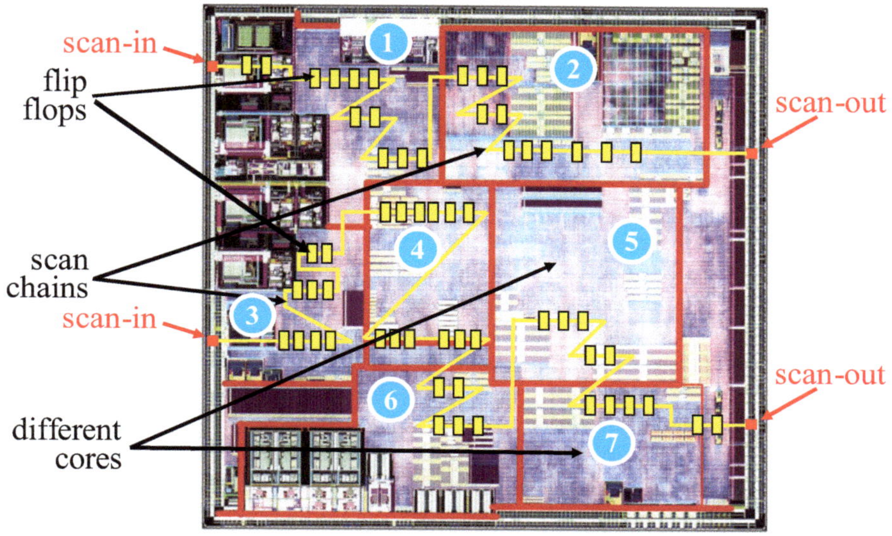

Fig. 9.3 Example of a scan chain in a complex VLSI circuit

flip-flops, so that they form a serial scan chain to propagate test signals from input pads (scan-in) to the logic core that is to be tested. In Fig. 9.3, a flip-flop of logic block 1 is directly connected to a flip-flop of logic block 2.

With a direct connection, the propagation time of the data between these flip-flops can be very short. As the clock signal line is routed through these blocks automatically, its time of arrival at the first flip-flop in the scan chain of logic block 2 can be later than the arrival time of the data signal (from block 1) at this flip-flop. This will result in a race, which can also occur in registers. Therefore, each (scan) register should be carefully checked with respect to the above critical timing situation. If necessary, additional delay by using several inverters must be included at these critical positions in the scan chain at the transition between these cores. This is a frequently occurring problem and results in a failing test (scan chain test), while the chip may work properly.

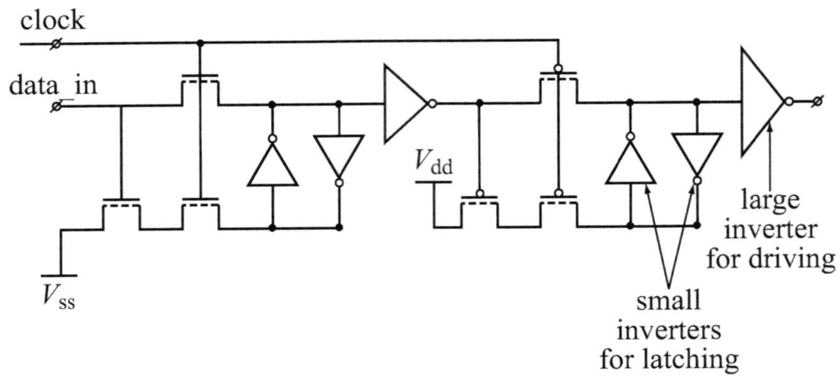

Fig. 9.4 Efficient single-phase clock flip-flop

Generally, there is a variety of single-phase clocked flip-flops in a library. As many of these flip-flops need two clock phases, one or both are generated inside the flip-flop by means of inverters.

Figure 9.4 shows a flip-flop [1] that can also be used in a low-power design: it has a small number of transistors, it is controlled by a single-phase clock, and it has a relatively low clock load. Here, the clock may also be generated locally in the latch by means of an additional inverter.

Clock Skew and Clock Jitter

Clocks are routed locally in IP cores and globally across the chip, which may consist of different clock domains. For proper operation, it is required that the clock signal arrives at all flip-flops within the same clock domain almost exactly at the same time. However, due to many different causes, this cannot always be sufficiently guaranteed. The difference in clock arrival times at these flip-flops is called *clock skew*.

There are several causes of clock skew:

- Different lengths of clock paths
- Different loads of the clock drivers
- Variations in clock driver delay due to:
  - Process variations
  - Supply noise and IR drop
  - Temperature variations

The actual clock skew between two flip-flops is different from one chip to another and varies with the time.

The probability of occurrence of clock skew is higher, when there is direct communication between flip-flops within digital cores or between flip-flops in different cores. An example of clock skew is demonstrated in Fig. 9.3, where the clock signal in logic block 1 has to propagate through a different number of clock drivers in the clock tree than the clock signal in logic block 2, before it arrives at the respective flip flops. Clock skew is also introduced in ICs that include cores whose clock may temporarily be switched off (*gated clocks*) to reduce power consumption when its functionality is not needed.

Particularly in this era of IP reuse, there is a large variety of digital cores to be integrated onto a single SoC. These logic and memory IP cores may originate from different design sites within one company or even from different IP vendors. Also, only one type of flip-flop is preferred in these different cores, particularly when there is data interaction: e.g. positive edge-triggered flip-flop. Clock and further timing parameters must very well be specified in order to have these cores operated correctly with one another.

When the clock skew exceeds a certain limit, it may cause *timing violations*. *Set-up and hold-time violations* can play a dominant role in the operating frequency of the circuit. *Set-up time* is the time a flip-flop requires its input data to be valid before its clock-signal capture edge. *Hold time* is the time a flip-flop requires its input data to be stable after its clock-signal capture edge. A hold-time violation is caused by a so-called short-path delay fault.

Usually, the set-up time is defined by the worst-case behaviour of the design, while the minimum hold time is defined by the best-case situation. Set-up time violations can be recovered if the application allows a reduction of the frequency. A hold-time violation means that you have to throw away the chip.

Within the cores, the clock skew is usually limited by applying a well-balanced clock-tree approach. It is extremely important that the different branches in the tree are equally loaded (same number of flip-flops and same lengths of the clock wires). This must be verified by tools, particularly in high-performance complex circuits. Current tools offer a well-balanced *clock-tree synthesis*, which enhances the quality of clock timing. An important advantage of this clock-tree approach is the distribution of the different small clock drivers over the logic blocks. The use of distributed clock drivers also puts the clock drivers right there where they are needed. Distributed clock drivers keep the current loops short, and they also do not switch simultaneously but distributed over a small time frame. Moreover, they can use the intrinsic decoupling capacitance which is available in a logic standard cell block. This reduces the $\frac{di}{dt}$ fluctuations, which are responsible for most of the supply/ground bounce in VLSI designs.

In many synchronous designs, the total dissipation of the clock-related circuitry may vary from 10% to even more than 60% of the total IC dissipation. It is obvious, then, that the clock system will also generate a large part of the total supply bounce.

Today's semiconductor fabrication processes allow us to integrate complete systems onto one single die. Such an IC may contain a large variety of functions which may operate more or less independent from one another. The corresponding large currents introduce voltage drop across the on-chip power distribution network, which negatively affect the timing behaviour. Also, other physical effects, such as cross talk, supply and substrate noise, variability, etc., which are discussed in this chapter, all affect the local and global timing behaviour.

Moreover, because power consumption has become one of the biggest concerns in the design of these systems, they need to support more and more state-of-the-art power reduction techniques: clock gating, power switching, voltage reduction, (dynamic) voltage and frequency scaling techniques, etc. Signals that cross different voltage domain boundaries require *level shifters* in their paths, which create additional delay. This additional delay may introduce timing errors if it is not included in the simulations. Reduction of the supply voltage causes a complete change in timing behaviour of the connected cores. In other words, it is not only the increasing IC functionality and performance that pose a serious threat to a secure and reliable timing closure; it is also these additional design measures to reduce power in both active and standby modes that create new timing constraints.

Next to clock skew, there is another major problem related to the propagation of the clock signal. Particularly as a result of supply voltage changes, e.g. due to supply noise, IR drop or temperature variations, the clock period may vary from one clock cycle to another. This is called *clock jitter*. While clock skew represents the difference in clock-edge arrival times at different flip-flops in the same clock cycle, clock jitter is the difference in clock-edge arrival times and/or clock period at the same flip-flop in different clock cycles.

It will be clear by now that both clock skew and clock jitter may have serious impact on the overall timing and functional behaviour of an IC. This section is only meant to review some of the most potential problems related to the clock, which is the most important signal on a synchronous chip and must be handled with care. More information on clock distribution, clock skew and clock jitter can be found in [2] and [3], respectively, and also in [4].

Other Timing Problems

Particularly in low-power CMOS ICs, some logic blocks (or sometimes even the complete chip) may often be inactive for certain periods of time. Such a chip may contain different clock domains, of which the mode of operation (active or standby) is controlled by a *gated clock* . In many cases then, the main clock is used as input to a logic gate which performs a logic operation on the clock signal (gated clock). Figure 9.5 shows an example:

When the delay between the clock ϕ and then enabled clock ϕ' is longer than the data delay between the output Q_1 of one flip-flop in a certain core and the input D_2 of the next flip-flop in another core, this "new" data sample will be clocked into this flip-flop by the "old" clock, and a race will occur.

Such clock-enabled signals are also often used in the design of memory address decoding circuits and are very critical with respect to timing margins.

Finally, timing problems could also occur when the data delay (caused by the logic and interconnection delay) between two successive latches or flip-flops becomes equal to or larger than one clock period. Figure 9.6 shows an example. When the total propagation time through the logic from Q_1 to D_2 exceeds the clock period, the data at D_2 can arrive after the sample period of flip-flop 2 has been terminated. It will then be sampled in the next clock period, resulting in incorrect

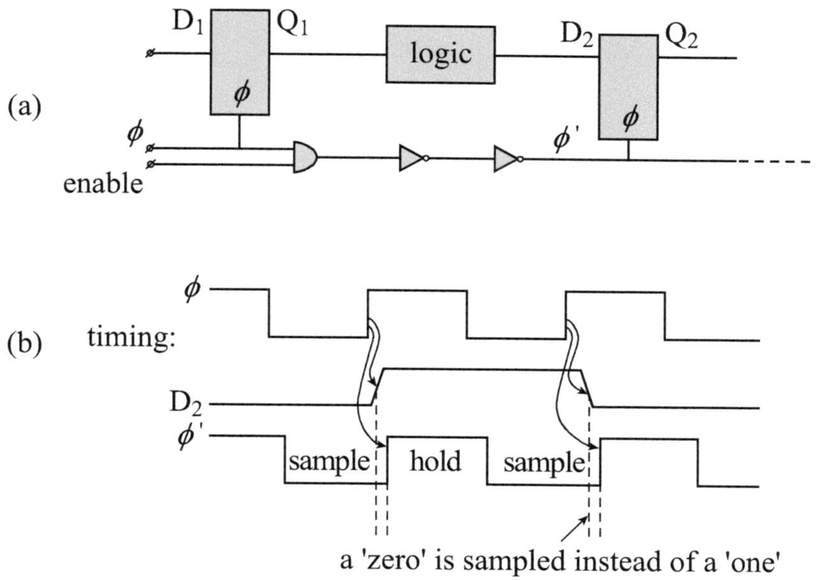

Fig. 9.5 (**a**) Example of a local clock-enabled circuit and (**b**) the corresponding timing diagram

output data. Timing simulation to find critical delay paths is therefore a must in CMOS VLSI design and is part of the design flow.

With accurate library cell characterisation (for worst-case, typical and best-case parameters) and "perfect" wire load models and wire-length estimation, the RTL synthesis tools used by the front end designer should create the correct logic network, operating fully according the specifications, using minimum power and area. In most IC designs, today, certain logic-path delays are much more dependent on propagation delay across the wires than on the cell timing. *Timing closure* is an iterative process, which eventually results in a close match between the front end timing numbers from synthesis and the back end timing after place and route. A good predictability is essential to an efficient timing closure in which the number of iterations is brought to a minimum.

Slack Borrowing and Time Stealing

When a data path uses more than a full clock cycle in a single clock system, this is referred to as *cycle stealing*.

Slack borrowing refers to the case where a logical partition utilises time left over (slack time) by the previous partition [5]. Important to note is that it can be used without the adjustment of circuitry and/or clock arrival times. This precludes the use of edge-triggered circuitry (dynamic logic and flip-flops). *Time stealing* refers to the case where a logical partition steals a portion of the time allotted to the next partition. This can only be obtained by adjusting the clock arrival time(s). Using one

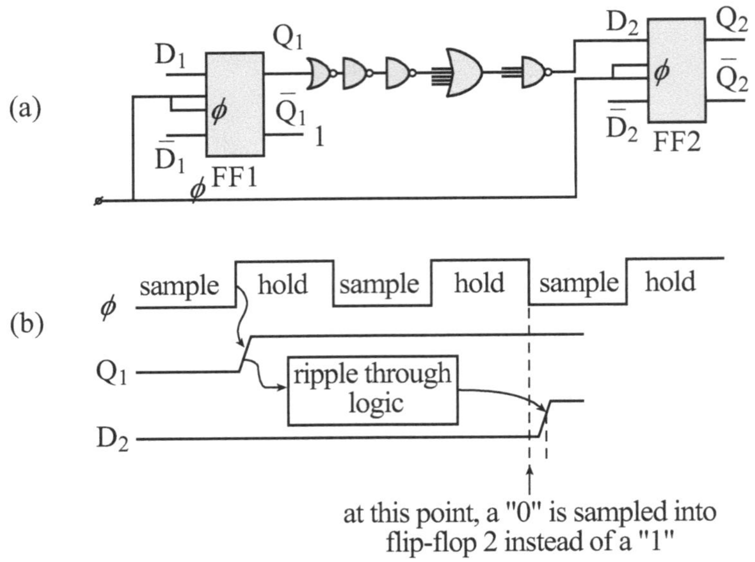

Fig. 9.6 (**a**) Example in which the data delay exceeds a clock period and (**b**) its corresponding timing diagram

of these concepts to solve timing problems in (ultra-)high-speed designs forces the designer to match certain design rule requirements. A well-documented list of such design rules can be found in [5].

Source-Synchronous Timing (Clock Forwarding)

In a source-synchronous interface, data and clock signal propagation between transmitter and receiver are matched. This technique is not only currently used in high-performance microprocessors and SDRAM interfaces [6, 7] but is also a potential candidate for on-chip time-of-flight compensation.

9.2.3 Clock Generation and Synchronisation in Different (Clock) Domains on a Chip

With IC complexities exceeding hundreds of millions of transistors, the total effort required to complete such complex VLSI designs is immense. This stimulates the reuse (IP) of certain logic blocks (cores) and memories. Current heterogeneous systems on chip may not only incorporate many clock domains but can be built from cores, which are designed at different sites, with different specifications. Because each core has a different clock skew from the core's clock input terminal to the farthest away flip-flop, the clock phase of each core has to be synchronised with the main clock. This subsection discusses the generation of multiple clocks and the

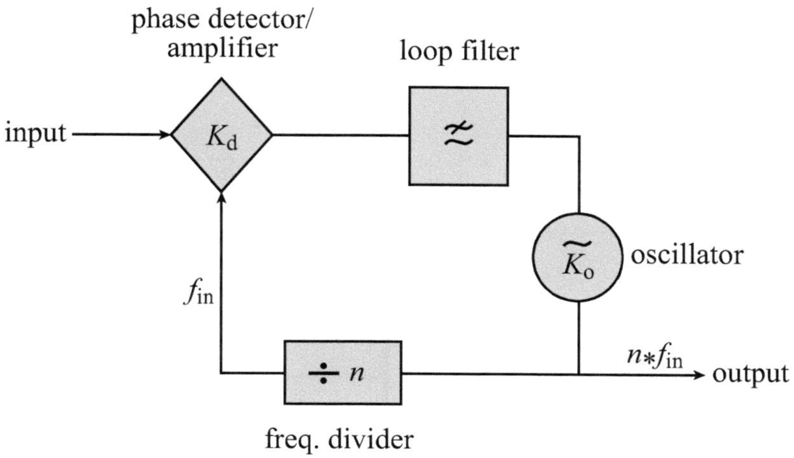

Fig. 9.7 Basic concept for a phase-locked loop

synchronisation of clocks in systems that use different cores running at different clock frequencies.

On-Chip Multiple-Clock Generation

On-chip multiples of the clock can be generated by phase-locked loops (PLLs). Figure 9.7 shows a basic phase-locked loop concept.

The voltage-controlled oscillator (VCO) (current-controlled oscillators (CCOs) are also used) is basically an oscillator whose frequency is determined by an externally applied voltage. Its output frequency is a multiple of its input frequency. The phase detector is sensitive to differences in phase of the input and VCO signals. A small shift in the frequency of the input signal changes the control voltage of the VCO, which then controls the VCO frequency back to the same value as that on the input signal. Thus, the VCO remains locked to the input. Based on this principle, a PLL can be used to generate an output frequency which is a multiple of the input frequency. The output frequency equals n times the input frequency. A frequency divider ($\div\, n$) is then used to create a feedback signal with the same frequency (f_{in}) as the input signal.

As current complex ICs require many different clock domains, multiple frequencies must be generated on chip. Figure 9.8 shows an example of a multi-clock generator based on PLL.

In this example, the PLL output frequency equals $n \times m_1 \times f_{\text{in}}$. Using different divisions (m_i), many different clocks can be generated. The PLL, by nature, automatically locks these clocks in phase with the input.

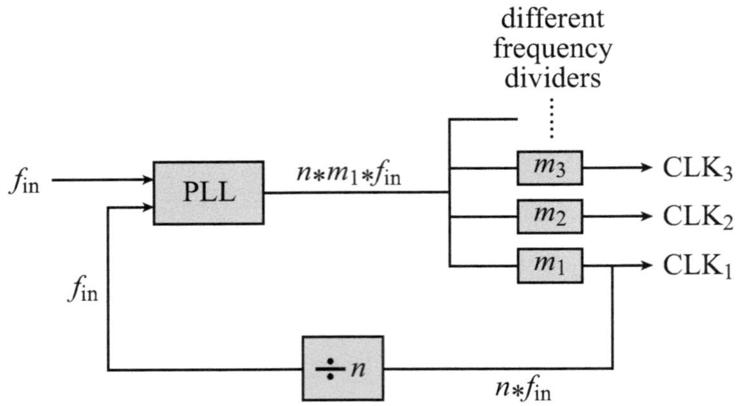

Fig. 9.8 Multi-clock generator, based on PLL

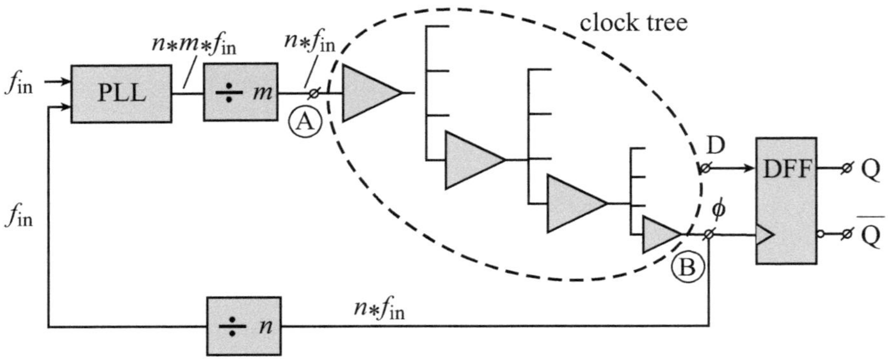

Fig. 9.9 Clock-phase synchronisation to compensate for different clock skews inside different cores

Clock-Phase Synchronisation in Multiple Core Environments

Because of differences in the clock arrival times at the flip-flops of different cores, these delays must be compensated for to allow proper communication between different cores. There are several methods of synchronising the clock phase at the actual flip-flops in each core.

The first method is *adaptive skew control*. In this approach, the clock network of each core (domain) is extensively simulated. The clock skew in each core is then made equal to the worst-case clock skew by using a chain of inverters. The length of this inverter chain is then adapted to the required additional delay in the specific core clock path.

The second method uses the PLL concept. The PLL property of locking one signal phase to the phase of another reference signal makes the PLL also suitable for the compensation of clock skew in different cores; see Fig. 9.9.

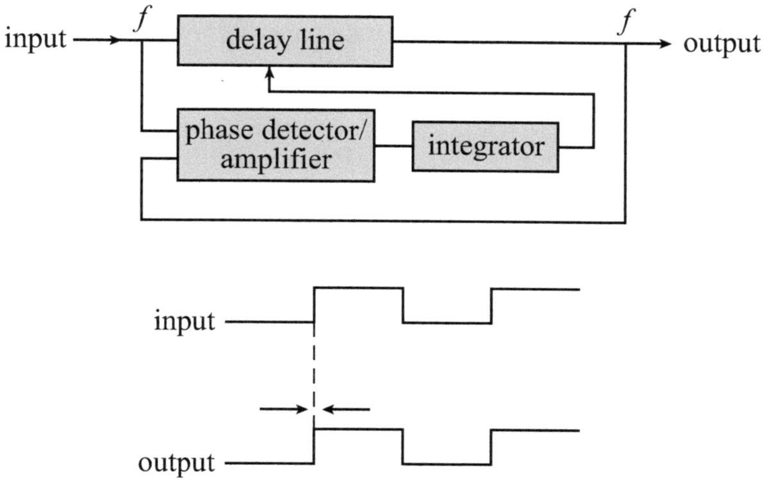

Fig. 9.10 Basic concept of a delay-locked loop and its timing

Node A represents the clock terminal of the core, and node B represents the clock terminal of the actual flip-flop in that core. The clock phase at the flip-flop will then be locked to the input reference signal, which is usually the chip's main clock. In this way, the clock-tree delay (which might be different in all cores) can be compensated for. Moreover, when the frequency dividers in Fig. 9.9 are made programmable, then the same PLL can be used in all cores, even when they run at different frequencies.

Sometimes, reusable cores are only available with fixed instances and only in GDSII (layout description) format. In these cases, the clock tree must be thoroughly simulated, and a delay chain, which mimics the core's internal clock delay path, replaces the clock tree between nodes A and B (Fig. 9.9) in the feedback path. The PLL must be placed outside the core. Disadvantages with the use of PLLs are:

- Because of high internal frequencies, PLLs can consume relatively high power.
- PLLs are difficult to start and stop.
- Especially the startup takes a relatively long time.
- Multiple-clock concepts and the use of PLLs for clock generation and synchronisation make testing very difficult. During testing, such PLLs must be set to the right mode first before the test procedure can be started.

Finally, to synchronise the clock phases to compensate for the different clock skews in different cores, delay-locked loops (DLLs) can also be used; (see Fig. 9.10).

The delay of the delay line can be controlled by the output voltage of the integrator. In this concept, the output signal is delayed over one complete clock period with respect to the input. If the delay is less, then the phase detector produces a signal which increases the delay of the delay line, via the integrator. The output signal in such a DLL has the same frequency as the input, and this concept of the DLL cannot be used to multiply the frequency.

Because the VCO or CCO in a PLL generates frequencies that depend on the supply voltage, clock jitter can occur when there is supply noise. Also, the delay in a DLL is susceptible to supply noise. Control of the clock jitter is therefore one of the most important constraints in the design of a PLL and DLL. For the synchronisation of the clock phases of all cores in a heterogeneous chip, each core needs its own PLL (DLL).

9.3 Signal Integrity

Signal integrity indicates how well a signal maintains its original shape when propagating through a combination of circuits and interconnections. On-chip effects from different origin may influence this shape. Signals can be influenced by switching of nearby neighbours (cross talk, interference), by voltage changes on the supply lines (voltage drop and supply noise), by local voltage changes in the substrate (substrate noise) or when the signal node is hit by radioactive or cosmic particles (soft error). In addition, the speed at which a signal propagates through bus lines is heavily affected by the switching behaviour of neighbouring bus lines.

The next subsections will focus on each of these signal integrity topics individually and also present ways to limit the noise level or the influence of the potential noise sources that threaten the signal integrity.

9.3.1 Cross Talk and Signal Propagation

Due to the scaling of the transistors, their density has almost doubled every new technology node for more than five decades already. This forced the metal lines (width and spacing) to be scaled in the same order to be able to connect this increasing number of devices per unit of area. Per unit of area, however, the total length of the interconnections in one metal layer only increased with a factor of 1.4. This means that additional metal layers were needed to allow a high-density connection of all logic gates. The metal layers are also used to distribute the current from the top metal layer all the way down to the individual devices. As will be discussed in the subsection on electromigration, the current density also increased with a factor of 1.4 every new technology node, meaning that the thickness of the metal layers could not be scaled at the same pace as the width and spacing. Consequently, the mutual capacitance between neighbouring signal lines has dramatically increased.

Figure 9.11 shows two cross sections of three parallel metal lines: one in a conventional 120-nm CMOS technology and the other one in a 22-nm process. It clearly shows that the bottom (C_b) and top capacitances (C_t) reduce while the mutual capacitances (C_m) increase. This increase in mutual capacitance has dramatic effects on the performance and robustness of integrated circuits. The first

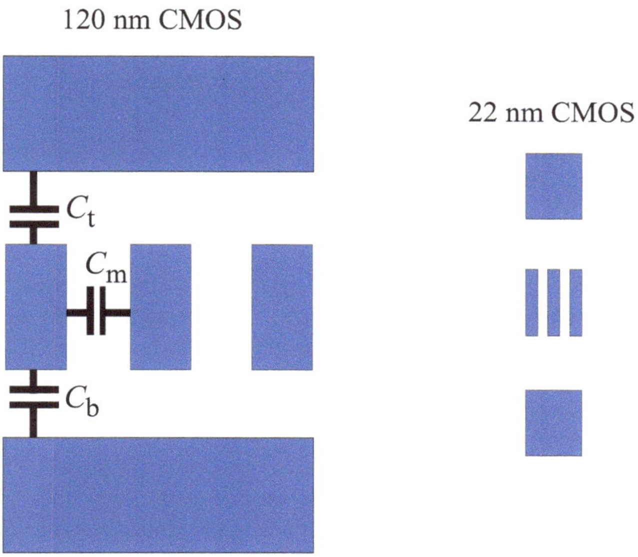

Fig. 9.11 Expected scaling of metal track width and spacing

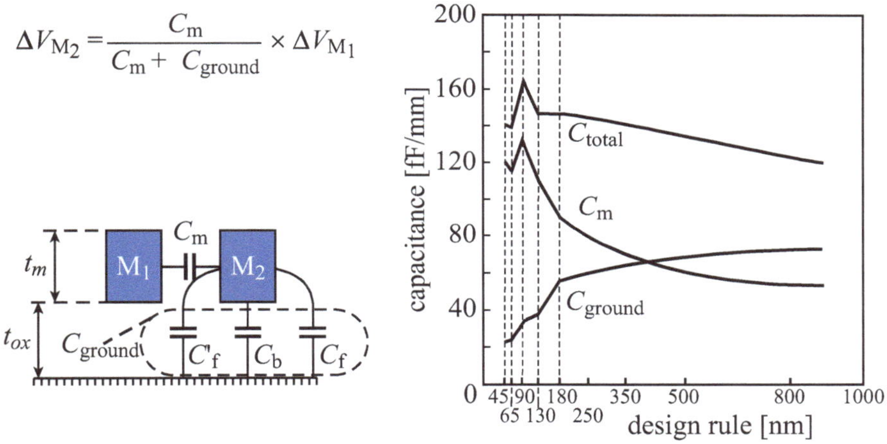

Fig. 9.12 Interconnect capacitances across various technology nodes

one is the growing interference between two neighbouring interconnect lines, which is usually referred to as *cross talk*. The second one is the growing signal *propagation delay* across the interconnect because of its increasing *RC* times. Third, the increased interconnect capacitances also affect the overall IC's power consumption. We'll discuss each one of these effects in more details now. Figure 9.12 depicts the trend in the cross talk over several technology nodes.

Table 9.1 Capacitance values for second metal layer in different CMOS technologies

Node	180 nm	130 nm	90 nm	65 nm	45 nm	32 nm	22 nm
C_m	89 fF	110 fF	132 fF	115 fF	120 fF	126 fF	123 fF
C_{ground}	58 fF	36 fF	32 fF	21 fF	18 fF	15 fF	13 fF
C_{total}	147 fF	146 fF	164 fF	136 fF	138 fF	141 fF	136 fF
$\Delta V_{M2}/\Delta V_{M1}$	0.6	0.7	0.8	0.84	0.86	0.89	0.90

The used model refers to two minimum-spaced interconnect wires, M1 and M2, in the same metal layer. A signal swing ΔV_{M1} on metal track M1 causes a noise pulse ΔV_{M2} on the floating neighbouring metal track M2, as defined by:

$$\Delta V_{M2} = \Delta V_{M1} \cdot \frac{C_m}{C_m + C_{ground}} \tag{9.2}$$

Table 9.1 shows the capacitance values for various technology nodes. The bottom line in this table presents the amount that one signal propagates into the other one through cross talk. For the 22 nm node this means that 90% of the switching signal propagates into its floating neighbours. For smaller process nodes, this percentage remains close to that 90%, as the transistor gate capacitance of FinFET and GAA transistors relatively increases compared to planar transistors, because of their 3D structure.

Because of this cross-talk effect, all floating lines (e.g. precharged bit lines in a memory and tri state buses) are very susceptible to cross-talk noise.

Even non-floating (driven) lines in digital cores are becoming increasingly susceptible to cross talk causing spurious voltage spikes in the interconnect wires. Traditional design flows only deal with top-level cross-talk analysis in the back end part, to repair the violations with manual effort, after the chip layout is completed. Because timing and cross talk are closely related, they need to be executed concurrently with the place-and-route tools. The introduction of multi-V_{dd} and multi-V_{T} poses a challenge for the physical synthesis and verification tools because both design parameters affect timing and signal integrity.

In memory design, scaling poses other challenges to maintain design robustness. The layout of a static random-access memory (SRAM), for example, includes many parallel bit lines and word lines at minimum spacing in different metal layers. It is clear that these will represent many parasitic capacitances with an increasing contribution of mutual capacitances between the various contacts and vias (pillars) (Fig. 11.4 in Chap. 11).

Memories in nanometer technologies therefore require very accurate 3D extraction tools in order to prevent that the silicon will, unexpectedly, run much slower than derived from circuit simulations.

Next to the cross talk between metal wires, the signal propagation across metal wires is also heavily affected by scaling. In a 32-bit bus, for example, most

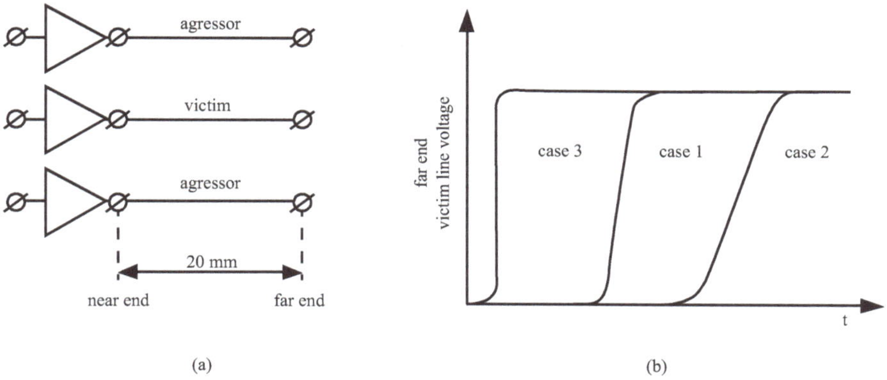

Fig. 9.13 Model for three neighbouring metal lines of an on-chip bus

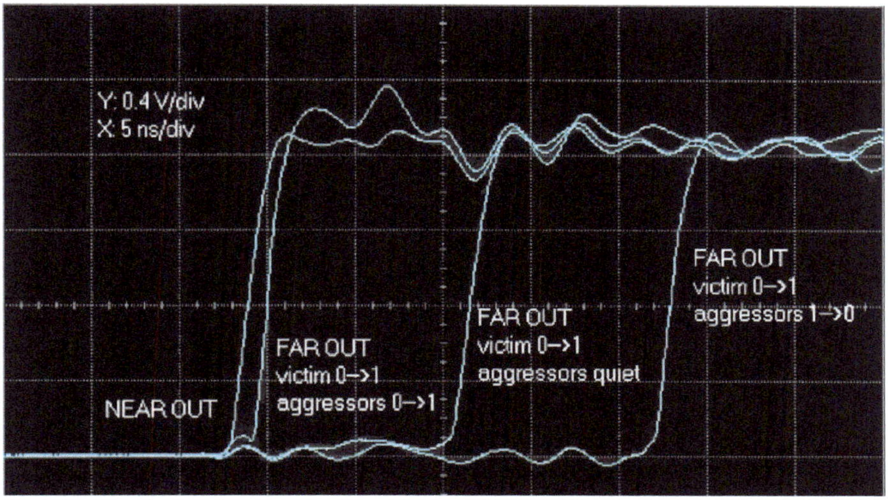

Fig. 9.14 Signal propagation across a 20-mm bus line in 180-nm CMOS technology in relation with the switching activity of both of its neighbours (at minimum distance)

internal bus lines (victims) are embedded between two minimum-spaced neighbours (aggressors) (Fig. 9.13).

The switching behaviour of both aggressors with respect to the victim causes a large dynamic range in signal propagation across the victim line. In case both aggressors switch opposite to the victim (case 2), the signal propagation across the victim lasts about 16 times longer than in case the aggressors and victim all switch in the same direction (case 3). Figure 9.14 shows this effect for 20-mm-long bus lines in a 180-nm CMOS technology.

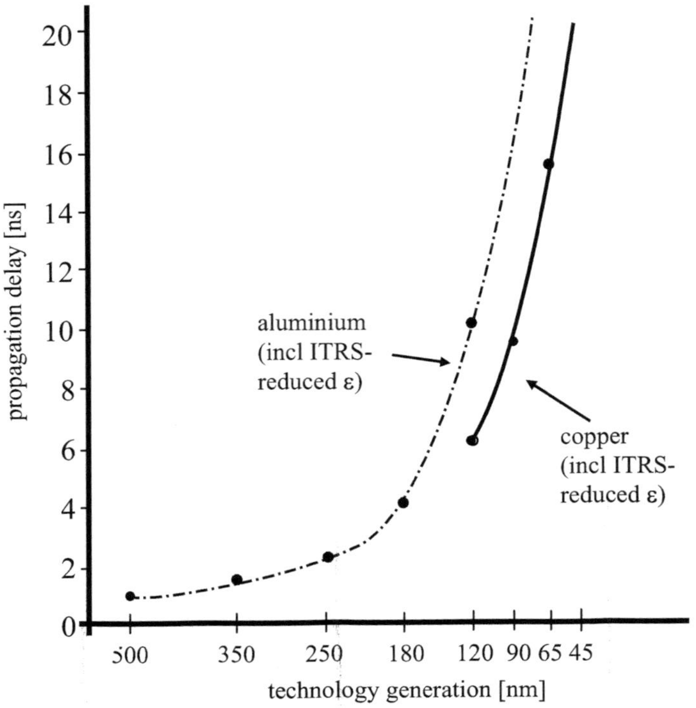

Fig. 9.15 Propagation delay versus technology node in case aggressors is quiet

Figure 9.15 plots the increasing *propagation delay* (in nanoseconds) with the technology node for a 20-mm-long bus line, embedded between two quiet (non-switching) aggressors.

Although the introduction of copper with the 120-nm node shows some relief in the increase of the propagation delay, it only helped for about one technology node. This means that in the 120-nm node, with an aluminium back end, the interconnect propagation delay would have reached the same order of magnitude as the 90-nm node with a copper back end. The diagram also shows that the propagation delay will further increase. This requires different design architectures, in which the high-speed signals are kept local. Such architectures must allow latency in the global communication or communicate these global signals asynchronously (i.e. *islands of synchronicity, globally asynchronous, locally synchronous*(*GALS*).

In the preceding discussions, self-inductances and mutual inductances were not taken into account. However, with the advances in speed and clock frequencies, the influence of these inductances becomes increasingly pronounced. The resistances of the metal lines in most of today's ICs still exceed the values of *inductance* by more than one order of magnitude. For one reason, this is due to the fact that the *resistance*

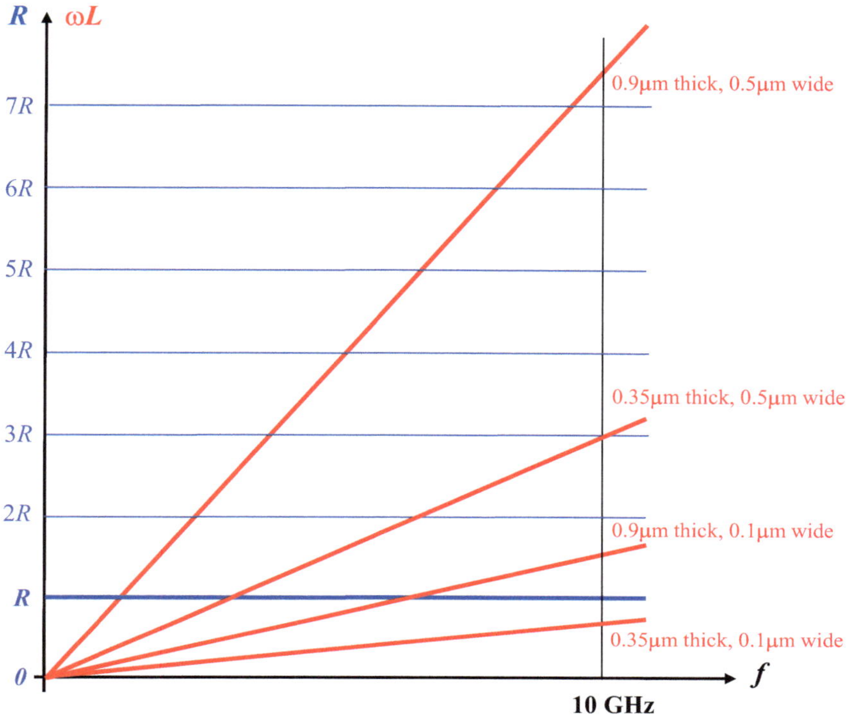

Fig. 9.16 Influence of the frequency on the inductance of 500 μm-long on-chip metal lines [8]

increases every technology node. The second reason is that the inductance is linearly proportional to the frequency (Fig. 9.16; [8]).

At a frequency of 10 GHz, the inductance contribution (of a 350-nm-thick signal line in a 65-nm CMOS process) to the total impedance of a metal wire reaches about two third of the resistance contribution. This means that we need to change from an *RC* interconnect model to an *RLC* model for designs that exceed 1 GHz (at this frequency, the inductance value is about 10% of the resistance value and can thus no longer be neglected). Generally, there are two effects determining the difference in accuracy between an *RC* and an *RLC* model: the damping factor and the ratio between the input signal rise time and the signal propagation speed across the line. Therefore, even in designs that do not yet reach 1 GHz, the wider metal lines, with lower resistance (e.g. in clock distribution networks and upper metal layers), can exhibit significant inductive effects. The lines in Fig. 9.16 represent the relative ωL values with respect to their resistance values. So, the ωL value of a 0.9-μm-thick and 0.5-μm-wide metal track reaches the level of its resistance value already at a frequency of close to 1.5 GHz.

Because the rise times of signals on interconnect lines are also reducing with the advance of the technologies, *RLC* models need to be included in CAD tools soon, in order to avoid inaccurate performance predictions or underestimate signal integrity effects, which may also reduce the operating margins.

Finally, a number of methods, depending on the requirements of the application, exist to reduce cross talk and/or improve signal propagation. We will summarise them here, without discussing them in more details:

- Use fat wires to reduce track resistance.
- Increase spacing to reduce mutual capacitance.
- Use shielding between the individual bus lines.
- Use staggered repeaters to compensate noise.
- Use tools which can detect, replace and reroute critical nodes.
- Use current sensing or differential signalling for improved speed and noise compensation.

Many of these solutions are described in various publications, which can easily be found on the internet.

9.3.2 Power Integrity, Supply and Ground Bounce

Every new technology node allows us to almost double the number of transistors. Next to this, the bus widths have also gradually grown over the last couple of decades: from 4 bits in the mid-1970s to 64 bits, or even 128 bits, today. The interface to a 1 Gb DDR-1 SDRAM, for instance, requires to communicate 64 data bits, about 30 address and control bits, totally adding up to some 96 parallel bits. In addition, due to the increased speed requirements, more flip-flops/pipelines are used within the logic blocks. All these individual trends contribute to a dramatic increase of simultaneously *switching activity* in an IC causing huge currents (i) and current peaks ($\mathrm{d}i$). These currents cause dynamic voltage drop across the resistance (R) of on-chip supply network, while the current peaks cause relatively large voltage dips and peaks across the self-inductances (L) in the supply path. As is discussed in the previous subsection, most of the *self-inductance* is still in the bond wires and the package leads, instead of in the on-chip metal supply lines.

Another trend that keeps pace with technology advances is the reduction in switching times ($\mathrm{d}t$) of the logic gates and driver circuits. The combination of these two trends leads to a dramatic increase of $\mathrm{d}i/\mathrm{d}t$, which term is mainly responsible for the *supply* and *ground bounce* generated on chip.

Referring to Fig. 9.17, we define the *dynamic voltage drop* (ΔV_{dyn}) as:

$$\Delta V_{\mathrm{dyn}} = i \cdot R + L \cdot \frac{di}{dt}. \tag{9.3}$$

The impact of this voltage drop on the behaviour of the chip is twofold: First, the average supply voltage $V_{\mathrm{dd}_{\mathrm{average}}}$ throughout the complete clock period determines the speed of a circuit. Let $V_{\mathrm{dd}_{\mathrm{nom}}}$ be the nominal supply voltage of a chip. Most

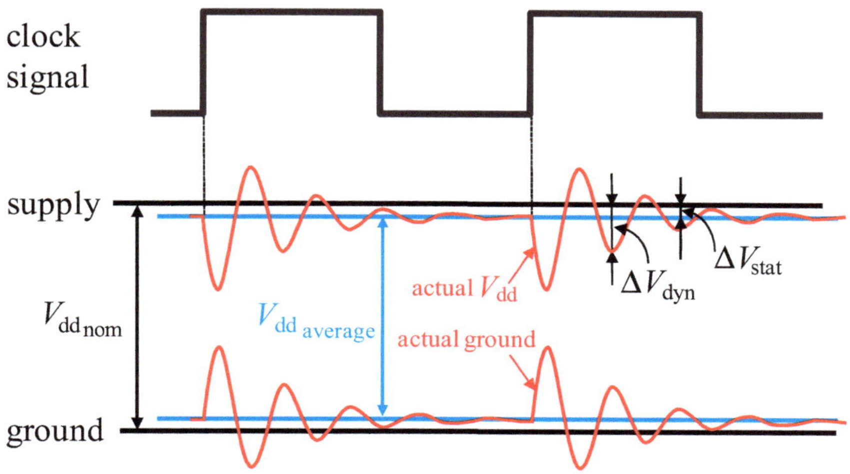

Fig. 9.17 Supply noise and ground bounce in a synchronous chip

commonly this means that the chip is specified to operate within 10% margin in this supply voltage. Assume a 1 V 32-nm CMOS design; this means that it should operate between 0.9 and 1.1 V. So, in the application, the IC should operate correctly, even at 0.9 V. Because the logic synthesis is done using the gate delays specified at this lower voltage, an additional IR drop within the chip could be disastrous for proper functionality. In other words, the designer should keep the total average voltage drop within stringent limits to assure the circuit operates according to the required frequency spec.

It is commonly accepted that this *static voltage drop* $\Delta V_{stat} = I \cdot R$, which is also called *static* IR *drop*, with I being the average current through the supply lines, is limited to just a small percentage of the supply voltage (around a few per cent). Second, ΔV_{dyn} refers to the noise in the supply line (*supply noise*). The current is supplied through the V_{dd} supply lines and leaves the circuit through the V_{ss} ground lines. When the impedances of the supply and ground lines are identical, which is most commonly the case, the introduced bounce on the respective lines show complementary behaviour and are identical in level. This means that the ground line faces the same dynamic and static voltage drop. So, the average supply voltage is then:

$$V_{dd_{average}} = V_{dd_{nom}} - 2 \cdot \Delta V_{stat}. \tag{9.4}$$

The total inductance (L) consists of on-chip contributions of the supply and ground networks and off-chip contributions of the bond wires, package leads and board wires. Usually, the damping effect of high resistive narrow signal wires reduces the effect of on-chip inductive coupling. To reduce the dynamic iR drop and static voltage drop in the above expression, however, the supply and ground

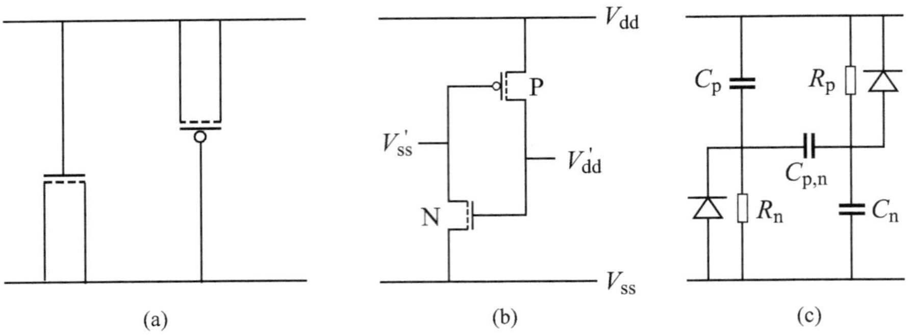

Fig. 9.18 Normal decoupling capacitor (**a**), tie-off cell decoupling capacitor (**b**) and equivalent circuit (**c**)

networks require wide metal tracks in the upper metal layers with very low sheet resistance. Particularly for designs operating at GHz frequencies, inductance in IC interconnects is therefore becoming increasingly significant.

The supply noise can be reduced in several ways. When using n supply pads for the supply connection, which are more or less homogeneously distributed across the IC periphery, the self-inductance will reduce to L/n. The use of both a low-resistive supply network and multiple supply pads, however, contributes to a reduction of the overall impedance of the supply network. Because the bond wires, package leads and board wiring all act as antennae, the resulting increase of the current peaks ($\mathrm{d}i/\mathrm{d}t$) leads to a dramatic rise of interference with neighbouring ICs on the board and may cause EMC problems in the system. Therefore, it is also required to keep the peak currents local within the different cores on the IC. In other words, it is necessary to lower the global $\mathrm{d}i/\mathrm{d}t$ contribution in the preceding equation as well. The use of staggered driver turn-on, to limit the amount of simultaneous switching activity as well as to encourage the use of "slow" clock transients, will directly contribute to a lower $\mathrm{d}i/\mathrm{d}t$. Another measure to limit the global $\mathrm{d}i/\mathrm{d}t$ is the use of decoupling capacitors within each of the different cores.

Figure 9.18 depicts two implementations of *decoupling capacitor* cells. Figure 9.18a is a complementary set of transistors connected as an nMOS and pMOS capacitor, directly between V_{dd} and V_{ss}. Because the supply voltage in this cell is directly across the thin gate oxides, this cell needs some additional resistances either in the gate connection or in the source/drain connections, to limit the chance for ESD damage. Figure 9.18b is a *tie-off* cell used as decoupling capacitor.

In several applications, a tie-off cell supplies dummy $V_{dd}^{'}$ and $V_{ss}^{'}$ potentials to inputs of circuits, which, for reasons of electrostatic discharge (ESD), are not allowed to be directly connected to the V_{dd} and V_{ss} rails. The channel resistances R_n and R_p (Fig. 9.18c) of the nMOSt and pMOSt, respectively, serve as additional ESD protection for the transistor gates connected to the $V_{ss}^{'}$ and $V_{dd}^{'}$. This advantage can also be exploited when we use this cell only as a capacitor cell between V_{dd} and V_{ss}, without using the dummy $V_{dd}^{'}$ and $V_{ss}^{'}$ terminals. When a supply dip occurs, the charge stored on the gate capacitance C_n (C_p) of the nMOSt (pMOSt) must be supplied to the V_{dd} (V_{ss}) in a relatively short time, which puts some constraints to the value of R_n (R_p). Therefore, *decoupling capacitor* cell *b* shows a better ESD behaviour compared to cell *a*.

These decoupling capacitors are charged during steady state, e.g. at the end of the clock period when the total switching activity has almost or completely come to an end. The additional charge, stored in these capacitors, is then redistributed to the supply network during moments of intense switching, particularly at the clock transient that initiates the next signal propagation through the logic paths. These decoupling capacitor cells are designed as standard cells and are usually available in different sizes. The amount of decoupling capacitance that needs to be added in each core depends on the number of flip-flops in it and on the switching activity of its logic. The switching activity α is defined as the average number of gates that switch during a clock cycle. When a logic core has an activity factor of $\alpha = 1/3$, it means that the average gate switches one out of every three clock periods. Different algorithms require different logic implementations, which show different switching activities. It is known that average telecom and audio algorithms show less switching activity ($0.05 < \alpha < 0.15$) than an average video algorithm ($0.1 < \alpha < 0.4$), for example. These activity factor values are only meant as illustration, rather than an exact range for an application. As an example, the total additional decoupling capacitance in a logic block, performing a video algorithm, running above a gigahertz in a 28-nm CMOS core in a digital chip, may occupy about 10–20% of its total area. When the standard-cell block utilisation is less than 85%, most of this decoupling capacitance fits within the empty locations inside a standard-cell core. In certain mixed analog/digital ICs, however, this amount could grow dramatically, since the noise in these ICs is much more restricted by the sensitivity of the analog circuits.

Because of further scaling, i, R and di (in Eq. 9.3) will increase, while dt will just do the opposite, potentially requiring an increasing number of design measures to limit the dynamic as well as the static voltage drop across the power network. Therefore, the power(-grid) integrity must be sufficiently guaranteed in order to enable correct chip behaviour. This power integrity must also be supported by the tools that can analyse (dynamic) voltage drop early in the design flow and will reduce the cost of chip debug and failure analysis and prevent the need for a respin. Static voltage-drop analysis targets at the average current (I) and on the resistive nature of the power supply network. Nanometer CMOS designs, however, also require a dynamic voltage drop analysis, which focuses on the impact of instantaneous currents (i) and current peaks (di), early in the design cycle to avoid dangerous compromises between power grid design, on the one hand, and power integrity, noise and timing requirements, on the other.

Backside Power Distribution

With increasing power consumption and current, it is expected that for process nodes below 3 nm the *power distribution network (PDN)* may be implemented at the backside of the chip [9, 10]. The active circuitry on the frontside can then be packed more dense, which enables higher transistor counts and increased functionality within the same chip footprint. This is particularly beneficial in modern integrated circuits, where the demand for higher integration and more complex functionality is continuously increasing. Moreover, the separation of the PDN from the frontside also mitigates the interference between the power delivery network and the active circuitry. *Backside power distribution* (*BPD*) decouples power noise from the frontside, resulting in improved signal integrity and reduced power supply noise. It is also claimed [11] that the heat dissipation capability of the chip may be enhanced. Heat generated by the active circuitry can be efficiently conducted away from the chip through the backside metal layers and attached heat sink, improving overall thermal management and reliability. Despite the various benefits, the backside power grid also introduces a lot of challenges and considerations. For example, the additional manufacturing steps and complexities in integrating the backside power grid may lead to increased design and manufacturing costs. Figure 9.19 shows an example of a possible BPD architecture [12].

For realizing a backside power delivery network, a dedicated wafer thinning process is necessary with the ability to process *nano-through silicon vias (nTSVs)* that connect the backside and the frontside of the device wafer [11, 13]. The top drawing shows the wafer with a standard back end-of-line (BEOL) process. The lower drawing shows the revised BEOL "flow". After creation of the (FinFET) transistors and the first local interconnect layers, identical to a normal CMOS (FinFET) process, the wafer is thinned. Next, it is flipped and then bonded to a silicon carrier through a dielectric bonding layer. The drawing shows the flipped state. Subsequently, the backside power distribution network is produced, after the nano-through silicon vias are created. These nTSVs connect the backside BEOL (or BPD) to the transistors and to the original standard BEOL, which now only consists of the signal routing layers. As stated, this is just one example of a possible BPD architecture. There are currently (2024) several other BPD approaches. Intel, Samsung and TSMC plan to implement BPD in some form at around the 2-nm node. In this node, for example, Intel plans to combine their nano-ribbon transistors with PowerVias [13].

In summary, BPD introduces a new process-integration scheme with many reliability, robustness and yield challenges. As the nTSVs require holes to be etched in the thinned substrate and then filled with a conducting material (Cu), they will cause stress and strain in the substrate which may influence transistor performance and reliability. The difference in the thermal coefficients of Cu and Si may also cause stress to its surrounding during the annealing process. Therefore, they may require some distance to transistors. Another important issue with BPD is how the I/O bonding pads on the backside are connected to the input/output signal terminals at the frontside. Still a lot of R&D is required to overcome these challenges and determine the most efficient BPD solution for future chip designs.

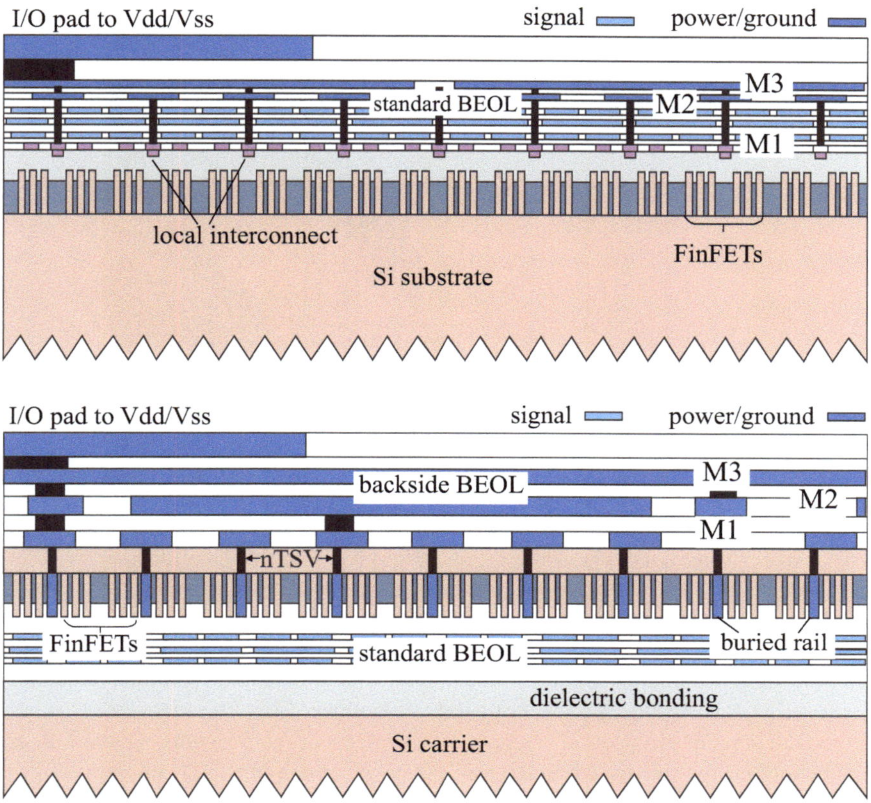

Fig. 9.19 Die placement and BEOL formations for a conventional frontside PD configuration (top) and backside PD configuration (bottom). (Source: [12])

9.3.3 Substrate Bounce

Substrate bounce is closely related to the ground bounce. On a mixed analog/digital IC, usually, the digital circuits are responsible for most of this bounce, while the analog and RF circuits are most sensitive to it (Fig. 9.20). The substrate bounce has several contributors. The transistor substrate current injection is responsible for only a few mV. Junction and interconnect capacitances account for several tens of mV. The highest noise levels (several hundred mV), however, are introduced through the current peaks in the supply network, also causing the previously discussed supply noise.

In most CMOS circuits, it is a common practice to connect the substrate to the V_{ss} rail, meaning that the ground bounce that is generated in the V_{ss} rail is directly coupled into the substrate. This is even a bigger problem, when the chip is realised on epitaxial wafers (see Sects. 3.2.2 and 9.5.5) with a low-ohmic substrate,

Fig. 9.20 Symbolic representation of a mixed analog/digital IC

because it propagates the noise through the substrate to the analog part almost instantaneously and with hardly any loss of amplitude. Because the noise margins reduce with reducing supply voltages, the use of high-ohmic substrates is becoming increasingly important. Triple-well technology allows improved isolation of analog and RF circuits from digital cores. The level of isolation also depends on the frequency of the RF circuits. The use of a silicon-on-insulator (SOI) technology allows even a complete separation of the analog and digital circuits. Several other measures exist to reduce the level of substrate bounce: First, the measures that help reduce the supply and ground bounce, as discussed in the previous subsection, are also beneficial for substrate bounce reduction. Second, a physical separation of the core and I/O supply nets from the analog supply net, according to Fig. 9.21, prevents the relatively large noise introduced in these nets to propagate directly into the analog net [14].

The figure also illustrates that most digital and analog circuits share the same ground (V_{ss}), because it also serves as a reference for the communicated signals. Usually, the impedance of the internal and external V_{dd} and V_{ss} networks is almost symmetrical, meaning that they have equal widths and the same number of bonding pads. An increase in the impedance of the V_{dd} network with respect to the impedance of the V_{ss} network would increase the bounce in the V_{dd} supply network while reducing it in the V_{ss} ground network. Because the analog and digital V_{dd}s were separated anyway, this additional digital supply bounce is not coupled into the analog V_{dd}. Due to the fact that the analog and digital circuits share the same ground, the lower V_{ss} ground bounce also reduces the substrate bounce. Therefore, to increase the margins and robustness of mixed analog/digital ICs, it may be advantageous to dedicate more supply pads to V_{ss} and less to the V_{dd}. Advanced CMOS technologies offer an additional deep-n-well (*triple-well technology*), which allows to isolate the p-wells in digital cores from those in analog cores. These cores may then also have separated V_{ss} pads.

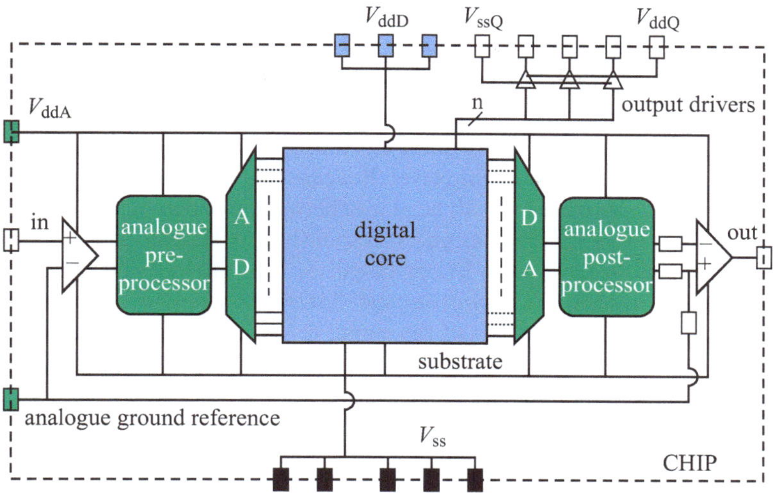

Fig. 9.21 Proposed supply connections in a mixed analog/digital IC

Finally, particularly in the case of high-ohmic substrates, circuits with the highest switching activities and driving strengths, e.g. I/O pads, clock drivers and drivers with a high fan-out, must be located as far away from the analog circuits as possible.

9.3.4 EMC

The problem of supply and ground bounce caused by large current changes is not restricted to on-chip circuits only. High current peaks may also introduce large electromagnetic disturbances on a printed circuit board (PCB) because of the electromotive force and threaten the off-chip signal integrity. Because bonding pads, package and board wiring act as antennae, they can "send" or "receive" an *electromagnetic pulse (EMP)*, which can dramatically affect the operation of neighbouring electronic circuits and systems [15].

When realising *electromagnetic compatible (EMC)* circuits and systems, the potential occurrence of EMPs must be prevented. The use of only one or a few pins for supply and ground connections of complex high-performance ICs is one source of EMC problems. Even the location of these pins is very important with respect to the total value of the self-inductance. The use of three neighbouring pins for V_{dd}, for instance, results in an electromagnetic noise pulse that is twice as large as when these supply pins were equally divided over the package. The best solution is to distribute the power and ground pins equally over the package in a sequence such as V_{dd}, V_{ss}, V_{dd} and V_{ss}. Bidirectional currents compensate each other's electromagnetic fields in the same way as twisted pairs do in cables. Another source of EMC problems is formed by the outputs. They can be many (about 96 I/O pins for the address, data and control signals in a 1 Gb DDR SDRAM interface), contain relatively large drivers with high current capabilities and often operate at higher voltages than the cores. Actually, each output requires a low-inductance current return path, such that the best position for an output is right between a pair of V_{dd} and V_{ss} pads. This results in the smallest electromagnetic disturbances at PCB level and reduces the supply noise at chip level. Because this is not very realistic in many designs, however, more outputs will be placed between one pair of supply pads. The limitation of this number is the designers' responsibility (simulation!) or defined by the characteristics of the library I/O or output cell. In this respect, the maximum number of *simultaneously switching outputs* (*SSOs*) per supply and ground pad is a combination of the characteristics of the output driver and the package. In addition, the $\mathrm{d}i/\mathrm{d}t$, generated by these outputs, must also be limited to what is really needed to fulfil the timing requirements. Finally, all measures that reduce on-chip supply and ground bounce also improve the electromagnetic compatibility of the chip and result in a more robust and reliable operation.

9.3.5 Soft Errors

Because of the continuous shrinking of devices on an IC, the involved charges on the circuit nodes have scaled dramatically. Ionising particles, independent of their origin, do have an increasing impact on the behaviour of these shrinking devices. At sea level, several categories of particles can be distinguished, which all generate free electron-hole pairs in the semiconductor bulk material [16]:

- *Alpha particles*, originating from radioactive impurities (mainly uranium and thorium) in materials. These materials can be anything in the vicinity of the chip: solder, package or even some of the materials used in the production process of an IC (metals or dielectrics). These so-called α-particles can create a lot of electron-hole pairs along their track.
- *High-energy cosmic particles*, particularly neutrons, can even fracture a silicon nucleus. The resulting fragments cause the liberation of large numbers of electron-hole pairs.
- *Low-energy cosmic neutrons*, interacting with boron-10 (^{10}B) nuclei. When a ^{10}B nucleus breaks apart, an α-particle and a lithium nucleus are emitted, which are both capable of generating soft errors. This is only an issue if BPSG (Chap. 3) is used in the processing, because it contains relatively large amounts of boron atoms.

In all cases, the generated electrons and holes can be collected by reversed-biased pn-junctions in the circuit. This charge collection causes a current pulse that can discharge capacitors (in dynamic logic and DRAMs) and can flip states in both dynamic and static storage circuits (memories, latches and flip-flops).

Figure 9.22 shows a cross section of an nMOS transistor which is part of an SRAM cell. Assume that the drain of this transistor is connected to the logic one

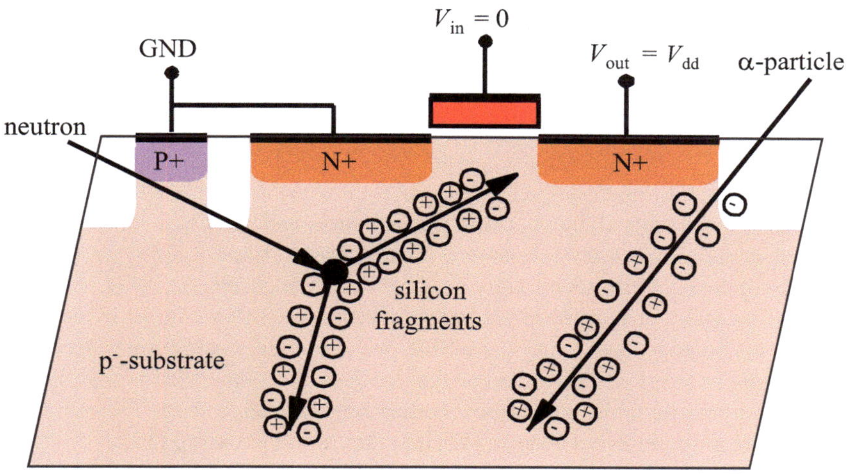

Fig. 9.22 Cross section of an nMOS transistor hit by an α-particle or neutron

(V_{dd}) side of the memory cell. An α-particle or neutron creates electrons and holes in the silicon. Electrons are attracted to the positive node (V_{dd}). If the number of collected electrons is large, it may discharge the node so much that the SRAM cell can flip its state.

The resulting incorrect state is called a *soft error*, also known as a *single-event upset (SEU)* because the flipped state has not caused permanent damage to any of the circuit devices. The rate at which a soft error occurs is called *soft-error rate (SER)* and is expressed in *failures in time (FIT)*. This rate reflects the number of expected failures in 1 billion operating hours. α-particles usually cause single-bit errors, while neutrons may cause both single- and multi-bit errors. The probability of a multi-bit upset is increasing with technology scaling. The total charge of a node is an important criterion for the probability of flipping its state after being hit by an ionising particle. The node charge Q_{node} representing one bit or one flip-flop equals:

$$Q_{\mathrm{node}} = C_{\mathrm{node}} \cdot V_{\mathrm{dd}} \tag{9.5}$$

where V_{dd} equals the supply voltage and C_{node} the total capacitance of the node.

In static storage cells (SRAM cell, latch or flip-flop), the critical charge is dependent not only on the capacitance of the nodes in these cells but also on the drive strengths of the transistors that try to maintain the logic state. In this case, the critical charge varies with the width of the transient current pulse induced by a particle hit.

As a first approximation, the critical charge (Q_{crit}) needed to generate a soft error at a specific circuit node is given by:

$$Q_{\mathrm{crit}} = Q_{\mathrm{node}} + I_{\mathrm{drive}} \cdot W_{\mathrm{pulse}} \tag{9.6}$$

where I_{drive} represents the transistor current needed for keeping the state and W_{pulse} is the width of the particle-induced current pulse.

Also, as a first approximation, the SER of a single bit (or cell) can be represented by:

$$\text{SER} \propto A_{\text{diff}} \cdot e^{(-Q_{\text{crit}}/<Q_{\text{coll}}>)} \tag{9.7}$$

where A_{diff} represents the sensitive diffusion area and $< Q_{\text{coll}} >$ the average collected charge. All three parameters in the above expression (A_{diff}, Q_{crit} and $< Q_{\text{coll}} >$) reduce with technology scaling. As a net result, the SER per Mb of SRAM is roughly constant. Because memory capacity (Mb per chip) is increasing with new technology generations, the SER per chip increases as well. Because of the shrinking memory cell sizes and distances, there is an increased probability that a high-energy particle hit may cause multiple bits to change state. The number of the so-called multiple-event upsets (MEU) is therefore increasing [17].

As already mentioned before, latches and flip-flops are also sensitive to soft errors. It turns out that their SER per cell is also fairly constant. This also leads to an increasing SER per chip, because the number of cells (flip-flops and/or latches) per chip increases with new technology generations. The average soft-error rate for both SRAM cells and flip-flops in the 0.18-μm CMOS technology node is about equal and close to 1000 FIT/Mb. Table 9.2 shows the relative trend in soft-error rates for SRAMs and flip-flops (latches). Generally, an IC contains much more SRAM cells than flip-flops. Therefore, when the SRAMs are not protected against soft errors, they will dominate the SER per chip. However, when the SRAMs are protected with ECC, as discussed below, flip-flops (latches) may dominate it. Reference [17] more or less confirms the above-described trend in SER sensitivity, although it is more optimistic regarding the trend in SER/bit for SRAMs, in that this rate even reduces every new technology node. It therefore claims that the SER at chip level is not expected to dramatically increase; but it also states that the SER remains an important point of focus for the quality of future semiconductor products.

Also logic gates become increasingly prone to soft errors, their contribution to the SER per chip is usually (much) less than 10–15% of the total SER.

DRAM vendors keep the storage capacitance per memory cell at a level between 25 and 50 fF. As a result, the DRAM SER per bit has decreased dramatically because of the reduction of the sensitive diffusion are, which reduces the $< Q_{\text{coll}} >$. However, at system level, the SER is almost constant because of the increasing memory capacity needs per system.

Finally, a high-energy particle hit may also introduce a *single-event latch-up (SEL)*. Once such a hit creates sufficient charge, it may trigger latch-up. This may be limited to a local region, but it may also propagate to larger parts of the chip, where the total induced current may become so high that it can even destroy the device. To recover the device operation after a non-destructive SEL condition, the power supply needs to be temporarily removed.

There are several measures to prevent or limit the occurrence of soft errors:

Table 9.2 Relative trend in soft-error rate for different circuits (source: Marc Derby (iRoC Technologies), IOLTS 2007) [18]

	Technology node					
	180 nm	130 nm	90 nm	65 nm	45 nm	comment
Integration level (Mtransistor/device)	48	97	193	386	773	Source: ITRS 2005
Non-protected memory						
Memory integration (Mbit/device)	6.4	12.9	25.7	51.5	103.1	Hyp: 80% of T → 6 – T mem cells
Memory SER per Mbit (norm. units)	1	0.78	0.91	1.03	1.10	Source: iRoC Technologies
Memory SER at chip level (norm. units)	1	1.58	3.66	8.28	17.71	**180 to 65 nm:** ′8.3
Non-protected sequential logic						
Flip-flop integration level (Mbit/device)	0.096	0.194	0.386	0.772	1.546	Hyp: 6% of T → 30 – T mem cells
Flip-flop SER per Mbit (norm. units)	1	0.93	0.73	0.69	0.63	Source: iRoC Technologies
Flip-flop SER at chip level (norm. units)	1	1.88	2.94	5.55	10.15	**180 to 65 nm:** ′5.6

SER numbers apply to high-energy-neutron and alpha-particle effects for an arbitrary device/technology (i.e. numbers do vary depending on source)
SER values given here include both single-bit upset (SBU) and multiple-cell upset (MCU) events; the SER values denote the bit-flip rate
45-nm SER values are estimates

- Careful selection of purified materials (package, solder, chip manufacture, etc.) with low α-emission rates.
- Usage of a shielding layer, most commonly polyimide. This layer must be sufficiently thick (20 μm) in order to achieve about three orders of magnitude reduction of the soft-error rate (SER) caused by α-particles. This measure does not help to reduce the soft-error rate caused by the high-energy cosmic particles because they can pass through even half a meter of concrete.
- SER hardening of the circuits by changing memory cells, latches and flip-flops. This may be achieved by increasing transistor sizes and/or adding an additional capacitance between the two (latching) nodes.
- Usage of process options or alternative technologies. The use of an additional *buried well* (*triple-well technology*) prevents the electron-hole pairs generated in the substrate below this buried well to reach the individual source and drain regions, which reduces the SER. Silicon on insulator (SOI) circuits exhibit even an order of magnitude reduction of the SER because charges that are generated along a particle track in the main part of the silicon chip are physically isolated from the circuits in the top layer.
- Inclusion of error detection/correction (ECC) circuits or making the designs fault-tolerant. ECC is a very effective method to protect SRAMs and DRAMs against soft errors. In combination with design techniques such as physical

interleaving of bit cells (scrambling), most multi-bit upsets may be considered as a multiple of individual single-bit upsets in multiple correction words and the ECC needs only to deal with single-bit errors.

Still, a lot of effort is being put into the evaluation, estimation and prevention of soft errors, particularly in aerospace systems [19] and systems containing large amounts of densely packed memories [20].

9.3.6 Signal Integrity Summary and Trends

From the previous subsections, it can be seen that all noise components increase because of scaling and integrating more devices onto the same die area. At the same time that noise levels in digital CMOS ICs increase with scaling, the noise margins reduce due to reducing supply voltages (Fig. 9.23). Because they deal with large current peaks, high-performance ICs such as the PowerPC (IBM), the alpha series (DEC), the Pentium series (Intel) and the Athlon (AMD) faced signal integrity effects already in the 1990s. Most application-specific integrated circuits (ASICs), however, consume one or more orders of magnitude less power (and current) and therefore faced these problems to a lesser extent or a couple of technology generations later in time.

When a certain noise level has reached a maximum limit, a design or technology measure is required to reduce the noise level.

Examples of technology measures are as follows:

- The use of copper instead of aluminium allowed a reduction of the metal height, thereby reducing the cross talk (see Sect. 9.3.1).
- The use of low-ϵ dielectrics in the back end of the technology has the same effect.

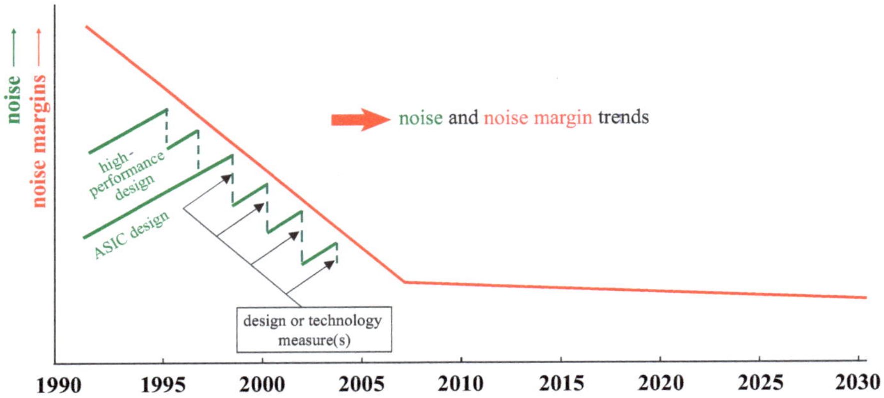

Fig. 9.23 Noise and noise margin trends over the last three decades

Examples of design measures are as follows:

- The increase of space between long signal lines (buses) also reduces the cross talk.
- The use of on-chip decoupling capacitors reduces supply, ground and substrate bounce.

Whatever technology or design measure is taken, it only fulfils the requirements in that technology node. The next technology node offers twice the number of transistors, which can intrinsically switch faster. This results in a huge increase in the noise levels. In addition, the noise margin has reduced. Therefore, in every new technology node, it becomes more difficult to limit the noise within shrinking boundaries. In other words, the line (in Fig. 9.23) that represents the increasing noise must be bended in the direction of the line that represents the reducing noise margins. This can only be obtained by applying more and more design and/or technology measures. For example, in today's high-performance ASIC designs, the decoupling capacitors occupy between 5% and 20% of the total area within a standard-cell block. It is expected that this number will further increase for such designs in the sub-10 nm nodes, which means that, for certain designs, a large part of all transistor equivalents on a chip is needed to support the other part in their functional and storage operations. This is yet another factor that adds up to the already rocketing semiconductor development costs.

Another increasingly important topic is the relation between signal integrity and test. Because noise has the tendency to increase, while noise margins reduce (again Fig. 9.23), there is not much room left for a reliable operation of an IC. Different operating vectors introduce different local and global switching activities. In many complex ICs, the operation and switching activity during testing are different from the operation and switching activity in the application. As a result, the noise, generated during a test, is different from the noise generated in the application. Because of the reducing noise margins, this increasing discrepancy between "test noise" and "application noise" causes products that were found correct during testing to operate incorrectly in the application. This is because, in many cases, scan tests are performed to verify the IC's functional operation. These tests are mostly performed locally and in many cases at different frequencies causing a lower overall switching activity and less noise than in the application. On the other hand, depending on the design, different scan chain tests may run in parallel, synchronous and at the same frequency, causing much more simultaneous switching and noise than in the application. These ICs may be found to operate incorrectly during testing while showing correct functional behaviour in the application. Because of this and because of the lack of access to most internal signals, debugging a system on chip (SoC) has become a very difficult and time-consuming task. In fact, first (and second) silicon debug has become a major component of time to market, as it may take up to 50% of designers' time.

A *signal integrity self-test (SIST)* architecture [21] allows real-time monitoring of different parameters (e.g. temperature, voltage drop, switching activity, supply noise, substrate noise, cross talk, process parameter spread, clock jitter and clock duty cycle) that characterise the quality of operation of an IC, during test and debug or in the application. Moreover, even when first silicon is functionally correct, this SIST architecture allows the monitoring of signal integrity margins, in order to anticipate potential operating failures due to technology parameter variation in future silicon batches.

Since the margins continue to decrease, the additional design for debug measures will not be enough, and provisions need therefore to be made in the designs to enhance operation robustness during the test as well as in the application. This poses additional challenges to the design, increases its complexity and also adds up to the total development costs.

9.4 Variability

This section presents a short overview on the increasing number and influence of different components of variability: process parameter spread, electrical variations, environmental variations, etc. The description starts by categorising these variability components by their nature in different ways. Next, their influence on the behaviour of analog and digital circuits is discussed.

9.4.1 Spatial vs. Time-Based Variations

Spatial variations are variations due to the fact that identical devices can and will have a different physical environment, caused by a different orientation, a different metal coverage or other *proximity effects*, such as mechanical stress (e.g. *STI stress*), the position of a well in the vicinity of a transistor (*well-proximity effect (WPE)*) [22] and/or pattern shape deviations as a result of imperfect lithographic imaging and pattern density variations.

Time-based variations include not only signal integrity effects, such as cross talk, supply noise, ground bounce and iR drop, but also temperature variations over time, due to variations in workload.

9.4.2 Global vs. Local Variations

Device parameters can vary from lot to lot, from wafer to wafer, from die to die and from device to device. The first three are usually referred to as global or *inter-die variations*, which are more or less systematic and common to all devices on the same die. A mask misalignment, a smaller processed channel length or a V_T-shift are all examples that cause global variation. Advances in both the lithography and diffusion process have led to a much tighter overall process control, such that in mature processes *global variations* are currently much better controlled than in the past. Most global variations are a result of systematic process imperfections. Nanometer CMOS technology nodes (e.g. 90-nm CMOS and beyond) show an increasing relation between design style and systematic yield. Symmetrical and regular layout styles, such as used in *litho-friendly design* [23], can be applied to make the design less sensitive to these variations.

Variations between devices on the same wafer are usually referred to as local or *intra-die variations* or *mismatch*. Most *local variations* (*random variations*) are caused by stochastic processes during fabrication. Polysilicon gate *line-edge roughness (LER)* and the channel doping statistics are examples of local variations. The resulting device mismatch is particularly a problem in analog circuits, e.g. circuits with a differential transistor pair, clock generating circuits, current mirrors, operational amplifiers, etc. A common design approach is to simulate a circuit with respect to best-case and worst-case process corners (see also Table 2.2 in Chap. 2), e.g. slow-nMOSt slow-pMOSt corner (snsp) [24]. The circuit in Fig. 9.24 is used for Monte Carlo simulations. These result in the 'diamond' in Fig. 9.25, which connects these corners and represents the global variations in the voltages across

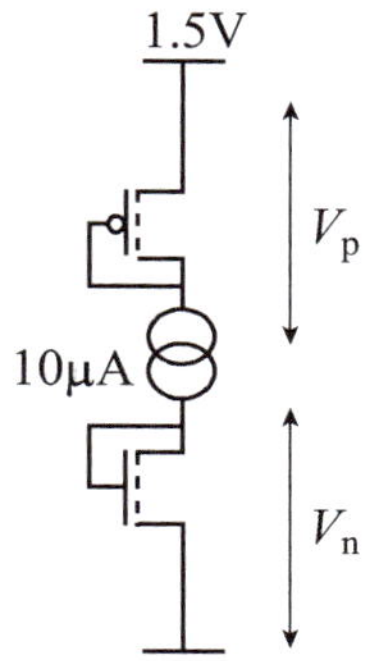

Fig. 9.24 Circuit used for the Monte Carlo simulation results of Fig. 9.25

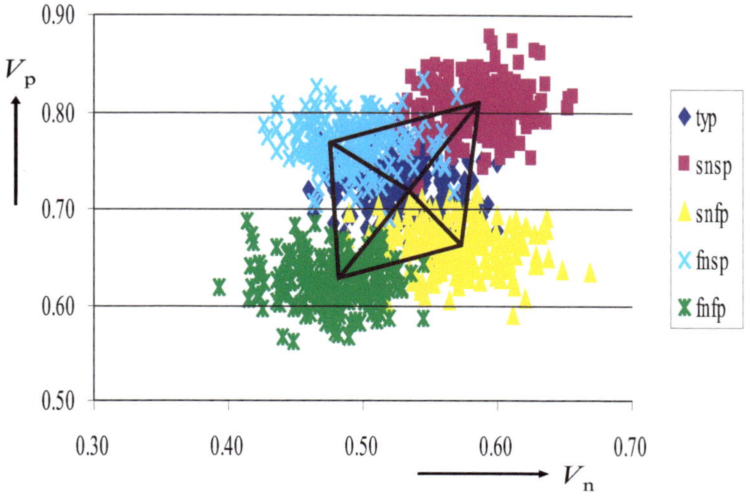

Fig. 9.25 Effect of global and local variations on the voltage across an individual device [25]

the devices. The clouds around every corner represent the random variations. The diagram clearly shows that, for individual devices, these random variations are in the same order of magnitude as the systematic variations.

Analog circuit performance is determined by the behaviour of the basic individual analog cells. Differential design eliminates the effect of the global variations in analog circuits, but they remain sensitive to local variations. The effects of global and local variations on the behaviour of a memory are comparable to that of analog circuits, because the behaviour of a memory is determined by the robustness of the individual cells.

In digital circuits, these random variations become increasingly important due to tighter timing restrictions and reducing noise margins.

Since most local variations are of random nature, their influence on the total delay of a logic path relatively averages out with the depth N of the logic path (Fig. 9.26). The figure shows that the spread in the logic path delay, due to global variations, increases linearly with N, while the delay spread due to local variations "only" increases with $\sqrt{N}$. (To be able to depict all three characteristics into one diagram, global and local variations have been scaled by the maximum length of the logic path (40) in the diagram.)

The effects of this relationship on the diagram of Fig. 9.25 are that the size of the diamond will increase linearly with N, while the size of the clouds will only increase with $\sqrt{N}$, showing a relative decrease of the local variability.

Figure 9.27 shows the influence of the local variability on the relative spread in logic path delay for different technology nodes. In this analysis, based on simulation results, the relative performance spread defined as sigma/mean (σ/μ) was used as figure of merit. It shows the increase in random variability with the scaling of the

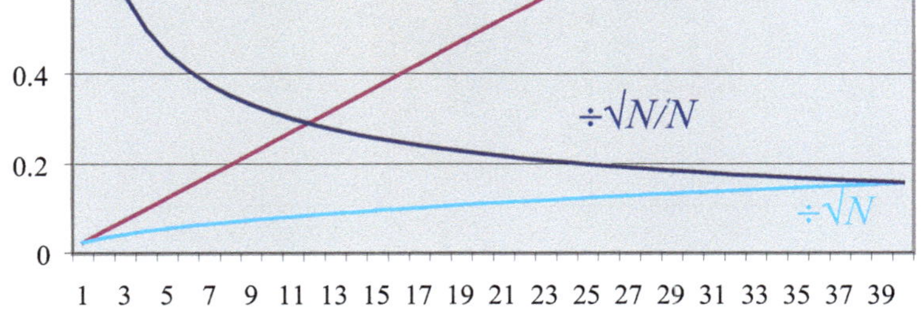

Fig. 9.26 Effect of global and local variations on the logic path delay vs. logic depth

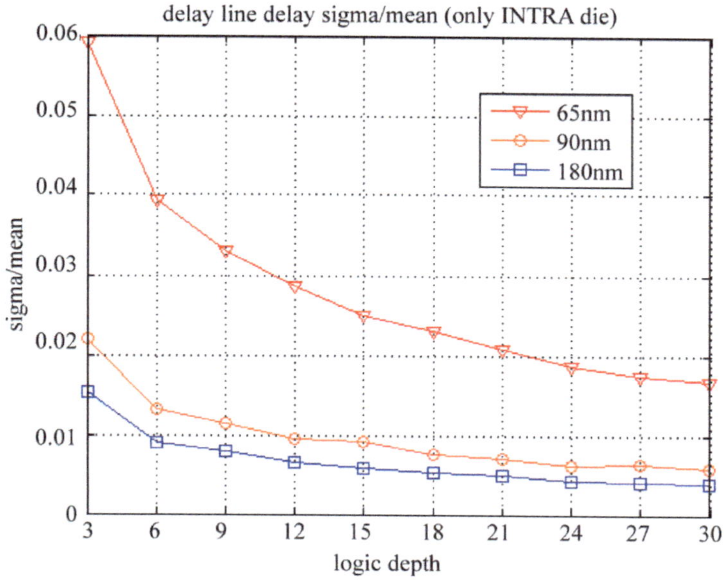

Fig. 9.27 Logic path delay (σ/μ) vs. logic depth

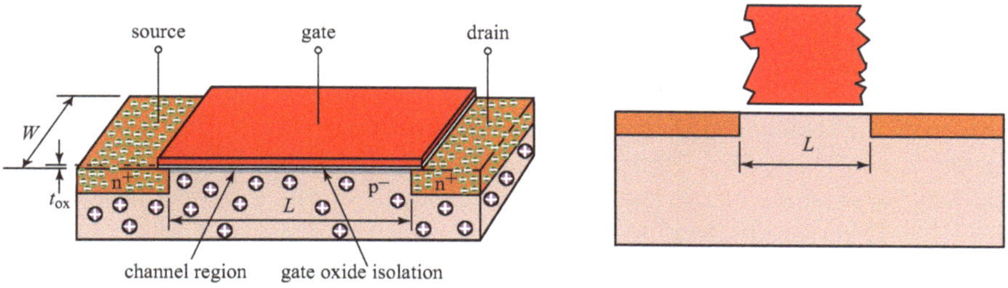

Fig. 9.28 Example of random doping fluctuations (left) and polysilicon gate line-edge roughness (right)

technologies. It also shows that high-speed circuits, which only have a limited logic depth, face a larger influence of the random variability.

In planar MOS transistors, much of the variability is caused by *random dopant fluctuations (RDF)* ΔN as a result of the continuous scaling of the channel length and channel width (Fig. 9.28). In an average transistor in 28-nm CMOS, for example, the average number of dopants in the channel has dropped between 50 and 100. The uncertainty in the number of dopants for such small transistor dimensions is in the order of 10%, which is dramatic when we integrate hundreds of millions to a few billion transistors on a single chip. The best solution would be to create transistor channels without dopants and make the threshold voltage only dependent on the work function of the gate stack. This is actually more and more the case with the introduction of (fully depleted) SOI, metal gate transistors and FinFETs. This means that, for these devices, the K-factor in expressions (1.16) and (1.17) in Chap. 1 is also low. However, with the advent of metal gates, another source of variability turns up. These so-called *random work function variations (RWF)*, which may lead to V_T variations between 20 and 30 mV, arise from the *metal grain granularity (MGG)* where relatively large metal grains (several nanometers) with different crystal orientations interact differently with the underlying gate dielectric and silicon, such that the charge carriers in the channel may face different gate potentials [26, 27]. The influence of these variations on the characteristics of FinFETs and circuits is also described in these references.

Another example of random variability is the so-called line-edge roughness (LER) of the polysilicon gate (Fig. 9.28). This is the local variation of the polysilicon gate edge along the width of the transistor. It is caused by the granularity of the photoresist, which creates a non-uniform polysilicon width resulting in channel length variations along the width of the transistor. This granularity is the result of stochastic effects during illumination, because EUV requires a lower number of photons for the same illumination energy. Less photons means less averaging and, so, more spread. Typical values for 3σ LER in a 32-nm process are around 5% of the nominal CD.

Although extreme ultraviolet (EUV) lithography enables 16-nm pitches, it still faces a number of challenges. Line-edge roughness (LER) is one of the critical issues that significantly affect critical dimension (CD) and device performance in the advanced process nodes because LER does not scale with the feature size [28].

The different expansion coefficients between silicon and STI cause permanent non-uniform compressive stress from the STI isolation into the active areas after processing, influencing both the device saturation current and the threshold voltage of the nearby transistors. Figure 9.29 (left) shows the influence of this stress on the behaviour of the charge carriers in the channel. In this example, the edge of the left STI region is closer to the transistor channel (distance a) than the edge of the right STI region, which is at a larger distance b from the channel. The left

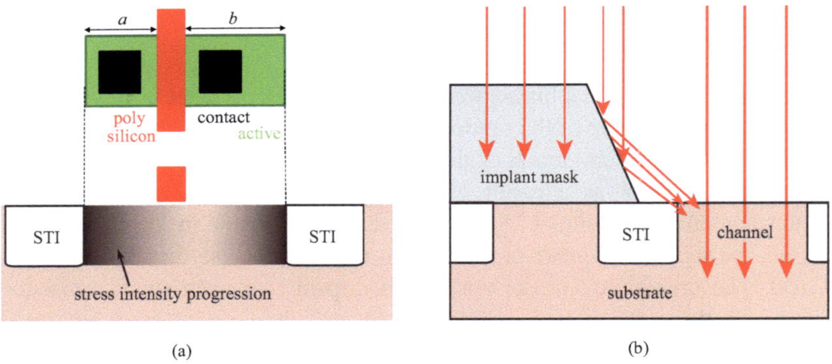

Fig. 9.29 STI stress (left) and well-proximity effect (right)

STI region has therefore more influence on the transistor current than the right STI region. In other words, transistors that are closer to an STI edge face more influence on their current behaviour than transistors that are further away, as in a series connection of transistors, for instance. During the retrograde-well implant (see Chap. 3, Sect. 3.9.3), doping atoms may scatter laterally from the photoresist or hard-mask into the silicon, close to the edge of a well as shown in Fig. 9.29 (right).

This causes a non-uniform well dope over a distance of one or more microns from the well-edge and a change in threshold voltage and (saturation) currents as well of the affected transistors. The influence of these effects is different for nMOS and pMOS transistors and requires optimised and symmetrical layout design methods in order to minimise their influence on analog circuit performance [29].

Timing analysis has been troubled by the growing circuit complexity in the number of functions on a chip, in combination with the diverging requirements for these functions when used in different operating modes (standby, low-power, high-speed, etc.) to support a variety of applications. The conventional *static timing analysis (STA)* is almost unable to handle within-die correlation, particularly when spatial correlation is included. *Statistical static timing analysis (SSTA)* can detect these correlations more easily and creates statistical distributions of the involved delays in order to mimic these variations. This is further discussed in Sect. 9.4.4. More details on various variability sources and characterisation can be found in [30, 31].

9.4.3 Transistor Matching

Matching of transistors means the extent to which two identical transistors, both in type, size and layout topology, show equal device parameters, such as β and V_T. Particularly in analog circuits (a memory is also an analog circuit) where transistor pairs are required to have a very high level of matching [32], the spread ($\sigma_{\Delta V_T}$) in the difference (ΔV_T) between the two threshold voltages of the pair results in inaccurate or even anomalous circuit behaviour. This spread is mainly caused by the variation in doping atoms ΔN, due to random doping fluctuations in the channel region of the MOS transistors. For bulk-CMOS devices, this is defined as [25]:

$$\sigma_{\Delta V_T} = \frac{A_{V_T}}{\sqrt{WL}} \text{ in mV} \tag{9.8}$$

in which the *matching coefficient* A_{V_T} is defined as:

$$A_{V_T} = \frac{q \cdot t_{ox}}{\epsilon_0 \cdot \epsilon_{r,ox}} \sqrt{2} N t_{depl} \tag{9.9}$$

where N equals the number of active doping atoms in the depletion layer. The expression shows the proportionality of A_{V_T} with t_{ox}. Technologies with a good matching behaviour have their A_{V_T}/t_{ox} ratio close to 1 V/mm^2. In [33], the following simple expression is derived:

$$\sigma_{\Delta V_{\mathrm{T}}} = constant \cdot t_{\mathrm{ox}} \frac{\sqrt[4]{N}}{\sqrt{WL}} \tag{9.10}$$

Until the 45-nm node, N increased every next (bulk-CMOS) technology node with a factor close to the reverse of the scaling factor s ($s \approx 0.7$), while beyond this node it is expected to be nearly constant with further scaling [33]. Therefore, its impact on the spread is only marginal, and we can simplify the relation to:

$$\sigma_{\Delta V_{\mathrm{T}}} \propto \frac{t_{\mathrm{ox}}}{\sqrt{WL}}. \tag{9.11}$$

It is clear from this relation that the ΔV_{T} spread is inversely proportional to the square root of the transistor area.

For minimum transistor sizes (area), the ΔV_{T} spread increases every new IC process generation, such that both the scaling of the physical size and the operating voltage of analog CMOS circuits lag one or two generations behind the digital CMOS circuits. Analog designs are more prone to the so-called proximity effects, such as *STI stress* and *well-proximity effect (WPE)* [29, 32], as discussed in the previous section. Differential analog circuits use "balanced transistor" pairs from which it is expected that both transistors show completely identical operation. Also, the rising demands of combined analog/digital systems require very high-performance A/D converters with an increasing resolution. Both requirements can only be implemented with good matching properties of the transistors. Particularly these types of analog circuits are more vulnerable to all kinds of random and systematic variations as discussed in the previous paragraph. These circuits therefore require the development of fully symmetrical layouts, in which the influence of a particular variability cause is preferably the same for all critical transistors. As this will never be the case, designers need to create circuits and layouts that are sufficiently immune to these variations, such that they are no longer a critical element of a larger electronic system.

Also, for logic circuits, matching of transistors is becoming an important issue, resulting in different propagation delays of identical logic circuits. Figure 9.30 presents two identical inverter chains (e.g. in a clock tree), but due to the V_{T} spread, they show different arrival times of the signals at their output nodes.

While the difference in arrival times at the second clock-tree stages may be 38 ps in a 130-nm CMOS technology, it can be as high as 110 ps in a 32-nm CMOS technology, which is in the order of several gate delays.

Table 9.3 shows the trend in this spread, simulated for a couple of technology nodes. It is right opposite from the ever-increasing timing requirements. For sub-20-nm process nodes, based on FinFET or GAA/nano-sheet/nano-ribbon transistors, this spread in arrival times relatively reduced, as their threshold voltages are much more dependent on the transistor work function than on the channel dope.

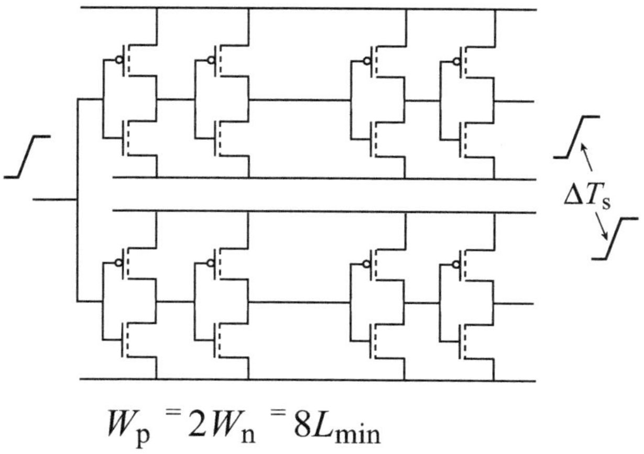

Fig. 9.30 Spread in signal arrival times due to transistor mismatch

Table 9.3 Spread in signal arrival times for different technology nodes based on [34]

Technology node	130 nm	90 nm	60 nm	40 nm	32 nm
$\sigma_{\Delta T_s}$ ($C_{load} = 50$ fF)	38 ps	68 ps	88 ps	100 ps	110 ps
$\sigma_{\Delta T_s}$ (50,35,25,20,15 fF)	22 ps	33 ps	32 ps	36 ps	37 ps
Clock period T	2 ns	1 ns	500 ps	450 ps	400 ps

Particularly for high-speed circuits, for which timing is a critical issue, transistor matching and its modelling are of extreme importance to maintain design robustness at a sufficiently high level.

9.4.4 From Deterministic to Probabilistic Design

In the above subsection, the influence of device parameter spread with respect to circuit performance is discussed. However, process-induced parameter spread in both the device and interconnect structures is also increasingly challenging chip-level timing behaviour and analysis. Transistors vary in relation to oxides, doping, V_T, width and length. Interconnects vary in relation to track width, spacing and thickness and dielectric thickness. So far, this spread was included in simulators in the so-called worst-case, nominal and best-case parameter sets in order to provide sufficient design margins. For example, in worst-case timing analysis, it is assumed that the worst-case path delay equals the sum of the worst-case delays of all individual logic gates from which it is built. This produces pessimistic results, incorrect critical paths and overdesign. *Static timing analysis (STA)* is a means to optimise and estimate timing across the chip. Current static timing analysis tools use the above-mentioned deterministic values for gate and wire delays, which is appropriate for inter-die parameter variations, but does not account for in-die variations. Particularly these in-die variations show significant impact on the overall timing behaviour. Delay faults caused by noise sources (cross talk, supply noise, etc.) are also unpredictable with respect to the induced delay. In STA, die-to-die variations are approximated by creating multiple corner files. However, within-die variations have become important contributors to the total variation. *Statistical static timing analysis (SSTA)* is therefore needed in order to cope with these local variations, which cause random gate and wire delays. These local variations are often dealt with by adding an artificial *on-chip variation (OCV)* derating factor. However, some companies use OCV for the compensation of unforeseen and unpredicted variability effects or parameter degradation over the lifetime of a chip. In fact, OCV is the discrepancy in timing properties of standard cells and/or their interconnections due to *PVT* (process, voltage and temperature) variations across the die. In standard OCV, the applied derating factor is the same for all cells globally over the die. This leads to pessimistic results, as it is not very likely that all cells will have the same worst-case *PVT*. This variation is often very local and also depends on the depth of the logic path. *Advanced OCV (AOCV)* use derating factors that are not flat and so better mimic the real variability picture [35].

An objective of SSTA is to find the probability density function of the signal arrival times at internal nodes and primary outputs. However, characterising libraries for SSTA is not easy, and the probability density functions are difficult to compute. Traditionally, SSTA has suffered from extreme run times. Related research is therefore focused to reduce run times [36, 37]. SSTA is just taking off. For the 45-nm technology node and below, SSTA is considered to be a must, particularly for the complex and higher-performance categories of ICs. Over the years, SSTA methods have been simplified and have become a standard component of the verification process, but there is research conducted to improve SSTA approaches and algorithms [38, 39].

9.4.5 Can the Variability Problem Be Solved?

A lack of modelling accuracy in current IC designs turns variability into uncertainty and increases the risk of functional failures and reduced yield. Statistical timing analysis can only predict the probability of a circuit operating at a given frequency. A hold-time violation, however, only shows up in the *process corner* that causes it and still needs a complete set of simulations for all possible corners. Various foundries offer the possibility to run *corner lots* to evaluate the circuit's behaviour in extreme process, temperature and voltage conditions. A limited amount of wafers is then processed with process variations that statistically can occur during wafer production over several years. In these process lots, also called split lots or skew lots, certain process parameters have been skewed by the fab to different corners. It helps to find out to what degree a design is immune to process (and litho) variations in the future. It requires production of lots in all *process corners*. There are five classic process corners (Fig. 9.25 shows an example); the first letter refers to the nMOS transistor and the second to the pMOS transistor:

- FF: Fast-fast corner batch represents the best-case scenario, which assumes the best-case process parameters yielding the highest performance for both types of transistor.
- FS: Fast-slow corner batch represents a scenario where the pMOS has degraded performance compared to a high-performance nMOS.
- SF: This corner batch is the compliment of the FS corner.
- SS: Slow-slow corner batch represents the worst-case scenario, which assumes the worst-case process parameters yielding the lowest performance for both types of transistors. This corner takes into account the potential impact of aging and other reliability factors.
- TT: Typical-typical corner batch represents the average or typical operating conditions. It assumes the nominal process parameters without any additional variations or performance enhancements.

These corner lots not only influence the performance of the circuit, but they may also explore leakage currents and uncover timing errors. Generally, when corner lots are used, they should be processed right after tapeout or before high-volume production as part of the product release process. Skewing of different process parameters can also be used to evaluate the sensitivity of certain circuits, e.g. memory or analog blocks, to specific process parameters. Many ICs, today, e.g. in a mobile phone, combine a mixture of high-speed and low-performance parts on the same multi-voltage domain chip. For these designs, it is not clear which process corners need to be explored. Designs in advanced process nodes may require the front end (transistor) corners to be combined with the parasitic *RC* corners of the back end (interconnect), e.g. the FF corner with the $R_{\text{min}}C_{\text{min}}$ and the SS corner with the $R_{\text{max}}C_{\text{max}}$ [40]. Particularly, designs fabricated with multi-patterning lithography require interconnect corners. It will be clear that designs made in advanced process nodes of 7 nm and below corner design become an increasing challenge.

There are several measures in the design that can reduce the effects of variability, but these are only effective for a limited category of circuits. As already mentioned before, in many analog circuits, the transistors do not use the smallest channel lengths and often do not operate at the same (low) voltages as supplied to the digital cores. Also, incorporating more regularity into the layouts of the library cells to support litho-friendly design will reduce the variability effects of the lithographic process. A fully regular library, built from only one type of transistor, would allow the technologists to optimise the transistor architecture such that it supports the lithography, reduces the variability and optimises the yield. The FinFET layout approach explained in Sect. 4.8 is an example of a fully regular layout style. It resembles a *sea-of-transistors* architecture that was introduced around the mid-1980s (see Sect. 7.6.6) for implementation of fast turnaround ASICs.

A lot of research is currently focused on methods and tools to reduce the impact of variability on the performance, to reduce design overkill and to shorten design time, since it may lead to solutions that no longer need extensive full process corner simulations. It will certainly take a while before these methods and tools will become mature instruments to effectively deal with the broad spectrum of causes of variability. But, even then, new technology nodes may introduce new variability sources requiring a continuous R&D effort to create appropriate methods and tools.

9.5 Reliability

The continuous scaling of both the devices and interconnect has severe consequences for a reliable operation of an IC. Reliability topics, such as electromigration, hot-carrier effects, negative bias temperature instability (NBTI), latch-up and ESD, are all influenced by a combination of geometrical, physical and electrical parameters: materials, sizes, dope, temperature, electrical field, current density, etc. Improving reliability therefore means choosing the right materials, the right sizes and the doping levels and preventing excessive electrical fields, temperatures and currents. Reliability issues are therefore an important part of chip design. Chip designers are even beginning to incorporate aging models in an attempt to increase reliability. This section will discuss the effects of scaling on each of the aforementioned reliability issues.

9.5.1 Punch-Through

The drain and source depletion regions of an MOS transistor may merge when a sufficiently large reverse-bias voltage is applied to the drain-to-substrate junction. This is particularly likely to occur in MOS transistors with very short channel lengths. The energy barrier, which keeps electrons in the source of an n-channel device, is lowered when the drain and source depletion regions merge. Consequently, many electrons start to flow from the source to the drain even when the gate voltage is below the threshold value and the transistor is supposedly not conducting. This effect is known as (subsurface) *punch-through*. The drain-source voltage V_{PT} at which punch-through occurs is approximated as follows:

$$V_{PT} = \frac{q}{2\epsilon_0\epsilon_r} \cdot N_A \cdot L^2 \tag{9.12}$$

where N_A represents the substrate dope, L represents the transistor channel length and q represents the charge of an electron. The effect of this leakage mechanism can be reduced during processing by increasing the doping level of the substrate with an *anti-punch-through (APT)* implantation. The associated increase in the threshold voltage of the transistor can be compensated by reducing the oxide thickness. Also, the retrograde well-doping profile formation helps to suppress punch-though. Punch-through is also regarded as a subsurface version of DIBL (see Chap. 2). It is obvious that punch-through will not occur in the devices of a well-defined CMOS technology.

9.5.2 Electromigration

The increase in current density associated with scaling may have detrimental impact, not only on circuit performance but also on the IC's reliability. High currents, flowing through the metal lines, may cause metal ions to be transported through the interconnection layers due to the exchange of sufficient momentum between electrons and the metal atoms. For this effect, which causes a material to physically migrate, many electrons are required to collide with its atoms. This physical migration of material from a certain location to another location creates open circuits or *voids* (Fig. 9.31a) on locations where the material is removed and *hillocks* (Fig. 9.31b) on locations where material is added. This *electromigration* effect damages the layer and results in the eventual failure of the circuit. Electromigration may therefore dramatically shorten the lifetime of an IC. The impact of electromigration is eliminated by preventing excessive current densities. Electromigration design rules are therefore part of every design kit. These rules specify the minimum required metal track width for the respective metal (e.g. aluminium or copper) for a certain desired current flow at given temperatures. Electromigration effects increase with temperature because of the temperature dependence of the diffusion coefficient. This causes a reduction of the maximum allowed current density (J_{max}) at higher temperatures in on-chip interconnect. The required metal width for electromigration roughly doubles for every 10 °C increase in temperature. Since many IC data sheets show a maximum ambient temperature of around 70 °C or higher, the real

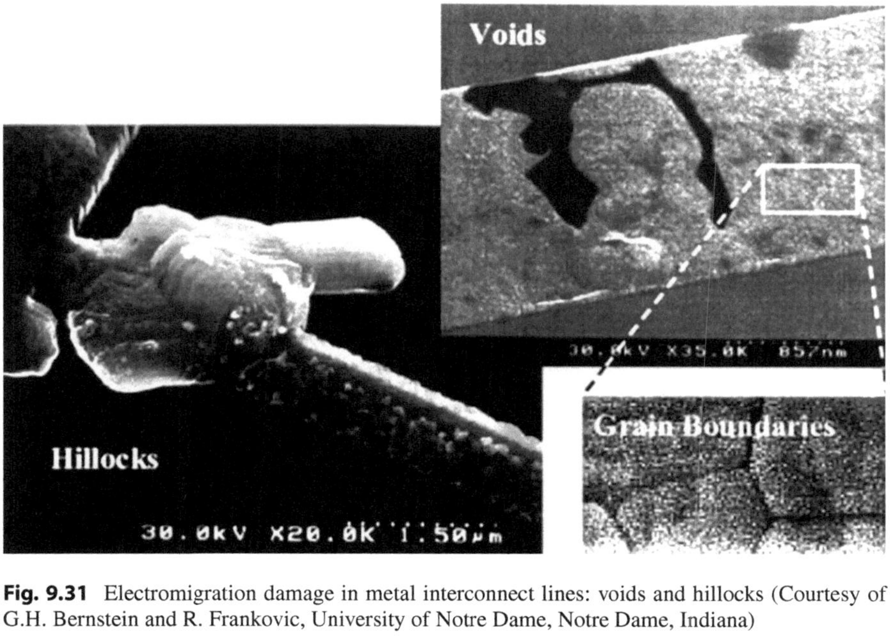

Fig. 9.31 Electromigration damage in metal interconnect lines: voids and hillocks (Courtesy of G.H. Bernstein and R. Frankovic, University of Notre Dame, Notre Dame, Indiana)

Table 9.4 Metal characteristics for a 65-nm bulk-CMOS technology

Technology node and metal layer	R_{sheet}	H	$\mathbf{J_{max}}$@125 °C
Lower metal layer (copper)	85 m$\Omega/\square$	350 nm	3.6 mA/μm^2
Top metal layer (copper)	26 m$\Omega/\square$	900 nm	3.6 mA/μm^2

Table 9.5 Metal characteristics for a 5-nm FinFET technology

Metal layer	$\mathbf{J_{max}}$@125 °C
Middle metal layer (copper)	5 mA/μm
Top metal layer (copper)	15 mA/μm

worst-case junction temperature of the silicon itself may exceed 100 °C in many applications. Therefore, it is common design practice to use the value for J_{max} at 125 °C.

The minimum allowed width W_{em} of a metal wire with height H, to carry a current I, according to this electromigration requirement, is then equal to:

$$W_{em} = \frac{I}{J_{max} \cdot H}. \tag{9.13}$$

Tables 9.4 and 9.5 show some parameter values, which are characteristic for metal layers in a 65 and 5-nm CMOS technology, respectively.

As can be seen from these tables, electromigration rules very much depend on the related process. Also, the impact of the operating temperature on electromigration increases with smaller process nodes. For example, in a 5-nm FinFET process, the maximum current through a top metal layer copper line at 125 °C may only be 1/4 of the current allowed at 105 °C. At 150 °C, that ratio is even 1/20. Particularly for medical, aviation and automotive products, this has severe impact on the design of the power distribution network, as these products often require higher-temperature specs. It is therefore very important for designers to pay serious attention to the electromigration rules in the design-rule manual of the specific process. Since most of the currents in an IC flow through the supply lines, it is obvious that these are often implemented in the upper metal layer(s), which usually have a larger height. Because AC currents flow in both directions through a wire, the maximum value of these currents with respect to electromigration is about one order of magnitude larger than the maximum values for the (average) currents mentioned above. Similarly, currents through contact holes and vias must also be limited to eliminate electromigration-induced damage of the contact conductor. A typical maximum current density value for a $0.1 \times 0.1\ \mu$m^2 contact or via in a 28-nm CMOS technology is around 0.16 mA/μm^2 at 125 °C. The increase in the aspect ratios of the contacts and vias, in combination with a reduction of maximum currents through them, makes them an incremental part of the overall IC reliability.

The continuous scaling of feature sizes and voltages (constant-field scaling) by about a factor of 0.7, for every new technology node, did not change the intrinsic power density of most standard-cell designs. However, due to the reduction in supply voltage, the supply current per unit area of logic increased with about a factor of 1.4 every generation. This has put severe constraints to maintaining electromigration reliability across complex designs.

Because of the expected increase in currents through the metal layers, more Joule heating is expected in these layers. This, in combination with low-ϵ dielectrics, which show a higher thermal resistance, made designers start worrying about the so-called *wire self-heating* mechanism. However, the width of a metal wire is specified not only by the appropriate electromigration requirements but also by the maximum allowed voltage drop across the wire in order to limit speed loss of the connected circuit(s). Suppose an active logic block draws an average supply current of 100 mA. When this block is located nearby the supply pads of the chip, the width of the supply lines is determined only by the electromigration requirement for this 100-mA current. When this block is near the centre of the chip, say at 5 mm distance from the supply pads, the supply lines must be much wider in order to limit the voltage drop across it. So, above a certain distance from the supply pads, the width of the metal (and thus its cooling area) grows with its length, keeping the voltage drop across the line constant. As a result also, the resistance of the line (and thus its total I^2R *Joule heating*) will then be constant. In other words, the maximum wire self-heating occurs in wires with length equal to a crossover length L_{co}, which is defined to be the length at which the metal width required by electromigration is identical to the width required by the maximum allowed voltage drop. In [41], it is shown that for 0.18 and 0.12-μm bulk-CMOS technologies, wire self-heating in supply lines causes only a limited temperature rise of the wires of just a few degrees. Also, for the 65 and 45-nm technology nodes, this temperature rise is by far negligible compared to the temperature rise due to the power consumption of the silicon part of the chip. From this result, it can be concluded that wire self-heating in supply lines should not be a real issue in current (and near future) properly designed bulk-CMOS VLSI chips.

9.5.3 Hot-Carrier Injection (HCI)

When carriers in the MOS transistor channel are given enough energy, they collide with the substrate atoms and generate electron-hole pairs. These, in turn, will also be accelerated and may also collide with substrate atoms. This so-called *impact ionisation* may cause large substrate currents, device breakdown and/or degradation of the silicon-to-gate oxide interface. Electrons actually collide with the gate oxide. When electrons achieve sufficient energy, they may cross this silicon to silicon dioxide (Si/SiO_2) interface barrier (with a barrier energy of about 3.1 eV for electrons and 4.7 eV for holes) and are injected into the gate oxide. Injected

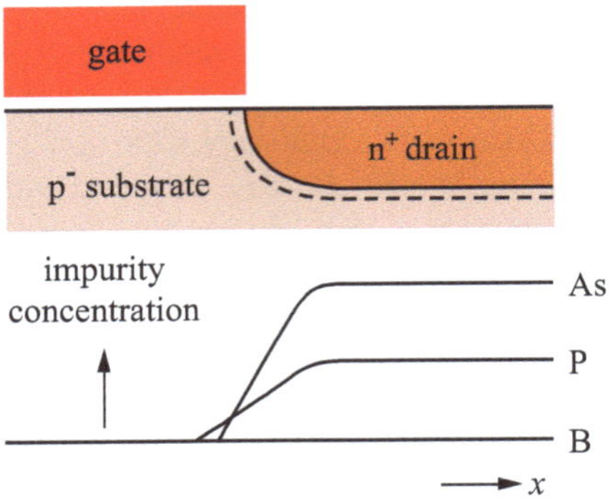

Fig. 9.32 Phosphorous (P) halo around arsenic (As) in the cross section of a graded drain transistor and the concentrations as a function of the position x

carriers lead to the degradation of the Si/SiO_2 interface (electrically active interface defects are generated), to the generation of defects in the gate oxide film and to charge trapping in the oxide interface (both pre-existing and newly generated). Oxide charge trapping and interface state generation induce a shift of the transistor threshold voltage and cause a degradation of the device drive current. This effect is called *hot-carrier injection (HCI)* and leads to degraded device performance and reliability problems. Due to the lower mobility of holes with respect to electrons in the transistor channel, impact ionisation in p-channel MOSFETs is less. Therefore, the hot-carrier effect is more severe in n-type MOSFETs.

Graded drain and lightly doped drain structures are used to reduce the maximum value of the electric field in small transistors and thus prevent hot-carrier degradation. The *graded drain* transistor is a very simple adaptation of the conventional transistor. The junction between the drain and the substrate is made much more gradual by simply implanting phosphorous with a relatively low concentration in the highly concentrated (arsenic) n^+ area. The phosphorous has a much higher diffusion coefficient than the arsenic in this area and therefore diffuses much further. This results in a donor profile with a low gradient; an example is shown in Fig. 9.32. The graded drain reduces the maximum electric field by about 30%. This implies that the operating voltage can be increased by 50% for given transistor dimensions.

The *lightly doped drain* (*LDD*) method is a more difficult means of reducing the drain-substrate concentration gradient. It was introduced in CMOS processes with channel lengths of around 1 μm, which exhibited extremely large peaks in the electric field in the channel close to the drain. The maximum electric field obtained by using LDD is lower than that achieved with the graded drain. The various LDD process steps are explained with the aid of Fig. 9.33. A conventional 0.35-μm nMOS transistor with a gate oxide thickness of about 7 nm thickness is shown in Fig. 9.33a.

Conventional CMOS processing, which is described in Chap. 3, is used to create the gate oxide. Phosphorous with a concentration that varies from 1×10^{18} to 4×10^{18} atoms per cm^3 is subsequently implanted. An oxide layer of about 0.35 μm

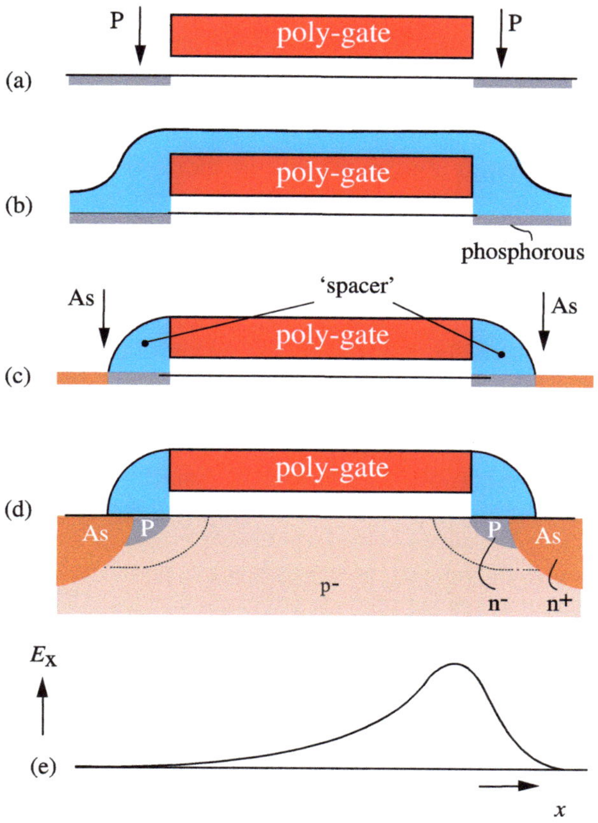

Fig. 9.33 Process steps for the creation of an LDD transistor and the resulting reduced horizontal electric field distribution

thickness shown in Fig. 9.33b is then deposited. This is followed by an anisotropic etch, which leaves the *oxide spacers* shown on both sides of the gate in Fig. 9.33c. A subsequent highly concentrated implantation of arsenic and a drive-in diffusion produce the resulting n^- and n^+ areas shown in Fig. 9.33d. The magnitude of the transistor's horizontal electric field as a function of the channel position x is shown in Fig. 9.33e. Its maximum value is 50% of that obtained in a comparable transistor with conventional arsenic drain and source areas. Two factors account for this significant reduction. The first is the relatively long region with a low donor (n^-) concentration. A depletion area will form much sooner in this area than in the n^+ area. A large proportion of the drain-source voltage drop is distributed over this area. The second factor is the extra separation between the gate and the n^+ drain area. This also reduces the influence of the second-order effects as discussed in Chap. 2.

The LDD transistor was difficult to create and has the added disadvantage of possible increased series resistance in the source and drain as a result of the n^- areas. Careful optimisation, however, yields small transistors with high operational voltages that can deliver high currents. LDD implants were included in CMOS technologies from about 1 µm down to 0.35-µm channel lengths. As discussed before, the required energy for carriers to cross the Si-SiO_2 interface barrier is at least 3.1 eV for electrons and 4.7 eV for holes. As supply voltages reduce with the advent of new process generations, these carriers can hardly ever reach such energies when the supply voltage is 2.5 V or less.

Theoretically, in a 0.18 µm CMOS technology with a supply voltage of 1.8 V, an electron can only get an energy level of 1.8 eV during its flow through the channel from source to drain. This is less than the above-mentioned barrier energy to create hot electrons. However, due to multiple collisions, some electrons may acquire more energy than the required barrier energy and become "hot". From these considerations, it was generally accepted that when supply voltages are reduced, the chance to generate hot carriers in the transistor channel would reduce as well and the hot-carrier effect was expected to eventually disappear totally. As a result, LDD implants are no longer required in 0.25 µm CMOS processes and below. These are then replaced by a (lightly doped) drain extension, as discussed in Sect. 3.9.3.

With the continuous scaling process, critical dimension (CD) control becomes more difficult, leading to transistors with different channel lengths showing different hot-carrier behaviour. Shorter channel lengths easily introduce punch-through. Both punch-through prevention and short-channel effect suppression require different doping profiles around sources and drains, with increased doping levels. This has some negative effects on the hot-carrier behaviour.

When voltages across the transistor are scaled at the same pace as the transistor feature sizes, the electrical fields remain almost constant, and the chance for impact ionisation would hardly change. However, particularly with 90 nm and smaller CMOS technologies, the effective channel length is scaling faster than the supply voltage, so that the increase in electrical field may lead to increased impact ionisation. Although these carriers do not acquire sufficient energy to cross the silicon-to-silicon dioxide barrier, they will still cause substrate currents. Hot-carrier effects may therefore manifest themselves again more in sub-100 nm technologies than in the technology nodes just above 100 nm, especially in the early process development phase due to bad transistor drain engineering. Assuming that the transistor is stressed under worst-case condition ($V_{\mathrm{ds}} = V_{\mathrm{dd}}$ and $V_{\mathrm{gs}} = V_{\mathrm{dd}}/2$), such that the substrate current is maximal, the hot-carrier lifetime is described by a well-accepted empirical expression (Takeda) as:

$$\tau_{drift} = A \cdot {L_{eff}}^{C} \cdot e^{B/V_{\mathrm{ds}}} \tag{9.14}$$

where τ_{drift} represents the lifetime (usually at 10% degradation), L_{eff} represents the effective channel length and A, B and C are process-related coefficients. It is clear that the hot-carrier lifetime reduces with decreasing channel length and increasing voltage. So, when we scale the supply voltage with the same factor as the feature sizes, still this lifetime may increase, dependent on the constants A, B and C.

An additional effect is that for future technologies the silicon dioxide will be replaced by high-ε dielectrics. Most of them, however, have a significantly lower barrier [42], and the hot-carrier effects are not just slowly fading away because of reducing supply voltages below the barrier. Results from literature [43, 44] stress the importance of a continuous attention for hot-carrier degradation in nanometer technologies, in order to maintain functional reliability at a sufficiently high level.

Finally, the temperature behaviour of *hot-carrier degradation (HCD)* changes with reducing channel lengths: conventional long-channel MOS devices show less degradation at higher temperatures, while short-channel devices (particularly in metal gate processes) show increasing degradation at higher temperatures [45].

9.5.4 Bias Temperature Instability (BTI, NBTI and PBTI)

Negative bias temperature instability (NBTI) is a result of a negative bias applied to the gate of a p-channel MOS transistor with respect to the bulk. The mechanism is temperature-activated. NBTI results in the degradation of many transistor parameters (threshold voltage, drive current and transconductance), but the threshold voltage appears to be the most degrading one. NBTI was first reported in 1967, but the attention devoted to this mechanism has been escalating since the millennium, due to the introduction of gate oxide nitridation [46] that enhances NBTI and the fact that other oxide wear-out mechanisms, such as HCE and oxide breakdown, were expected to become less severe as the gate oxide scales down. NBTI is strongly process-dependent. It has been reported that a higher nitrogen concentration in the oxide [46], boron penetration [47] and plasma processing can enhance NBTI, while fluorine incorporation in the gate dielectric is beneficial against NBTI [48]. The physical nature of the wear-out mechanism induced by NBTI is very difficult to identify. The most accepted models imply positive charge build-up in the oxide bulk and at the Si/SiO_2 interface (donor-like interface states) [48, 49].

While hot-carrier injection mostly affects n-channel MOSFETs and depends on the transistor channel length, NBTI mostly affects the pMOS transistor and is only slightly dependent on the transistor geometry, although it has also been reported that in shorter channel devices NBTI can be more severe [50]. Furthermore, the NBTI does not imply a current flow in the transistor channel and can occur at zero drain to source bias. This would mean that NBTI stress could even occur in the standby mode. Design configurations in which matched p-channel MOSFET pairs are subjected to unbalanced stress are reported as most sensitive to NBTI degradation, since the threshold voltages of the transistor pair change differently with the stress [51]. Also, matched p-channel MOSFET pairs operated symmetrically can lead to reliability fails due to NBTI when the transistors are subjected to different biases in power-down mode. Burn-in can also be a source of NBTI-induced circuit fails, due to the involved high temperature. The NBTI effect is more severe for pMOS transistors than for nMOS transistors (PBTI) because of the difference between holes and electrons in interacting with oxide states. The NBTI (PBTI) degradation almost always recovers after the stress is removed. This requires a quick engineering test to demonstrate the impact of this reliability mechanism.

Even when an IC is produced in different fabs that run the same process, it may perform differently with respect to NBTI, because not all individual processing steps are completely identical. NBTI is therefore a technology issue, but critical design configurations, such as matched p-channel MOSFET pairs subjected to unbalanced stress, either in operation or power-down mode, should be avoided.

The physical understanding of BTI is continuously improving, leading to the development of various models. A description is beyond the scope of this book, but many of them can be found on the internet. Assuming a power-law dependence on the stress voltage (field), then the change in V_{T} is proportional to:

$$\Delta V_{\mathrm{T}} = D \cdot {E_{ox}}^{m} \tag{9.15}$$

where D is a process-dependent parameter, E_{ox} represents the electrical field across the oxide, and m is a coefficient dependent on, for example, the dielectric material and the dielectric thickness (an approximate value is $m \approx 4$).

V_{T} shifts of 50 mV and more have been reported, so designers need to be convinced to build enough tolerance in their designs. The occurrence of NBTI can be lowered when a device (chip) is not subjected to voltage overshoot and/or high temperatures, either from its own heat dissipation or from its application environment. Therefore, a reduced power consumption would also be beneficial to reduce the chance for NBTI stress.

With shrinking devices, the spread in V_{T} due to BTI stress has increased. Larger devices show a kind of average ΔV_{T} shift, while ΔV_{T} in small devices varies widely. This leads to the fact that some devices fail quickly, while others maintain their performance over long periods of time. As explained before, the stress creates interface traps. After removing the stress conditions, these interface traps are partially annealed (repaired), such that some devices recover very fast (in less than a millisecond), while others still fail after an hour. The (partial) recovery is different for varying device dimensions, materials and structures.

A lot of research is currently still be carried out on various aspects of BTI for different devices and circuits. Also, embedded SRAMs are affected by NBTI, particularly when they are integrated on ICs whose temperature may dramatically increase during active modes of operation [52]. BTI also has an increasing impact on the reliability of scaled DRAMs [53]. The problems manifest themselves particularly in the peripheral MOSFETs. Some DRAM vendors use a special process flow to reduce the effects of BTI. More details about BTI fundamentals can be found in [54]. BTI is not just a reliability issue in planar MOS devices; also FinFETs [55] and high-ϵ, metal gate transistors [56] are affected. Even today, BTI is still seen as a dominant degradation effect (aging), in advanced CMOS process technologies [57, 58]. The reader is advised to search the internet for more details on all kinds of BTI aspects, since it is too broad and too specialised to be included in this section.

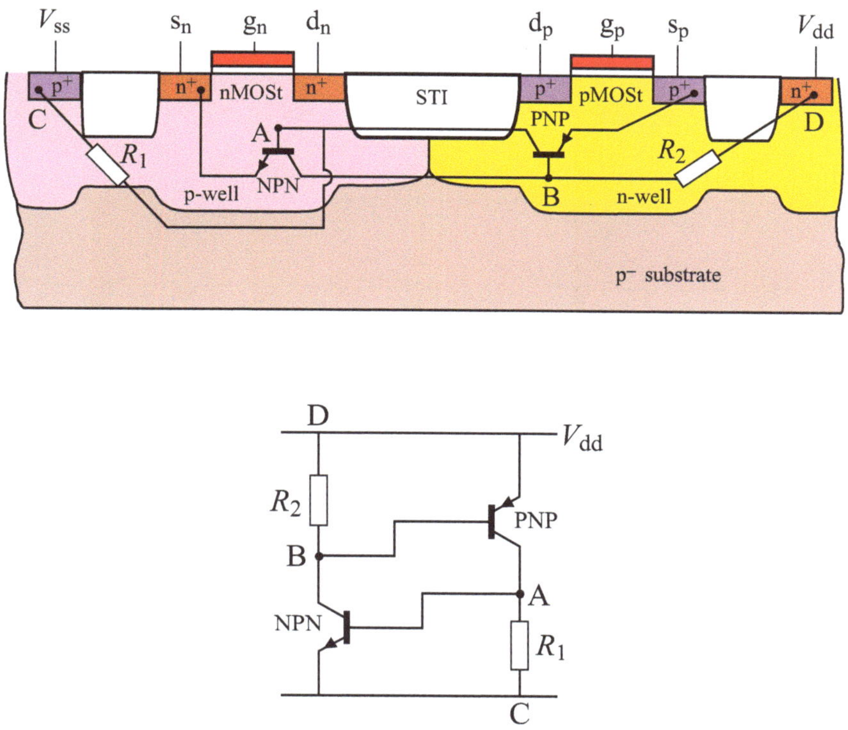

Fig. 9.34 Parasitic thyristor in CMOS and its equivalent circuit diagram

9.5.5 Latch-Up

The presence of nMOS and pMOS transistors in a CMOS process leads to the creation of parasitic thyristors, as shown in Fig. 9.34. In this figure, R_1 and R_2 represent p-well and n-well resistances, respectively.

Relatively high currents through the bipolar transistors will create relatively high voltages in the substrate and/or n-well. When a sufficiently high positive voltage is present somewhere in the p-well (e.g. at position A), it will turn on the parasitic NPN transistor, or when a local voltage (e.g. at position B) within the n-well that is sufficiently lower than the V_{dd}, it will turn on the parasitic PNP transistor. When both bipolar transistors conduct, they are connected into a feed-forward loop, which means that they enhance each other's conduction state, which will finally be latched (maintained) in the thyristor. This state can only be recovered when the supply is completely switched off. This undesirable effect is called *latch-up* and leads to incorrect circuit behaviour or even damage. Also, inductive effects or coupling capacitances may cause the node connected to the drain to have overshoots and/or undershoots, thus forward biasing the drain to p-well junction, which may initiate

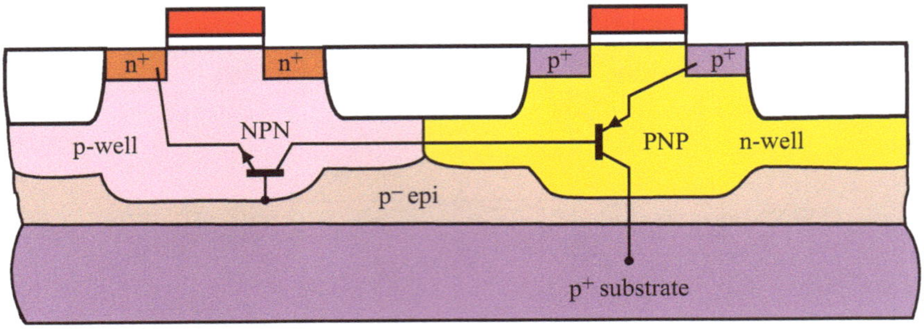

Fig. 9.35 Cross section of a wafer with a thin p− epi-layer on a thick p+ substrate

latch-up. This requires a controlled startup of ICs. Certain low-standby power applications require the use of additional back-bias voltages to reduce subthreshold leakage: a negative voltage (often between −1 and −2 V) for the p-well and a positive voltage for the n-well (often between 1 and 2 V above the supply voltage). These back-bias voltages can be generated on-chip or externally supplied through additional supply pads. In either case, one must guarantee that these well voltages are always switched on before the nominal supply is switched on. This is to prevent that the wells are floating when the supply is switched on. Floating wells are very susceptible to noise and may initiate latch-up already during startup of the chip (system). This is a real burden for those ICs that have on-chip back-bias generation, because then the nominal supply need to be switched on before the back-bias voltage can be generated by the on-chip charge pumps.

Latch-up in CMOS circuits can be avoided by applying the following technological and/or design remedies:

- Minimise the substrate (p-well) and/or n-well resistances. This can be done in two ways: One is the use of many p-well and n-well contacts in the design, which will reduce the values for R_1 and R_2, respectively. These p-well and n-well contacts are part of the library cell design. There are libraries with fixed substrate and n-well contacts. Most CMOS libraries in 60 nm CMOS and beyond also contain separate *well-tap cells* that can be placed in the standard-cell rows as required by the corresponding design rule manual. Section 4.6.4 describes these library aspects in more details. So, a reduction of the substrate (or p-well) resistances then requires to place a larger number of these well-tap cells in the standard-cell rows. The parasitic thyristor is then unlikely to turn on. Reducing both resistances by increasing the substrate and n-well doping is not an option, since it also changes the threshold voltages and overall transistor behaviour. A good alternative is the use of the so-called epitaxial wafers (Fig. 9.35).

  Epitaxy is a layer of single-crystalline silicon deposited/grown onto a single-crystalline silicon wafer (see Sect. 3.2.2). The crystalline structure of the substrate is reproduced in the growing material. This epitaxial layer, in which the devices are formed and whose thickness is usually between 1 and 5 μm, can be doped, as it is deposited, to the required doping type and concentration (usually with a resistivity of ≈10–20 Ωcm) while continuing the substrate's crystalline structure. So, we can create a thin p^--epitaxial layer on top of a p^+-substrate. Because the current (unground) wafer thickness is typically 750 μm, the p^+-substrate is relatively thick and has a low resistivity (≈5–10 mΩcm). Such low-ohmic substrates show very low values for R_1. A large part of the PNP collector current will therefore flow through this substrate, and only a small part will flow into the base of the NPN transistor. This transistor can no longer be turned on easily and is then largely excluded from the latch circuit. *Epitaxial wafers* with low-ohmic substrates have been massively used for CMOS products in 0.18-μm technologies and above. Due to decreasing supply voltages and increasing noise levels, the combination of analog and digital circuits onto one single chip has made its design a difficult and cumbersome task. Particularly the substrate noise sensitivity of analog circuits requires a good isolation from the digital noise "generators", which is why a high-ohmic substrate is preferred for mixed analog digital circuits. Because the latch-up effect decreases with reducing voltages, CMOS in 120 nm and beyond are most commonly processed on high-ohmic wafers (Sect. 3.2.2).

- The use of *guard rings* (Sect. 9.5.7) is another way to make strong (low-ohmic) connections of local substrate and/or n-well areas to V_{ss} and V_{dd} respectively. Moreover, the distance between n-type and p-type areas is also a matter of concern during the design phase and is particularly of interest in I/O circuits, which are usually supplied by higher voltages. Guard rings are more effective on high-ohmic substrates.
- Apply a back-bias voltage to the substrate. When the p-well in Fig. 9.34 is connected to a negative voltage instead of to V_{ss}, the base voltage V_A of the NPN transistor will be lowered. Therefore, this transistor can no longer be turned on easily. This technique is more a theoretical option and is not frequently used for latch-up prevention.
- Use silicon-on-insulator technology to completely isolate the nMOS transistors from the pMOSTs. In this technology, the NPN and PNP transistors are completely isolated from one another, and so the connections to create latching thyristor circuits are missing. The amount of "electrical isolation" depends heavily on the applied frequencies.

The application of one or more of the above remedies has increased latch-up immunity to a very high level. The highest chance of occurrence for latch-up is during testing. Standard testing requirements include immunity to 100 mA or more, depending on what the IC can and should withstand from an application point of view. This means that with epi-wafer material, 100 mA can be supplied to the output of an output buffer (driver) even though no output transistor is conducting. This current, then, directly flows into the substrate, thereby raising the substrate voltage and possibly turning the thyristor on (Fig. 9.34). In practice, some latch-up tests are done with 150–200 mA at a maximum ambient rated temperature for the device, depending on the target application area.

In future technologies, the latch-up phenomenon is likely to disappear inside electronic circuits, as the supply voltages will be reduced every new technology node. However, at the chip I/Os, the requirements on latch-up remain relatively high, since many applications still require a higher interface voltage (1.8, 2.5, 3.3 V). Also, non-volatile memories are more prone to latch-up since they use high program and erase voltages and sometimes even higher voltages during reading. The higher a voltage somewhere on a chip, the bigger the chance to trigger the thyristor, and the higher the probability of latch-up occurrence. More on latch-up basics can be found in [59, 60].

9.5.6 Electrostatic Discharge (ESD)

Integrated circuits are exposed to many possible sources of damage, both during and after the manufacturing process. The principal cause of damage is *electrostatic discharge (ESD)*, due to the transfer of charge between bodies at different electrical potentials. ESD pulse durations are very short and normally range from 1 to 200 ns, but they may introduce very large power spikes. The high impedance of MOS input circuits makes them particularly vulnerable to physical damage when they are exposed to these spikes. This may result from operations during the fabrication process or from handling (un)packaged dies and bonding. It may also occur during testing and maintenance or in the application. While only a few devices or connections may be severely damaged, many more may suffer damage that is not immediately apparent. These latent failures will result in customer returns, which is one of the biggest worries of semiconductor vendors. Thus, ESD is one of the most important factors that determine the reliability of an IC. It may also trigger the parasitic thyristor in Fig. 9.34 resulting in the occurrence of latch-up.

The damage caused by ESD is irreversible. The human body is one of the main sources responsible for ESD. Just by walking on a carpet on a low-humidity day, for instance, a person wearing shoes with highly insulating soles can build up a voltage in excess of 30.000 V. The resulting charge can then be transferred via an ESD to an electronic circuit during touching. It is also very important that precautions need to be taken to prevent ESD damage during IC fabrication. In addition, protective measures must be included in an IC's design to ensure that it can withstand acceptably large ESD pulses. On-chip MOS protection circuits are used to increase the immunity of an IC to ESD pulses. These circuits are designed to provide input and output circuits with low-impedance shunt paths, which prevent excessive voltages to arrive at the IC's input, output and core transistors.

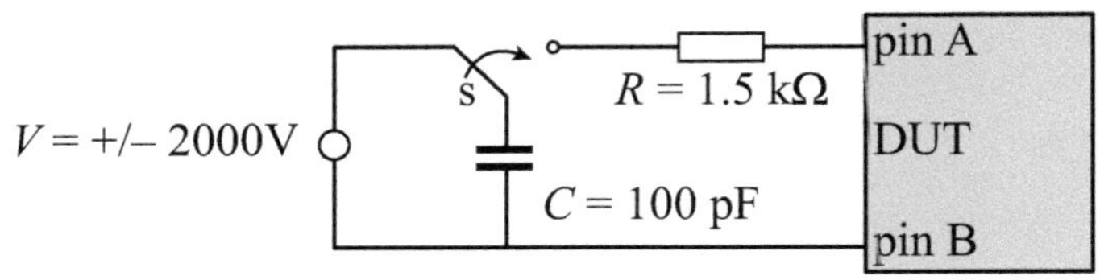

Fig. 9.36 A typical equivalent circuit based on the human-body model

Table 9.6 Different ESD test states

State	DUT Pin A	Pin B
1	Input	V_{ss}
2	V_{ss}	Input
3	Input	V_{dd}
4	V_{dd}	Input
5	Output	V_{ss}
6	V_{ss}	Output
7	Output	V_{dd}
8	V_{dd}	Output
9	Input	Output
10	Output	Input
11	Input	Input
12	Output	Output
13	V_{dd}	V_{ss}
14	V_{ss}	V_{dd}

ESD Test Models and Procedures

ESD sources are emulated in several different ways. The *human-body model* is currently the most popular industry model and simulates the direct transfer of electrostatic charge from the human body to a test device. It is internationally accepted as a standard (JEDEC Standard No.22-A114-B). Figure 9.36 shows a human-body test set-up. The basic requirement for this model, in combination with the parasitics (L) of the tester interface cables, is to generate ESD pulses with rise times between 10 and 15 ns.

The test is normally done on an ESD tester. This human-body model has not changed much over the last few decades. Basically, a 100- pF capacitor is charged to the test voltage and then discharged through a 1.5- kΩ resistor across any combination of pins A and B (Table 9.6) of the DUT (device under test). The chip may consist of several supply (V_{dd}) and ground (V_{ss}) domains. Each domain may be supplied by more than one pin. The V_{ss} and V_{dd} in the table below refer to just one of the respective pins of a supply domain. In other words, each pin is then tested with respect to all grounded V_{ss} and V_{dd} domains and not to all grounded V_{ss} and V_{dd} pins, to save test time. Each signal pin is also tested with respect to all other grounded signal pins. The maximum test voltage ranges from 2 to 8 kV and depends on the application area of the chip. Since production environments are well controlled, a maximum voltage of 2 kV is usually required. However, because more and more IC pins can be touched in daily life (plug-ins like USB ports, chip cards, SIM cards, memory sticks, flash cards, etc.), the ESD test requirements tend to increase. The 8- kV requirement is therefore no exception anymore. The devices are classified when meeting a particular sensitivity criterion. A class-2 device, for instance, has passed the 2 kV but fails after exposure to an ESD pulse of 4 kV (see the above-mentioned standard: www.jedec.org). There are also discussions led by the Industry Council on ESD Target Levels to target for a safe human-body model of 1 kV for many applications (e.g. automotive and consumer), because the 2- kV level is really overdesign for many (but not all) of these applications.

The first ESD tests start at 100 V. Generally, three to five positive and negative pulses are applied at 300 ms intervals in all test states. Stressed pins are tested after application of each ESD pulse series. If no failure is observed for a sequence through the pins, then the ESD voltage level is increased by 100 V, and the sequence is repeated. The ESD test is complete when a failure is observed or when all pins on the DUT have been stressed until the required maximum voltage is reached. Generally, the following (example) criteria may be used to determine failure:

- Incorrect functional operation or a violation of the device specifications
- A change of more than 5% in the forward voltage drop and breakdown voltage in the diode characteristic
- An increase of more than 10% in the I_{ddq} leakage current (see Chap. 10)

Another standardised and popular ESD test model is the *machine model*, which emulates the rapid direct transfer of electrostatic charge, from a charged conductive object (tool or equipment) to a test device. Compared to the human-body model of Fig. 9.36, the machine model specifies a discharge of a 200-pF capacitor through a 0.5- μH inductor. Due to the absence of the current limiting resistor, this model was seen as more severe, and the tests are done at lower voltages. The *charged device model* is an alternative ESD test set-up, which is most commonly used to emulate rapid electrostatic charge transfer during, for example, packaging and assembly. More details on the latter two models can be found in [61] or directly from the JEDEC website: www.jedec.org.

On-Chip ESD Protection Circuits

Although much ESD and ESD protection knowledge has been built over the last couple of decades, the design of on-chip ESD protection circuits is both scientific and experimental. This is due to the fact that in every new semiconductor node, device architectures and feature sizes (e.g. width, spacing, oxide thickness, etc.) have changed with respect to the previous node, which requires new protection solutions. Usually, several alternative protection circuits are explored in each new technology node, and often semiconductor process development goes hand in hand with ESD protection development.

The purpose of a protection circuit is that it provides a low-ohmic shunt path in parallel with the MOS input and output transistors during the occurrence of an ESD pulse. In its simplest form, a protection circuit consists of a spike filter and a set of diodes (Fig. 9.37).

Resistor R_i, in combination with the parasitic capacitance at node A, forms the input voltage spike filter, while diodes D_1 and D_2 clamp the input voltage to a junction voltage above the V_{dd} and a junction voltage below the V_{ss}, respectively. R_o, D_3 and D_4 do the same but then for the output node. In many cases also, MOS

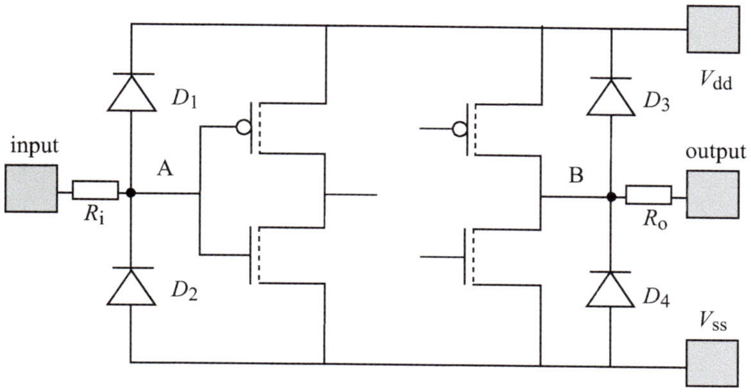

Fig. 9.37 Simplest form of an MOS input and output protection circuit

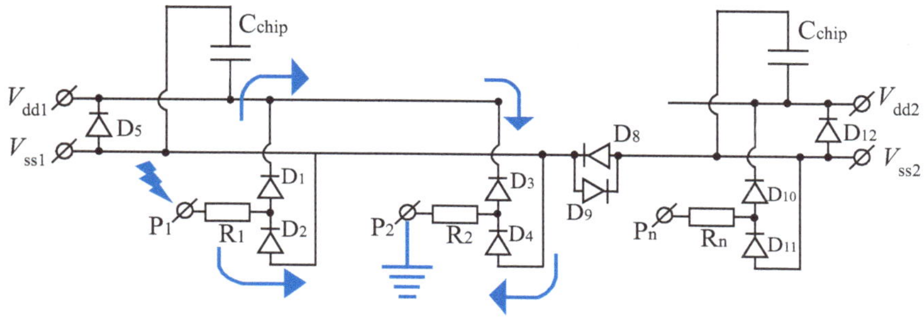

Fig. 9.38 Connection of ESD protection devices between multiple pins in the same voltage domain or in different voltage domains

diodes are used for clamping the input and output voltages. Values for the input and output resistors are in the order of 100 Ω and 5 kΩ, respectively.

The idea behind ESD protection is to protect the internal core and I/O circuits of the chip from overvoltage during an ESD event by guiding the ESD current through a defined current path. ESD tests are performed with floating supply pins. Therefore, the diodes in Fig. 9.36 do not really clamp their voltages to the Vdd and Vss terminals of the circuit. Figure 9.38 shows a generic protection architecture as it may be found in modern ICs with multiple supply domains. In fact, when performing an ESD test between two pins, e.g. a positive pulse between P_1 and P_2 in the figure, then there are always two parallel current paths that clamp the voltage. As indicated by the arrows, one path is formed by the forward-biased diode D_1 and the reverse-biased diode D_3, while the other path is formed by the reverse-biased diode D_2 and the forward-biased diode D_4.

To manage an ESD event between the supply pins, e.g. between V_{ss1} and V_{dd1}, the protection diode D_5 is placed between the supply rails which will clamp the voltage to a diode voltage above V_{dd1} in the case of a positive spike on V_{ss1}. In the

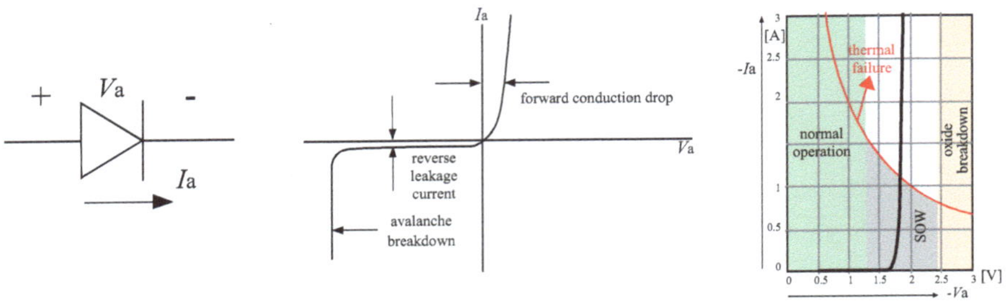

Fig. 9.39 Symbol of a diode, its current-voltage characteristic and its safe operating window (SOW)

case of a negative spike on V_{ss1}, D_5 will go into breakdown and thereby limit the internal voltage spike.

In the case of multiple supply domains, attention should be paid to managing ESD events between supply pins of the different domains. For this purpose, the back-to-back diodes D_8/D_9 are placed between the ground lines of the two V_{ss} domains. These ensure that the voltage between ground lines is always limited to just one diode voltage drop.

In this drawing, it is assumed that the V_{dd1} (e.g. voltage of the digital cores) and V_{dd2} (e.g. voltage of the I/Os) domains run at different voltages. Therefore, there is not a set of cross-coupled diodes placed in between them. When the level of these voltages is the same, then this set of diodes must be present.

Figure 9.39 shows a conventional ESD device: a diode and its current-voltage characteristics.

When, for example, the voltage across the reverse-biased diode D_3 in Fig. 9.38 exceeds a certain reverse voltage, it will go into avalanche breakdown, as shown in the middle diagram in Fig. 9.39, thereby clamping its voltage to the corresponding breakdown voltage.

The right diagram in this figure shows a graphical representation of the safe operating window (SOW) for an ESD protection device. The left part of the window covers normal IC operation. In this part, the current through the protection device should be very low. The right part of the window covers the area in which the applied voltage could cause damage to the internal devices. The top of the operating window is limited by the maximum energy and covers the maximum current the protection should be able to handle while clamping the voltage during an ESD event. Note that in the case of the HBM a protection that is able to cope with an event of a 2 kV requires to sink a current of 1.3 A! From this requirement, the size of the device can be derived.

In current ESD protection circuits, however, the diode has often been replaced by a grounded-gate nMOS transistor operating as a snapback device (Fig. 9.40).

Such a device is compact and can be relatively easily implemented, without additional process steps. It uses the parasitic bipolar junction transistor (BJT) to perform a snapback clamp after exceeding the first breakdown voltage.

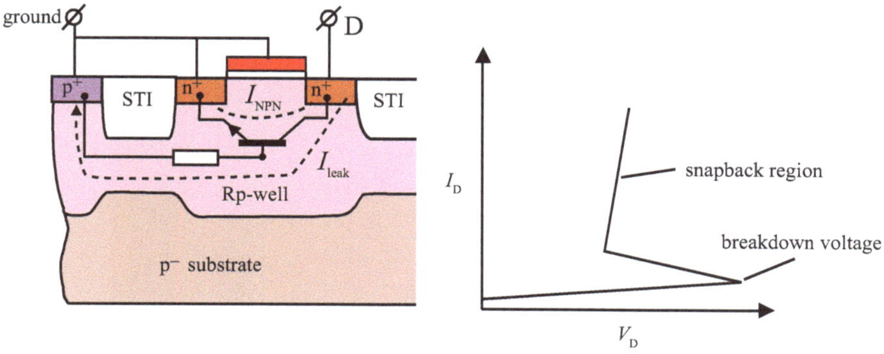

Fig. 9.40 Grounded-gate nMOS transistor operating as a snapback device, with its current-voltage characteristic

Since MOS inputs are connected to high-ohmic transistor gates, the protection of input circuits is more critical than that of output circuits. Output pads are connected to drain areas. Usually, these drain areas are relatively large, because outputs usually have to drive large capacitances (10–50 pF) and the complementary drain junctions act as intrinsically available diode clamps. Of course also, the outputs must fulfil ESD design rules.

The behaviour of MOS protection circuits depends very much on their size and layout and on various process parameters. Each manufacturing process has its own specific design rules for ESD protection circuits. Therefore, the design of such circuits is done in co-operation with specialists in the field of protection devices. As an ESD event can occur between any combination of two pins on the IC, it requires to include all possible current paths during an ESD event. For systematic analysis of ESD current paths, software is available from various vendors, some of which provide a full-chip ESD analysis solution.

Advanced technologies require different transistor architectures (such as FinFETs) and/or different substrates, such as SOI and/or silicon germanium (SiGe). ESD protection for FinFET technology is very challenging as FinFET devices show only a limited available silicon volume to carry the required ESD currents and dissipate the corresponding ESD power. Many of the initial roadblocks for ESD protection in FinFETs have already been removed [62]. SOI technologies need a different approach for the development of ESD protection devices, since their devices are built on an isolating substrate. The implementation of ESD protection diodes on SOI needs to change from the high-perimeter bulk-CMOS diodes to an SOI lateral-gated diode structure. SiGe technology has become another important alternative for high-speed communications and wireless applications. Since the change in material and mobility will also influence ESD, developing an ESD strategy for SiGe circuits will be very challenging. More about general ESD and ESD models can be found in [63] and [64]. Today, not only the adoption of 3D packaging solutions, such as stacked chiplets and HBM, but also the implementation

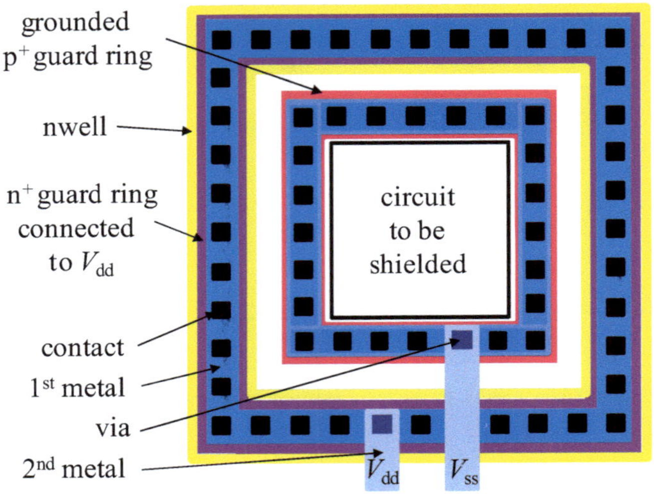

Fig. 9.41 Layout example of a guard ring to shield circuits from substrate noise, latch-up or other physical phenomena

of backside power distribution will have a serious impact on ESD behaviour. These topics are discussed in [65].

9.5.7 The Use of Guard Rings

All robustness phenomena discussed in this chapter, so far, require their own design strategy and solutions to prevent related problems and failures. There is one particular solution that can be used to solve a couple of design problems or at least reduces them. The *guard ring* was already mentioned during our discussions on the prevention of latch-up. Figure 9.41 shows an example of a guard ring. The target circuit is first surrounded with a p^+ guard ring, which makes a low-ohmic connection between the p-well and the ground (V_{ss}) net. It means that the p-wells in the circuit are now better shielded against substrate noise from other regions of the chip. Next, this p^+ guard ring is surrounded by an n^+ guard ring, which connects a surrounding n-well stripe to the supply (V_{dd}) net.

The minority carriers (electrons) are attracted to the n^+ ring, while the majority carriers (holes) flow to the p^+ ring. Together, they form a shield to noise coupling. Guard rings can be placed around the noise generating circuit as well as around the circuit that needs to be shielded. With a p^+ doping depth of several nanometers and an n-well of less than a micron deep, guard rings do not completely solve the problem. Parts of the substrate current and noise propagate under the rings and can still reach the circuit. It is therefore important to put the rings in close proximity of the noise sources on the chip. Most CMOS processes, today, support the use of a *triple-well (deep n-well)*. This allows putting each core on the chip in its own deep n-well, which also results in a better isolation between noise sources and noise-sensitive circuits. A model that predicts the noise suppression performance is described in [66].

9.5.8 Charge Injection During the Fabrication Process

Many IC processing steps use plasma-etching, sputter-etching or deposition techniques, which introduce a flow of charged particles towards the wafer surface. These charge particles are then collected by conducting surface materials (polysilicon, metals). This so-called antenna effect can create significant electrical fields across the thin gate oxides which can be stressed to such an extent that the transistor's reliability can no longer be guaranteed. It can also cause a threshold voltage shift, which affects the matching behaviour of transistor pairs in analog functions. It is industry practice to introduce additional "antenna design rules" to limit the ratio of antenna area to gate oxide area (see Chap. 3). The back end design tools can handle these design rules by limiting the maximum wire (antenna) length in the different metal layers. Also, protection diodes can be used in the library cells to shunt the transistor gates. Due to the trend in gate oxide thickness scaling, the appearance of the antenna effect is expected to have greater consequences for the design. The use of high-ε gate dielectrics in building the transistor stack would therefore also be beneficial to reduce this antenna effect.

9.5.9 Reliability Summary and Trends

Most of the previously discussed reliability topics depend on size, doping profiles and levels, voltages, temperatures and device materials. Scaling requires a change in many of these parameters and will therefore have severe effects on the reliability of CMOS devices and circuits. Moreover, in technologies with channel lengths below 45 nm, the transistors are increasingly built from a completely different stack of materials, compared to the traditional poly-Si (gate)-SiO_2 (gate oxide)-mono-Si (channel) stack. In some applications, the bulk silicon substrate will probably be replaced by SOI and/or SiGe. Due to the high leakage current, the SiO_2 gate oxide is being/has been replaced by a high-ε dielectric and because of gate depletion the polysilicon gate has already been replaced by a metal gate. This has an additional impact on the reliability of the devices and vice versa. The introduction of FinFETs has completely changed the transistor architecture: from planar to three dimensional. In fact, it looks like the transistor now lies on top of, rather than inside, the wafer, which increases the thermal resistance from the transistor to its substrate. Moreover, also the FinFET current density per unit area is more than two times higher than in lateral MOS devices. This may introduce *self-heating* effects in the transistor channels as FinFETs may locally show larger temperature extremes around the average die temperature, compared to lateral MOS devices. This creates problems with thermal reliability and requires new simulation models. While the generated heat in planar transistors is predominantly transferred through the bulk, the largest part of the FinFET-induced heat must be transferred through the top interconnect layers. The mean time to failures (MTBF) due to electromigration in metal lines on a die is inversely proportional to the temperature, meaning that a small temperature increase may dramatically reduce the lifetime of a chip. The FinFET architecture also affects other reliability issues, such as bias temperature instability (BTI) and hot-carrier injection (HCI). More detailed discussions on FinFET reliability issues are addressed in [67, 68] but are beyond the scope of this book.

Maintaining reliability at a sufficiently high level will put severe demands on this new transistor stack and makes the choice for the right materials a very difficult and cumbersome one.

9.6 Design Organisation

Robustness of nanometer CMOS designs not only impacts the level to which the chip operates with respect to the functional and electrical requirements. A robust design intrinsically includes a decent organisation of its *database* [69].

A very important requirement for increasing the integrity of the database is to enable quick design changes (be it in the final design stage or during a redesign). There are two requirements for the database with respect to design changes:

- It should take minimum effort.
- It should not introduce new errors caused by:
  - Unintended modifications
  - Forgetting something

These requirements even hold for design changes after one or more years. Current VLSI chips often reuse existing building blocks, such as multipliers, memories or microprocessor cores. This requires a structured and well-documented database set-up and design documentation.

Ten to thirty per cent of test engineer's time is lost as a result of incomplete documentation of the design. What really is required is:

- Good and complete specification.
- Complete (sufficient) test vectors.
- Mixed signal ICs: The test engineer must understand the complete IC.

Generally, the best solution for a database set-up is a hierarchical approach, in which one has:

- Directory hierarchy = design hierarchy (easy to find your way through)

- Good version management; what object (netlist and layout, etc.) is generated from which source (e.g. VHDL or Verilog description) and which one is the latest; use of make files!

A first-time-right design requires a full hundred per cent discipline, not only to fulfil the requirements of the specification but also to create a structured and logical database in order to reduce the chance of failures during (re-)design or during the creation of design derivatives.

9.7 Conclusions

After almost six decades of scaling, the robustness of integrated circuits has dramatically been affected by the consequences of the continuous increase of device and interconnect density. There are three major effects that contribute to the negative trend in the robustness of operation. The first one is the increase in IR drop, supply, substrate and ground bounce due to the increase of average current and current peaks and cross talk as a result of the increasing interference of the signal wires. The second one is the fact that the reduction of physical sizes of the devices and of the supply voltages has reduced the operating margins of the ICs and made them more sensitive to both internal and external influences. The third one is the increasing parameter variation, which is another source of nonideal circuit behaviour and also has a negative impact on the robustness of operation. The reduction of the device sizes in combination with the increased density has also made ICs to become more vulnerable to soft errors. All these effects tend to move in the wrong direction and have severe consequences for the design methods and flow to keep robustness at a sufficiently high level.

Until the change of the millennium, the reliability of ICs has not suffered dramatically from the continuous scaling of the technologies. This, however, has changed since the semiconductor industry has moved to new device concepts and new interconnect strategies. The impact of hot-carrier effects, BTI and ESD is very much dependent on the physical sizes and material properties of the transistor stack. In this respect, the alternative MOS transistor architectures, such as multi-gate devices and FinFETs, are very challenging regarding reliability topics. An accurate prediction of device behaviour with reliability models turns out to be very difficult. The combination of new materials and device stacks introduces new reliability problems, for which the models frequently need to be adapted and further improved.

Higher device densities and increased frequencies have led to a continuously increasing demand in IC current. At the same time, the metal widths and heights reduce, particularly of the lower metal layers. The corresponding expected increase in current density requires the permanent attention of the designers, to prevent electromigration to occur anywhere in the chip. Power estimation tools, capable of calculating the required currents, are therefore needed early in the design flow, to create a proper global supply network, while also paying attention to the local supply wire widths. Also, the adoption of 3D packaging solutions, such as stacked chiplets and HBM, as well as the implementation of backside power distribution will have severe consequences for the reliability of integrated systems.

From these considerations, it is clear that it will be a continuously challenging task for the designers to maintain a high robustness and reliability level in a "shrinking environment".

A robust design not only requires the integrity of the electrical and physical operation of the chip; it also includes the set-up of a very well-organised database. This allows easy, correct and rapid design modifications when redesigns or different versions of the design are required.

9.8 Exercises

1. Explain why the internal chip latch-up sensitivity will decrease every new process generation.
2. What are the main causes of supply noise inside a VLSI chip?
3. Explain why the power supply lines to a large driver circuit (e.g. clock driver or output driver) should be wider than the output signal track.
4. When we decide to reduce the thickness of the copper wires in a CMOS process, explain what would be the advantages and disadvantages in terms of signal integrity and reliability if we would not adapt our design to this reduction.
5. What is generally the best place to position the clock drivers and why?
6. What are the main causes of clock skew and what are the measures to reduce it?
7. Explain how the back end of the manufacturing process is dominating the IC behaviour.
8. Mention several reasons for increasing $\mathrm{d}i/\mathrm{d}t$. What are the consequences?
9. If you are facing IR drop problem, during power and signal integrity analysis, then how can you prevent/reduce it?
10. Explain why the use of a good database management system is required during the design of a VLSI chip.
11. Why would the implementation of an ESD protection be more problematic on SOI than on bulk-CMOS?
12. Explain the use of decoupling capacitors and why they are needed.
13. Assume a certain IP core is consuming 100 mA at 1 V. Assume also that the maximum allowed combined average voltage drop (IR drop) across the V_{dd} and V_{ss} supply lines in the top-level metal layer to this core is only 2% of the supply voltage. What would be the distance of the block to the supply bonding pads at which both the electromigration and the voltage drop requirements would exactly be fulfilled? Use the electromigration number of copper in the section on electromigration at 125 °C. The square resistance of copper is $R_{\square} = 22\,m\Omega/\square$ for the top-level metal ($R = R_{\square} \cdot (L/W)$).
14. What would be the difference between a synchronous and an asynchronous implementation of a logic core in terms of signal integrity?

15. What would be the effect of a small forward body bias on the performance of a transistor? What would be the risks of using forward body bias with respect to reliability?
16. Suppose that a chip requires an additional on-chip decoupling capacitance. How would you implement this capacitance to achieve the largest capacitance density (i.e. capacitance value per unit area)?
17. Explain why a company may first order several corner lots before they ask for high-volume production.

References

1. B. Barton et al., *ESSCIRC, low-power workshop 1997*, Southampton
2. P.J. Resle et al., A clock distribution network for microprocessors. IEEE J. Solid-State Circuits **36**(5), 792–799 (2001)
3. S. Rusu, Clock generation and distribution for high-performance processors. SoC 2004, http://www.tkt.cs.tut.fi/kurssit/8404941/S04/chapter5.pdf
4. S. Henzler, Digital system clocking, in *High-Speed Digital CMOS Circuits*, Technical University, Munich, Summer Term 2015
5. K. Bernstein et al., *High-Speed CMOS Design Styles* (Kluwer Academic Publishers, Dordrecht, The Netherlands, 1999)
6. S. Rusu, Circuit design challenges for integrated systems, in *Workshop on Integrated Systems, European Solid-State Circuits Conference*, Sep 1999
7. C. Kim et al., *High Bandwidth Memory Interface*. Springer Briefs in Electrical and Computer Engineering (Springer, NewYork, 2014)
8. H. Basit et al., Practical multi-gigahertz clocks for ASIC and COT designs, in *DesignCon*, 2004
9. L. Peeters, Challenges in backside power delivery, in *Semiconductor Engineering*, 17 Nov 2022
10. D. Patel, Imec demonstrates backside power delivery network for enhanced system performance, in *EEPOWER*, 29 June 2021
11. M.O. Hossen et al., Power delivery network (PDN) modeling for backside-PDN configurations with buried power rails and μTSVs. IEEE Trans. Electron Devices, **67**(1), January 2020
12. A. Jourdain et al., Buried power rails and nano-scale TSV: technology boosters for backside power delivery network and 3D heterogeneous integration, in *2022 IEEE 72nd Electronic Components and Technology Conference*
13. S.K. More, Intel is all-in on backside power delivery, in *IEEE Spectrum*, June 2023
14. B. Nauta, G. Hoogzaad, How to deal with substrate noise in analog CMOS circuits, in *European Conference on Circuit Theory and Design*, Budapest, Sep 1997
15. H.B. Bakoglu, *Circuits, Interconnections and Packaging for VLSI* (Addison-Wesley, Boston, USA, 1990)
16. E. Dupont et al., Embedded robustness IPs for transient-error-free ICs. IEEE Design Test Comput. **19**(3), 56–70 (2002)
17. T. Heijmen et al., Soft-error rate testing of deep-submicron integrated circuits, in *Test Symposium*, ETS'06, 2006
18. M. Derby, Soft-error impacts on design for reliability technologies, in *Keynote talk at IOLTS*, July 2007
19. D. Lee et al., Enhanced soft error rate estimation technique for aerospace electronics safety design via emulation fault injection. J. Appl. Sci. **14**(4) (2024)
20. Y. Ko, Survey of software-implemented soft error protection. J. Electron. **11**(3) (2022)
21. V. Petrescu et al., A Signal Integrity Self Test (SIST) concept for the debug of nanometer CMOS ICs, in *ISSCC 2006*, Digest of Technical Papers, session 29 (2006)

22. P. Drennan et al., Implications of proximity effects for analog design, in *IEEE 2006 CICC conference*
23. J.M. Brunet, Modelling process variability in the design flow, in *Chip Design Magazine*, Dec 2005/Jan 2006
24. M. Vertregt, The analog challenge of nanometer CMOS, in *IEDM 2006*, Digest of Technical Papers, Dec 2006, pp. 11–18
25. M. Pelgrom et al., Transistor matching in analog CMOS applications, in *International Electron Device Meeting (IEDM)* 1998, pp. 915–918
26. V. Moroz, FinFET structure design and variability analysis enabled by TCAD, *EE|Times*, 10 Aug 2012
27. A.R. Brown et al., Impact of metal gate granularity on threshold voltage variability A full-scale three-dimensional statistical simulation study. IEEE Electron Device Lett. **31**(11), 1199–1201 (2010)
28. S.-K. Kim, Line-edge roughness from extreme ultraviolet lithography to fin-field-effect-transistor: computational study. Micromachines (Basel), 30 Nov 2021
29. T. Kanamoto et al., Impact of well edge proximity effect on timing, in *ESSCIRC 2007*, Digest of technical papers, 2007, pp. 115–118
30. K. Qian, Variability modeling and statistical parameter extraction for CMOS devices. Dissertation No. UCB/EECS-2015-165, Electrical Engineering and Computer Sciences, University of California at Berkeley, June 2015
31. N. Damrongplasit, *Study of Variability in Advanced Transistor Technologies*, Electrical Engineering and Computer Sciences, University of California at Berkeley, fall 2014
32. M. Vertregt, Embedded analog technology, in *IEDM short course on system-on-a-chip technology*, 5 Dec 1999
33. P. Stolk et al., Modeling statistical dopant fluctuations in MOS transistors. IEEE Trans. Electron Devices **45**(9) (1998)
34. M. Pelgrom, et al., Digital circuit insights from analog experiences, in *ISSCC 2007*, Special Topic Evening Sessions, 2007
35. A. Dadheech et al., Leakage power optimization for 28 nm and beyond, in *EDN*, 7 Apr 2014
36. A. Agarwal et al., Statistical timing analysis using bounds, in *Proceedings of the Design, Automation and Test in Europe Conference and Exhibition (DATE'03)* (2003), pp. 62–67
37. J.-J. Liou et al., Fast statistical timing analysis by probabilistic event propagation, in *DAC 2001*, Las Vegas, June 2001
38. A.M. Baker, Y. Jiang, Modeling and architectural simulations of the statistical static timing analysis of the non-Gaussian variation sources for VLSI circuits. Int. J. Sci. Res. Publ. **3**(1), 1 (2013). ISSN 2250-3153
39. J. Chen, M. Tehranipoor, Critical paths selection and test cost reduction considering process variations, in *2013 22nd Asian Test Symposium*, 18–21 Nov 2013, pp. 259–264
40. A. Mutschler, Process corner explosion, in *Semiconductor Engineering*, 13 Sep 2018. https://semiengineering.com/process-corner-explosion/
41. H.J.M. Veendrick, Wire self-heating in supply lines on bulk-CMOS ICs, in *ESSCIRC 2002*, Digest of Technical Papers, Sep 2002, pp. 199–202
42. G.D. Wilk, et al., High-k dielectrics: current status and materials properties considerations. J. Appl. Phys. **89**(10), 5243–5275 (2001)
43. A. Kottantharayil, Low-Voltage Hot-Carrier Issues in Deep-sub-micron MOSFETs. Thesis, University of Munich, 2001, http://137.193.200.177/ediss/kottantharayil-anil/inhalt.pdf
44. S. Mahaptra et al., Device scaling effects on hot-carrier induced interface and oxide-trapped charge distributions in MOSFETs. IEEE Trans. Electron Devices **47**(4) (2000)
45. S. Tyaginov et al., Understanding and modeling the temperature behavior of hot-carrier degradation in SiON nMOSFETs. IEEE Electron Device Lett. **37**(1), 84–87 (2016)
46. K. Kushida-Abdelghafar et al., Effect of nitrogen at SiO_2-Si interface on reliability issues negative bias temperature instability and Fowler-Nordheim stress degradation. Appl. Phys. Lett. **81**(23) (2002)

47. Y. Hiruta et al., Interface state generation under long-term positive-bias temperature stress for a p+ poly gate MOS structure. IEEE TED **36**, 1732 (1989)
48. T.B. Hook et al., The effect of fluorine on parametric and reliability in a 0.18 μm, 3.5/6.8 nm dual gate oxide CMOS technology. IEEE TED **48**(7), 1346 (2001)
49. Ogawa et al., Interface-trap generartion at ultrathin (4–6 nm) interfaces during negative-bias temperature aging. JAP **77**(3) (1995)
50. A. Scarpa et al., Effect of the process flow on negative-bias-temperature-instability, in *Proceedings of 8th International Symposium on Process- and Plasma-Induced Damage*, 2003, p. 142
51. P. Chaparala et al., NBTI in dual gate oxide PMOSFETs, in *Proceedings of 8th International Symposium on Process- and Plasma-Induced Damage*, 2003, p. 138
52. P. Rani et al., Impact of negative bias temperature instability on 6T CMOS SRAM cell performance. Int. J. Comput. Appl. (0975 – 8887) **128**(12) (2015)
53. S. Han et al., In-depth analysis of NBTI at 2X nm node DRAM, in *2016 IEEE 8th International Memory Workshop (IMW)*
54. S. Mahapatra (ed.), *Fundamentals of Bias Temperature Instability in MOS Transistors*. Springer Series in Advanced Microelectronics (Springer, 2016), ISBN 978-81-322-2507-2
55. C.D. Young et al., Investigation of negative bias temperature instability dependence on fin width of silicon-on-insulator-fin-based field effect transistors. J. Appl. Phys. **117**, 034501 (2015)
56. B. Linder et al., Process optimizations for NBTI/PBTI for future replacement metal gate technologies, in *International Reliability Physics Symposium*, 2016
57. S. Mahapatra et al., Modeling of NBTI using BAT framework: DC-AC stress-recovery kinetics, material, and process dependence. IEEE Trans. Device Mater. Reliab. **20**(1), 4–23 (2020)
58. P.R. Genssler et al., Modeling and predicting transistor aging under workload dependency using machine Learning. IEEE Trans. Circuits Syst. I: Regul. Pap. **70**(9) (2023)
59. R.R. Troutman, *Latchup in CMOS Technology* (Kluwer Academic Publishers, 1986), ISBN 0-89838-215-7
60. F. Farbiz, Modeling and suppression of latch-up. PhD dissertation, Department of Electrical and Computer Engineering, University of Illinois at Urbana-Champaign, Urbana, 2010
61. http://www.esdlab.com/others.htm
62. J.-H. Lee et al., ESD in FinFET technologies: Past learning and emerging challenges, in *IEEE International Reliability Physics Symposium (IRPS)*, 2013
63. A. Ameraskera, C. Duvvury, *ESD in Silicon Integrated Circuits* (Wiley, 2002), ISBN 0-471-95481-0
64. M.D. Ker et al., ESD test methods on integrated circuits; An overview, in *8th IEEE International Conference on Electronics, Circuits and Systems*, vol. 2 (2001), pp. 1011–1014
65. W.-C. Chen et al., How to protect advanced CMOS technologies with thin Si substrates against ESD events, https://www.imec-int.com/en/articles/how-protect-advanced-cmos-technologies-thin-si-substrates-against-esd-events, 8 Feb 2024
66. M. Shen et al., Modeling and design Guidelines for P guard rings in Lightly doped CMOS substrates. IEEE Trans. Electron Devices **60**(9), 2854–2861 (2013)
67. S. Uppal, FinFET reliability, in *2015 IEEE International Integrated Reliability Workshop (IIRW)*, Oct 2015
68. M.I. Khan, Self-heating and reliability issues in FinFET and 3D ICs, in *12th IEEE International Conference on Solid-State and Integrated Circuit Technology (ICSICT)*, Oct 2014
69. C.A. Chami, Keeping your design files organized, SemiWiki.com, 07 Sep 2016

Testing, Yield, Packaging, Debug and Failure Analysis

10

10.1 Introduction

Although this is almost the final chapter in this book, it does not mean that the topics discussed here are less important than those of the previous chapters.

Testing, debugging, yield and packaging have become a substantial part of the ultimate costs and quality of a chip. An overview of the most important issues related to these topics is therefore included in this chapter.

An integrated circuit can fall victim to a large variety of failure mechanisms. Ideally, the related problems are detected early in the manufacturing process. However, some only show up during the final tests, or even worse, they might not be identified before the chip is soldered on a customer's board, leading to customer returns.

The next paragraph starts with an overview of different test methods currently in use and continues with the measures that a designer can implement to improve the testability and support the debug of his design.

The engineering and evaluation of first silicon until it is considered to be 'error-free' happens to be a tough job. Programmable processors, for example, may be used in an almost unlimited number of different applications. It is almost impossible to give a 100% operation guarantee, even for the fifth respin of silicon for these kinds of ICs.

Even when a failure is detected during the testing of first silicon, it might take a considerable time before the cause of failure is located and proven. This is because some complex ICs contain more than a hundred billion transistors and over a thousand I/O pins. It is therefore very complex to locate an internal failure via a limited number of external pins (I/Os). Moreover, because of the increased number of interconnection (metal) layers, physical probing of signals has almost become impossible. Design for test and debug is therefore adopted as a general design approach, to enhance controllability and observability and to ease the detection of design bugs and other failure mechanisms during the engineering phase of first silicon.

H. Veendrick, *Nanometer CMOS ICs*,
https://doi.org/10.1007/978-3-031-64249-4_10

Eventually, the results of all tests determine which of the dies pass and which fail. The related yield is a function of the test coverage. A flavour of the most important yield topics is therefore part of the discussions.

Many of the chip characteristics in terms of performance, robustness and reliability are co-defined by the way the chip is designed and packaged. This chapter, therefore, also contains a summary of the most commonly used packages. It includes a few presentations that describe the influence of the package on some of the electrical and thermal characteristics of the chip. Also trends in SoC and SiP integration solutions are identified.

Finally, this chapter is concluded by a presentation of failure analysis methods that support the detection of failures and their diagnosis to enable a fast identification of the failure mechanism during first silicon debug. This should prevent customer returns and will shorten time to market.

10.2 Testing

Testing is done to bridge the gap between customer requirements and the quality of the design in combination with the manufacturing process. Testing thus helps to increase the quality of an IC. The yield is determined by testing and can be influenced by the complexity of the test: a simple test may lead to a higher yield but can lead to more *customer returns*. The yield for large and complex ICs can be relatively low and can dominate the ultimate costs. How extensive a design needs to be tested depends on many different factors. Three major different test categories can be distinguished.

Characterisation tests, which are mostly executed manually, are developed for characterising the chip operation and/or operation area with respect to different operation conditions. These tests are more focused on the accuracy of the test than on the speed of testing. These tests are very much related to which parameters and what type of circuits are to be characterised:

- Design errors, design margins, manufacturing defects.
- The on-chip circuits: pure logic, pure analogue, pure memory, static circuit, dynamic circuit or a mix of analogue, RF and digital circuits.
- An increasing number of chip failures are related to the dynamic effects such as cross-talk, charge sharing, critical timing and noise. Most automatically generated tests detect the 'stuck-at-one' and 'stuck-at-zero' faults, which cause circuit nodes to remain at '0' and '1', respectively. Nanometer CMOS technologies, however, result in lower supply voltages and, consequently, reduced noise margins. This will produce faults that are much more difficult to classify than the traditional stuck-at faults.

Production tests, which are performed in an automated mode, provide a way to reject those ICs that do not meet the required specification criteria or performance limits. Production tests include a large number of different tests to achieve the best possible test coverage and depend on the quality requirements of the target product and/or target application area:

- Consumer
- Computer
- Aviation: aircraft and spacecraft
- Automotive

Due to safety requirements it will be clear that the latter two require higher test coverage.

Reliability tests, which are mostly performed manually, challenge the chip operation during and after the exposure to extreme electrical and environmental conditions:

- Electrical stress, burn-in.
- Temperature cycles, thermal shock and high-temperature storage.
- Increased humidity levels.
- Mechanical vibrations and shocks.
- Other reliability tests include electrostatic discharge (ESD) and latch-up, but since these are very much design related, they are discussed in Chap. 9.
- Pressure pot (PPOT) (autoclave); testing moisture resistance of plastic encapsulated devices.
- (Bond) wire pull.

Because of safety requirements, aviation, navigation and medical applications usually require very exhaustive testing particularly with respect to reliability standards. For the same reason, also automotive products require exhaustive testing. Moreover, they operate in more 'hostile' environments which may include large supply transients or interference caused by switching of heavy or inductive loads such as lamps and starter motors and require specific protection and more stringent reliability tests.

When all tests are executed properly, only a few of the chips that pass all tests still may be returned by the customer (*customer returns; escapes*) because of a failure, which displayed itself in the application either directly or after a while (day, week, month, year). The number of customer returns is expressed in *ppm* (parts per million), which represents the ratio of customer returns per million supplied chips. This ppm level has become representative for the quality of a delivered product. Ppm acceptance levels are related to the quality requirements of the application domain. While typical automotive applications allow between 1 and 5 ppm, consumer applications and microprocessors may show ppm levels of 100 and 300, respectively.

Tests can be performed both on the wafer and on the final packaged product. Pretests, also known as *e-sort* (electrical sort), are usually performed directly on a wafer to prevent unnecessary assembly costs. The final tests are performed on the packaged die. There is often a lot of overlap between the pretests and the final tests.

Fig. 10.1 Example of a probe card (Source: MICRAM Microelectronic GmbH)

As a result of the associated additional costs, the number of redundant tests must be limited. During pretest, the individual ICs are tested on the wafer by probing the bond pads of the chip. Figure 10.1 shows a photograph of an example of a probe card with more than 120 probes.

A probe station brings these small needles into contact with the IC's bond pads. During probing, debris from the bond pads may contaminate the probe tips, which requires minor, on-line, cleaning in combination with a complete thorough cleaning and recalibration operation typically after very 50,000 touchdowns. To reduce test time for stand-alone memories, *multi-DUT probe cards* are used to simultaneously test, e.g. 32 memory dies on the wafer. With the recent accelerated transition to the D1y and D1α nanometer process nodes (see second bullet below Fig. 6.22), die count per wafer is increasing rapidly. As a result, full-wafer DRAM probe cards that simultaneously test every die on the wafer must keep pace with this increase. Formfactor's Matrix^{TM} architecture provides test parallelism in excess of 3000 dies per wafer on a single touchdown [1]. The company also produces full-wafer probing solutions for NAND and NOR flash memories.

A test system provides predetermined stimuli for the IC and compares actual output signals to expected responses (Fig. 10.2). The stimuli should ensure that a large percentage of possible faults will result in discrepancies. This percentage is called fault coverage with respect to the applied fault model and is most commonly targeted at above 99%.

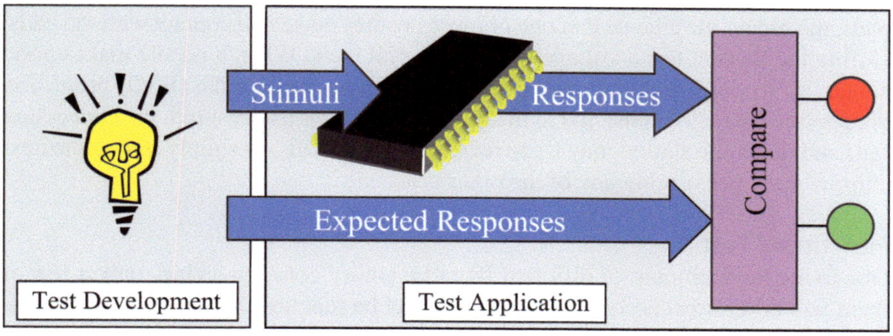

Fig. 10.2 Basic principle of IC testing

The fault coverage, however, is always related to the fault model used (stuck-at, bridging, stuck-open, transition [only once], gate delay and path delay). Redundancy can also be reason for reduced fault coverage. The test stimuli and response signals are transferred through a connector that provides a bidirectional link between the probes and the test system. The test system, *automatic test equipment (ATE)*, can also be used to control the wafer prober or handler to automatically step from one circuit to another so that a number of ICs can be tested in rapid sequence.

It was relatively easy to manually determine test stimuli (vectors) for complete SSI (small-scale integration) circuits. For VLSI circuits, however, this is impracticable and has led to the development of computer programs that generate test vectors. A complete test program may consist of several subtests, of which the most important ones are discussed in the next paragraph.

The quest for high bit or gate densities consumes much design effort aimed at the realisation of a maximum amount of electronics on a minimum area. However, designers must ensure that their circuits are testable. For VLSI circuits, an increase in testability may, for instance, result in a chip area 'sacrifice' of 5% and a 50% reduction in test costs.

10.2.1 Basic IC Tests

This section discusses the most important different tests that are applied to a chip to achieve sufficient test coverage. Some of these tests are done at chip level, some at core level, using so-called test protocols. A test protocol is a detailed description of how the test stimuli must be supplied (to which terminals of the core and in which time slots) and how the responses of the core to these stimuli must be captured.

Contact Test

As discussed before, pretests are performed directly on the wafer to prevent unnecessary assembly costs. To perform the test, the needles of the probe card are brought into contact with the bond pads of the chip. The larger the number of bond pads, the bigger the chance that one or more needles make bad contact with the pads. During the contact test a voltage is applied to all pads. When a needle makes good contact to its corresponding pad, a current will flow through the (ESD) protection diodes (see Chap. 9) connected to that pad. If no current is measured, the connection fails and the probe station may try to reconnect to the pads, or simply step to the next chip, without performing any of the other tests.

Functional Test

Due to the large amount of different IP and memory cores on a chip, only a few of them will have direct access to the pins and can be functionally tested. A functional test reuses stimuli from the design simulation phase which exercise the function of the circuit. Preferably, the whole function of the IC should be tested similar to the way it is used in the application. This full functional test should be performed at application speed, since only then the IC will show its real application behaviour. Some IC problems only manifest themselves if all circuits operate simultaneously and identical to the application:

- Voltage drop along supply lines
- Supply and ground bounce due to peak currents through the supply lines bond Wires and package leads
- Cross-talk between neighbouring signals
- EMC behaviour

The on-chip noise is maximal when all circuits are active, just like in the application. When a circuit fails, it may not directly be an indication for a bad chip or bad design. It may also lead to the conclusion that the simulation is not correct. In many cases, however, the malfunction may be caused by a timing-critical failure, a noise-related operation failure, a process-related defect causing (resistive) shorts or opens or a design error. Debug and failure analysis techniques are used to locate the failure and support the diagnosis. These are discussed in the final subsection in this chapter.

A test is not only used as a method to verify the correct behaviour of an IC. It is also used as a quality measure for its operating area. A Shmoo plot can also be used for this purpose. A Shmoo plot is a graph that represents when a certain test passes or fails with respect to a large number of parameter settings at which the chip is repeatedly tested. Figure 10.3 shows an example of a Shmoo plot in which both the supply voltage and frequency are varied over certain ranges around the spec area. The shape of the Shmoo plot may contain information about the cause of the failure and will be discussed in more detail in Sect. 10.6.3. The Shmoo plot must also be measured for the worst- and best-case temperatures and the way it changes its shape is representative for the operation of the chip (see section on failure analysis).

Let us assume that the Shmoo plot in this figure represents that of first silicon. When the margins of the operating area, with respect to the spec area, are small, there is a big chance that one or more of the following batches will show failing circuits. In other words, this kind of test allows you to anticipate already on what problems may show up in future wafer batches, and so, debug and failure analysis

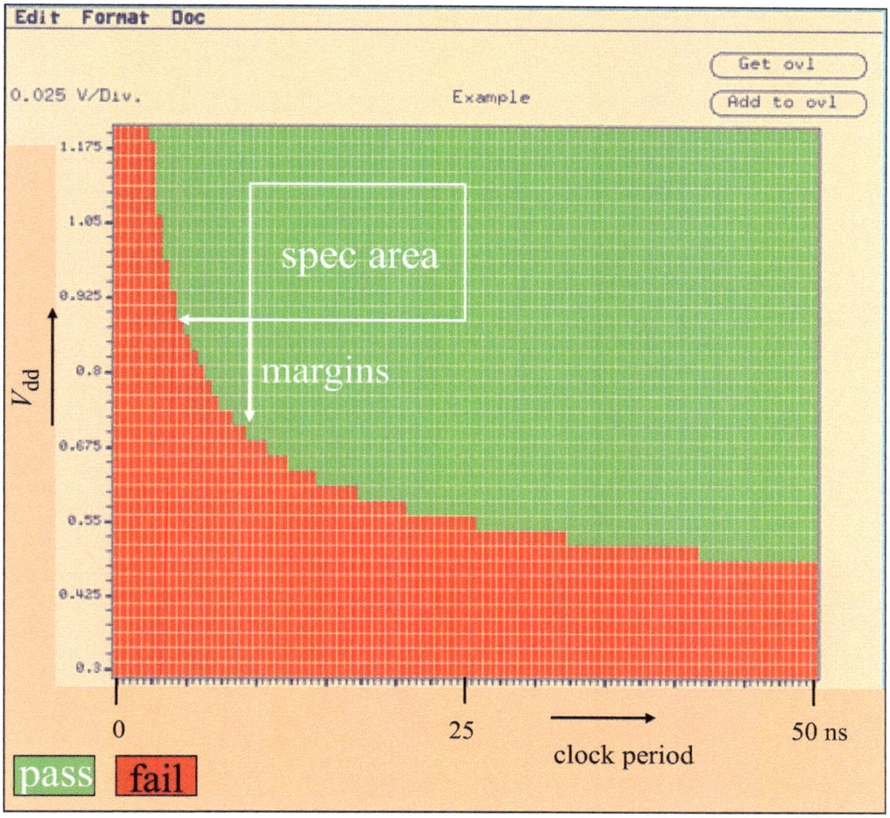

Fig. 10.3 Shmoo plot example, showing the margins between the measured operating and the spec area

can already be started to determine the cause and location of the of the circuit that causes the small margin.

Delay Fault Test

For a digital circuit to operate correctly, it is required that the propagation delay along all of its logic paths is more than, or less than, a specified limit. Random process parameter variations and/or defects may cause propagation delays to exceed these specified limits, leading to operation faults. These faults can only be detected by a form of so-called at-speed testing. At-speed tests were only very scarcely applied and usually carried out with functional test patterns. The increasing IC complexity, in terms of both components and speed, has made at-speed functional testing unattractive mainly due to the test development costs and the cost of a complex test system with a very high timing accuracy. *Structural test* techniques, such as scan tests, which were originally developed for relatively slow stuck-at fault testing, are forming today a viable cost-efficient alternative to at-speed testing. Delay fault testing offers a structural approach to at-speed timing tests while keeping test hardware costs limited.

A *delay fault* usually refers to a single logic gate or a logic path exceeding its maximum specified propagation delay. The output of a logic gate responds to a transition of one or more of its inputs. The time such a response takes is called *gate delay*. *Path delay* represents the accumulated gate delay plus the interconnect delays within that path.

There are two fault models related to delay fault testing: the transition fault model and the path delay model. Transition faults, also called gate delay faults, model defects which occur at the inputs or outputs of a logic gate and which lead to a gate delay outside its specified range. Path delay faults model defects that cause signal propagation along the path that takes longer than the specified maximum time (usually defined by the clock period).

A *transition delay fault* is caused by random variations occurring both in the devices and interconnections. Examples are as follows: threshold voltage shifts, CMOS opens, highly resistive vias, narrow metal lines, signal integrity causes such as cross-talk, supply noise and IR drop.

Test pattern generation for delay fault testing (for transition delay and path delay faults) is based on stuck-at procedures and requires two tests. The first test puts the targeted faulty circuit path in a certain state, while the second test introduces an input transition, such that it propagates to one or more primary outputs or scan flip-flops. Scan chains are normally used to guide the stimulus patterns to the faulty path, which may have been identified through static timing analysis. The responses of the faulty path to these input stimuli are captured by the output scan chain. It is virtually impossible to detect all possible delay faults. Therefore, several techniques have been proposed to enhance delay fault coverage for standard scan designs [2]. However, a discussion of these techniques is beyond the scope of this book.

A complete functional test of an IC or of parts of the IC is often complex and time-consuming. Therefore, a structural test approach is applied to most of today's ICs. This type of test is also called scan test.

Scan Test (Structural Test)

As already stated before, the complexity of today's ICs is so high that most cores have no direct access to the pins. To be able to test whether these cores show correct behaviour, we have to create access to them by artificial means. A synchronous design is built from logic gates and flip-flops, which are used to control the data flow through a chip. The flip-flops are controlled by a clock signal and temporarily store a logic value and let it go on clock demand. So, in normal operation, the flip-flops are an essential part of the total function of a core. But, during test mode, these flip-flops are put in series, thereby bypassing the logic path in between them. As such, the flip-flops form a so-called *scan chain*, which operates as a shift register to guide the data to target nodes in a logic core. Each *scannable flip-flop* is therefore equipped with an additional multiplexer to enable connection to the logic path

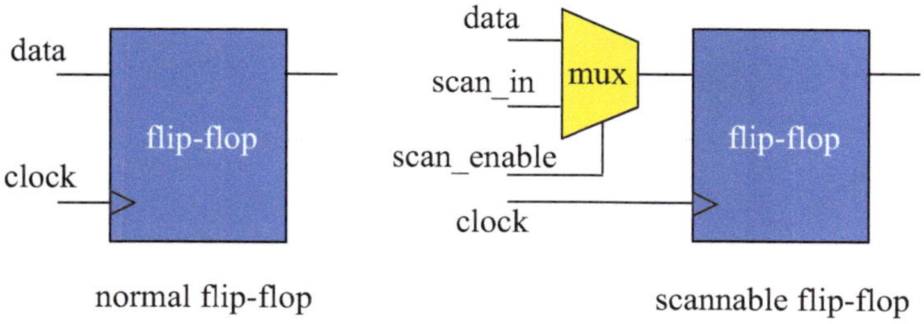

Fig. 10.4 Example of the use of a scannable flip-flop

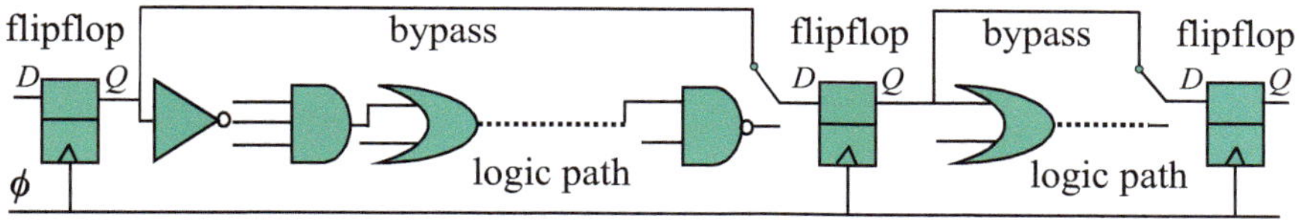

Fig. 10.5 Example of how a scan chain architecture is implemented

during normal operation and connection to the previous flip-flop in the chain during the test (Fig. 10.4).

This multiplexer is controlled by the *scan_enable* signal, which needs to be routed through the chip as well. Figure 10.5 shows how flip-flops in the same scan chain are connected. In normal operation mode the flip-flop input selector is connected to the logic gate. In the scan mode, this input selector is connected to the previous flip-flop, whereby the logic is bypassed.

Different cores can be connected to the same scan chain (see Fig. 10.6 for a scan chain example). Scan tests are performed to detect manufacturing defects or design errors in combinatorial logic cores. An *automatic test pattern generator (ATPG)* tool tries to excite every node within the logic core by generating input test vectors at the inputs of the scan chains.

To test one of the cores in a chain, the following sequence of operations is executed:

- Put the chip in the scan-test mode: *scan_enable* goes high to form the scan chain.
- Guide data through the scan chain to the flip-flops in the target core.
- Put the core in functional mode for performing a single clock cycle.
- Put the chip in test mode again.
- Scan out the results of the functional operation.
- Compare these results with the expected data (most commonly from simulations).

A test may activate many scan chains (10–200) in parallel to reduce test time. Their number is limited by the number of available (reused) I/O pads of the application. Overall accessibility is ensured because each flip-flop in each logic block is part of

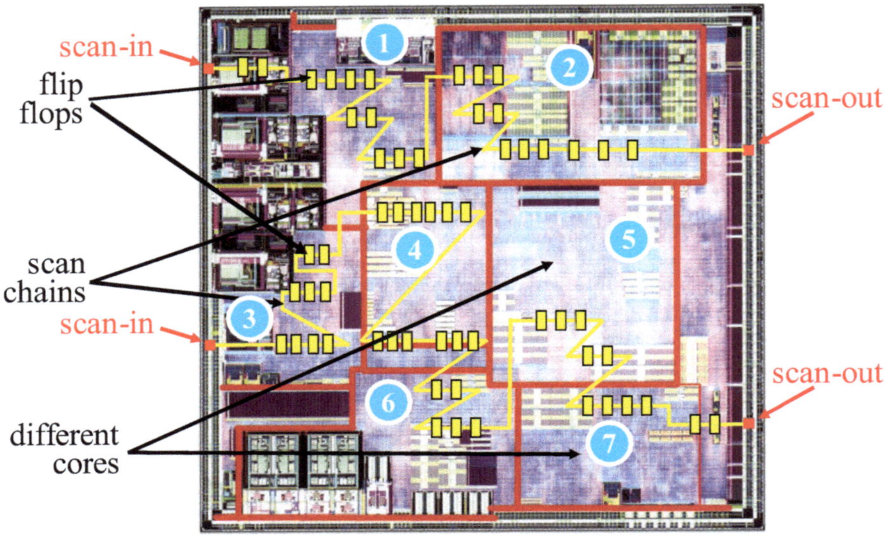

Fig. 10.6 Example of a scan chain in a complex VLSI circuit

a scan chain. The total chip area overhead to support the scan test (the multiplexer in each flip-flop, the routing of the scan data, *scan* and *scanb* signals and the test-control block [TCB]) is typically less than 5%. Today, most scan tests are packetised and execution is controlled by a scheduler (no longer synchronous to the outside world).

The I_{ddq} test, which is discussed next, uses the scan-test infrastructure to put the chip in different states, in order to detect defects or faults.

I_{ddq} and ΔI_{ddq} Test

During the eighties, the IC testing was based on *stuck-at fault* models, which could detect failures at logic gates and flip-flops when their outputs were short-circuited to V_{dd} or ground (*stuck-at fault*: stuck-at-one or stuck-at-zero, respectively), or floating (*stuck-open fault*). However, with these simple models, it was not possible to cover all process-oriented defects. I_{ddq} tests are particularly good at detecting bridging faults, power supply short circuits and punch-through failures.

In a normal static CMOS logic gate, either the pMOS pull-up network is conducting, keeping the output at high level (logic '1'), or the nMOS pull-down network is conducting, keeping the output at low level (logic '0'). In the steady state, no current usually flows through such a logic gate, except for a negligibly small subthreshold leakage current of the logic gates. Above 0.25 μm CMOS technologies, the magnitude of this leakage current was usually below 1 μA. At such a level of background current, larger steady-state currents, caused by different process-defect mechanisms, can easily be detected by measuring, as these currents are several orders of magnitude higher than the leakage current.

For example, common gate oxide defects may result in current values in the order of microamperes to several milli-amperes, depending on the size of the defect and on the size of the transistor involved. A drain-source bridging defect can easily cause steady-state currents up to several milli-amperes as well.

However, a *defect* is not always leading to a fault. It may cause a *structural fault* when it is large enough to connect two neighbouring conductors or disconnect a continuous pattern. Only faults lead to yield loss. Some defects do not lead to structural faults, but only to *parametric faults* which may affect the circuit performance.

During the measurement of the *steady-state current*, the chip has to be put in the steady-state mode. In many CMOS ICs, this state can be achieved by just switching off the clock. However, the chip often has to be put in a special I_{ddq} test mode before switching off the clock. In this way, defects are detected by the level of the supply current during the steady state. This is called I_{ddq} testing. I_{ddq} test pattern generation is only needed to put the chip or different parts of the chip in a certain mode (controllability). Observability need not be supported, as results of the test are simply measured via I_{ddq} currents. Because the current needs to settle during the measurement, I_{ddq} testing is a relatively slow process.

Especially in circuits that contain non-static CMOS circuits, such as PLLs, A/D, D/A and other analogue circuits and floating nodes (e.g. tristate buses), dynamic and pseudo-nMOS circuits need additional attention during the design to make the total chip I_{ddq} testable.

At which I_{ddq} level the chip should be considered defect depends on many things. The number of gates is one important parameter, while the level of the threshold voltages of the nMOS and pMOS transistors is also dominant in determining the critical I_{ddq} level. Because of scaling, the threshold voltage is reduced every process generation, to maintain or increase the speed of each new generation of ICs. As shown in Chap. 8, the subthreshold current in a transistor is defined by the *subthreshold slope* (Chap. 2) of the device. A typical value of 80 mV/dec, for bulk-CMOS devices, leads to an increase of about a factor eighteen for each threshold voltage reduction of 100 mV.

CMOS technologies beyond 120 nm exhibit even more leakage mechanisms. Next to subthreshold leakage also gate oxide and junction leakage start playing a role. Because of the increased leakage levels, I_{ddq} testability is therefore no longer possible for most ICs made in 120-nm CMOS technologies and beyond. An alternative, in this respect, is the ΔI_{ddq} test. During this test, the chip is put into several different states, by scanning a variety of test vectors through the scan chain. After every new test vector, the chip is put into the corresponding quiescent state and the I_{ddq} is measured. Next, all I_{ddq} values are compared with each other and so the ΔI_{ddq} is determined. Figure 10.7 shows some measurement results for three ICs.

These ICs may originate from different batches. A small threshold-voltage shift can cause a large change in leakage current. Different test vectors are needed to put the chip in such a state that a defect is detected. In Fig. 10.7a the short will only lead to a defect-oriented current when the logic levels across the short are different.

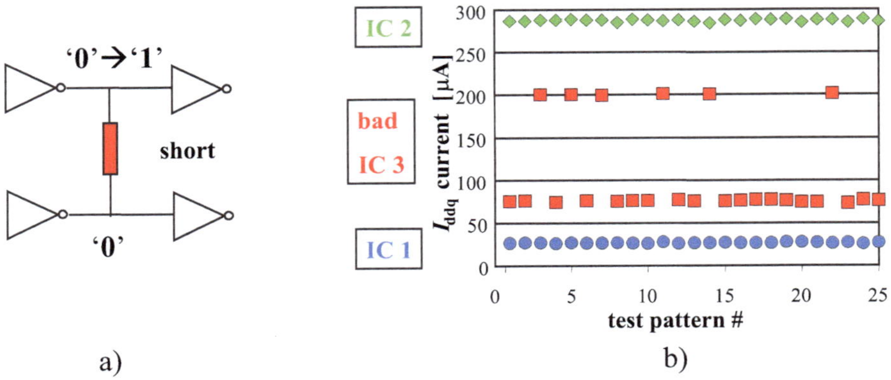

Fig. 10.7 Leakage measurements for different test patterns in different ICs

Although IC 1 and IC 2 show a relatively large difference between their I_{ddq} values, these values are independent of the state (test vector) of the chip. The bad IC 3, however, shows different values of the I_{ddq} current, which means that certain test vectors bring the chip in a state in which it manifests a short. The rejection criterion is not the absolute value of I_{ddq}, but the difference between I_{ddq} values of several measurements. Therefore, this test is called ΔI_{ddq}.

Very-Low-Voltage (VLV) Testing

Bridging, *gate-delay* and *path-delay* faults may not always be detected. These *delay faults* are mostly caused by opens (e.g. a bad via, which causes too much contact resistance). In noncritical delay paths, these faults may not be detected, but, then, they may not lead to functional errors and can thus be tolerated. When we define the *golden device* to be a product that operates perfectly according to the specification and even at voltages well below the worst-case specified levels, then *VLV (very-low-voltage)* tests can be performed on the other devices to detect weak (high-ohmic) shorts. At such low-voltage levels, these shorts can easily cause a much longer delay, leading to a detectable fault. Figure 10.8 shows an example of voltage-based testing of shorts.

Assume that, due to a manufacturing defect, there is a high-ohmic short (or bridge) between the drain and source of the transistor as indicated in the figure. When switching the gate of this transistor from a logic 'one' to a logic 'zero', it will not switch completely off. With no bridge, the output of the two-input NAND gate would switch to the supply voltage, indicated as 'normal' in the upper diagram. However, dependent on the resistance of the bridge, this output will not reach this level. Shorts with more than 2-kΩ resistance will only manifest themselves as additional gate delay at the output of the inverter connected to the NAND (lower diagram). The fault will manifest itself as an additional path delay.

Figure 10.9 shows an experimental Shmoo plot measurement using VLV testing of a resistive short in an inverter.

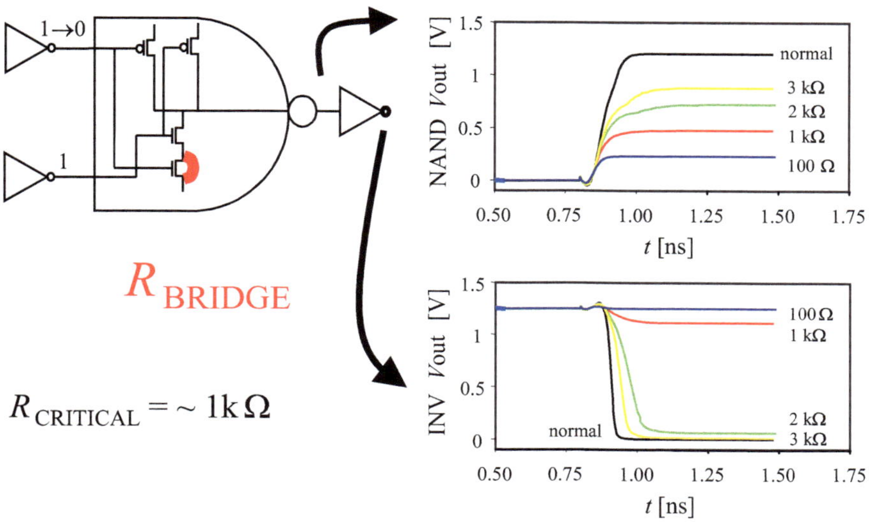

Fig. 10.8 Example of voltage-based testing of shorts

The Shmoo plot clearly shows that the operating area reduces when the short becomes stronger (less resistive). At a resistive value of 300 Ω, the circuit does not operate at voltages below 0.525 V, even at frequencies of only 20 MHz.

BIST

The costs of testing will dramatically increase as a result of the increase in the speed of the circuits, the reduction of the voltages (smaller noise margins) and the increase in the number of bond pads. The cost of a tester will increase from a few million to more than 10 million US$ in the next decade. *Built-in self-test (BIST)* techniques are currently used in several (embedded) memories. Figure 10.10 shows an example of BIST in an embedded memory: *memory BIST*. To reduce the cost of overall chip testing, BIST techniques must also be included in the design of digital and analogue blocks.

Although the embedded memories in a VLSI chip contain most of its transistors, it is relatively easy to achieve a large memory test coverage, because of the regular architecture of its memory array. To be able to detect defects between neighbouring bit lines or word lines, they have to be set in different logic states. First, the complete memory is loaded with '1's and then read. Next, it is loaded with '0's and then read. Then a checkerboard (1 0 1 0 1 0 1) pattern is loaded into the memory, such that every '1' is surrounded by '0's and then read. Next the inverse checkerboard pattern is loaded and read. These tests only contain very regular patterns of '1's and '0's, which can easily be generated by a finite state machine (FSM) and an address

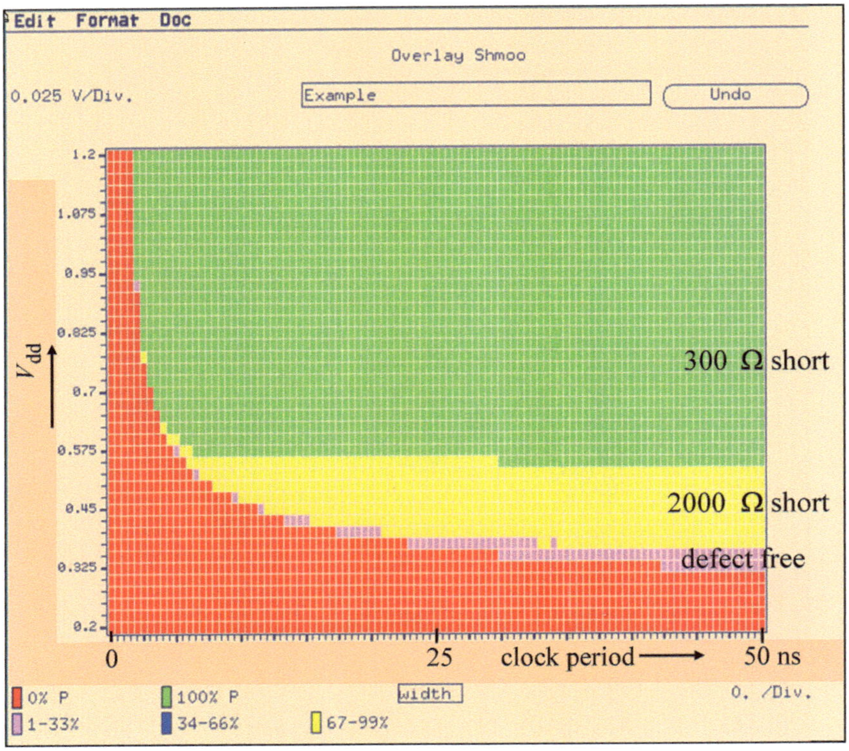

Fig. 10.9 Shmoo plot measurement used during a VLV test of an inverter

sequencer. All output data (read back data) is sequentially stored in a *multiple-input signature register (MISR)*. A MISR basically contains a shift register with a built-in linear feedback loop. It generates a signature which is dependent on all bits that are fed into it. If one or more bits are wrong, the signature does not match its expected value, meaning that it has detected a fault. A MISR actually compresses the output data to a single signature to save test time.

Because of its simplicity in both the on-chip generation and comparison of the test vectors, memory BIST has already been incorporated in many IC designs and has become more or less standard practice, today. Currently, BIST is often combined with a repair action. This technique is called *built-in self-test and repair* (*BISTAR*) in which faulty columns are replaced by correctly operating redundant columns, by blowing polysilicon fuses.

Introducing self-test into logic cores is much more complicated. This so-called *logic BIST* (LBIST) measures the response to random test patterns, fed to the different scan chains in the logic core. Figure 10.11 shows the basic architecture of LBIST.

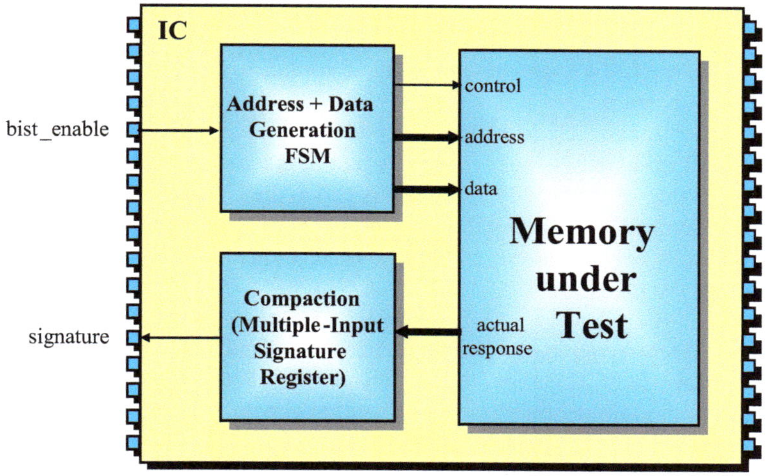

Fig. 10.10 Example of built-in self-test in an embedded memory

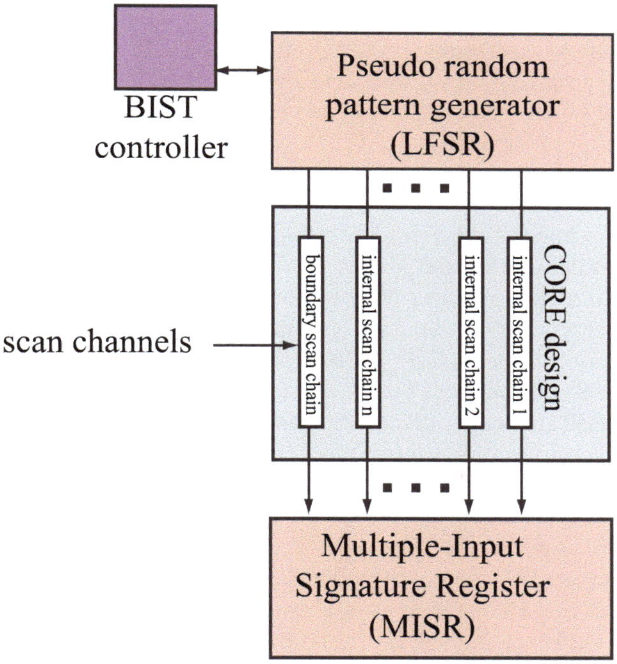

Fig. 10.11 Basic architecture of logic BIST

The BIST controller generates a sequence of activities to perform the execution of BIST:

- The MISR is first put into a known state.
- A pseudo/random pattern generator (PRPG), implemented as a linear feedback shift register (LSFR), generates the input stimuli to the different internal scan chains on the chip.
- Then a functional cycle is started to capture the responses to the input stimuli.
- The MISR compresses these responses and at the end of a BIST execution, the final state that is captured in the MISR is called the signature.
- This signature is compared with the expected signature, known from simulation. A mismatch between these signatures is an indication for the occurrence of defects in the logic circuit.

There are a few remarks to be made here. A disadvantage of LBIST is the associated low fault coverage. *Deterministic LBIST (DLBIST)*, which uses an LBIST decompression/compression architecture capable of applying deterministic test patterns, shows a reasonable test coverage, however at the cost of a relatively large area overhead.

LBIST has therefore not yet become a mainstream test solution. It was and still is not an integral part of the synthesis tools and design flow. However, with the rapidly growing test cost, LBIST may become more generally accepted as a standard design for testability methodology. LBIST is already in use for some time in special applications, particularly in security applications where a scan test would enable unwanted read-out of the security key and in applications that require field tests and where there is no tester nearby.

Boundary Scan Test

Advances in semiconductor and packaging technologies lead to such densely integrated modules that overall system accessibility is reduced. Also, the need for shorter time to market requires flexible and fast in-system testability. In 1990, a breakthrough in system test methods was made with the standardisation of the so-called *boundary scan test (BST; IEEE 1149.1, JTAG)* method. BST reduces the overall test costs and simplifies board- and system-level testing.

Although BST increases chip and board costs (additional area dedicated to *design-for-testability* circuits), this is recovered by the advantages mentioned in this section. BST also supports system production efficiency and in-field serviceability. With BST, interconnection failures during the assembly of ICs and in between ICs on a board, such as the open circuits, short circuits and stuck-at faults, can be detected. In the BST approach, a boundary cell, which contains a flip-flop, is positioned between every pin to core connection. Each cell is also connected to its two neighbours; see Fig. 10.12. In the BST test mode, these cells form a scan register, which is able to serially scan in and scan out test data.

Independently of the cores, such a scan chain can drive and monitor the pin connection of each chip in the system. A test clock and an additional test pin control the BST test mode of the system. BST supports three basic tests: interconnection tests between BST chips, IC core tests and function monitoring during normal circuit operation.

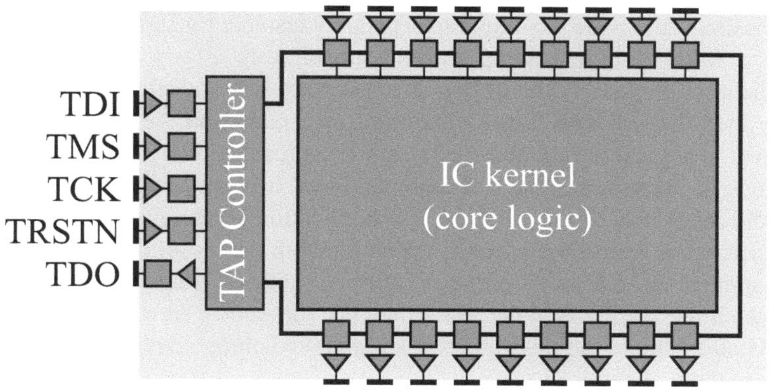

Fig. 10.12 The boundary scan-test approach

Because of the BST standard, ICs from different vendors supporting BST can be placed on the same board in a system to allow overall system testability. BST replaces the conventional 'bed of nails' test technique, in which a tester is connected through numerous wires to an array of pins on an acrylic substrate, whose footprint matches that of the test points on the PCB to be tested. Ideally, all components on a board are equipped with BST. However, even if some components do not have BST, there are still substantial benefits. IEEE 1149.1 mandates a few instructions to support board-level interconnection testing, but is open for private instructions. Many companies make dozens of such private instructions, e.g. for IC production testing, silicon debugging, emulation and application debugging, etc.

10.2.2 Design for Testability

The previous subsection discussed the different tests that are currently applied to achieve the best possible test coverage. This section discusses what a design team needs to do to support the test—*design for testability* (DfT):

- Make the design I_{ddq} or ΔI_{ddq} testable.
  There are several generally accepted guidelines for this test. Usually test patterns are created at core level or at chip level by an *automatic test pattern generator (ATPG)* tool (e.g. Tetramax, Fastscan, EncounterTest and AMSAL). These test patterns are capable of putting all cores into various different states for making defects visible through varying I_{ddq} currents. The amount of test vectors needed to create a sufficient number of different states depends on the application domain of the chip. For certain IC categories, no I_{ddq} test is performed at all, because of test-cost savings. Other categories may require around ten to twenty well-chosen test vectors, which may create 98% I_{ddq} test coverage. Some ICs might even require several hundreds to a thousand test vectors. These then require the usage of an I_{ddq} monitor on the load board, to speed up the test.

- If possible, subdivide the chip into separately testable functional blocks.
  The possibility of executing full functional tests allows to mimic the real application. Preferably the whole chip should be fully functional tested, because only then the real application conditions are created on the chip, including all current peaks, cross-talk and supply noise sources. Although the complexity of today's ICs does not allow full functional test, it is very important that the test conditions match closely with the real application conditions, because there are an increasing number of cases in which the chip passes all tests, but still fail in the application, due to the different noise conditions. There are even examples in which the chip fails in the test mode, but still shows correct behaviour in the application due to the possibility of a higher switching activity in the test mode than in the real application!
- Add self-test logic to suitable cores of the chip.
  This is sufficiently discussed in the previous subsection and need no additional guidelines here.
- Make the design scan-testable.
  The scan test improves accessibility and observability and enables to guide signals to the relevant cores on the chip that are not directly accessible through the pins of the chip. This holds for almost all logic blocks on the chip, today. Compressed scan data, combined with on-chip decompression techniques, may lead to a reduction of five to ten times in test time at the costs of only 5% in area overhead.
- Include boundary scan test (BST) for enhanced system testability.
  Most of the current PCBs are very densely packed with a lot of components and interconnect, which makes direct test access to the relevant areas on the PCB impossible. BST verifies the operation at pin level of every device in the system and checks the connection from the device pads through the leads of package to interconnections on the board.

Methods for testability improvement are meant for production testing of ICs. Prior to the computer test phase, however, design problems may appear during *IC characterisation*, debug and *engineering*. On-chip *waveform measurements* are essential when timing errors, noise margin problems or other non-stuck-at errors are suspected. These measurements facilitate functional checking of different IC parts and local verification of timing specifications.

10.3 Yield

The current diameter of wafers used in modern IC production is mostly 8 to 12 inches. The size of an IC determines the number of dies per wafer. Most IC sizes range between 25 mm^2 and 200 mm^2 and their number per wafer therefore ranges from a few thousand to a few hundred, respectively. The ultimate price of an IC is determined by the number of *functionally good dies per wafer* (FGDW). This number is not only dependent on the number of dies per wafer but also on the *yield*.

Quite a lot of dies on a wafer do not meet their specified requirements during testing. An additional number of dies is lost during packaging. The yield observed during wafer probing depends on the quality of the manufacturing and on the sensitivity of the design to process-induced defects. The production of nanometer CMOS ICs places very high demands on the factory building, the production environment and the chemicals. Disturbances in the production environment may be attributed to the following parameters:

- *Temperature*: Fluctuations in temperature may cause the projected image of the mask on the wafer to exceed the required tolerances. Also several processing steps are done at elevated temperatures.
- *Humidity*: High humidity results in a poor bond between the photoresist layer and wafer. This may result in under-etching during the subsequent processing step (delamination).
- *Vibrations*: Vibrations that occur during a photolithographic step may lead to inaccurate pattern images on the wafer and result in open or short circuits.
- *Light*: The photolithographic process is sensitive to UV light. Light filters are therefore used to protect wafers during photolithographic steps. The photolithographic environment is often called the *'yellow room'* because of the specially coated lamps used in it. The long wavelength of this yellow light does not interfere with the chip's process, as photoresist materials are most sensitive to short wavelengths.
- *Airflow*: In a CMOS cleanroom, airflow plays a critical role in maintaining the cleanliness and integrity of the environment. The airflow is carefully designed to create positive air pressure, which ensures that any particles or contaminants are pushed away from the cleanroom area. To achieve the desired airflow, various techniques are employed, such as the use of *high-efficiency particle air (HEPA)* filters to filter the incoming air and clean the recirculated air. Additionally, the cleanroom airflow is carefully controlled to create specific airflow patterns. A commonly used pattern is the laminar airflow, where the filtered air moves in a continuous, unidirectional flow from the ceiling to the floor. This ensures that any particles, airborne microorganisms or contaminants generated within the cleanroom are consistently carried away and do not settle on surfaces or contaminate the manufacturing process.
- *Process-induced or dust particles*: Particles that contaminate the wafer during a processing step may damage the actual layer or disturb a photolithographic step. This can eventually lead to incorrect circuit performance or to a complete circuit failure. A lot of effort is done to keep the contamination level as low as possible. Cleanroom operators need to wear special suits to maintain high-quality standards of the *cleanroom* with respect to contamination. Silicon wafers are subjected to many process steps to build a complete circuit. Each step requires physical treatment performed with a dedicated tool. Feature size reduction has constantly increased the requirements with respect to the purity of the chemicals, gases and environments that contact the wafers during processing. The exposure of the wafer surface to the less pure cleanroom environment

Table 10.1 US FED STD 209E cleanroom standards [3]

Class	Maximum Particles/ft^3					ISO equivalent
	≥0.1 um	≥0.2 um	≥0.3 um	≥0.5 um	≥5 um	
1	35	7	3	1		ISO3
10	350	75	30	10		ISO4
100				100		ISO5
1000				1000	7	ISO6
10,000				10,000	70	ISO7
100,000				100,000	700	ISO8

introduces defects and results in yield loss. Modern cleanrooms have class 10–100 for the overall environment. A mini environment, with controlled airflow, pressure and much less particles (e.g. better than class 1), is used to transport the wafer to the various process tools. For wafers of 200 mm and smaller, such a mini environment is called a standard mechanical interface pod (*SMIF pod*). It protects the wafers from particle contamination and provides an automated and standardised interface to the process tools. The wafers remain either in the SMIF pod or in the tool and are no longer exposed to the surrounding airflow. The class that is assigned to a cleanroom is based on the level of cleanliness required within the cleanroom area. There are two standards that define that class: the Federal Standard 209e and ISO standards. Table 10.1 shows these standards. In the Fed St 209e a class 1 cleanroom may contain no more than 1 particle of size 0.5 μm or larger per cubic foot, while class 100 may contain no more than 100 of these particles. The ISO 14644-1 standard is based on the logarithm of the maximum number of particles of 0.1 μm or larger per cubic metre of air. An ISO5 cleanroom may contain no more than 10^5 particles per cubic metre. As $1\text{m}^3 \approx 35\ \text{ft}^3$, both standards are similar when measuring 0.5 μm particles, although the testing standards differ [3].

The above mentioned mini environment class 1 and cleanroom class 100 refer to Fed St 209e. Although the maximum permitted particle count in the cleanroom is a hundred times higher than in the mini environment, it still requires a complex, highly controlled and automated production line and a very expensive fab construction to prevent contamination, as even a small particle or impurity can significantly impact the performance, yield and quality of semiconductor devices.

The front opening unified pod (also with cleanliness better than class 1) or *FOUP mini environment* (Fig. 10.13) was particularly developed for the constraints of the 300-mm wafer generation. FOUPs may have RF-identification tags for automatic handling in the cleanroom. It may be clear that the class 1 requirement not only holds for the FOUP (or SMIF pod), but also for all installed tools in the cleanroom for lithography, deposition, etching, planarisation, etc.

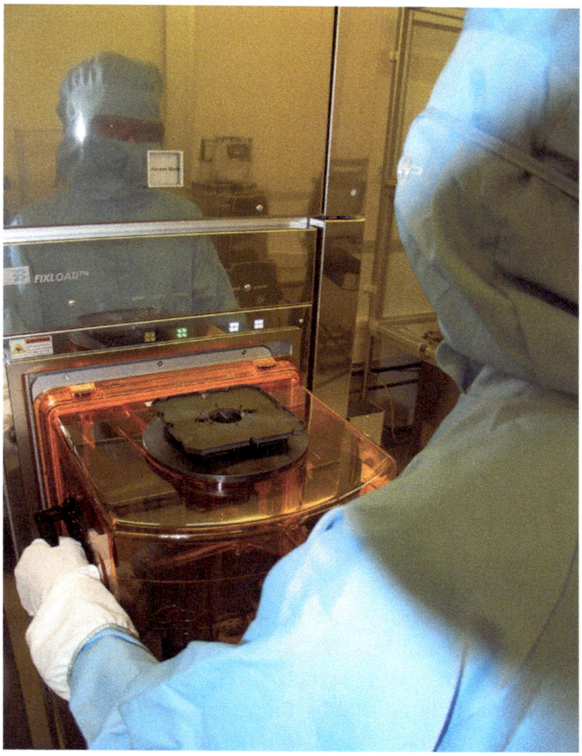

Fig. 10.13 Example of the use of a FOUP mini environment in a modern cleanroom (Source: Entegris)

- *Electrostatic charge*: Electrostatic charge attracts small dust particles. Very high charge accumulation may occur at a low humidity. This can lead to a discharge which damages the electronic circuits on ICs.
- *The purity of the chemicals*: The chemicals used must be extremely pure to guarantee the high grade of reproducibility and reliability required for ICs.

10.3.1 A Simple Yield Model and Yield Control

The above parameters, the complexity of the process and the size of an IC determine the yield. Disturbances anywhere during wafer processing may cause defects. In order to control the production costs and predict the product's performance, yield loss mechanisms must be very well understood and accurately modelled. The basic cause of yield loss can be threefold. *Systematic yield loss* is usually caused by the sensitivity of process variations, process or lithography steps to certain pattern topographies in the layout. These are usually spatially or temporally correlated. *Parametric yield loss* is often caused by marginal operation of the design, e.g. critical timing, too much switching noise or small noise margins. Finally, *random yield loss*, which is typically associated with physical mechanisms, such as metal shorts and opens due to defects (particles) or contaminants, or open contacts and vias due to misalignment or formation defects. These are usually characterised by the absence of any kind of correlation.

There exists several yield models today. Each model assumes a particular defect density distribution: exponential in the Seeds model, triangular in the Murphy model, gamma in the negative binomial model and random in the Poisson model. For a given process, IC manufacturers compare wafer yield (versus die size) with simulation results based on the selected yield model, to achieve the best fit.

The overall die yield can generally be described as a product of parametric/systematic limited yield Y_s and random-defect limited yield Y_r. To keep the explanations simple, we will use the Poisson model. According to this model, the yield Y is expressed as:

$$Y = Y_s \cdot Y_r = Y_s \cdot e^{-D_0 A} \tag{10.1}$$

where Y represents the pretest yield, D_0 the *defect density* (#defects/cm^2) in diffusion and the product defect susceptibility and A the chip area. The yield Y_s incorporates the wafer area usage factor, stepper wafer layout definition, stepper alignment mark areas or other drop-in structures (if applicable) and systematic and parametric yield loss. Today's production lines use electrostatic clamping devices for wafer handling, which offers significant advantages over the conventional mechanical clamp ring by increasing wafer edge utilisation and yield.

The parametric yield is determined by the match of the product design and process window. Especially in the early phase of process development, yield loss is dominated by parametric/systematic issues. Such defects are the result of structural failure mechanisms, which may be caused either by physical process defects or by an incorrect or process-sensitive design, and are relatively easy to find.

Most nonuniformly distributed defects originate from 'critical' processing steps. Particularly the steps that involve masks with very dense patterns are considered to be potentially critical. These masks include those used to define patterns in thin oxide regions, polysilicon layers and metal layers.

The factor Y_S, which is area independent, does not include the unusable wafer area close to the wafer edge. The usable wafer area (see Fig. 10.14) is defined by the total area occupied by complete dies, with the exclusion of a circular edge area (with a width of several millimetres) and a bottom flat side. Current wafers (8" wafers and larger) no longer contain a flat side, but only a notch (Sect. 3.2.3). The total number of dies within this usable area is called *potential good dies per wafer* ($PGDW$).

The defect density D_0 in equation (10.1) represents the density of defects causing uniformly distributed failures. These are uncorrelated and randomly distributed over the wafer. Examples include dust particles which may affect each process step.

The number of *functionally good dies per wafer* ($FGDW$) is:

$$FGDW = PGDW \cdot Y \tag{10.2}$$

Fig. 10.14 Useful wafer area for PGDW

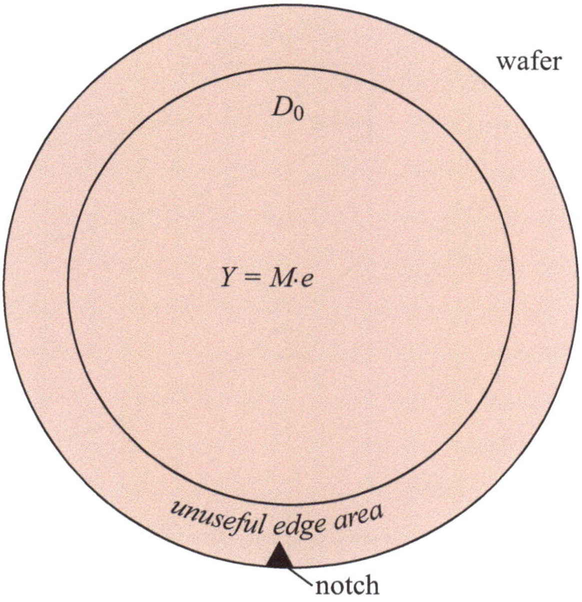

The eventual production cost of a chip is determined by the cost of a fully processed wafer and FGDW:

$$Cost/chip = wafercost/FGDW \tag{10.3}$$

Clearly, the best way to reduce the fabrication cost per chip is to increase the yield.

Particularly in the early phase of process development, Y_s will be relatively low and D_0 will be relatively high. Figure 10.15 shows an example of the yield Y according to equation (10.1) as a function of the die area A for two cases for a 65-nm CMOS process. Case 1 shows the situation during an early development stage of a new process, when $Y_s = 0.6$ and $D_0 = 2$ [defects/cm^2]. Case 2 may represent the situation after a year ($Y_s = 0.85$ and $D_0 = 0.5$ [defects/cm^2]). For more mature processes, typical values for $Y_s = 0.97$ and $D_0 = 0.25$ [defects/cm^2] (case 3).

Traditionally, during a wafer test, an ink dot was deposited on every die that failed the test. Today, the distribution of correct and failing dies across the wafer, a so-called *wafer map* or *wafer bin map*, is stored in the tester's memory. Some prober control tools allow real-time monitoring of the wafer map during testing, with the X-Y coordinates displayed relative to the reference die. The test results are put in bins, presented by colour-coded dies on the wafer. Figure 10.16 shows an example of such a wafer bin map.

All white dies pass all tests. The red dies fail the scan-continuity test, which test the correct behaviour of all scan chains. The pink dies fail the full scan test of the logic cores on the die. The light-blue dies have at least one failure in the SRAM.

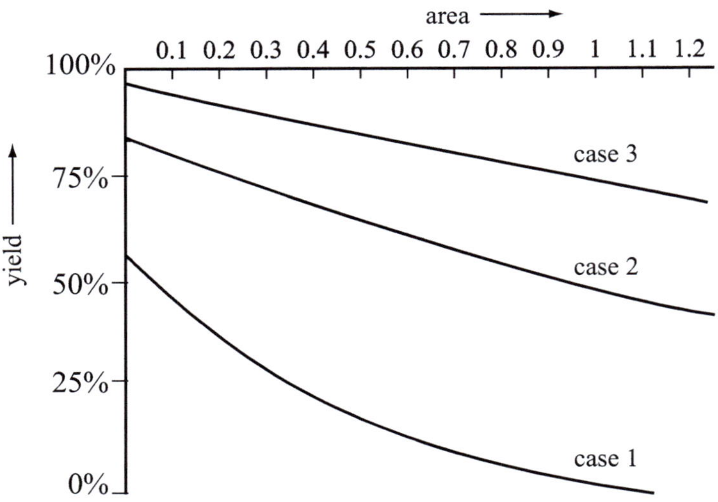

Fig. 10.15 Yield curves at different stages of process maturity

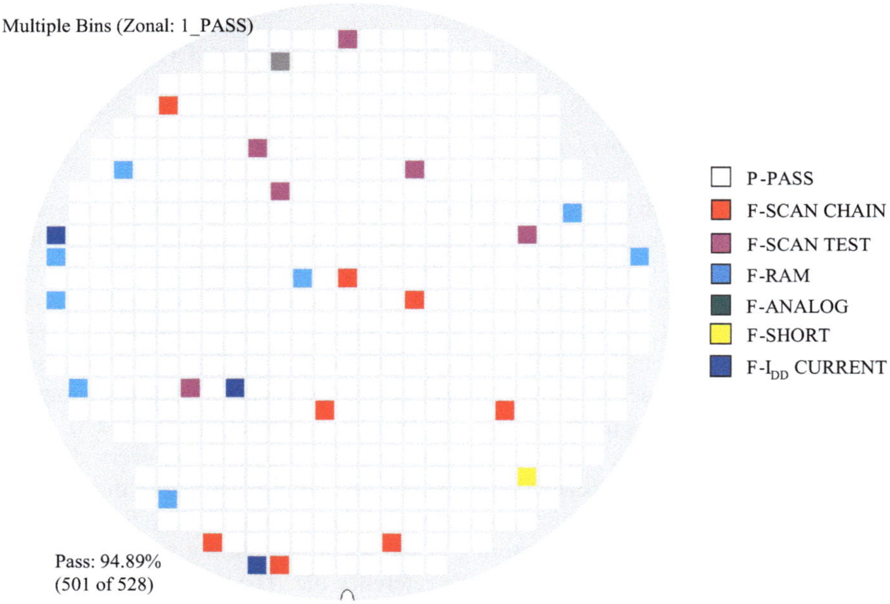

Fig. 10.16 Example of a wafer bin map showing correct and failing dies (Source: NXP Semiconductors)

Dies with a failure in the analogue circuits are coloured green. Finally, the dark-blue dies represent dies, which fail the supply current I_{ddq} test. Wafer bin maps can also be used to aggregate data from multiple wafers and stack them for cross wafer or lot analysis. Specific patterns in a bin map are usually an indication for equipment problems or process variations. Several tools exist that can automatically recognise wafer bin map patterns and can provide valuable information for the diagnosis of failure causes. This supports the designers and the foundries to ramp up yields in shorter time. Figure 10.16 shows that the yield near the wafer edge is less than in the centre area. This is a common phenomenon in semiconductor manufacturing. This is because the wafer edge is a zone of discontinuities and inhomogeneities which cause various defect and failure mechanisms. Additional *wafer bevel* etch, polish and clean process steps are required to reduce the influence of these mechanisms (see Sect. 3.2 and [4]). Even though the wafers are bevelled, the wafer edge is still leading to a lot of imperfections during the semiconductor processing:

- Potential lithographic wafer-edge problems
  - Defocus, CD and overlay variation; contamination by residues and particles; *wafer edge exposure** adaptations, e.g. to compensate warpage
- Potential layer deposition wafer-edge problems
  - Layers wrap around the edge, bevel area and apex; thickness variation near the wafer edge; warpage due to multilayer deposition
- Potential etching wafer-edge problems
  - Edge roughness particles; bevel edge process; etch residues
- Potential CMP wafer-edge problems
  - Chipping during bevel edge process; CMP (deep-) scratch damage; film peeling
- Potential cleaning wafer-edge problems
  - Particles, peeling, residues
- Other potential wafer-edge problems
  - Warpage; edge profile issues

* The *wafer edge exposure (WEE)* tool must carry out the necessary metrology to determine with great precision where the edge of the wafer is so that the exposure means can be properly manoeuvred to create the desired exposure at the wafer's edge [5].

Figure 10.17a shows the variation in yield from the centre towards the edge of the wafer, while Fig. 10.17b represents the yield improvement at both the wafer centre and edge, from ramp-up to a more mature process.

In process nodes at and beyond 20 nm the wafer stress is worsening due to multi-film stacking, particularly in 3D NAND flash processes. Manufacturers use various techniques, such as modifying process chemistry or optimising equipment placement, to create a more uniform processing environment across the entire wafer and improve the quality of the final product while reducing manufacturing costs. This can also be concluded from Fig. 10.17, which shows that when the process becomes more mature, the wafer-edge yield gets closer to the centre yield, because

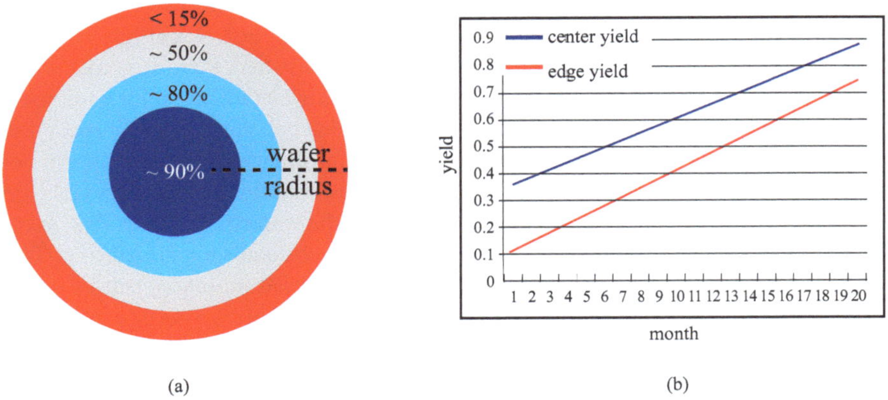

Fig. 10.17 (**a**) Example of how the yield may vary from the wafer centre to the edge and (**b**) example of how the yield may progress over time, starting at ramp-up of the wafer fab (Source: Hynix Semiconductor Inc and KLA-Tencor Corp [6])

the edge-yield line is steeper than the centre-yield line. Still the die failure rate at the edge remains higher than in the centre (e.g. *isolated vias*) and, as such, wafer-edge inspection and metrology are critical tools to determine the root cause of failures. Particular in SOI, where one wafer needs to be bonded to a carrier wafer, the bonding properties near the wafer edge lead to even more profound yield problems.

For the purpose of manufacturing process monitoring and yield control, *process control modules (PCMs)* are included on wafers. For effective monitoring, PCM structures should be distributed across the entire wafer. Traditionally, a wafer contained about five PCMs, located in the centre, edge, north, south, east and west positions on the wafer. In Fig. 3.4 in Chap. 3, some of the PCMs can still be distinguished on the one- and two-inch wafers. Today, these PCMs are positioned within the *scribe lanes* between the dies (Fig. 10.18). There will be many of them on a 300-mm wafer, but usually still only a limited number per wafer are measured. As explained in Sect. 3.3.1, the scribe lanes are also used to position the alignment marks during the lithographic process, which means that the available scribe lane space is limited.

A PCM often includes transistors of various sizes (W, L) for the electrical characterisation of parameters such as β and V_T. PCMs often also contain relatively large structures that facilitate the measurement of layer resistances and capacitances, via resistances, possible shorts and opens through meander structures, etc., for example. They also help to monitor the consistency of key parameters across the wafer and distinguish between acceptable and defective wafers based on specified parameters. These structures are sampled during and at the completion of the wafer fabrication process. Often, several hundred parameters can be measured on a PCM. As advanced CMOS processes moved to the sub-5-nm regime, both the number of device and interconnect structures on a chip increase dramatically. This requires

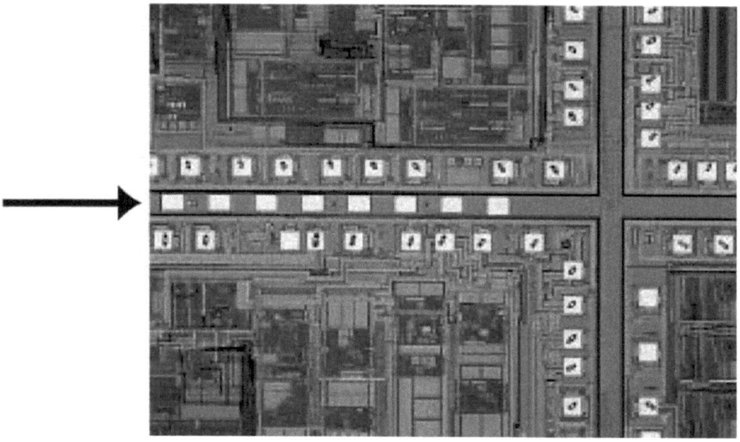

Fig. 10.18 Example of process control modules within the scribe lane between dies

much more devices and structures to be included in the PCMs to still yield a good coverage of the process parameters.

During the introduction of a new process, the PCMs on all wafers are often measured. When a process becomes mature, usually a few wafers per lot and a few devices per wafer are measured. The measurement results are also used as an early feedback to control the process and to provide reliable *wafer acceptance test* (*WAT*) results. The wafer customer typically sees a selected subset of the PCM data.

Finally, when the correct dies are packaged, the final tests are done, which, besides functional, structural and reliability tests, also check the connections between package and die. These final tests, in combination with the pretest (wafer test), must limit the number of customer returns to a minimum.

10.3.2 Design for Manufacturability

Over the last decade, design costs for a complex (not leading-edge) ASIC have started to explode, from approximately $1million in 1998 up to $100 million in 2024. This, combined with reducing product life cycles and manufacturing yields, has increased the drive to reduce the number of respins and to ramp up the yield in shorter time, to meet time-to-market, quality and cost targets. Design rules form the real link between process technology and design. In conventional CMOS technologies, 'absolute' design rules *(DRC-rules)* were sufficient to create circuits with relatively high yields. From 90 nm downwards, these absolute design rules are no longer sufficient. Additional rules *(DfM-rules)* are required to make the designs tolerant to photolithography and process deficiencies, in order to maintain a sufficiently high yield level. In current nanometer CMOS technologies, extensive yield evaluation must be performed before a design is sent to the fab.

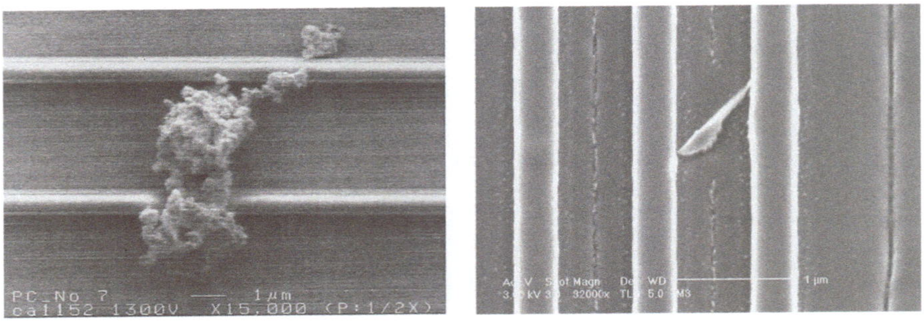

Fig. 10.19 Example of random failures: particles causing a potential short (Source: NXP Semiconductors)

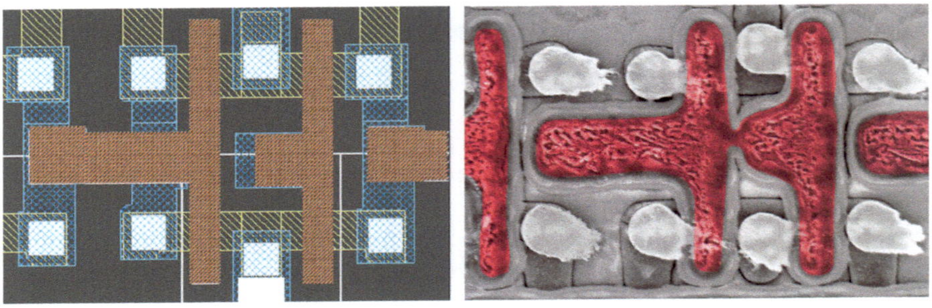

Fig. 10.20 Example of a systematic failure, showing the layout and a photograph of a polysilicon short

Particularly layouts are adapted to increase this yield. This so-called *design for manufacturability (DfM)* can reduce the design sensitivity to defects (opens or shorts), but it may also support the lithographic process (litho-friendly design; Chap. 3). Figure 10.19 depicts some examples of random failure.

There exists no uniform definition for DfM. Some include all effects that are potential candidates to reduce the yield: defects, shorts and opens, lithographic variations, process variations, power integrity, substrate noise, electromigration, leakage currents, reducing noise margins, etc. Many of these effects are discussed in the previous chapter, since they also influence the design robustness and product reliability [7].

DfM includes a set of guidelines to make designs more robust against systematic, parametric and random yield loss and create more easy producible products. DfM is a way of anticipating on critical features or critical areas in the layout early in the design phase. Figure 10.20 shows an example of a systematic failure: a short between the two polysilicon areas. The photo has been taken after de-processing of the metal and via layers.

Particularly at product introduction, when the design rules and process are not yet mature, the operating margins can be low and may cause parametric yield loss.

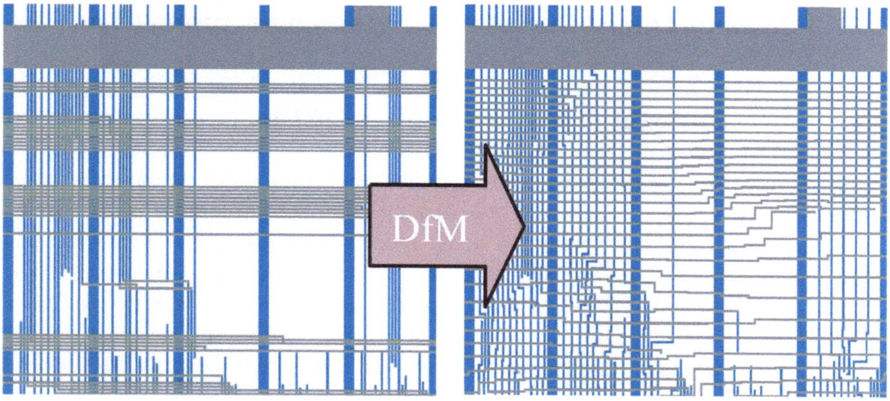

Fig. 10.21 Example of wire spreading to improve yield (Source: NXP Semiconductors)

Consequently, DfM rules may change as the process technology becomes more mature [8].

A few DfM rules have already become commonplace, such as antenna rules and rules for tiles (area fills to improve the CMP planarisation process (Chap. 3). Rules for wire widening and improved wire distribution (*wire spreading*; Fig. 10.21) were introduced around the turn of the millennium.

Wire spreading was particularly an issue in those areas of the chip, where many wires were routed at minimum width and spacing, while there is ample room for wider wires at (much) larger than minimum spacing. Over the last couple of years *via doubling* has entered the scene to reduce the number of opens on a chip. This is due to the fact that the number of contacts and vias with minimum metal overlap has dramatically increased. Doubling every via in a design is not possible, since it would have a serious area impact. The current approach is to double only the vias that have sufficient white space around them and which causes no area increase. Particularly at the wafer edge, isolated vias are an increasing source of concern, as they can have a severe impact on the overall yield.

These additional DfM rules require the development of intelligent tools. An important requirement for applying these tools to achieve maximum yield improvements is that they are supported with adequate yield models that have the potential to recognise and fix critical layout areas without area increase. Approaching the end of Moore's law has forced the designers to also deal with many nanometer CMOS effects such as shrinking process windows, increasing process variability, changing defect mechanisms, increasing lithographic effects, increasing noise levels, reducing noise margins, etc, which are not all covered by DfM. It is better to use the term *design for anything* or *DfX*, which includes DfT, DfM, design for robustness, design for reliability, litho-friendly design, design for debug, design for failure analysis, etc. Each of these 'design for' topics requires additional design resources and increases the design complexity and costs.

10.3.3 Towards Chiplet Design to Reduce Manufacturing Costs

When the semiconductor fabs moved from the 7-nm and 5-nm nodes towards the 3-nm and 2-nm nodes, both the printing (lithography) and production process of the related minimum feature sizes of CMOS circuits became extremely complex. As a result, it became much harder to produce more transistors on the same piece of silicon. As we have seen in the memory chapter, DRAMs and flash memories increasingly use the 3rd dimension to be able to continuously increase the cell capacitance (DRAM) and memory capacity (flash), respectively. For current logic chips, however, the 3rd dimension is used for increased transistor current drive (FinFET and gate-all-around transistors) as well as increased transistor density (nano-ribbon, nano-sheet, fork-sheet, nano-wire, CFET transistors). For several applications the fabrication of large monolithic dies has become less financially attractive (Fig. 10.22).

As transistor sizes continue to shrink, maintaining high yields and achieving the desired performance becomes increasingly challenging. To ease the economics of manufacturing chips with a large amount of transistors, the industry has introduced *chiplet*-based design whereby a large ASIC is partitioned into multiple dies (chiplets), which are then interconnected with very advanced packaging techniques to build a heterogenous system. Let us assume a simple yield model in which defects scatter randomly across a wafer and that a defect anywhere on the die will cause a failure. Wafers with large dies will then show lower yields than wafers with small dies (Sect. 10.3.1 and [9]). Figure 10.23 shows a plot of the wafer yield for a die of various sizes versus the same die function built from two, three and four chiplets [10]. An additional 10% overhead for the cross-die communication has been added to the chiplet-based designs. According to the graph, a monolithic die with an area of 200 mm^2 will only have a yield of 24%. However, a 4-chiplet design with each chiplet having an area of 50 mm^2 will have a yield of 49%, more than twice

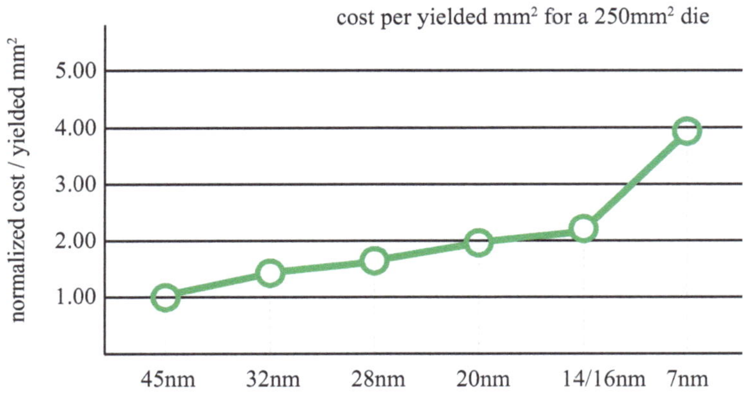

Fig. 10.22 Trends of normalized cost in advanced process nodes (Source: [9] and AMD)

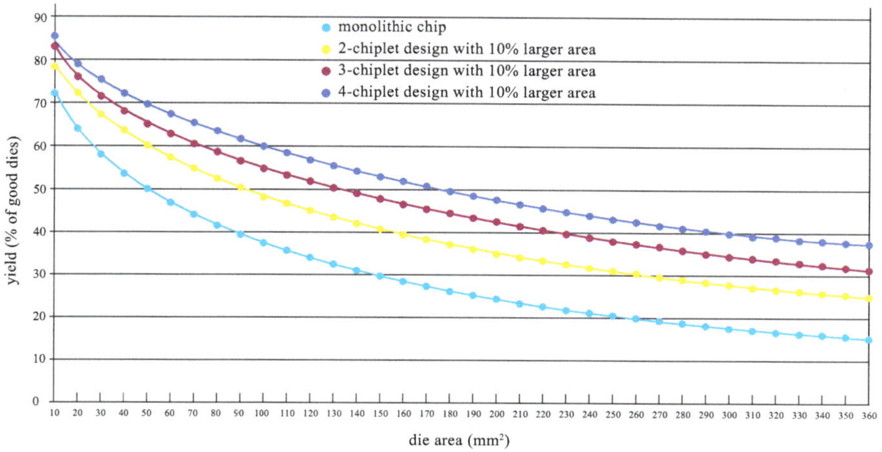

Fig. 10.23 Comparison of yield of various chiplet implementations versus the yield the monolithic design of the same function (Source [10])

the yield of the monolithic die. Although the total die area of the 4-chiplet design is slightly larger due to an additional 10% area penalty (20 mm^2 for a combined silicon area of 200 mm^2), the noticeable improvement in yield directly results in lower costs, which justifies the additional area penalty.

In addition to the above, chiplets offer several other advantages over traditional monolithic chips, the main one being the ability to leverage different manufacturing technologies and processes. This can result in cost savings, as each chiplet can be optimised for its specific function, and the combination of several chiplets can result in a more capable and efficient overall system. Additionally, the modularity of chiplets means that they can be easily swapped out or added to a system, making upgrading or repairing electronic devices much easier. High-bandwidth memory (HBM) (see Sect. 6.4.4) is the first example of heterogenous integration that is being widely used. HBM is built from various dies from different process technologies which are integrated using dense parallel package interconnects, often *through-silicon vias*, to enable efficient communications between the chiplets.

One of the main challenges associated with chiplets is the complexity of integration. The process of designing, testing and integrating chiplets can be time-consuming and costly. Additionally, the different processes used to manufacture chiplets can result in variations in terms of performance, functionality and reliability, which can complicate the integration process even further. The 3D stacking of dies often requires an additional critical process step: the *backside grinding* of the wafer to less than 100 μm, before the dies are cut from the wafer. Despite these challenges, chiplet technology is enjoying growing popularity in the semiconductor industry. It offers a way to continue to increase the capabilities of electronic devices while also providing benefits in terms of cost and flexibility. Examples of current chiplet usage can be found in [11], where researchers at UCLA and the University of Illinois in the USA integrated a large number of chiplets on a wafer-scale interconnect substrate. TSMC developed a 3D logic-on-logic and memory-on-logic chiplet stacking technology platform [12, 13]. Another example of the increasing popularity of the use of chiplets is the Intel Meteor Lake™ architecture which can run various AI workloads [14]. It consists of four separate chiplets: a CPU die processed in 'Intel 4', a GPU die processed in 'TSMC N5' and a SoC and I/O die, both processed in 'TSMC N6'. These four dies are mounted on a passive interposer that connects them with high bandwidth and low latency so that its functionality is very close to a monolithic chip implementation. In 2023, TSMC unveiled its plans to deliver chip packages containing one trillion transistors, similar to Intel's announcement in that year. These high-capacity systems will be achieved by utilising 3D-packaged chiplets in a single chip package. However, TSMC is also working towards developing chips with 200 billion transistors on a monolithic chip. To achieve this, the company is actively working on the development of 2-nm process nodes, as well as 1.4-nm and 1-nm fabrication processes. The last two are expected to be ready by the year 2030. TSMC also anticipates advancements in packaging technologies such as CoWoS, InFO, SoIC and others, enabling them to create large-scale multi-chiplet solutions that can house more than a trillion transistors by around 2030 [15]. As chiplets continue to be developed and refined, it will be interesting to see how they are incorporated into the devices we use every day.

Finally, one remark must be made here: in the explanations above, when dies are stacked on top of each other, or placed on an interposer, we used the term 3D. However, in literature we often see that people distinguish between 2.5D and 3D solutions, which are both semiconductor packaging technologies that offer enhanced performance, size reduction and improved power efficiency to different extents [16].

2.5D packaging is an intermediate step between traditional 2D packaging and full-fledged 3D packaging. In 2.5D packaging, multiple semiconductor dies, typically from different process technologies, are placed next to each other on a silicon interposer. The interposer acts as a bridge, connecting the individual dies and providing a high-speed communication interface [17]. This arrangement allows for greater flexibility in combining different functionalities on a single package. 2.5D packaging is advantageous for combining various components and reducing footprints. It suits applications in high-performance computing and AI accelerators.

In *3D packaging*, on the other hand, multiple dies, which are often ground to sub-100-nm thickness, are stacked vertically on top of each other, with (micro-) through-silicon vias (TSVs) connecting the dies. The distance between the successive dies is small, which provides a higher level of integration and enables improved performance and reduced power consumption. 3D packaging is beneficial in applications that require high performance and small footprint. It is often used in advanced memory technologies like high-bandwidth memory (HBM) and advanced processors for high-end gaming gadgets and smartphones and in specific high-speed computing applications.

10.3.4 Redundancy in Logic

In Sect. 6.8 we discussed how we can include redundancy in memory blocks. Memories, whether these are stand-alone memories or embedded memories, usually contain a large number of transistors packed with extremely high densities. They are therefore highly prone to failures due to process defects or small operating margins. It is common practice to include redundancy in these memories. They often have a number of additional (redundant) rows or columns. Let us take row redundancy as an example. When, during testing, a cell on a word line shows a failure, the faulty word line is replaced by a redundant word line, by storing the faulty row address in an OTP memory. Every time, during its application, when the faulty row is addressed, the redundancy hardware guides it automatically to the corresponding redundant row. This redundancy technique, which only adds a limited amount of on-chip hardware, is described in Sect. 6.8. However, including redundant core(s) on a multicore processor chip would lead to a relatively large area increase and a lower number of dies on the wafer, resulting in an increase of chip costs. Modern multicore micro-, graphics and AI processor chips often contain several billion up to a hundred billion or more transistors. The probability that they all operate perfectly is low. Therefore, the chip is designed such that cores which do not work can be disabled. Disabled cores are turned off for a simple reason: they failed factory tests. These processors can then be sold as a lower-end model. This is known as part binning. It is not only commonly applied in the semiconductor industry but also in many other industries where poor quality or incomplete parts can still be sold.

10.4 Packaging

10.4.1 Introduction

The development of the IC package is a dynamic technology. Applications that were unattainable only a few years ago are now common place thanks to advances in package design. Moreover, the increasing demand for smaller, faster and cheaper products is forcing the packaging technology to keep pace with the progress in semiconductor technology. The huge diversity of application areas, e.g. automotive, identification, mobile communications, medical, consumer and military, to name a few, combined with an exponentially growing device complexity and the continuous demand for increased performance has generated a real explosion of advanced packaging techniques.

Packaging is no longer a final step in the total development chain of a semiconductor product, and as such, it has become an integral and differentiating part of the IC design and fabrication process.

The package supports various important functions:

- Allow an IC to be handled for PCB assembly and protect it during further PCB production

- Mechanical and chemical protection against the environment
- Mechanical interface to the PCB
- Good electrical connection (signals and power supply) between PCB and chip
- Enhance thermal properties to improve heat transport from IC to environment
- Allow standardisation

Currently, ICs may contain hundreds of millions to a hundred billion transistors. With such high integration densities, the IC package has become increasingly important in determining not only the size of the component, but also its overall performance and price. Higher lead count, smaller pitch, minimum footprint area and reduced component volume all contribute to a more compact system implementation. As the package directly affects factors such as heat dissipation and frequency dependency, choosing the right package is essential in optimising IC performance.

10.4.2 Package Categories

Packages can be classified into different categories: board-mounting techniques, construction form and power handling capabilities. The packages in these 'power' categories offer a high thermal capability, enabling IC usage in some of the most power-demanding application areas. Four major package categories that depend on different board-mounting techniques can be distinguished:

- Through-hole packages, whereby the pins are inserted and soldered into plated through-holes in a PCB. These packages are widely used in cheap electronic equipment where board space is less important. Example packages (Fig. 10.24) in this category are single-in-line (SIL), dual-in-line packages (DIPs) and pin grid arrays (PGAs).
- Surface-mount dual/quad packages are relatively thin packages (see quad flat pack [QFP] example in Fig. 10.24) whereby the leads are directly soldered on corresponding metal lands at the PCB surface. This allows smaller dimensions and components at both sides of the PCB.
- Surface-mount area array packages, which have an array of balls, or lands that are soldered on a PCB as well.

SIL DIL PGA QFP MLF/QFN FLIP-CHIP

Fig. 10.24 Various package images

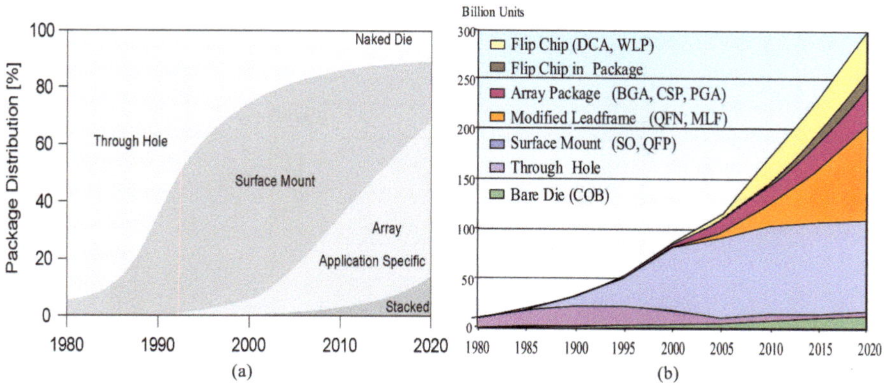

Fig. 10.25 (**a**) Market penetration of the different package categories and (**b**) IC shipments by package category (Source: Prismark Partners)

- Surface-mount wafer-level packages (bumped dies). In *wafer-level packaging (WLP)* most or all packaging steps are carried out directly at the wafer level instead of the traditional assembly of each individual die after wafer dicing. Because the final package is almost of the same size as the chip itself, WLP is a true *chip-scale packaging (CSP)* technique. The IC production process is then extended and includes additional interconnection (redistribution) and protection steps.

The package choice depends on the expected priority in properties of the application area, e.g. high density (very small package), high bandwidth (many pins and low self-inductance), high power (good thermal behaviour), etc. Figure 10.25a shows an overview of the (expected) market penetration of the major package categories.

Figure 10.25b shows the changing use of various package technologies over the last four decades. It is expected that surface-mount packages remain popular for at least another decade and that flip-chip and micro lead frame (MLF and QFN [quad flat no-lead]) packages become more widely used. They show an improved heat transfer path because they use perimeter lands on the bottom of the package and can be soldered directly to a PCB.

Figure 10.26 presents an overview of the different package categories.

Different versions have been developed for each of the package categories. The ceramic versions show better thermal behaviour and are meant for ICs that consume more power. The increase in power density, combined with a limited temperature budget, requires creative approaches to thermal management. Dependant on the application demand, several alternative technologies can be applied to uniformly cool an IC, e.g. normal airflow, heat sink, heat spreader, thermally enhanced interface material and fan.

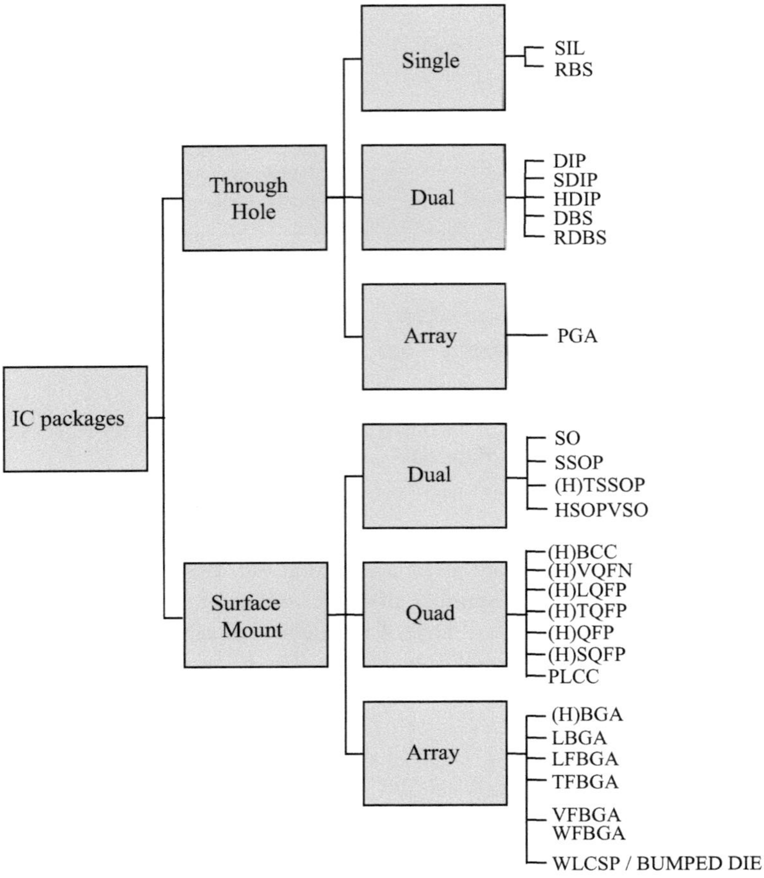

Fig. 10.26 Overview of the most commonly used packages

These versions can be categorised with three characteristics: overall package height (L, T, V, W), lead/ball pitch (S, F) and thermally enhanced (H). This naming convention is standardised by JEDEC, an industrial standardisation committee.

Figure 10.27 shows the evolution in use of the various package categories, with a focus on density increase. Through-hole packages (e.g. TO3P, DIP and PGA) were the most preferred ones during the 1960s and 1970s, while surface-mount device (SMD) technology (e.g. TSOP, QFP, flip chip) became very popular during the 1980s and 1990s at the cost of the through-hole packages.

Products like mobile and smart phones get smaller and thinner every year, which automatically requires the same shrink for the components they are built from. This means that conventional leaded parts, such as quad flat packs (QFPs), will increasingly be substituted by leadless parts like QFNs or even bare dies (wafer-level *chip-scale package [CSP]* [WLCSP]).

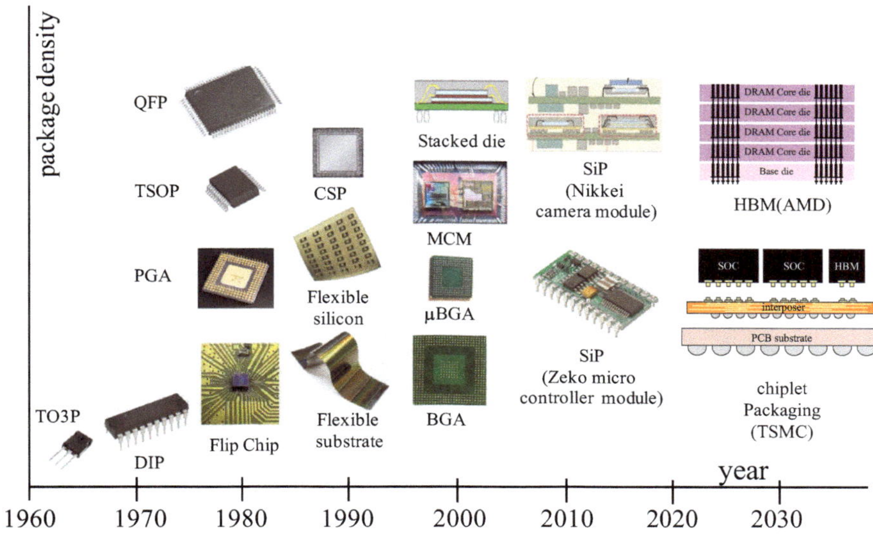

Fig. 10.27 Worldwide package evolution

The combination of increased computing power and memory capacity with reduced footprint sizes of all kinds of electronic gadgets has led to an increasing attention to 3D package techniques, which are discussed in the next paragraphs.

10.4.3 Packaging Process Flow

When a wafer leaves the waferfab, it first needs to be functionally and electrically tested before it can be assembled. This electrical testing, which is often referred to as probing, is done by means of metal needles that physically contact the bond pads on each die. Three main technologies can be identified: cantilever probing, as depicted in Fig. 10.28; membrane probing, often used for RF solutions; and vertical probing, the preferred technology for bumped dies. ICs that do not pass the functional and electrical tests were marked by an ink dot, in conventional processes. During the 1990s, the red dots have been replaced by an electronic wafer bin map (Fig. 10.16), which ensures that they will not be used during the die-attach/die-placement process for packaging.

Backgrinding and Sawing

Once the design of a CMOS chip is completed, it must be fully surrounded by a *seal ring* for reliability reasons. The metallised seal ring serves as a hermetic seal from all external 'bad influences'. It encloses the whole chip and is used to isolate the active silicon circuits from the external environment, protect it from die-

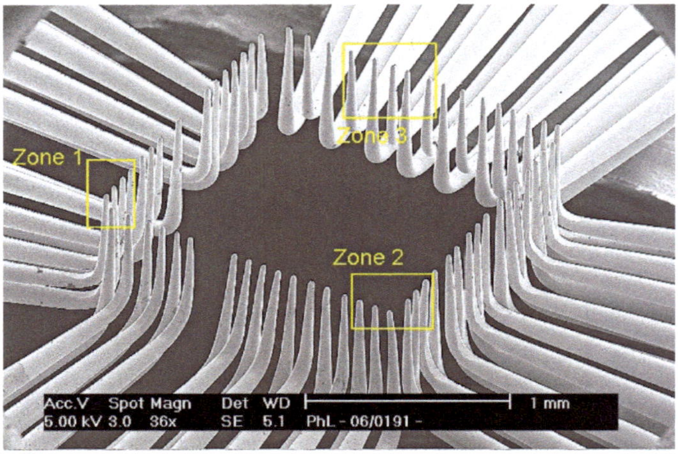

Fig. 10.28 Picture of cantilever probe card (Source unknown)

sawing stress and prevent penetration of contaminants into the chip circuits. Seal ring dimensions and architectures vary from manufacturer to manufacturer and must therefore be designed according to the specific foundry design rules. It often consists of a stack of metal layers and via bars from the top metal all the way down to the silicon substrate. The porosity of the high-ϵ dielectrics used in between the metal lines and metal layers enables penetration of contaminants and moisture into the die, such that they can reach the integrated circuits. Therefore, the seal ring must completely seal the chip circuits, meaning that all via layers in the seal ring consist of a few parallel continuous via lines all along the seal ring (Fig. 10.29).

As dies are configured into arrays on a wafer, they need to be separated from each other after processing and testing. This can either be done by a diamond saw, with a width between 30 and 90 μm, or by a laser. Therefore, the seal ring itself is again completely surrounded by a *scribe lane* (also called scribe line), through which the saw or laser can cut the wafer into separate dies. The width of the scribe lane varies from 30 to 100 μm. Figure 10.29 shows the structure and an example cross section of the seal ring and scribe lane. It also shows the various assembly design rules for the chip pads to the scribe lane.

Most 300-mm wafers have a thickness of about 775 microns ($\approx$ 30 mils). As discussed in Chap. 3, the circuits on a chip only use the top few micron of the wafer substrate. This leaves a lot of unused excess substrate material at the backside of the wafer. Thinning of the wafers by backgrinding, in combination with polishing and de-stress steps, supports advanced packaging techniques which increasingly uses stacked dies to reduce the overall system footprint. So, before actual packaging starts, each wafer will have to be backgrinded to the optimal thickness. Typical thicknesses are 280 μm and 380 μm, while 100 μm is more common for very thin packages.

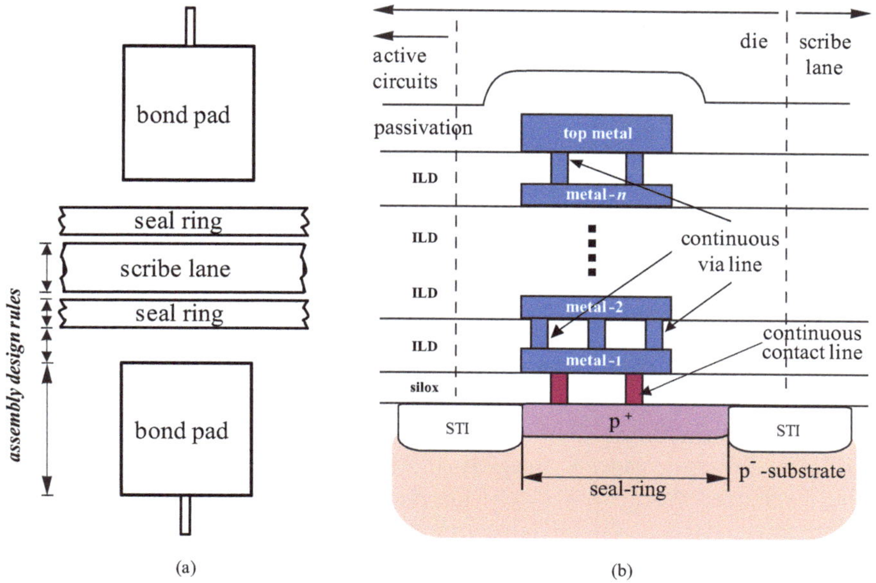

Fig. 10.29 Structure (**a**) and cross section (**b**) of seal ring and scribe lane around the chip and a representation of the corresponding assembly design rules

The physical backgrinding is typically a two-step process, in which the wafer is first backgrinded to about 20 μm above the required thickness, with a coarse grinding wheel. The remaining last 20 μm are then grinded with a much finer grinding wheel. For very thin wafers (< 150 μm) an etch step can be added for stress relief.

Once backgrinded to the right thickness, the wafer has to be separated into individual dies. This is typically done by means of a diamond saw, although laser separation is an upcoming trend. *Laser dicing* has a couple of advantages compared to diamond sawing. It is faster, it causes less material stress, it requires a smaller scribe lane and it is able to die devices with different form factors on the same wafer [18]. The conventional laser dicing used laser wavelengths that are highly absorbed by the materials to be diced. The introduced heat and debris led to reliability issues. As the laser energy was absorbed in the top of the wafer, it created a lot of stress and strain at the die edges, which influenced transistor behaviour in the nearby electronic circuits. *Stealth dicing* uses (higher) laser wavelengths that transmit through the wafer and where the beam can be focused at different depths of the wafer. The position of this focal point enables selective localised reactions within the wafer. It avoids damaging of the wafer surface, where the transistors and interconnections are located. To allow this dicing scribe lanes of 30 μm to 100 μm are designed around each die.

Dicing with the conventional diamond saw can be performed in two different techniques. In the first technique, a single cut is made directly through the complete

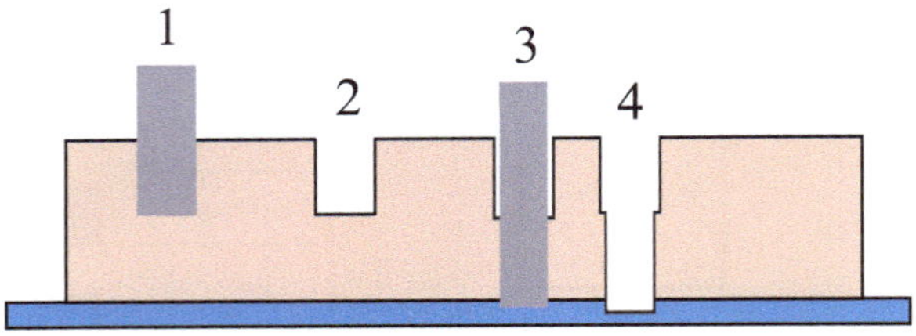

Fig. 10.30 The step-cut process for singulation of dies (dicing)

wafer. This process is typically used for conventional production processes. The more advanced processes, e.g. the ones that use low-ϵ dielectrics, use a so-called step-cut process. This process first uses a wide blade to only cut through the active area (1 and 2 in Fig. 10.30) into the bulk silicon. Next, a finer blade in the second step will separate the dies (3 and 4).

Since the demand for thinner wafers is increasingly enabling 3D stacking of dies in different electronic gadgets' integration, the risk of wafer-level breakage during dicing the thinned wafers has become a big challenge. This has led to another way of creating individual dies: the so-called *dice before grind (DBG)*, which refers to the order of operations in the process of cutting a semiconductor wafer into individual dies. In this context, first, a half-cut is performed on the wafer with a dicing saw. Then the wafer is mounted upside down, with a special adhesive film, to a substrate. Next, the wafer is ground to its desired thickness, whereby the individual chips become "diced".

Packaging

Once the wafers are backgrinded and diced, the actual packaging can start. The package choice is very much related to the electrical, thermal and size requirements dictated by the application domain. Two main interconnect technologies can be identified to realise the electrical connections. The most common one is *wire bonding (WB)*, which is still responsible for about 90% of all chip interconnects. Before the actual wire bonding can take place, the die is first mounted on a carrier (lead frame or substrate) in a process called *die attachment*. During this die attachment, an adhesive is deposited on the carrier and the functional good dies are picked from the wafer and placed in the adhesive. Which dies are being picked is determined by the wafer map that has been generated during probing (see Fig. 10.16), or, in conventional processes, by the ink dots placed during the wafer test. Adhesive materials are typically a mixture of epoxy and a metal (aluminium or silver) to ensure a low-electrical and low-thermal resistance between the die and package. For thermally enhanced applications also solder can be used to attach the die to the carrier.

IC reliability is strongly influenced by the quality of bonding wires. Diameters of the wires range from 15 μm for fine- pitch applications to 150 μm for high-power devices.

There are two common wire bond processes, depending on the applied wire material: Au ball bonding and Al wedge bonding. However, copper is gaining more popularity because of its reduced electrical and thermal resistances. But copper is more readily oxidised than aluminium and copper oxidation may cause reliability problems, in the form of poor adhesion or in the introduction of cracks at the bond interfaces (see Sect. 10.4.6). To prevent corrosion of the copper wire and make it more flexible for bending during the bonding process, palladium-coated copper (pdCu) wires are used. All copper wire bonding varieties must be performed in the presence of so-called forming gas (95% nitrogen and 5% hydrogen) or a similar anoxic gas in order to prevent corrosion. More details on the physical and material properties of copper bonding processes can be found in [19]. It provides good electrical performance and is cheaper than gold (Au) wires.

During wire bonding the bonding tool is guided to the bond pad on the die. This so-called first bond (Fig. 10.31) is achieved by using thermal and ultrasonic energy. Next, the wire is stretched to the corresponding finger of the lead frame on the carrier and, again by using pressure and ultrasonic energy, the opposite end is welded to the leadfinger to form a stitch bond (also known as wedge bond). At the formation of this second bond the wire is also automatically cut in preparation for the next bond. After all the pads have been bonded, the die is encapsulated. Figure 10.31 shows an overview of the wire bonding process.

Because the number of transistors grows quadratically with the scaling factor, while the number of pad positions only grows linearly with it, there is an increasing demand for multi-row bond pad (staggered bond pads) connections to support a variety of applications. Figure 10.32 shows examples of a double and a triple-row wire bond interconnections.

Very high current variations in high-speed VLSI circuits can cause an increased voltage drop ΔV across the bonding wires. This is because of the inductance (L) of

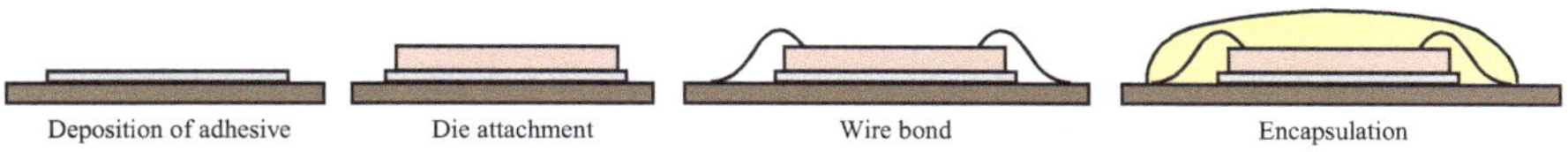

Fig. 10.31 Overview of the wire bonding process

Fig. 10.32 Examples of a double- and a triple-row wire bond interconnects (Source: NXP Semiconductors)

the wires and is expressed as follows:

$$\Delta V = L\frac{di}{dt} \tag{10.4}$$

The above voltage drop may become critical in analogue/mixed-signal and high-speed digital circuits, unless suitable design measures are taken. This topic is addressed in Chap. 9.

In *flip-chip bonding (FCB)*, which is the second interconnect technology, the die is assembled face down directly onto the circuit board (Fig. 10.24) with solder, Au or Au/Ni bumps. Compared to wire bond, this technology comes with less area overhead, because there is no additional area needed for contacts on the sides of a chip. It enables the final packaged chip to be only marginally larger than the original die (*chip-scale package*).

In the example of the *controlled-collapse chip connection (C4)* soldering process, first, solder bumps are deposited on the die bond pads (Fig. 10.34), usually when they are still on the wafer and at the corresponding locations on the substrate. Figure 10.33 shows a wafer-level CSP with two rows of dies sawn from the wafer. The zoom-in shows that also a redistribution layer is used, which is covered with a polyimide layer for passivation and stress relief. Then holes are etched in this passivation layer and a direct ball drop method produces the balls needed for connection to the board.

During placement, the array of balls on the die is aligned with the array on the substrate. Depending on the FCB technology, it is then either pressed and reflowed (melted) or the complete embodiment is reflowed (melted) in a furnace, to create all electrical connections. During this reflow step the chip is self-aligned to its exact position on the substrate (Fig. 10.34). If no material was deposited before, there will always be a gap between the die and substrate. In the next step, the die is encapsulated with good isolating material (epoxy) to fill this gap (underfill). This underfill serves to protect the chip from humidity and impurities, but also improves reliability in terms of mechanical and thermal stress.

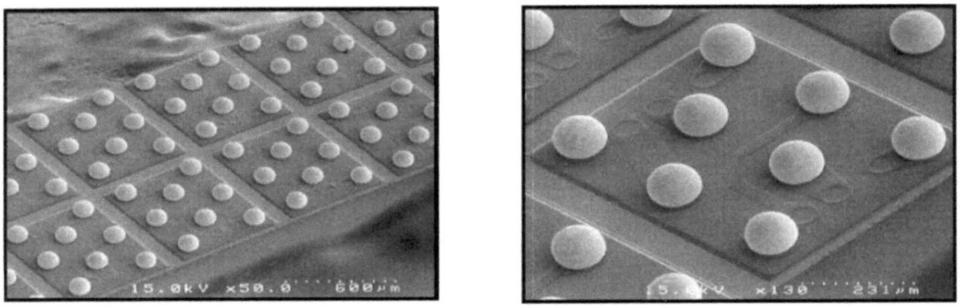

Fig. 10.33 A wafer-level chip-scale package with direct ball drop

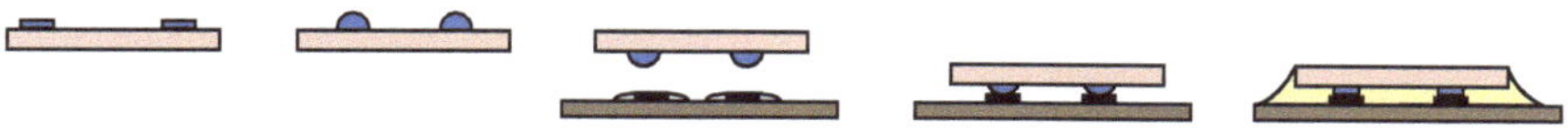

Fig. 10.34 Overview of flip-chip bonding process

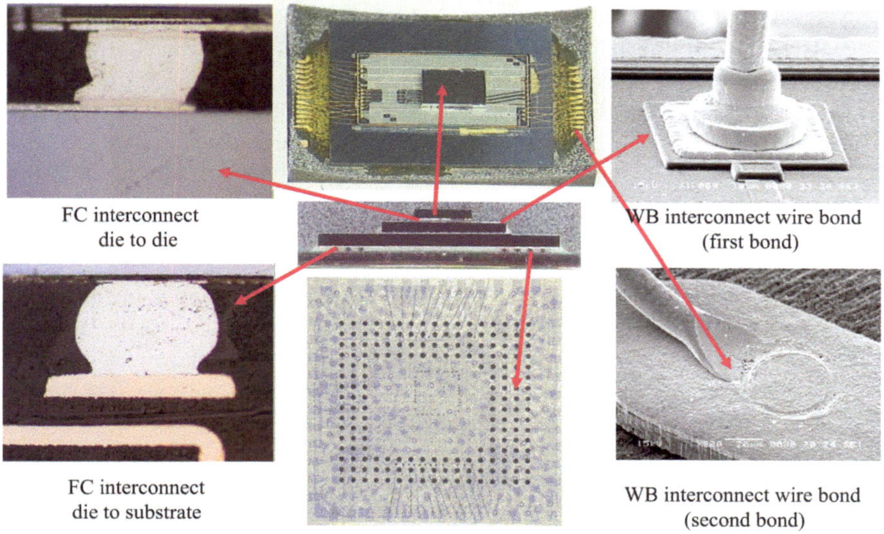

Fig. 10.35 Package with both flip-chip (FC) and wire bond (WB) interconnections (Source: NXP Semiconductors)

Finally, the FCB process is completed by curing (hardening) the underfill material by using heat or light at a certain wavelength, depending on the underfill material.

Flip-chip bonding accommodates dies that may have several hundred bond pads placed anywhere on their top surface. In many cases an additional redistribution metal layer is required to transfer periphery wire bonding pad connections into an area array of connections for flip-chip bonding. Compared to other interconnection techniques, FCB results in very short connections and exhibits improved performance in high-speed applications.

Flip-chip connection can be found both in a silicon to substrate and silicon to silicon bumping. Figure 10.35 shows a combination of wire bonding and flip-chip bonding in a stacked-die application.

The substrate with the array of balls, in the lower middle of Fig. 10.35, shows such a redistribution layer. Currently, the redistribution layer is part of the top aluminium layer, which is needed for coverage of the bond pads to support reliable bonding.

10.4.4 Electrical Aspects of Packaging

The drive for higher bandwidths and the resulting increase in signal speed, combined with the ever present demand for area and cost reduction, has caused the package development to become a significant and integral part of the total development process of integrated circuits. Package costs are mainly driven by the size of the package and is closely related to the number of package connections. A package with poorly controlled electrical characteristics (such as resistance R, inductance L, capacitance C and impedance Z) needs more pins than a package that meets the electrical signal interface design criteria.

The electrical characteristics of a package are determined by its construction. The type of signal interface indicates the desired $RLCZ$ of the package interconnects. A high-speed differential interface may want $100 - \Omega$ differential impedance between metal tracks, while minimising the inductance of the power and ground pins is more important for a single-ended interface. The signal type will determine the desired electrical characteristics of the package interconnect.

Generally, a small value of L is advantageous for both signal integrity and speed. Fast current changes may introduce large voltage changes (ΔV); see also Chap. 9, according to expression (10.4). Certain package types, however, can only support certain ranges of electrical parameters. Conventional package types using lead frames have relatively large lead inductances (1–50 nH), because of longer lead lengths. They also tend to have a high mutual coupling. Ceramic multilayer packages, such as pin grid arrays (PGAs), have better performance due to the presence of power or ground planes, but are relatively expensive. As previously discussed, the use of flip-chip bonding can also improve electrical performance by minimising the lengths of the connections between the die and the substrate, resulting in inductances of 0.5–1 nH. Recent developments in package technology, however, allow more flexibility in the design of package interconnects to meet cost targets. Some package types allow the design of specific impedances while others try to minimise them altogether. *Ball grid array (BGA)* packages use inexpensive laminate substrates allowing the inclusion of power and ground planes and therefore the design of transmission line structures. On the other hand, so-called no-lead packages, such as QFNs, have virtually no pins and sometimes not even bond wires thereby minimising the influence of the package on the overall electrical behaviour of the chip. It is therefore extremely important to understand that the requirements of the interconnect are dictated by the application. Particularly in high-speed applications, a reasonably accurate electrical model of the package is an essential part of the final performance simulations of the integrated circuit.

In the packaging of ICs, we distinguish several hierarchy levels of interconnections:

- first level of interconnection: chip to package connection
- second level of interconnection: package to PCB connection
- third level of interconnection: PCB wires
- fourth level of interconnection: PCB to system (back-planes) connection

It should be clear that not only the package-to-board connections must be optimised for high-performance ICs, but to achieve maximum system performance, all levels of interconnections must be optimised and adapted to each other. This becomes even more important with the state-of-the-art packaging technologies, such as multi-chip modules (MCMs) and stacked dies (Fig. 10.40), system in a package (SiP) and system on a package (SoP) (Fig. 10.42). These packaging technologies are discussed in Sect. 10.4.8.

10.4.5 Thermal Aspects of Packaging

Another dominating parameter in the performance and reliability of an integrated circuit is the physical temperature of the die inside the package, which is determined by the power consumption of the IC in combination with the *thermal behaviour* of the package. This requires a strong interaction between the IC, the package, the system design and its application. The most commonly used, but simple model for IC packages includes two thermal resistance parameters. For a given power dissipation P, the junction-to-air thermal resistance R_{JA} represents the ability of a package to conduct heat from junction (die) to ambient and is expressed as follows:

$$R_{JA} = \frac{(Temp_J - Temp_A)}{P} [°\text{C/W}] \tag{10.5}$$

where $Temp_J - Temp_A$ represents the temperature difference between the chip (junction) and its environment (ambient).

R_{JA} is often determined corresponding to the JEDEC [20] requirements for standard test boards and in different air conditions, including still air. Table 10.2 shows some values for R_{JA} under still-air conditions.

In many applications the maximum junction temperature is defined as 125 °C. If we assume a consumer application with an ambient temperature of 70 °C, the maximum allowed power consumption of an IC, packaged with a 48-pin SSOP (see Table 10.2) under still-air conditions (on a reference board of the supplier), is then equal to:

$$P = (Temp_J - Temp_A)/R_{JA} = (125 - 70)/88 = 625\,\text{mW}$$

If the power consumption is more than this calculated maximum, either a heat spreader is required or an airflow must be introduced, using a fan.

The other parameter is defined as the junction-to-case thermal resistance R_{JC} and represents the ability of a package to conduct heat from the junction (die) to the surface (top or bottom) of the case (package) and is expressed as follows:

$$R_{JC} = \frac{(Temp_J - Temp_C)}{P} [°\text{C/W}] \tag{10.6}$$

Table 10.2 Thermal resistance values for different packages under still-air conditions (Source: NXP Semiconductors [21])

Package pins	Package designator	Package outline code	RTH(J-A)$^{o}C/W$	RTH(J-C)$^{o}C/W$
DQFN-14	BQ	SOT762	61	32
DQFN-20	BQ	SOT764	50	23
HVQFN-16	BS	SOT629	40	18
HVQFN-24	BS	SOT616	40	18
LFBGA-96	EC	SOT536	60	16
LFBGA-114	EC	SOT537	55	15
PLCC-20	A	SOT380	80	32
PLCC-28	A	SOT261	73	29
PLCC-52	A	SOT238	48	13
SSOP-20	DB	SOT339	136	40
SSOP-24	DB	SOT340	125	37
SSOP-28	DB	SOT341	98	35
SSOP-48	DL	SOT370	88	25
SSOP-56	DL	SOT371	84	24
TSSOP-8	DP	SOT505	120	30
TSSOP-16	PW	SOT403	160	39
TSSOP-24	PW	SOT355	128	32
TSSOP-48	DGG	SOT362	104	23
TSSOP-64	DGG	SOT646	110	26
QFP-52	BB	SOT379	62	15
VFBGA-56	EV	SOT702-1	80	21

This parameter is only applicable if an external heat sink is used and the heat is only conducted through that surface that connects to the heat sink.

If we assume an IC, consuming 1 W, which exceeds the above calculated maximum allowed power under the same conditions, then the required thermal resistance of the device must be equal to:

$$R_{JA} = (Temp_J - Temp_A)/P = (125 - 70)/1 = 55\,°\mathrm{C/W}$$

This can either be achieved by introducing an airflow or by using an external heat sink.

This model, however, only describes the steady-state heat conduction capability and does not account for the dynamics in power behaviour of the product in a real application. Heat flows are rarely one-dimensional. Different application boards, or stacking packages, change the environment of the product and can have a huge impact on its thermal behaviour. However, the value for R_{JA} can very well be used to compare the thermal capabilities of different packages.

Previous discussions were focused towards the average die temperature. However, although the average die temperature may not exceed the maximum specified temperature (e.g. 150 °C), the momentary temperature gradient across the die can vary a lot with the workload, such that locally this maximum can be exceeded. An example of such simulated temperature gradients is given in [22].

To obtain an accurate model for a particular thermal situation including two- and three-dimensional heat conduction paths therefore soon leads to a complex network. This has led to the development of compact thermal models, describing the thermal behaviour with an accuracy of 5% by using a thermal network with seven or more nodes connected by thermal resistances. A discussion of such compact models is beyond the scope of this book. An example of such a compact model is described in [23].

10.4.6 Reliability Aspects of Packaging

There are a few packaging aspects that are related to the reliability of the chip. First of all, the trend to reduce the dielectric constant of the *inter-level dielectric (ILD)* layers (low-ϵ dielectrics) in semiconductor fabrication processes makes these dielectrics more porous, less robust and more sensitive to physical pressure during test (probing) and bonding. A combination of adhesion problems and poor mechanical strength limits a fast integration of alternative low-ϵ dielectrics. Second, in a copper-backend CMOS process, copper is used for all metal layers including the one(s) used to create the bond pads. However, copper oxidises quickly and the oxidation prevents the creation of a good and reliable contact between the bond wire and the pad. Therefore, during an additional re-metallisation step, a so-called *Al cap (aluminium cap)* is formed above the pad area to create a good electrical contact with the bond wire. But this does not solve all reliability aspects. Particularly the drive for finer pad pitches and smaller pads requires probe cards with smaller and sharper needles, which increases the probability to punch through the Al cap and expose the underlying copper. Also these exposed copper areas oxidise quickly, showing the same problems as described above. A solution to this problem is to increase the Al cap area such that the probe needles do not land in the wire bond region (Fig. 10.36) and can no longer damage the underlying copper layer because it is separated by the passivation layer.

The increasing number of pads, combined with the drive for smaller chip areas, has forced the semiconductor industry to create *bond-over-active (BOA)* layout techniques, in which bond pads are not only located at the chip's periphery but also on top of active silicon areas at the periphery of the die core area on top of diodes, power and ground lines, I/O transistors and ESD protection circuits. BOA may lead to a significant reduction in die size. However, it also has consequences for the reliability of the integrated circuit [24].

In summary, any change in pad-related design and technology concepts may have severe consequences for the reliability of the bonding process and for an overall reliable chip operation.

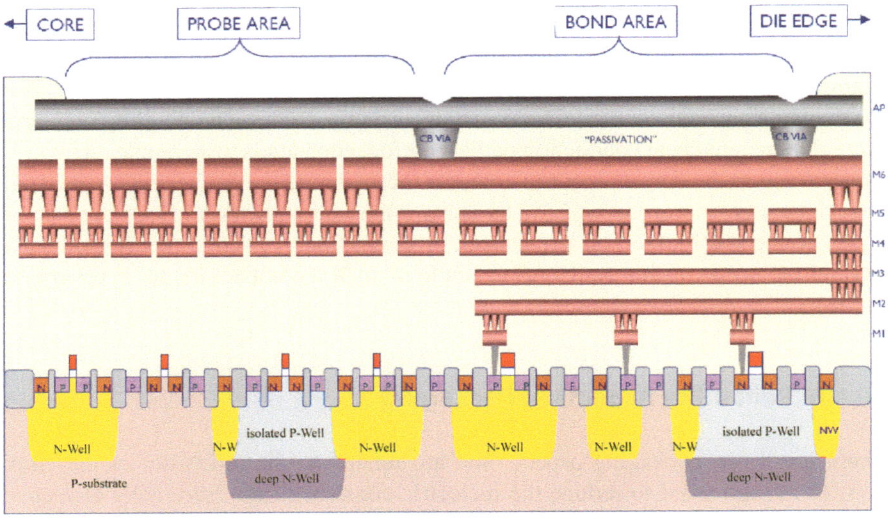

Fig. 10.36 Example of extended Al cap area to prevent pad damaging during probing (Source: NXP Semiconductors)

10.4.7 Future Trends in Packaging Technology

While during the 1980s and 1990s *surface-mount device* (*SMD*) technology became very popular at the cost of through-hole packages, a new trend towards miniaturisation is observed. Products like mobile phones get smaller and thinner every year, which automatically requires the same shrink for the components they are built from. This means that conventional leaded parts, such as quad flat packs (QFPs), will increasingly be substituted by leadless parts like QFNs or even bare dies (wafer-level *chip-scale package* [*CSP*] [WLCSP]).

Because the complexity of nanometer ICs has reached an incredibly high level and will continue to increase, it puts severe demands to the density of die pads and package connections (pins or balls). This drives the trend towards area array packages, e.g. ball grid arrays (BGAs), in which the connection between the package and the application board is formed by solder balls (Fig. 10.37).

Figure 10.38 shows a variety of BGA packages for different applications. They mainly differ in the ball-pitch, package height and/or thermal capability. Also a differentiation can be made between the bonding technologies used within the BGA packages, e.g. wire bonding and flip-chip bonding.

In literature BGAs are frequently combined with CSPs, which are usually referred to as packages whose sizes are less than 20% larger than the die itself. Most CSPs are wafer-level packages, where the peripheral bond pads are redistributed and rerouted to an area array of pads, using a thin-film-like technology. This can be executed as an extension of the wafer fab process, or the wafer can be sent to a bump supplier who creates the redistribution directly on the wafer. On those rerouted pads

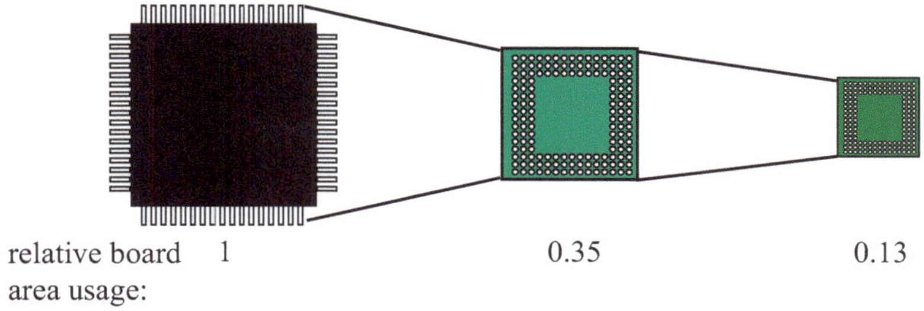

Fig. 10.37 Miniaturisation in packaging (reduced board areas) (Source: OKI)

Fig. 10.38 Different flavours of BGA packages

Fig. 10.39 Rerouted Wafer-Level CSP

balls are placed, by means of direct ball attach, to create the CSP (Fig. 10.39). Next, flip-chip bonding technology is applied, when the CSP needs to be attached to a kind of laminate carrier.

10.4.8 System on a Chip (SoC) Versus System in a Package (SiP)

Over the last couple of years, bulk CMOS processes have enabled the integration of digital, RF and mixed-signal functions onto one single die. Time has come to make a trade-off between traditional board design with discrete passive devices and a fully integrated system solution like a *system on a chip (SoC)*. A SoC is an extremely integrated single chip solution built from in-house and/or external IP. It contains the computing engine (e.g. μproc and/or DSP core), logic and various memories on the same chip. The key benefits of a SoC realisation are:

- Better performance due to the smaller on-chip delays, compared to the delays caused by going on- and off-chip
- Small physical size
- Reduced overall system costs, due to a reduction in the number of components
- Less power consumption
- Increased reliability due to a reduced number of system components

However, there are also some critical remarks to be made here. For many applications, the time between inception and high-volume production of a SoC may take several years. Most SoCs are therefore expensive, custom-designed products for high-volume market segments with a relatively long lifetime expectancy. The increasing diversity of the system's applications requires the development of more sophisticated IP.

Today, system complexity is growing at a faster rate than that of a SoC and a printed-circuit board (PCB). In many applications, Moore's law only deals with the integrated fraction of the system, leaving the largest part to relatively large discrete passive components such as antennas, filters, capacitors, inductors, resistors and switches. In the example of a cell phone, only one-tenth of the system consists of ICs, with the remaining part being passives, boards and interconnections and switches. An alternative to SoC integration is the use of *system-in-a-package (SiP)* technology, which usually refers to a single package that includes a multiple of interconnected integrated circuits and/or passive devices. SiP technology enables hybrid systems built from subfunctions that may have been created by different design methods in different technologies. Some people see a SiP and *multi-chip module (MCM)* as the same system solution; however, an MCM is usually referred to as the integration of different dies on the same plane on the same substrate in one single package, while SiP also refers to stacked dies and/or passives in one single package. SiPs may use a combination of different packaging technologies including wire bond, flip-chip, wafer-level packages, CSPs, stacked dies and/or stacked packages. Figure 10.40 shows examples of both an MCM and wire-bonded stacked dies.

Compared to a SoC realisation, a SiP approach offers much more flexibility in adding new functions and features to the system. As is the case with all packaging technologies that combine one or more naked dies, also SiP technology faces the

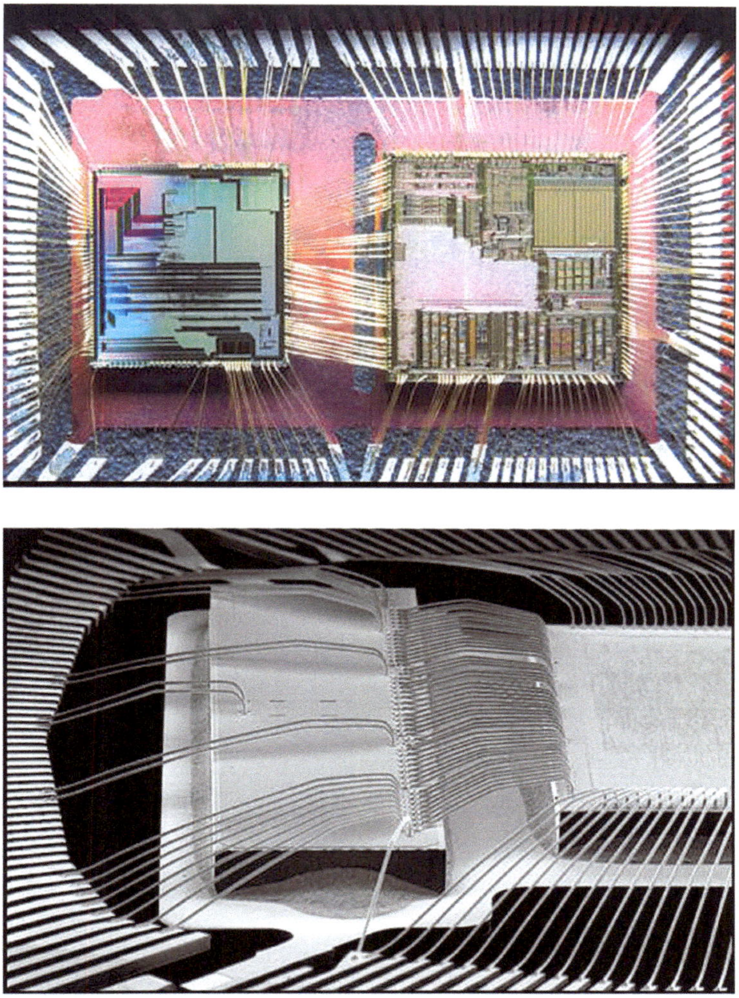

Fig. 10.40 Example of an MCM and of wire-bonded stacked dies

challenge of *known good die (KGD)*, which is a chip that has been extensively tested before being placed into its package. When an expensive processor is to be combined with a cheap peripheral chip onto one substrate or into one package, an almost 100% guarantee is required that this peripheral chip will operate fully according to its spec. This is to prevent to throw away the total substrate, including the expensive processor, if only the cheap peripheral chip does not work properly. To avoid this problem, a new upcoming trend can be identified: *package on a package (PoP)*. In this concept (Fig. 10.41) the expensive processor/ASIC is separately packaged in the POP bottom package, while the *stacked memory dies*

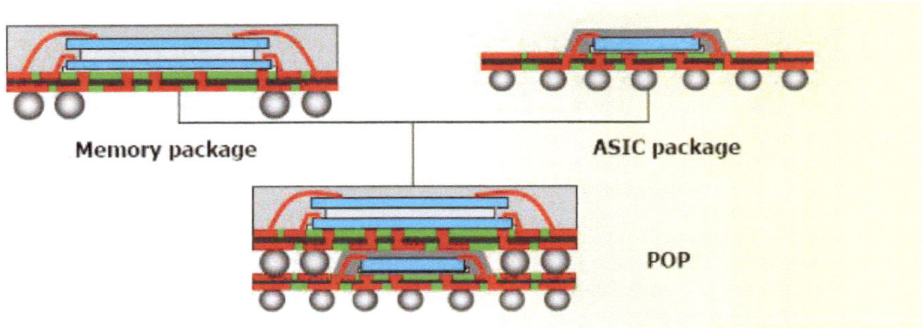

Fig. 10.41 PoP concept (Source: Amkor technology)

are encapsulated in the top package. Each of these two can be tested separately, while for the memory even burn-in can be applied. Once both are proven to be fully functional, the parts can be mounted on the application board.

Future systems, however, will incorporate features and functional complexity that will be even beyond today's imagination. They will combine the potentials of physics, optics, biology and chemistry with analogue and digital signal processing and storage capabilities packed onto one composite substrate. This is usually referred to as *'more than Moore'*. A target application may be a device that could be encapsulated below the human skin to permanently monitor a person's health. This could be done by checking vital organs through the monitoring of breath, heartbeat, blood pressure, blood glucose level, etc. The results could then wirelessly be communicated through the Internet to a medical advisor or physician, which can then propose the appropriate medical treatment, when necessary. In this case the sensors (monitors) may be attached on top of a substrate. Figure 10.42 shows an example of such a *system on a package (SoP)*, which may combine optical circuits with passives, MEMs, SoCs and SiPs inside or attached to a composite substrate [25].

In conclusion: SoC and SiP are competing technologies, but this does not mean that there will be a winner in the end. SoC and SiP systems will live together, but each in its specific application domains, where their properties are exploited to their full advantage.

In certain applications the traditional monolithic system on chip (SoC), particularly in the latest process nodes, is replaced by advanced, more cost-effective 3D multi-chiplet architectures. High-bandwidth memories (HBM) and the stacking of chiplets are examples of this trend. *Chiplets* split a monolithic chip design into multiple functional blocks and implement these into separate chiplets. These are then reassembled at the package level through dense, fast and high-bandwidth interconnections.

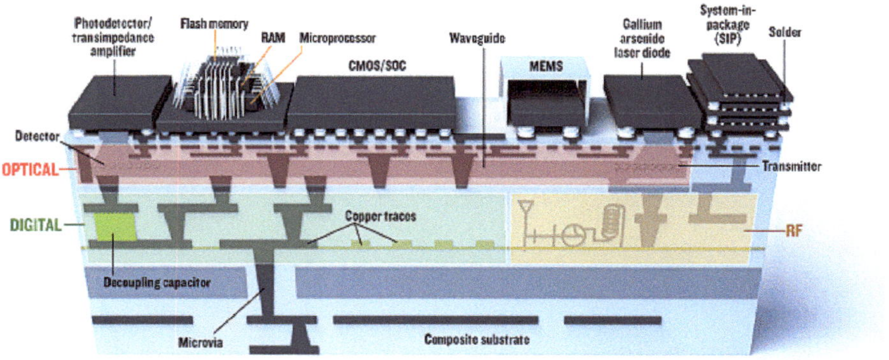

Fig. 10.42 Example of a system on a package (SOP), consisting of optical circuits and devices, resistors, inductors, antennas, decoupling capacitors within a layered substrate and a combination of bare and packaged such as flash memory, CMOS/SoC, optical circuits, laser diode and SiP devices attached to the top layer [25] (Graphics design by www.bryanchristiedesign.com)

10.4.9 Quality and Reliability of Packaged Dies

Various quality and reliability tests are applied to packaged ICs before they are approved for sale or for application in high-volume production. Many of these tests are standardised. An insight into the background to these tests and their implementations is provided below.

Quality

Vulnerability to *electrostatic discharge (ESD)* and sensitivity to *latch-up* are two important quality criteria on which chips are tested.

Parasitic bipolar devices in all CMOS chips form a *thyristor* between its supply and ground terminals. Activation of this thyristor results in latch-up. The result is a dramatic increase in current consumption and a chip malfunction. A chip's latch-up sensitivity can be tested by sequentially applying a voltage of one-and-a-half times the maximum specified voltage to each pin, while limiting the available current to, for example, 500 mA. The actual current consumption is observed for signs of latch-up.

Since ESD and latch-up sensitivity can be influenced by the design, these topics are discussed in detail in Chap. 9. In addition, Chap. 9 describes technological and design measures which can be taken to reduce the chances of failure in the associated tests. ESD tests and the related quality requirements are also discussed in that chapter.

Reliability

The increasing complexity of ICs means that their reliability has a considerable effect on the reliability of electronic products in which they are applied. Reliability is

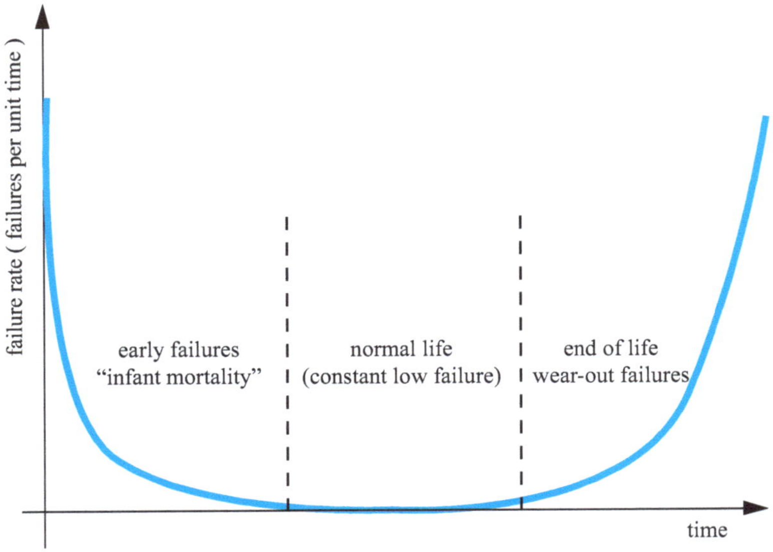

Fig. 10.43 Example of a bath tub curve, showing the failure rate of an electronic product over its life cycle

therefore an important property of an IC and receives considerable attention from IC manufacturers. Traditionally, the failure rate of an electronic product is represented by the well-known *bath tub curve*, shown in Fig. 10.43.

In the early life, the failure rate starts at a high level, but is rapidly reducing as failing products are quickly detected and rejected and failure mechanisms due to immature production steps, material defects or design errors are investigated and eliminated. During normal life, when the failure rate is constantly low, the product lifetime is expected to last till wearout, due to ageing effects. Normal life failures are usually caused by random stress phenomena, such as the occurrence of a soft error. Wearout failures become more likely when the product becomes older. Ageing is a physical process that causes deterioration, fatigue or depletion of material. The useful lifetime of a product is determined by the worst-case component, the one that lives shortest. Manufacturers must guarantee that their products function correctly over the intended lifetime. Therefore, many different reliability tests have been developed to characterise the product quality to achieve a sufficiently long normal life for their products.

Related tests subject an IC in active and non-active states to various *stress conditions*. This facilitates rapid evaluation of the IC's sensitivity to external factors such as temperature changes and humidity. The most important *reliability tests* are as follows:

- *Electrical endurance test*: This test exposes an IC to a high temperature (125 °C to 150 °C), while its supply voltage exceeds the specified maximum. Constant and varying signals are applied during the test, which may last for 1000 hours. These stress conditions will make the weak devices fail, which is detected by applying normal tests that monitor whether the ICs still show correct functional behaviour.

The electrical endurance test reveals the following:

- *Infant mortality*, i.e. faults which are likely to arise in the early months of an IC's normal application
- *Early failure rate*, i.e. faults which are likely to arise after half a year
- *Intrinsic failure rate*, i.e. the probability of a fault occurring during a specified number of years
- *Wearout*, i.e. effects of prolonged use on the product

Faults that are observed during the electrical endurance test can usually be traced to errors in the manufacturing process which preceded IC packaging.

- *Burn-in test*: This is a type of stress test that is performed to ensure the stability, reliability and performance of a chip over an extended period of time. The chip is then subjected to a high workload for a set period, under certain test conditions, e.g. several test cycles during 160 hrs at higher junction temperatures (up to 175 °C) and/or higher voltages with many (15,000) components tested simultaneously. It should lead to the detection and identification of any potential issues, such as hardware failures, faults or errors that may arise during normal use. The burn-in test involves running a variety of intensive, resource-heavy applications or programs, such as benchmarks, simulations or games, that push the system to its limits. The purpose of this is to simulate the demands of real-world usage scenarios and force the components to operate at their maximum capacity, generating heat, voltage fluctuations and other stresses that the hardware may encounter during heavy usage. By performing a burn-in test, hardware manufacturers and testers can assess the quality and reliability of the system or component under extreme conditions and ensure that it meets the required standards and specifications. This type of testing is commonly used in the production and quality control of computer systems, servers, CPUs, GPUs, hard drives, RAM modules, consumer, automotive and AI chips and other hardware components.
- *High-temperature operating life (HTOL) test*: This reliability test is performed on microchips to evaluate their long-term performance and durability under high-temperature conditions. This test is designed to simulate and predict the vulnerability of the chip to thermal stress. During the HTOL test, the chip is subjected to a high temperature for an extended period, usually between 1000 and 2000 hours. The temperature is typically set to 125–150 degrees Celsius, which is higher than the normal operating temperature of most chips, to accelerate the ageing process and expose any weaknesses or defects in the chip's structure. The chip is monitored during the test to detect any changes in its electrical characteristics, such as increased leakage current, noise, or reduced functionality. If any faults are detected, they are documented and analysed to help improve the design and manufacturing process.

- *Time-dependent dielectric breakdown (TDDB) test*: TDDB may result from exposing a dielectric over a certain period of time to a constant electric field strength, which is less than its breakdown field strength. Aggressive scaling of the gate dielectric in MOSFETs for higher electrical performance led to a continuous increased R&D interest in TDDB. As a complex VLSI chip may carry billions of transistors, its reliability is not just determined by average characteristics. The *ppm levels* (part-per-million failures) due to this type of failures can only be predicted from extrapolation of accelerated measurement results. Often a çonstant-stress test is used in which the leakage current is monitored while the device is subjected to somewhat higher temperature and voltage conditions than in normal operation. This is done for many devices leading to a distribution diagram of time-to-failure of these devices. There is no general procedure for testing TDDB, as it heavily depends on the used dielectric material and architecture, e.g. silicon dioxide or high-ϵ gate dielectric (hafnium oxide, zirconium oxide and others). More on TDDB can be found in [26].
- *Temperature-cycle test (TCT)*: This test emulates practical temperature changes by exposing the product to rapid and extreme temperature variation cycles. The minimum temperature in each cycle is between $-55\,^{\circ}$C and $-65\,^{\circ}$C. The maximum temperature is $150\,^{\circ}$C. The number of cycles used is typically five hundred. The test is carried out in an inert gas and/or an inert liquid. The main purpose of the temperature-cycle test is to check the robustness of the package and the robustness of the connections between the package and its die. The test should reveal possible incompatibilities between the temperature expansion coefficients of the various parts of an IC, e.g. the die, the lead frame and the package material.
- *Humidity test*: This test exposes an IC to a relative humidity of 85% in a hot environment ($85\,^{\circ}$C to $135\,^{\circ}$C). The test reveals the effects of corrosion on the package and provides an indication of the quality of the *scratch-protection layer*. Usually, the corrosion process is accelerated by applying different voltages to consecutive pairs of pins, with 0 V on one pin and V_{dd} on the other. Most humidity tests last 1000 hours.
- *Highly accelerated temperature and humidity stress (HAST)*: This test is a type of reliability test used to evaluate the performance of electronic chips and devices under harsh environmental conditions. The test subjects the chips to a combination of high temperature and high humidity levels in order to simulate the effects of extreme environments. During a HAST test, the chips are placed in a specially designed chamber with controlled temperature and humidity levels. The temperature is typically set to around $130\,^{\circ}$C and the relative humidity is set between 85% and 100%. These conditions accelerate the diffusion of moisture into the chip's internal components, which can lead to various types of failure, such as corrosion, electrochemical migration or delamination. The test typically lasts for a few hundred hours, during which the chips are subjected to a range of different stresses, such as temperature cycling and high temperature/humidity soak. The performance of the chips is monitored and recorded throughout the test, and any failures or malfunctions are noted.

- *Thermal shock test (TST)*: It is a type of stress test that is used to determine the ability of electronic chips to withstand sudden changes in temperature. The test involves subjecting the electronic chip to a series of extremely high and low temperatures in rapid succession. This can be done using specialised equipment that rapidly cycles the chip between two environmental chambers, one set to a high temperature (e.g. 150 °C) and the other set to a low temperature (e.g. −65 °C). The magnitude of the temperature change and the number of cycles used in the test depend on the specific requirements of the chip being tested. For example, ten of these temperature cycles may be required for lot acceptance, while a complete qualification test may take more than a thousand cycles. During the test, the chip is monitored for any changes in performance, such as changes in electrical conductivity or resistance that could indicate the presence of cracks or other defects in the chip.
- *High-voltage stress test (HVST)*: During this test the chip is subjected to extremely high voltages at wafer probe in order to activate any latent fab defects so that they can be screened out by later tests. A higher than usual voltage is applied to the chip, well above the normal operating range, for a specific period of time. In order to determine the duration for the stress, an HVST loop study is performed. The loop study consists of repeating a short HVST stress a number of times and then retesting the population of samples. The total HVST duration for the final test program is determined when adding shorter-duration HVST loops no longer cause additional test failures. This result is then checked by running the passing samples through appropriate burn-in to ensure that no more latent fab defects remained in the sample population. A successful HVST stress test implementation at wafer probe will eliminate the requirement for sample burn-in in order to achieve the desired quality levels. This is particularly important in industries such as the automotive, medical, servers and aerospace industries, where electronic systems must perform flawlessly under extreme conditions. While 10 Dppm (defect parts per million) was a traditional solid benchmark, the trend is now to move towards 10 Dppb (defect parts per billion) in these application domains.
- *High-temperature storage test (HTST)*: This is a test which is used to determine how a chip performs when subjected to high temperatures over an extended period of time. This test is conducted in order to ensure that the chip can withstand a range of operating conditions, including high temperatures that might be encountered during normal use. During this test the chip is typically stored for 1000hrs at a temperature of between 125°C and 175°C. The chip is then tested to ensure that it still operates correctly and that its electrical characteristics have not degraded as a result of the exposure to high temperature. The purpose of this test is to simulate the effects of long-term exposure to high temperatures that a chip may experience during use in real-world applications. This type of testing is particularly important for chips that are used in high-temperature environments, such as those found in automotive and aerospace applications.
- *Vibration test*: It is a type of mechanical testing that is performed to assess the ability of a semiconductor chip to withstand mechanical vibrations that it may

encounter during use in real-world applications. In this test, the chip is subjected to a series of mechanical vibrations at various frequencies and amplitudes. The vibrations are typically applied using a shaker table or vibration testing machine. The chip is then tested to ensure that it still operates correctly and that its electrical characteristics have not degraded as a result of the exposure to mechanical vibrations. The purpose of this test is to simulate the effects of mechanical vibrations that a chip may experience during its lifetime, such as those encountered in automotive and aerospace applications. The vibration test helps to ensure that the chip can operate reliably under such conditions without failing prematurely. The vibration test is typically performed along with other mechanical tests such as temperature cycling, mechanical shock and bending test to ensure that the chip is robust enough to withstand the various mechanical stresses it may encounter during its lifetime.

The required reliability specifications of an IC depend on its application field, the process node and process maturity, the package and the expected circuit ageing. It can therefore take a relatively long time before the quality and reliability of a new IC in a new manufacturing process reaches an acceptable level.

10.4.10 Conclusions

While packaging, in conventional IC designs, was seen as a necessity to be tackled at the end of the design process, today it is more and more a critical factor towards the success of an IC development. The combination of increased IC complexity, the drive towards miniaturisation and the continuous pressure on cost and power reduction will not make the design process easier. It is expected that packaging technology will get closer to its limitations. For wire bonding, reductions in bond pad pitches will slow down or maybe even stagnate. Alternative bond pad architectures and new die to package connections will have to be developed. To follow the fab technology miniaturisation, flip chip will gain in popularity, while, on the other hand, the clear distinction between fab and assembly will fade. Within semiconductors vertical integration architectures are increasingly being applied and new packaging trends will appear, where fab technologies will be used in combination with assembly techniques, as well as substrate manufacturing technologies will be combined with assembly techniques. Examples of vertical integration have already been discussed before and include high-bandwidth memory and the 3D stacking of chiplets.

SoC and SiP systems will live together, but each in its specific application domains, where their properties are exploited to their full advantage. One thing will be sure; SiP will be extended towards nonconventional technologies integrating MEMS-based applications, biosensors and/or optics.

Particularly in the latest process nodes, monolithic system-on-a-chip solutions are more and more replaced by advanced, more cost-effective 3D multi-chiplet architectures. High-bandwidth memories (HBM) and the stacking of chiplets are examples of this trend.

Finally, the package choice has a huge impact on the overall performance, quality and reliability of the chip.

10.5 Potential First Silicon Problems

When first silicon, either on a wafer or mounted in a package, is subjected to the first tests, one or even all tests might fail. Passing a test means that everything must be correct: the technology must be within specification, the tester operation must be correct, the test software (vectors and timing) must be right, connections between tester and chip (interface and probe card) must be proper and, finally, the design must be right. Therefore, passing a test means the logical AND of correct processing, correct tester and interface operation, correct software and, finally, correct design.

Especially in the beginning of the engineering phase of first silicon, problems may occur with the tester, its interface or the test software. Also, problems may arise from marginal processing or marginal design.

The following subsections discuss each of the different categories of failure causes.

10.5.1 Problems with Testing

Very complex ICs contain hundreds of millions to several billions of transistors and can have several hundreds to more than a thousand bond pads. It is therefore a tough job to locate the failure somewhere in the chip, when, for instance, one output signal fails. The relation between an incorrect signal on one of the output pins and the location of an internal failure is very vague. Dedicated advanced testing techniques are already included in the design to support testing. Because not all functional blocks have (direct) access to output pins, they will be part of a scan chain (see Sect. 10.2.1). In many cases, these scan chains run (and are tested) at lower frequencies. A potential problem is that such blocks are found to operate correctly on the tester (at a lower frequency) but may show failures when the chip is put in the application (board; speed check). Therefore, the chip should run at the same speed during scan test as in the application.

Test data, such as test vectors and expected output data from simulations, are also subjected to failures. Testing of complex high-performance VLSI chips requires a lot of different test vectors to be applied to the chip at the right time. Normally, the test response is compared with the "expected data", most of which is generated during the simulation of the silicon at the verification phase of the design. To reduce the number of test pins and test time, large parts of the chip are simultaneously tested via scan chains.

A reduction of the number of test vectors is often achieved by the implementation of *multiple-input signature registers (MISRs)* which allow compression of data over a number of clock cycles. The final data is then scanned out. Because such tests are not functional tests, they may not yet have been simulated thoroughly during the design phase, leading to incorrect test pattern generation or incorrect comparison data. Moreover, when a bit failure occurs in compressed test data (signature), it is very difficult to locate the cause of the failure. This requires a lot of simulation. Data compression techniques during testing must only be used if other techniques are not satisfactory.

Other causes of test errors are timing errors. Sometimes, the switch from a functional test to a scan test or vice versa may take more time on the chip for the multiplexers to adopt the new state. Waits must then be included in the test programs to properly test the chip. Even setup and hold times for input pins or the amount of load that the tester offers to a chip output pin must be thoroughly verified. In some cases, even the tester hardware might show problems.

An important, but not yet discussed, source of initial test failures is the probe card, which is used in the initial test phase during failure analysis on the wafer instead of on packaged dies. In such a test environment, limited ground bounce can only be achieved by taking several measures. These measures are all related to preventing or limiting current slew rates (di/dt). Placing decoupling capacitances close to the supply pads is one measure. Another measure is to prevent large (ground) current loops. This can be achieved by using star grounds instead of serial grounds; see Fig. 10.44.

Especially outputs can generate large current slew rates. The measurement of V_{OL} and V_{OH}, for instance, will often be done sequentially instead of testing simultaneously for all outputs. In conclusion, failures may arise during the development of the tests, during the development of the test boards and during the testing itself. Passing these test phases carefully can save a lot of time and frustration during the evaluation of first silicon.

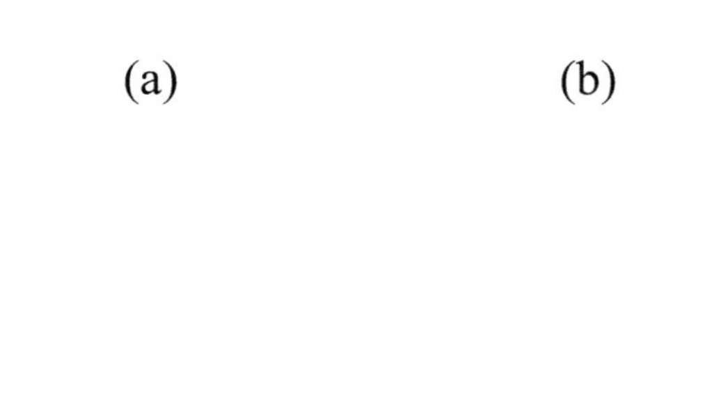

Fig. 10.44 Limiting large (ground) current loops by using (**a**) star grounds instead of (**b**) serial grounds

10.5.2 Problems Caused by Marginal or Out-of-Specification Processing

Each batch of wafers is processed under different environmental conditions: dust, temperature, humidity, implanter energy, etching time and doping levels, etc. This means that dies from different batches may show different electrical behaviours. The number of dust particles, for example, is one of the dominating factors that determines the yield; see Sect. 10.3. In the following, we describe the influence of the most important technology parameters on the electrical behaviour of the chip.

Gate Oxide Thickness

The gate oxide thickness is the smallest dimension in the manufacture of MOS devices. It controls the gain factor β and the threshold voltage V_T, and it can also affect the IC's reliability.

When the gate oxide is thin, β will be high and an increased current capability of the transistors will be the result. In some circuit blocks, especially in memories, signals have to arrive in a certain sequence and they therefore propagate through different delay paths. However, when transistors become faster, the difference in delay paths may change, or may even become negative. This may cause a race, resulting in malfunctioning of the circuit.

Thin gate oxide may also lead to *pinholes*. These are oxide imperfections at locations where the oxide thickness is locally reduced (Fig. 10.45).

Sometimes, the oxide thickness at such a pinhole may be so thin that the voltage across it might cause carriers to tunnel through this oxide. The resulting leakage current increases slowly over time and eventually, as a result of this oxide breakdown mechanism, the chip no longer functions correctly. This process might take an hour, a week, a month or even a year. The sooner it is detected, the better. However, detection after shipping the device to customers will be disastrous and lead to customer returns. Therefore, a bad gate oxide reduces reliability and can often be detected by means of I_{ddq} testing. Today most semiconductor fabs use an *atomic layer deposition (ALD)* step for the formation of very thin layers with atomic-size thicknesses. ALD is capable of a pinhole free gate oxide deposition (with variety

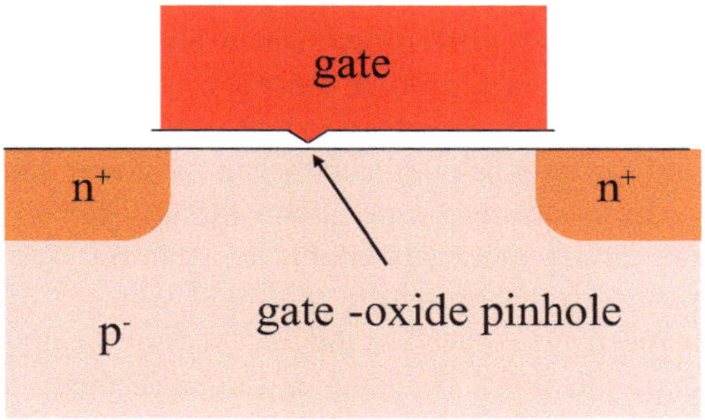

Fig. 10.45 Example of a pin hole in the gate oxide

of dielectric constants) with an extremely well-controlled thickness; see Sect. 3.5 and [27].

Polysilicon Width

The distance between the source and drain of a transistor (called the channel length) is determined by the polysilicon width, forming the gate of the transistor. The wider the polysilicon, the larger the transistor channel lengths will be and the slower the transistor becomes. Signals that propagate through a combination of metal tracks and transistors will show different timing diagrams when polysilicon widths are smaller or larger than expected, which could have been caused by imperfections during the lithographic process, or due to line-edge roughness, for example. This may lead to timing problems as a result of slowly operating circuits. On the other hand, narrow polysilicon leads to fast transistors. This may again result in timing problems such as races.

Threshold Voltage

A change in threshold voltage can have different effects on the electrical behaviour of the chip.

A high threshold voltage, caused by a different channel dope, a thicker gate oxide or a larger body factor (K-factor), results in slower operation of the transistors. Especially a high body factor may lead to problems in pass-transistor logic and latches that use pass transistors. This may cause these circuits to operate significantly slower.

In contrast, a low threshold voltage results in somewhat faster circuits. Subthreshold currents, which increase by a factor of about 18 for every 100-mV decrease of the threshold voltage, may cause larger standby currents. This is an important consideration in battery-driven applications.

A variation in the threshold voltage can have severe impact on the performance of analogue, RF and digital circuits, while it also determines their noise margins.

Substrate (p-Well) and/or n-Well Dope

All together the n^+ diffusions of an nMOS transistor, the p^- substrate (or p-well), the p^+ diffusions of the pMOS transistors and the n-well form parasitic thyristors.

When the p^- substrate is pulled to more than a junction voltage ($\approx$ 0.6–0.7 V) above the n^+ diffusion, such a thyristor might switch on; see also Sect. 9.5.5. Because of the positive feedback in such a thyristor, it operates like a latch and the current may increase to unacceptably large values. This effect is called latch-up and can only be eliminated when the power supply is switched off.

Low substrate dope allows the thyristor to switch on much earlier and makes the circuit more susceptible to latch-up. The doping levels of substrate and n-well also determine the threshold voltages of the nMOS and pMOS transistors, respectively, as well as the thickness of the depletion layers across their source and drain junctions. The latter, in turn, determines the parasitic junction capacitances.

Next to these examples of how process technology can impact circuit behaviour and reliability, there are several physical mechanisms (such as STI stress, hot-carrier injection, well-proximity effects and NBTI) that will also cause variations in β and V_T. These effects are modelled quite well starting from the 65-nm technology node. These also form potential causes of performance reduction, which may lead to incorrect chip behaviour.

10.5.3 Problems Caused by Marginal Design

Currently, verification software for integrated circuits has evolved to mature tools that are part of every design flow. Especially the verification on register transfer level (RTL) and logic level (gate level) offers the potential of designing chips in which no logic error can occur. These tools almost guarantee that everything on the chip is connected correctly according to the specification. It is therefore important to first verify the specification, either by simulation or by emulation. Sometimes, in an application, the chip does not perform the function it was supposed to execute. In several cases it turned out that the specification was insufficiently verified.

A hardware failure in very complex programmable chips can sometimes only be detected during very dedicated application tests. The number of different applications (and thus programs) of such chips is almost unlimited and extremely hard to simulate within an acceptable time.

Currently, most ASICs are designed in a mature process via a mature design flow and run at medium clock frequencies. First-time-right ASICs therefore should be the rule rather than the exception. However, modern logic processes (10-nm CMOS and below) offer high transistor densities and thus the ability to integrate up to a hundred billion transistors on one single chip. This, combined with the trend of increasing chip area, the use of multi-voltage and multifrequency domains and/or adaptive voltage scaling techniques, challenges the designer with many potential electronic problems that are, although supported by the tools, not easy to deal with. Chapter 9 focuses on the underlying physical effects and on the measures that a designer can take to maintain the IC's reliability and signal integrity at a sufficiently high level.

10.6 First-Silicon Debug and Failure Analysis

10.6.1 Introduction

Current VLSI chips may contain hundreds of millions to several billion transistors, with only several hundred I/O pins. This means that hardly any logic block has direct access to output pins.

Without a direct access to the output pins, the other blocks must be accessed through a scan chain and tested as such (see Sect. 10.2). In many cases, these scan tests run at a lower speed. This might lead to problems that show up only when the blocks are used in the real application because only then are all circuits running at full speed.

Logical (design) errors are easy to locate, both in scan test and in full functional test. On the other hand, the identification of timing errors is much more complex!

When failures show up during the debug and engineering phase of an IC, it is important to know their source: whether it is logical, short circuit, latch-up or timing failure, etc. I_{ddq} testing is a means to quickly detect leakage currents and floating nodes, etc.

For circuits that can be tested at full functional speed, Shmoo plots can be drawn to gather information about the behaviour of the IC. Afterwards, different *failure analysis (FA)* techniques can be applied to locate the failure. The conventional failure analysis techniques such as picoprobing, liquid crystal and electron beam usually need access to the circuits from the frontside of the wafer or chip. Current FA techniques also use the backside to get access to the circuits, e.g. time-resolved photoemission and many *scanning optical beam (SOM)* techniques. There are several other techniques that support these analysis tools and allow a quick repair of only a few samples.

10.6.2 I_{ddq} and ΔI_{ddq} Testing

I_{ddq} and ΔI_{ddq} tests are described in Sect. 10.2. In the following text with I_{ddq} we intrinsically mean I_{ddq} and ΔI_{ddq}. When a synchronous chip has been completely designed in static CMOS, hardly any current should flow when the clock is switched off. The only currents that flow are leakage currents caused by subthreshold and gate leakage mechanisms. However, in some cases, local higher-amplitude currents can flow. I_{ddq} testing, which is extensively discussed in Sect. 10.2, is therefore a good means of locating certain defects or unusual behaviour which cause increased current levels during steady state.

10.6.3 Traditional Debug, Diagnosis and Failure Analysis (FA) Techniques

This section discusses debug and FA techniques that were already in place during the last century. They may still be in use, either for designs in conventional CMOS technologies with only a few metal layers or for designs in which special arrangements have been made, e.g. taking critical nodes up to the top metal so they can still be probed from the frontside of the wafer or chip or after de-processing of the IC.

Diagnosis via Shmoo Plots

When a complete chip or part of a chip can be functionally tested, and an insight about operating margins with respect to the specification is required, then a *Shmoo plot* can be made. The name Shmoo plot is after a character introduced by Al Capp in his comic strip Li'l Abner in 1948. It could adopt all kinds of shapes.

A Shmoo plot shows the operating area of the software, the tester, the interface between tester and chip and the chip itself, with respect to different parameters. When a Shmoo plot is not according to expectation (specification), the failure does not necessarily need to be in the IC design. It can also be in the technology, tester software or interface or the tester itself.

A Shmoo plot, which shows the operating area of a chip, is, in fact, a quality measure. It shows whether the chip is marginal with respect to its specification (see Fig. 10.3). Measurements of Shmoo plots can be repeated at different temperatures to see how the margins shift.

Once the environment (tester, tester interface, connections, etc.) has proven to be correct, then, if the small operating areas of the chip are found to be too small, several different Shmoo plots must be measured to find dependencies—supply voltage, frequency, setup time, temperature and I/O levels, etc.:

- If delay paths between flip-flops are too long:
  → frequency versus supply voltage Shmoo plot: lower frequency → better operation and higher voltage → faster circuits
  Conclusion: ⇒ use frequency versus supply voltage Shmoo plot at a fixed temperature.
- If races, which are independent of the frequency, occur:
  → supply voltage versus temperature Shmoo plot: higher voltage → faster circuits and higher temperature → slower circuits
  Conclusion: use supply voltage versus temperature Shmoo plot at a fixed frequency. For this test, often Shmoo plots with error count are made. Such Shmoo plots show bands of errors, which are independent of the frequency (Fig. 10.46).
- A Shmoo plot diagnosis may take a lot of time. Once a diagnosis has been made, it must always be verified by other techniques (such as probing). This is shown by the following example. Although this example refers to a chip in a conventional 0.25 μm CMOS technology, it is still very well suited for educational purposes.

Example Although this example discusses a rather old 0.25-μm CMOS design, it is still a perfect example of what could happen today when designers are too quick to draw conclusions.

A certain signal processor contained two separated V_{dd} supply connections, V_{dd_1} and V_{dd_2}, which should have been connected together on the chip, but they were not. Figure 10.47 shows the Shmoo plot of the operating area of the memory on that chip:

Both supplies are connected to the same V_{dd} on the board. The 45° line in the plot shows the operating points for which V_{dd_1} equals V_{dd_2}. The Shmoo plot shows that when there is only a small on-chip supply noise in one of the supply domains, the chip would no longer operate. After a first inspection, the input registers of the memory were suspected. Because the ϕ and $\overline{\phi}$ clocks were generated by a clock

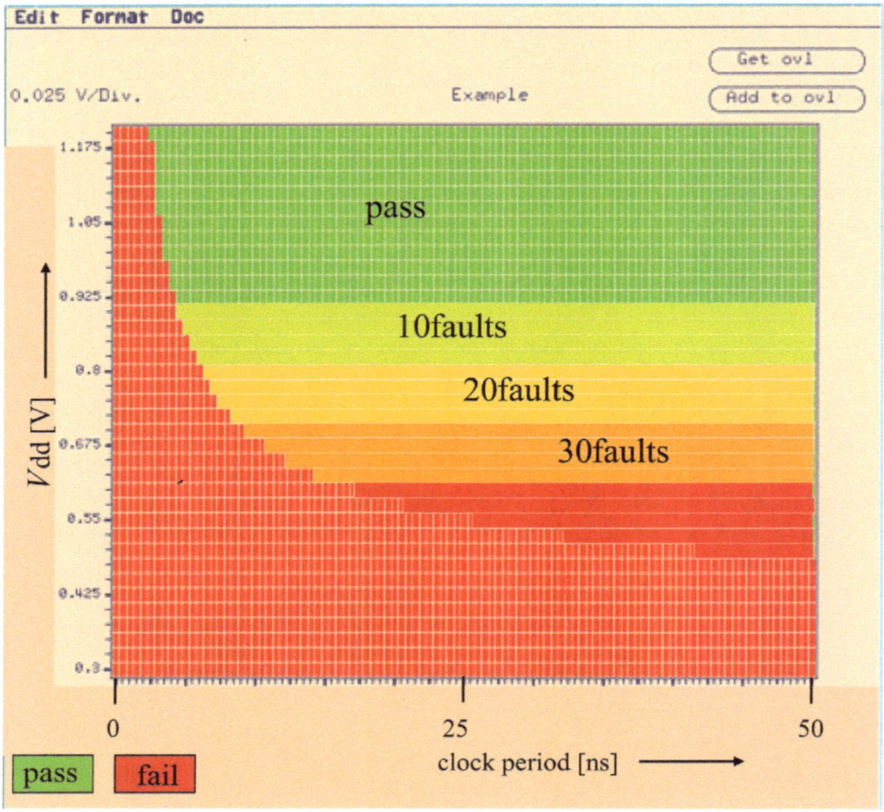

Fig. 10.46 Shmoo plot showing bands of faults, which are independent of the frequency

buffer which was supplied by V_{dd_1} and the latches (Fig. 10.48) of the input register were supplied by V_{dd_2}, these latches could be the cause of the problem.

Inverters 1 and 3 have a switching point equal to about $V_{dd_2}/2$, because the pMOS transistor width is three times the nMOS width (so $\beta_n \approx \beta_p$). Clocks ϕ and $\overline{\phi}$ are supplied via power supply V_{dd_1}. When $\overline{\phi}$ is high (V_{dd_1}), the voltage on node A will not be higher than: $V_{dd_1} - V_{T_a}$.

Because of the back-bias effect, V_{T_a} will be relatively high. Therefore, if $V_{dd_1} - V_{T_a} < V_{dd_2}/2$, the flip-flop will fail to switch to a logic '1'. The results of a circuit simulation using worst-case process parameters are shown in Fig. 10.49.

Below the solid line, the circuit operates correctly; above the line, it does not. If we compare this with the Shmoo plot of Fig. 10.47, we see that they are almost completely identical, and one would believe that the flip-flop is the real cause of the problem.

However, before changing the flip-flop design, the outputs of the flip-flops were probed to check the diagnosis. These flip-flops happened to operate much better

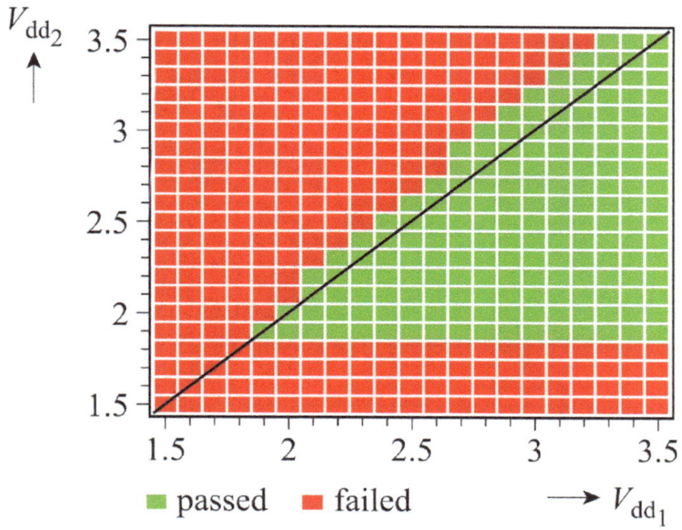

Fig. 10.47 Example of a very critical Shmoo plot

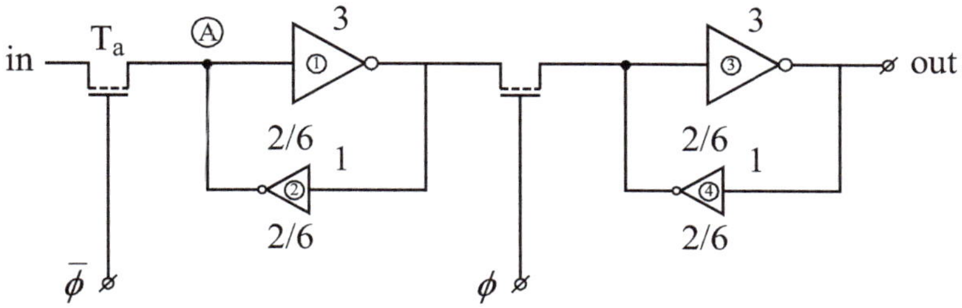

Fig. 10.48 Circuit for potential cause of the problem in the video signal processor

than simulated and thus the real cause of the problem had to be found elsewhere. The process was certainly not in the worst-case corner.

Although a legacy technique with picoprobes was used for further analysis, this example is still representative for the necessity to check a potential cause of a failure.

Diagnosis via Probing

Probing is a method that allows us to measure any node that is available at the top-level metal when there is still no passivation layer (scratch protection) on the wafer, or when this layer has been removed locally (by etching, laser cutting, or focused ion beam [FIB]).

Conventionally, picoprobes were used, Fig. 10.50. They consist of needles as thick as a hair and with a very thin tip, less than several tens of a micron. This needle is connected to the input of a FET to reduce the capacitance. Values of 5

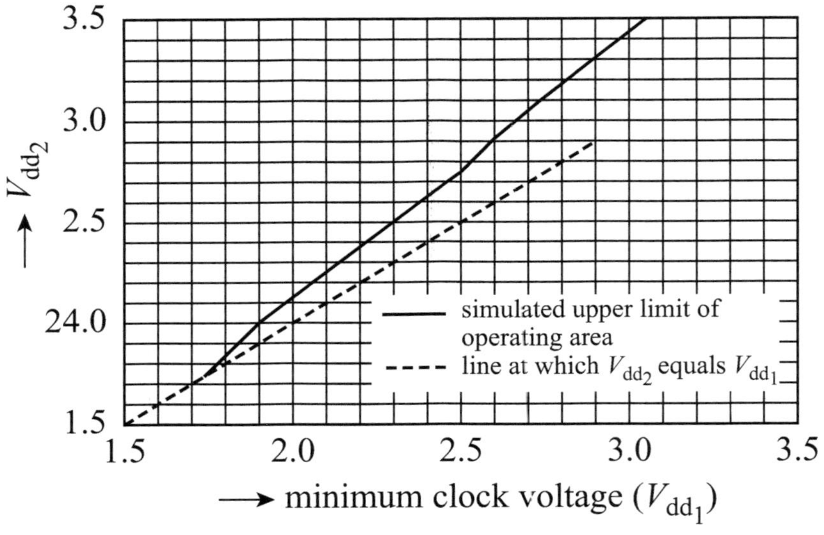

Fig. 10.49 Circuit simulation of the operating area of the latch of Fig. 10.48

to 100 fF for such a FET probe are available and are so low that they can be used within a digital IC almost without affecting the probed signal itself.

This technique was a reliable method for analysing incorrectly operating VLSI chips in semiconductor technologies with up to three metal layers. It can therefore only be applied in products that are still made in such conventional IC processes and in certain advanced memory products that also use only three metal layers. However, with the advent of multi-level metal technologies, it became increasingly difficult to probe a signal that is only available in the lowest metal layer(s). During the design phase, additional metal stacks could be placed at the critical nodes to create probe pads. Another way to cope with this problem is to adapt the design style to design for debug; see Sect. 10.6.7.

In the previous example, picoprobes were also used to locate the failure, which resulted in finding the real cause of the failure. Eventually it turned out to be a timing error in the word line decoder circuitry. Figure 10.51 shows the corresponding schematics. When a new read address is given, it takes a while before the selected word line signal at node B becomes active (high). This signal should be stable before the clock signal at node A arrives.

However, the one-shot pulse on node A came 100 ps too early. At that moment the row address of the previous clock cycle was still available at the address decoder pins. This resulted in reading the wrong word from the memory. By cutting track A with a laser, probing the one-shot signal right before the cut and forcing it back via a pulse generator with variable delay right after the cut, the correct pulse could be found and a new Shmoo plot was measured. Figure 10.52 shows the result.

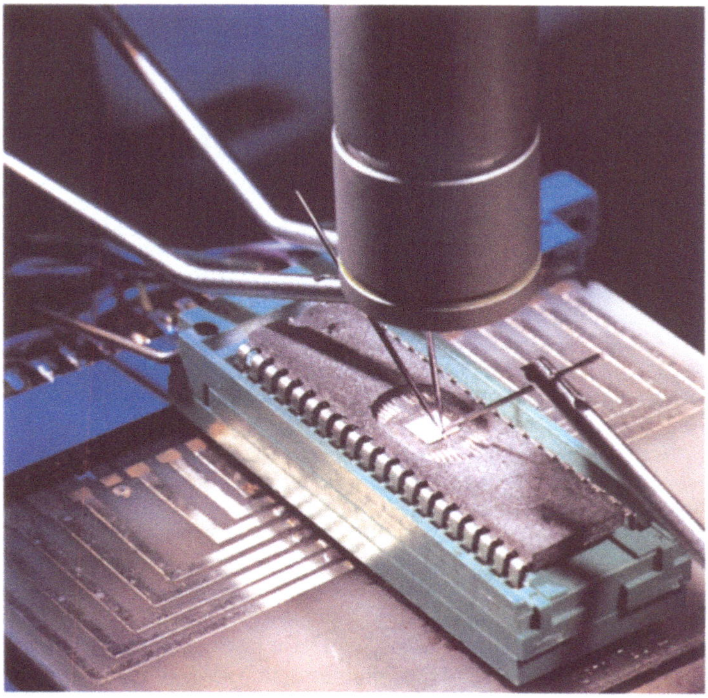

Fig. 10.50 Picoprobes were used to measure a chip's internal signals (Source: NXP Semiconductors)

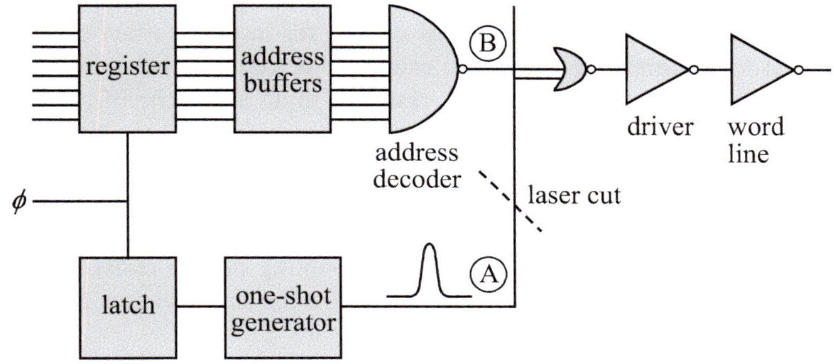

Fig. 10.51 Schematic of the real cause of failure

Thus, probing identified the exact location of the failure and also a way to solve the problem. A redesign (a one-mask change only) was made and the devices in the next batch operated correctly.

Besides picoprobing, there were several other conventional techniques to accommodate failure analysis, such as liquid crystal and electron beam. Because of the

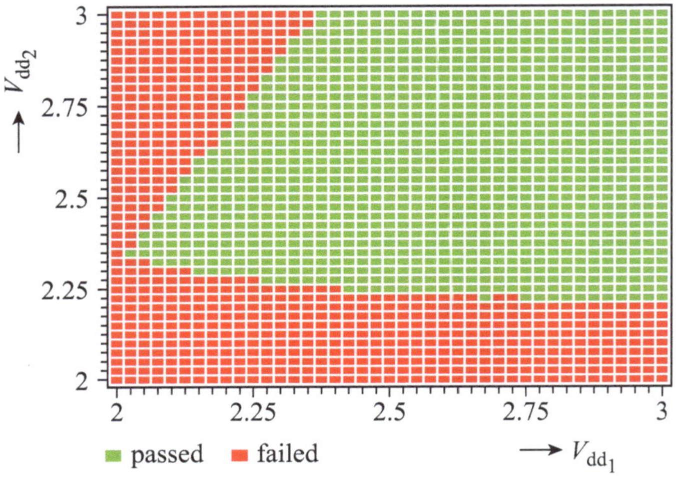

Fig. 10.52 Shmoo plot after correcting the one-shot pulse by forcing it via picoprobes

large number of metal layers which shield the lower signal lines and devices, these techniques are only applied occasionally and have been replaced by such techniques as laser scanning and time-resolved photoemission. We will therefore no longer focus on these legacy techniques.

Diagnosis by Photon Emission Microscopy (PEM)

When charge carriers decay to a lower state of energy, the energy surplus is converted to photon emission (PE). This occurs, for instance, when electrons are accelerated in the transistor channel to excessive velocities, when carriers cross a potential barrier, or during breakdown, resulting in an avalanche of carriers [28]. The *photon emission microscopy (PEM)* tool, also referred to as *light emission microscopy*, utilises a high-gain camera to detect the extremely small amounts of light emitted by some semiconductor devices and defects. It is therefore a good tool in identifying hot electrons. The light that is given off by operating ICs is captured by a microscope and used for imaging. The resulting photon emission image is overlaid with the GDS-II layout to easily determine the fault location. Figure 10.53 shows several hot spots on a CMOS chip layout.

The key performance parameter for PEMs is the overall sensitivity of the system, which is a function of the wavelength, but also depends on the optical system. Currents well below a microampere can be visualised at spatial resolutions less than 1 μm. In this way, latch-up, gate oxide defects, saturated MOS transistors, degraded (avalanching) junctions and unwanted forward-biased junctions can be detected.

The emitted photons can have both visible and near-infrared (IR) wavelengths. With the large numbers of metal layers, frontside analysis faces severe limitations. Because silicon is transparent to near-infrared light, IR PEM can also be used for die backside analysis. With highly sensitive cameras (LN2-cooled InGaAs camera,

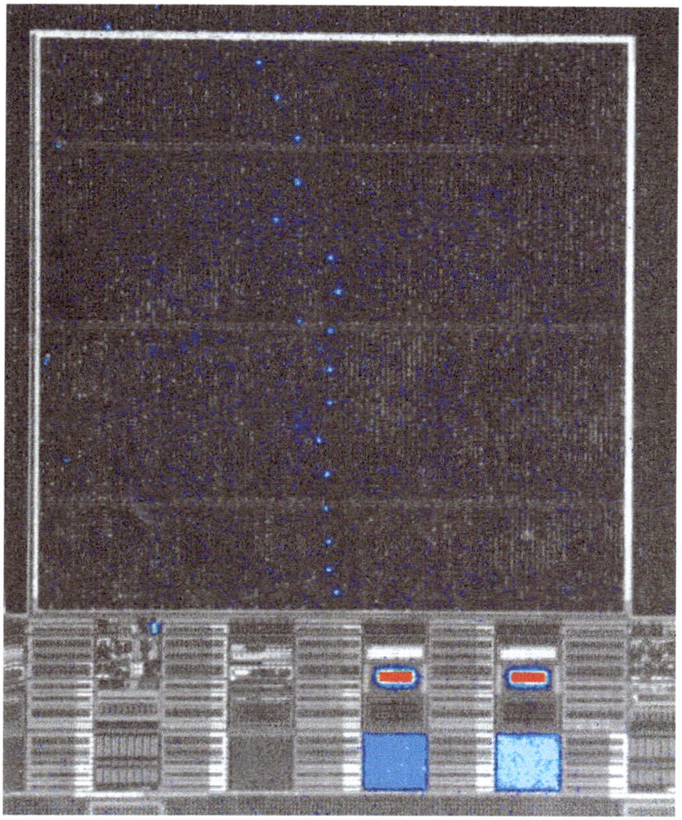

Fig. 10.53 Photoemission image of operating ARM microcontroller core in 90-nm CMOS, taken through the backside of the silicon. Field of view is about 300x300 microns. The 16 strong emission points in the core are part of the clock tree (Source: NXP Semiconductors)

900–1600 nm wavelength), this greatly reduces image capture time and prevents optical obstruction by multi-level metal layers and flip-chip packaging.

10.6.4 More Recent Debug and Failure Analysis Techniques

The continuously growing complexity and density of integrated circuits, both in terms of number of transistors and timing requirements, have increased the variety of failure mechanisms. These failures can be originated by manufacturing defects or by design-related failure mechanisms. Section 10.3.2 discussed DfM basics to improve yield and to reduce the number of manufacturing defects. Still many ICs are not first-time-right products and require a lot of support to reduce the debug, failure analysis and diagnosis time, in order to reduce time to market. *Design for*

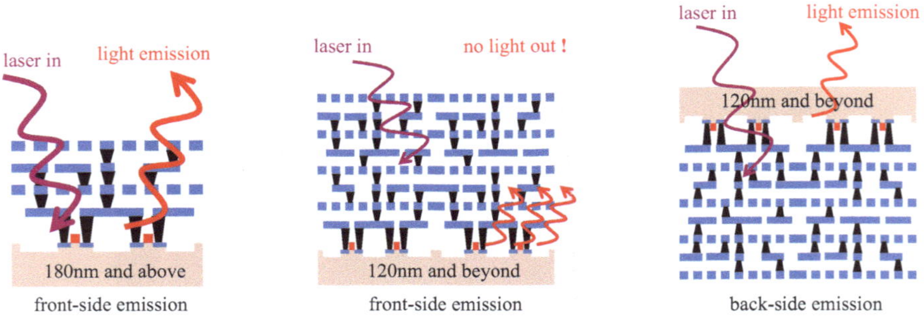

Fig. 10.54 Frontside FA techniques could be used in conventional technologies, but beyond 120 nm, backside techniques are required

debug is a step in that direction and helps to improve observability, both at system level and at the physical level.

Developments in failure analysis (FA) techniques have enabled the way to access critical nodes from the backside of the wafer or chip, because of the above-described inability to observe the transistors through analysis techniques from the frontside (Fig. 10.54).

One technique, which is based on photon emission microscopy but with improved sensitivity for backside usage, is the time-resolved PEM technique. The others are based on stimulating the circuit with either laser beam or electron beam. The following subsections present a flavour of state-of the art failure analysis techniques. For a more detailed summary on electron and laser beam failure analysis techniques, the reader is referred to references [29] and [30]. The chapter ends with a short discussion on techniques that can be applied already during the design phase to support the debug and failure analysis phase.

Time-Resolved Photoemission Microscopy (TR-PEM)

The basics of PEM are already discussed in the previous section, for the detection of breakdown, hot carriers, latch-up, gate oxide defects, degraded junctions and even saturated transistors.

Picosecond imaging circuit analysis (PICA) is a form of *time-resolved photon emission microscopy* developed by IBM. During switching of MOS transistors, light pulses are generated due to hot-carrier injection, Fig. 10.55.

When the substrate is thinned, a very sensitive, high-resolution, time-resolved camera can capture the integrated IR component of these light pulses through the backside of the IC. The pulses are a measure for the switching moment and can be used to measure the timing behaviour of any node in a digital circuit. The PICA technique images the switching activity of several circuits in a relatively large field. Because not every switch generates a photon, it takes many hours to a day to aggregate sufficient photons for the creation of the switching activity image in the scanned field.

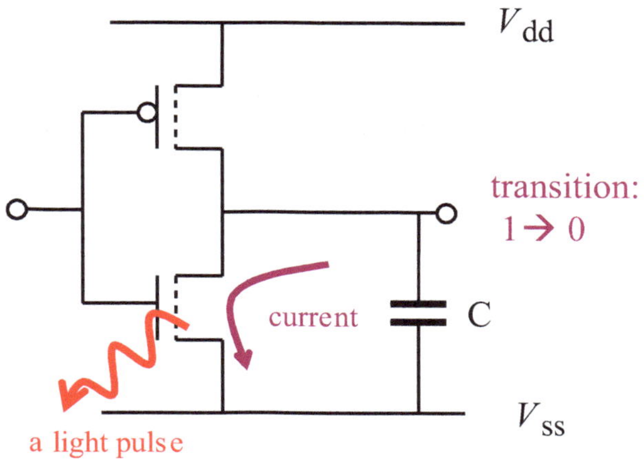

Fig. 10.55 The emission of a photon during switching transition of a logic gate

The technology is further developed by Credence Systems in their Emiscope tool. It is based on single-point detection, rather than on field imaging. It uses an avalanche photo diode to capture single photons from a single switching node over a certain collection time and transports it via an optical fibre to the time-resolving equipment. It can create a measured waveform in a few minutes to an hour, much less than was needed by the original PICA system. The number of photons generated by a switch is much less than one and is dependent on the voltage swing. In a 90-nm technology the number of photons per detected switch is in the order of 10^{-5}. For smaller technologies, with reduced supply voltages, the aggregation of sufficient photons will become more time-consuming. Hot electrons generate one to two orders of magnitude more photons than hot holes, which makes this technique better suited for the timing analysis of nMOS transistor switching.

Individual light pulses of transistors can be visualised by a histogram of detected photons versus time (Fig. 10.56). The signal is the result of the integration of photons accumulated by the microscope objective.

However, the silicon substrate has a relatively high refraction index ($n = 3.5$), which means that most rays are strongly refracted at the silicon-to-air interface and rays with angles more than seven degrees stay completely within the substrate. The result is a loss of 90% of the captured light. The solution to this problem is to bring a **silicon** *solid immersion lens (SIL)* in contact with the silicon substrate (Fig. 10.57).

In this way the rays are no longer refracted and the system can capture about ten times more light and, due to the higher NA, its improved resolution enables visualisation of even the smallest 90-nm CMOS devices (Fig. 10.58b). The SIL is also used to improve resolution of other laser-based FA techniques.

Scanning Optical Beam (SOM) Techniques (or Laser Signal Injection Microscopy [LSIM])

Basically we can divide scanning optical beam FA techniques into two categories. The first SOM category does not use a tester to generate test stimuli, but only applies

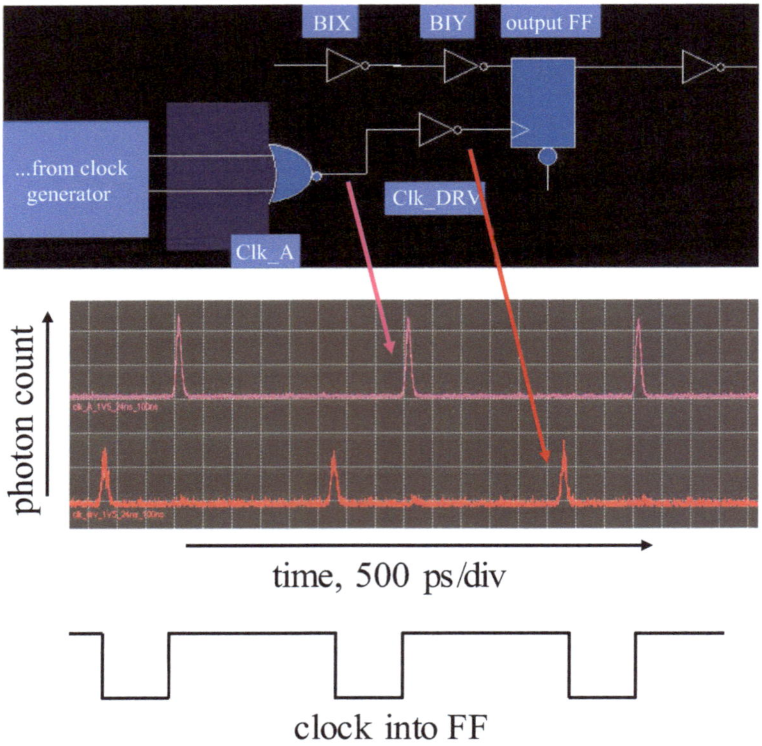

Fig. 10.56 A typical time-resolved PEM measurement result

a constant voltage source across the supply terminals to sense current changes, or a constant-current source to sense voltage changes, both as a result of internal circuit stimulation by thermal or charge induction.

The second SOM category requires an IC tester to create the required input stimuli and generate the optimum operating conditions to enable detection of even the smallest change in electrical performance. The operating point is then set at the edge of the Shmoo plot (Fig. 10.59).

In this point the chip is marginally operating and the circuit that causes the failure will be very sensitive to any injection of heat and/or charge into its critical node. Next, the chip is scanned with a laser beam, which induces thermal heating of the material (silicon, metal, etc.) in the laser spot or injects charge into a device. Once the spot has arrived at the critical node, it will introduce a change of its electrical behaviour, either in amplitude or in timing. The example Shmoo plot refers to a chip that exhibits too much delay in one of its critical paths. When the laser strikes this critical path, it will increase the delay because of the local thermal heating that it causes in the spot of the beam. When we keep the laser focused at this failing critical path and we measure the Shmoo plot again, it shows that the edge of the Shmoo plot

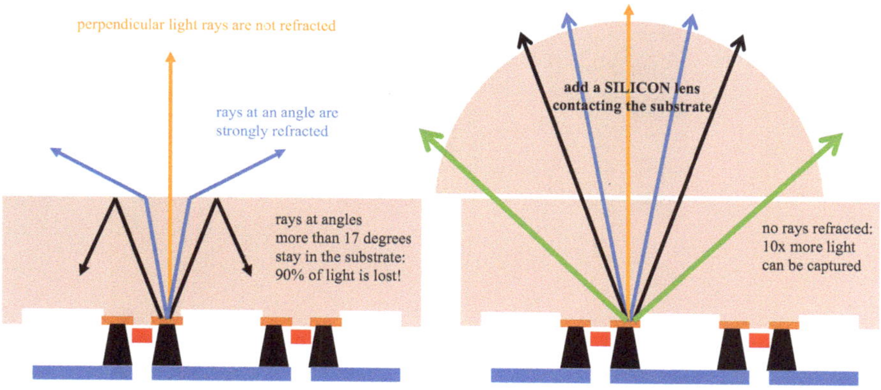

Fig. 10.57 A 90-nm CMOS IC viewed through the (backside) silicon with a conventional microscope objective (**a**) and with a solid immersion lens (SIL) (**b**). The size of the image is about 80 × 50 microns

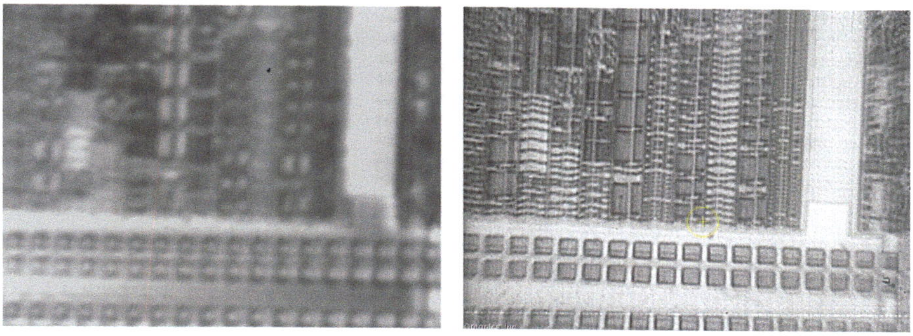

Fig. 10.58 (**a**) Image captured with a normal lens and (**b**) same image captured by a solid-immersion lens (SIL)

has moved, resulting in a smaller operating area of the chip. In this way it allows the detection of critical paths in a logic design.

Many FA optical techniques can be used both from the top and backside of the chip, depending on what layer needs to be analysed. If the failure happens to be in one of the in-between metal layers, it becomes very difficult to create access to that layer. Optical backside analysis exploits the relative transparency of silicon to (near) infrared light. Backside analysis requires thinning and polishing of the substrate since the transmission of IR light through silicon decreases exponentially with its thickness (Fig. 10.60). Particularly heavily doped substrates, which are much less transparent, often need to be mechanically ground down until a thickness in the order of 50 μm is reached and then further polished to achieve an adequate optical backside surface quality for proper light injection and propagation during laser beam stimulation. Many CMOS circuits, today, employ lightly doped substrates (see

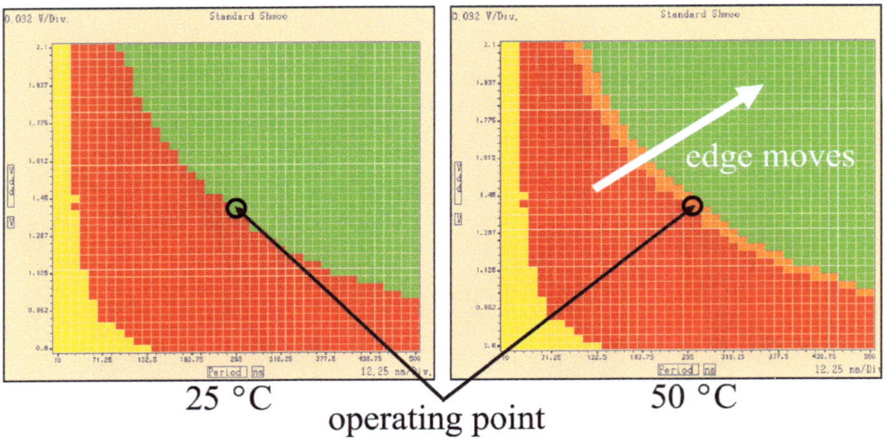

Fig. 10.59 Preferable operating point for failure analysis test

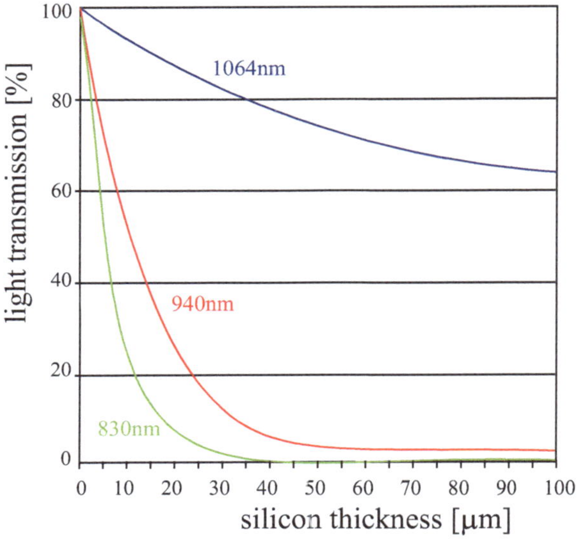

Fig. 10.60 Behaviour of light transmission vs silicon thickness at different wavelengths of the light

Chap. 3) and do not require thinning at all. The package has a large influence on the ease of use of backside failure analysis. Needless to say, that backside analysis is easier for flip-chip packaged devices.

Many of the laser beam stimulation techniques are based on the same principle of scanning a laser beam across the chip and measure potential changes in its electrical or logic behaviour.

The applied laser wavelength depends on its usage. Systems meant for thermal stimulation use a laser wavelength of 1340 nm and are most commonly used for detection of resistive changes in the backend of the process, e.g. the metal layers and

Table 10.3 Different categories of SOM failure analysis techniques

	Thermal stimulation	Photonic stimulation
Wavelength of applied laser	1340 nm	1064 nm
Application of constant-voltage source or constant-current source to sense current or voltage changes, resp.	OBIRCH TIVA, XIVA SEI (Seebeck effect)	OBIC LIVA
Application of IC tester to observe test results	SDL RIL	LADA Laser voltage probing

vias, or for the local heating of transistors to measure changes in timing behaviour. A laser wavelength of 1064 nm is used for photonic stimulation to create electron-hole pairs in the p-n junction, which may also change local timing behaviour.

There exist many alternative laser-based FA techniques. Based on the above discussion we can categorise them as shown in Table 10.3.

In most failure analysis techniques, the response is visualised by an image of (part of) the circuit or of the chip. It uses a *confocal laser-scanning microscope (CLSM)*, in which the laser beam is guided through a combination of mirrors across a chip. It creates a first image, which is based on the intensity of the reflected light. Next, an identical scan is performed, but now the current (or voltage, dependent on the specific FA technique) is measured with respect to the position of the chip. These current (or voltage) levels can be transferred into different grey levels, or colours.

When we define all current changes above 100 μA to become red, then the current measurement image shows these red spots only on the positions where the response to a laser strike results in a larger current. This image is then used as an overlay on top of the chip layout, so that the defect location can be easily matched with the position in the chip. The spatial resolution of conventional SOM techniques is limited to 200 to 300 nm (for visible light) by the applied wavelength and numerical aperture of the system. This was the primary motivation for the development of *near-field scanning optical microscopy (NSOM or SNOM)* with improved lateral resolution up to 20 nm. An example application of the use of NSOM in integrated circuit failure analysis can be found in [31].

Optical-Beam-Induced Resistive Change (OBIRCH), Thermally Induced Voltage Alteration (TIVA), Externally Induced Voltage Alteration (XIVA)
These analysis techniques are very similar and based on the generation of changes in IC power demands because of changes in the resistance of a short (*OBIRCH* and *TIVA*) when illuminated by a scanning laser beam which induces heat into the spot area of the beam. TIVA systems use a constant-current source at the supply terminals of the chip and measure the voltage changes. OBIRCH systems apply a constant voltage to the supply terminals and measure the current changes. Both constant-voltage and constant-current systems often offer about the same sensitivity. It depends on the impedance of the sample, which of the systems is the best one to use.

XIVA systems claim the same high sensitivity as TIVA by enabling constant-current sensing by supplying a constant voltage to the terminals of the device [32]. Actually, XIVA can be applied to detect all defect mechanisms, junction defects, opens, shorts, defective vias, etc., because it may use thermal or photonic stimulation, depending on the applied wavelength of the laser. Application of the technique is more difficult than TIVA and OBIRCH.

Seebeck Effect Imaging (SEI)
The Seebeck effect imaging is based on the fact that electrical potential gradients are generated in conductors because of the creation of thermal gradients. In correct functioning ICs, the potential gradient in a metal line, produced by local laser beam heating, is compensated by a current in a transistor driving that line. However, if a void in the metal line, or an open via, is isolating this line from a driver transistor, the potential of the open line will change because of the Seebeck effect, causing a change in the gate voltage of the connected transistors. These transistors will change their conducting states and cause a change in power demands of the chip. The SEI technique uses a constant current source applied to the supply terminals of the IC.

The laser may generate thermal gradients in a conductor of several tens of degrees Celsius, causing voltage changes in the order of a few tens of microvolts. The technique works from both the frontside and backside. For SEI no external bias is needed to observe a signal. SEI is today not very often used. One reason might be that the resulting voltage changes are very difficult to interpret.

Soft Defect Localisation (SDL) and Resistive-Interconnect Localisation (RIL)
This technique can be applied to both the front and backside of the chip to detect soft defects (e.g. gate oxide leakage, resistive vias, process spread, marginal timing). It is based on the fact that when a 1.3-μm laser is targeted at a soft defect, the defect changes its electrical behaviour during testing. As discussed previously, the IC is operated at the edge of the Shmoo plot, such that the device only functions marginally. When the laser reaches a sensitive position in the chip, it might change the pass/fail status, e.g. by changing the resistance of a defect or by changing the conductive state of a transistor to change timing behaviour of critical signals.

Light-Induced Voltage Alteration (LIVA), Optical-Beam-Induced Current (OBIC) and Laser-Assisted Device Alteration (LADA)
LIVA analysis can be performed either from the frontside using a visible laser or from the backside using a photon beam generated by an infrared laser. Next to junction defects, this technique can also be used for a localisation of opens in interconnections, contacts and vias, by creating an image of the entire chip. Such an image is the result of the monitoring of the voltage fluctuations of a constant-current power supply during a laser beam scan across the chip, using a scanning optical microscope (SOM), which allows zooming in to the physical defect location. Both LIVA and OBIC analysis are based on photon-induced electron-hole pairs, which are representative for the material to which the laser spot is focused. Recombination of the electron-hole pairs causes a change in the power demands of the chip, leading

to voltage fluctuations that are dependent on the amount of generated electron-hole pairs. When, for example, the spot is focused onto a transistor junction which is connected to an open conductor, it will result in a change in conductive state of the transistor(s) connected to that junction and causes a change in the supply voltage. A photomultiplier detector is used for the visible-light laser, while a germanium-diode detector is used for the infrared laser(s). The LIVA system produces an image of the voltage fluctuations in relation with the position on the chip. When a laser spot arrives at a junction, the junction will always cause a supply voltage fluctuation. However, it is difficult to interpret the measurement data, because one has to discriminate between the fluctuation caused by a correct (connected) junction and that of a defective junction.

The resolution of the OBIC technique is relatively low and can be used for analysing junction sizes of 0.18-μm technologies and larger. For smaller technologies its usage is limited to large transistor junctions and wells. For nanometer ICs this technique will lose more of its popularity.

Laser Voltage Probing and Laser Voltage Imaging
Both laser voltage probing and laser voltage imaging are noninvasive and destruction-free FA techniques that are based on optical measuring of transistor activity through a thinned die (backside stimulation/probing), very similar to the above LIVA technique. Laser voltage probing (LVP) is used to measure electrical signals in a chip. This technique uses a laser beam which is focused at a specific point in the chip where it creates a pulse of light, which induces electron-hole pairs in the silicon. A photodetector then measures both the amplitude and phase of the reflected light, which are measures for the voltage behaviour of that specific device. The technique is capable of detecting voltage variations of less than a millivolt. It is therefore very useful for analysing the electrical behaviour of a chip during operation. It can help in identifying defects or performance issues in the chip and can also be used for debugging and characterisation of the IC. The laser voltage probing requires specialised equipment and expertise to operate. Its implementation involves several steps, including grinding the chip, controlling the laser beam and analysing the output signals. Laser voltage imaging (LVI) is a technique used to analyse the electrical properties of semiconductor chips by using a laser beam that follows a raster pattern across the chip to create a 2D image. It obtains voltage waveform and timing information by monitoring the reflected light whose amplitude is proportional with the changing node voltages. These voltage changes across the chip can be visualised and analysed to determine the distribution of electrical properties and potential hot spots that could cause device failure. The technique is extremely useful in localisation of defects in circuits generating specific frequencies within modern integrated circuits and serial chains, such as scan chains. Different causes of scan chain failures can be visualised:

- Timing problems between scan chains of different cores: clock skew or data timing
- Opens or bridges in the scan chain data path

- Defects in the clock tree or scan-enabled signal
- Defects in the scan chain control logic

By mapping the voltage distribution across a chip, LVI can provide insights into circuit performance, power consumption, timing and signal integrity. One of the advantages of LVI is that it is nondestructive, meaning that the chip can be imaged repeatedly without damaging it. This enables the monitoring of the voltage distribution across a chip over time, providing valuable insights into the circuit's degradation or ageing. Ageing can cause transistors to slow down, switch on and off more erratically or fail completely. Additionally, LVI is a fast and accurate technique, with image acquisition times ranging from a few milliseconds to a few microseconds.

Scanning Electron Beam Microscopy (SEM) Techniques

Most, if not all, scanning electron beam techniques are used during FA after the location of the failing mechanism or defect has already been determined by one of the previously discussed FA techniques, due to the about two orders of magnitude better resolution. It also has a large depth of focus, which allows that during imaging a large part of the circuit is within focus. The basic idea behind the use of electron beam microscopy for failure analysis is analogous to optical (laser) beam techniques, but with the difference that an electron beam is used as stimulus instead of an optical beam. SEM operation is based on the capture of backscattered and secondary electrons produced by the sample. *Backscattered electrons* are electrons from the original beam that are attracted by the gravitational force of the positive nucleus, such that it will partly circle around the nucleus and then leave it with a different angle, without losing speed. *Secondary electrons* are electrons that are physically hit by the strike of an electron of the original beam, due to the fact the beam electron repels the device electron. If this repulsion is large enough, this secondary electron may be pushed out of the atom and may exit the device through the surface of the die, but much slower than the backscattered electrons. A positive charge on the detector must physically attract these secondary electrons to create an image.

The amount of secondary emission depends on the voltage on the scanned node (metal line). Metal lines at V_{dd} level absorb most of the secondary electrons and provide dark fields in the voltage-contrast image. Active voltage-contrast SEM techniques require the chip to operate to create the image. In passive voltage-contrast SEM techniques, the contrast is created by the beam. When the beam strikes an open metal, it will charge the metal line to a different potential than when it strikes a normal connected metal line.

Similar to SOM FA techniques, with SEM techniques the sample also needs to be prepared (grinded) to create access for the beam to the point or layer of interest. In many cases also FIB tools are used for the deposition of additional test points (probe pads) on the metal connections to the suspicious node. In literature, a couple of scanning electron beam FA techniques are described, particularly *(low-energy) charge-induced voltage alteration ((LE)CIVA)* and *electron-beam-induced current*

(EBIC). However, these techniques are generally not very extensively used and a detailed discussion on their application is therefore beyond the scope of this book.

Electron beam or e-beam technology is also an important tool in testing and failure analysis of *stacked dies*. Stacked dies refer to multiple semiconductor dies that are vertically stacked and interconnected to form a single device. The use of stacked dies is becoming increasingly popular in the semiconductor industry due to its ability to increase device functionality and reduce size and cost (see chiplets in Sect. 10.3.3). However, the manufacturing and testing of stacked dies presents several challenges, including the need to create high-quality interconnects between the stacked dies. Moreover, stacked-die technology is also prone to reliability issues like thermal stress and performance degradation due to the high-density packaging. This is where e-beam technology comes in.

10.6.5 Observing the Failure

After the test has identified a failing chip, the debug and failure analysis techniques support the identification of the failure location. If the failure is caused by a manufacturing defect, such as a short circuit or an open circuit, it is necessary to make a detailed material analysis of the defect to understand its cause.

The next example shows the failure analysis process, starting with the Shmoo plot from a failing device all the way to a TEM cross section of the physical defect. The example is related to a device, which has passed all tests, except for a delay fault test (see Sect. 10.2.1). Shmoo plots (Fig. 10.61) of the digital tests confirmed that a certain critical path suffered from an additional propagation delay of around 14 ns (compare Shmoo plots [a] and [b]) at a supply voltage of 1.8 V. The Shmoo plots show the operating area (green) as a function of the reducing clock period (= increasing frequency) (horizontal) and supply voltage (vertical). The first plot shows the Shmoo plot of the delay fault test of a correct operating reference device, while the second one represents the Shmoo plot of the same test for the failing device.

By using a software tool that combines logged test data with an image of the IC, the particular failing net could be localised in the same way as described below in Fig. 10.59. The result is depicted in Fig. 10.62, with the failing net highlighted.

Next, backside SDL is used to visualise the response to the laser scan. Figure 10.63(left) shows a zoom-in backside image with GDS-II layout overlay and the failing net. No spots were visible at the driver side of the net, but all three gates on the receiver side of the net showed a response to the laser scan. This suggests that there exists a relatively high resistance somewhere between the driver and the receiving gates.

Because backside SDL did not show further responses to the laser scan, the device was opened from the top and a frontside laser scan was performed. This yielded a clear response from only one VIA-5 between the fifth and sixth metal layer. Because this via was in the top of the metal stack, it was not directly visible with the backside SDL analysis. Figure 10.63(right) shows an overlay of the failing

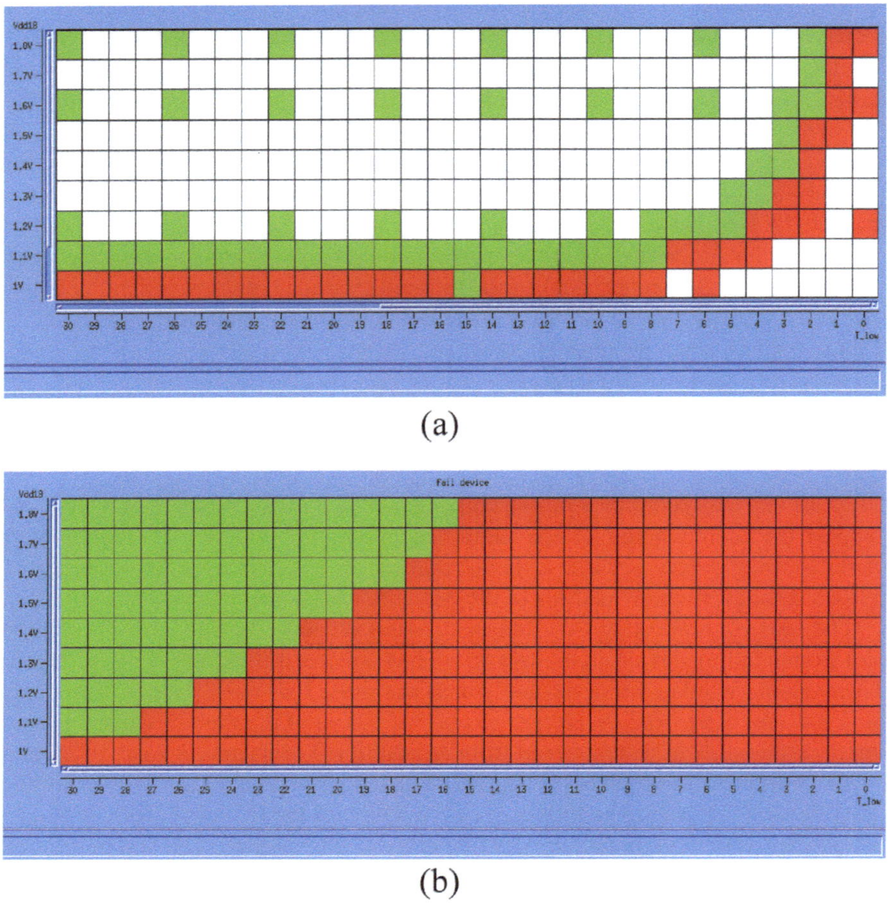

Fig. 10.61 Shmoo plots of a correct device (**a**) and of a failing device (**b**)

net on a frontside image of the IC. Because the response is not clearly visible in the figure, a more detailed view is presented in Fig. 10.64a.

The GDS-2 overlay helped to locate the coordinates of the defective via. Once the location is known, the failure analysis is continued to find the real cause of the failure. A thin cross-section slice was cut from the chip by a focused ion beam (FIB). Figure 10.64b shows the cross section as viewed in a transmission electron microscope (TEM). It clearly shows the defect in the via-5. The via-etching process step did not stop at the top of the titanium nitride capping layer. Finally, the enlarged image of the defective via showed that there was only partly contact between the via and the metal. This caused the increased resistance in the delay path, which was the cause of the original delay fault test.

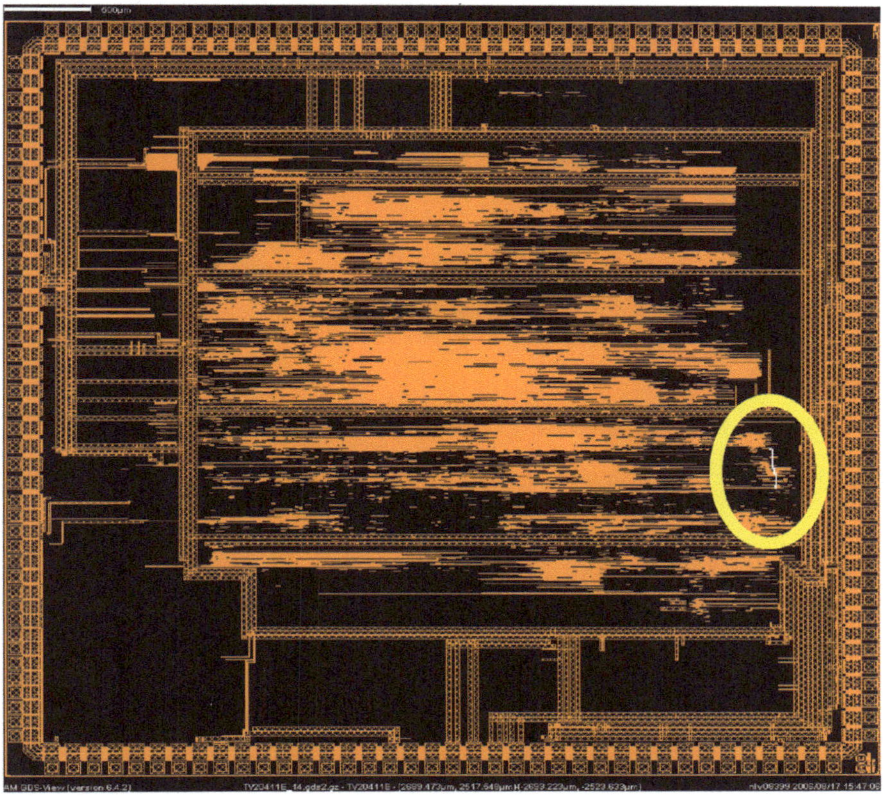

Fig. 10.62 Image of the die with the failing net highlighted

This example shows that the complete failure analysis process, from a failing test, down to the defect, is a costly, time-consuming and nontrivial effort, which may dramatically increase time to market of the final product. Therefore, a well-equipped failure analysis lab, combined with highly skilled FA engineers, is an essential part of the process of bringing a new design to volume production.

10.6.6 Circuit Editing Techniques

Once the diagnosis has been made, the chip can be repaired directly by making and breaking techniques. The ability to physically edit an IC *(circuit editing)* may reduce the number of respins and helps in reducing time to market.

Traditionally, laser beams were used for both cutting lines and laser-induced liquid-phase metal deposition to create interconnections on top of the scratch protection. A disadvantage of laser systems is that their resolution is limited, which

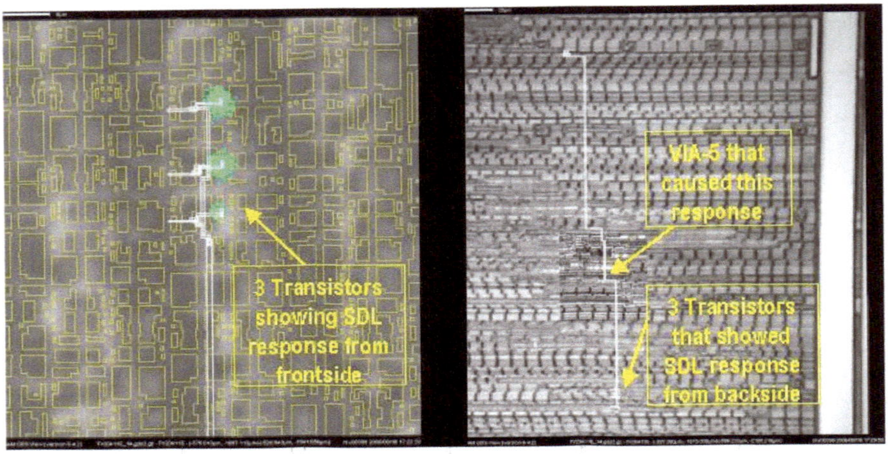

Fig. 10.63 Zoom in of a GDS-II layout backside image with an overlay of the SDL laser response (left) and and an overlay of failing track on a frontside image of the IC (right)

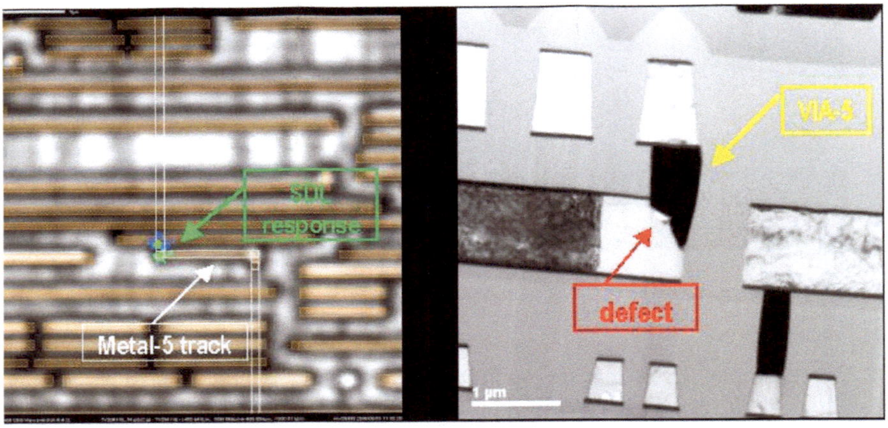

Fig. 10.64 (**a**) Zoom in of the SDL response of the failing VIA-5 in the overlay of the GDS-2 and (**b**) a TEM image of the defect

has made them much less popular for circuit editing techniques in deep-submicron and nanometer ICs.

Focused ion beam (FIB) systems show better resolution: they can achieve imaging resolution better than 5 nm, while for cutting conductors, spatial resolutions of less than 10 nm have already been demonstrated. Operation of a FIB system is similar to that of a SEM. Instead of an electron beam, a FIB system uses a focused beam of gallium ions, which is scanned across the chip, to image the sample in a vacuum chamber. At the location where the beam strikes the chip,

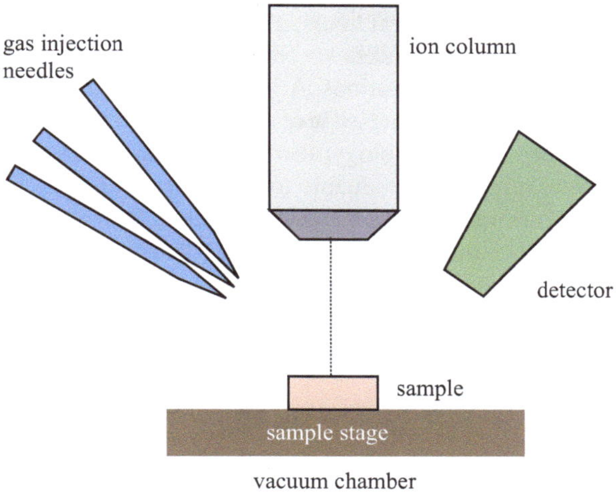

Fig. 10.65 Basic schematic diagram of a FIB system [33]

ions and secondary electrons are emitted. The secondary electrons are captured and their intensity is used to create an image of the surface of the chip. An important advantage of a FIB system is that it can be used to remove material from the surface of a chip *(milling)* or to deposit dielectric or metal layers *(deposition)*. FIB is often used to physically edit a circuit on a chip with nanometer precision. It can cut unwanted connections and deposit metal to change or add connections on top of the passivation layer or to create additional probe pads. Because holes can be made with high accuracy, even connections between different metal layers can be made, providing the capability to rewire ICs directly on a chip with nanometer precision. These 'design modifications' may fix design errors or implement spec changes.

Additionally, it enables the connection of an internal circuit node to a FIB-deposited metal area on top of the scratch protection. This will increase its load capacitance and can be used to correct timing violations. Figure 10.65 shows a schematic diagram of a basic FIB system [33].

It is equipped with a computer-controlled gas injection system that can handle various different gases for the deposition of metal and dielectric material or for enhanced and selective etching.

A modern FIB system consists of complex and expensive equipment, which is capable of removing and depositing material (metal and dielectrics) and making smooth cross sections for SEM or TEM analysis. It is sometimes combined with a SEM column for high-resolution imaging. To allow faster material removal at lower beam intensities, advanced FIB systems use gas-assisted etching. With this technique, holes can be etched down to the first metal layer. In this respect, holes with aspect ratios of up to 30 with a minimum feature size below 25 nm can be created [34].

The deposition of the conductive material on top of the scratch protection is easy but time-consuming. But the turnaround time of the modified chips is only in the order of several hours. The combination of a new mask and fabrication respin is very expensive, takes several months and introduces additional risks, since it is only based on simulations. A FIB circuit change allows the customer to perform all system-level tests and assures the next respin will include all the necessary changes.

Since FIB technology also allows small holes to be accurately cut through the wafer, it may be a valuable tool in inspecting flip-chip packaging. Advanced FIB machines also support backside editing. It then requires an additional IR microscope and bulk silicon trenching package that supports accelerated silicon etching for faster wafer-level cross sectioning [35]. Backside FIB circuit editing can also be used to introduce a resistor across two nodes [36].

10.6.7 Design for Debug and Design for Failure Analysis

The enormous density of VLSI chips with nanometer feature sizes and small operating voltages have reduced the margins in the electrical operation of the digital circuits. Voltage drop, temperature variations, cross-talk, supply and substrate noise, all depend on unforeseen operational conditions and, consequently, circuits may fail although they passed structural test procedures. There are even examples of ICs that fail during test, but pass the application. Particularly for nanometer CMOS ICs, the large number of metal layers with the increasing metal densities (metal fill (tiles), to support the CMP planarisation process), prevents physical probing of the signals for debug purposes. To enhance observation of important design and technology parameters, such as supply noise, capacitances, temperature, threshold voltage, etc., monitors can be embedded within the functional cores. Dedicated monitor circuits have been proposed in literature [37–39]. Another article [40] describes a fully integrated signal integrity self-test (SIST) system that combines a variety of on-chip signal integrity and performance sensors with a very simple digital readout mechanism. This technique supports debug by enabling additional observability by monitoring important electrical signals, which represent certain operating conditions of (different parts) of the IC. This type of *design for debug* techniques may also complement the physical failure analysis tools. Design for debug also includes support for fault observation at higher levels of design, e.g. software fault isolation. It will enable all flip-flops to be monitored and controlled at full functional speed. This requires additional on-chip hardware, which supports the debug software running separately from the digital tester.

Design for failure analysis is the process of intentionally designing a product or system to enable easy identification and diagnosis of potential failures. The concept is to design the product in a way that allows quick and efficient troubleshooting and problem-solving in the event of a failure. This requires a thorough understanding of the physical and electrical properties of the chip, as well as careful consideration of potential sources of failure and the best strategies for mitigating these risks. Some key aspects of design for failure analysis include:

1. Fault isolation: this involves designing the product in a way that makes it possible to isolate specific areas that may cause a problem. This can be achieved by the design-in of additional test point for probing (through careful design of interfaces, connectors and diagnostic tools.
2. Standardisation: standardising components and interfaces makes it easier to diagnose and replace failed components, as well as to ensure that replacement components are compatible with the existing system.
3. Include redundancy: designing a chip with multiple identical circuits or components so that if one fails, the others can still function, maintaining overall chip functionality. For example, in a multicore processor or one with multiple cache memories within a single chip, a failing block may be disconnected, so the chip may still be sold as a lower-end device.
4. Include fault tolerance mechanisms, which are designed to detect and correct errors that may occur during operation. These mechanisms include error-correcting codes, which are algorithms designed to detect and correct errors in data storage or data transmission.
5. Improve diagnostics: designing an easy-to-test product allows engineers and technicians to quickly identify and diagnose potential problems. This includes designing products that include additional test points (for e-beam, physical or other probing techniques) and ensuring that diagnostic tools are easy to use and widely available (see also Sect. 10.2.2).

Overall, design for failure analysis (DFA) is a methodology for designing ICs that are easier to analyse and diagnose if they fail. The goal of DFA is to reduce the time and effort required to identify the root cause of a failure, which makes it easier to fix the problem and get the system or product back up and running.

10.7 Conclusions

The general requirement of a high fault coverage during the test of an IC is being challenged by an ever-increasing design complexity. Advanced test methods have been developed to maintain high fault coverages, both during IC testing and board testing. Additional hardware is included in the design to support these methods. This also reduces test time, which can be a severe impact on the ultimate price of an IC.

The purpose of IC testing is not only to separate the good from the bad dies, but the test results can also be used for feedback on the operating margins that the chip has with respect to its specification. Testing is also closely related to yield. Therefore, a single yield model served to present a flavour of the most important aspects that determine the yield. For several applications the fabrication of large monolithic dies has become less financially attractive. Therefore, the industry has introduced chiplet-based design whereby a large ASIC is partitioned into multiple dies (chiplets), which are then interconnected with very advanced packaging techniques to build a heterogenous system. They offer a way to continue to increase the capabilities of electronic devices while also providing benefits in terms of cost and flexibility.

Although packaging is not a typical CMOS issue only, the overview here shows the importance of choosing the right package. The temperature of the packaged die and the self-inductance of the package pins are important parameters which may dominate the performance and the operation integrity of an integrated circuit.

The increased device complexity, combined with more levels of metal, reduces the fault observability. A design must therefore be supported by a design-for-debug approach to support both a rapid identification of the failure and the failure mechanism.

When the cause of a failure cannot be traced with the combination of test/debug software and on-chip test/debug hardware, various techniques can be used to further analyse the failure. Because of the increasing complexity of integrated circuits, e.g. smaller feature sizes, increasing number of devices and metal layers and higher densities, semiconductor companies have installed very advanced failure analysis (FA) labs with complex FA tools, which offer sufficient observability both from the frontside and the backside of the die with very high resolution.

Finally, circuit editing techniques, performed with a focused ion beam tool, support the debug and failure analysis. It allows to physically remove and/or add dielectric and metal species on top of the passivation layer to enable circuit changes before mask changes are made and a respin is started.

10.8 Exercises

1. Why is design for testability an increasingly important design requirement?
2. A given CMOS manufacturing process has the following parameters at a certain point in time: $Y_s = 0.75\,\text{cm}^{-2}$, $D_0 = 1\,\text{cm}^{-2}$.
   (a) Express your opinion about this process.
   (b) Calculate the expected yield for a chip with a die area of $80\,\text{mm}^2$.
   (c) Three months later, an $80\,\text{mm}^2$ chip has a yield of 60.3% and a $120\,\text{mm}^2$ chip can be produced with a yield of 49.4%. Calculate Y_s and D_0, assuming that the class of the cleanroom has not changed.
3. An IC dissipates 300 mW while its junction temperature is 32 °C. If the thermal resistance of its package is 30 °C/W, then what is the IC's ambient temperature?
4. What are the major differences between through-hole and SMD area array packages? Specify their respective advantages and disadvantages.
5. Explain the differences between test, debug and failure analysis of first silicon.
6. Why is design for debug a must for current and future ICs?
7. What is the drive for reducing the self-inductance (L) of the package pins?
8. What kind of tests are required to determine the quality and reliability of packaged dies?
9. Explain what is meant by observability and discuss its trend with respect to future process generations.

10. In current and future processes, the transistors and lower metal layers are shielded by the upper ones. What could be done during the design phase to support failure analysis in this respect?
11. Explain how FIB supports failure analysis.

References

1. High Volume Memory Test, DRAM-Full-Wafer Contact Testing, https://www.formfactor.com/applications/high-volume-test-on-wafer/memory-test/
2. S. Wang, et al., A Scalable Scan-Path Test Point Insertion Technique to Enhance Delay Fault Coverage for Standard Scan Designs, ITC 2003, pp. 574–583
3. American Cleanroom Systems, https://www.americancleanrooms.com/cleanroom-classifications/
4. P. Nanja, *Challenges and Solutions for Silicon Wafer Bevel Defects during 3D NAND Flash Manufacturing* (A Conventor Inc. publication, May 21), 2019
5. S-CUBED, https://www.s-cubed.com/wafer-edge-exposure-wee-process/
6. T.H. Kim, et al., Criticality Of Wafer Edge Inspection And Metrology Data To All-Surface Defectivity Root Cause And Yield Analysis, in *Proceedings of the SEMI Advanced Semiconductors Manufacturing Conference (ASMC 2016)*, Saratoga Springs, New York, May 16–19, 2016
7. L. Peters, *DFM: Worlds Collide, Then Cooperate* (Semiconductor International, June 2005). www.semiconductor.net
8. P. Rabkin, *DFM for Advanced Technology Nodes: Fabless View* (Future Fab International, Issue 20, January 2006)
9. W.T. Beyene, *Chiplet Technology and Heterogeneous Integration* (IEEE Electronics Packaging Society, April 2022) https://eps.ieee.org/images/files/enews/EDMS_TC_-_Chiplet_Technology_and_Heterogenous_Integration.pdf
10. What is a chiplet? WikiChips, Chips & Semi; Semiconductor and Computer Engineering. https://en.wikichip.org/wiki/chiplet?utm_content=cmp-true
11. N. Flaherty, 14,336 ARM cores in chiplet-based waferscale AI engine, Technology News, October 15, 2021
12. D. Yu, TSMC Packaging Technologies for Chiplets and 3D, in *HotChips, conference tutorials*, 2021
13. New World, New Opportunities, in *TSMC Technology Symposium*, 2022. https://pr.tsmc.com/static/campaign/2022july/tsmc_eu/EU_Special_Edition.html
14. P. Alcorn, Intel Details Core Ultra 'Meteor Lake' Architecture, Launches December 14, tom'sHARDWARE, September 20, 2023
15. A. Shilov, TSMC charts a course to trillion-transistor chips, eyes 1 nm monolithic chips with 200 billion transistors, tom'sHARDWARE, December 27, 2023
16. 2.5D and 3D ICs: New Paradigms in ASIC, https://www.einfochips.com/blog/2-5d-3d-ics-new-paradigms-in-asic/, January 16, 2024
17. 2.5D vs. 3D Packaging, https://resources.pcb.cadence.com/blog/2023-2-5d-vs-3d-packaging, October, 2023
18. T. Lizotte, Laser Dicing of Chip Scale and Silicon Wafer Scale Packages, in *2003 IEEE/CPMT/SEMI Int'l Electronics Manufacturing Technology Symposium*, 2003
19. E. Rutkowski, M.J. Sampson, Body of Knowledge (BOK) for Copper Wire Bonds, https://nepp.nasa.gov/files/26611/2015-370-Rutkowski-Final-Paper-NEPPweb-Copper-Wire-Bonds-TN26444.pdf, August 2015
20. JEDEC, Joint Electron Devices Engineering Councils, www.jedec.org
21. G.Q.Zhang, et al., *Mechanics of Microelectronics* (Springer, 2006). www.springer.com

22. J.H.J. Janssen, H.J.M. Veendrick, *Spatial and Temporal Temperature Variations in CMOS Designs* (EDA Publishing/THERMINIC, 2009)
23. H. Pape, et al., Thermal transient modeling and experimental validation in the European project PROFIT. IEEE Trans. Compon. Packag. Technol. **27**(3), 530–538 (2004)
24. K.J. Hess, et al., Reliability of bond over active pad structures for 0.13 μm CMOS technology, in *Proceedings of Electronic Components and Technology Conference*, January 2003
25. R. Tummala, *Moore's Law Meets its Match*, (IEEE Spectrum, June 2006), pp. 38–43
26. Time-dependent gate oxide breakdown, Wikipedia, 2024
27. K.J. Kuhn, Considerations for Ultimate CMOS Scaling. IEEE Trans. Electron Dev. **59**(7), (July 2012)
28. K. de Kort, *Techniques for Characterization and Failure Analysis of Integrated Circuits*, Analysis of Microelectronic Materials and Devices (Wiley, 1991)
29. E.I. Cole, *Beam-based defect localization methods*, Microelectronics Failure Analysis Desk Reference, 5th edn., 2005, pp. 406–416
30. F. Beaudoin, et al., *Principles of thermal laser stimulation techniques*, Microelectronics Failure Analysis Desk Reference, 5th edn., 2005, pp. 417–425
31. R. Giridharagopal, Near-Field Scanning Optical Microscopy for Through-Silicon Imaging and Fault Isolation of Integrated Circuits, in *ISTFA 2014: Proceedings of the 40th International Symposium for Testing and Failure Analysis* (November 9–13, 2014, Houston, Texas, USA)
32. R. Aaron Falk, Advanced LIVA/TIVA Techniques, in *Proceedings of the 27th International Symposium for Testing and Failure Analysis*, 2001, pp. 59–65
33. *Introduction to Focused Ion Beams; Instrumentation, Theory, Techniques and Practice*, in ed. by L.A. Giannuzzi, F.A. Stevie (Springer Science + Business Media, Inc., Boston, 2005)
34. H.D. Wanzenboeck, S. Waid, *Focused Ion Beam Lithography*, Lithography, vol. 1 (InTech Open Access Publishers, Vienna University of Technology Institute for Solid State Electronics, Austria, 2012)
35. FEI, The V400ACE Focused Ion Beam (FIB) system (2015). www.fei.com/products
36. Focused Ion Beam (FIB), Anysilicon, June 19, 2015. http://anysilicon.com/focused-ion-beam-fib/
37. M. Nourani, et al., Detecting signal-overshoots for reliability analysis in high-speed system-on-chips. IEEE Trans. Reliab. **51**(4), 494–504 (2002)
38. E. Alon, et al., Circuits and techniques for high-resolution measurement of on-chip power supply noise. IEEE J. Solid State Circuits **40**(4), 820–828 (2005)
39. D. Schinkel, et al., A 1-V 15μW High-Precision Temperature Switch, in *Proc. ESSCIRC*, Sept. 2001, pp. 77–80
40. V. Petrescu, et al., A Signal-Integrity Self-Test Concept for Debugging Nanometer CMOS ICs, ISSCC Digest of Technical Papers, February 2006, pp. 544–545

Effects of Scaling on MOS IC Design and Consequences for the Roadmap

11

11.1 Introduction

The prolonged scaling of CMOS devices according to Moore's law has brought the design complexity in terms of number of transistors and performance requirements to such a high level that design styles and methods continuously need to be changed in order to manage this complexity and to enable full exploitation of the potentials of advanced and future CMOS technologies. A prediction of these potentials was presented in the *International Technology Roadmap for Semiconductors (ITRS)* [1], launched in 1991 by the *Semiconductor Industrial Association (SIA)*. Traditionally, this ITRS roadmap was updated every two years. Because the semiconductor industry is rapidly approaching a brick wall, leaving only a few players who can afford the huge investments needed for further scaling, the continuation of the ITRS roadmap has become questionable. The ITRS 2013 roadmap was the last version. The IRDS™ (International Roadmap for Devices and Systems) is a set of predictions that serves as the successor to the ITRS (International Technology Roadmap for Semiconductors). The first edition of the IRDS was published in 2017. The 2022 update of the More than Moore roadmap of the IRDS can be found in [2].

This chapter discusses the consequences of the scaling process for nanometer IC design, with the focus on future trends of power, speed, reliability and signal integrity. Nanometer CMOS design requires more focus on the physical design and on the consequences of further scaling. This will certainly have an impact on the semiconductor technology roadmap. In the race towards a tera-transistor heterogeneous *system on a chip (SOC)* (see Fig. 11.1) design methods and tools not only have to be changed to make the design manageable (system design aspects) but also to make a functional design (physical design aspects). Note the difference with Fig. 7.8, which only shows the system design aspects.

H. Veendrick, *Nanometer CMOS ICs*,
https://doi.org/10.1007/978-3-031-64249-4_11

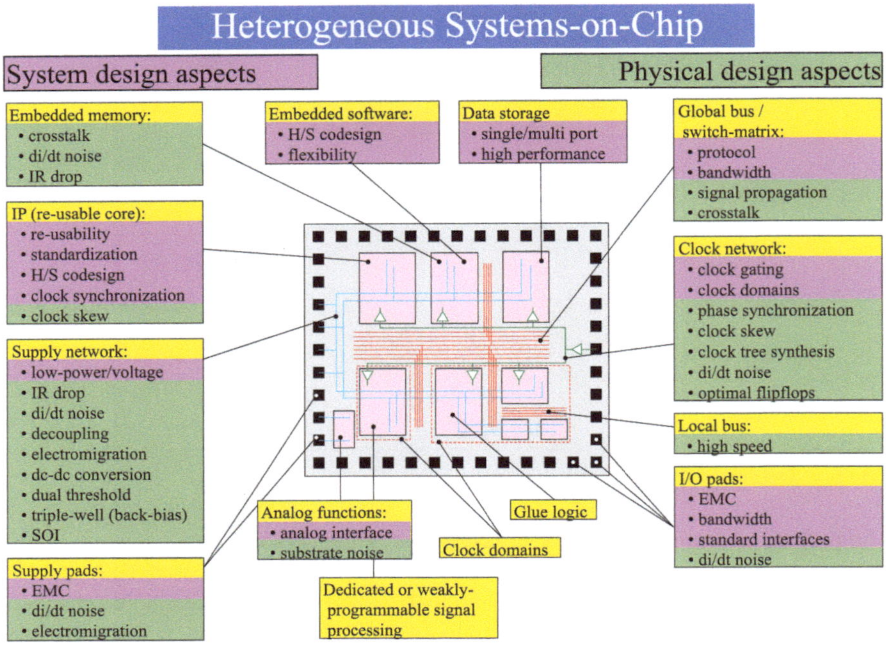

Fig. 11.1 Important aspects of a (heterogeneous) system on a chip

The complexity of such an SoC can only be managed by applying:

- A platform with integrated hardware/software architecture and application development tools
- System-level synthesis to improve design efficiency
- Design reuse
- Increased design resources per chip

The first three items deal with system-level design aspects; see Chap. 7. The increased design resources, however, are not only required to manage the SoC design complexity, they are also needed to cope with the increasing number of physical design aspects as depicted in the figure.

For a long time, only analogue circuits were susceptible to these physical effects. The scalability of analogue circuits is discussed in Chap. 9 and in [3]. Analogue effects in digital circuits, such as cross-talk, supply noise and clock jitter, have become important design considerations from 250-nm CMOS onwards. In current and future process generations, these effects dominate the SoC's performance, signal integrity and reliability. VLSI design therefore requires an increasingly analogue approach. Design is no longer about switches and ones and zeros only, but also about resistors, capacitors, inductors, noise, interference and radiation.

Basically, a VLSI chip is just a bunch of transistors that perform a certain function by the way they are interconnected. The next sections focus on the influence of scaling on the basic elements, the transistor and the interconnections and the consequences for the overall performance, reliability and signal integrity of nanometer IC designs. This increasing design complexity has caused a paradigm shift in the cost components of an IC: from fabrication costs towards design costs. Section 11.5 will therefore discuss trends in design complexity and costs in some more detail.

11.2 Transistor Scaling Effects

Over many technology generations, both the feature sizes and the supply voltage have scaled with the same factor. This type of scaling is called *constant-field scaling* and involves technology nodes from 0.8 µm (5 V) through 0.5 µm and 0.35 µm (both 3.3 V), 0.25 µm (2.5 V) and 180 nm (1.8 V) to 120 nm (1.2 V). When scaling transistor sizes and bias voltages by a factor of s ($s \approx 0.7$), the transistor current scales by the same factor. To maintain performance, the threshold voltage is also required to scale with s. The threshold-dependent leakage current is an important factor that limits the pace of scaling. This subthreshold leakage current can be estimated by the following relation, which means that this current increases by about a factor of 18 for every 100-mV (depending on the subthreshold slope) decrease in V_T:

$$I_{\text{subthreshold}}(scaled) = 18^{10(1-s)V_T} \cdot I_{\text{subthreshold}}$$

For large SoCs, this background leakage current will be larger than a gate oxide short current, for example, which will dramatically limit the potentials of I_{ddq} testing. The eventual reduction of the threshold voltage requires alternative techniques, such as the dual-V_T concept [4] and the triple-well concept [5], to limit the subthreshold power consumption during standby and test modes. Both concepts are discussed in Chap. 8.

Another result of transistor scaling is the increased channel dope, caused by V_T correction and by the implants (drain extensions and halos) needed to suppress short-channel effects (Chap. 2). As a consequence, the thinner depletion layers cause higher parasitic junction capacitances. This particularly occurred in scaling planar MOS transistors towards the 28-nm node. In high-ϵ-metal-gate and FinFET transistors, the value of the threshold voltage is mainly determined by the transistor work function, rather than by its channel dope.

In combination with gate oxide thickness (t_{ox}) scaling, the higher channel dope mirrors depletion charge into the gate (gate depletion), which reduces the effective current control by the gate. Below a t_{ox} of about 2 nm [6], quantum-mechanical tunnelling of charge through the gate oxide may occur, resulting in additional standby currents and possibly a reliability problem. Finally, scaling of the channel length will increase the mismatch between 'equal' transistors, as a result of increased spread of the number of dopants in the transistor channel. Following Moore's law beyond the 10-nm node has severe consequences for transistor performance and requires alternative architectures to improve the transistor efficiency in terms of driving current per unit area. This particular section need not further discuss the scaling consequences for FinFET transistors and the introduction of alternative 3D transistor architectures such as gate-all-around, nano-wire, nano-sheet, nano-ribbon, fork-sheet or CFET transistors, as they have already been discussed in previous chapters, along with their usage in various logic and memory circuits.

11.3 Interconnection Scaling Effects

Scaling of widths and spacings has caused the metal interconnections to start dominating the IC's performance, reliability and signal integrity. The output load of a logic gate is equal to the total of the fan-in capacitances of its connecting gates and the total wire load of the interconnections. Table 11.1 shows the increase in the average ratio between wire load and fan-in, for average standard cell blocks, caused by scaling. These numbers represent average values; for each individual chip, this ratio may be different from the table.

The relative reduction in this ratio for the 16-nm FinFET node and beyond is due to the 3D architecture of the transistor channel in these devices.

The increasing resistance values of the on-chip interconnections lead to larger voltage drops and the increasing mutual capacitance values cause more cross-talk, while the combination leads to larger signal propagation delays. At 1 GHz, the required signal rise and fall times should be less than 50 ps to perform some computational tasks within the available 1-ns timeframe. Even on-chip wires then cause interference with other modules. For such signal edges, line lengths of 3 mm and above become critical and require transmission line modelling. The numbers

Table 11.1 Increasing interconnect dominance

Technology	Ratio: wire load/fan-in
350 nm	30/70
250 nm	33/67
180 nm	36/64
130 nm	45/55
90 nm	54/46
65 nm	66/34
45 nm	75/25
32 nm	81/19
28 nm	83/17
22 nm	85/15
16 nm (FinFET)	78/22
12 nm and beyond	80/20

in the table for 32-nm CMOS and beyond only represent an order of magnitude, since they are not average for each process in these respective nodes. This is due to the fact that most of these technologies do not have a generic backend (metal layer architecture) and this is also different for different semiconductor vendors. For 16-nm FinFET technologies, the ratio reduces, because of their relatively larger transistor width compared to lateral devices in this technology node. But it is clear that the trend of increasing wire load continues again with further FinFET process scaling. The ratio in the table also depends on the thickness of the metal lines/layers in the used process node. Since, again, different vendors use different backend architectures, it is difficult to come up with a ratio beyond the 16-nm node. Therefore, a minimum value for this ratio is given for process nodes of 12 nm and beyond.

Figure 11.2 shows the *propagation delay* of an embedded metal track (metal track embedded between two minimum-spaced neighbours) in different technologies.

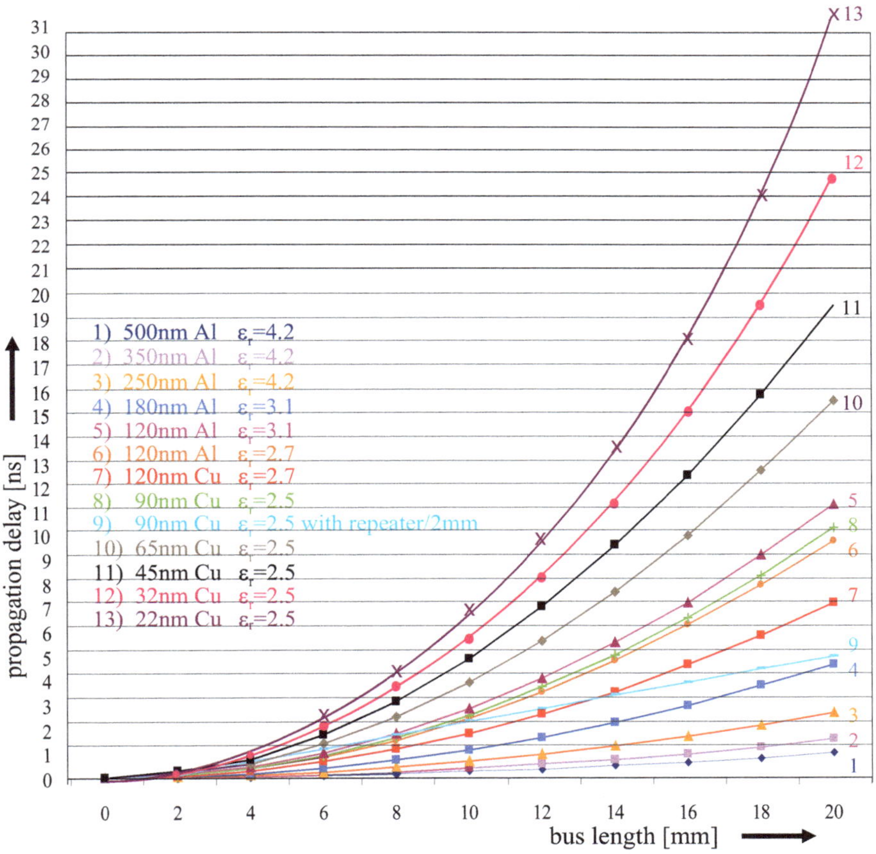

Fig. 11.2 Propagation delay of an embedded track in different technologies

There are several approaches to reduce the negative effects of scaled interconnections. One is to reduce the capacitance, which is expressed as $C = \epsilon_0\epsilon_r A/t_{\text{dielectric}}$. Until the mid-1990s, the *interlevel dielectric (ILD)* coefficient ϵ_r was around 3.9 (for silicon dioxide). That value has steadily been reduced to below 2.8 by making the oxide more porous. Alternative *low-ϵ dielectrics*, such as SiCOH, for instance, have reduced the dielectric constant to below 2.5 [7]. In the ITRS roadmap, edition 2006, values around $\epsilon_r \approx 2$ were already predicted for the years 2016/2017. With the introduction of air gaps, as discussed in Sect. 3.9.6, the dielectric constant would be further reduced. However, according to the IRDS roadmap (see Fig. 11.5) this dielectric constant is expected to remain between 2.2 and 2.5 until 2035. The increased porosity of these low-ϵ materials threatens the integrity of these ILD layers. The integration of organic or hybrid dielectrics with lower dielectric constants than that of silicon-dioxide, however, all failed because of the lower mechanical strength, the reduced chemical robustness [8] and the increased electromigration effect. This is particularly an issue at higher metal layers, which also suffer from thermal stress caused by the difference in the thermal expansion coefficients of the silicon die, encapsulation material and package. This may result in delamination issues of the BEOL metal layers [7]. The second approach is to reduce the resistance. The sheet resistance of conventional aluminium alloys is around 3 $\mu\Omega$cm, while that of copper is about 1.8 $\mu\Omega$cm. However, the potentials of the reduced copper resistance cannot fully be exploited. Because copper diffuses through oxides, it cannot be deposited and etched like aluminium. By applying a damascene backend flow, copper had to be completely encapsulated within a barrier material, as shown in Fig. 3.38 (Chap. 3). The effective sheet resistance of copper wiring depends on the barrier material and, for a metal layer used for global wiring, increased to between 3.5 and 6 $\mu\Omega$cm. This value is even more than that of the original aluminium metal wiring of the 180-nm CMOS nodes and above. Section 3.9.5 discussed the scaling trends in the BEOL process for advanced sub-7-nm nodes.

Figure 11.2 also shows the individual influence of copper and low-ϵ dielectrics. The signal propagation delay over a metal wire is proportional to the square of its length. The use of repeaters, however, reduces the propagation delay to a linear dependency on length. Particularly for longer wires, this may reduce the propagation delay by more than a factor of two (compare curves 8 and 9) The increasing clock skew and propagation delay for global signal wires are in direct contrast to the reducing clock period. Therefore, there will be an increased drive to limit the size of the standard cell blocks (between one and several square millimetres), which will also limit local interconnect lengths and clock skew. Simulation results beyond 22 nm are not available. However, it will be clear that the trend of increasing signal propagation delays across and signal interferences between metal lines will continue with further scaling. This results in an increasing impact on the timing and signal integrity of long signal lines within the logic blocks and it requires extensive top-level chip validation and verification of the signal lines between these logic blocks.

To further relief the propagation delay problems, pipelines could be built into the global interconnects, but the bus latency will then become an important design parameter.

Figure 11.3 shows an example cross section of an advanced CMOS process. The figure clearly demonstrates the increased dominance of the interconnect in current and future nanometer CMOS processes.

For CMOS technologies beyond 20 nm, the metal dominance will be even stronger, as the number of metal layers may be well above ten and the spacing between metal lines in the same layer reduces.

As an example, in nanometer CMOS SRAMs not only the increase of the mutual capacitance between two minimum spaced wires is important, but also the increase of the mutual capacitance between two minimum spaced contacts or two minimum spaced vias. Figure 11.4 shows a 3D cross section of an SRAM array. It requires an accurate 3D extraction tool to enable proper SRAM timing simulation.

11.4 Scaling Consequences for Overall Chip Performance and Robustness

For many technology generations in the past the supply voltage has been constant and equal to 5 V. The scaling process over that period of time was called *constant-voltage scaling*. Figure 11.5 shows the evolution of the voltage scaling over the last couple of decades.

During the 1990s, the advances in CMOS technology were not just related to scaling of the devices and the minimum features sizes, but also of the supply voltages. This is called *constant-field scaling*. Because of the shrinking voltages, the threshold voltages also reduced with about the same factor. At a certain moment in time, the subthreshold and gate oxide leakage currents reached levels that were unacceptable for many mobile applications. Particularly these applications forced the semiconductor manufacturers to differentiate between three roadmap variants: high-performance (HP), low operating power (LoP) and low standby power (LSTP). The different target applications in each variant are already discussed in Sect. 4.7.

In the period from the early 1990s to the early 2000s the nominal supply voltage has scaled to values close to 1 V. For processes beyond the 120-nm node, the supply voltage only marginally reduced, such that this period is also considered as a period of constant-voltage scaling. There are several reasons for this very limited voltage scaling. A further reduction of the supply voltage would also require a further reduction of the threshold voltage leading to a disproportionally high increase of the subthreshold leakage currents. Also the noise margins of an integrated circuit reduce with reducing supply voltage. Particularly the embedded SRAMs are affected. A lower supply voltage would reduce their static noise margin and threatens a stable SRAM operation. Moreover, the SRAM cell delay would considerably increase, leading to a dramatic loss of speed. For analogue circuits, a further reduction of the supply voltage will reduce their performance as well as their noise margins while

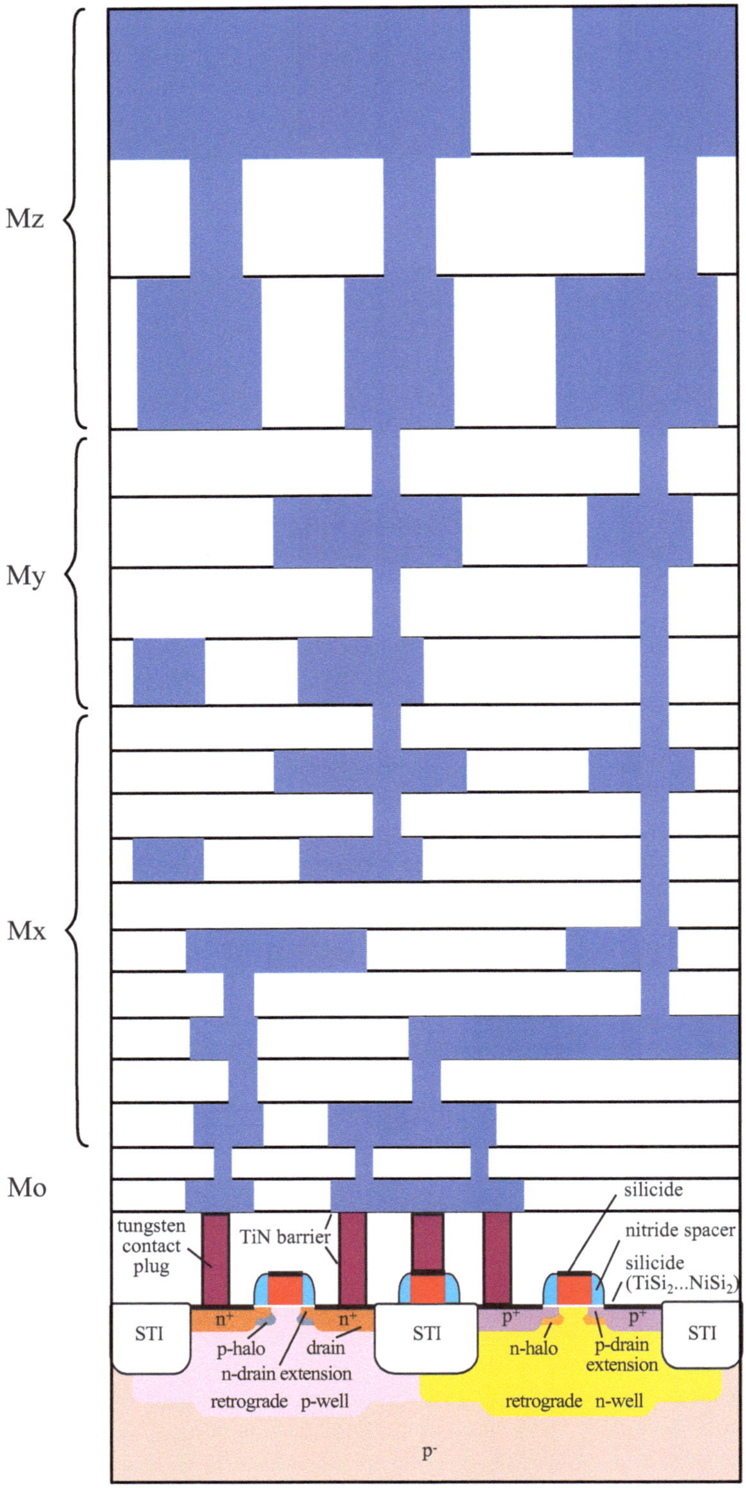

Fig. 11.3 Cross section to show the various metal layers in the backend of an advanced CMOS process

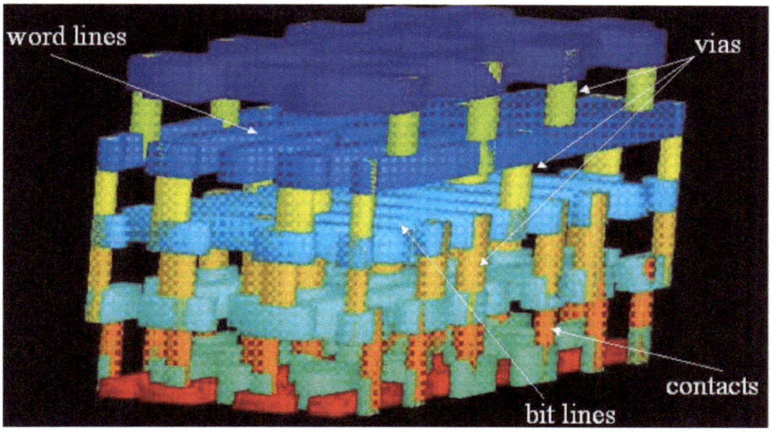

Fig. 11.4 3D cross section of a nanometer CMOS SRAM array

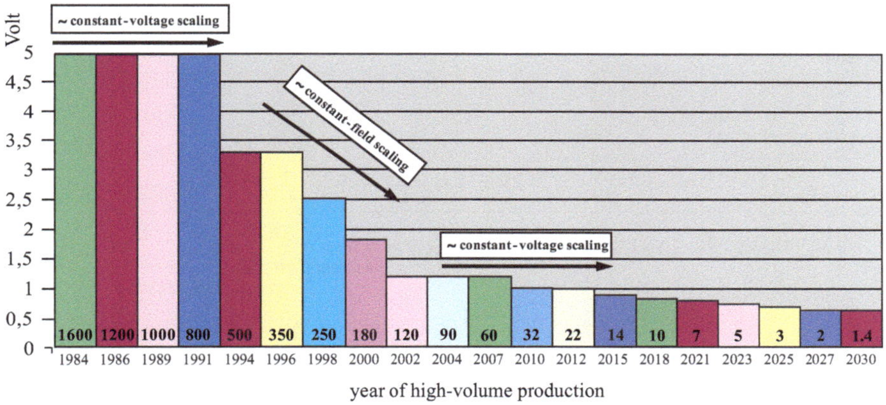

Fig. 11.5 Process node versus year of high-volume production

increasing their susceptibility to process variations at the same time. These problems are the drive for alternative design approaches for embedded SRAMS and analogue circuits while continuing Moore's law.

It is obvious that these different scaling scenarios in Fig. 11.5 have a different impact on the basic transistor parameters and on the performance and robustness of CMOS ICs. Table 11.2 shows how the transistor performance, reliability and signal integrity parameters depend on the scaling factor s ($s \approx 0.7$) and the impact of the voltage scaling scenarios over the last three decades, assuming bulk-CMOS designs using lateral MOS transistors. Also the next three diagrams are based on the same assumptions.

Table 11.2 Different scaling scenarios

Different scaling scenarios

topic	relation	Scaling factor ($s \approx 0.7$)			
		$p \neq s \neq q$	$p=1$	$p=s$	$s=q=1$
Voltages	V	p	1	s	p
	V_t	q	1	q	1
Feature sizes	W, L, t_{is}, dist, t_{ox}(EOT)	s	s	s	1
Devices per unit area	$\div 1/A$	$1/s^2$	$1/s^2$	$1/s^2$	1
Transistor bias current	$I = \beta/2(V_{gs}-V_t)^2$	$\sqrt{p}\sqrt{s}$ [1]	$\sqrt{s}$	s	$\sqrt{p}$
Capacitance	$C=\varepsilon_0\varepsilon_r A/t_{is}$	s	s	s	1
Metal resist. (top metals)	$R=\rho\ell/(t_m W)$ ($t_m \approx$ const)	1	1	1	1
Gate delay τ ($\div 1/f$)	CV/i	$\sqrt{p}\sqrt{s}$	$\sqrt{s}$	s	$\sqrt{p}$
Average current/unit area	$I = CVf/A$	$\sqrt{p}/s\sqrt{s}$	$1/s\sqrt{s}$	$1/s$	$\sqrt{p}$
Power dissipation/gate D	CV^2f	$p\sqrt{p}\sqrt{s}$	$\sqrt{s}$	s^2	$p\sqrt{p}$
Power-delay product τD	$\div CV^2$	p^2s	s	s^3	p^2
Power density P	CV^2f/A	$p\sqrt{p}/s\sqrt{s}$	$1/s\sqrt{s}$	1	$p\sqrt{p}$
Subthr. leakage current	Expon. with V_t and V	$s^{-1}18^{10(1-q)Vt+0.1(p-1)V}$	$s^{-1}18^{10(1-q)Vt}$	$s^{-1}18^{10(1-q)Vt+0.1(s-1)V}$	$18^{0.1(p-1)V}$
Gate leakage current	Expon. with V and t_{ox}	$s^2.10^{5(1-s)t_{ox}+{}^2\log p}$	$s^2.10^{5(1-s)t_{ox}}$	$s^2.10^{5(1-s)t_{ox}+{}^2\log s}$	$10^{{}^2\log p}$
Electromigr. (cur.dens.)	$I=P/V$	$\sqrt{p}/s\sqrt{s}$	$1/s\sqrt{s}$	$1/s$	$\sqrt{p}$
Latch-up (for Vdd >>1V)	$\div V/\text{dist}$	p/s	$1/s$	~1	p
ESD susceptibility	$\div 1/t_{ox}$	$1/s$	$1/s$	$1/s$	1
Hot-carrier lifetime	f(V,V/distance)	$s^C.e^{B/V(1/p-1)}$	s^C	$s^C.e^{B/V(1/s-1)}$	$e^{B/V(1/p-1)}$
NBTI V_t-shift	$\delta Vt \div f(V,L,t_{ox})$	$(p/s)^m$	$(1/s)^m$	1	p^m
Cross-talk / unit length	$\div 1/\text{dist.}$	$1/s$	$1/s$	$1/s$	1
Induct. noise/unit area	$(di/dt)/A$	p/s^3	$1/s^2$	$1/s^2$	p
Voltage drop/unit length	$I.\sigma_{metal}.W_{metal}/L_{metal}$	p/s^3	$1/s^2\sqrt{s}$	$1/s^2$	p
SER per Mb or Mflip-flop	$A_{diff}\cdot e^{(-Qcrit/\langle Qcoll \rangle)}$	1	1	1	$1/p$
Noise margin	V_{dd} and V_t	p and q	1 and q	s and q	p and 1

With velocity saturation — *constant-voltage scaling* ($p=1$) — *constant-field scaling* ($p=s$) — *constant-size 'scaling'* ($s=q=1$)

Let us assume that the transistor sizes are scaled with a factor of s, the voltages with a factor of p and both the threshold voltage and gate oxide thickness with a factor of q. The first scaling column ($p \neq s \neq q$) shows how a parameter scales when the voltages scale with a different factor than the sizes. In the first constant-voltage scaling column ($p = q = 1$), only the sizes scale, while the voltages are kept constant. In the constant-field scaling column ($p = q = s$), both the sizes and voltages scale with the same factor, keeping the field in the channel constant. Finally, in the second constant-voltage scaling column (after yr 2000), the physical effects that cause mobility reduction (Chap. 2), such as velocity saturation, are included in the relations. These are therefore different from the first constant-voltage scaling column (before yr 1990).

The table (column $p \neq s \neq q$) shows that signal integrity is more affected by scaling of the sizes, while the performance is more affected by voltage scaling. For several parameters the relation with the scaling factors s and p is not completely clear. The hot-carrier lifetime is described by the well-accepted empirical expression from Takeda, as discussed in Chap. 9. For NBTI there is not such a well-accepted model. In the table the V_T shift due to NBTI is taken as a parameter. Also here, lifetime issues are involved, but due to the complex behaviour, it is not included in the table.

When we combine this table with the voltage scaling evolution as shown in Fig. 11.5, we are able to visualise the trends in performance, variability and signal integrity and reliability. The following figures show these trends, assuming that we continue to use bulk-CMOS wafers and that both the transistor architecture and the supply voltage do not change dramatically. Figure 11.6 shows the improvements in various performance parameters over the last couple of decades and their expected improvements. The figure also assumes that the first year for volume production for the 32-nm and 22-nm nodes are delayed with respect to the two-year cycle with which technology nodes were introduced before.

The figure shows that the constant-field scaling era has been particularly beneficial for power-efficiency improvement, due to the combined scaling of the sizes and the supply voltage. This allowed new generations of electronic devices to exhibit about two-and-a-half times more functionality for the same power needs, compared to their previous generation. In that same period of time, the subthreshold leakage power has increased by more than three orders of magnitude, which was a major drive to limit further supply and threshold-voltage scaling. The diagram also clearly shows that below 100 nm the improvements both in power efficiency and speed are only limited and that it will remain so in the future, while the fab costs are still expected to increase at least with the same trend.

The above-discussed trends have severe impact on the design of integrated circuits. For high-performance microprocessors, this has led to moving away from higher-frequency architectures towards multicore architectures. For vendors of general VLSI and ASIC ICs, not only the performance and functionality of an IC are differentiators, but certainly also the power consumption, both in active and standby

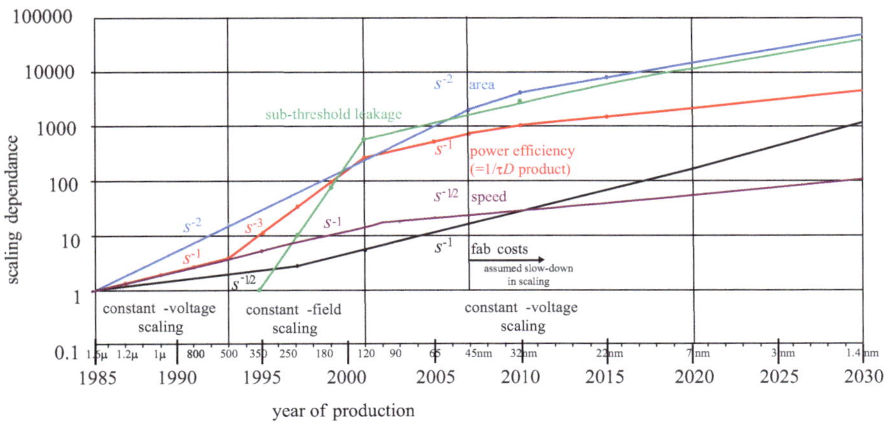

Fig. 11.6 Experienced and expected trends in various performance parameters in relation with the fabrication cost

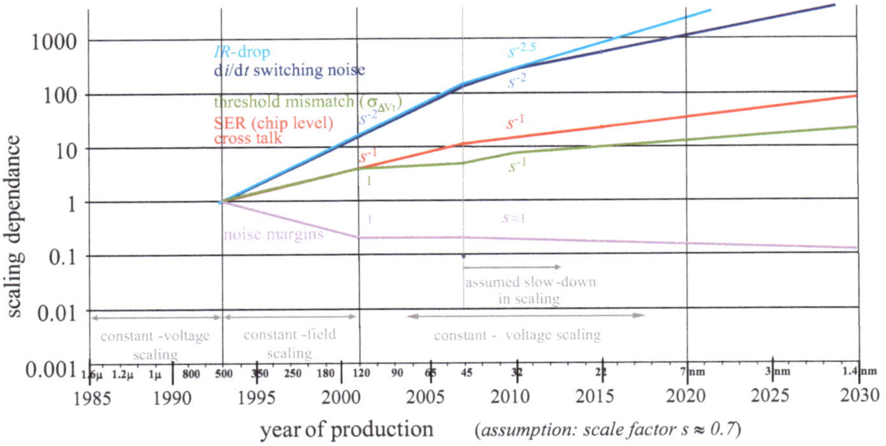

Fig. 11.7 Experienced and expected trends in variability and signal integrity parameters

modes, because it has severe consequences for the size and cost of the package for most of the products, as well as for the battery lifetime of mobile products. Power management at all hierarchy levels of design has therefore become a necessity for a successful introduction of a product into the market.

Figure 11.7 shows the variability and signal integrity trends, starting with the 500-nm CMOS technology node as a reference. Before that node the impact of most of these parameters on the behaviour of digital circuits was hardly visible.

The diagram shows that the noise, IR-drop, cross-talk and variability parameters are all increasing, while the noise margins (V_{dd} and V_T) reduce or remain almost constant. This requires the design styles and design flows to continuously adapt to the increasing impact of these parameters. More decoupling capacitance, larger wire spacing in buses, limited di/dt in clock, bus and I/O drivers and *variability-aware design* are examples of how to deal with these effects. But the trend shows that what seems to be a sufficient solution in one technology node will certainly be not sufficient in the next. Design solutions and tool development must therefore be targeted to bend the positive slopes of the parameters into the direction of the noise-margin slope.

The relation between the reliability parameters and the scaling factors, as presented in the table, is shown in the diagram in Fig. 11.8. This diagram is only meant to show the trends rather than representing an accurate estimation. Some lines therefore show a 'question-mark' relation with the scaling factor.

It shows that almost all parameters are getting worse, again if we assume that both the transistor architecture and the supply voltage would not dramatically change. Generally, reliability parameters are very difficult to model and predict, because they depend on more factors than just the scaling of the sizes and/or voltages. Huge R&D efforts are required to understand all physical mechanisms that underlay these

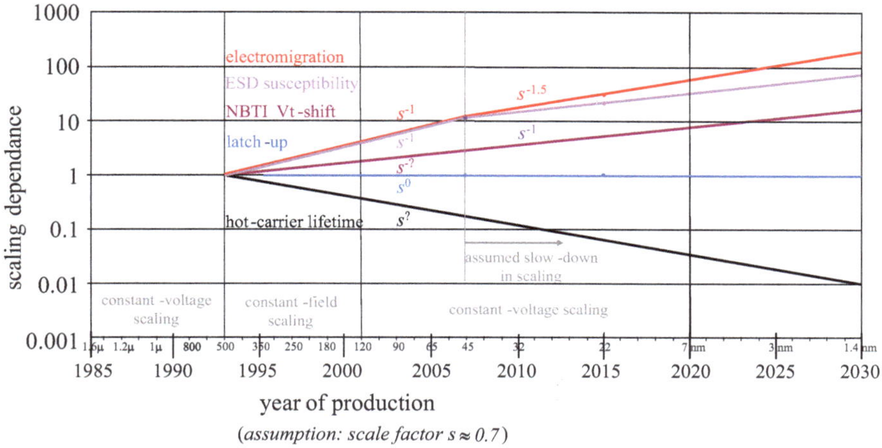

Fig. 11.8 Experienced and expected trends in reliability parameters

reliability parameters in order to build an accurate model that can be used to predict the behaviour of individual MOS transistors as well as of the ICs built from them.

Creative solutions, both in technology and design, are needed to keep the IC's robustness at a sufficiently high level in order to extend Moore's law for yet another lustrum. However, this will lead to a further increase of the complexity and total development and production costs of an IC.

11.5 Potential Limitations of the Pace of Scaling

Originally, Moore's law predicted a doubling of the transistor density every 18 months. During the last decade, this has reduced to a doubling every two years. In this decade the pace has reduced to a doubling every three to four years. It is sometimes called a self-fulfilling prophecy and was seen as a measure for trends and set the pace of innovation. Almost according to this law, the Semiconductor Industrial Association has based its ITRS roadmap [1] on the continuation of the law. Tables 11.4 and 11.5 show several important parameters of this IRDS roadmap [2].

The previously discussed scaling trends show that there are potentially several key factors that may limit the pace of scaling. As can be seen from the table, the growth in complexity of logic CMOS ICs is expected to be much less than according to Moore's law (please refer to the line: NAND2-eq gate count). However, the complexity of the design and test tasks is accelerated and forms a potential barrier to obtaining full exploitation of the available manufacture potentials. The overall success of the semiconductor industry will be increasingly dominated by how the complex design, engineering and test challenges will be addressed [1]:

- System complexity
  - Huge amount of transistors on one single chip (10 million to more than 100 billion).
  - Convergence of consumer, computing and communications domains, which accelerates the introduction of new features on a single chip. This makes a design more heterogeneous, with a large variety of domain-specific, general-purpose IP and memory cores.
  - Development of application software to handle specific user tasks.
  - Validation of the total system through extensive (hybrid) simulation and emulation.
  - Different performance demands on a single chip, e.g. high performance and low power, which require multi-threshold voltages, multi-gate oxides, multi-clock and multi-voltage domains.
- Silicon complexity
  - Huge amount of transistors on one single chip (10 million to more than 100 billion).
  - Increasing manifestation of deep-sub-micron/nanometer physical effects like Cross-talk, voltage drop, supply noise, electromigration, variability, stress, leakage, etc.
  - Performance increase is no longer an implicit benefit of further scaling (beyond 90 nm CMOS) and we are approaching the frequency barrier. This results in a complex and very time-consuming timing closure process.
  - Imperfect lithography.
  - Changing process defect mechanisms.
- Design flow complexity
  - Development of an application-domain-specific design platform.
  - Increasingly complex design flow to cover all previous additional design tasks (deep-submicron/nanometer physical effects, power switches, multi-voltage/frequency domains, adaptive voltage/frequency scaling, etc.).
  - Validation, verification and timing closure are increasingly complex.
  - 3D extraction tools required for memory and analogue circuit design.
  - Design closure, which is the process of a (slow) convergence to a fully functional design that meets all constraints.
  - Test development and test coverage, also dealing with multi-clock and multi-voltage domains.
- Fabrication complexity
  - Increasingly complex and expensive lithography
  - Mask cost explosion
  - Wafer fab cost explosion
- Package complexity
  - Increasing number of power and I/O pads/balls
  - MCM, SiP and SoP solutions
  - Limited thermal conductivity improvement
- Debug and failure analysis complexity

- Increasing variety of failure causes: defects, stress, proximity effects, process spread, noise, temperature.
- Less noise margins.
- Less first-time-right products.
- Metals shield access from top side.
- Complex backside stimulation and analysis tools.

An important factor that limits the pace of scaling is the *design productivity gap*, as shown in Chap. 6, Fig. 6.51.

The diagram shows the increase in available logic transistors per chip as a result of the continuous scaling of transistor dimensions, as well as the increase in design productivity in transistors per man month. The fact that IC design teams continuously use an increasing amount of smarter CAD design tools (see Chap. 7), also at higher abstraction levels of design, does not increase the pace of design productivity. The main reason is that the complexity of a design is not just related to correctly connect an increasing number of transistors that switch at higher frequencies. More transistors also mean more current, more power, more noise, more timing problems, reduced reliability and signal integrity, which together cause a dramatic increase in the number of design tasks and their complexity. One of the major design productivity bottlenecks, however, is the validation and verification process of a design. In advanced designs, it may take about 30–40% of the total development time!

Figure 11.9 shows a summary of increasing design tasks. It also shows the exponential increase in leading-edge monolithic chip design costs because of the rapidly increasing design complexity. These costs are just an order of magnitude and representative for new SoCs at the leading-edge node in the respective year. For some SoCs, these costs may be higher, while for others, these may be lower, depending on the amount of available reusable IP.

The design will therefore have an increasing impact on the total price setting of an integrated circuit. As said, the costs in the diagram reflect a leading-edge monolithic chip design. However, no chip has ever cost that amount, because many chip designs are based on the reuse of IP. This also holds for the software which is built on huge libraries of code [9]. For a 28-nm SoC, these costs (including the complete system, the architecture and software development and the total costs for validation and verification) has jumped to about 50 M US$. In 10 nm a leading-edge design costs about 140 MUS$. Assuming that the chip is meant for a consumer application and that the profit per chip is expected to be in the order of a few US$, then a simple calculation learns that at least a total volume of about 50 million products is required to reach breakeven with respect to the development costs. There are not so many applications that generate market volumes for a single supplier in this order of magnitude. For the smaller nodes, the total development costs will only increase further.

The total costs of a 1-cm^2 chip, designed to run at the maximum frequency achievable with a standard-cell design flow (say 1 GHz for a 16-nm SoC), including the costs of software (for both platform and application), are so high that the required

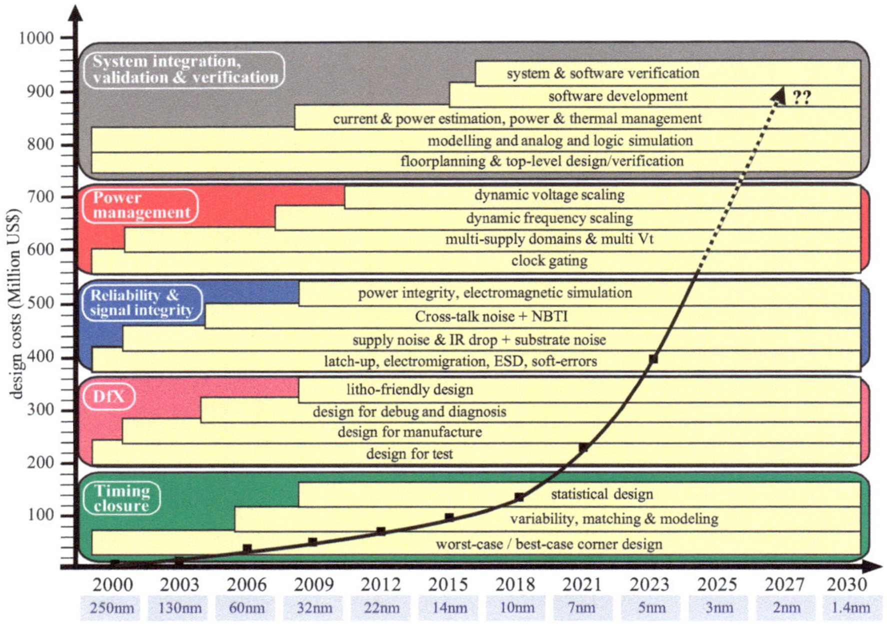

Fig. 11.9 Increasing number of design tasks and growing design costs for a leading-edge monolithic SoC design

volume to recover the development and fabrication costs about equals the number of transistors in unique cores on that chip. What is meant here is that current chips (re)use multiple, or even large arrays of identical cores. In that case the design costs per transistor are less. For several application areas, we may expect that the move to the next technology node will no longer be economically attractive. When the fabrication costs have only become a minor part of the total costs of an IC, scaling to the next technology node will hardly reduce the price. Therefore, it may be possible that the 16-nm node may be the last economically viable technology for many applications. For some cheap consumer products, the 28-nm or 22-nm node may be the final one, while this may be the 7-nm or 5-nm node for certain high-performance processors. Both the price and profit per chip in this category is at least an order of magnitude larger than that of an average consumer chip, which allows to recover the huge development and production costs.

The ability to completely verify, test, debug and diagnose future complex designs will reduce dramatically. It is therefore likely that current design styles with fixed and dedicated logic will be replaced by design styles that allow flexibility and configurability. This flexibility can be enhanced by software solutions (programmability) as well as hardware solutions (*reconfigurable computing* such as embedded *FPGA* and/or sea-of-gates architectures). Remaining bugs can then be bypassed by changing the program or by remapping the function, respectively. Another solution

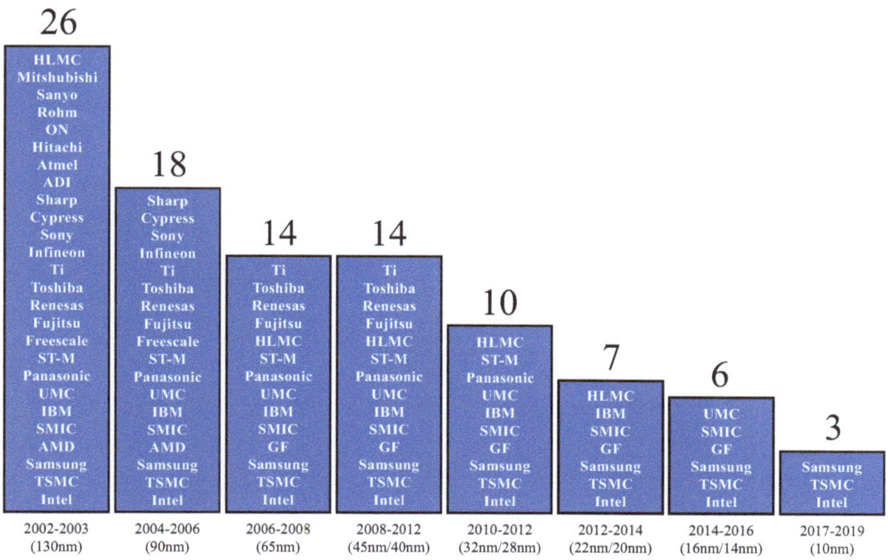

Fig. 11.10 Semiconductor companies with a leading edge fab in various process nodes (Source: Intel/Bits on Chips)

that tackles the complexity problem is to move away from a monolithic chip design solution towards chiplet design. This is extensively discussed in Sect. 10.3.3.

Another potential key factor in lowering the pace of process innovation, which is already discussed, is formed by the economics of the production facilities. From 1966 to 2024, the costs of a wafer factory increased by roughly three orders of magnitude, from about $15 million to $20 billion [9]. A 2-nm fab with a capacity of roughly 50,000 *wafer starts per month* (*WSPM*) is expected to cost around $28 billion, up from around $20 billion for a 3-nm fab with a similar production capacity [10].

These investments can only be raised by a few individual large semiconductor companies and several semiconductor alliances. For 45-nm CMOS and beyond we saw already an increasing number of semiconductor companies that outsource chip fabrication and will become fabless: "only the elite few will be able to afford it" [11]. Figure 11.10 shows the number of semiconductor houses with a leading edge fab in process nodes down to 3 nm. As can be seen, beyond the 10-nm node, there are only three semiconductor houses, who can still bear the huge investments costs, left.

It has already been discussed before that, when the design costs become significantly larger than the fabrication costs, the drive for scaling an application to the next technology node will reduce. Moreover, since electrons run close to their saturation speed in the 65-nm technology node and beyond, only limited circuit performance improvements had been gained from this scaling. This has changed the focus from GHz to multicore designs. It is doubtful whether the cost

Table 11.3 Some key factors and cost per gate trend towards 16-/14-nm nodes [12]

technology	gates/mm^2 (KU)	gate utilisation (%)	used gates/mm^2 (KU)	parametric yield impact (Δ from D_0 yield)	actual used gates/mm^2 (KU)	gates/wafer (MU)	wafer cost ($)	wafer cost (Δ)	cost per million gate ($)
90nm	637	86	546	97	532	33,831	1,357.62	-	0.0401
65nm	1,109	83	919	96	885	56,330	1,585.71	16.8	0.0282
40nm	2,139	78	1,677	92	1,538	97,842	1,898.83	19.7	0.0194
28nm	3,946	76	3,011	87	2,610	166,086	2,326.12	22.5	0.0140
20nm	6,992	65	4,524	73	3,293	209,541	2,981.75	28.2	0.0142
16/14nm	12,391	54	6,716	61	4,090	260,228	4,205.37	41.0	0.0162

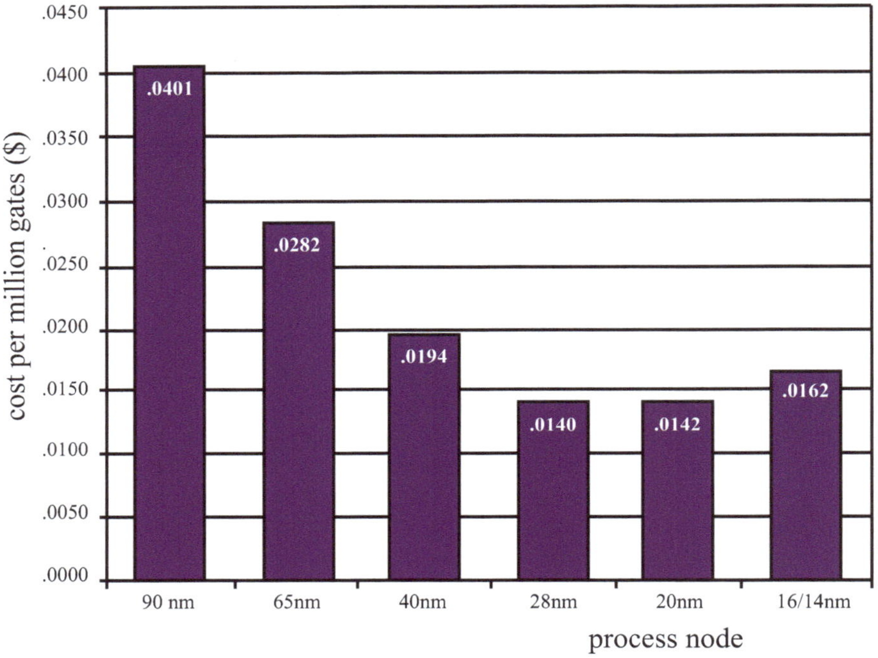

Fig. 11.11 Expected trend in reduction of cost per million gates, which seem to saturate at the 28-nm node

per transistor will further reduce with scaling. An analysis shows that, although the number of gates per unit area still increases with further scaling, the cost per gate saturates around the 28-nm node (Table 11.3 and Fig. 11.11 [12]), as it was one of the last planar (2D) nodes. FinFETs, which have replaced planar transistors, use a 3D structure and require more complex process steps. Therefore, it was expected that the cost per gate for the 16-/14-nm FinFET nodes is higher than for 20 nm and 28 nm. The expected transition from FinFET to gate-all-around (GAA), nano-sheet/nano-fork/nano-wire and *CFET (complementary FET)* transistors for the sub-3-nm nodes leads to a further cost increase. These devices require even more 3D-intensive structures at the cost of a lot more complex process steps.

In the table, the parametric yield is assumed to reduce due to the expected increase of problems with doping uniformity, line-edge roughness and other physical effects [12].

From this table, the following cost diagram is derived (Fig. 11.11 [12]).

In 2014, Intel presented a cost-per-transistor diagram [13] which still shows a continuation of the cost reduction to the 14-nm node. The cost per transistor is the result of the chip architecture, its design cost and its fabrication cost and the package cost. In process nodes below 10 nm we experience an explosion of both design costs (see Fig. 11.9) and manufacturing cost due to approaching the physical limits in device operation and of the production tools in a wafer fab. As an example, the estimated TSMC processed-wafer prices in the year 2020 for the 7-nm and 5-nm nodes are US$9500 and US$17,000 [14], respectively.

A typical cleanroom may include more than 1200 multimillion-dollar factory tools and over 1500 pieces of utility equipment [15], e.g. pumps, filters and airflow control, as well as to supply electricity, gases and chemical liquids. The following table gives an impression of some well-known tools used in a logic wafer fab. The first four lines also include some indication of how many of these tools may be installed in a cleanroom. These numbers depend heavily on the process node and on the type of fab, e.g. logic, NAND flash or DRAM fab:

- 20–60 scanners/steppers plus wafer tracks
- 20–40 deposition tools
- 25–50 etch machines
- 15–30 ion implanters
- Annealing/diffusion tools
- Polishing and grinding tools
- Baking tools
- Cleaning tools
- Metrology tools

A *front-opening unified pod* (*FOUP*) can contain a batch of 25 wafers. An extremely well-controlled wafer track system is used to automatically move these sealed FOUPs between the tools efficiently and reliably, minimizing the risk of contamination or damage and minimizing the wafer cycle time. A typical fab is capable of 50-k to 100-k wafer starts per month.

The above discussions, in combination with expected lower yield numbers, show that it is very difficult to present a generic cost per transistor per node. Even when we would have that, it would certainly not be applicable for many chip designs as they may show completely different architectures, e.g. microprocessor versus graphics processor versus AI designs. Therefore, the number of products for which scaling to the next process node is still economically attractive is continuously reducing. In this respect, particularly for large die sizes, chiplets (see Sect. 10.3.3) offer an alternative solution to reduce costs.

The third key factor that may limit the pace of scaling is represented by the increased manifestation of physical and electrical effects in deep-submicron technologies. Larger current slew rates (di/dt) and mutual signal track capacitances will bring the circuit noise to unacceptable levels. In addition to this, the noise margins of future processes will further decrease due to the continuous drive for further (although marginal) reduction of the supply and threshold voltages (Fig. 11.7).

Figure 9.23 (Chap. 9) shows the voltage-scaling evolution over four decades: from 5 V at the end of the 1980s to around 0.7 V in 2030.

As high-performance designs, such as the Intel and AMD processors for PC and server applications, already experienced problems with process variation and switching noise in the early 1990s at relatively high supply voltages, ASIC designs faced about the same problems, however, a couple of process generations later at somewhat scaled supply voltages (and reduced noise margins). Particularly during the period from 1990 to 2007, where the supply voltage (= noise margin) scaling was the same as the feature-size scaling, every new design in the next process node required additional design and/or technology measures to reduce the noise and increase the gap between the noise and the noise margin. However, after scaling to the next node, the problem is the same again and new measures were required. Relatively large additional chip areas need therefore be devoted to on-chip measures like decoupling capacitances and to more widely spaced buses and other global signal interconnections etc. These physical effects, which are extensively discussed in Chap. 9, reduce the chance of fully exploiting the potentials of the new process generations. Design for test, design for robustness, design for debug, litho-friendly design, design for manufacturability, etc., all add up to an increased design complexity and chip area. The level to which these additional measures will limit the efficient use of chip area cannot be predicted because it also depends on the creative design alternatives that will be developed in the near future.

The above discussions regarding the cost trend per transistor were focused towards the leading-edge process fabs. Today, however, we also face increasing demand at the lagging-edge fabs, which means that the demand for chips in the more mature technologies is growing strongly. This is particularly driven by the automotive and IoT markets, which can be served by conventional, more mature technologies. These show higher yields, lower costs and increased reliability. The problem is that there are hardly investments in these older fabs resulting in capacity shortage and price increase. The 'old' chips are becoming just as important as new chips [16].

Finally, as already stated before, the ITRS 2013 roadmap was the last version of this roadmap. There are several reasons that drove the change towards a new roadmap set up [17]:

- Moore's law is approaching its limits.
- Driver applications have changed from mainframes and PCs to mobile, wireless, Internet of Things (IoT), artificial intelligence (AI) and other applications that

require processing of huge amounts of data, e.g. weather forecast, and to the cloud for data storage, which require other technology options and solutions.
- As discussed above, the cost per gate is saturating, so the drive for further scaling is changing towards other system requirements, such as small available footprint or limited available volume, or less power, due to an increasing number of features and interfaces on the same footprint. Also the move to 3D solutions, both on-chip and in the package, had its influence on the roadmap.
- Since there are only a very few companies that can afford to scale further and who are willing to make the required investments, there is less willingness and determination to share competitive data.

Traditionally, the technology node was named after the minimum width of the polysilicon track, which forms the transistor channel length. In a conventional 120-nm CMOS process the channel length was close to 120 nm. However, the further we scale, the more the channel length deviates from the process node name. In advanced CMOS process nodes, the physical gate length is quite different from the node name. Below 20 nm there is less of standardisation regarding features sizes and process nodes. This means that different IC vendors can have different process *node names* for processes that contain the same feature sizes. In other words: a 7-nm FinFET process of one IC vendor can use quite different minimum feature sizes than a 7-nm FinFET process of another vendor. This makes further area and performance comparisons complicated. Current process node names refer to a transistor density that corresponds to an equivalent channel length. In a 2-nm CMOS process node the transistor density is such, as if it would have a channel length of 2 nm.

The IEEE International Roadmap for Devices and Systems (IRDS) is the continuation and extension of the International Technology Roadmap for Semiconductors (ITRS). This ITRS showed a record of uninterrupted publications spanning from 1992 to 2015. From 2016, the IRDS has followed several of the successful methodologies demonstrated by the ITRS with additional expansion to system integration, data centres and Internet communication requirements as well as quantum computing and quantum communications. The IRDS includes 11 chapters. The most important ones for further CMOS integration for the target audience of this book are summarized below:

- More Moore
- Yield enhancement
- Lithography
- Metrology

The following two tables present the roadmap for several device and interconnect parameters, which may give the reader a more detailed insight into what is to be expected until the year 2034. For background behind and explanation of the parameters and numbers in these tables, the reader is referred to reference [2].

Table 11.4 IRDS roadmap for selected logic device and core characteristics (Source: IEEE IRDS More Moore Roadmap edition) [2] (for explanation of the abbreviations and the meaning of the used colours, please refer to the reference)

YEAR OF PRODUCTION	2022	2025	2028	2031	2034
	G48M24	G45M20	G42M16	G40M16/T2	G38M16/T4
Logic industry "Node Range" Labeling	"3nm"	"2nm"	"1.5nm"	"1.0nm eq"	"0.7nm eq"
Fine-pitch 3D integration scheme	Stacking	Stacking	Stacking	3DVLSI	3DVLSI
Logic device structure options	finFET LGAA	LGAA	LGAA CFET-SRAM	LGAA-3D CFET-SRAM	LGAA-3D CFET-SRAM
Platform device for logic	finFET	LGAA	LGAA CFET-SRAM	LGAA-3D CFET-SRAM-3D	LGAA-3D CFET-SRAM-3D
			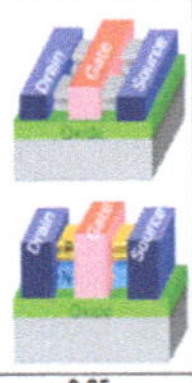	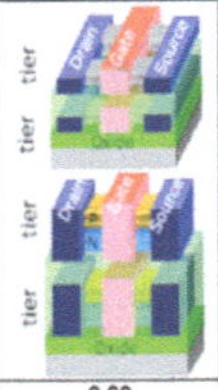	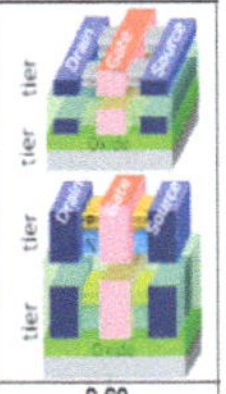
Vdd (V)	0,70	0,65	0,65	0,60	0,60
Gate length (nm)	16	14	12	12	12
Number of stacked tiers [1]	1	1	1	2	4
Number of stacked nanosheets in logic device [1]	1	3	3	4	4
Number of stacked nanosheets in SRAM device [1]	1	3	6	8	8
Maximum number of stacked nanosheets in one device [1]	8	8	16	16	32
Digital block area scaling	1,00	0,74	0,55	0,26	0,13
Digital block energy scaling	1,00	0,81	0,72	0,56	0,50
#MAC units in SoC - based on integration capacity	8192	11038	14980	30966	65191
Cell height (nm) - HD	144	114	90	80	80
CPU frequency (GHz)	3,18	3,28	3,36	3,42	3,47
CPU frequency at constant power density (GHz)	3,18	3,17	2,79	1,49	0,71
Power density scaling	1,00	1,03	1,20	2,29	4,85
TOPS/mm2 scaling	1,00	1,39	1,93	4,07	8,68
TOPS/W scaling	1,00	1,23	1,39	1,79	1,99
TOPS/mm2 * TOPS/W	1,00	1,71	2,70	7,29	17,24
LOGIC DEVICE GROUND RULES					
Mx pitch (nm)	32	24	20	16	16
M1 pitch (nm)	32	23	21	20	19
M0 pitch (nm)	24	20	16	16	16
Gate pitch (nm)	48	45	42	40	38
Lg: Gate Length - HP (nm)	16	14	12	12	12
Lg: Gate Length - HD (nm)	18	14	12	12	12
Channel overlap ratio - two-sided	0,20	0,20	0,20	0,20	0,20
Spacer width (nm)	6	6	5	5	4
Spacer k value	3,5	3,3	3,0	3,0	2,7
Contact CD (nm) - finFET, LGAA	20	19	20	18	18
Device architecture key ground rules					
Device lateral pitch (nm)	24	26	24	24	23
Device height (nm)	48	52	48	64	60
FinFET Fin width (nm)	5,0				
Footprint drive efficiency - finFET	4,21				
Lateral GAA vertical pitch (nm)		18,0	16,0	16,0	15,0
Lateral GAA (nanosheet) thickness (nm)		6,0	6,0	6,0	5,0
Number of vertically stacked nanosheets on one device		3	3	4	4
LGAA width (nm) - HP		30	30	20	15
LGAA width (nm) - HD		15	10	10	6
LGAA width (nm) - SRAM		7	6	6	6
Footprint drive efficiency - lateral GAA - HP		4,41	4,50	5,47	5,00
Device effective width (nm) - HP	101,0	216,0	216,0	208,0	160,0
Device effective width (nm) - HD	101,0	126,0	96,0	128,0	88,0
PN seperation width (nm)	45	40	20	15	15

Table 11.5 IRDS roadmap for selected logic and analogue device specs and interconnect technology (Source: IEEE IRDS More Moore Roadmap edition) [2] (for explanation of the abbreviations and the meaning of the used colours, please refer to the reference)

YEAR OF PRODUCTION	2022	2025	2028	2031	2034
	G48M24	G45M20	G42M16	G40M16/T2	G38M16/T4
Logic industry "Node Range" Labeling	"3nm"	"2nm"	"1.5nm"	"1.0nm eq"	"0.7nm eq"
Fine-pitch 3D integration scheme	Stacking	Stacking	Stacking	3DVLSI	3DVLSI
Logic device structure options	finFET LGAA	LGAA	LGAA CFET-SRAM	LGAA-3D CFET-SRAM	LGAA-3D CFET-SRAM
Platform device for logic	finFET	LGAA	LGAA CFET-SRAM	LGAA-3D CFET-SRAM-3D	LGAA-3D CFET-SRAM-3D
LOGIC DEVICE ELECTRICAL SPECS					
Power Supply Voltage - Vdd (V)	0,70	0,65	0,65	0,60	0,60
Effective mobility (cm2/V.s)	125	100	80	60	40
Ion (uA/um) at Ioff=10nA/um - HP [6]	874	787	851	753	737
Ion (uA/device) at Ioff=10nA/um - HP [7]	88	170	184	157	118
Ion (uA/um) at Ioff=100pA/um - HD [8]	644	602	656	551	532
Ion (uA/device) at Ioff=100pA/um - HD [9]	65	130	142	115	85
Energy per switching [CV2] (fj/switch) - FO3 load, HP	0,65	0,49	0,47	0,40	0,40
Gate height over fin (nm)	20	15	10	10	10
INTERCONNECT TECHNOLOGY					
Mx pitch (nm)	32	24	20	16	16
M1 pitch (nm)	32	23	21	20	19
M0 pitch (nm)	24	20	16	16	16
Aspect ratio - M0, M1, Mx, MP80, MP720	1.5-2.5	1.5-2.5	1.5-2.5	1.5-2.5	1.5-2.5
Power rail layer	M0	Buried Rail	Buried Rail	Buried Rail	Buried Rail
Power rail material	Co, W, Ru	W, Ru	W, Ru	W, Ru	W, Ru
Metallization - M0	Co, W, Ru	Co, W, Ru	Co, W, Ru	Co, W, Ru	Co, W, Ru
Barrier - M0	0.5nm TiN+WC	0.5nm TiN+WC	0.5nm TiN+WC	0.5nm TiN+WC	0.5nm TiN+WC
Metallization - M1, Mx	Cu	Cu, Co, Ru	Cu, Co, Ru	Cu, Co, Ru	Cu, Co, Ru
Barrier metal - M1,Mx	1.5nm TaNCo	0.5nm TiN+WC	0.5nm TiN+WC	0.5nm TiN+WC	0.5nm TiN+WC
Di-electrics k value - M0,M1,Mx	SiCOH (2.70-3.20)	SiCOH (2.70-3.20)	SiCOH (2.70-3.20)	SiCOH (2.70-3.20)	SiCOH (2.70-3.20)
Metallization - >MP40	Cu	Cu	Cu	Cu	Cu
Di-electrics k value - >MP40	SiCOH (2.40-2.55) Airgap (1.0)	SiCOH (2.20-2.55) Airgap (1.0)	SiCOH (2.20-2.55) Airgap (1.0)	SiCOH (2.20-2.55) Airgap (1.0)	SiCOH (2.20-2.55) Airgap (1.0)
ANALOG SPECIFICATIONS OF LOGIC DEVICE					
Transconductance - gm (μS/μm)	1605	1621	1751	1725	1653
High-current gain cut-off frequency - fT (GHz)	261	304	358	411	403
Maximum oscillation frequency - fmax (GHz)	169	175	166	210	233
1/f-noise (μV²·μm²/Hz)	16	16	13	13	13
Analog gain (dB)	42	40	36	38	39

11.6 Conclusions

Traditionally, there are three ingredients that drove the continuous scaling of integrated circuits: cost reduction, size reduction and performance improvement. However, scaling not only influences the system sizes and performance positively, it also has major negative effects on the reliability and signal integrity of nanometer

ICs. These effects have increased to such an extent that digital ICs can no longer be regarded as circuits that process and store ones and zeros in a certain order to execute a certain function. The design of digital circuits increasingly requires an analogue approach to maintain reliability and signal integrity at a sufficiently high level. The manifestation of the responsible physical effects, which increase with scaling of the feature sizes, will be a further challenge, if not a threat to the reliability and signal integrity of future VLSI designs.

These effects place high demands on the design and test strategies used for modern ICs and systems. Additional measures in the design are needed to maintain testability, observability, reliability and signal integrity at a sufficiently high level. In combination with power management solutions and DfX requirements, these measures all require additional chip area, which limit an efficient exploitation of the potentials of the new process generations. Moreover, they also contribute to the rising cost of IC design.

Already today, we face the fact that a move to the next process generation will no longer be commercially attractive for various categories of products. Physically, we may scale transistor channel lengths to about 7 nm at extremely high fabrication cost, but what does it mean in practice? It means that the channel is only about 35 silicon atoms long. Devices at that scale will certainly no longer operate reliably, since their behaviour largely depends on increasing quantum physics uncertainties and they are not cost-effective. Several publications stated that the minimum cost per transistor has already been reached at the 28-nm node, although not everybody agrees with it. But whatever the cost-per-transistor trend is, certain applications (e.g. PCs, laptops, AI, IoT, mobiles, etc.) still require more performance, less power or smaller footprints, which drive scaling of related products to the next process nodes. An alternative to the continuous scaling is to use the third dimension by stacking the devices on top of each other. This third dimension is already used in NAND flash memories, where hundreds of memory-cell layers are stacked on top of each other with rather relaxed design rules. DRAMs not only use the third dimension in the 3D capacitors in their memory cells, but also by stacking eight, sixteen or more individual dies on top of each other in a single package to create high-bandwidth memories. The future cannot be predicted, so nobody knows what the last semiconductor process node will be. There is also no generic definition of a process node anymore. Until the 28–32-nm node, the node name was more or less related to the transistor channel length. However, in a 16-nm process, there is not any physical size equal to 16 nm! Current *node names* refer to a transistor density that corresponds to an equivalent channel length. In a 2-nm CMOS process node the transistor density is such, as if it would have a channel length of 2 nm. Currently the IDRS roadmap predicts a constant channel length of 12 nm for the next ten years (up to 2035). But the node names will then be expressed in Å, rather than in nm. Finally, there is also a limitation in scaling the width of the fin. At, but certainly below, a *fin width* of 5 nm, the FinFET transistor shows extremely large V_T-variations and leakage currents, while also electrostatics reduce the FinFET's current drive, which require different transistor architectures to continue further scaling. These are discussed in the chapter on manufacturing.

11.7 Exercises

1. Explain the differences between the constant-voltage scaling and the constant-field scaling process. How did they influence the main driving force behind the scaling process?
2. Why was copper not used in the early MOS processes? What is the result of using copper instead of aluminium for the interconnection patterns of an IC?
3. Suppose that an IC with channel lengths of 40 nm is manufactured in a 40-nm CMOS process with $Vdd = 1.1$ V and used in a particular application. Suppose this IC is scaled by a factor of 0.7 and manufactured in a 28-nm process with $Vdd = 1.1$ V. You may assume that it obeys all 28-nm process design rules. What would happen to the following parameters when this IC is used in the same application:
   (a) The transistor gain factors β_n and β_p
   (b) The threshold voltages V_{T_n} and V_{T_p}
   (c) The chip's power dissipation
   (d) The chip's power density
   (e) The noise on the chip's supply and ground lines
4. What would be the four biggest threats for the pace of scaling? Motivate your answer.
5. When you look at the propagation-delay trend in Fig. 11.2, what measures, both in design and process technology, can be taken to reduce the propagation delay?
6. Give three reasons why the metal interconnect capacitance should be as low as possible?

References

1. Semiconductors Industrial Associations, ITRS roadmap, 2013 update, http://www.itrs.net
2. International Roadmap for Devices and Systems (IRDS™) 2022 Update: More Moore. https://irds.ieee.org/images/files/pdf/2022/2022IRDS_MM.pdf
3. M. Vertregt, et al., *Scalable high-speed analog circuit design, 2001 AACD* (Kluwer Academic Publishers, 2002), pp. 3–21
4. M. Izumikawa, et al., A 0.25 μm 0.9 V 100 MHz DSP Core (IEEE-JSSC, 1997), pp. 52–61
5. T. Kuroda, et al., A 0.9 V, 150 MHz, 10 mW, 4 mm^2, 2-D DCT Core Processor with Variable Threshold voltage Scheme (IEEE-JSSC, 1996), pp. 1770–1779
6. S.H. Lo, et al., Quantum-mechanical modelling of electron tunnelling current from the inversion layer of ultra-thin-oxide in MOSFET's. IEEE Electron Dev. Lett. **18**(5), 209–211 (1997)
7. T. Dillinger, The Latest in Dielectrics for Advanced Process Nodes, https://semiwiki.com/semiconductor-manufacturers/intel/294379-the-latest-in-dielectrics-for-advanced-process-nodes/, December 2021

8. A. Grill, et al., Progress in the development and understanding of advanced low k and ultralow k dielectrics for very large-scale integrated interconnects-State of the art. Appl. Phys. Rev. **1**, 011306 (2014)
9. B. Bailey, What Will That Chip Cost? SemiconductorEngineering, https://semiengineering.com/what-will-that-chip-cost/, October 30, 2023
10. A. Shilov, Firm predicts it will cost $28 billion to build a 2 nm fab and $30,000 per wafer, a 50% increase in chipmaking costs as complexity rises, Toms Hardware, December 2023
11. B. Johnson, et al., Market Trends: Rising Costs of Production Limit Availability of Leading-Edge Fabs, Gartner report, 17 September 2012
12. H. Jones, Why migration to 20 nm bulk CMOS and 16/14 nm FinFETs is not a best approach for Semiconductor Industry, White Paper, International Business Strategies, Inc., January 2014
13. H. Mujtaba, Intel 14 nm Broadwell CPU Architecture Analyzed – 5% IPC Increase Over Haswell, 2nd Generation FIVR, 20% More Compute Units, wccftech.com, August 2014
14. A. Shilov, TSMC's Estimated Wafer Prices Revealed: 300 mm Wafer at 5nm Is Nearly $17,000, tom'sHARDWARE, September 18, 2020
15. What does it take to build a fab, https://www.intel.com/content/dam/www/central-libraries/us/en/documents/what-does-it-take-to-build-a-fab.pdf
16. D. O'Laughlin, The Rising Tide of Semiconductor Cost, Fabricated Knowledge, https://www.fabricatedknowledge.com/p/the-rising-tide-of-semiconductor, November 22, 2021
17. F. von Trapp, Executive Interview: Bill Bottoms Talks about Revamping the ITRS Roadmap, http://www.3dincites.com/2015/03/executive-interview-bill-bottoms-talks-revamping-itrs-roadmap/, March 12, 2015

Index

Numbers and Symbols
NA, 94
V_T-implant, 419
(MLC), 327
(NVDRAM), 295
(PDN), 483
(PLC), 326, 327
(QLC), 327
(TLC), 327
(triple-level cell (TLC)), 318
1T cell, 285
2.5D packaging, 562
3D DRAM, 294
3D NAND flash, 78
3D Xpoint memory, 328
3D die stacking, 78, 115
3D packaging, 562
3 T cell, 344
6-σ design, 281

A
ABB, 417
ABCD process, 266
Abstraction level, 358
Accelerating voltage, 135
Acceptor, 7
Access time, 271, 277
Accumulation
  capacitor, 9
  layer, 31
  process, 9
ACTIVE
  mask, 141
  mask programmed ROM cell, 306
Active area, 145, 224
Activity
  factor, 414
  reduction, 441
Adaptive
  body bias, 417
  skew control, 472
  voltage scaling, 417, 432
Address buffer, 275
Advanced OCV, 502
AE, 309
AID PSM, 114
Airflow, 549
Air gaps, 170, 421
Al cap, 577
ALD, 107, 128, 130, 288, 591
ALE, 133
Algorithmic level, 357
Alignment
  marks, 90, 155
  sensors, 90
  sequence, 90
  tree, 90
Allocation, 371
Alpha particles, 488
  radiation, 305
Alternating phase-shift mask, 98
AltPSM, 98
Aluminium, 4, 626
  cap, 577
  gate process, 145
Amorphous silicon, 143
Anamorphic projection, 112
AND
  function, 189
  matrix, 387
Anisotropic etch, 131, 509
Annealing, 85
Anneal step, 135
Antenna effect, 133, 140, 523
Anti-irradiation damage PSM, 114
Anti-punch-through, 504
  implant, 150
Anti-reflective coating, 152

H. Veendrick, *Nanometer CMOS ICs*, https://doi.org/10.1007/978-3-031-64249-4

AOCV, 502
APCVD, 126
Application-specific
IC, 257
integrated circuit, 350
standard product, 350, 351
APS, 250
APT, 504
implant, 150
ARC, 152
Area efficiency, 309
ASIC, xvi, 257, 350, 381
turnaround time, 350, 352
Aspect ratio, 182, 194
Assist circuits, 284
Associative memory, 273
ASSP, 350, 351
Asynchronous
circuits, 446
design, 446, 447
ATE, 535
Atom, xxix
Atomic-scale terms, xxix
Atomic layer
deposition, 107, 128, 130, 288
etching, 133
ATPG, 539, 547
Attenuated phase-shift mask, 97
AttPSM, 97
Automatic test equipment, 535
Automatic test pattern generator, 539
AVS, 417, 432
A_{V_T}, 499

B
Back-bias, 416
controlled V_T, 416
effect, 21, 192, 417
Back-end design, 373
Back-end-of-line, 153
Back-gate effect, 21, 23
Back-scattered electrons, 610
Backside grinding, 561
Backside power distribution, 484
Ball grid array, 574
Bandgap, 6
Basic
CMOS process, 145
complementary MOS process, 145
MOS technologies, 140
silicon-gate nMOS process, 141
Bath tub curve, 584
Battery, 412
energy, 412
memory effect, 413
RAM, 329
BCCD, 246
surface-state immunity, 249
BCD
counter, 445
process, 263
Behavioural simulation, 373
BEOL, 153
Best-case corner, 281
BGA, 574
BICMOS
characteristics, 255
circuit performance, 256
digital circuits, 253
NAND gate, 256
performance, 257
technology, 253
Bipolar
CMOS-DMOS process, 263
gain factor, 255
noise, 255
BIRA, 341
Bird's beak, 122
suppression, 122
BIST, 543
BISTAR, 544
Bit line, 275
select, 274
Bit-parallel operation, 363
Bit-slice layout, 384
Block, 390
β_n, 197
BOA, 577
Body
bias, 417
effect, 21, 85, 592
factor, 22
Bond-over-active, 170, 577
Bootstrap-capacitance, 185
Bootstrapped load, 185
Boundary scan test, 546
Bow, 78
BOX layer, 86
β_p, 199
BPD, 484
BPSG, 126
BRAM, 329
Breakdown
mechanism, 591
voltage, 259
Bridging faults, 542

BST, 288, 546
Bubbles, 100
Buffer circuits, 199
Built-in repair analysis, 341
Built-in self-test and repair, 543, 544
Bulk silicon, 77
Buried-channel CCD, 246
Buried-oxide layer, 86
Buried well, 416, 491
Burn-in test, 289, 585
Burst mode, 277
Bus latency, 627

C
C4, 572
Cache memory, 302
Cache register, 314
CAD tools, 350
CAM, 269, 273
Capacitances, 34
Capacitor-under-bit line, 290
Carrier mobility, 143
  reduction, 47
Cascode voltage swing logic, 219
CBRAM, 333
CCD, 245
  cell, 246
  operating frequency, 249
  shift register, 246
CCO, 471
CD, 94, 102, 106
  uniformity, 94, 105
Cell-based IC design, 352
Cell abutment, 440
CFET, 168
Channel
  conductance, 26
  dope, 4
  hot electron injection, 311, 314
  length, 144, 157
  length modulation, 51
  stopper, 141, 150
  stopper implant, 28
Channel-free gate array, 390
Channel-less gate array, 390
Characterisation tests, 532
Charge
  bucket, 246
  characteristic, 197
  coupled device, 245
  distribution, 9
  pump, 187
  sharing, 210, 216
  transfer, 247
  transfer gate, 320
  trap flash, 316
Charged device model, 518
CHEI, 311
Chemical mechanical polishing, 138
Chemical vapour deposition, 126
Chip, vii
  select, 275
Chip-scale
  package, 565, 566, 572, 578
Chiplet, ix, xiv, 78, 379, 422, 560, 582
Choice of logic implementation, 219
CiM, 338
Circuit
  density, 147
  editing techniques, 613
  simulation, 38, 373
Circuit-analysis program, 188
Cleanroom, 549
Clock
  activity, 451
  generation, 471
  jitter, 468, 474
  -phase synchronisation, 472
  signals, 211
  skew, 213, 216, 220, 466
  -tree synthesis, 467
Clocked CMOS circuits, 211
Clocking strategies, 220
CMOS, 192
  buffer design, 203
  buffer optimisation, 203
  driver, 222
  image sensors, 245
  inverter, 193
  inverter design, 197
  inverter dissipation, 199
  inverter transfer characteristic, 195
  latch, 211
  NAND gate, 256
  output buffer, 222
  parasitic bipolar device, 583
  process, 145, 223
  transmission gate, 208
CMP, 138
Column decoder, 274
Compiled cell, 352
Complementary FET (CFET), 168
Complementary pass-transistor logic, 437
Complex PLD, 399
Compressive strain, 159
Computational lithography, 108

Compute-in-memory (CiM), 338
Computing power, 455
Conduction band, 6
Conductive bridging memory, 333
Conductivity, 4
Constant-field scaling, 627
Constant-voltage scaling, 623, 627
Contact filling, 152
CONTACT mask, 144
CONTACT-mask programmed ROM cell, 306
Content-addressable memory, 269, 273
Continuous array, 392
Control
  bus, 357
  path, 360
Controlled-collapse chip connection, 572
Copper, 626
Core, 351, 390
Corner lot, 502
Correct by design, 461
Cosmic particles, 305
Cost, 80
  of a wafer factory, 637
  of interconnect, 334
CPL, 437
CPLD, 399
Critical
  delay, 469
  dimension, 94, 102, 106
  dimension uniformity, 105
  module, 368
Cross-over capacitance, 218
Cross talk, 218, 475
Crystal
  growth, 76
  ingot, 81
Crystal-oriented particles, 82
CSP, 565, 566, 578
CTF, 316
CTG, 320
CUB, 290
Current-controlled oscillator, 471
Current density, 2, 130
Customer returns, 532
Custom IC, 351
Customisation, 350, 390
Cut mask, 107
CVD, 126
CVSL, 219
  logic gate, 219
Cycle
  stealing, 469
  time, 271
Czochralski, 76, 82

D

D-type flip-flop, 212, 247, 393
Damascene
  backend flow, 626
  patterning, 153
Damascene process flow, 128
Dark current, 248
Data
  bus, 357
  input buffer, 275
  output buffer, 275
  path, 360
  retention time, 269, 312
Database, 89, 524
DBG, 570
DDR, 294, 299
Decision tree, 367
Decoupling capacitor, 482, 483
Deep
  n-well, 416, 523
  ultra violet, 103
Deep neural networks (DNN), 339
DEF, 236
Defect density, 552
Delay-locked loop, 473
Delay fault, 538, 542
$\Delta I_{\rm ddq}$ test, 542
Depletion
  layer, 9
  layer thickness, 53
  load, 187
  process, 9
  transistor, 27
Deposition, 126, 615
Depth of focus, 78, 94, 97
Design
  for anything, 559
  for debug, 602, 616
  documentation, 524
  efficiency, 622
  for failure analysis, 616
  hierarchy, 524
  for lithography, 102
  for manufacturabilty, 169, 223
  organisation, 524
  productivity, 334
  productivity gap, 635
  resources, 622
  rule check program, 384
  rule manual, 223
  rules, 224
  style, 636
  for testability, 547
  verification, 373

Designing a CMOS inverter, 197
Destructive read-out, 286, 331
Developer, 120
DfL, 102
DfM, 169, 223, 373, 558
  rules, 557
DfT, 547
DfX, 559
DIBL, 54, 62
Dice before grind (DBG), 570
Dicing, 569
Dielectric relaxation time, 32
Differential split level logic, 219
Diffusion, 134
  coefficient, 134
Digital
  CMOS circuits, 206
  ICs, 351
  mixer, 374
Direct slice writing, 351
Direct-writing techniques, 117
Discharge characteristic, 197
Dishing, 140
Dislocations, 82
Disturbances in the production environment, 549
DLL, 473
DMOS transistor, 259
DOF, 78, 94, 97
DOMINO-CMOS, 214
Donor, 7
Dope profile, 135
Double-diffused MOS transistor, 258
Double-flavoured polysilicon, 125, 147, 194
Double-gate transistor, 161
Double-patterning techniques, 103
Double-sided polishing, 78
Double data rate, 299
Double pass-transistor logic, 438
DPL, 438
DPT, 103
Drain, 4
  extension, 144, 151
  series resistance, 510
Drain-induced barrier lowering effect, 62
DRAM, 62, 271, 285
  architectures, 295
  crown capacitor, 290
  pillar cell, 291
  process node name, 294
  refresh rate, 292
DRC, 384
  rules, 557
Drive
  current, 49
  strength, 235
Driver transistor, 181
  characteristic, 181
DRM, 223
DRO, 286, 331
DSL, 163, 219
DSW, 351
Dual-damascene, 153
Dual-dope polysilicon, 194
Dual-edge triggered flip-flops, 451
Dual-port memory, 304
Dual-V_T process, 23, 623
Dual polysilicon, 147
Dual stress liner, 163
Dummy
  metal, 140
  polysilicon, 232
DUV, 103
DVFS, 432, 434
Dynamic
  CMOS, 214
  CMOS circuits, 214
  CMOS latch, 215
  CMOS shift register, 215
  D-type flip-flop, 216
  flip-flop, 215
  memory, 271
  power consumption, 414
  power dissipation, 200
  RAM, 271, 285
  shift register cell, 215
  voltage and frequency scaling, 432, 434
  voltage drop, 480

E

Early failure rate, 585
E-beam, 117
  lithography, 114
EBL, 114
EBPG, 90
ECC (error-correction code), 303, 316, 319, 341, 491
EDA, 374
EDO, 297
EDO DRAM, 298
eDRAM, 285
E/D technology, 187
EEPLD, 354
Effective
  channel length, 51
  transistor channel length, 144

Electric
  field, 9
  potential, 9
Electrical endurance test, 584
Electromagnetic
  compatibility, 487
  pulse, 487
Electromigration, 130, 505
Electron, xxix
  beam, 117
  mobility, 1, 194
  valves, 1
Electron-beam pattern generator, 90
Electronic design automation, 374
Electronic disks, 316
Electrostatic
  charge, 551
  discharge, 516, 583
  potential difference, 11
Elevated S/D, 157
Embedded
  arrays, 400
  FPGA, 636
  logic, 336
  memory, 269, 334
  non-volatile memory, 336
  software, 371
  SRAM, 284
EMC, 487
EMP, 487
EMRAM, 332
Emulation, 369, 372
Endurance, 312, 319, 320
  characteristic, 312
Energy
  band, 6
  band diagram, 11
  band theory, 5
  delay product, 423
  gap, 6
Enhancement transistor, 27
EOT, 124, 158
Epi-layer, 81
Epitaxial
  film, 126
  wafer, 81, 515
EPLD, 354
EPROM, 310
Equivalence checking, 373
Equivalent oxide thickness, 124, 158
Erosion, 140
Error-correction code, 303
Error detection and correction codes, 319
ESD, 85, 516, 583
e-sort, 533
eSRAM, 284
ET-SOI, 166
Etching, 131
EUV, 109
Exclusive OR, 209
EXOR gate, 209, 450, 453
Extended data out, 297
  DRAM, 298
Externally-induced voltage alteration, 607
Extreme data rate RDRAM, 301
Extreme-UV lithography, 109
Extremely-Thin SOI, 166

F
FA, 594
Fab-lite, 173, 407
Fab cost, 80
Fabless, 173, 407
Faceted S/D, 163
Failure analysis, 593, 594
Fast page mode, 297
  DRAM, 298
Fat zero, 248
FD-SOI, 86
FeDRAM, 331
FeFET, 331
FeNAND, 331
FEOL, 153
Fermi level, 8
Ferroelectric
  DRAM, 331
  FET, 331
  NAND, 331
  RAM, 330
FIB, 614
Field-effect principle, 1
Field-programmable device, 354, 394
Field-programmable gate array, 352
Field oxide isolation, 390
FIFO, 271, 272
Filler cells, 401
Fill factor, 250
Fin
  height, 163
  width, 163, 644
Fin aspect ratio, 238
FinFET, 162
  layout, 237
  SRAM, 284
Firm cores, 353
First-silicon debug, 593

First-time-right silicon, 461
Flash, 320
  memory, 313
Flat-band
  condition, 11
  voltage, 11
Flip-chip bonding, 572
Flip-flop, 211
Floating gate, 310
Floor-planning, 373, 379
Focused ion beam, 614
Fork-sheet transistor, 168
Formal verification, 373
Forward-bias effect, 25
FOUP, 639
  mini environment, 550
Four-transistor SRAM cell, 278
Fowler–Nordheim tunnelling, 311
FPGA, 352, 394, 636
FPM, 297
FPM DRAM, 298
FRAM, 330
Front-end-of-line, 153
Front-opening unified pod, 639
Full adder, 190, 365
Full-CMOS SRAM cell, 278
Full-custom IC, 351
Full-featured EEPROM, 311
Fully regular library, 102
Fully silicided, 125
Functional level, 360
FUSI, 125
  gate, 156

G
GAA, 168
GAA FinFET, 167
Gain factor, 45
Gajski-Kuhn chart, 356
GALS, 478
Gate, 4
  array, 390
  delay, 538
  depletion, 125, 155, 623
  forest, 390
  inversion, 155
  oxidation, 142
  oxide, 123
  oxide tunnelling, 624
Gate-all-around (GAA), 167, 168
Gated clock, 452, 467, 468
Gate-drain overlap capacitance, 145
Gate-first process, 157
Gate-induced drain leakage, 66
Gate-isolation technique, 393
Gate-last process, 158
Gate-oxide
  leakage, 63
  leakage current, 420
  thickness, 123
Gate-source overlap capacitance, 145
GDDR, 300
GDSII, 407
General-purpose process, 124
Geometric layout
  description language, 407
  representation, 407
GIDL, 66
GLDL, 407
Glitches, 450
Globally synchronous, locally asynchronous, 478
Global variations, 494
Glue logic, 359
Golden device, 542
Graded-drain transistor, 508
Gray code counter, 445
Ground bounce, 480
Guard ring, 515, 522

H
Half pitch, 94
Halo, 53, 54
Handcrafted layout, 384
Handshake circuits, 449
Hard cores, 353
Hard mask, 118
Hardware
  accelerator, 372
  description language, 352, 364
  software codesign, 370
Haze effect, 92, 116
Hazing, 116
HBM, 78, 336
HBT, 254
HCD, 511
HCI, 508
HDD, 151
HDGA, 390
HDL, 352, 364
HDP, 132
Hemispherical grain, 287
HEPA, 549
Hetero-epitaxy, 126
Heterogeneous system, 359
  on a chip, 470, 621

Hierarchical
  design approach, 403
  layout, 403
High bandwidth memory (HBM), 294, 302, 336
High-density gate array, 390
  layout, 213
High-density plasma, 132
High-ϵ dielectric, 75, 158
High-efficiency particle air, 549
High-energy cosmic particles, 488
High-k dielectric, 75
High-level synthesis, 371
Highly doped drain, 151
Highly accelerated temperature & humidity stress (HAST), 586
High-NA EUV, 112
High-ϵ metal-gate (HKMG), 294, 419
High-performance process, 423
High temperature operating life, 585
High temperature storage test (HTST), 587
High-voltage CMOS, 260
High voltage stress test (HVST), 587
Hillocks, 505
HLS, 371
Hold-time violation, 467
Hole mobility, 194
Holes, 6
Homo-epitaxy, 126
Homogeneous sytem, 359
Hot-carrier
  degradation, 511
  injection, 66, 151, 508
Hot-electron effect, 314
Hot carrier, 66
Hot electron, 311
HP-CMOS, 234
HP process, 124, 423
HSG, 287
HTOL, 585
Human-body model, 517
Humidity
  sensitivity, 584
  test, 586
HV-CMOS, 260
Hyper-NA EUV, 112
HV_T, 124

I

IC, vii
  characterisation, 548
  customisation, 355
  database, 90
  design path, 358
  design rule check, 89
  early failure rate, 585
  electrical check, 89
  engineering, 548
  floor-plan, 368
  functional check, 89
  infant mortality, 585
  intrinsic failure rate, 585
  layout, 89, 357
  lifetime, 130
  package corrosion, 586
  package robustness, 586
  quality, 583
  reliability, 583
  reliability tests, 584
  wearout, 585
I_{ddq} testing, 540
IGZO-based DRAM, 304
ILD, 128, 170, 577, 626
Image sensor, 250
IMC, 338
Immersion lithography, 99, 103
Impact ionisation, 66, 507
Implantation duration, 135
Improved DRAM access time, 296
In-memory computing (IMC), 338
In-system programmability, 394
Inductance, 478
Inert
  gas, 586
  liquid, 586
Infant mortality, 585
Ingot, 76
Input protection, 85
Insulated gate bipolar transistor (IGBT), 264
Integrated circuit, vii
Intellectual property, 352, 371
Inter-die variations, 494
Interconnect sheet resistance, 626
Interlevel dielectric, 128, 170, 577, 626
Intermediate focus point, 110
Internet of Things, 435
Interstitial dope atoms, 135
Interstitials, 82
Intra-die variations, 495
Intrinsic
  failure rate, 585
  silicon, 8
Inverse curvilinear masks (ILM), 108
Inversion-layer transistor, 3, 14
Inverse lithography technology (ILT), 108
Inverse narrow-width effect, 56

Inverter, 180
chain, 203
DC behaviour, 181
INWE, 56
Ion, xxx
acceleration, 134
implantation, 134
implanter, 134
Ionisation energy, 7
IoT, 435
IP, 352, 371, 470
IR drop, 466
Islands of synchronicity, 478
Isolated vias, 556
Isotropic, 131
ISP, 394
ISPP, 318
Iterative multiplier, 363
ITRS, 621

J
Joule heating, 507
Junction spiking, 137

K
K-factor, 22, 24, 193
Known-good-die, 581

L
LADA, 608
Laser
beam, 117
beam pattern generator, 90
dicing, 569
fusing, 339
produced plasma, 109
signal injection microscopy, 603
Laser-assisted device alteration, 608
Laser voltage probing (LVP), 609
Latch, 211
Latch-up, 84, 253, 513, 583, 592
sensitivity, 583
thyristor, 583
Lateral
diffusion, 144
electric field, 47
Law for conservation of charge, 18, 32
Layout
description, 473
implementation form, 384, 405
level, 359, 368
process, 223
LBPG, 90
LDD, 508
transistor, 510
LDO, 432
Leakage, 67
current, 62, 216, 416
power, 417
power consumption, 414
Lean integration with pass-transistor, 439
LEAP, 439
LEF, 236
LELE, 103
LER, 105, 110, 164, 495, 498
Level shifter, 221, 433, 468
Library, 234
cell, 224
cell height, 235
LIFO, 272
Light-emission microscopy, 600
Light-induced voltage alteration, 608
Lightly doped drain, 508
LIL, 237
Line-edge roughness (LER), 105, 110, 164, 495, 498
Line-width roughness (LWR), 110
Linear region, 15, 16
Lithium-air batteries, 413
Litho-friendly design, 102, 232, 495
Lithography, 89
LIVA, 608
Load
elements, 180, 181
lines, 181
transistor, 182, 183
Loadless SRAM cell, 278
Local interconnect layer, 237
Local oxidation of silicon, 122
Local variations, 495
LOCOS, 141
oxide, 145
process, 122
Logic-gate level, 364
Logic simulation, 373
Long-channel devices, 420
Lookup table, 394
LOP, 423
CMOS, 234
process, 124
Low-dropout regulator, 432
Low-end IC market, 349
Low-energy cosmic neutrons, 488
Low-operating power, 423
process, 124

Low-power
CMOS, 411
library, 436
Low-voltage design, 429
Low-ϵ dielectric, 170, 626
Low standby power process, 124, 423
LPCVD, 127
LPP, 109
LSIM, 603
LSTP, 68
CMOS, 234
process, 124, 423
LUT, 394
LWR, 110
LV_T, 124

M
Machine model, 518
Macro, 390
cell, 352
Magnetic tunnel junction, 331
Magneto-resistive RAM, 331
Majority charge carrier, 9
Manhattan shapes, 109
Manhattan skyline effect, 404
Mapping, 371
Mask, 76, 224
costs, 116
ROM, 305
writers, 117
Mask-less lithography, 118
Mask-programmable
gate array, 390
ROM, 305, 388
Mass action law, 4
Master cell, 390
Matching, 499
coeffient, 499
of transistors, 624
Maximum storage time, 248
MCM, 315, 580
Mean free path, 171
Meet-in-the-middle strategy, 404
Mega cell, 352
Memory
address, 271
array, 269
bank, 275, 300
cell, 269
controller, 299
matrix, 269
word, 271
MEOL, 153, 171
Metal-oxide-semiconductor (MOS) capacitor, 8
Metal gate, 156
Metal grain granularity, 498
METAL mask, 144
Metrology, 115
MGG, 498
Microcode instruction, 387
Microcontrol unit, 360
Micro defects, 82
Microprocessor core, 371
Mid-end-of-line, 153
Military specifications, 3
Milling, 615
Minority carrier, 14
Mismatch, 495
MISR, 544, 590
Mixed signal CMOS process, 262
ML2, 118
MLC, 315
MLL, 118
MLR, 117
Mobility, 27, 45, 143
Module generator, 404
Molecule, xxix
Molybdenum, 4
gate, 145
Monocrystalline silicon, 143
More than Moore, 582
MOS, 1
capacitance, 31, 34
formulae, 18
transistor, 5
transistor leakage mechanisms, 59
transistor weak inversion operating region, 60
MPGA, 390
MPU, 349
MPW, 117
MRAM, 331
MTCMOS, 419
MTJ, 331
MTP, 338
MUGFET, 165
Multi-chip module, 580
Multi-DUT probe card, 534
Multi-gate FET, 165
Multi-level cell (MLC), 315
Multi-level flash memory, 315
Multi-patterning, 104
Multi-port memory, 304
Multi-project wafers, 117
Multi-time programmable memory, 338
Multi-voltage domain, 422

Multi-V_T, 124
Multilayer reticle, 117
Multiple input signature register, 590
Multiple threshold CMOS, 418
Murphy's law, 85

N
NAND logic D-type flip-flop, 213
Nano-imprint lithography, 113
Nano-ribbon transistor, 168
Nano-sheet transistor, 168
Nano-through silicon vias (nTSVs), 484
Nano-wire FET, 167
Nano-wire transistor, 168
Narrow-channel effect, 55, 56
NBTI, 511
N-channel MOS transistor, 27
Near-field scanning optical microscopy, 607
Negative bias temperature instability, 511
Negative resist, 121
Netlist, 364, 375
Neutron, xxix
Next-generation lithography, 112
NGL, 112
NIL, 113
Nitrided gate oxide, 125
NMOS
  inverter, 180
  process, 141
  transistor, 4, 192
  transistor gain factor, 197
  transistor threshold voltage, 192
NMOS-mostly, 214
  circuit, 180
NMOSt, 4
Node name, 641, 644
Noise
  immunity, 220
  margin, 181, 205
Non-overlapping, 213
  clocks, 216
Non-rechargeable batteries, 412
Non-recurring engineering costs, 382
Non-saturated enhancement load, 184
Non-volatile
  memory, 269
  RAM, 329
Non-volatile DRAM, 295
Non-volatile RAMs, 295
Normally off transistor, 27
Normally on transistor, 27
Notch, 80, 84
NRE costs, 382
NSOM, 607
N-tub CMOS process, 145
N-type silicon, 3
Number of process steps, 173
Number representation, 442
Numerical aperture, 94
Numerical aperture (NA), 112
NVRAM, 295, 329
N-well CMOS process, 145, 180

O
OAI, 97
OBIC, 608
OBIRCH, 607
OCV, 502
Off-axis illumination, 97
On-chip variation, 502
One-time-programmable
  EPROM, 311
  memory, 309
ONO, 287
OPC, 101
Optical-beam
  induced current, 608
  induced resistive change, 607
Optical proximity correction, 101
OR-function, 189
OR matrix, 387
OTP, 309
  EPROM, 311
OUM, 332
Output
  buffer, 222
  conductance, 26
  enable, 275
  impedance, 21
  protection, 85
Overlay, 115
  problem, 290
Ovonic Unified Memory, 332
Oxidation, 118, 122
Oxide-nitride-oxide, 287
Oxide spacer, 509

P
Package-on-a-package, 581
Packaging, 563
Page, 297, 314
PAL, 388
Parallel
  connection of transistors, 189, 207
  multiplier, 363

Parallelism, 427
Parametric
  fault, 541
  yield loss, 551
Parasitic
  capacitances, 85
  extraction, 236
  MOS transistor, 28
  thyristor, 253
Partial product, 363
Pass-gate logic, 437
Passivation layer, 126, 144
Pass transistor, 208
  logic, 209, 437
Path delay, 538
P-channel MOS transistor, 27
PCM (Phase Change Memory), 332
PCM (Process Control Module), 91, 556
PCRAM, 332
PD-SOI, 86
PDK, 374
PECVD, 127
Pellicle, 92, 111, 116
PEM, 600
Penetration depth, 134
Penta-level cell, 326
Periodic system of elements, 7
Perovskite crystals, 330
Persistent memory, 295
Phase-change memory, 332
Phase-locked loop, 471
Phase-shift masks (PSM), 94, 97, 111
Photolithography, 89
Photon-emission microscopy, 600
Photoresist, 118
  layer, 92
Physical design aspects, 621
PICA, 602
Picosecond imaging circuit analysis, 602
Piller DRAM cell, 291
Pinch-off
  point, 17
  region, 52
Pinhole, 591
Pinned photodiode, 250
Pipelining, 427
PLA, 387, 388
Place and route, 375, 404
Placement and routing, 404
Planar
  DRAM cell, 287
  IC technology, 29
  silicon technology, 3
Planarisation, 138
Plasma, 127
  etching, 131
Platform ASIC, 402
PLD, 354, 394
PLL, 471
PMOS transistor, 192
  gain factor, 199
  threshold voltage, 192
Pocket implants, 53
Point defects, 82
Poisson's law, 9
Polycide process, 152
Polycrystalline silicon, 75, 143
  layer, 4
Poly fuse, 339
Polygon pusher, 384
POLY mask, 142
Polymide layer, 305
Polysilicon, 75, 143
  gate, 143
  interconnect, 144
PoP, 581
Positively charged ion, 6
Positive photoresist, 120
Positive resist, 121
Potential first silicon problem, 589
Power
  binning, 434
  delay product, 27, 422
  dissipation, 199, 219
  distribution network, 483
  down mode, 452
  grid integrity, 483
  integrity, 483
  reduction techniques, 415, 420, 424
  switch, 418
  transistor, 258
Power MOSFET, 245, 259
  applications, 261
  transistor, 258
PPD, 250
Ppm, 533, 586
P&R, 375
PRAM, 332
Pre-deposition, 134
Primary battery cells, 412
Probe card, 590
Probing, 597
Process
  control module, 91, 556
  corner, 502
  cross section, 232
  design kit, 374

node naming, 644
parameter variation, 69
Production tests, 532
Product term, 387
Programmable
array logic, 388
logic array, 387
logic device, 354
read-only memory, 309
Programmable logic devices, 394
PROM, 309
Propagation
delay, 475, 478, 625, 626
time, 468
Protection circuit, 220
Proton, xxix
Prototyping, 393
Proximity effects, 494
Pseudo-nMOS
circuit, 197, 207
logic, 426
Pseudo-static RAM, 271
PSG, 126
PSM, 94, 97
P-type substrate, 4
Punch-through, 54, 150, 504
PVT, 58, 71, 236

Q

QLC, 326
Quad-level cell, 326
Quadruple patterning, 104

R

Race, 213
Radiation hardness, 317
Raised S/D, 157
RAM, 269, 273
Rambus DRAM, 297
Rambus™, 301
Random
access, 271
dopant fluctuation, 165, 497
doping fluctuations, 293
variations, 495
work function variations, 498
yield loss, 552
Random-access memory, 269, 273
Rapid thermal anneal, 135, 157
Ratioed logic, 182
RCAT, 286
RDF, 165, 293, 497
RDR, 158
RDRAM, 297, 301
Reactive ion etching, 132
Read-only memory, 269, 305
Read-retry, 319
Recessed-channel array transistor, 286
Recessed source/drain, 159
Rechargeable batteries, 412
Recombination, 32
time, 47
Reconfigurable computing, 636
Reduced latency DRAM, 300
Reduced voltage swing, 432
Redundancy
in logic, 563
in memories, 339
Refresh
amplifier, 286
operation, 286
Register-transfer
language, 362, 372
level, 357
Reliability
stress conditions, 584
tests, 533
Repeaters, 626
Replacement gate, 158
Replicator circuit, 433
ReRAM, 333
Resistance, 478
Resistive
load, 188
RAM, 333
Resistive-interconnect localisation, 608
Resolution enhancement techniques, 94, 95
Restrictive design rules, 158
RET, 94, 95
Reticle, 90
alignment sequence, 155
Retrograde
implant, 137
profile, 135
well, 149
Reuse, 354, 371, 393, 470, 622
Reverse-bias junction leakage, 64
Reverse short-channel effect, 54
RIE, 132
RIL, 608
RLDRAM, 300
R-load SRAM cell, 278
Roadmap, 621
ROM, 269, 305
layout, 386
logic function, 386

Routing channel, 390
Row
  decoder, 274
  refresh, 286
RRAM, 333
RSCE, 54
RTA, 135, 157
RTL, 355, 362, 372
  description, 362
RWF, 498

S
Sacrificial pad oxide, 142
SACVD, 127
SADP, 103
Salicide, 152
SAQP, 107
Saturated enhancement load, 182
Saturation
  current, 17
  region, 15, 17
  velocity, 49
Scaling
  effects, 621
  limitations, 633
  properties, 150
Scan
  chain, 538
  test, 390, 393, 538
Scannable flip-flop, 538
Scanning
  electron beam microscopy, 610
  optical beam, 594, 603
SCCD, 246
Scheduling, 371
SCM, 295
Scratch-protection layer, 144, 586
Scribe lane, 90, 91, 556, 568
Scribe line, 90
SDL, 608
SDRAM, 297, 299
Sea-of-transistors, 392, 503
Seal ring, 567
Sea of gates, 390
Secondary
  battery cells, 412
  electrons, 610
Seebeck effect imaging (SEI), 608
SEG, 157, 163
SEL, 490
Selective cobalt deposition, 129
Selective epitaxial growth, 157, 163
Self-aligned
  double patterning, 103
  drain, 75, 144
  quadruple patterning, 107
  salicide, 152
  source, 75, 144
  source/drain implantation, 147
  via, 169
Self-discharge rate, 413
Self-heating, 87, 523
Self-inductance, 480
Self-test, 390
Self-timed circuits, 447
SEM, 610
Semi-custom IC, 355
Semiconductor
  doping, 7
  material, vii
Sense amplifier, 275, 286
Separation by implantation of oxygen, 85
SER, 417
Serial
  memory, 269, 272
  ROM, 308
Series connection of transistors, 189, 207
Set-up time violation, 467
SEU, 489
SGRAM, 301
Shadow RAM, 308, 329
Shallow-trench isolation, 122, 147
Sheet resistance, 144, 188
Shift-and-add operation, 363
Shift register, 215
Shmoo plot, 594
Short-channel effect, 53
Short-circuit
  current, 199, 222
  free output buffer, 222
  power consumption, 414
  power dissipation, 200
Short-circuit current, 195
SIA, 621
Sign-magnitude notation, 442
Signal
  integrity, 474
  integrity self-test, 493
  processor, 357
  propagation, 625, 626
Signature, 590
SIL, 603
Silicidation, 137
Silicides, 151, 232
Silicon
  atom, 5

crystal, 6
dioxide, 2
insulator-silicon, 288
on-insulator, 84
on-sapphire, 85
SIMOX, 85
Simulation, 372
Simultaneously switching outputs, 488
Single-edge triggered, 451
Single-event
latch-up, 490
upset, 489
Single-level cell, 315
Single-phase, 220
clocking, 463
Single-port memory, 304
SiO_2, 2
SiON, 125
SiP, 382, 580
SIS, 288
SIST, 493
Six-transistor SRAM cell, 278
Slack borrowing, 469
SLC, 315
Sleep mode, 452
SLI, 394
Slurry, 138
SMD, 578
SMIF pod, 550
SNM, 280
SNOM, 607
SoC, 334, 349, 369, 382, 580, 621
design platform, 351
Soft
core, 353
defect localisation, 608
error rate, 305, 417, 489
SOG, 138
SOI, 84, 166
SOI-CMOS, 84
Solid immersion lens, 603
Solid-state drives, 316
SOM, 594, 603
SONOS, 316
SoP, 582
SOS-CMOS process, 85
Source, 4
series resistance, 510
synchronous timing, 470
Source/drain capacitance, 85
Spacer, 509
lithography, 104
Specification, 524, 593
Speed and area, 219
Spin-on-glass, 138
Spin transfer torque, 332
Spurious transitions, 450
Sputter etching, 132
SRAF, 101
SRAM, 271, 273
memory cell, 277
SRAM applications, 284
SRB, 159
SRPL, 439
SSD, 316
SSO, 488
SSO driver, 223
SSTA, 374, 499, 502
STA, 373, 499, 501
Stacked capacitance cell, 287
Stacked CMOS image sensor, 252
Stacked die, 302, 327, 581, 611
Stand-alone memory, 269
Standard
cell, 234, 389
commodities, 355
IC, 381
logic IC, 355
product, 355
Standard-cell, 389
height, 437
layout, 389
library, 389
Standby
current, 62, 417
mode, 329, 416
power, 68
Static
CMOS circuits, 206
CMOS flip-flop, 212
CMOS inverter characteristic, 195
column access, 296
IR drop, 481
memory, 271
noise margin, 280
power consumption, 414
RAM, 271, 273, 277
voltage drop, 481
Static CMOS circuit, 240
Static timing analysis, 373, 499, 501
Statistical static timing analysis, 374, 499, 502
STC, 287
Steady-state current, 541
Stealth dicing, 569
Step-and-repeat, 93
Step-and-scan, 93
Step coverage, 130

STI, 51, 122, 147
  stress, 494, 500
Stick diagram, 227
Stitching, 112
Storage class memory, 295
Storage gate, 246
Strained silicon, 159
Strain-relaxed buffer, 159
Stress memorisation, 163
Strong inversion, 13
Structural
  fault, 541
  test, 537
Structured ASIC, 402
STT, 332
Stuck-at fault, 540
Stuck-open fault, 540
Subresolution assist feature, 101
Substrate, 76
  bounce, 485
  dope, 4
Subthreshold
  behaviour, 60, 62
  current, 62, 416
  leakage current, 60, 415, 623
  logic, 206, 434
  region, 60
  slope, 60, 88, 541
Supply
  bounce, 480
  noise, 481
Surface
  scattering, 47
  states, 247
Surface-channel CCD, 245
Surface-mount device, 578
Swing Restored Pass-Transistor Logic, 439
Switching activity, 441, 480
Symbolic layout, 405
Synchronous
  CMOS circuits, 211
  DRAMs, 297, 299
Synchronous Graphics RAM, 301
Synthesis, 373
  tools, 371
System
  design aspects, 621
  in a package, 382, 580, 582
  level, 357–359
  on a chip, 349, 369, 372, 382, 580
  on silicon, 371
Systematic yield loss, 551
Systems on silicon, 394
SV_T, 124

T
Tap-less cells, 232, 236
Tape-out, 89
Tapering factor, 203
Tapped cells, 232, 236
τD-product, 27
TDDB, 586
TDP, vii
Temperature
  cycle test, 586
  expansion coefficient, 586
  sensitivity, 584
  variation cycle, 586
Tensile strain, 160
TEOS, 127
Test vectors, 589
Thermal
  behaviour, 575
  design power, vii
  energy, 6
  generation, 32
  oxide, 122
Thermal activation, 135
Thermally-induced voltage alteration, 607
Thermal shock test (TST), 587
Thick oxide, 122, 141
Three-beam imaging, 94
Three-dimensional DRAM cell, 287
Threshold
  adjustment implantation, 5
  implant, 419
  loss, 184, 209
Threshold voltage, 13, 24, 192, 592
  adjustment implantation, 142
  loss, 209
  roll-off, 53
  temperature dependence, 57
Throughput time, 173
Through-silicon via, 132, 561
Tie-off cell, 482
Tiles, 140
Time-dependent dielectric breakdown, 586
Time-resolved photon-emission microscopy, 594, 602
Time stealing, 469
Timing
  closure, 469
  margins, 468
  problems, 463
  verification, 548
  violations, 467
Titanium nitride, 152
TIVA, 607

TLC, 325
Tomic layer deposition, 591
Top-down design process, 358
Total thickness variation (TTV), 78
Track system, 92
Transconductance, 25
Transfer
efficiency, 247
gate, 246
Transistor
level, 365
lifetime, 123
matching, 499, 624
scaling effects, 623
Transistor gain factor, 20, 190
temperature dependence, 57
Transition
delay fault, 538
region, 16
Transmission gate, 208, 211
Transparency, 213, 216
Trench capacitance cell, 287
Tri-gate device, 165
Tri-state buffer, 222
Triode region, 15
Triple-level cell (TLC), 325
Triple patterning, 104
Triple-well, 523
concept, 623
device, 416
technology, 416, 486, 491
TSV, 132
TTL
compatible, 221
input buffer, 221
TTL-CMOS input buffer, 221
Tunnelling, 125
Turnaround time, 308, 352, 393
Two's complement notation, 442
Two-beam imaging, 94

U
Ultra-low-power application, 434
Ultra thin body, 166
Uncommitted array, 392
Under-etch, 131, 549
Uptime of EUV tool, 111
Usable gates, 355
User-specific integrated circuit, 350
USIC, 350
UTBB-SOI, 166
Utilisation factor, 355

V
V-NAND, 322
flash, 320
Valence
band, 6
electron, 6
Variability-aware design, 632
VCO, 471
VDMOS, 259
Velocity saturation, 47
Verilog, 371
code, 372
Vertical
double-diffused MOS, 259
electric field, 47
NAND, 322
NAND flash, 320
pillar capacitors, 289
Very low voltage test, 542
VHDL, 371
code, 372
Via doubling, 559
Vibration test, 587
Video
memories, 272
RAM, 272, 300
Virtual
component, 353
drain, 17, 51
Virtually static RAM, 271
Virtual socket interface alliance, 353
VLV test, 542
Voids, 505
Volatile memory, 269
Voltage
drop, 480
peaks, 222
regulator, 431
Voltage-controlled oscillator, 471
VPT, 289
VRAM, 272, 300
VSI Alliance, 353

W
WAF, 316
Wafer, 76
acceptance test, 557
bevel, 77
cycle time, 173
diameter, 548
flatness, 78
level packaging, 565

Wafer (*cont.*)
  map, 553
  probing, 549
  starts per month, 637
Wafer edge exposure (WEE), 555
Wallace tree multiplier, 363, 450
Warp, 78
Warpage, 78
WAT, 557
Waveform measurements, 548
Wear-levelling, 316, 319
Wearout, 585
Well-bias, 417
Well-proximity, 500
  effect, 494
Well-tap, 232
  cells, 236, 514
Well biasing, 417
Wet-etching method, 131
Wire
  bonding, 570
  self-heating, 507
  spreading, 559
WLCSP, 566
WLP, 565
Word line, 274
Work function, 11, 22, 156
  layer, 419
Worst-case
  corner, 281
  delay path, 355, 368
WPE, 494, 500
Write amplification factor, 316
Write and verify programming, 318
Write enable, 275
WSPM, 637

X
X-decoder, 274
X-ray lithography, 112
XDR RDRAM, 301
XIVA, 607
XRL, 112

Y
Y-decoder, 275
Yellow room, 549
Yield, 548
  control, 556
Yield degradation
  chemical impurities, 551
  dust particles, 549
  electrostatic charge, 551
  humidity, 549
  temperature fluctuations, 549
  UV light, 549
  vibrations, 549

Z
Zero capacitor DRAM, 303
Zero-temperature-coefficient, 58
Z-RAM, 303
ZTC, 58

MIX
Papier aus verantwortungsvollen Quellen
Paper from responsible sources
FSC® C105338
www.fsc.org

If you have any concerns about our products,
you can contact us on
ProductSafety@springernature.com

In case Publisher is established outside the EU,
the EU authorized representative is:
Springer Nature Customer Service Center GmbH
Europaplatz 3, 69115 Heidelberg, Germany

Printed by Libri Plureos GmbH
in Hamburg, Germany